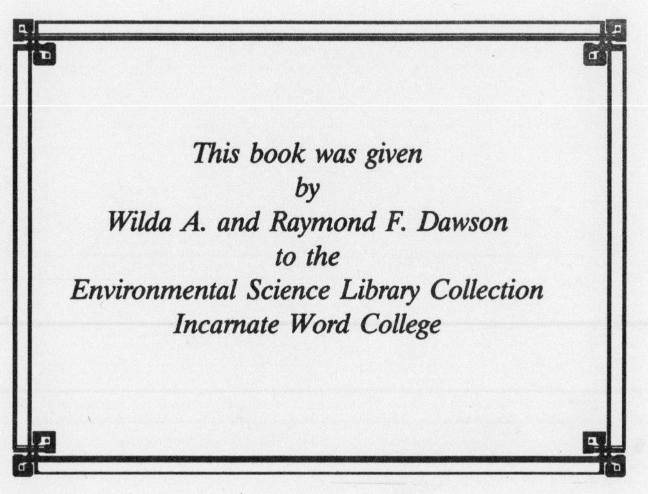

*Handbook of*

# Mass Spectra of Environmental Contaminants

*Second Edition*

**Ronald A. Hites**
School of Public and Environmental Affairs
and Department of Chemistry
Indiana University
Bloomington, Indiana

LEWIS PUBLISHERS
Boca Raton Ann Arbor London Tokyo

**Library of Congress Cataloging-in-Publication Data**

Catalog record is available from the Library of Congress.



Direct all inquiries to CRC Press, Inc., 2000 Corporate Blvd., N.W., Boca Raton, Florida 33431.

International Standard Book Number 0-87371-534-9

PRINTED IN THE UNITED STATES OF AMERICA
2 3 4 5 6 7 8 9 0

Printed on acid-free paper

# PREFACE

Mass spectrometry is the single most useful technique for the analysis of organic compounds in environmental samples (air, water, sediment, fish, etc.). Fortunately, many of the compounds encountered in environmental samples have known mass spectra. This handbook is a collection of the electron impact mass spectra of 533 commonly encountered environmental pollutants. This is an increase of 139 compounds over the first edition of this book.

The compounds which are included were selected by an analysis of several U.S. Environmental Protection Agency databases and the Merck Index, all of which were accessed through the Chemical Information System (Baltimore, MD).

The data were acquired from the Chemical Information System using a personal computer operating as a "smart terminal". In this way, data were written directly to diskettes, thus avoiding transcription errors. The spectra were edited and then plotted from these diskettes. Structures were drawn using Molecular Presentation Graphics and checked against the Merck Index. All of the spectra were carefully reviewed by two experienced mass spectrometrists. Numerous errors were detected and corrected. The most common error was spurious mass peaks, presumably due to instrumental contamination. This review also correlated the major fragment ions with the structure (see below).

The author thanks the following people for their assistance in the preparation of this book: Philip Anderson, Thomas Burgoyne, Anthony Borgerding, Louis Brzuzy, Kelly Dodson, Michael Howdeshell, Mark Krieger, Staci Massey, Voon Ong, Sandra Panshin, and Jeffrey Wallace. Veronica Hites acted as the graphics consultant, a task she performed on the first edition as well. The data are used with the permission of the copyright holder – the National Institute of Standards and Technology – and the author thanks Sharon Lias for this permission.

# THE AUTHOR

Ron Hites received a B.A. in Chemistry from Oakland University in 1964, and a Ph.D. in Analytical Chemistry with Klaus Biemann from the Massachusetts Institute of Technology (M.I.T.) in 1968. He remained on the staff and faculty of M.I.T. until 1979 when he became a Professor of Public and Environmental Affairs at Indiana University; he is also a Professor of Chemistry. In 1989, he was appointed to the special rank of Distinguished Professor. He was President of the American Society for Mass Spectrometry from 1988 to 1990. He is an Associate Editor of *Environmental Science and Technology*, and he is the winner of the 1991 American Chemical Society Award for Creative Advances in Environmental Science and Technology. His research focuses on the behavior of potentially toxic organic compounds in the environment.

# HOW TO USE THIS BOOK

The layout of the spectra is as follows: The *first line* gives the common name of the compound and the article number (if any) in the Merck Index (Eleventh Edition) where more information on the compound can usually be found. An asterisk after the name indicates that the name is a trade name. The *second line* gives the Chemical Abstract Service (CAS) Registry Number, molecular formula, and exact molecular weight of the compound. The *third line* lists the masses and intensities (in parentheses) of the four most abundant ions in the spectrum.

The spectrum is plotted as a bar graph always starting at M/z = 20. Several of the more intense peaks have been labeled automatically by the computer. In a few isotopic clusters, the first or second isotope peak has been labeled – be on guard for this. Some peaks have been labeled to indicate the source of the ion, e.g., *M-Cl* or *M-*$CH_3$; however, these assignments are "best guesses" and should be viewed accordingly. The structures have been marked to indicate the origin of selected fragment ions; again, these assignments are "best guesses". In marking the structures, hydrogen rearrangement ions are indicated by $+H = xx$ or $-H = xx$. The spectra are indexed at the end of the book by common chemical name, CAS Registry Number, exact molecular weight, and intense peaks.

*DEDICATED TO*

**BONNIE R. HITES**

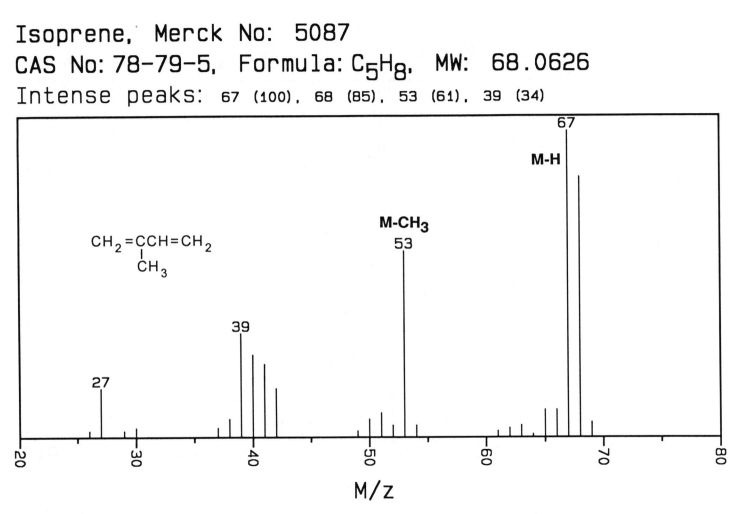

Isoprene, Merck No: 5087

CAS No: 78-79-5, Formula: $C_5H_8$, MW: 68.0626

Intense peaks: 67 (100), 68 (85), 53 (61), 39 (34)

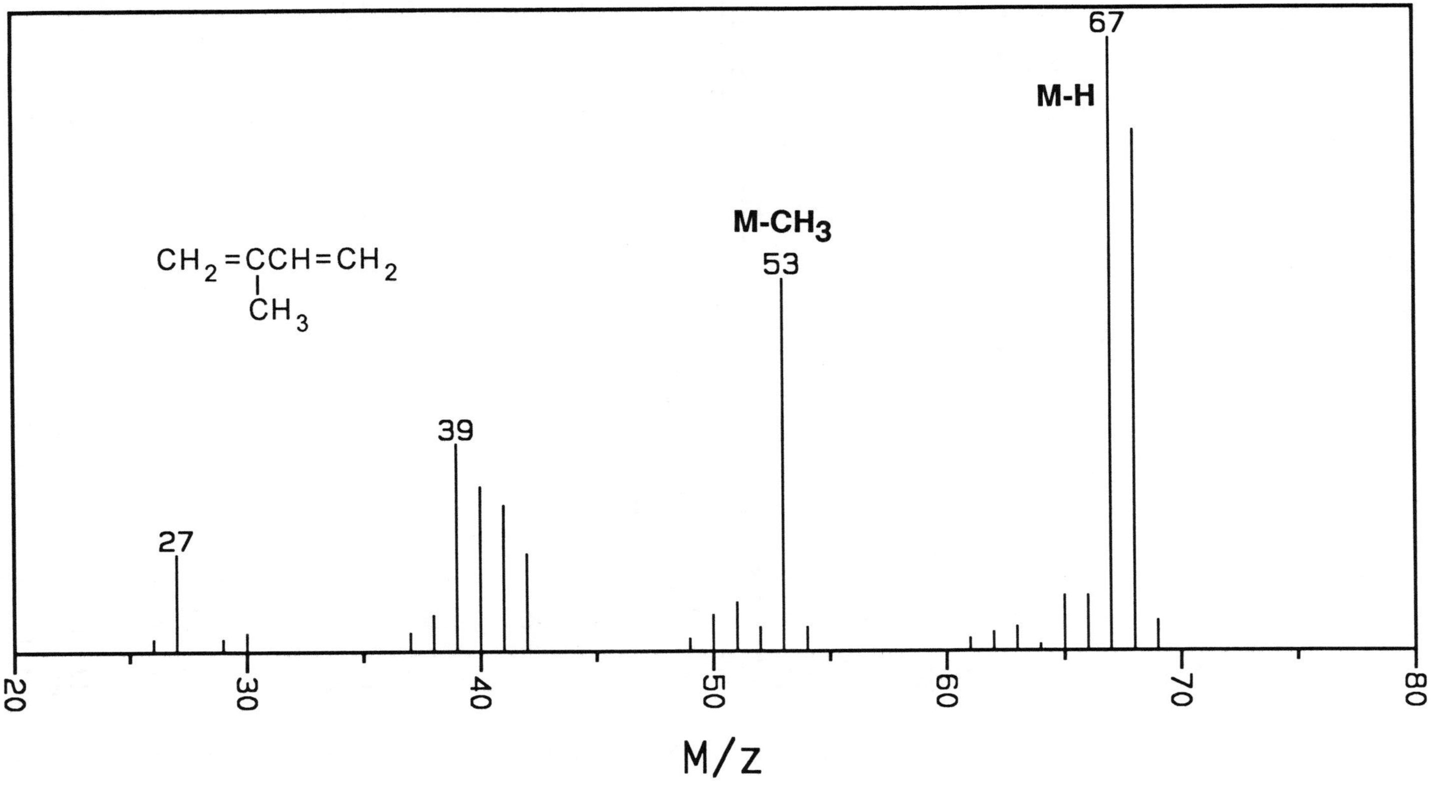

$CH_2{=}C(CH_3)CH{=}CH_2$

67

M-H

M-$CH_3$

53

39

27

20 30 40 50 60 70 80

M/Z

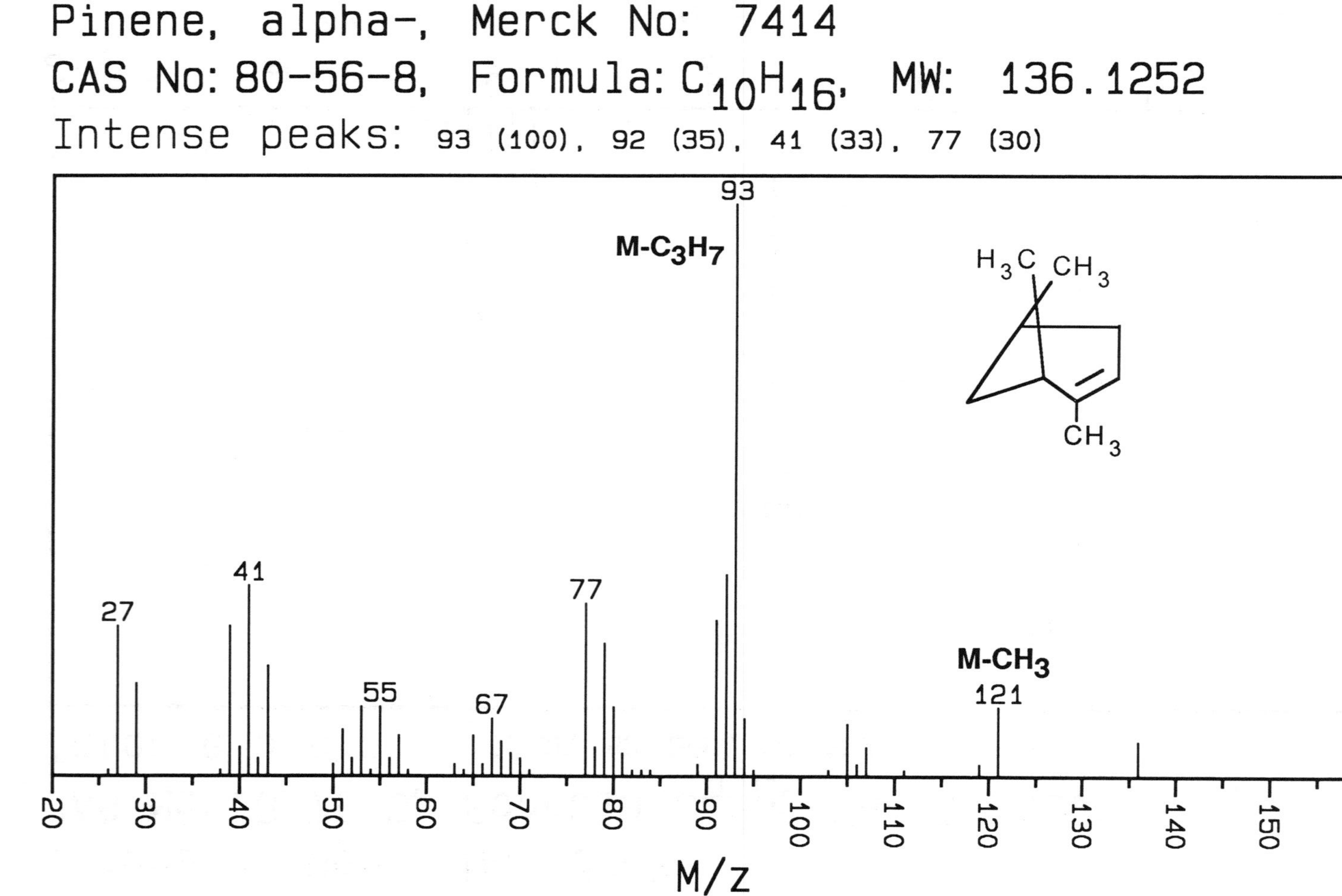
Pinene, alpha-, Merck No: 7414
CAS No: 80-56-8, Formula: C10H16, MW: 136.1252
Intense peaks: 93 (100), 92 (35), 41 (33), 77 (30)
93
M-C3H7
H3C CH3
CH3
27
41
55
67
77
M-CH3
121
20
30
40
50
60
70
80
90
100
110
120
130
140
150
160
M/Z

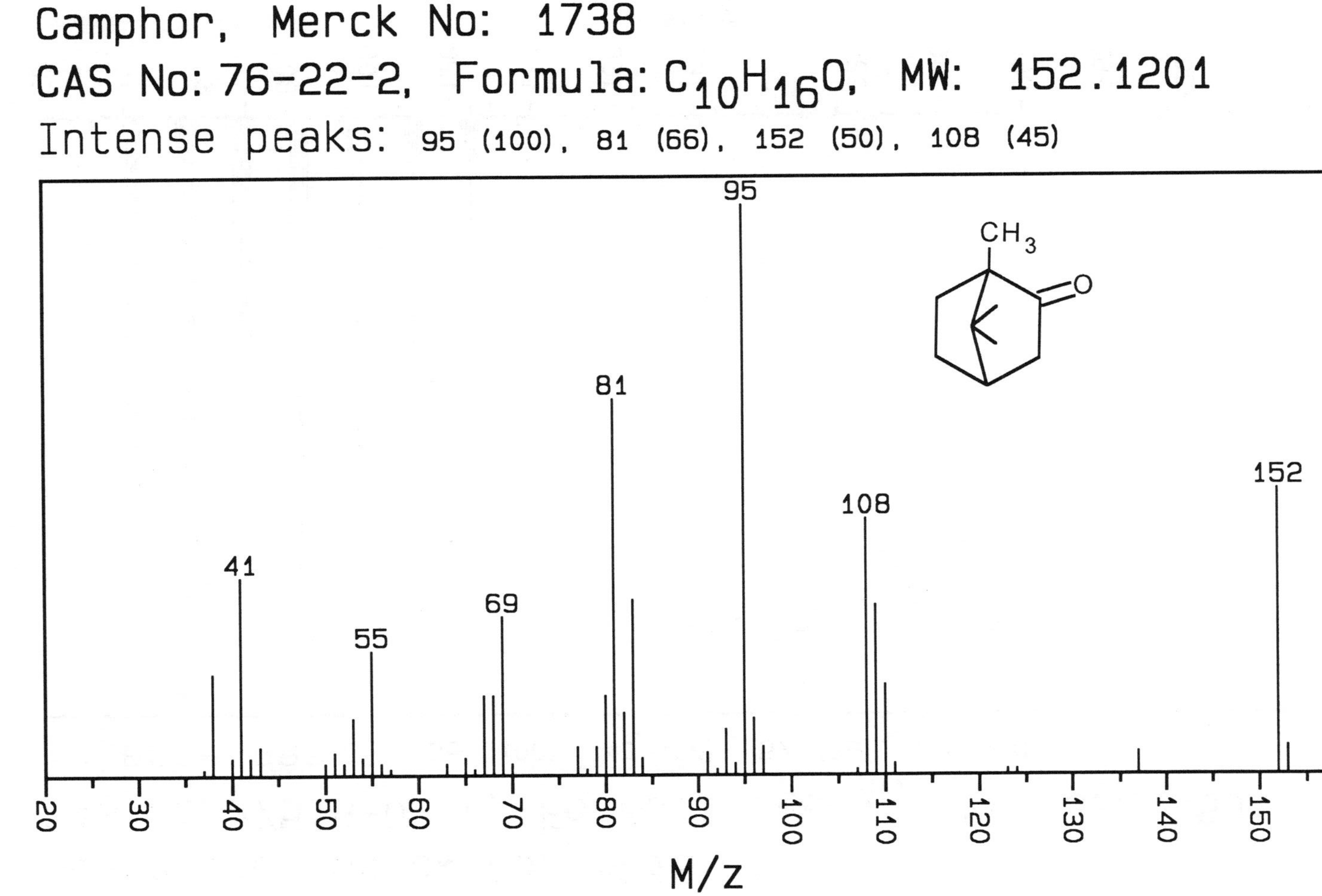
Camphor, Merck No: 1738
CAS No: 76-22-2, Formula: $C_{10}H_{16}O$, MW: 152.1201
Intense peaks: 95 (100), 81 (66), 152 (50), 108 (45)
95
81
152
108
41
69
55
$CH_3$
O
20
30
40
50
60
70
80
90
100
110
120
130
140
150
160
M/Z

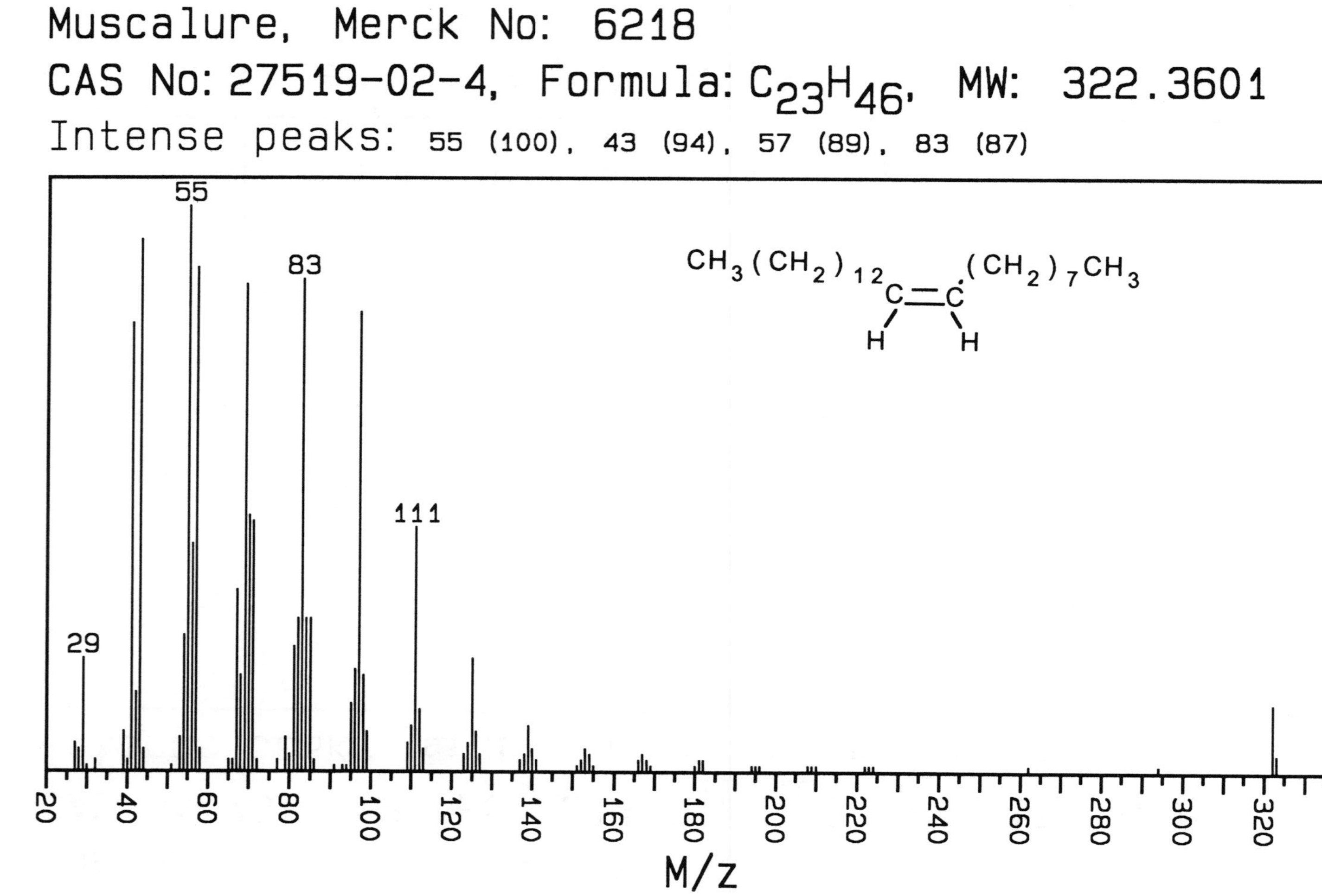
Muscalure, Merck No: 6218
CAS No: 27519-02-4, Formula: C23H46, MW: 322.3601
Intense peaks: 55 (100), 43 (94), 57 (89), 83 (87)
CH3(CH2)12C=C(CH2)7CH3
H H
55
83
111
29
20
40
60
80
100
120
140
160
180
200
220
240
260
280
300
320
340
M/Z

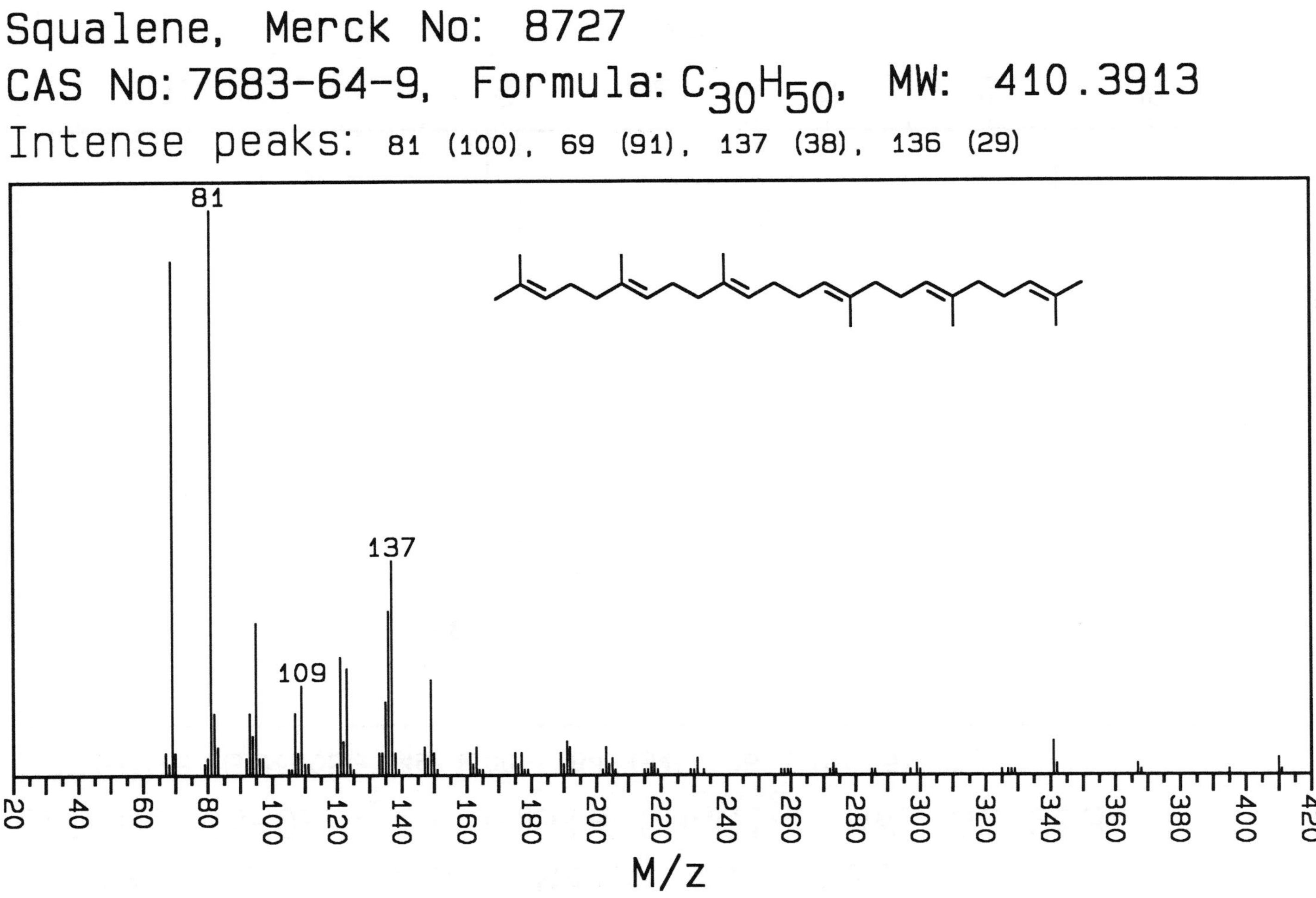

Squalene, Merck No: 8727
CAS No: 7683-64-9, Formula: C30H50, MW: 410.3913
Intense peaks: 81 (100), 69 (91), 137 (38), 136 (29)
81
109
137
20
40
60
80
100
120
140
160
180
200
220
240
260
280
300
320
340
360
380
400
420
M/Z

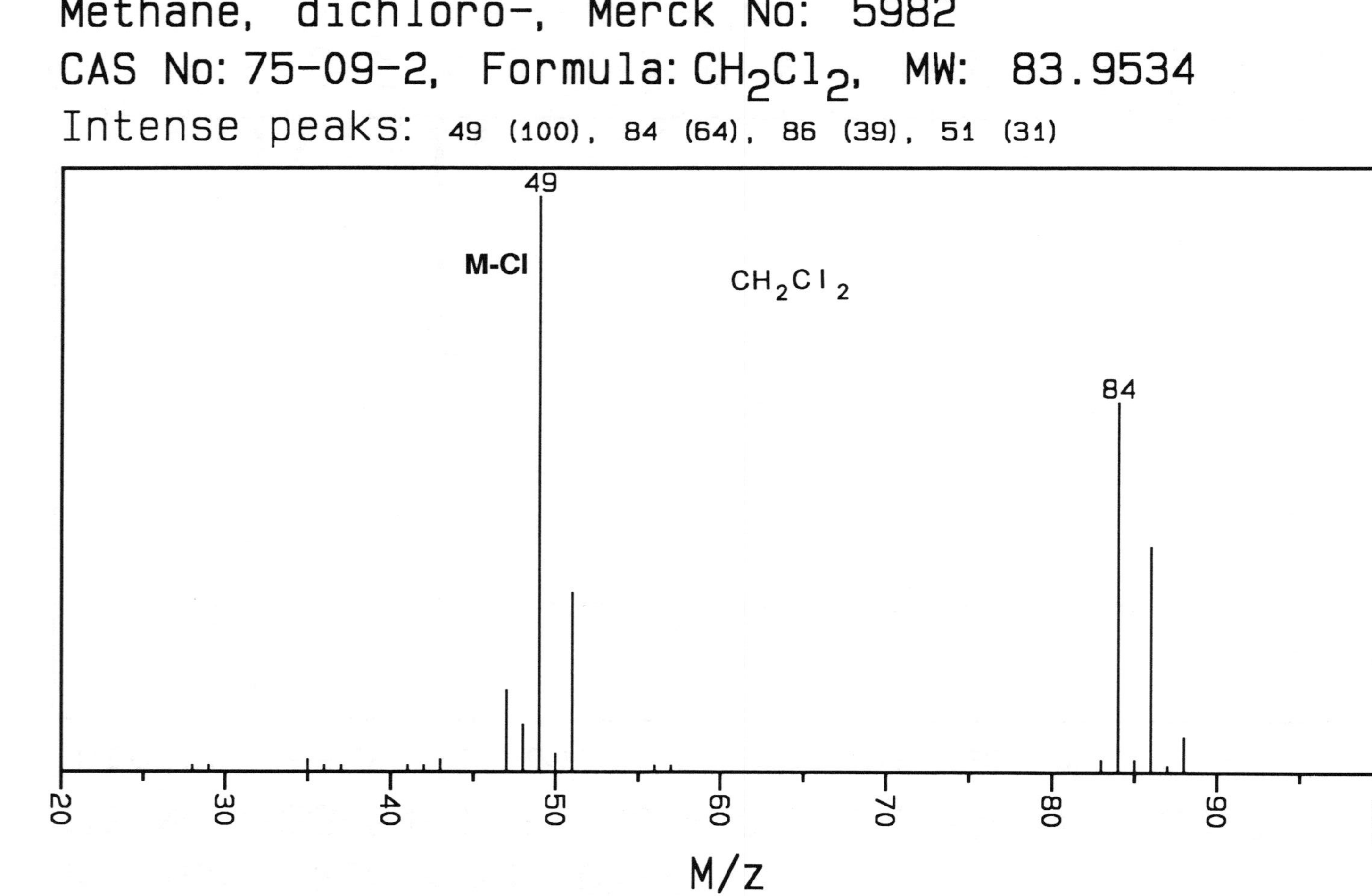
Methane, dichloro-, Merck No: 5982
CAS No: 75-09-2, Formula: CH2Cl2, MW: 83.9534
Intense peaks: 49 (100), 84 (64), 86 (39), 51 (31)
49
M-Cl
CH2Cl2
84
20
30
40
50
60
70
80
90
100
M/Z

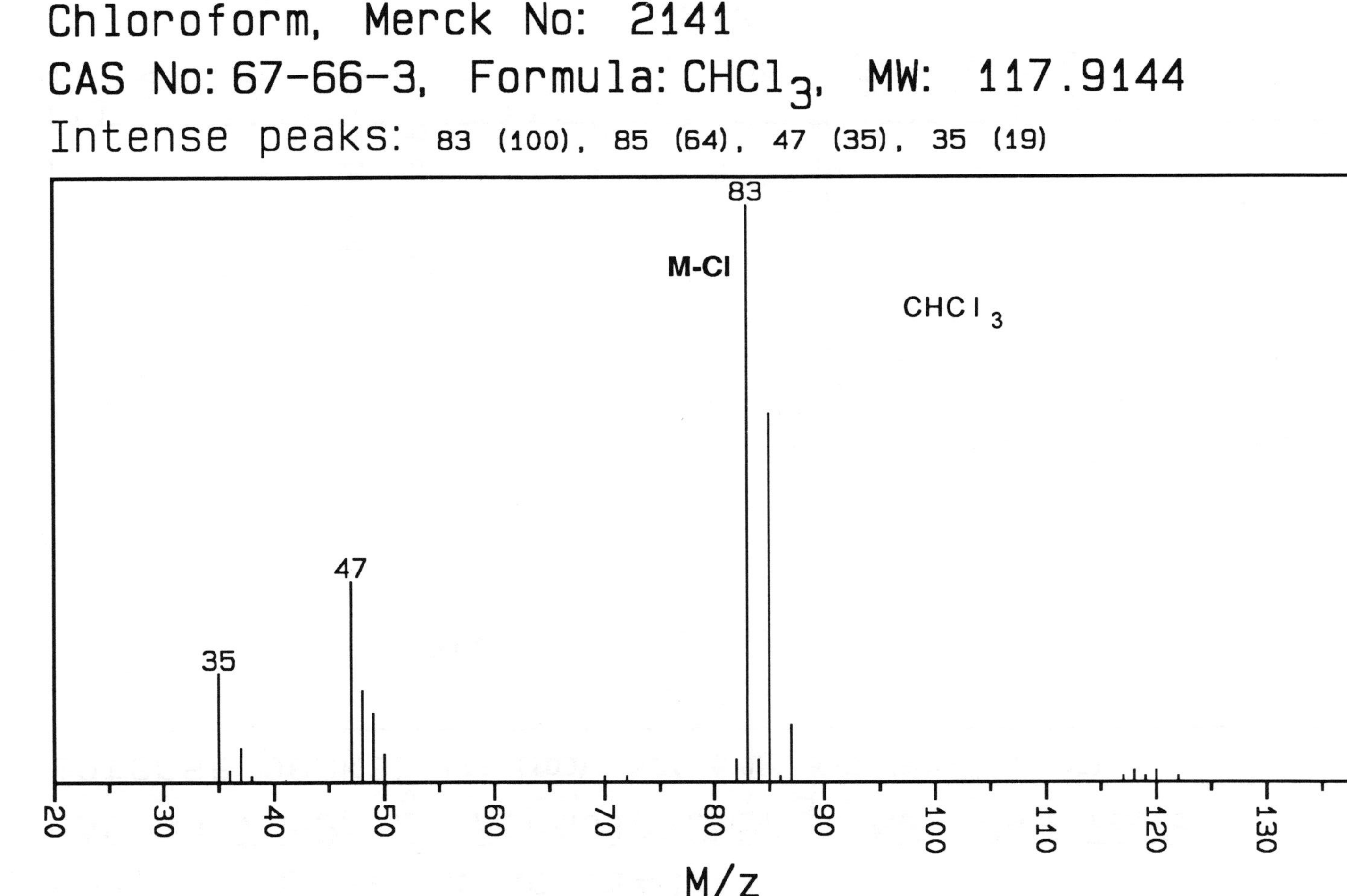
Chloroform, Merck No: 2141
CAS No: 67-66-3, Formula: CHCl3, MW: 117.9144
Intense peaks: 83 (100), 85 (64), 47 (35), 35 (19)
83
M-Cl
CHCl3
47
35
20
30
40
50
60
70
80
90
100
110
120
130
140
M/Z

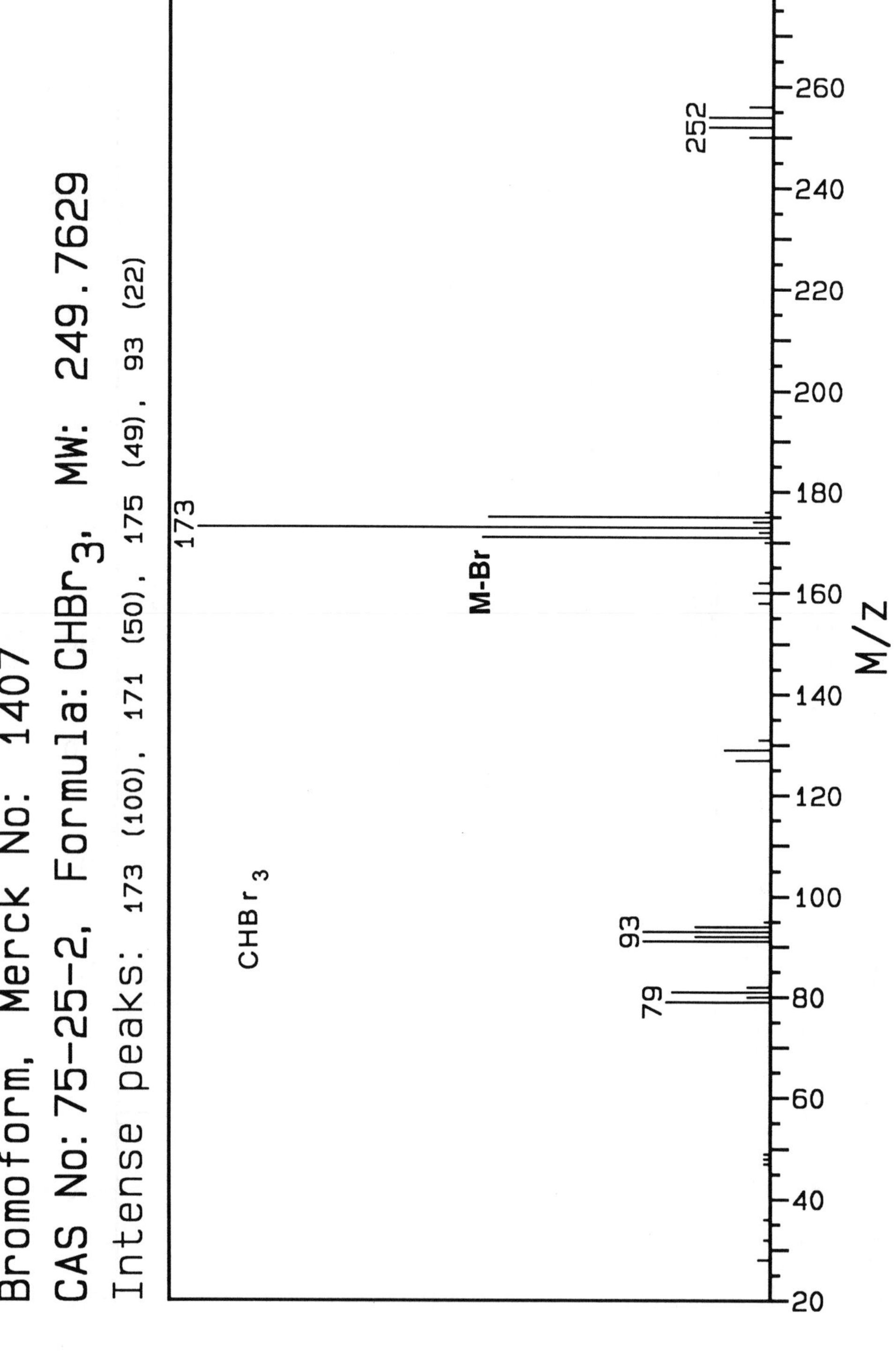
Bromoform, Merck No: 1407
CAS No: 75-25-2, Formula: $CHBr_3$, MW: 249.7629
Intense peaks: 173 (100), 171 (50), 175 (49), 93 (22)
$CHBr_3$
173
M-Br
252
93
79
M/Z
20
40
60
80
100
120
140
160
180
200
220
240
260
280

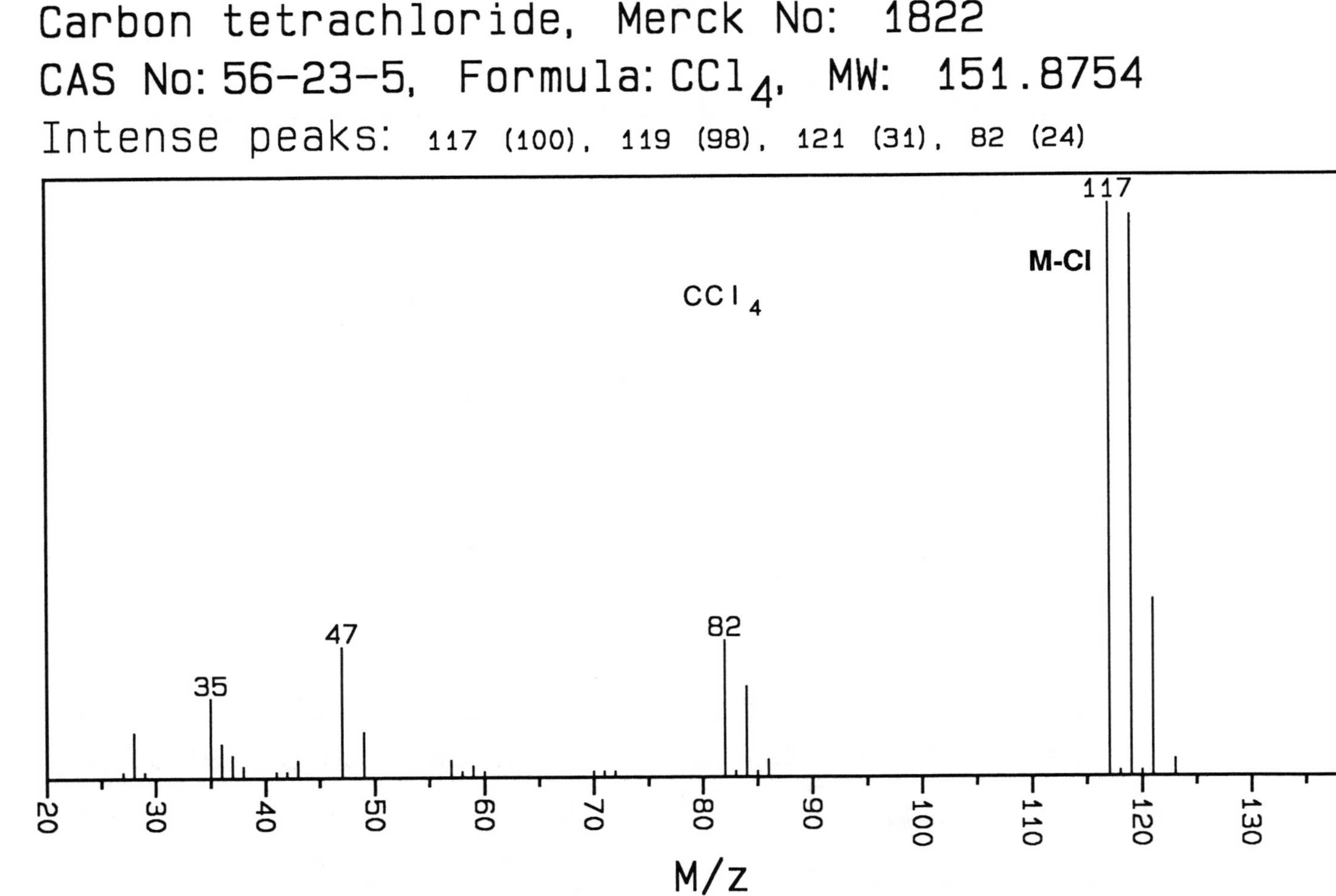
Carbon tetrachloride, Merck No: 1822
CAS No: 56-23-5, Formula: CCl4, MW: 151.8754
Intense peaks: 117 (100), 119 (98), 121 (31), 82 (24)
117
M-Cl
CCl4
82
47
35
20
30
40
50
60
70
80
90
100
110
120
130
140
M/Z

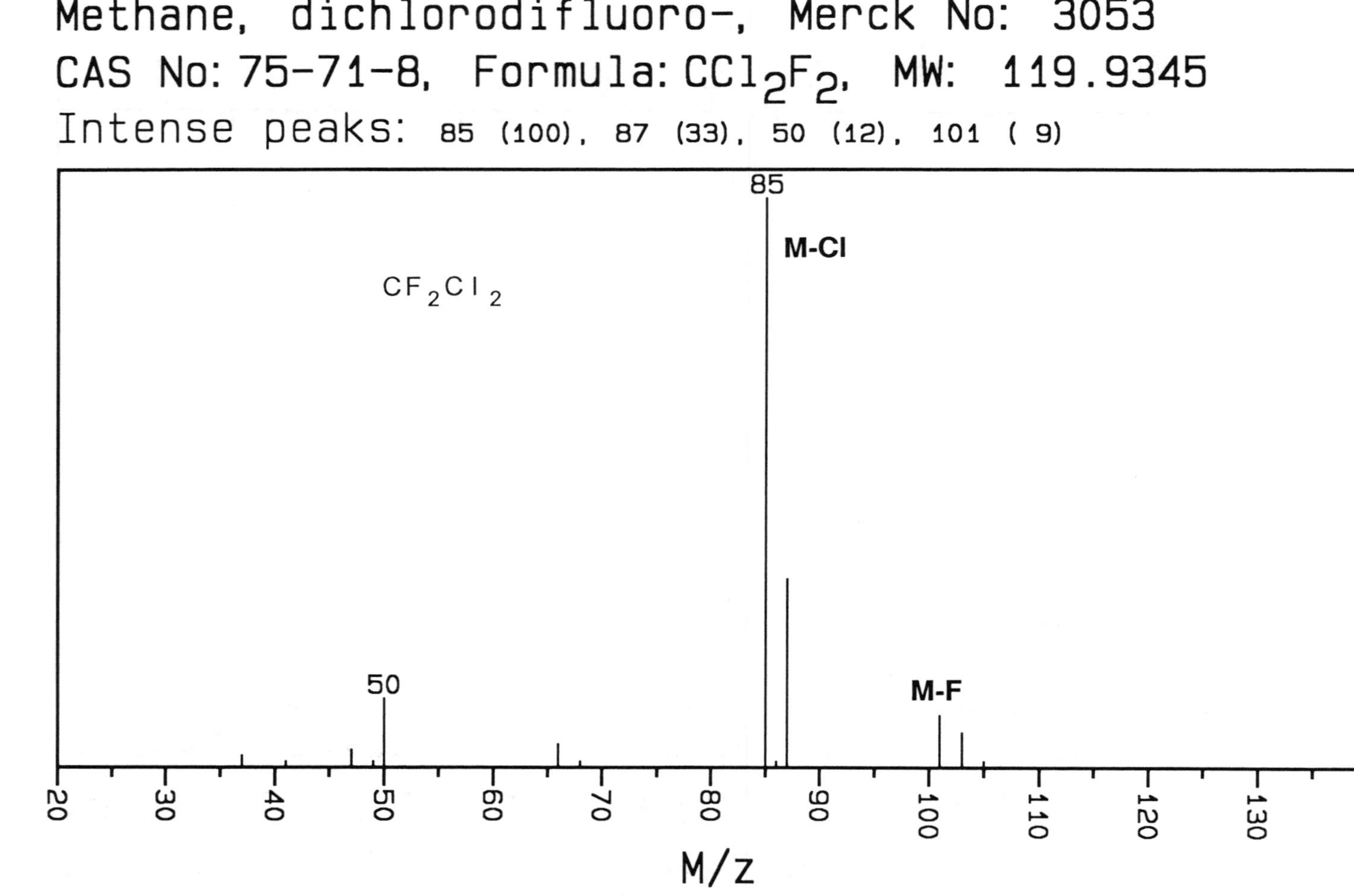
Methane, dichlorodifluoro-, Merck No: 3053
CAS No: 75-71-8, Formula: $CCl_2F_2$, MW: 119.9345
Intense peaks: 85 (100), 87 (33), 50 (12), 101 ( 9)
85
M-Cl
$CF_2Cl_2$
50
M-F
20
30
40
50
60
70
80
90
100
110
120
130
140
M/Z

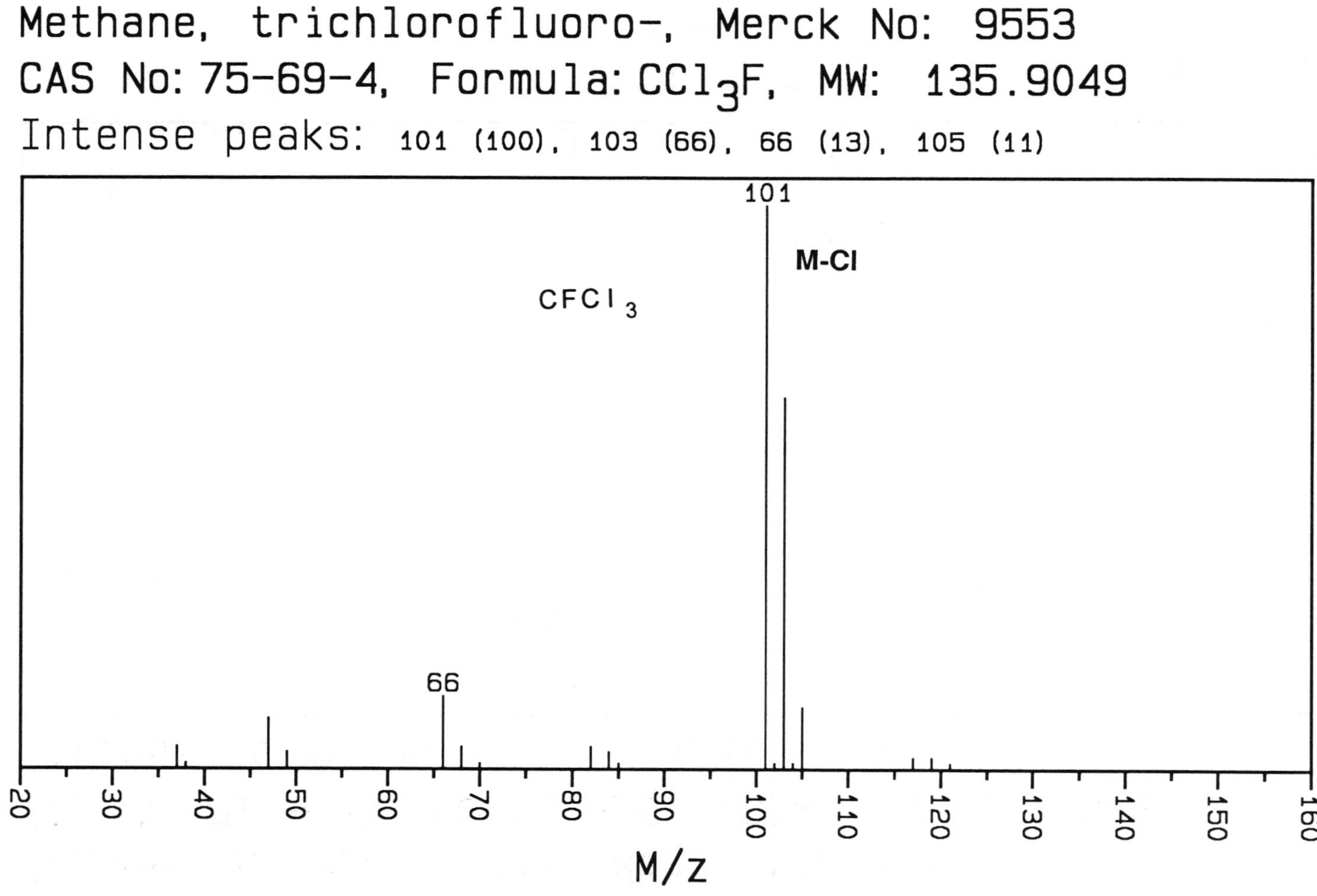
Methane, trichlorofluoro-, Merck No: 9553
CAS No: 75-69-4, Formula: CCl3F, MW: 135.9049
Intense peaks: 101 (100), 103 (66), 66 (13), 105 (11)
101
M-Cl
CFCl3
66
20
30
40
50
60
70
80
90
100
110
120
130
140
150
160
M/Z

Ethane, 1,1-dichloro-, Merck No: 3766

CAS No: 75-34-3, Formula: $C_2H_4Cl_2$, MW: 97.9689

Intense peaks: 63 (100), 27 (71), 65 (31), 26 (19)

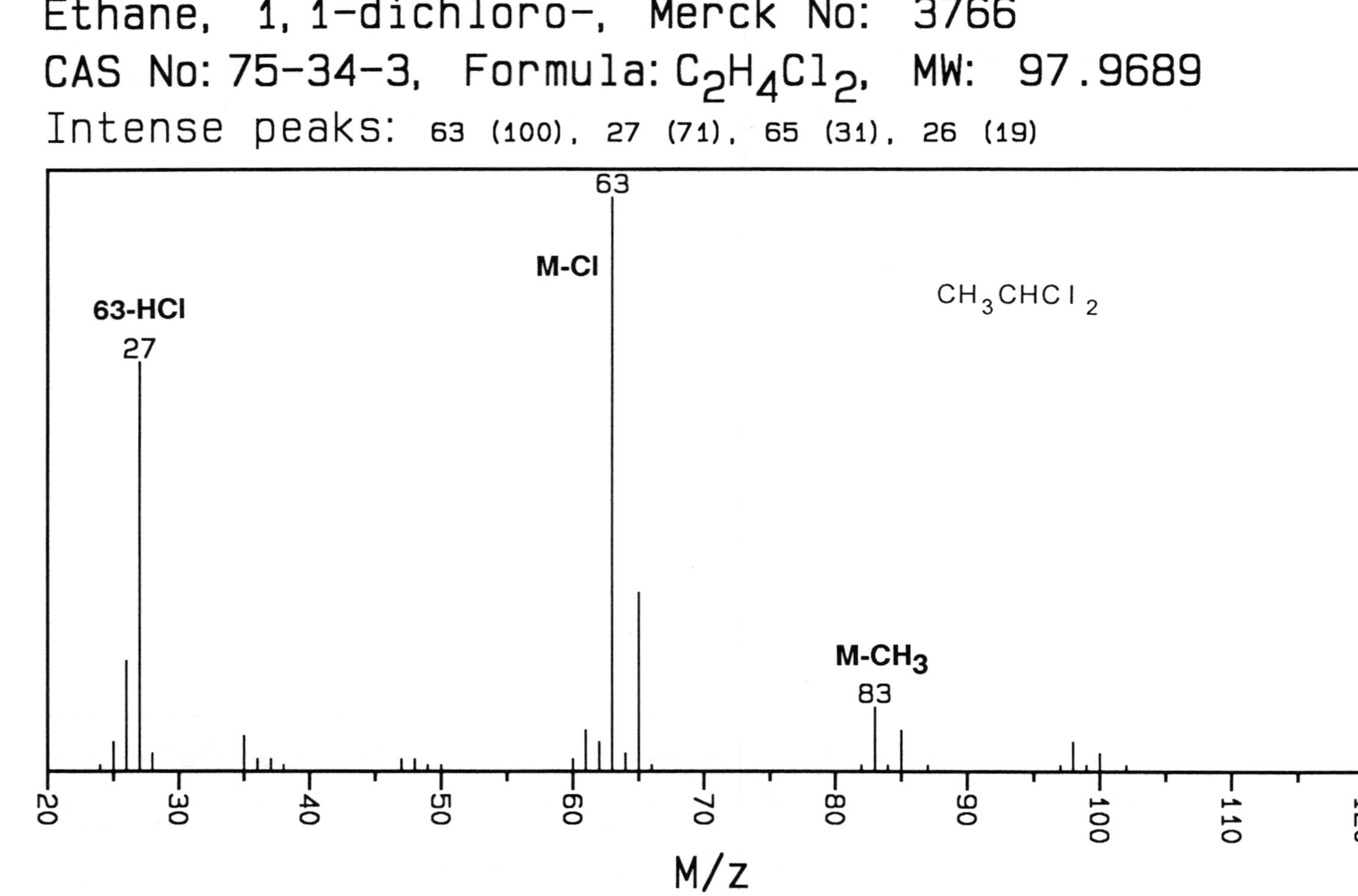

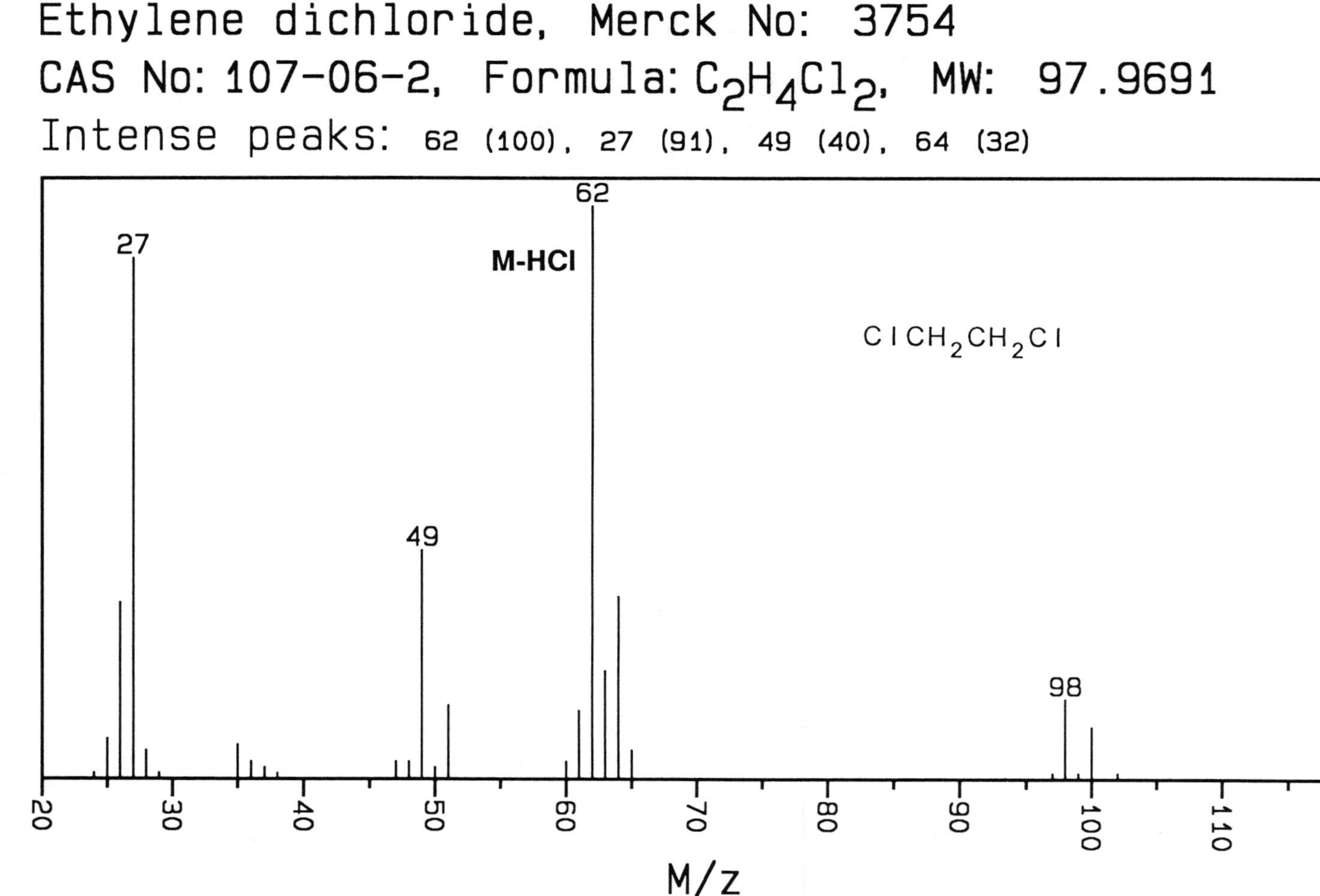

Ethylene dichloride, Merck No: 3754
CAS No: 107-06-2, Formula: $C_2H_4Cl_2$, MW: 97.9691
Intense peaks: 62 (100), 27 (91), 49 (40), 64 (32)
27
M-HCl
62
49
98
$ClCH_2CH_2Cl$
20
30
40
50
60
70
80
90
100
110
120
M/Z

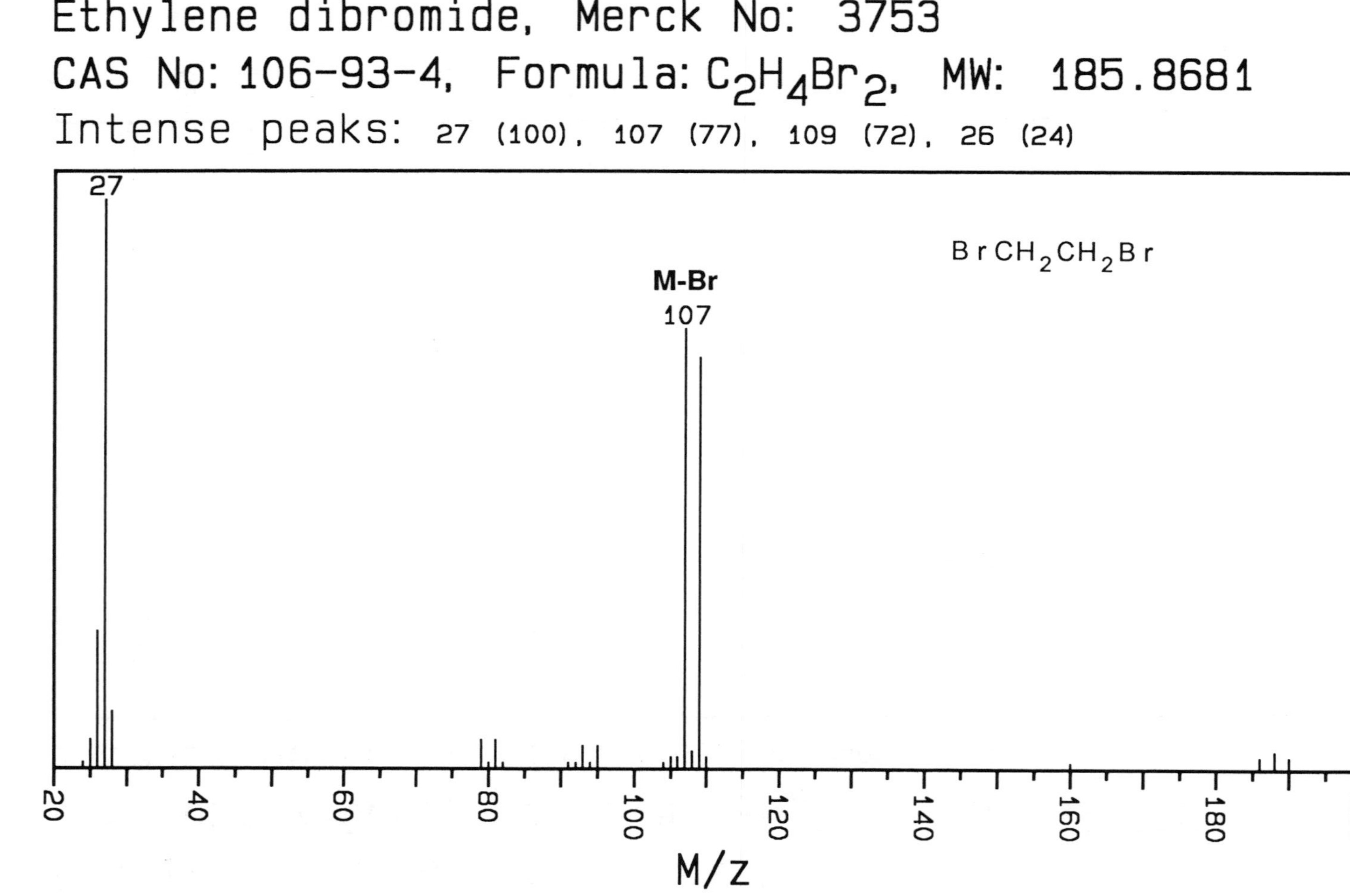

Ethylene dibromide, Merck No: 3753
CAS No: 106-93-4, Formula: $C_2H_4Br_2$, MW: 185.8681
Intense peaks: 27 (100), 107 (77), 109 (72), 26 (24)
27
M-Br
107
$BrCH_2CH_2Br$
20
40
60
80
100
120
140
160
180
200
M/Z

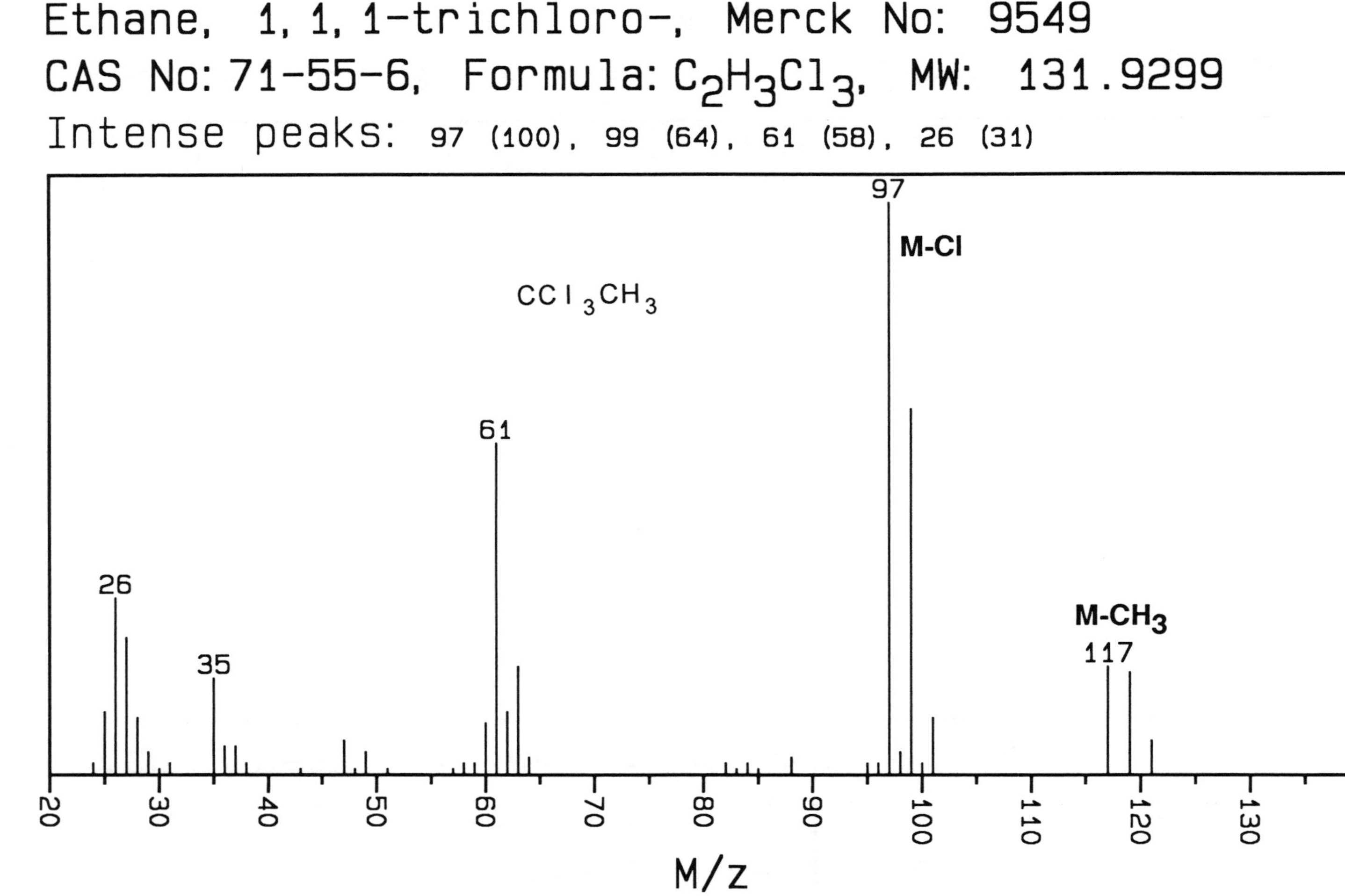
Ethane, 1, 1, 1-trichloro-, Merck No: 9549
CAS No: 71-55-6, Formula: C2H3Cl3, MW: 131.9299
Intense peaks: 97 (100), 99 (64), 61 (58), 26 (31)
CCl3CH3
97
M-Cl
61
26
35
M-CH3
117
20
30
40
50
60
70
80
90
100
110
120
130
140
M/Z

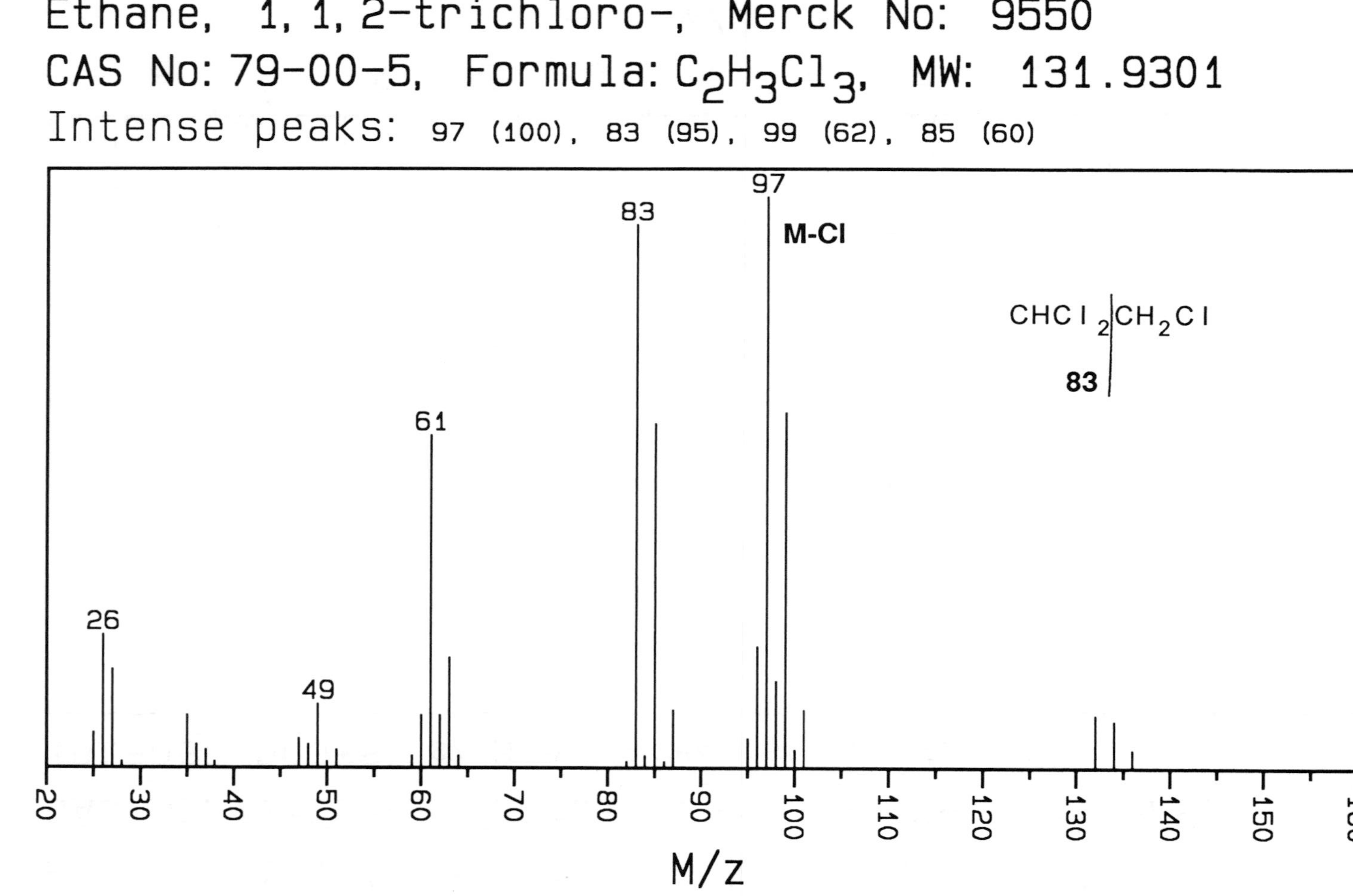

Ethane, 1,1,2-trichloro-, Merck No: 9550
CAS No: 79-00-5, Formula: $C_2H_3Cl_3$, MW: 131.9301
Intense peaks: 97 (100), 83 (95), 99 (62), 85 (60)
97
83
M-Cl
CHCl$_2$|CH$_2$Cl
83
61
26
49
20
30
40
50
60
70
80
90
100
110
120
130
140
150
160
M/Z

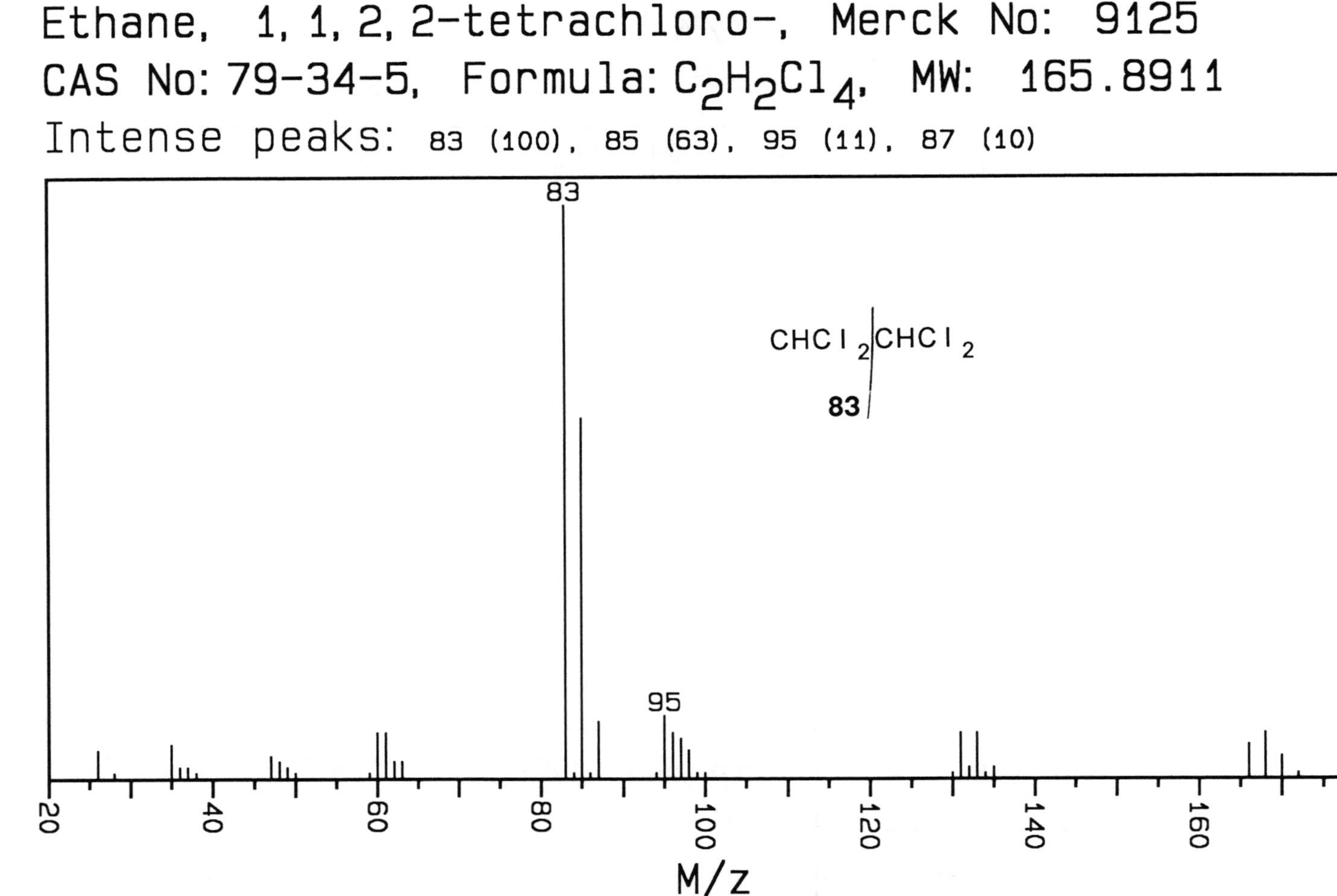

Ethane, 1,1,2,2-tetrachloro-, Merck No: 9125
CAS No: 79-34-5, Formula: $C_2H_2Cl_4$, MW: 165.8911
Intense peaks: 83 (100), 85 (63), 95 (11), 87 (10)
83
95
$CHCl_2$|$CHCl_2$
83
20
40
60
80
100
120
140
160
180
M/Z

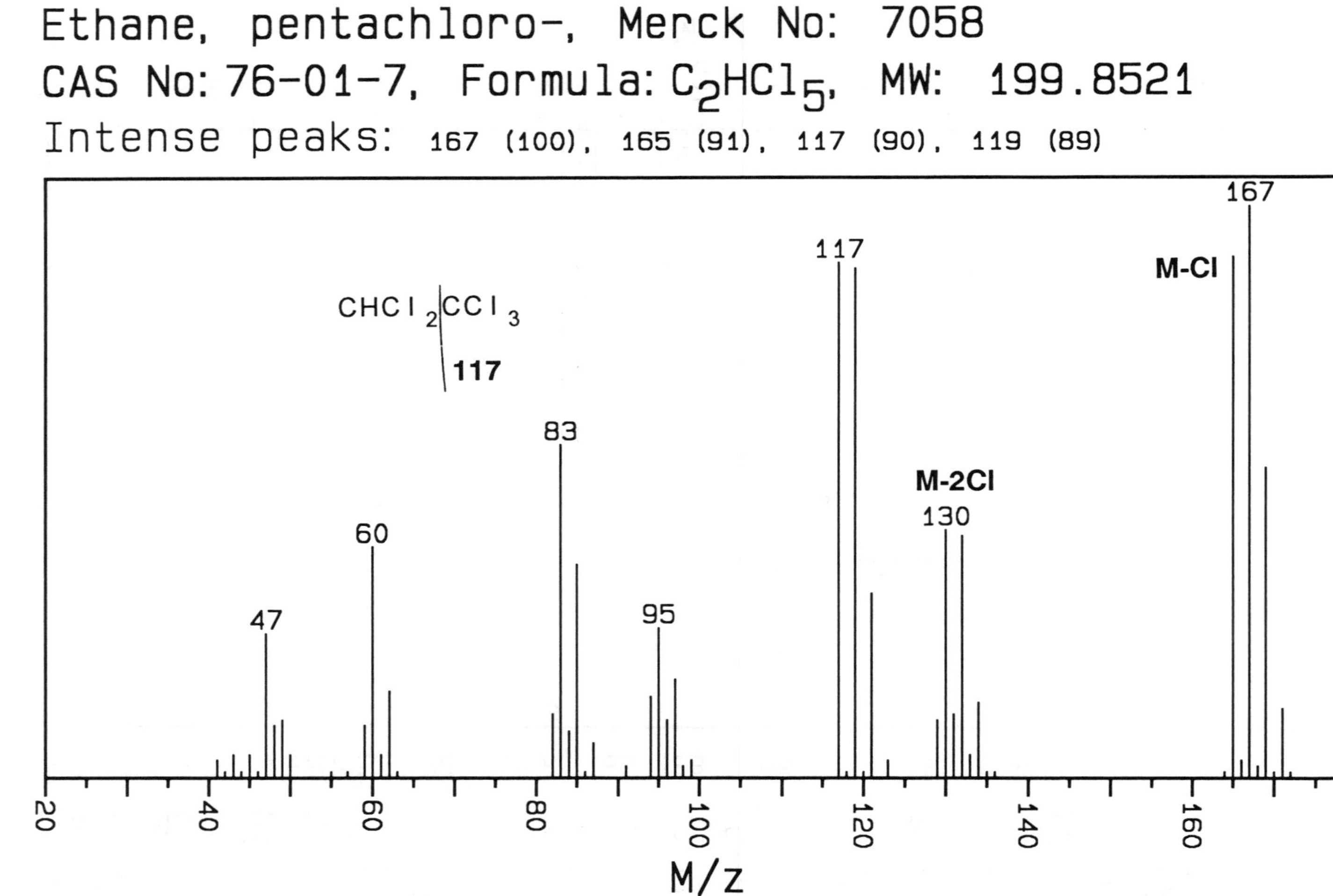
Ethane, pentachloro-, Merck No: 7058
CAS No: 76-01-7, Formula: $C_2HCl_5$, MW: 199.8521
Intense peaks: 167 (100), 165 (91), 117 (90), 119 (89)
$CHCl_2|CCl_3$
117
47
60
83
95
117
M-2Cl
130
M-Cl
167
20
40
60
80
100
120
140
160
180
M/Z

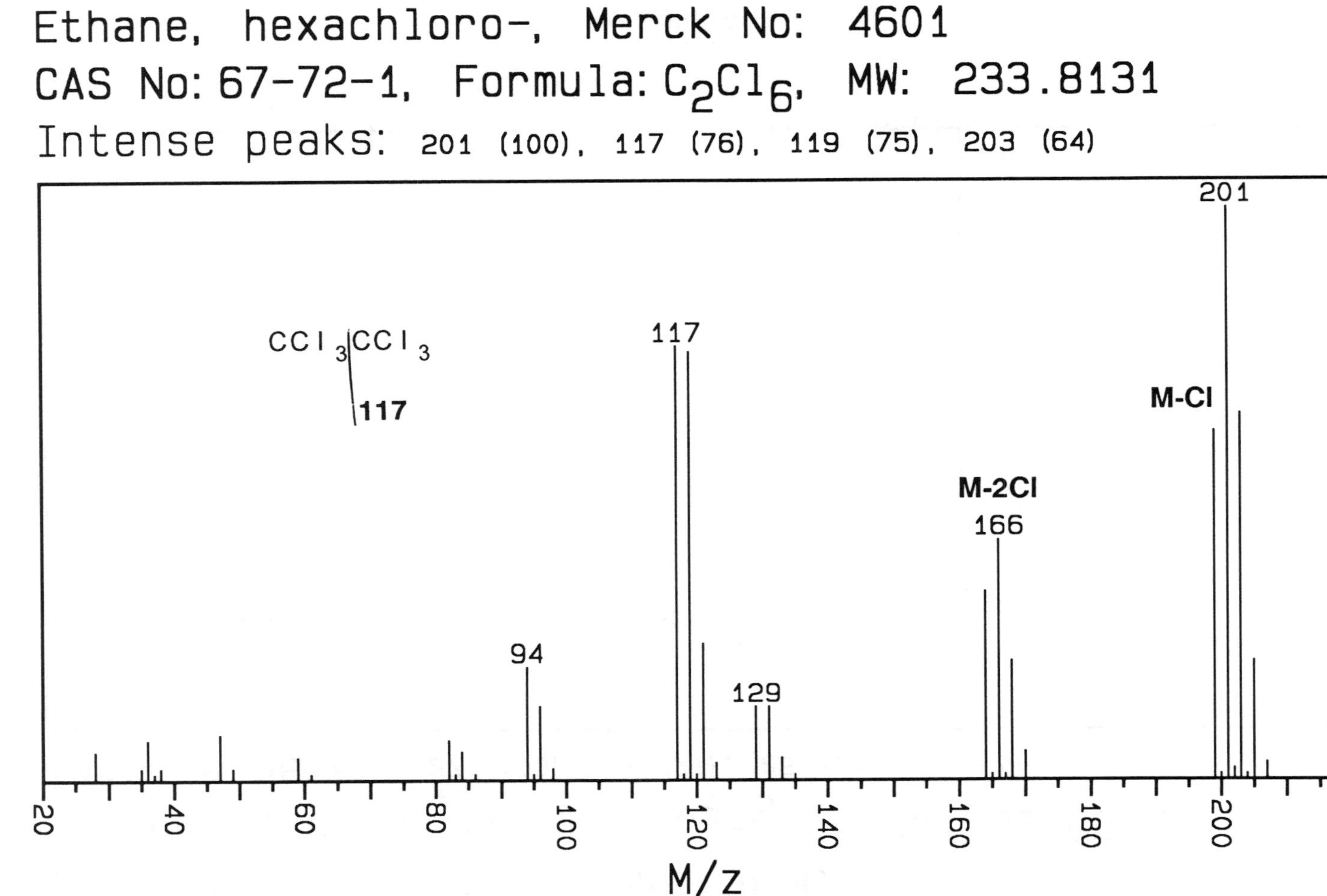
Ethane, hexachloro-, Merck No: 4601
CAS No: 67-72-1, Formula: $C_2Cl_6$, MW: 233.8131
Intense peaks: 201 (100), 117 (76), 119 (75), 203 (64)
201
M-Cl
M-2Cl
166
129
117
94
$CCl_3$|$CCl_3$
117
20
40
60
80
100
120
140
160
180
200
220
M/Z

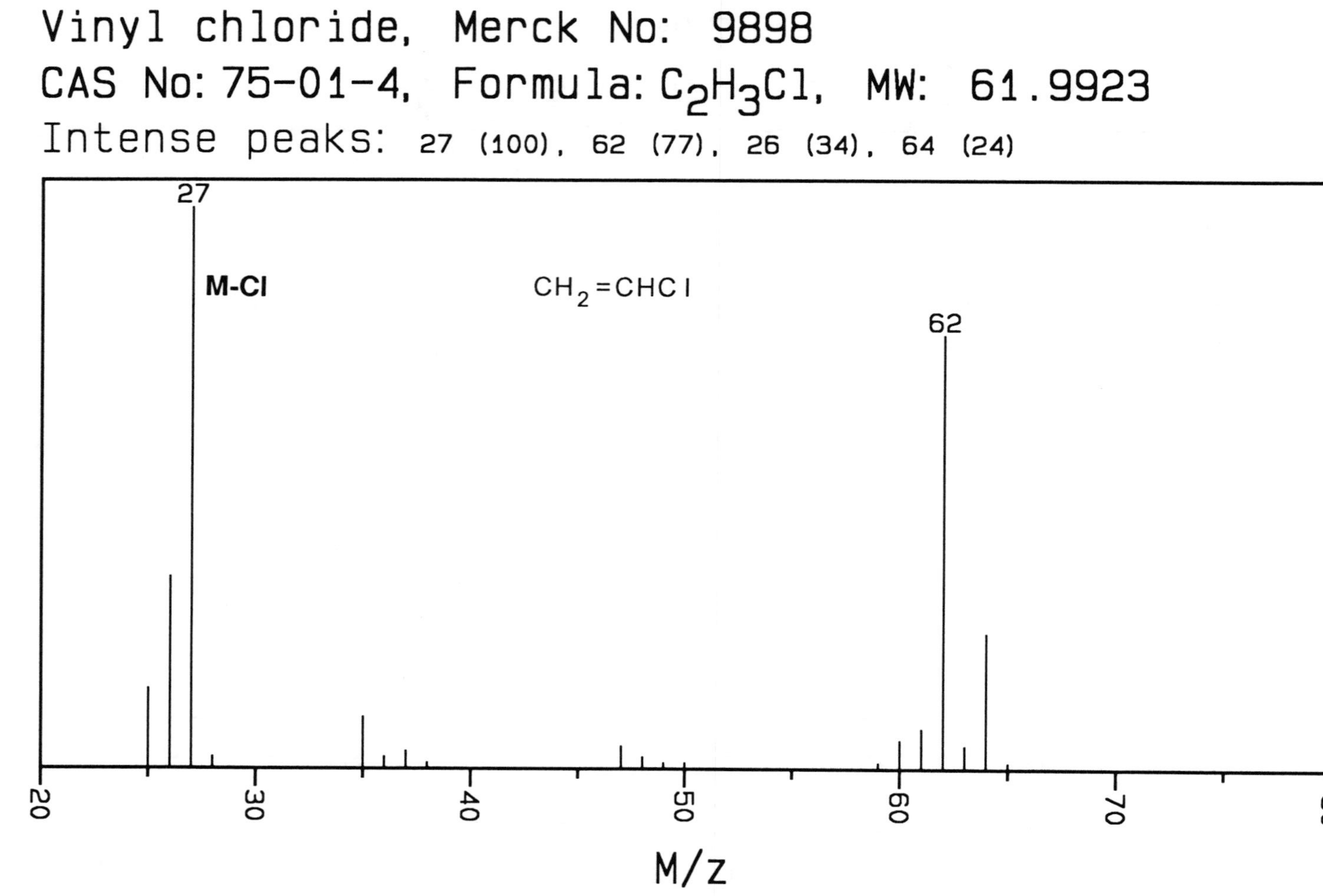
Vinyl chloride, Merck No: 9898
CAS No: 75-01-4, Formula: $C_2H_3Cl$, MW: 61.9923
Intense peaks: 27 (100), 62 (77), 26 (34), 64 (24)
27
M-Cl
$CH_2$=CHCl
62
20
30
40
50
60
70
80
M/z

Ethene, 1,1-dichloro-, Merck No: 9900

CAS No: 75-35-4, Formula: $C_2H_2Cl_2$, MW: 95.9534

Intense peaks: 61 (100), 96 (61), 98 (38), 63 (32)

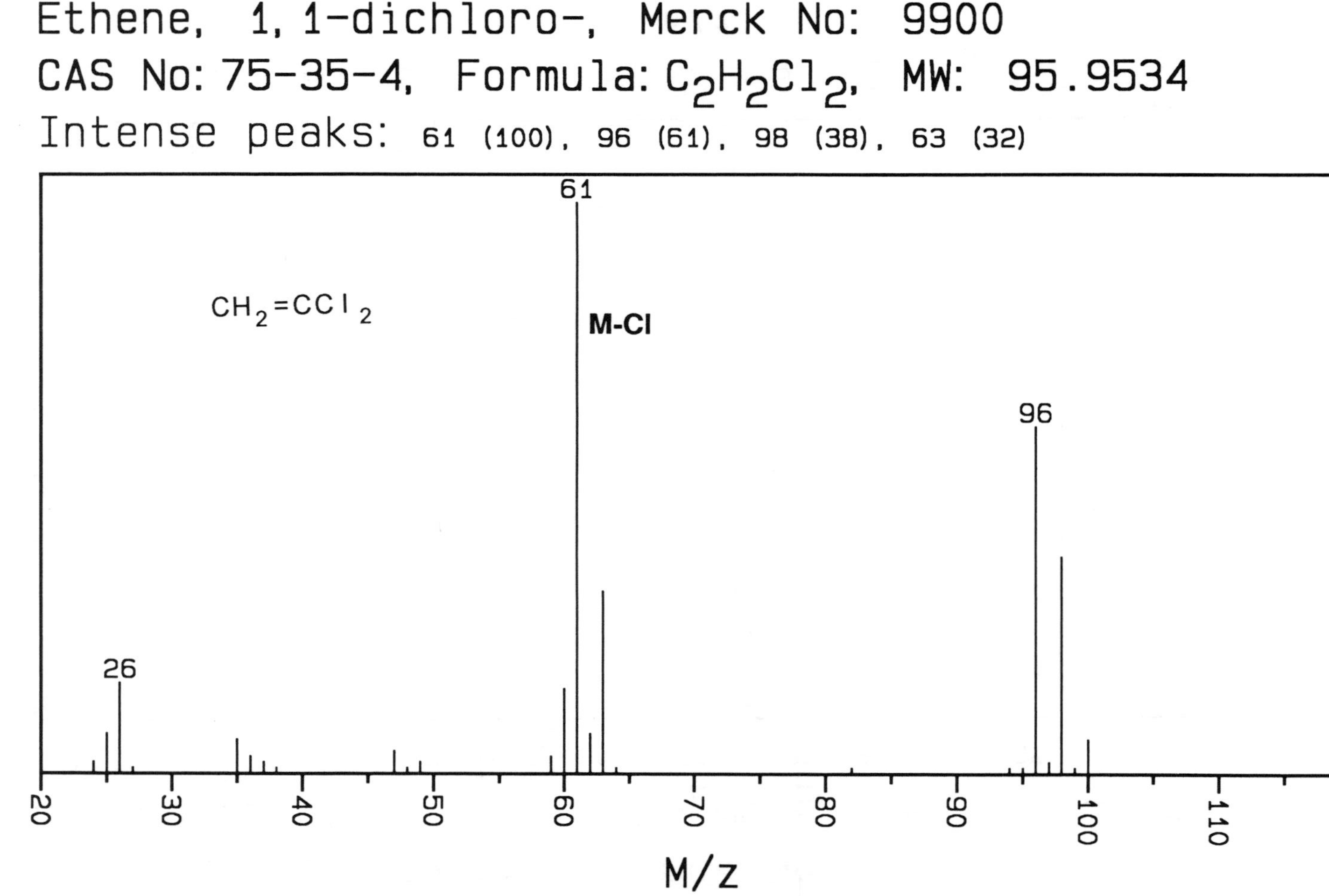

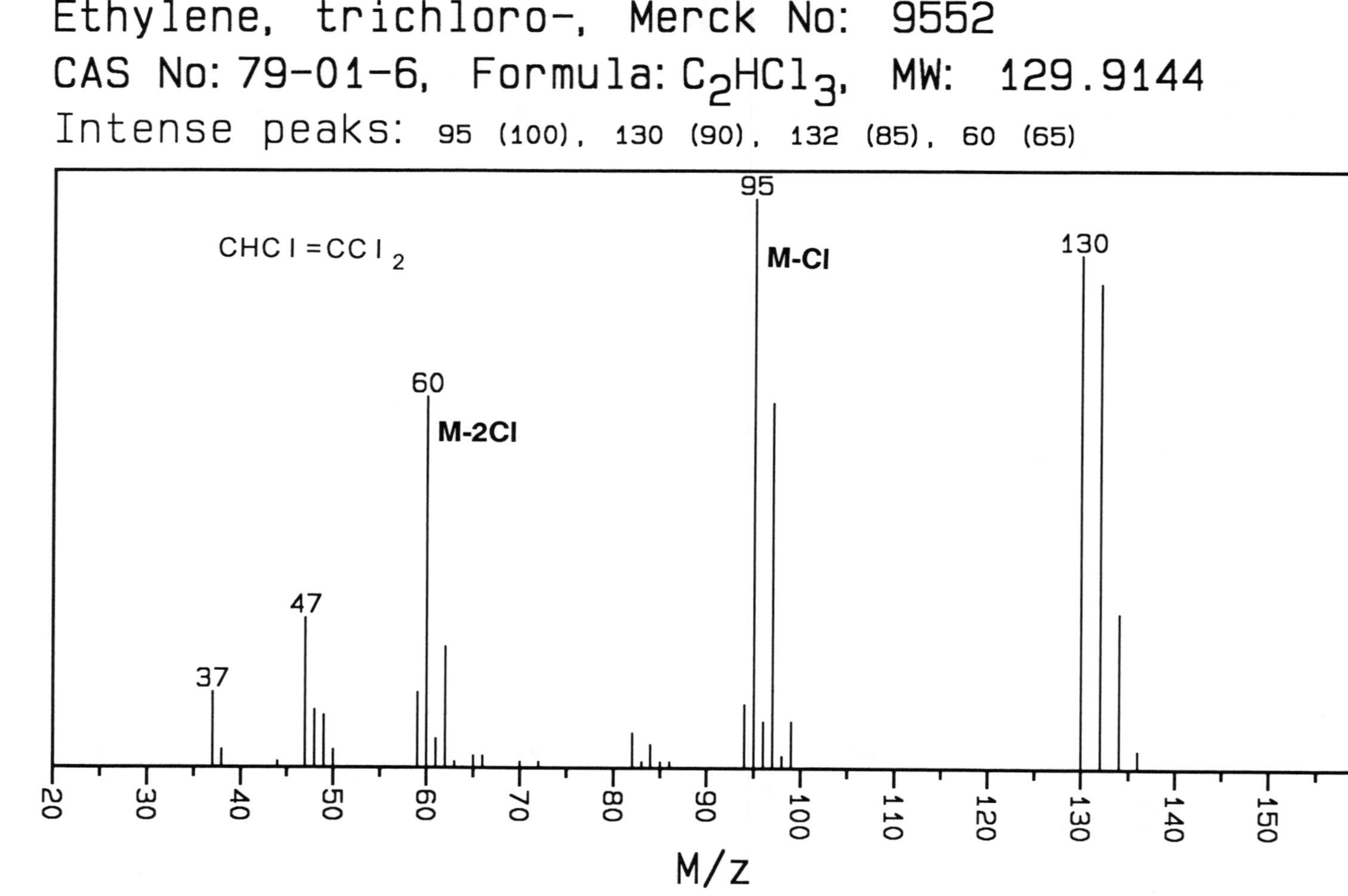

Ethylene, trichloro-, Merck No: 9552
CAS No: 79-01-6, Formula: $C_2HCl_3$, MW: 129.9144
Intense peaks: 95 (100), 130 (90), 132 (85), 60 (65)
CHCl=CCl2
95
M-Cl
130
60
M-2Cl
47
37
20
30
40
50
60
70
80
90
100
110
120
130
140
150
160
M/Z

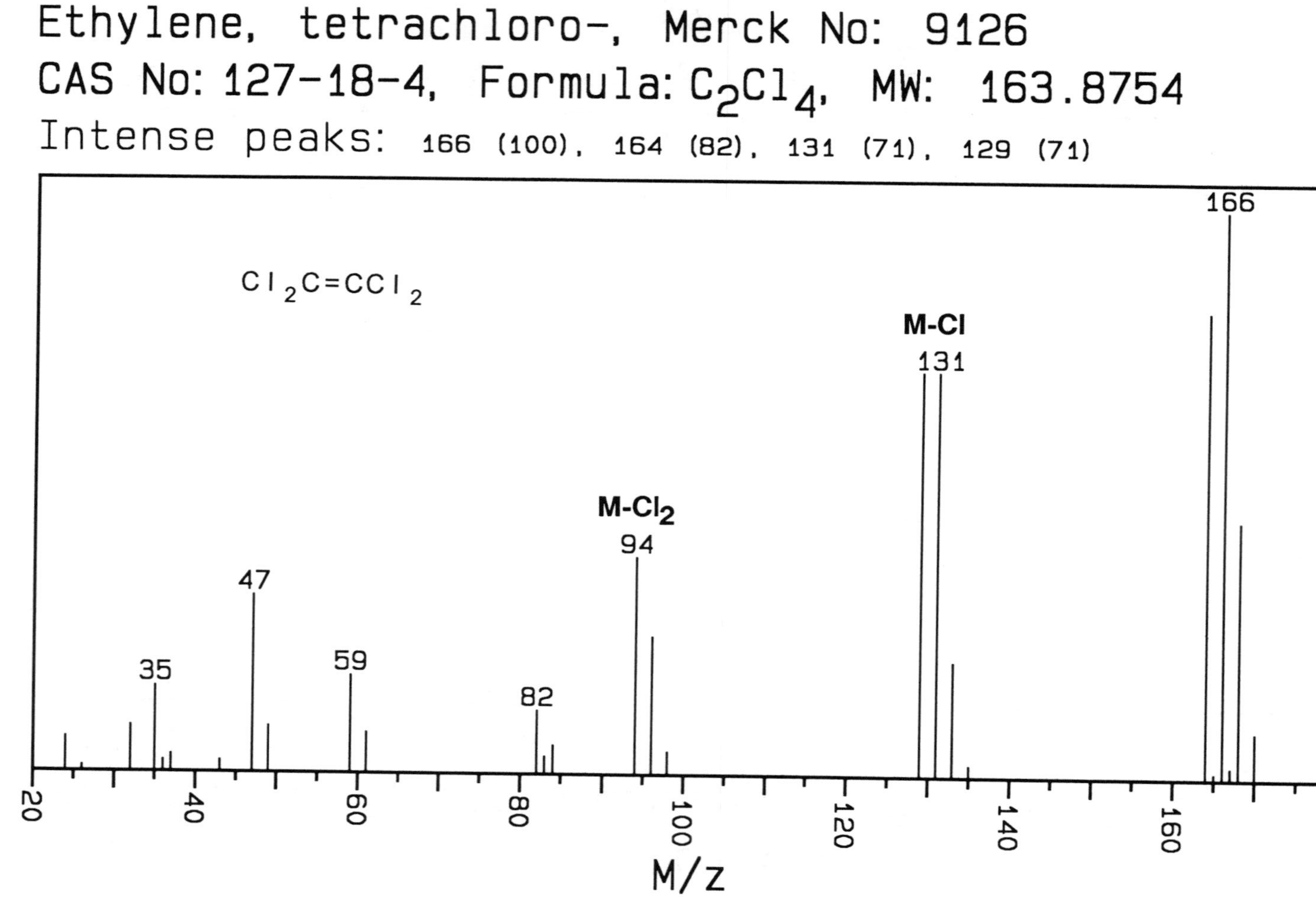

Ethylene, tetrachloro-, Merck No: 9126
CAS No: 127-18-4, Formula: $C_2Cl_4$, MW: 163.8754
Intense peaks: 166 (100), 164 (82), 131 (71), 129 (71)
$Cl_2C=CCl_2$
166
M-Cl
131
M-$Cl_2$
94
82
59
47
35
20
40
60
80
100
120
140
160
180
M/Z

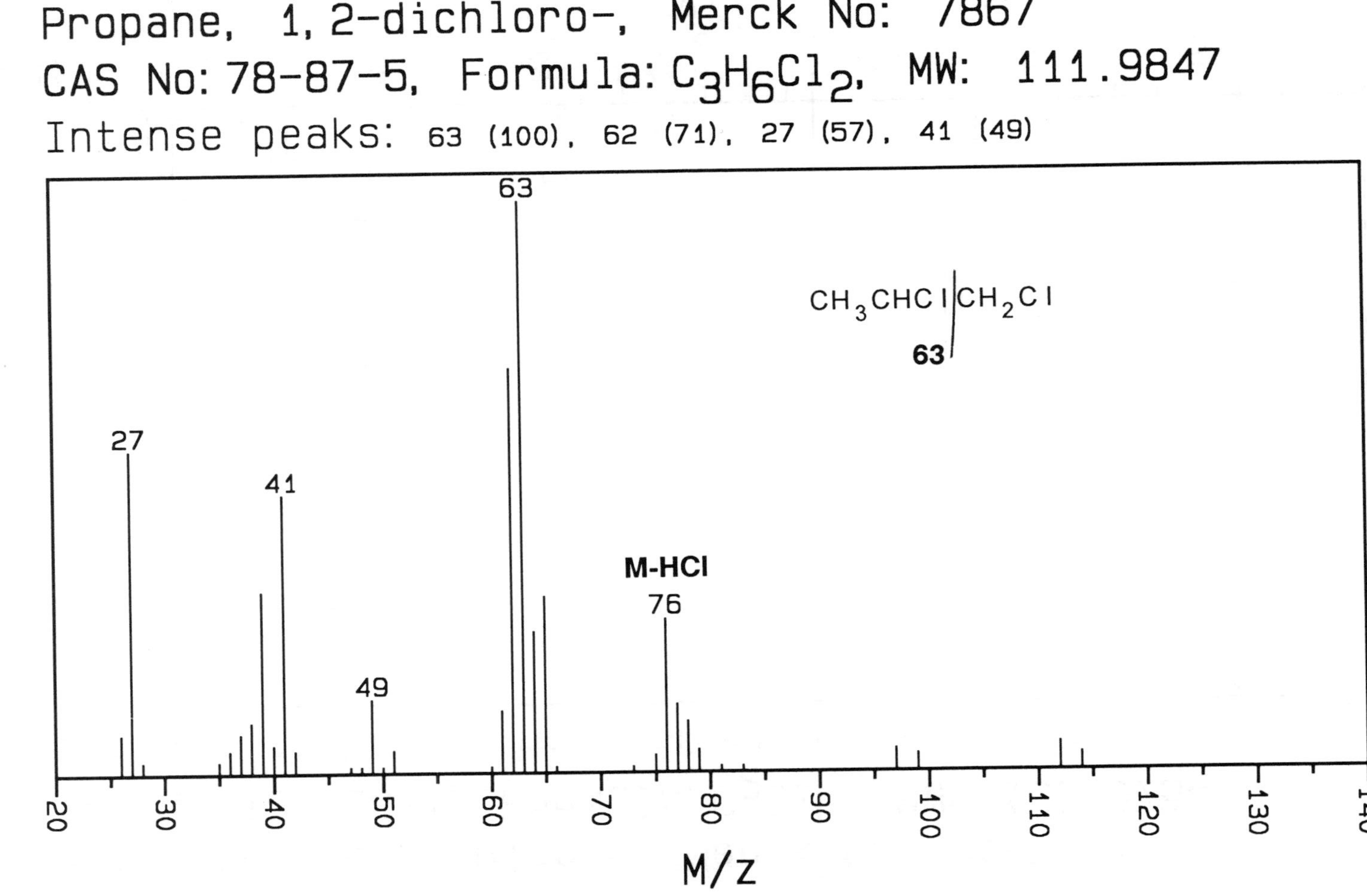
Propane, 1,2-dichloro-, Merck No: 7867
CAS No: 78-87-5, Formula: $C_3H_6Cl_2$, MW: 111.9847
Intense peaks: 63 (100), 62 (71), 27 (57), 41 (49)
63
$CH_3CHCl|CH_2Cl$
63
27
41
M-HCl
76
49
20
30
40
50
60
70
80
90
100
110
120
130
140
M/Z

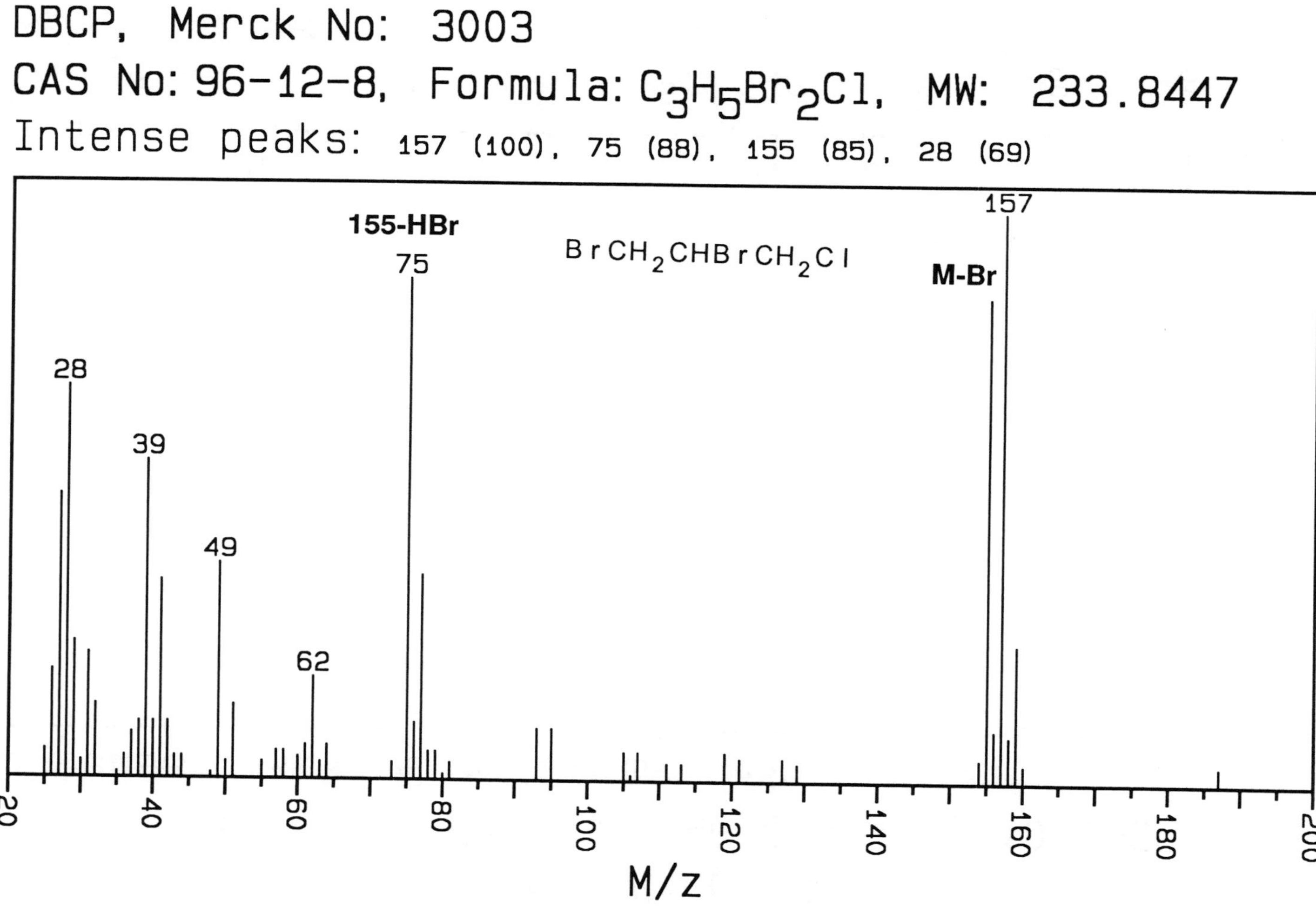
DBCP, Merck No: 3003
CAS No: 96-12-8, Formula: C3H5Br2Cl, MW: 233.8447
Intense peaks: 157 (100), 75 (88), 155 (85), 28 (69)
155-HBr
75
BrCH2CHBrCH2Cl
M-Br
157
28
39
49
62
20
40
60
80
100
120
140
160
180
200
M/Z

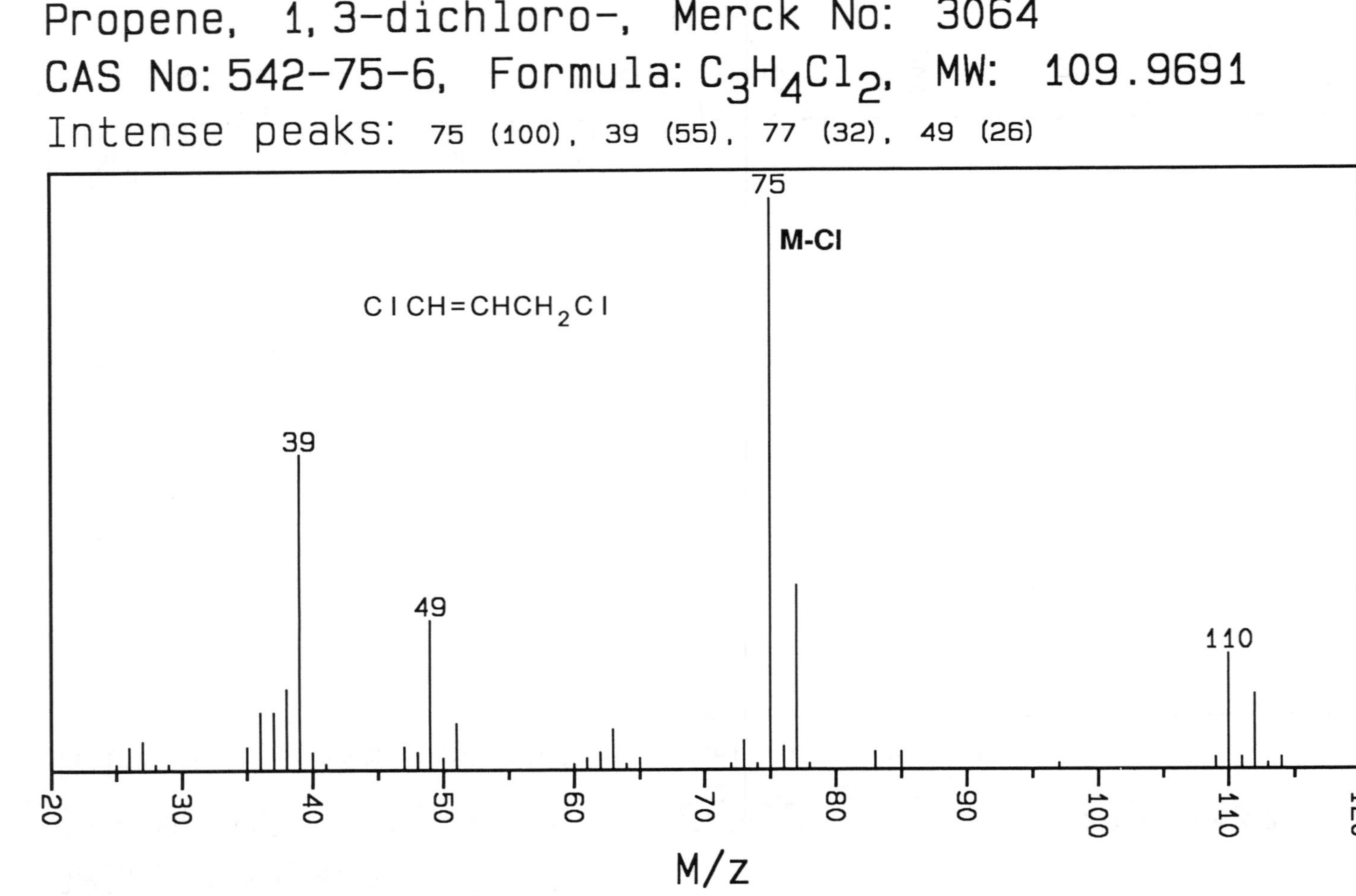
Propene, 1,3-dichloro-, Merck No: 3064
CAS No: 542-75-6, Formula: $C_3H_4Cl_2$, MW: 109.9691
Intense peaks: 75 (100), 39 (55), 77 (32), 49 (26)
75
M-Cl
ClCH=CHCH$_2$Cl
39
49
110
20
30
40
50
60
70
80
90
100
110
120
M/Z

Propene, 3-chloro-2-methyl-, Merck No: 2148

CAS No: 563-47-3, Formula: $C_4H_7Cl$, MW: 90.0236

Intense peaks: 55 (100), 39 (53), 29 (31), 90 (30)

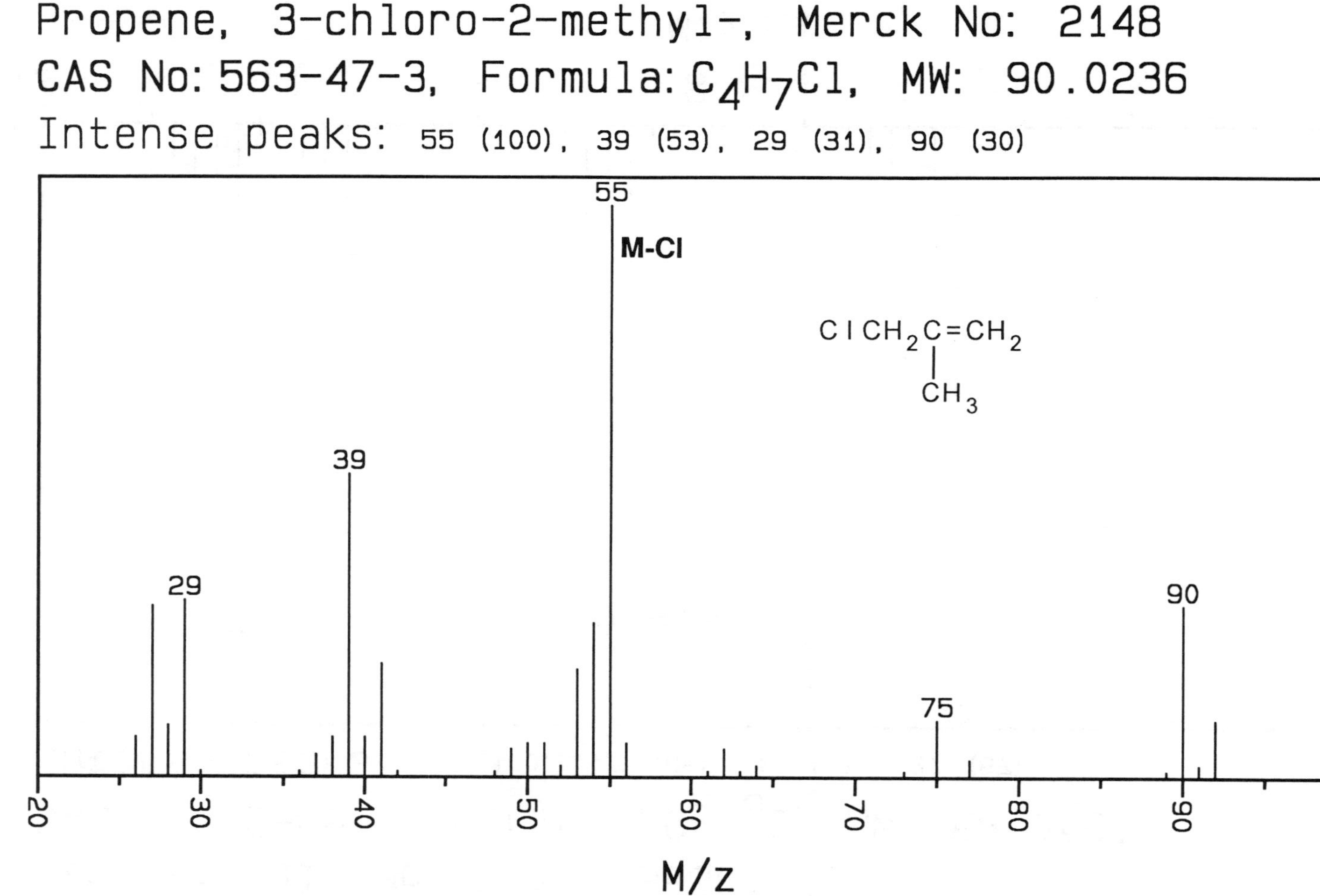

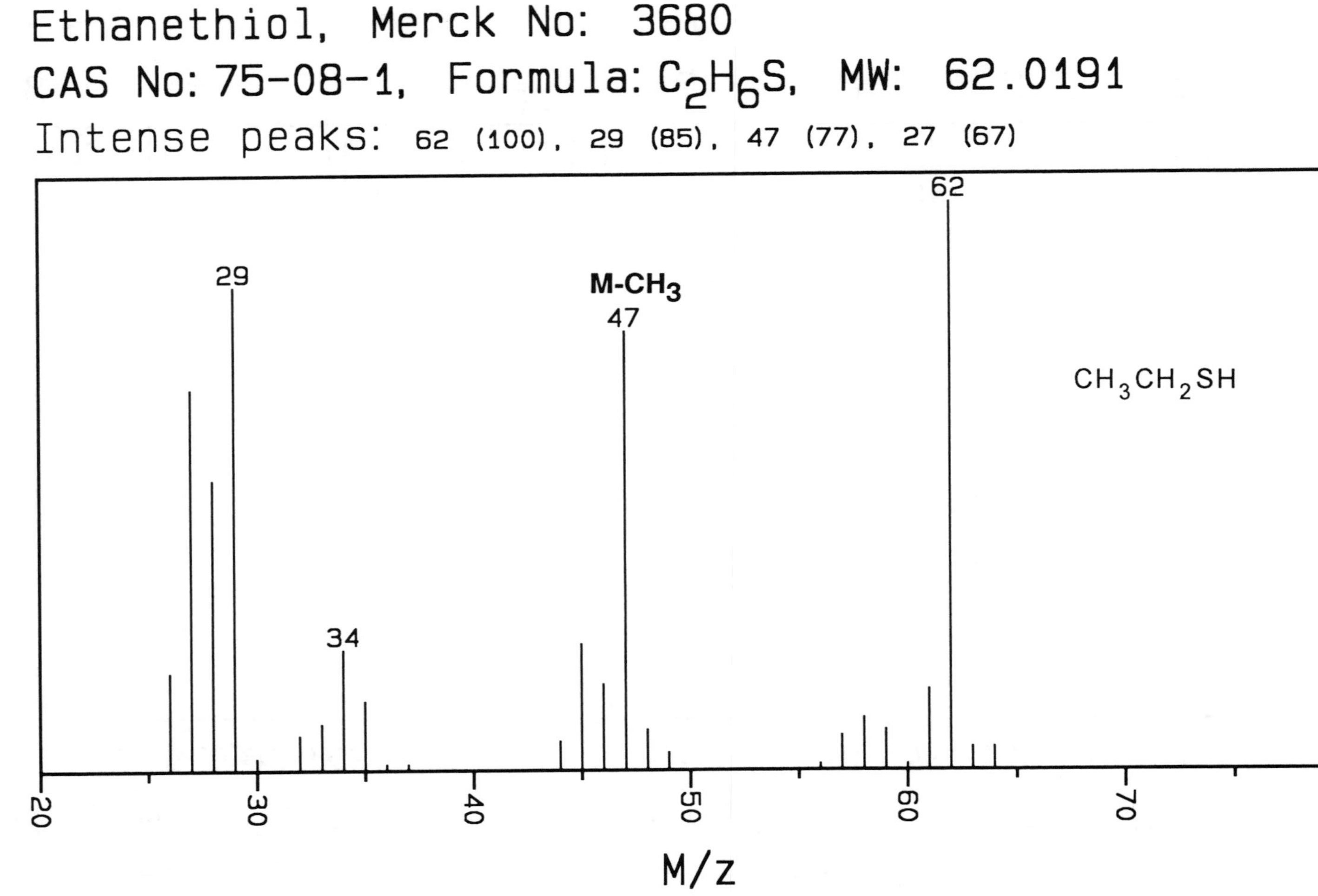
Ethanethiol, Merck No: 3680
CAS No: 75-08-1, Formula: C2H6S, MW: 62.0191
Intense peaks: 62 (100), 29 (85), 47 (77), 27 (67)
62
29
M-CH3
47
CH3CH2SH
34
20
30
40
50
60
70
80
M/Z

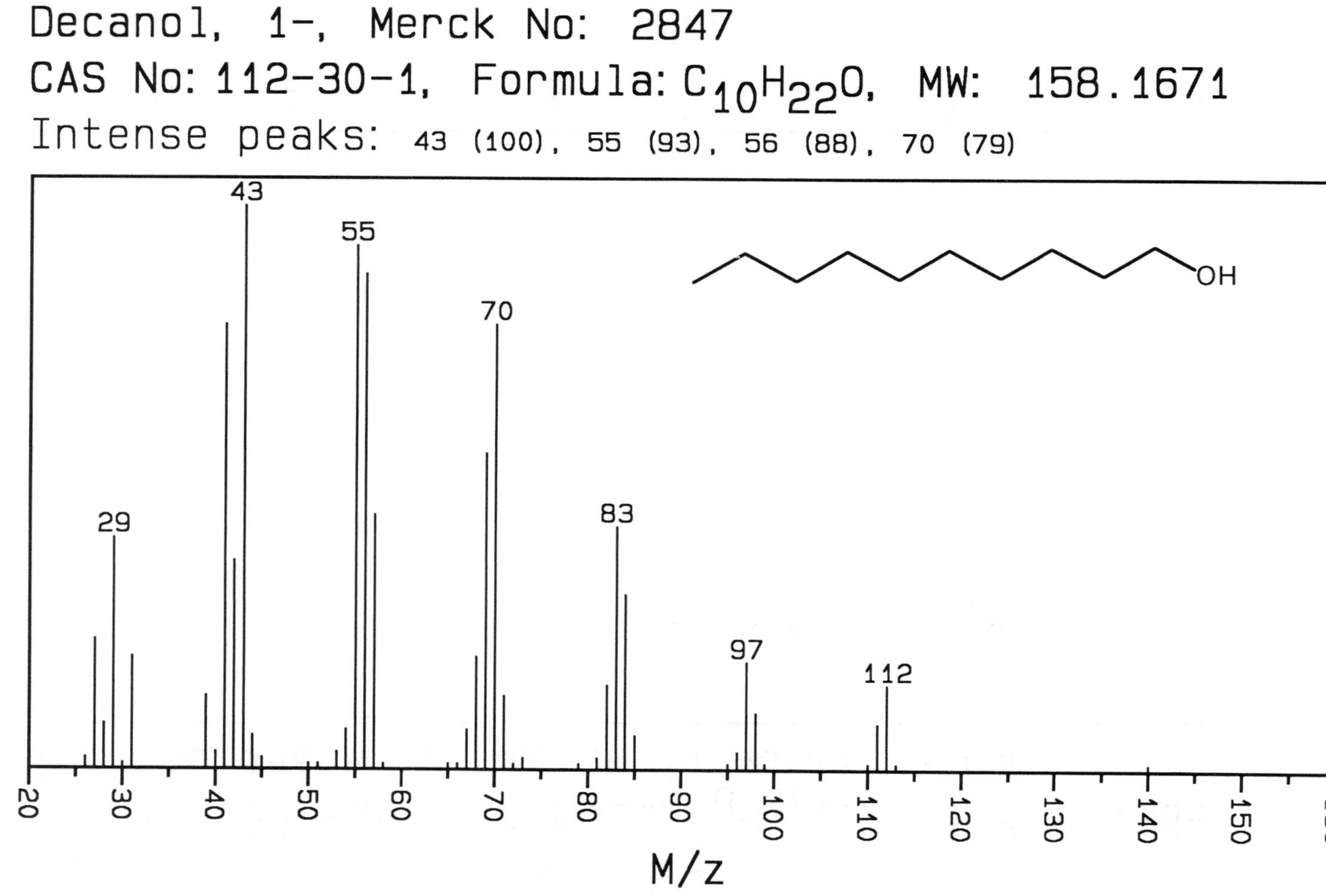

Decanol, 1-, Merck No: 2847
CAS No: 112-30-1, Formula: $C_{10}H_{22}O$, MW: 158.1671
Intense peaks: 43 (100), 55 (93), 56 (88), 70 (79)
OH
29
43
55
70
83
97
112
20
30
40
50
60
70
80
90
100
110
120
130
140
150
160
M/Z

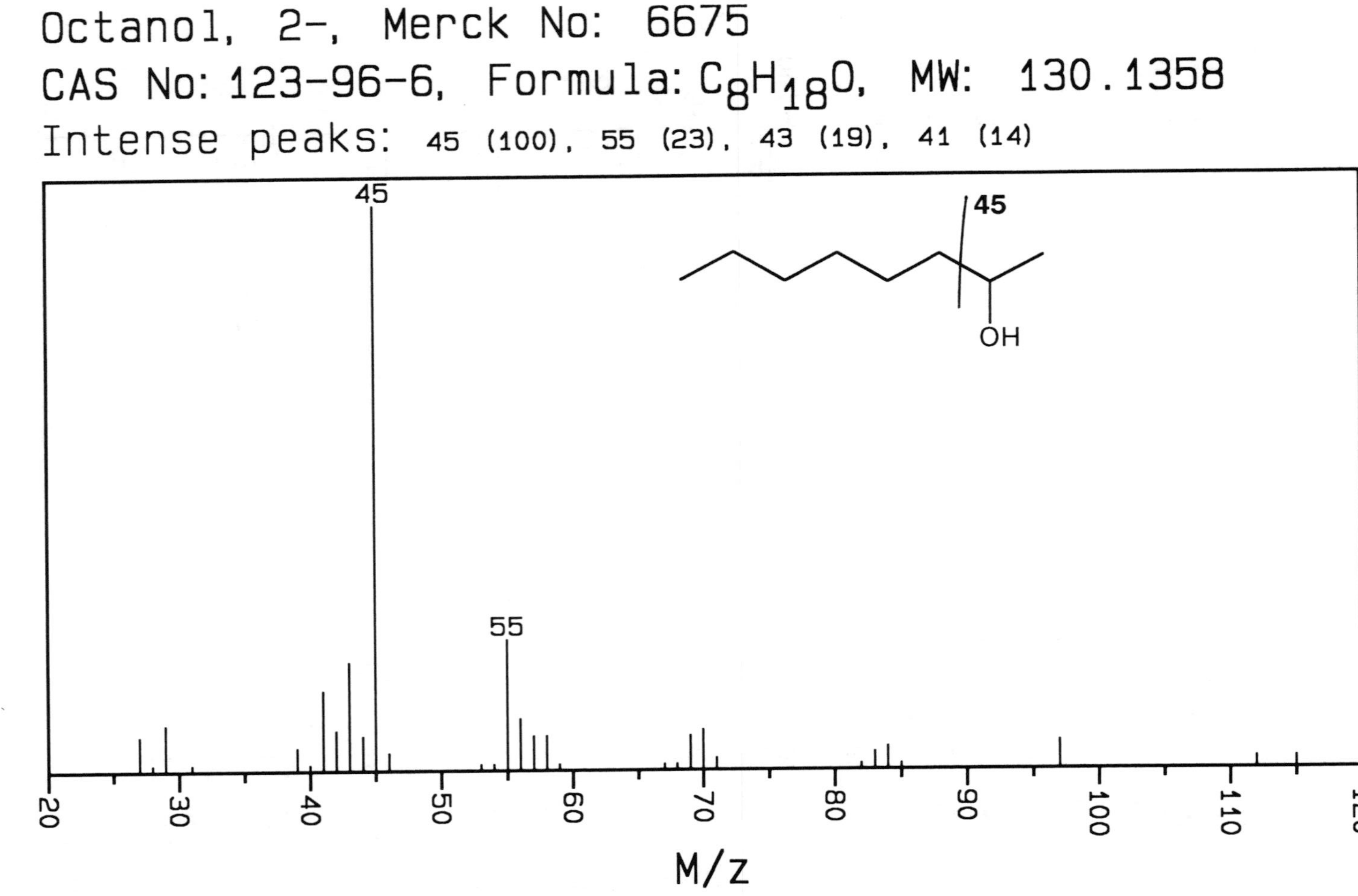
Octanol, 2-, Merck No: 6675
CAS No: 123-96-6, Formula: $C_8H_{18}O$, MW: 130.1358
Intense peaks: 45 (100), 55 (23), 43 (19), 41 (14)
45
45
OH
55
20
30
40
50
60
70
80
90
100
110
120
M/Z

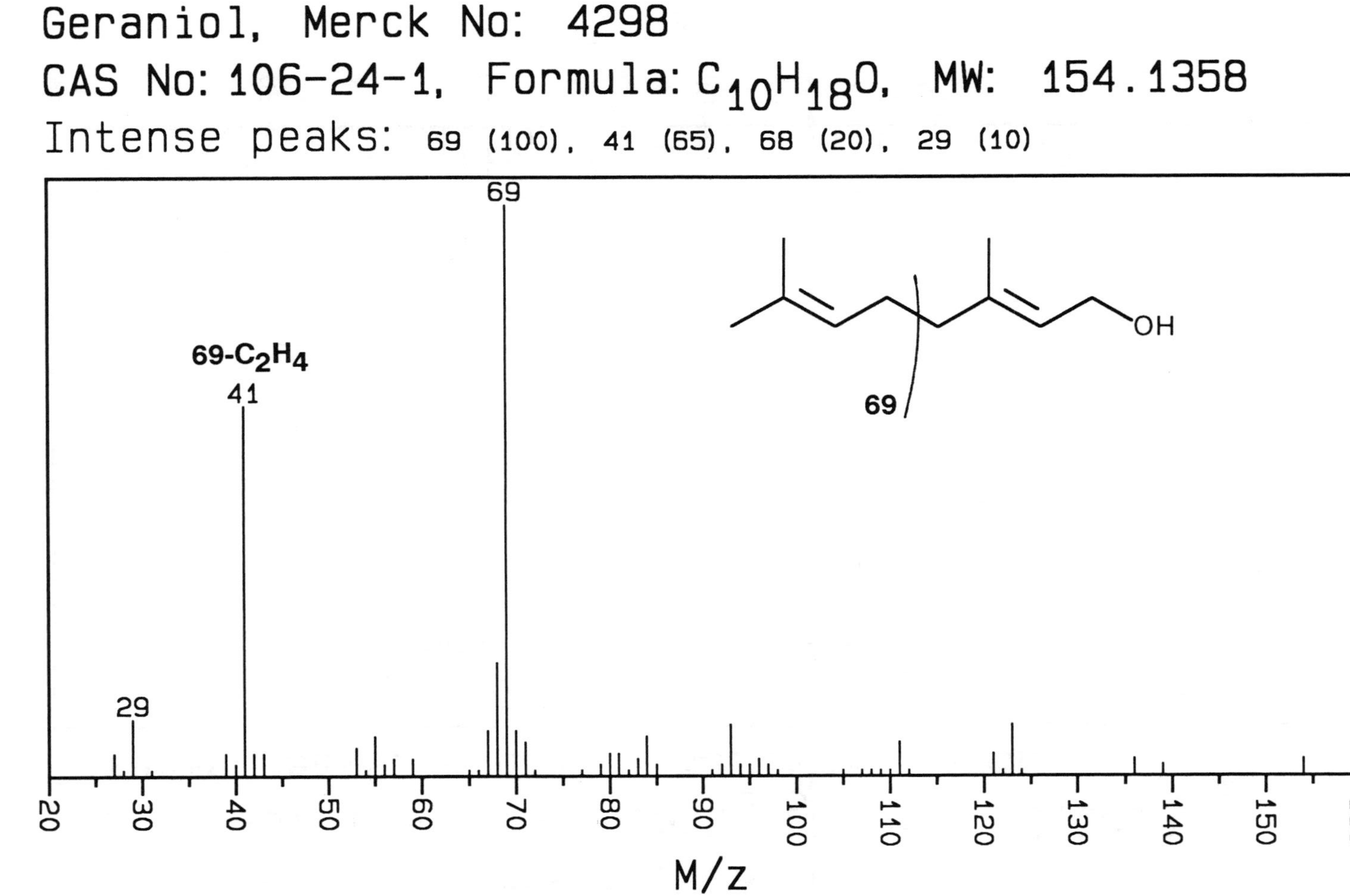
Geraniol, Merck No: 4298
CAS No: 106-24-1, Formula: $C_{10}H_{18}O$, MW: 154.1358
Intense peaks: 69 (100), 41 (65), 68 (20), 29 (10)
69
$69-C_2H_4$
41
29
69
OH
20
30
40
50
60
70
80
90
100
110
120
130
140
150
160
M/Z

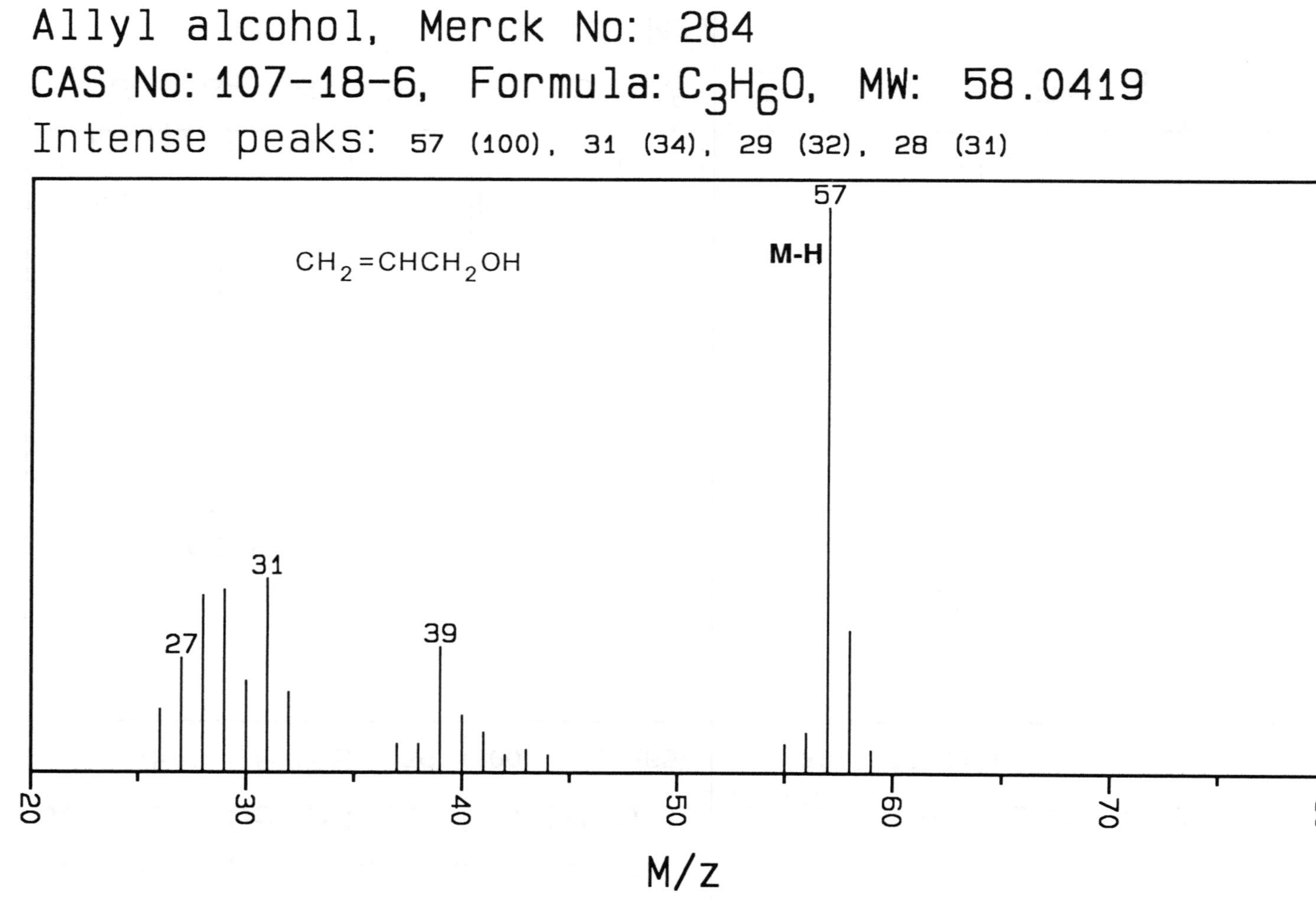
Allyl alcohol, Merck No: 284
CAS No: 107-18-6, Formula: $C_3H_6O$, MW: 58.0419
Intense peaks: 57 (100), 31 (34), 29 (32), 28 (31)
$CH_2=CHCH_2OH$
M-H
57
39
31
27
20
30
40
50
60
70
80
M/Z

Ethanol, 2-(ethylthio)-, Merck No: 3813

CAS No: 110-77-0, Formula: $C_4H_{10}OS$, MW: 106.0452

Intense peaks: 75 (100), 47 (77), 29 (43), 106 (43)

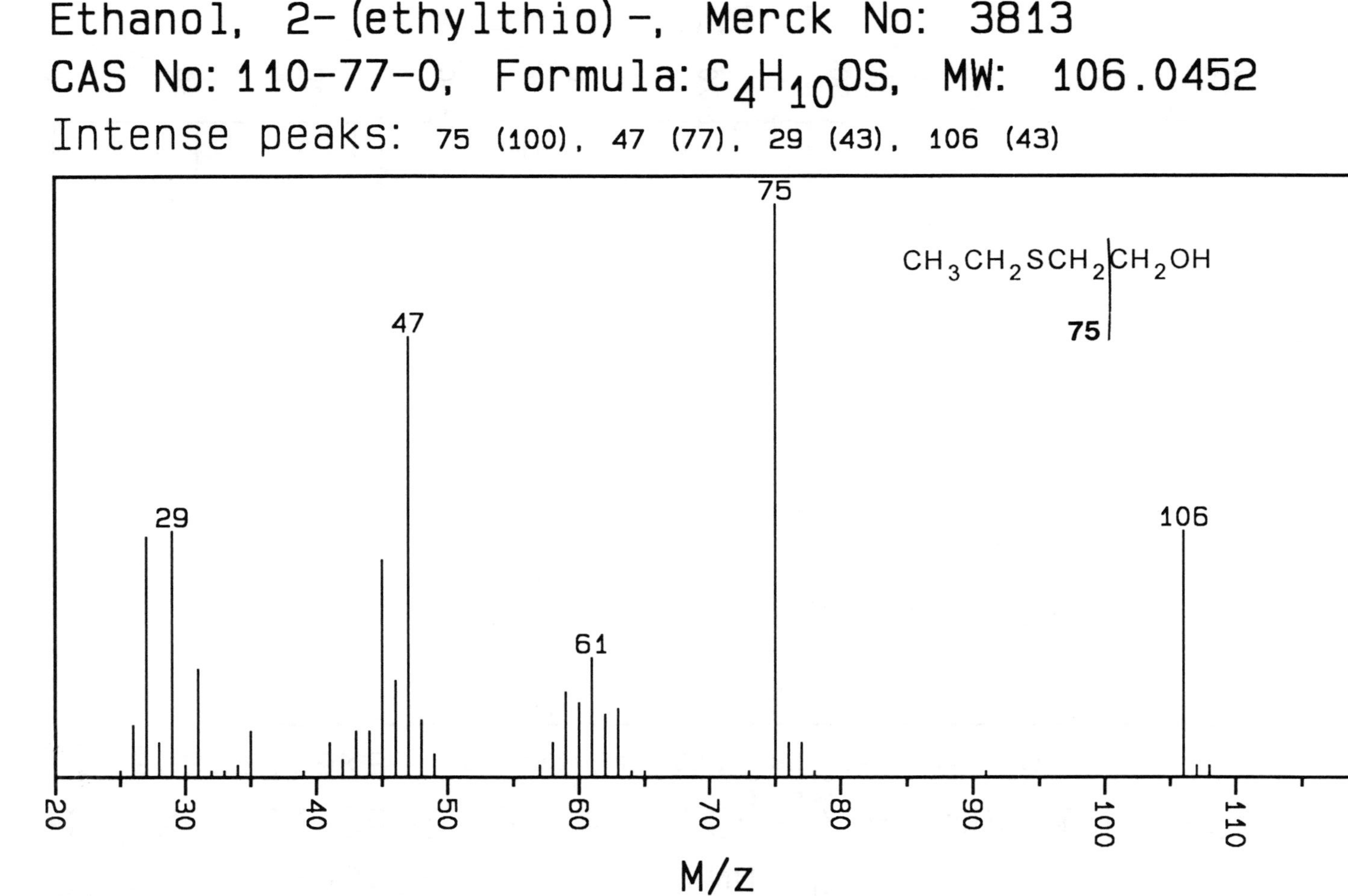

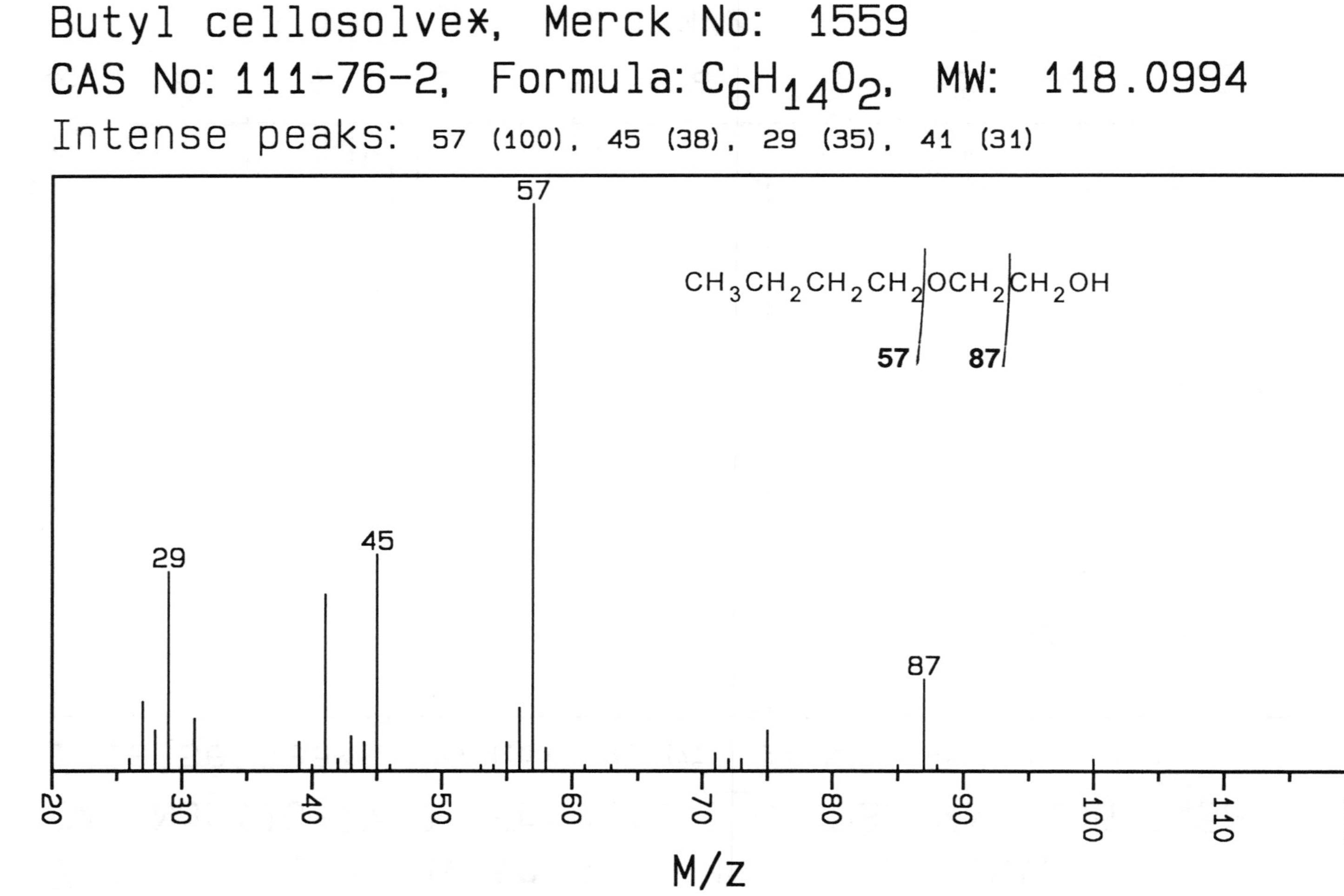
Butyl cellosolve*, Merck No: 1559
CAS No: 111-76-2, Formula: $C_6H_{14}O_2$, MW: 118.0994
Intense peaks: 57 (100), 45 (38), 29 (35), 41 (31)
57
$CH_3CH_2CH_2CH_2OCH_2CH_2OH$
57
87
29
45
87
20
30
40
50
60
70
80
90
100
110
120
M/z

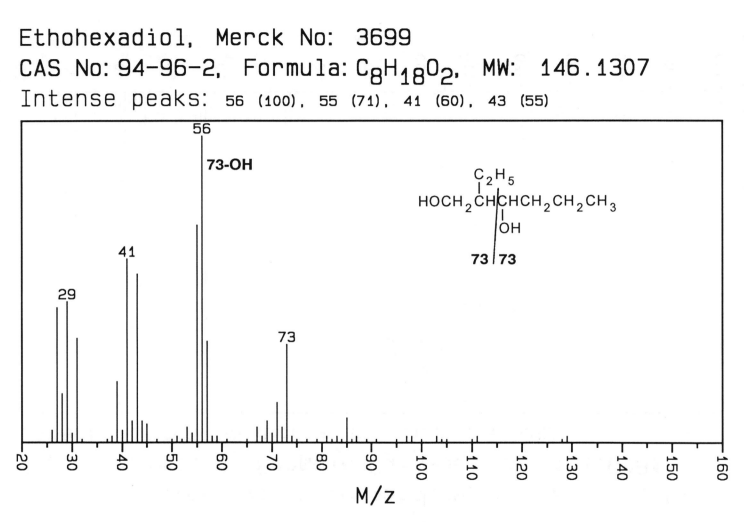

Ethohexadiol, Merck No: 3699

CAS No: 94-96-2, Formula: $C_8H_{18}O_2$, MW: 146.1307

Intense peaks: 56 (100), 55 (71), 41 (60), 43 (55)

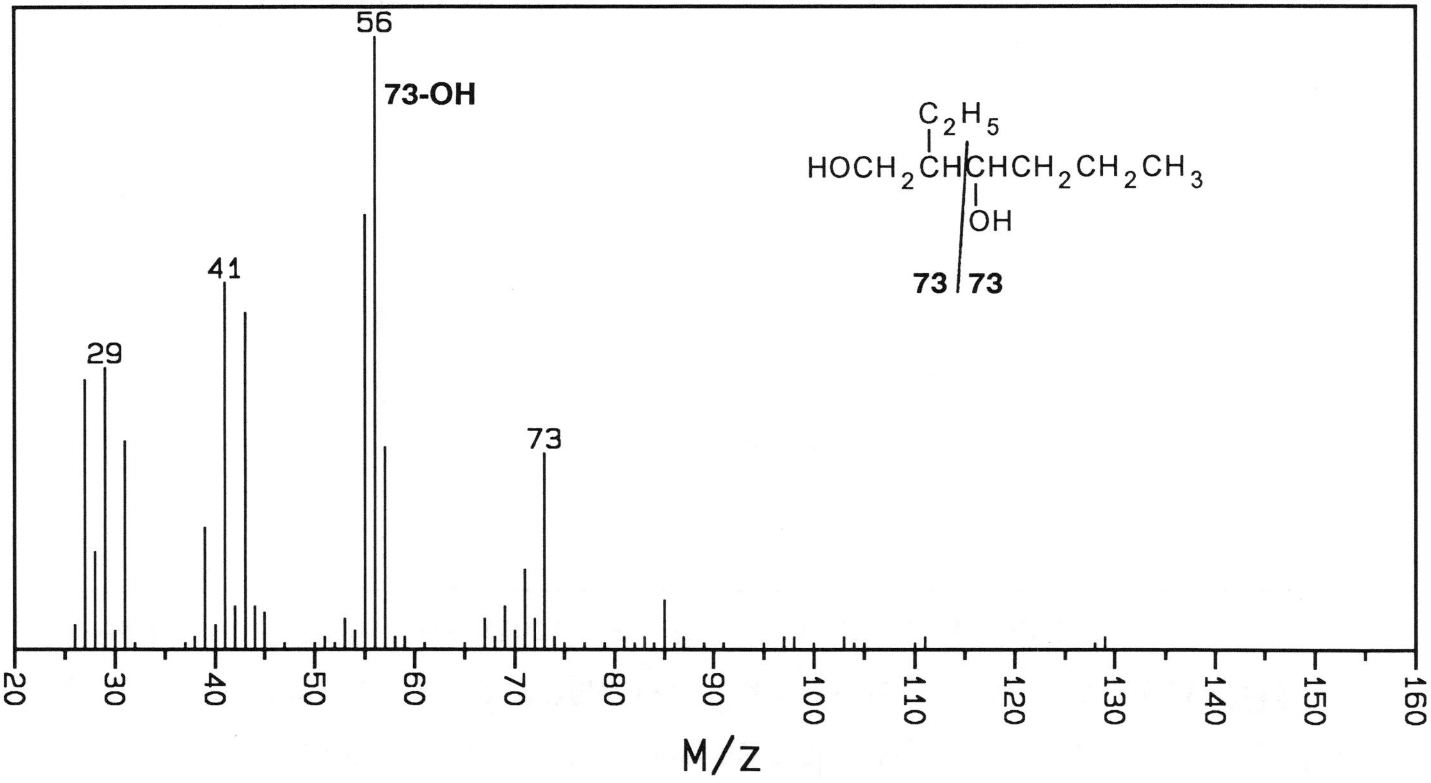

56

73-OH

41

29

73

$C_2H_5$

$HOCH_2CHCHCH_2CH_2CH_3$

OH

73 | 73

20 30 40 50 60 70 80 90 100 110 120 130 140 150 160

M/Z

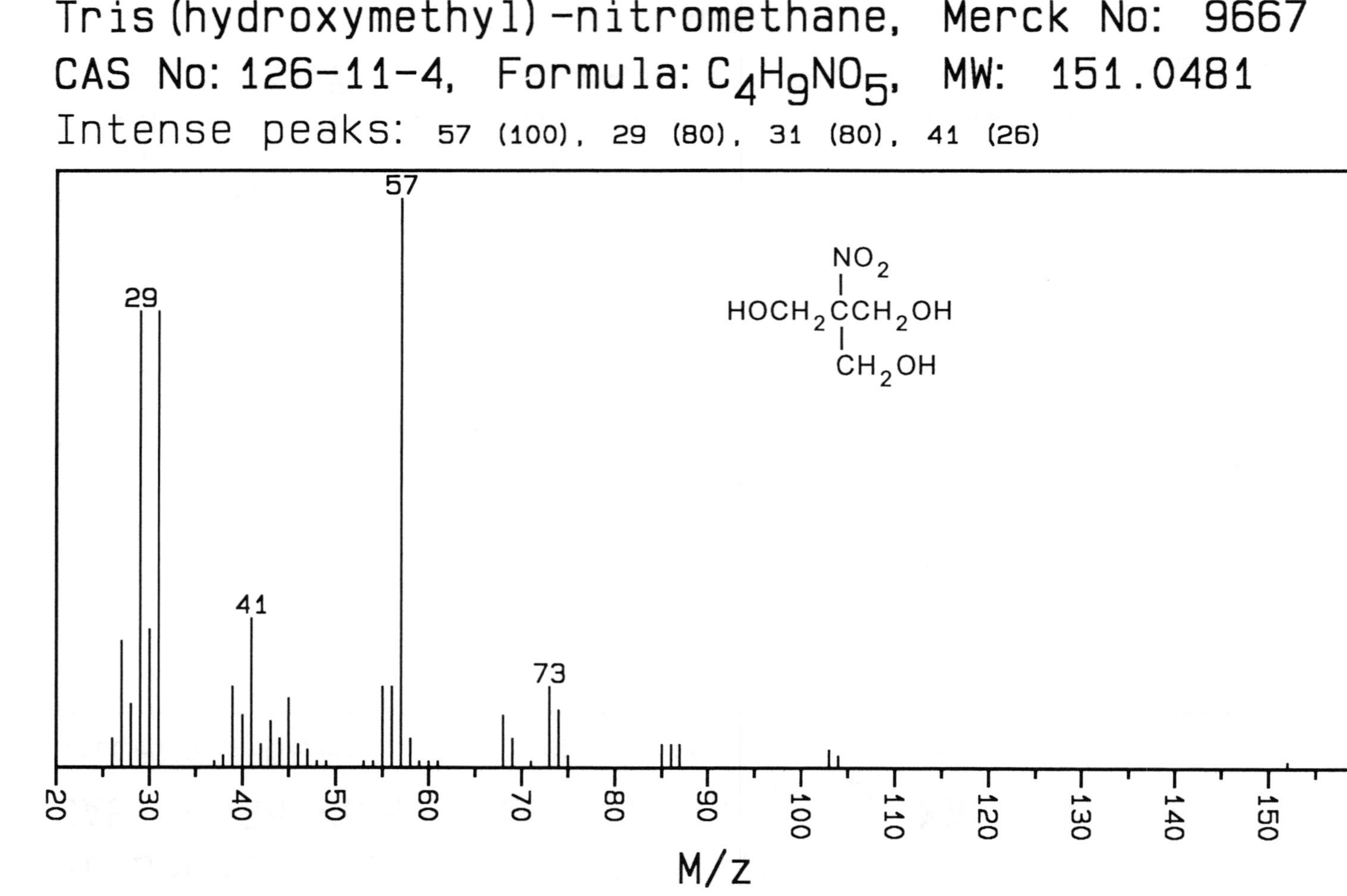
Tris(hydroxymethyl)-nitromethane, Merck No: 9667
CAS No: 126-11-4, Formula: $C_4H_9NO_5$, MW: 151.0481
Intense peaks: 57 (100), 29 (80), 31 (80), 41 (26)
57
29
41
73
NO2
HOCH2CCH2OH
CH2OH
20
30
40
50
60
70
80
90
100
110
120
130
140
150
160
M/Z

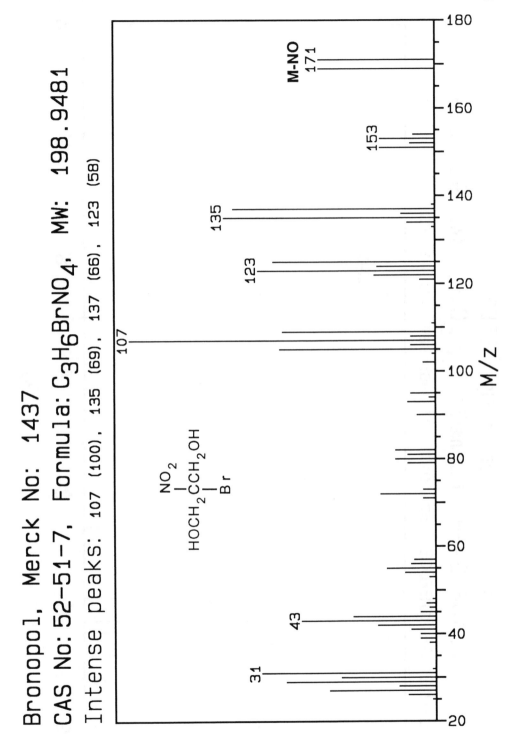

Bronopol, Merck No: 1437

CAS No: 52-51-7, Formula: $C_3H_6BrNO_4$, MW: 198.9481

Intense peaks: 107 (100), 135 (69), 137 (66), 123 (58)

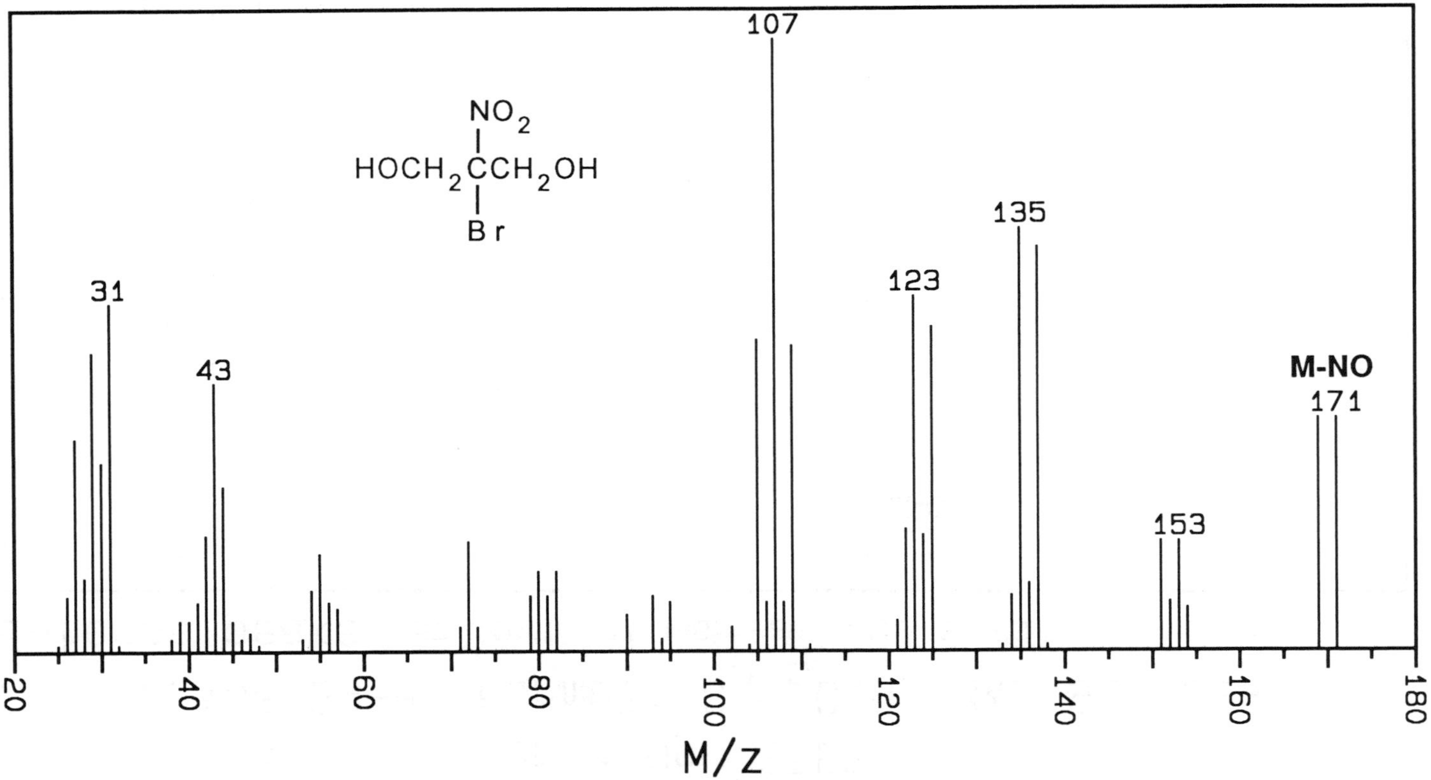

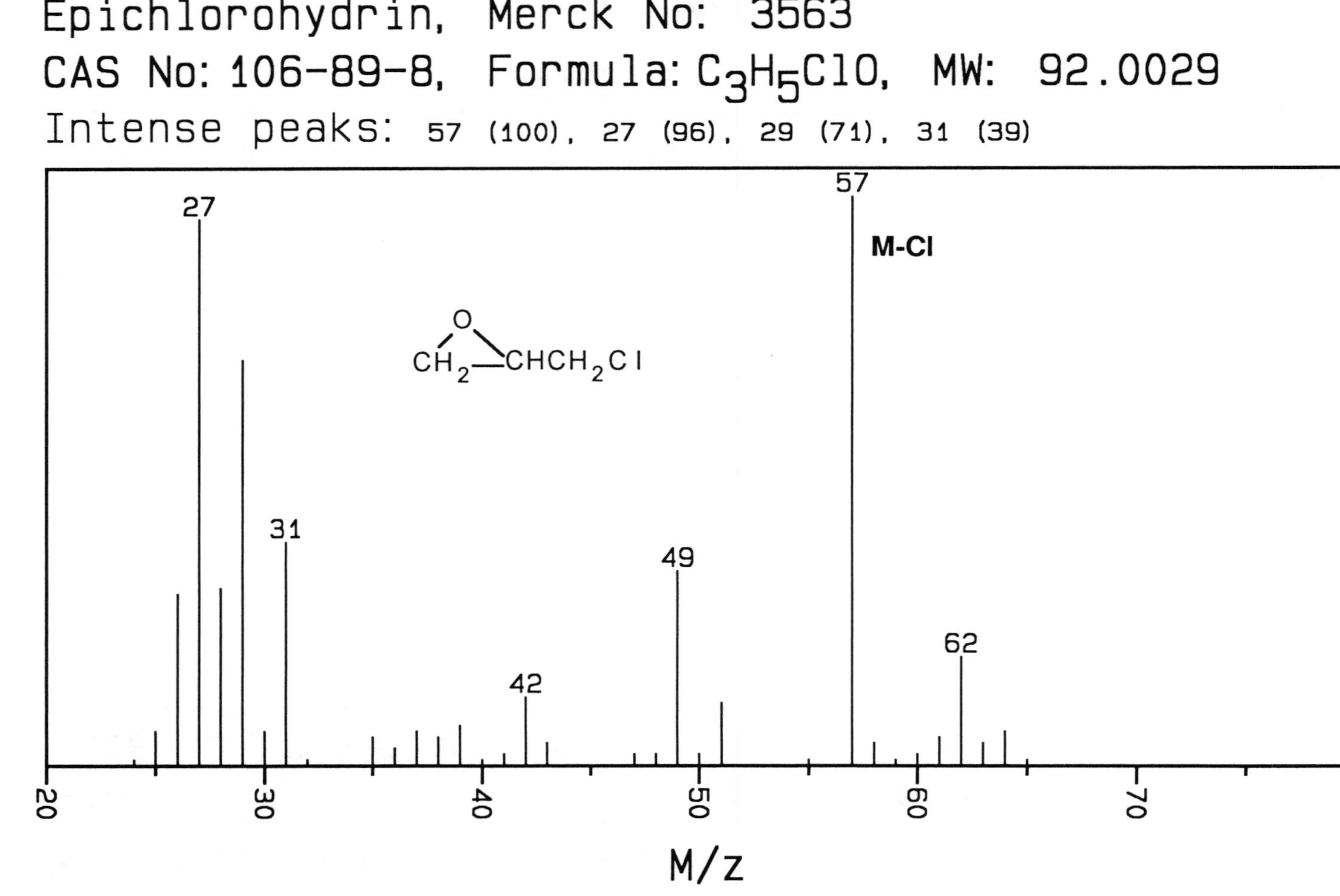
Epichlorohydrin, Merck No: 3563
CAS No: 106-89-8, Formula: $C_3H_5ClO$, MW: 92.0029
Intense peaks: 57 (100), 27 (96), 29 (71), 31 (39)
57
27
M-Cl
$CH_2$—CHCH$_2$Cl
O
31
49
62
42
20
30
40
50
60
70
80
M/Z

Dichloromethyl ether, sym-, Merck No: 3058

CAS No: 542-88-1, Formula: $C_2H_4Cl_2O$, MW: 113.9639

Intense peaks: 79 (100), 49 (62), 81 (32), 29 (23)

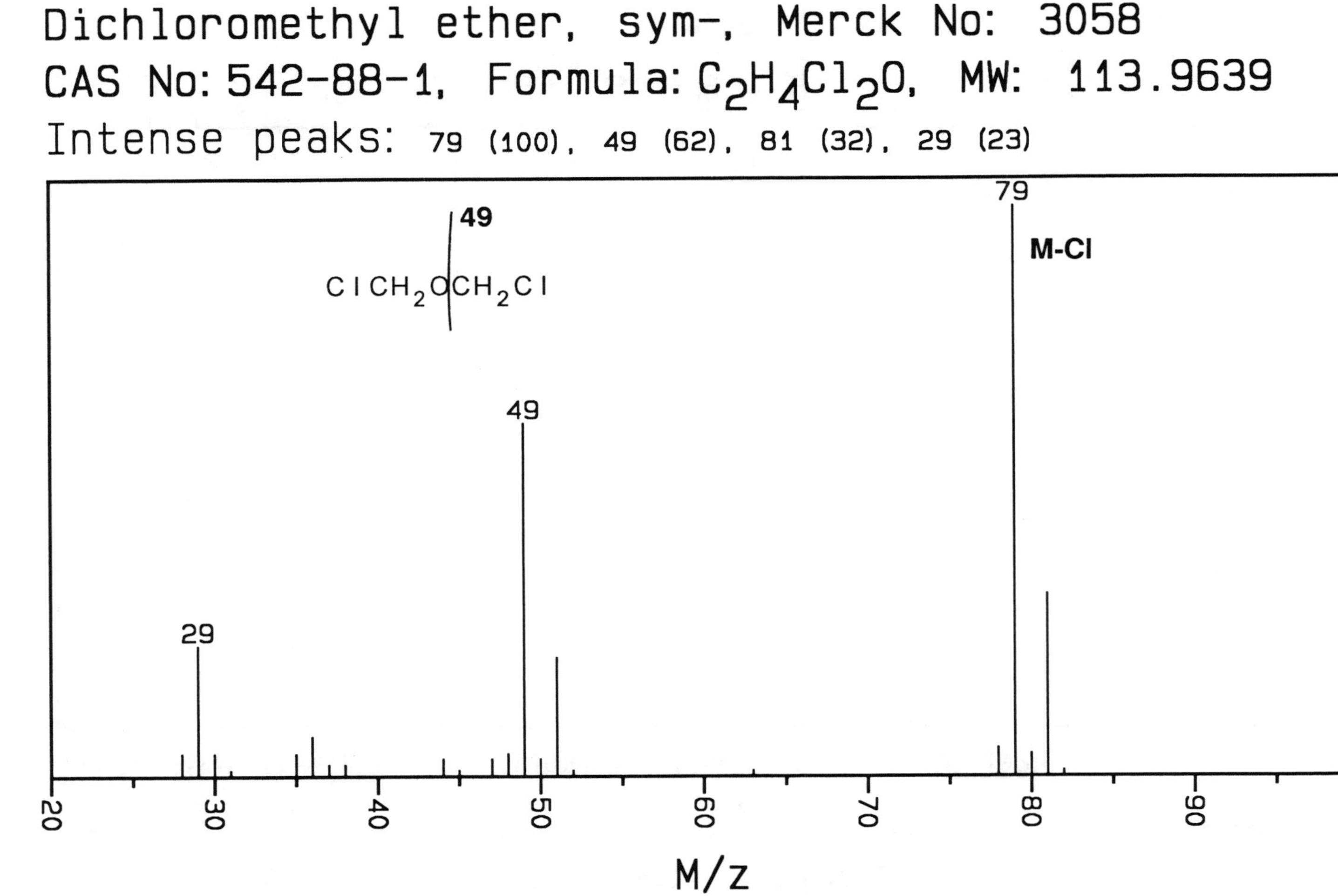

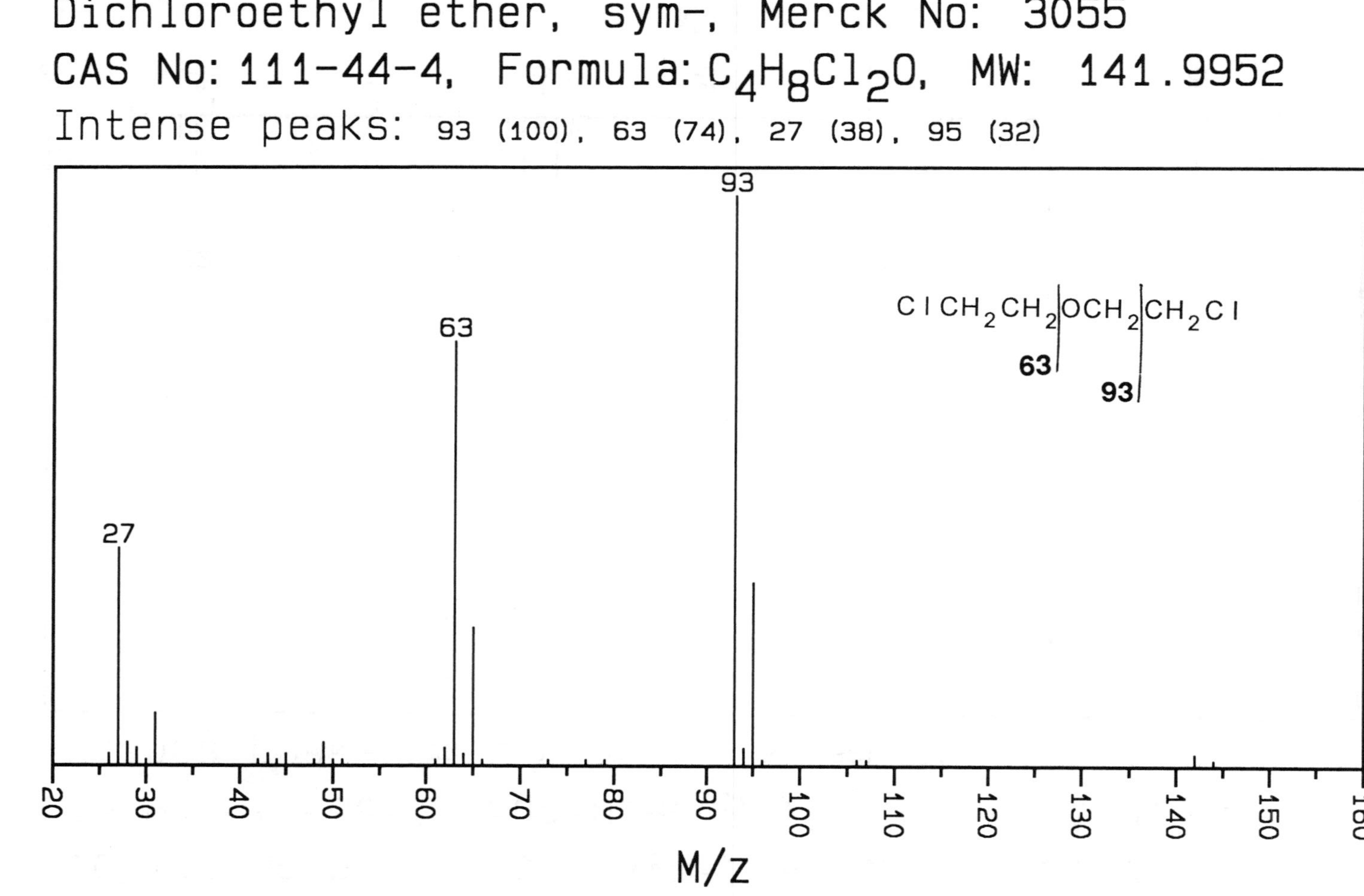

Dichloroethyl ether, sym-, Merck No: 3055
CAS No: 111-44-4, Formula: $C_4H_8Cl_2O$, MW: 141.9952
Intense peaks: 93 (100), 63 (74), 27 (38), 95 (32)
93
63
27
$ClCH_2CH_2OCH_2CH_2Cl$
63
93
20
30
40
50
60
70
80
90
100
110
120
130
140
150
160
M/Z

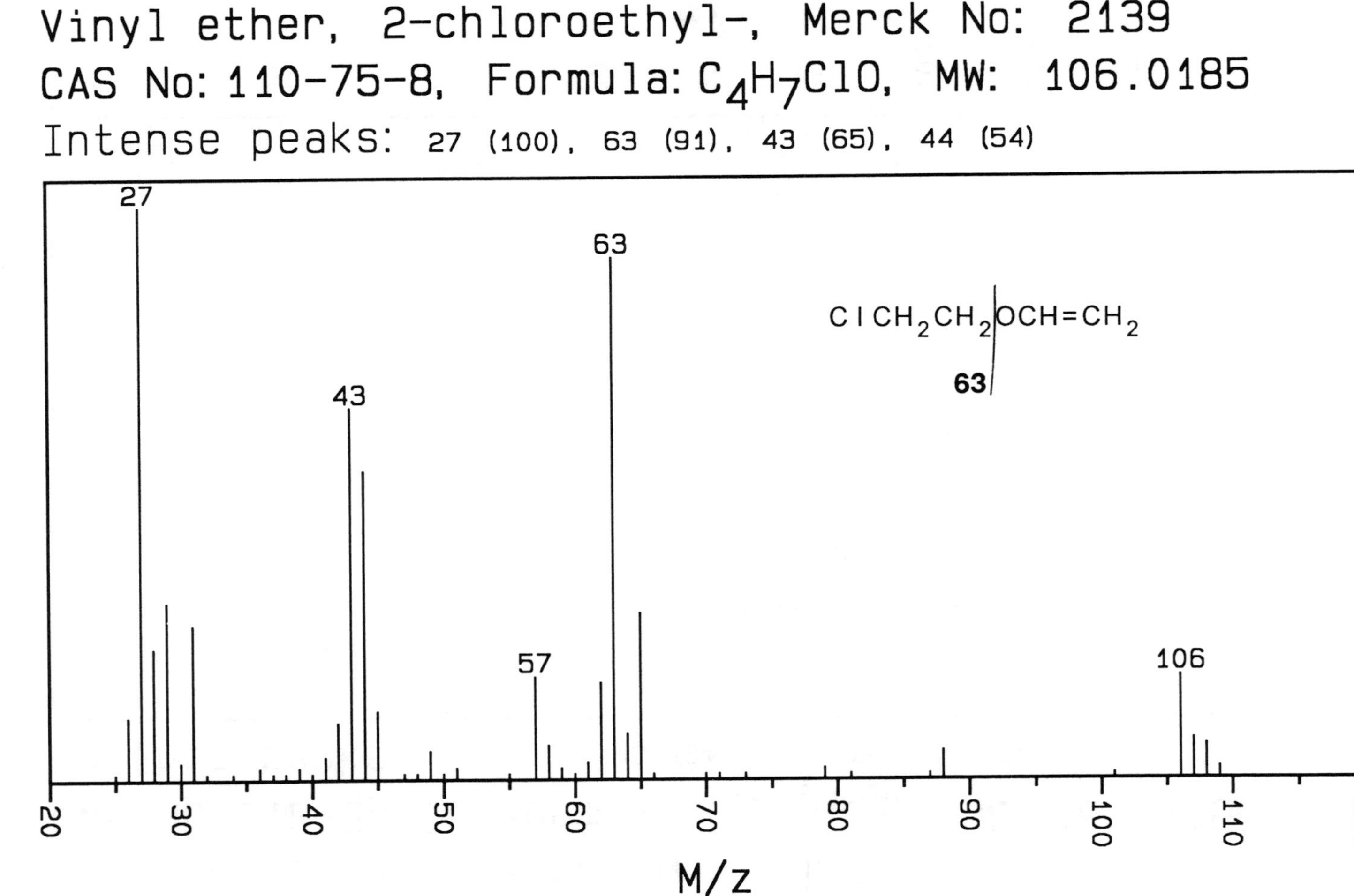
Vinyl ether, 2-chloroethyl-, Merck No: 2139
CAS No: 110-75-8, Formula: C4H7ClO, MW: 106.0185
Intense peaks: 27 (100), 63 (91), 43 (65), 44 (54)
27
63
43
57
106
ClCH2CH2OCH=CH2
63
20
30
40
50
60
70
80
90
100
110
120
M/Z

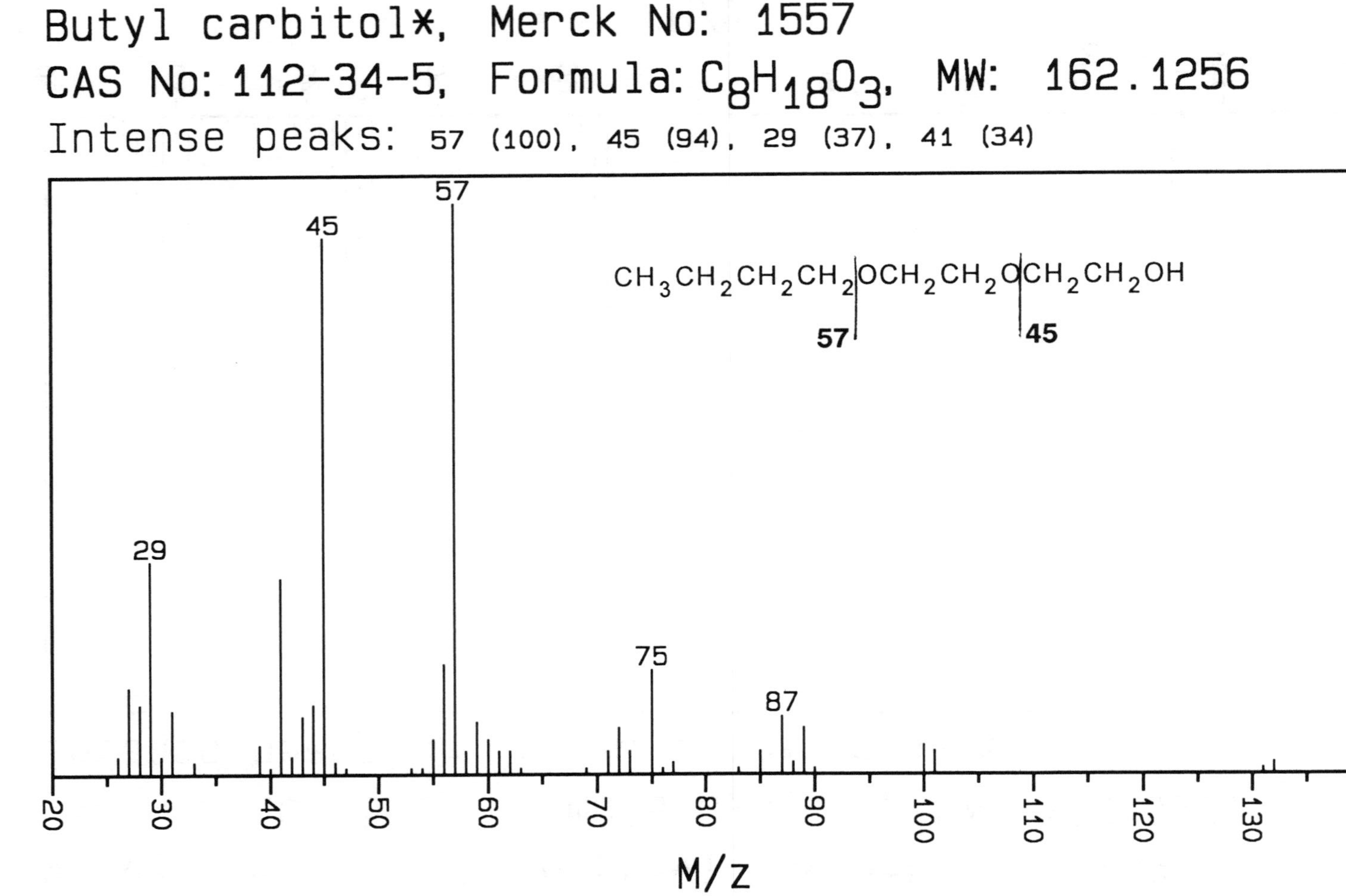
Butyl carbitol*, Merck No: 1557
CAS No: 112-34-5, Formula: $C_8H_{18}O_3$, MW: 162.1256
Intense peaks: 57 (100), 45 (94), 29 (37), 41 (34)
57
45
$CH_3CH_2CH_2CH_2OCH_2CH_2OCH_2CH_2OH$
57
45
29
75
87
20
30
40
50
60
70
80
90
100
110
120
130
140
M/Z

Lethane*, Merck No: 5329

CAS No: 112-56-1, Formula: $C_9H_{17}NO_2S$, MW: 203.0981

Intense peaks: 29 (100), 57 (95), 27 (84), 41 (67)

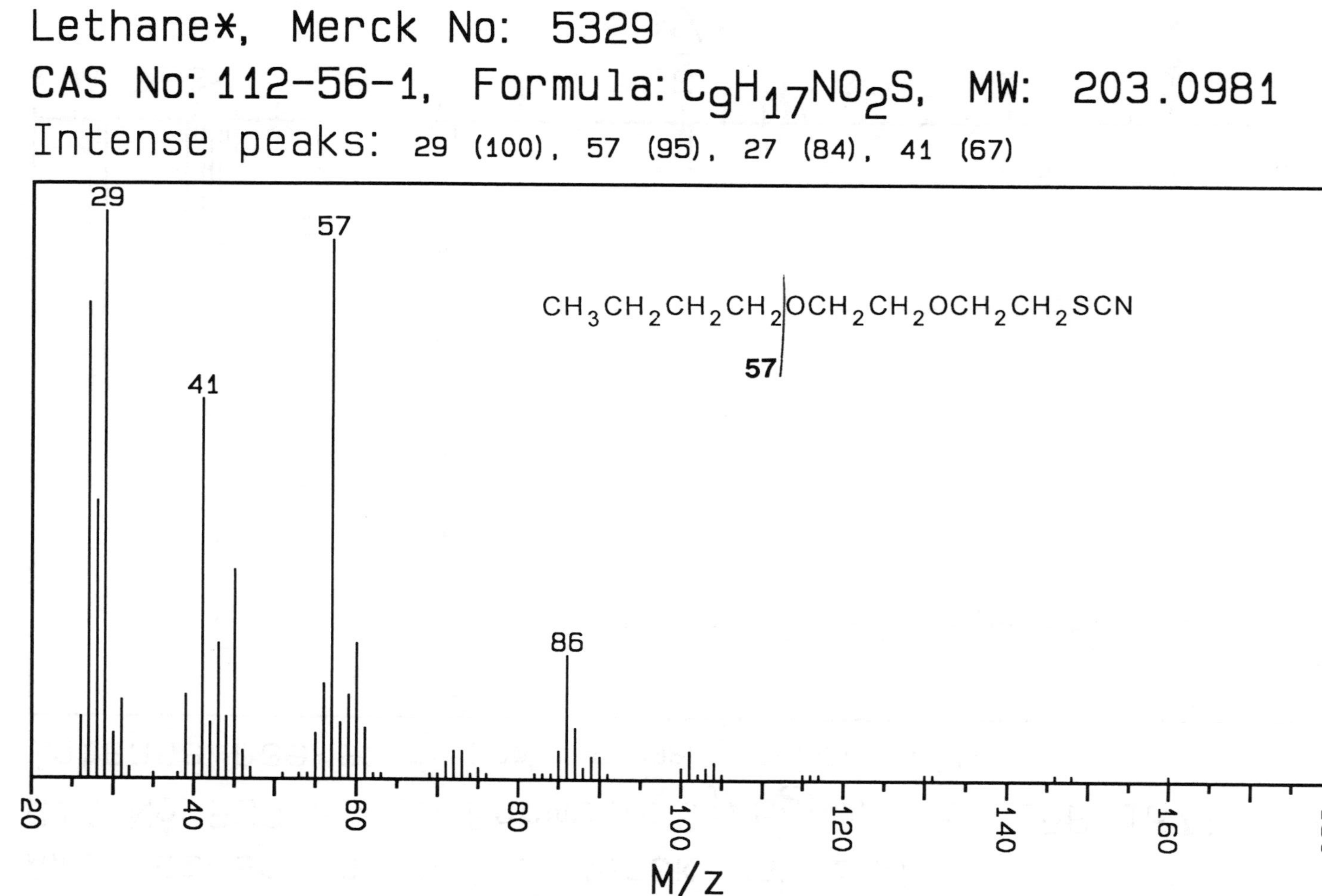

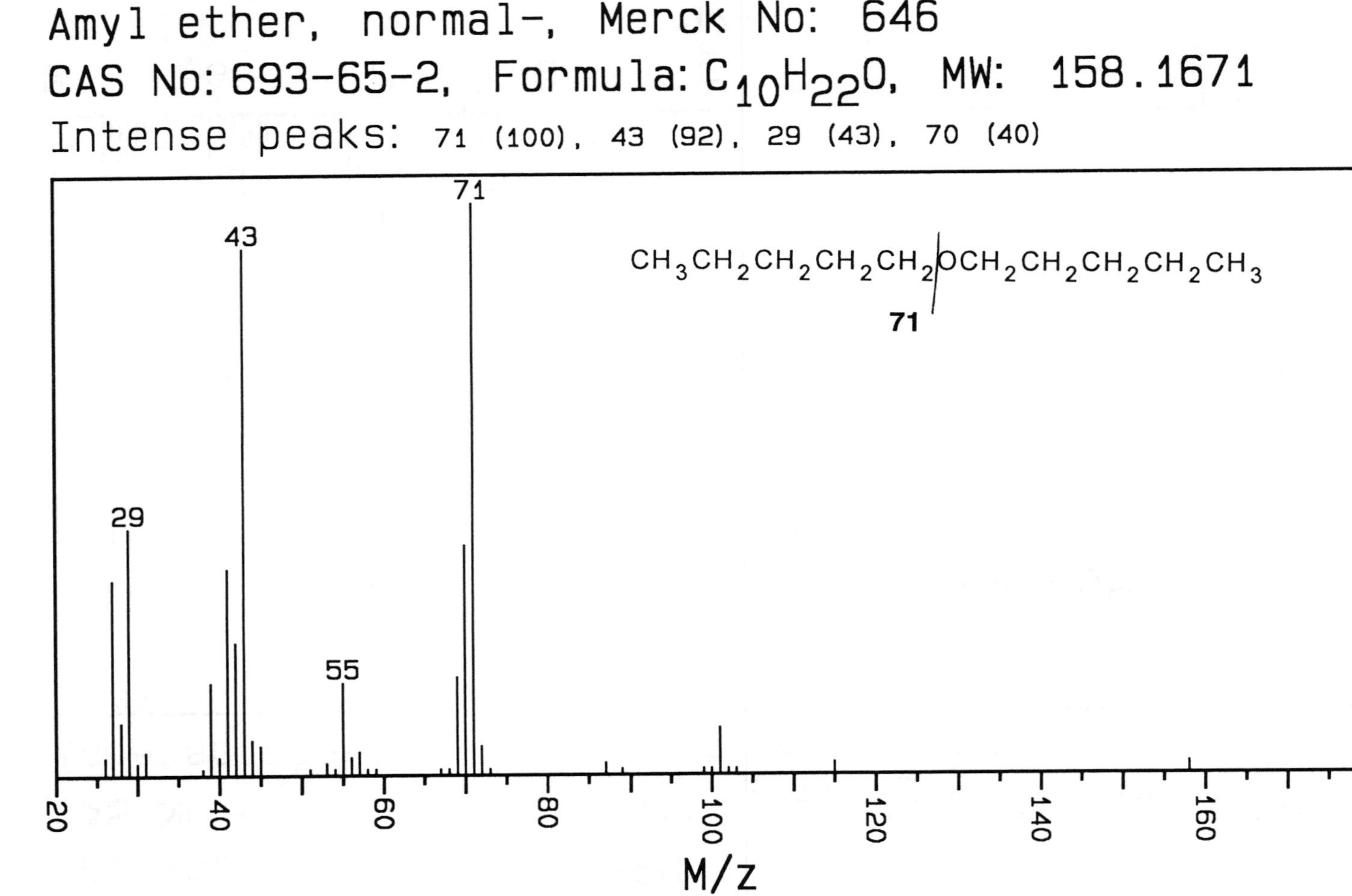
Amyl ether, normal-, Merck No: 646
CAS No: 693-65-2, Formula: $C_{10}H_{22}O$, MW: 158.1671
Intense peaks: 71 (100), 43 (92), 29 (43), 70 (40)
$CH_3CH_2CH_2CH_2CH_2$ | $OCH_2CH_2CH_2CH_2CH_3$
71
29
43
55
71
20
40
60
80
100
120
140
160
180
M/Z

Chloroacetone, Merck No: 2113

CAS No: 78-95-5, Formula: $C_3H_5ClO$, MW: 92.0029

Intense peaks: 27 (100), 49 (87), 42 (80), 29 (64)

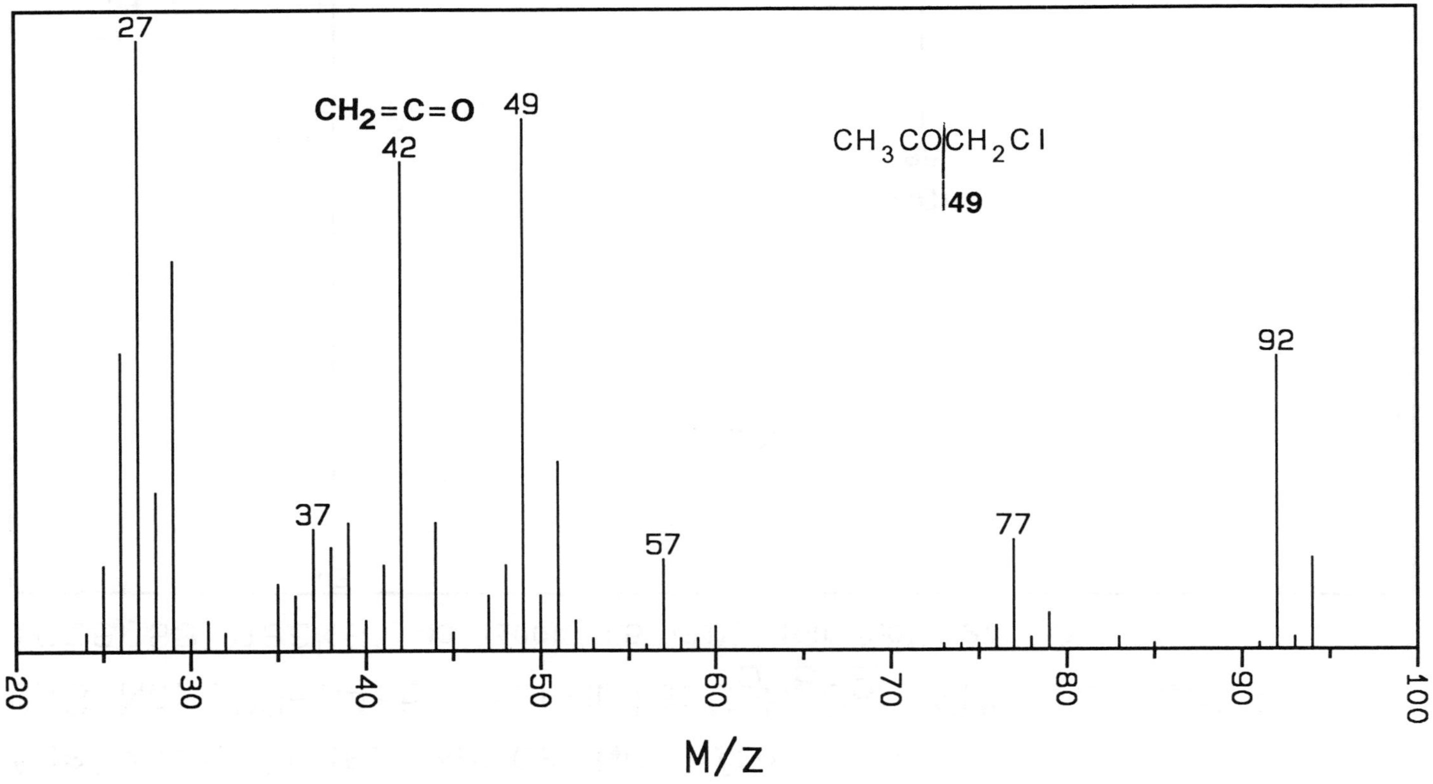

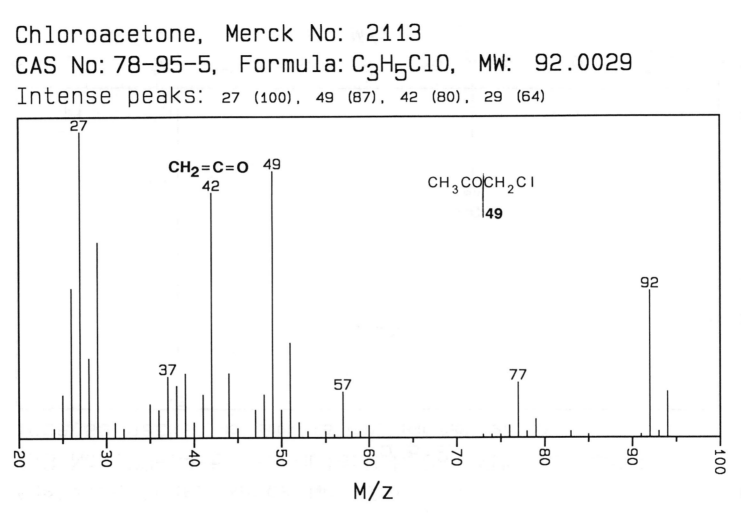

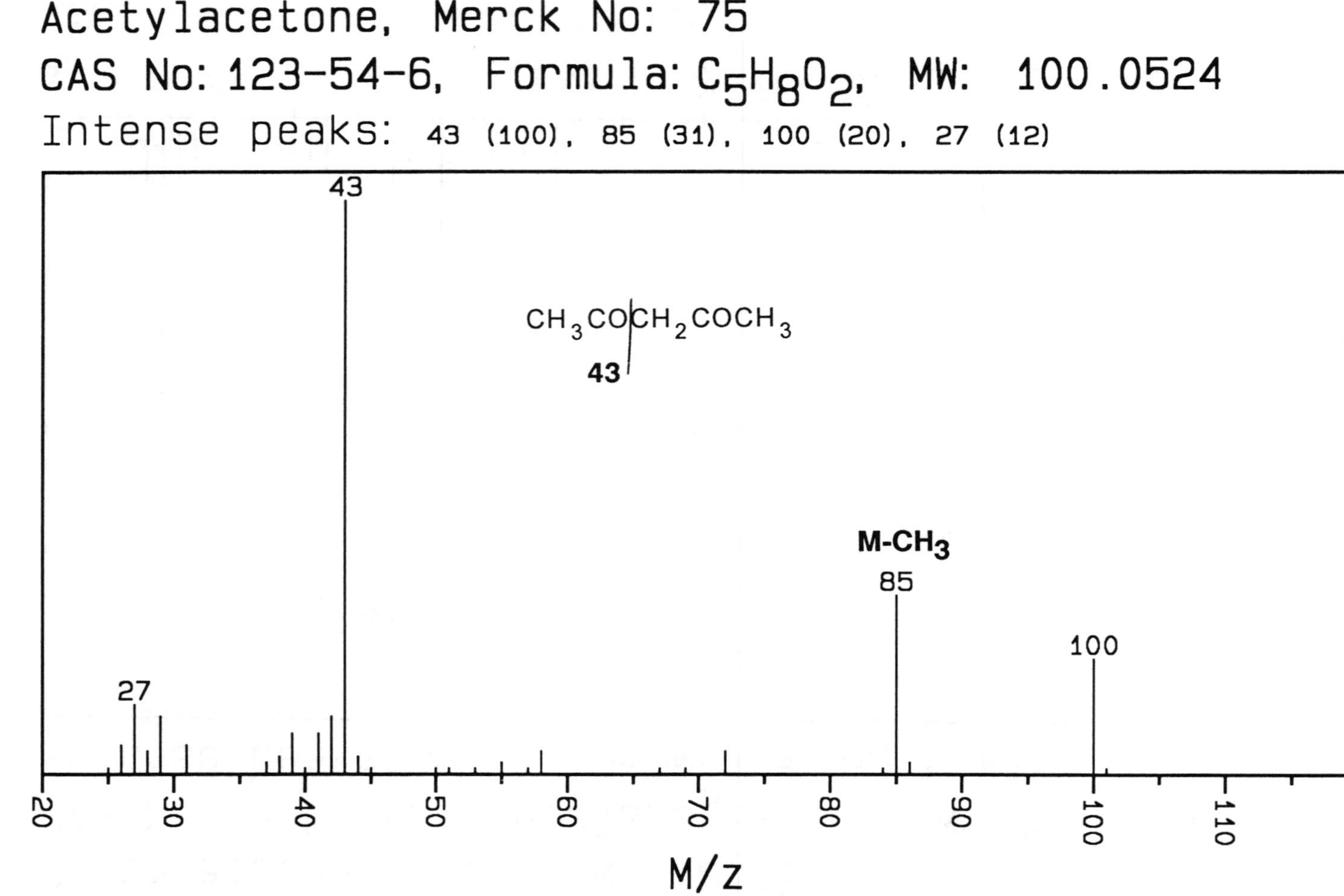
Acetylacetone, Merck No: 75
CAS No: 123-54-6, Formula: $C_5H_8O_2$, MW: 100.0524
Intense peaks: 43 (100), 85 (31), 100 (20), 27 (12)
43
$CH_3COCH_2COCH_3$
43
M-$CH_3$
85
100
27
20
30
40
50
60
70
80
90
100
110
120
M/Z

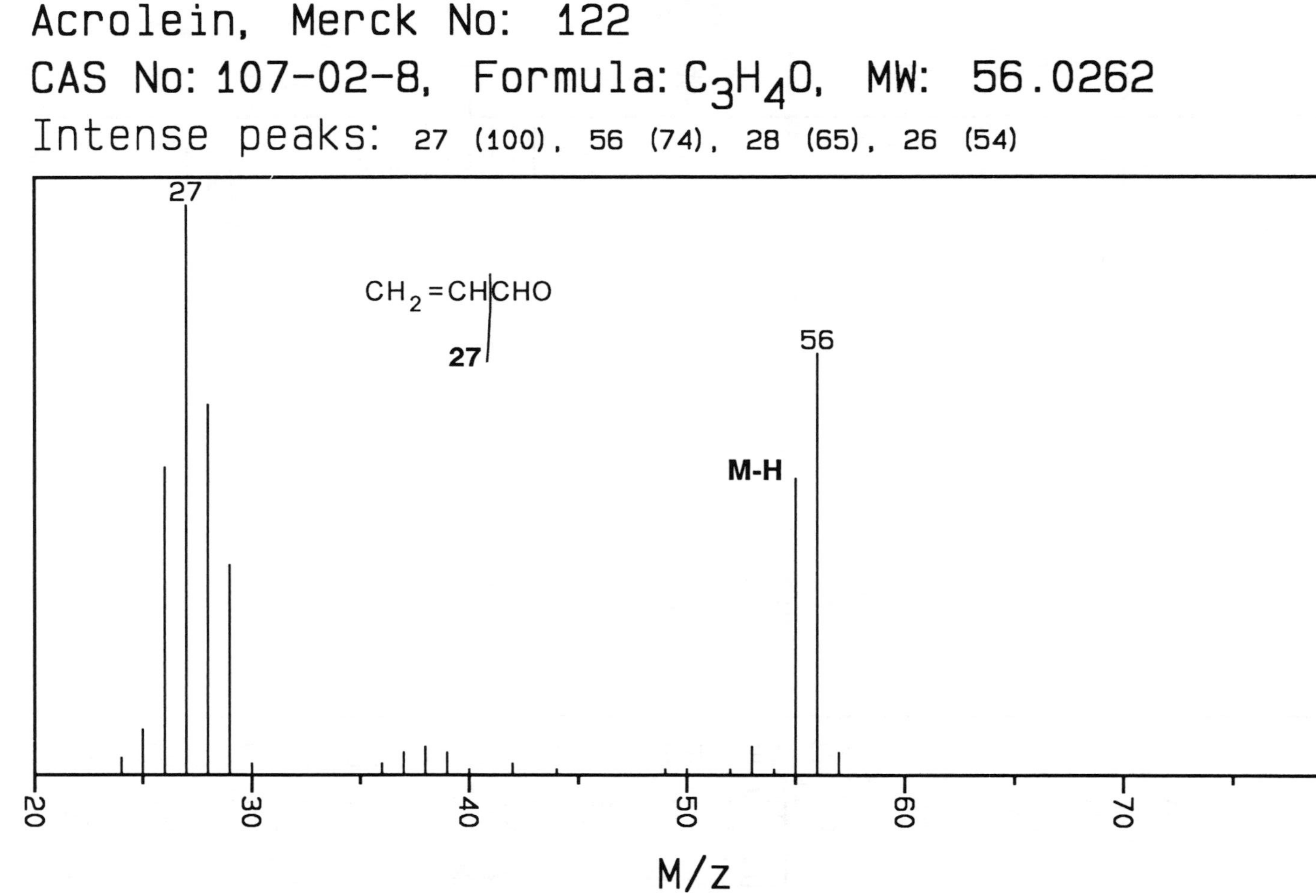
Acrolein, Merck No: 122
CAS No: 107-02-8, Formula: $C_3H_4O$, MW: 56.0262
Intense peaks: 27 (100), 56 (74), 28 (65), 26 (54)
27
$CH_2$=CH|CHO
27
56
M-H
20
30
40
50
60
70
80
M/Z

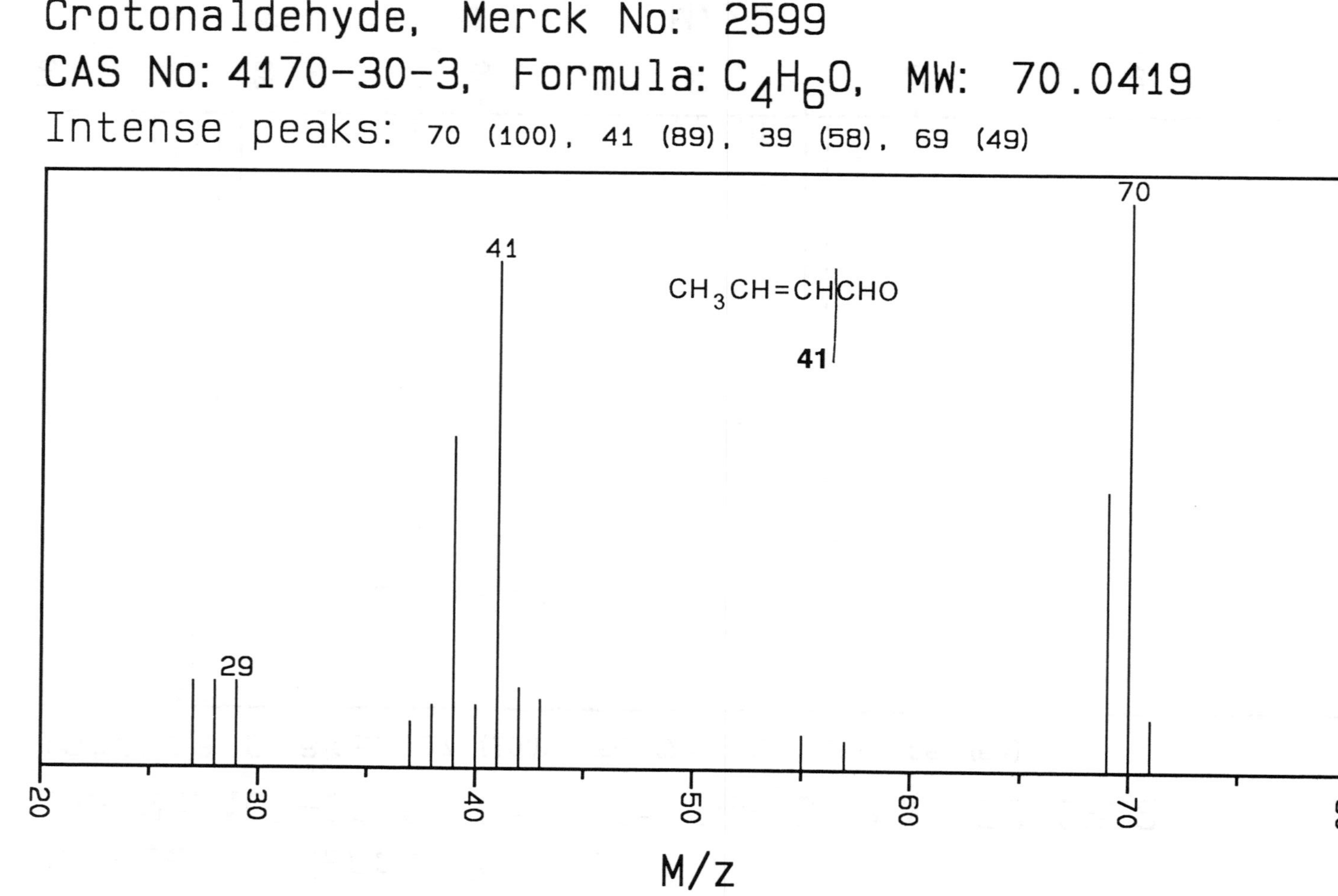
Crotonaldehyde, Merck No: 2599
CAS No: 4170-30-3, Formula: $C_4H_6O$, MW: 70.0419
Intense peaks: 70 (100), 41 (89), 39 (58), 69 (49)
70
41
$CH_3CH=CHCHO$
41
29
20
30
40
50
60
70
80
M/Z

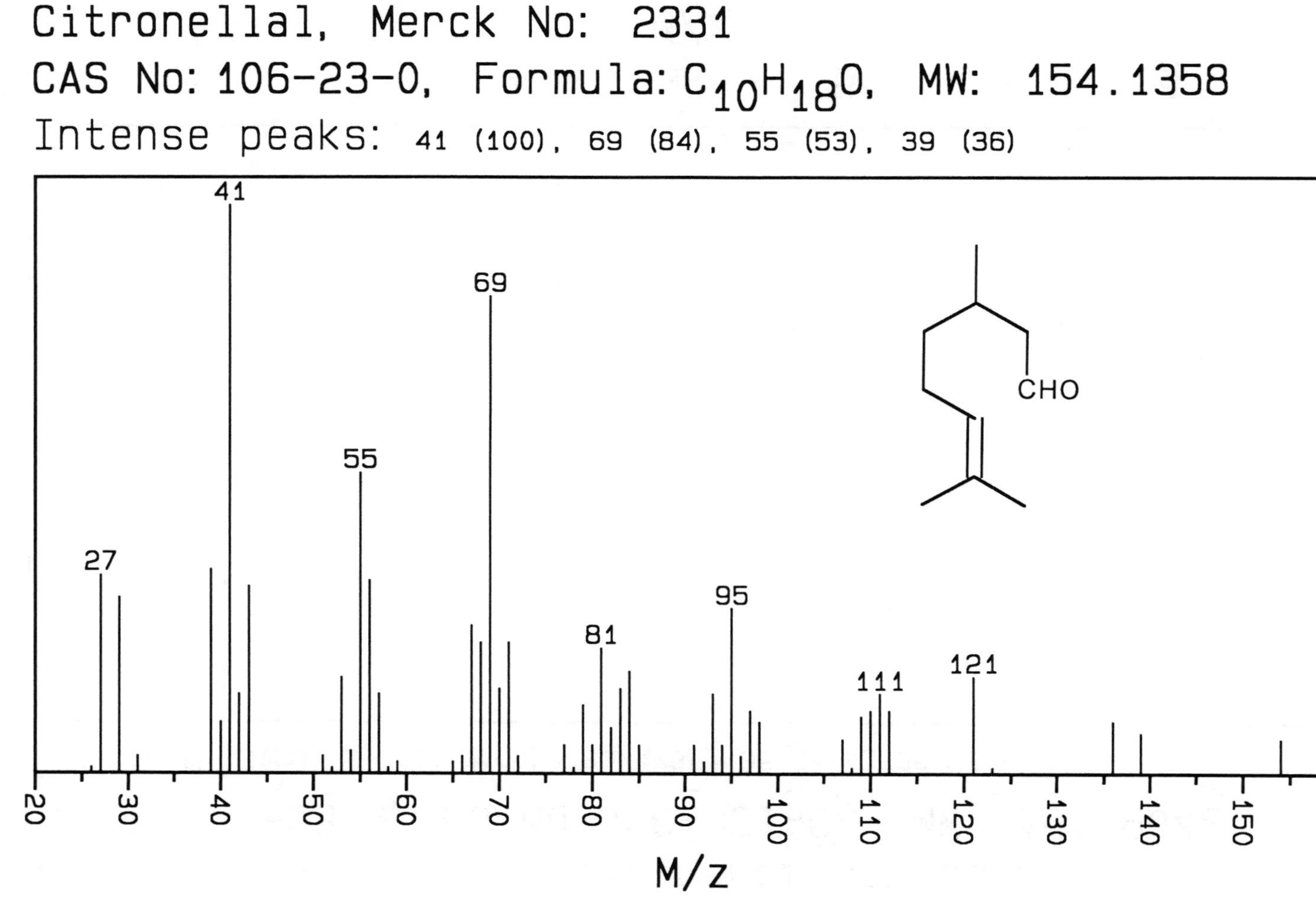
Citronellal, Merck No: 2331
CAS No: 106-23-0, Formula: C10H18O, MW: 154.1358
Intense peaks: 41 (100), 69 (84), 55 (53), 39 (36)
41
69
55
27
81
95
111
121
CHO
20
30
40
50
60
70
80
90
100
110
120
130
140
150
160
M/Z

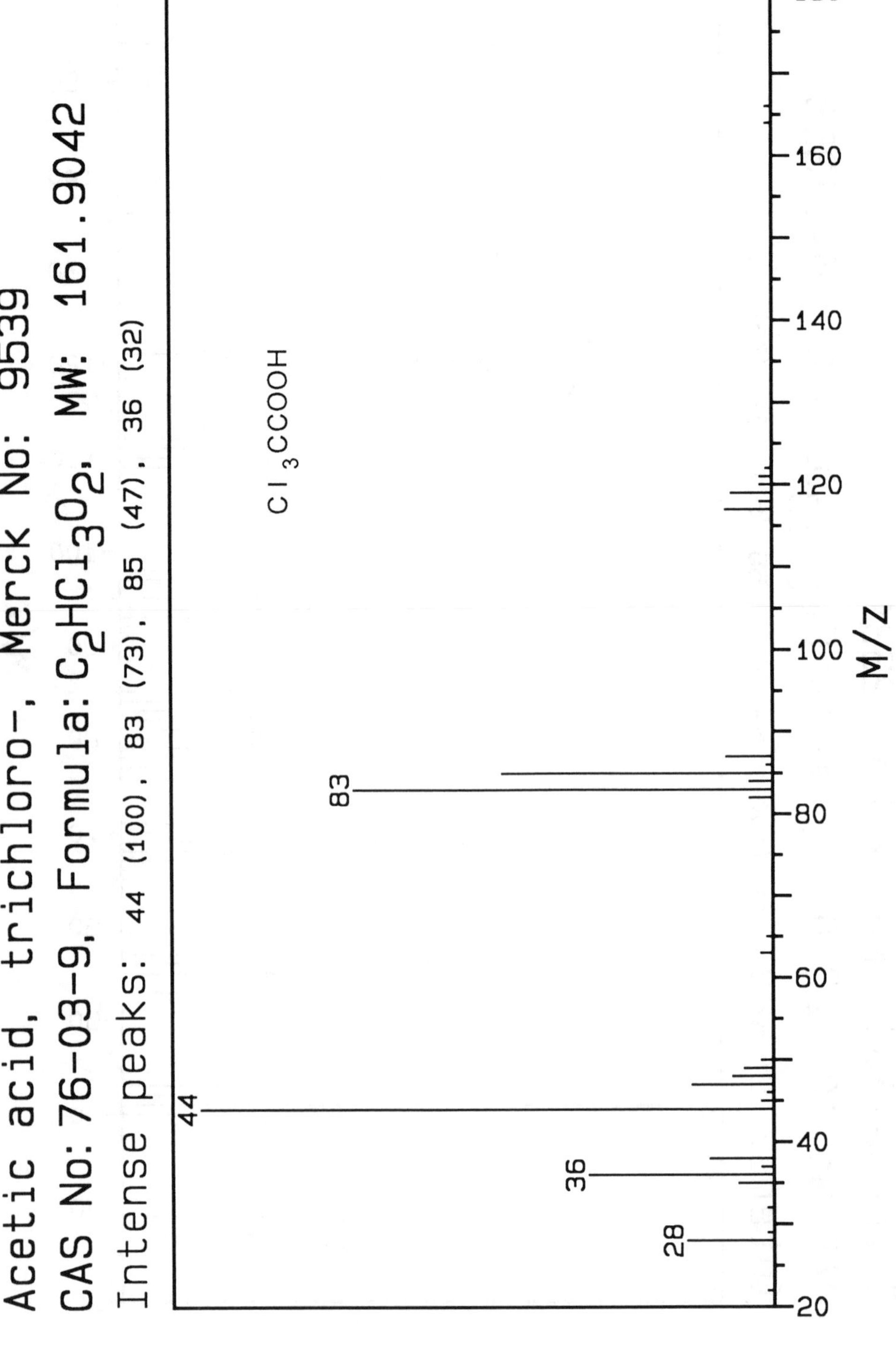
Acetic acid, trichloro-, Merck No: 9539
CAS No: 76-03-9, Formula: $C_2HCl_3O_2$, MW: 161.9042
Intense peaks: 44 (100), 83 (73), 85 (47), 36 (32)
$Cl_3CCOOH$
28
36
44
83
20
40
60
80
100
120
140
160
180
M/Z

Propanoic acid, Merck No: 7837

CAS No: 79-09-4, Formula: $C_3H_6O_2$, MW: 74.0368

Intense peaks: 28 (100), 29 (84), 74 (79), 27 (62)

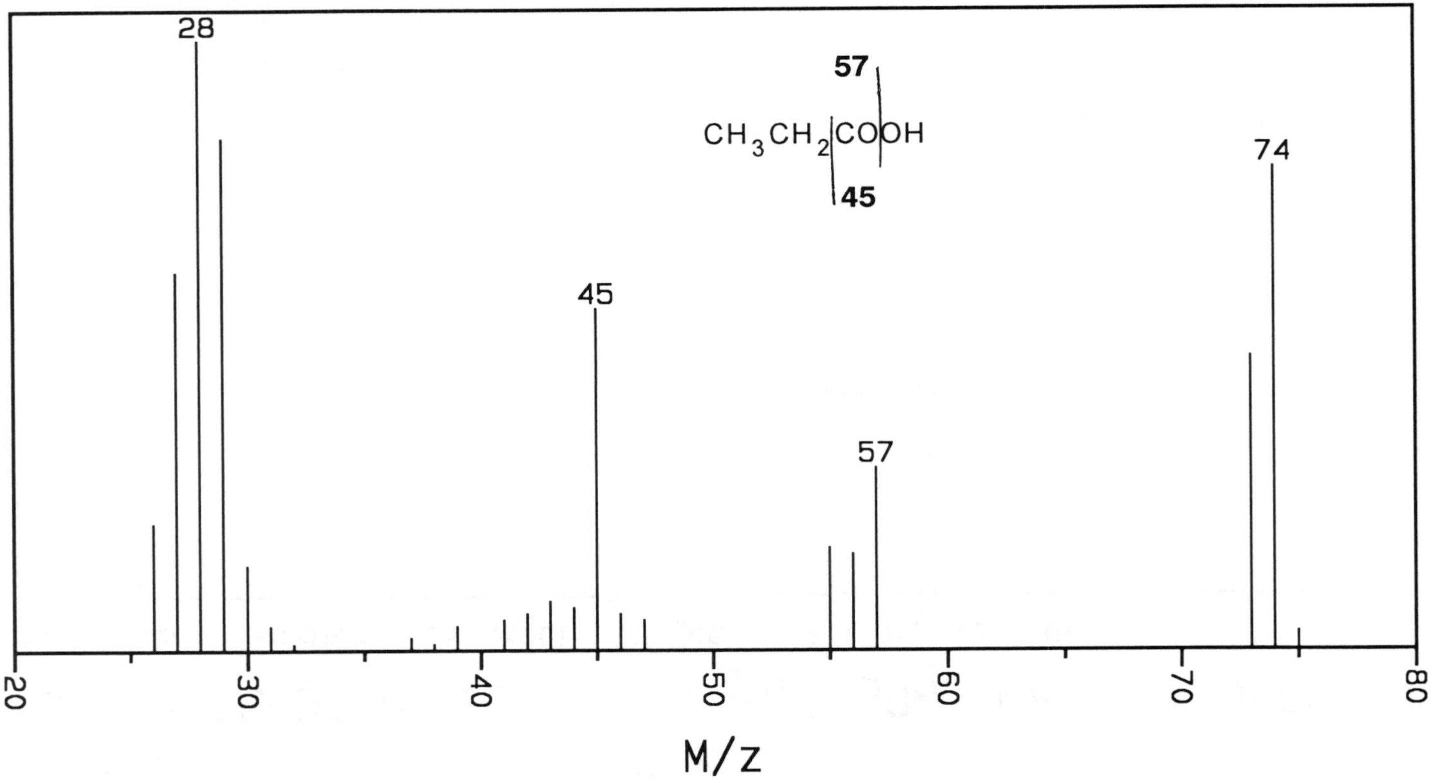

**57**
$CH_3CH_2$|CO|OH
**45**

28
45
57
74

20 30 40 50 60 70 80

M/Z

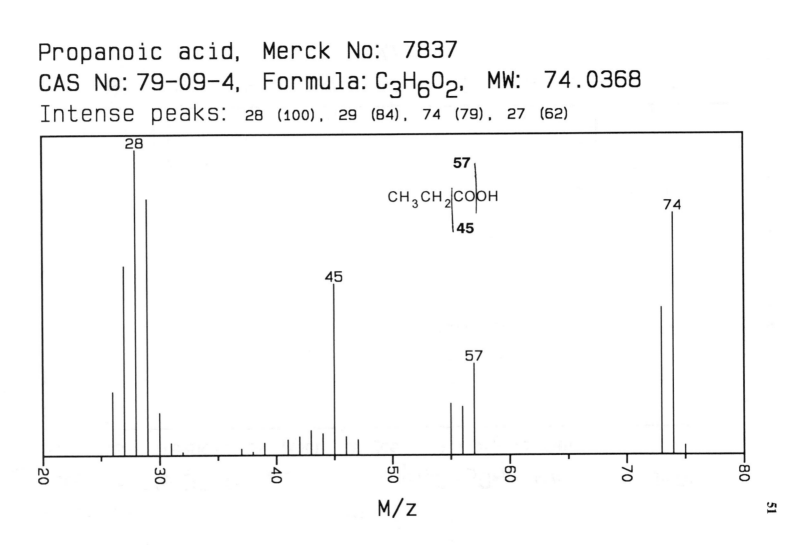

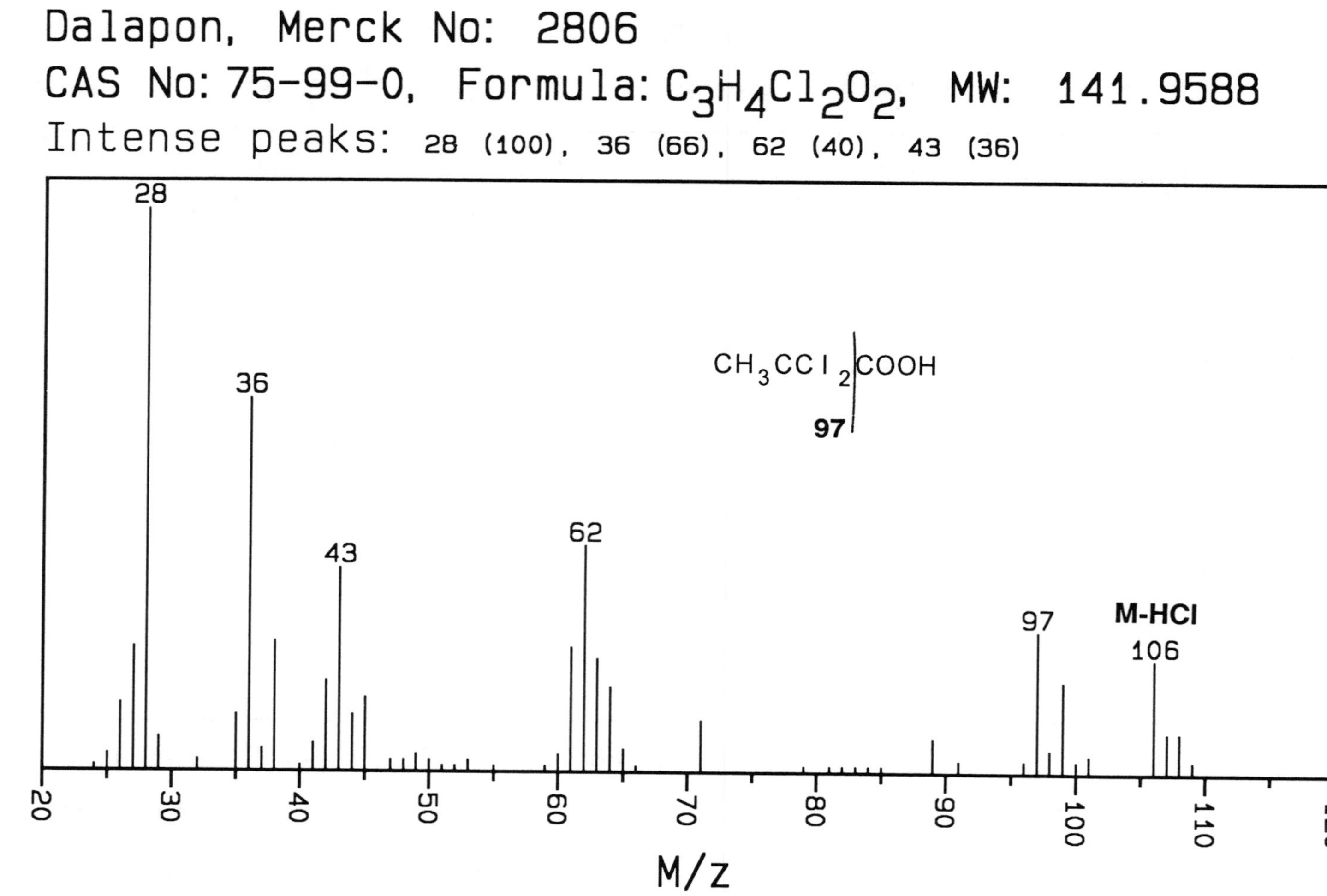
Dalapon, Merck No: 2806
CAS No: 75-99-0, Formula: $C_3H_4Cl_2O_2$, MW: 141.9588
Intense peaks: 28 (100), 36 (66), 62 (40), 43 (36)
28
36
43
62
$CH_3CCl_2COOH$
97
97
M-HCl
106
20
30
40
50
60
70
80
90
100
110
120
M/Z

Butyric acid, Merck No: 1593

CAS No: 107-92-6, Formula: $C_4H_8O_2$, MW: 88.0524

Intense peaks: 60 (100), 27 (50), 73 (27), 42 (25)

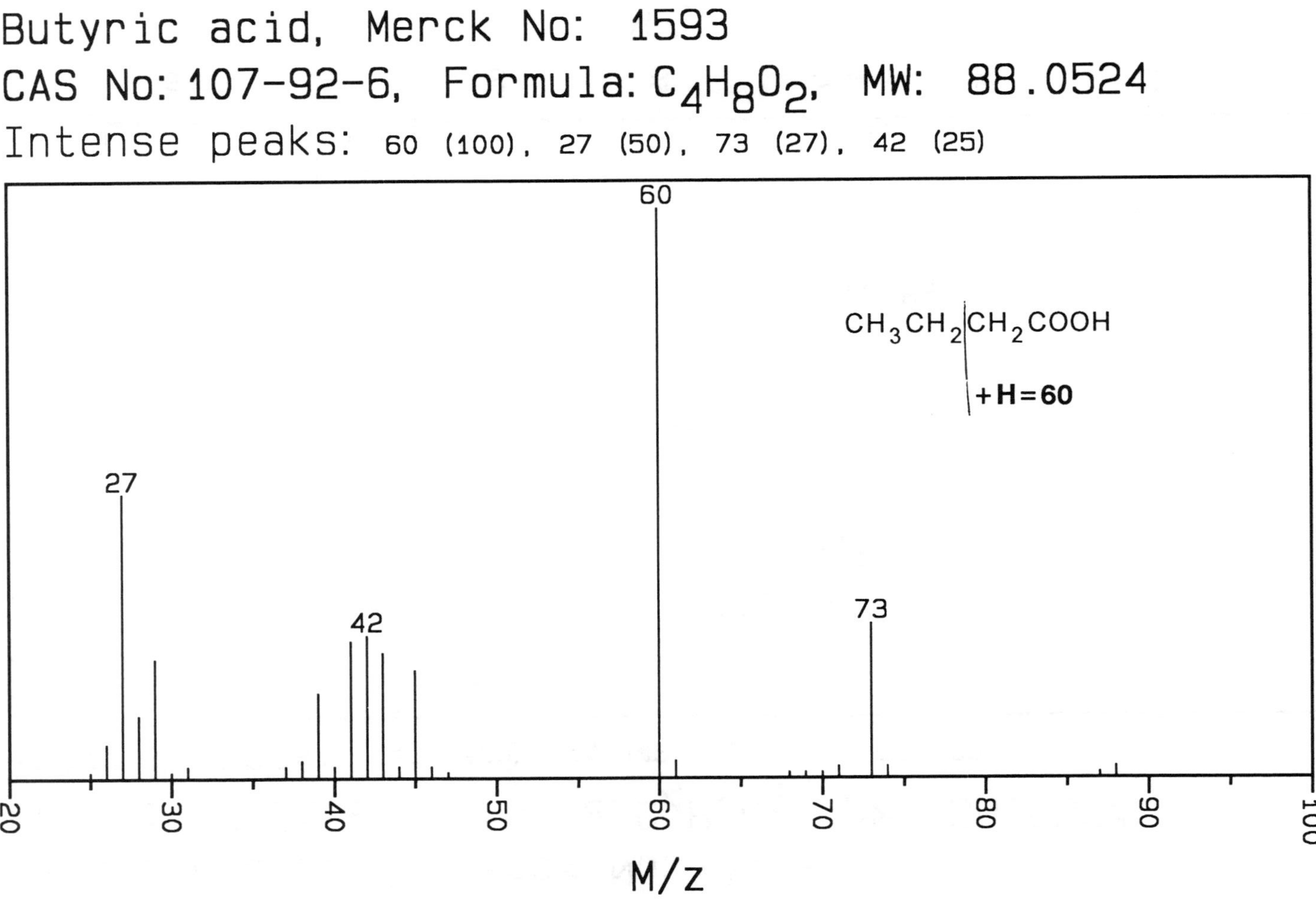

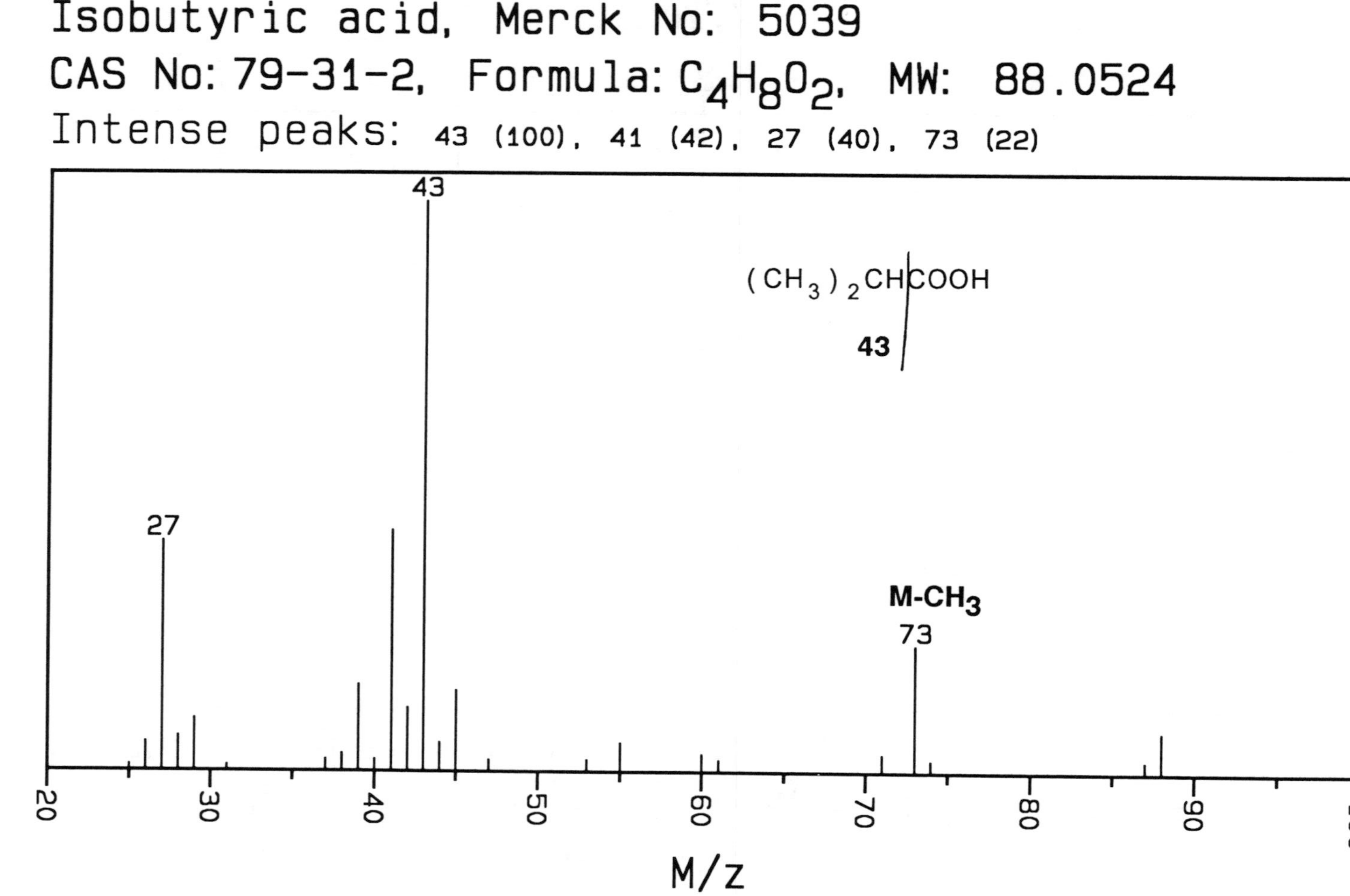
Isobutyric acid, Merck No: 5039
CAS No: 79-31-2, Formula: $C_4H_8O_2$, MW: 88.0524
Intense peaks: 43 (100), 41 (42), 27 (40), 73 (22)
43
$(CH_3)_2CH|COOH$
43
27
M-$CH_3$
73
20
30
40
50
60
70
80
90
100
M/Z

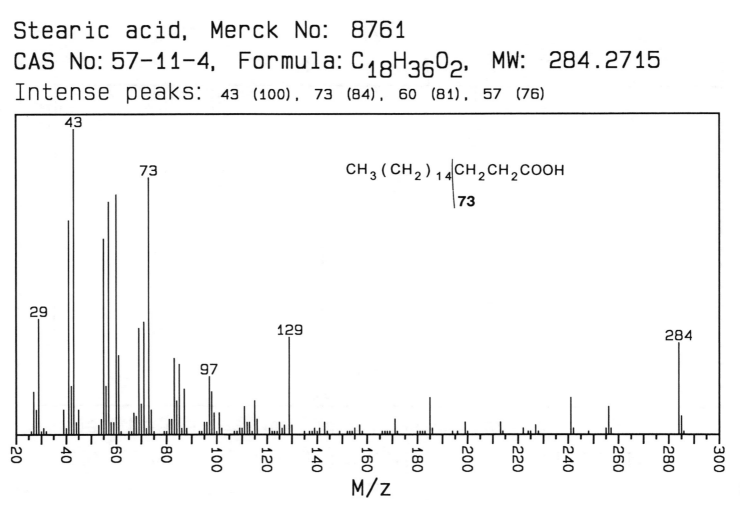

Stearic acid, Merck No: 8761

CAS No: 57-11-4, Formula: $C_{18}H_{36}O_2$, MW: 284.2715

Intense peaks: 43 (100), 73 (84), 60 (81), 57 (76)

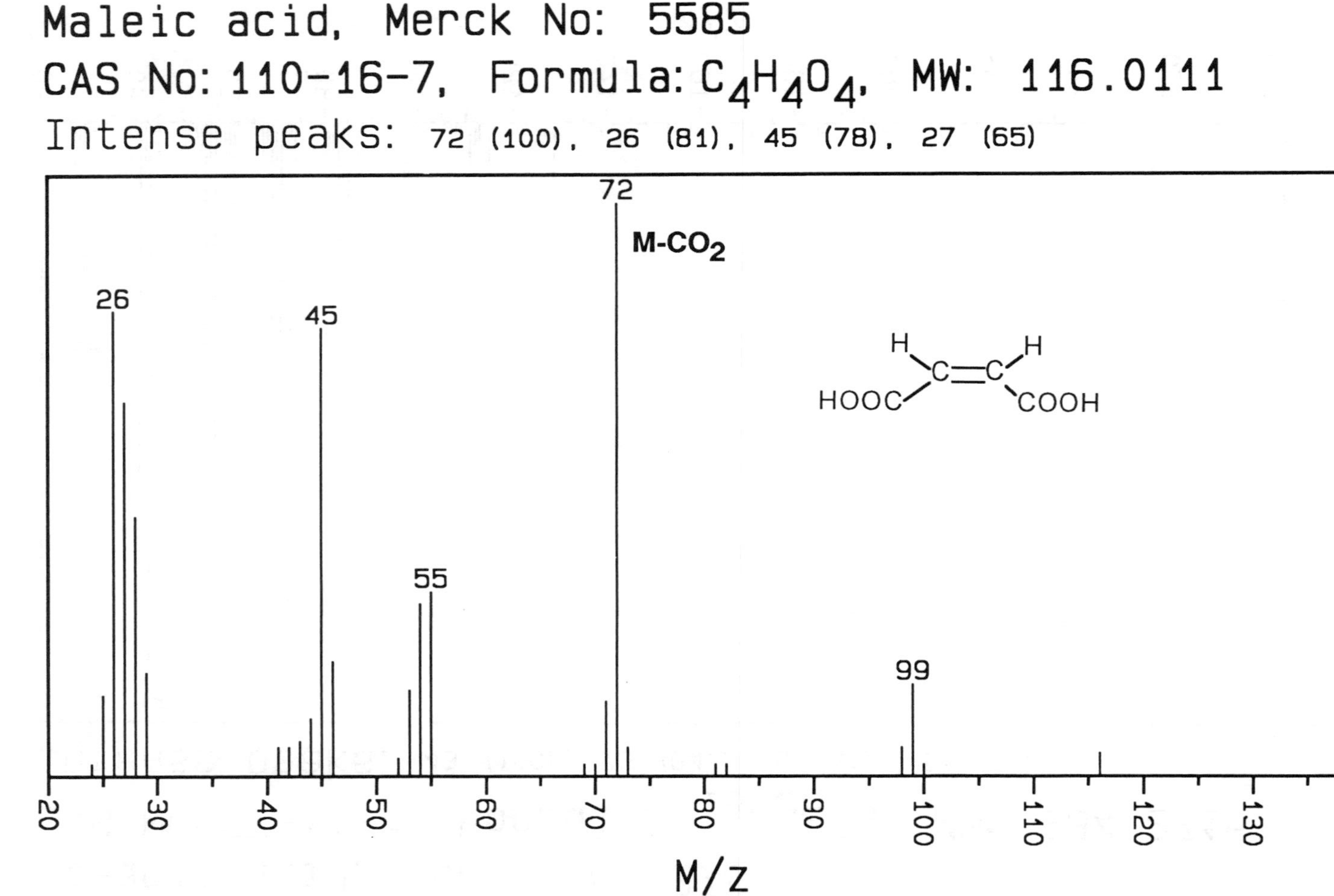
Maleic acid, Merck No: 5585
CAS No: 110-16-7, Formula: $C_4H_4O_4$, MW: 116.0111
Intense peaks: 72 (100), 26 (81), 45 (78), 27 (65)
72
M-$CO_2$
26
45
55
99
H
H
HOOC
COOH
20
30
40
50
60
70
80
90
100
110
120
130
140
M/Z

Fumaric acid, Merck No: 4200

CAS No: 110-17-8, Formula: $C_4H_4O_4$, MW: 116.0111

Intense peaks: 27 (100), 98 (93), 45 (93), 26 (77)

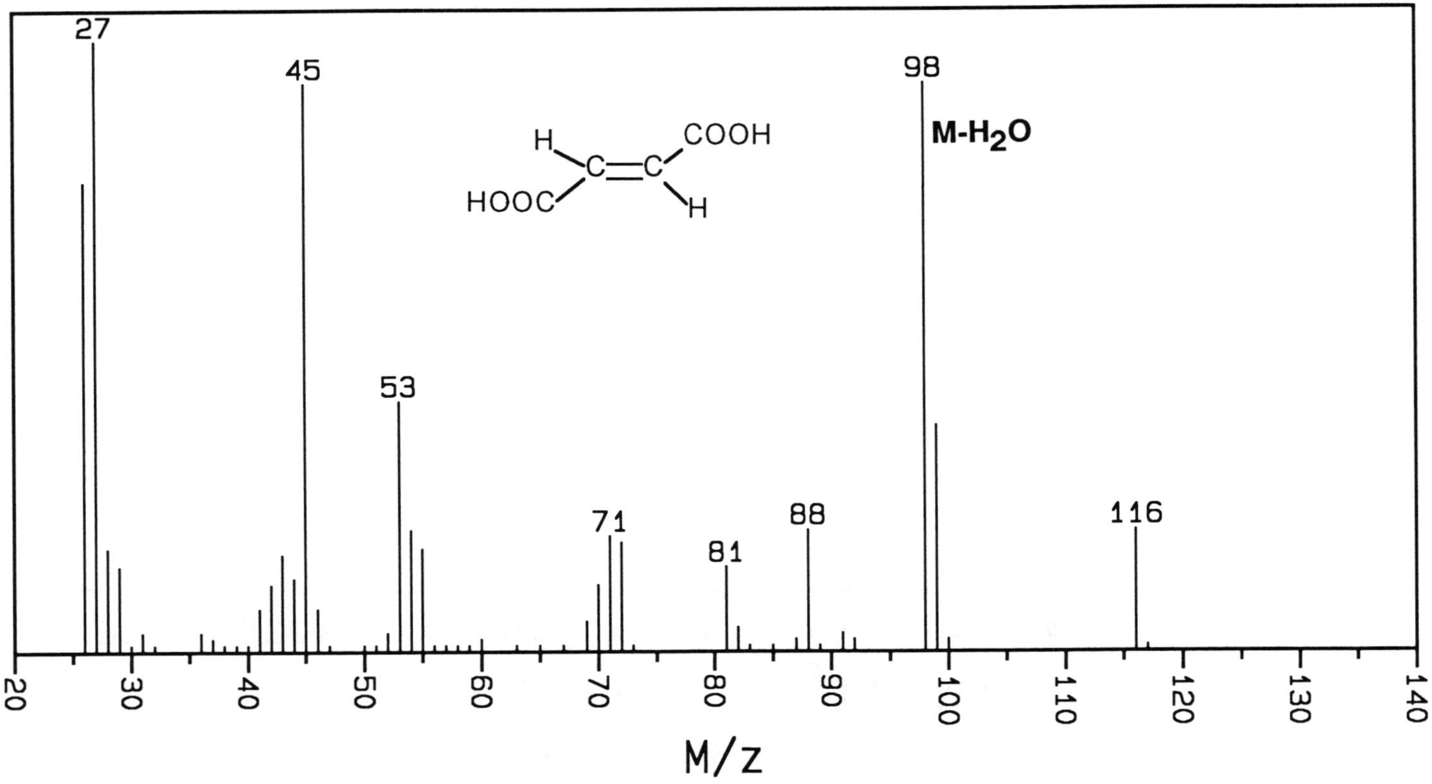

27
45
98
$M-H_2O$
53
71
81
88
116

H
COOH
HOOC
H

20 30 40 50 60 70 80 90 100 110 120 130 140

M/Z

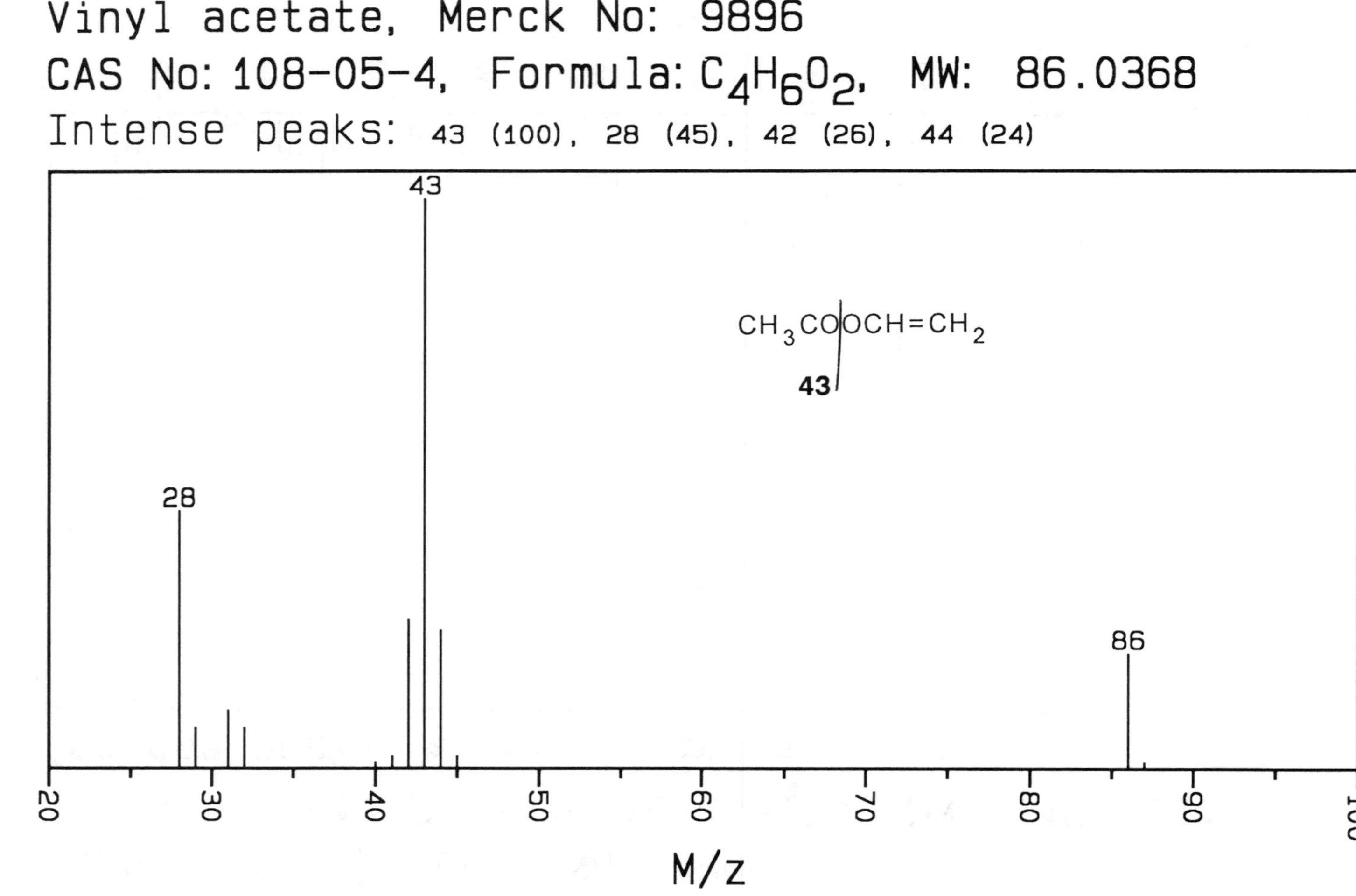
Vinyl acetate, Merck No: 9896
CAS No: 108-05-4, Formula: $C_4H_6O_2$, MW: 86.0368
Intense peaks: 43 (100), 28 (45), 42 (26), 44 (24)
43
$CH_3CO|OCH=CH_2$
43
28
86
20
30
40
50
60
70
80
90
100
M/Z

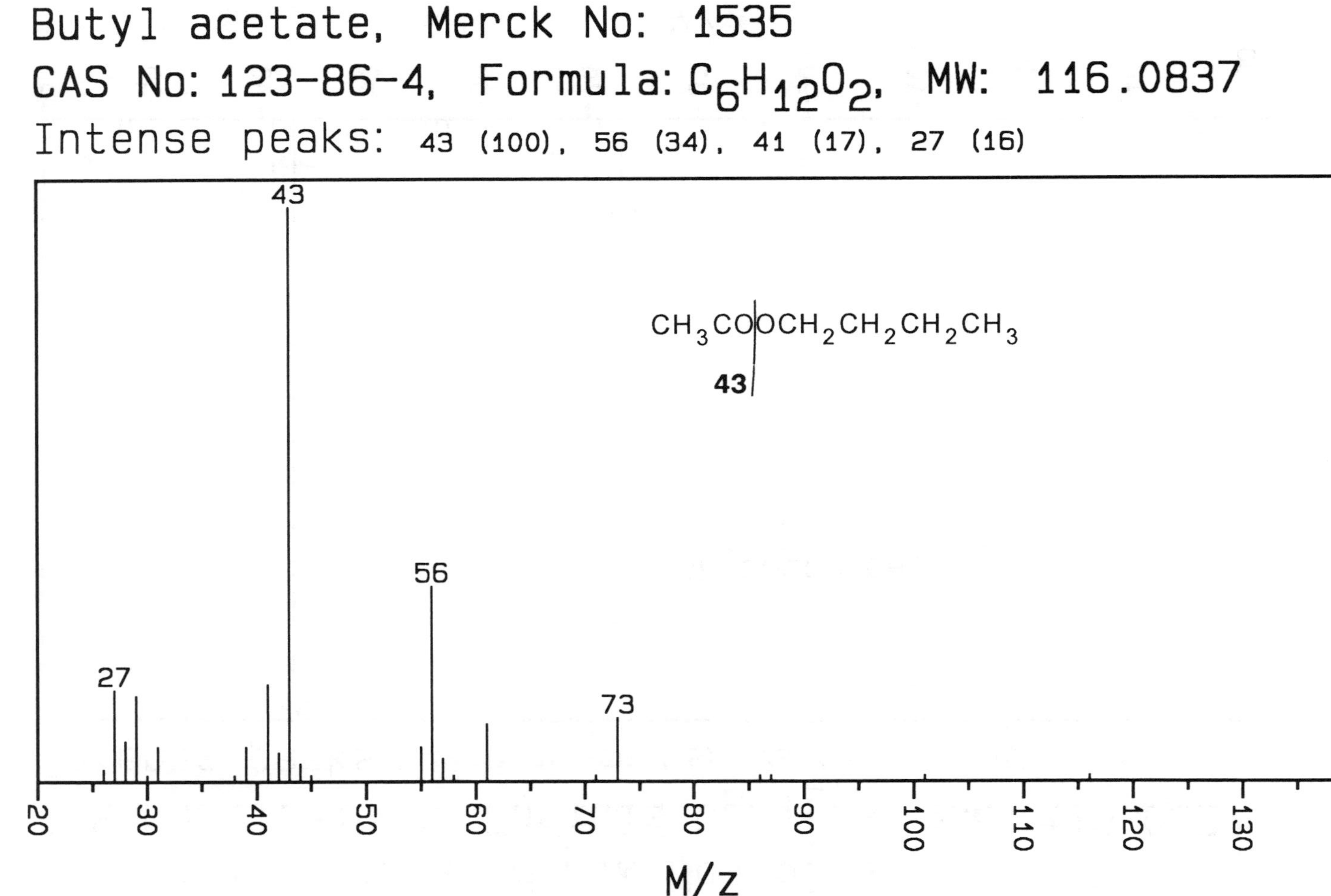
Butyl acetate, Merck No: 1535
CAS No: 123-86-4, Formula: $C_6H_{12}O_2$, MW: 116.0837
Intense peaks: 43 (100), 56 (34), 41 (17), 27 (16)
43
56
27
73
$CH_3CO|OCH_2CH_2CH_2CH_3$
43
20
30
40
50
60
70
80
90
100
110
120
130
140
M/Z

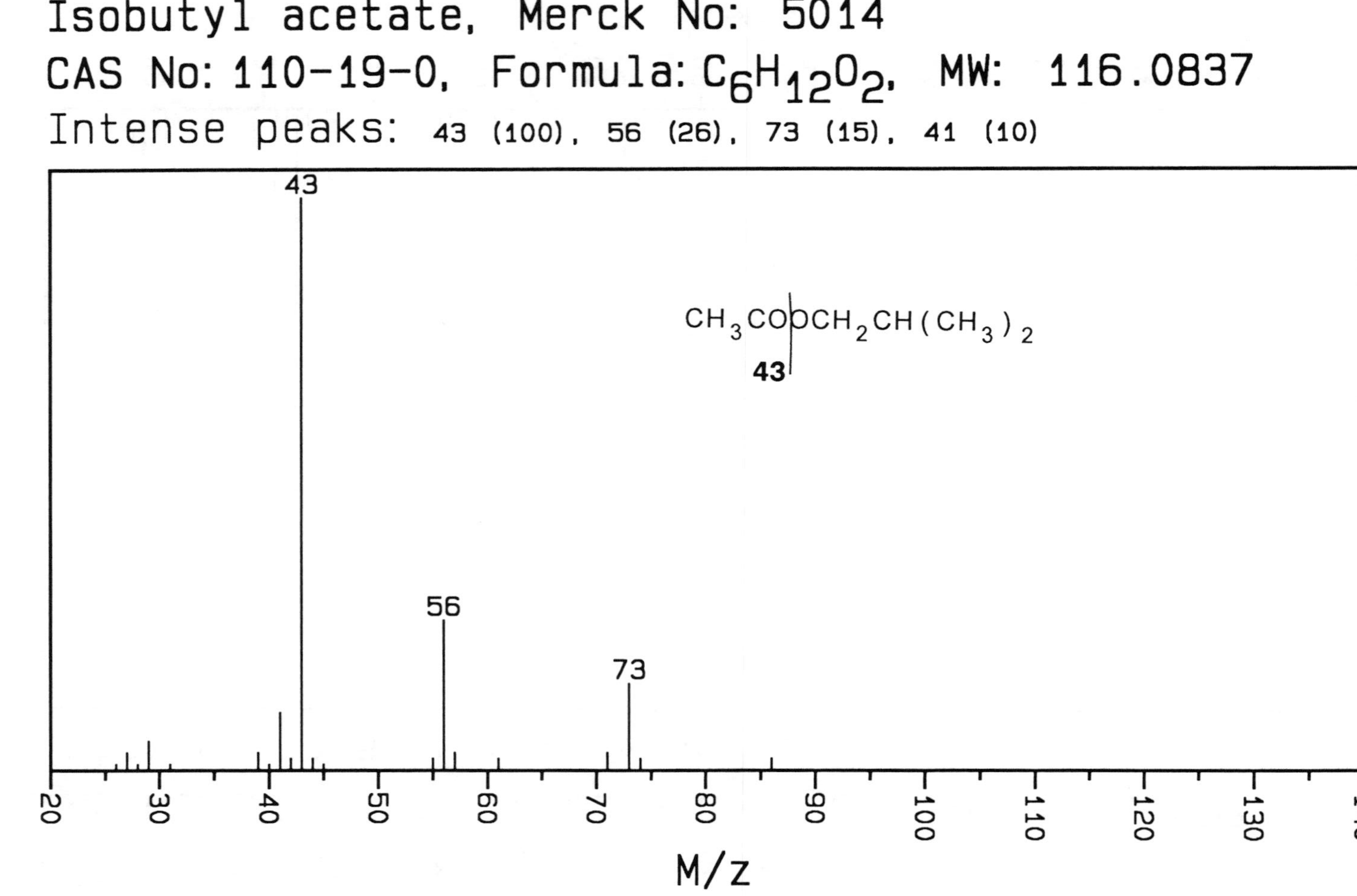

Isobutyl acetate, Merck No: 5014
CAS No: 110-19-0, Formula: C6H12O2, MW: 116.0837
Intense peaks: 43 (100), 56 (26), 73 (15), 41 (10)
43
CH3COOCH2CH(CH3)2
43
56
73
20
30
40
50
60
70
80
90
100
110
120
130
140
M/Z

Isoamyl acetate, Merck No: 4993

CAS No: 123-92-2, Formula: $C_7H_{14}O_2$, MW: 130.0994

Intense peaks: 43 (100), 70 (49), 55 (38), 61 (15)

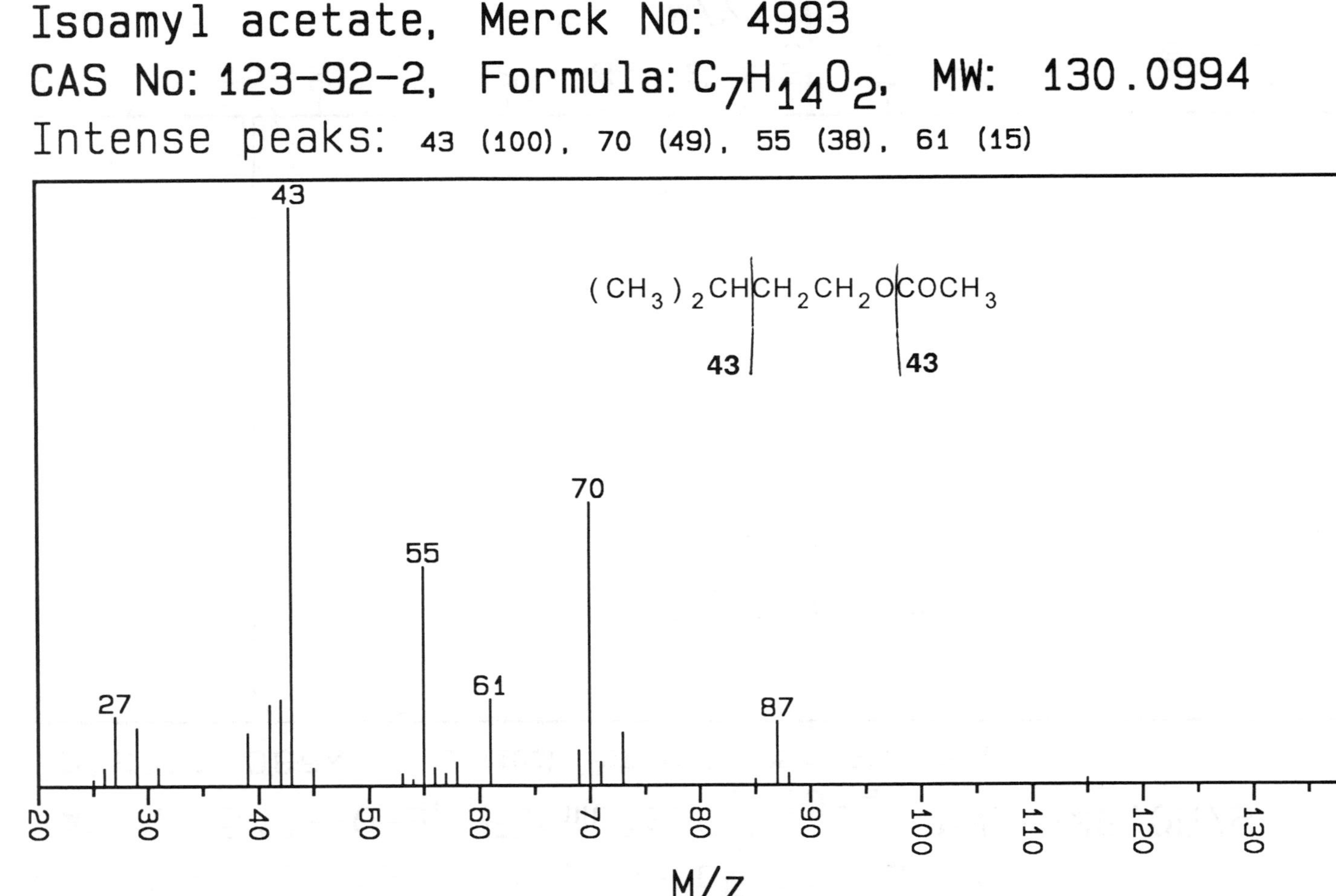

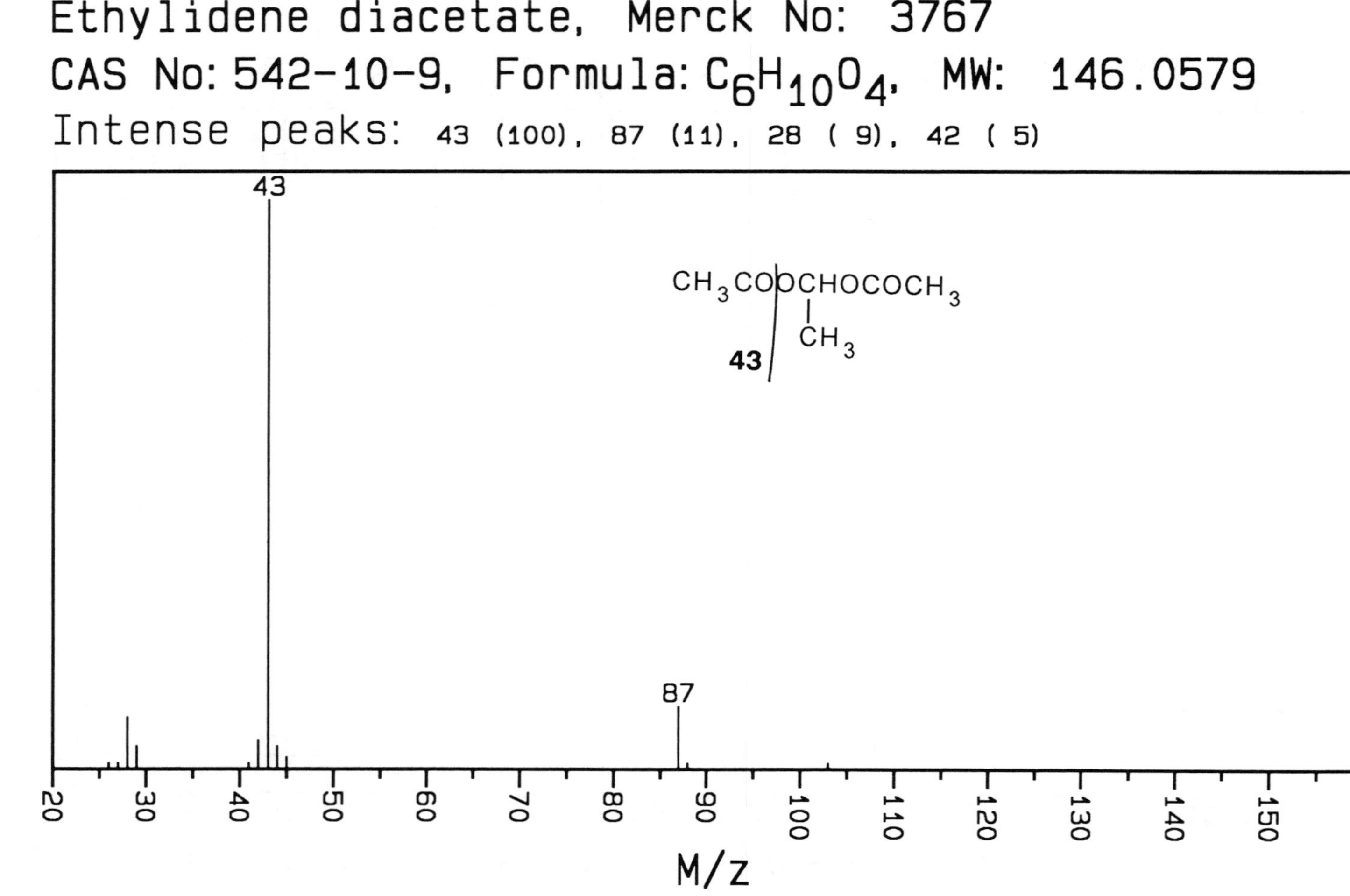
Ethylidene diacetate, Merck No: 3767
CAS No: 542-10-9, Formula: $C_6H_{10}O_4$, MW: 146.0579
Intense peaks: 43 (100), 87 (11), 28 ( 9), 42 ( 5)
43
$CH_3COOCHOCOCH_3$
$CH_3$
43
87
20
30
40
50
60
70
80
90
100
110
120
130
140
150
160
M/Z

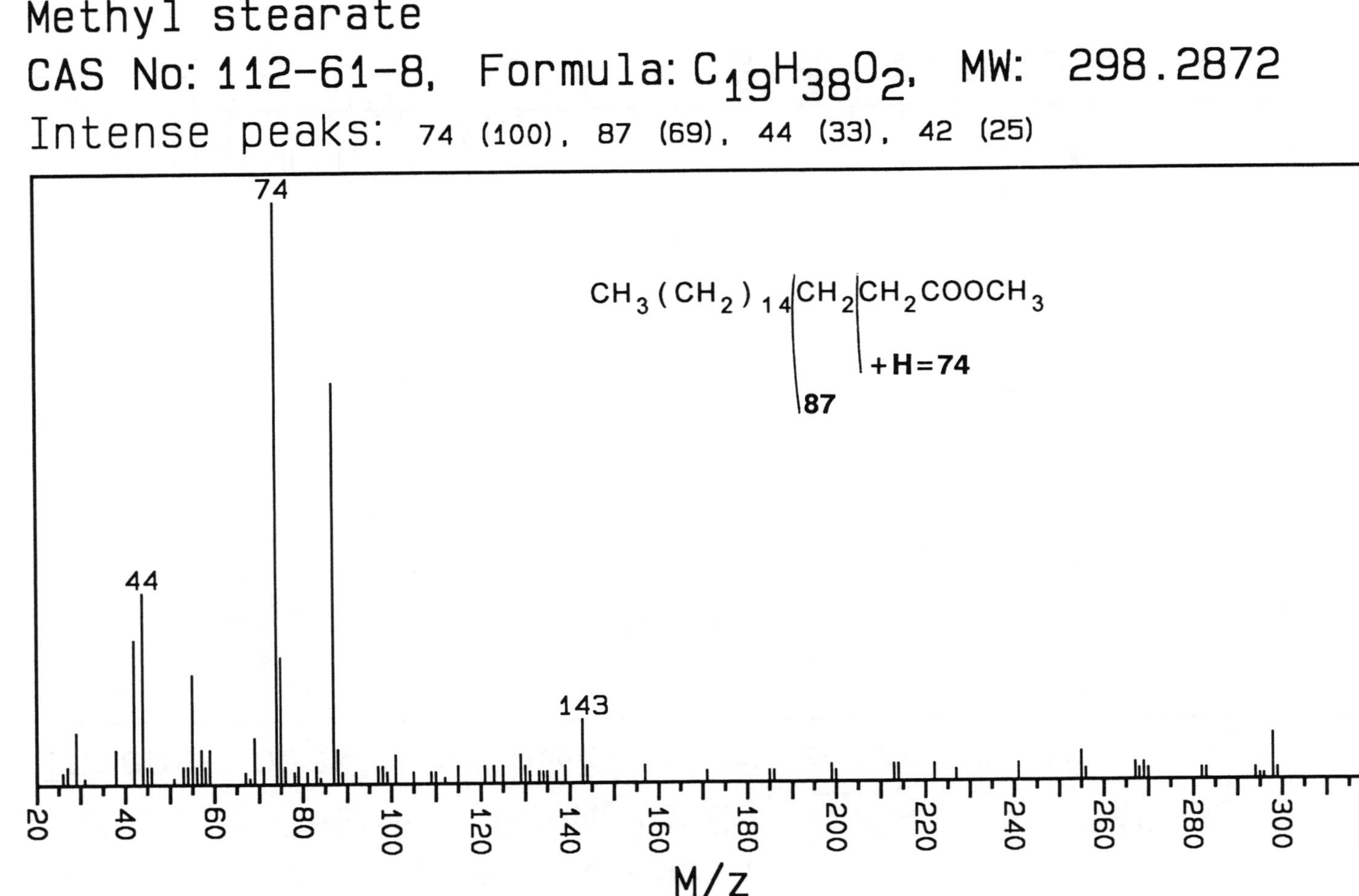
Methyl stearate
CAS No: 112-61-8, Formula: $C_{19}H_{38}O_2$, MW: 298.2872
Intense peaks: 74 (100), 87 (69), 44 (33), 42 (25)
74
44
143
$CH_3(CH_2)_{14}CH_2CH_2COOCH_3$
+H=74
87
20
40
60
80
100
120
140
160
180
200
220
240
260
280
300
320
M/Z

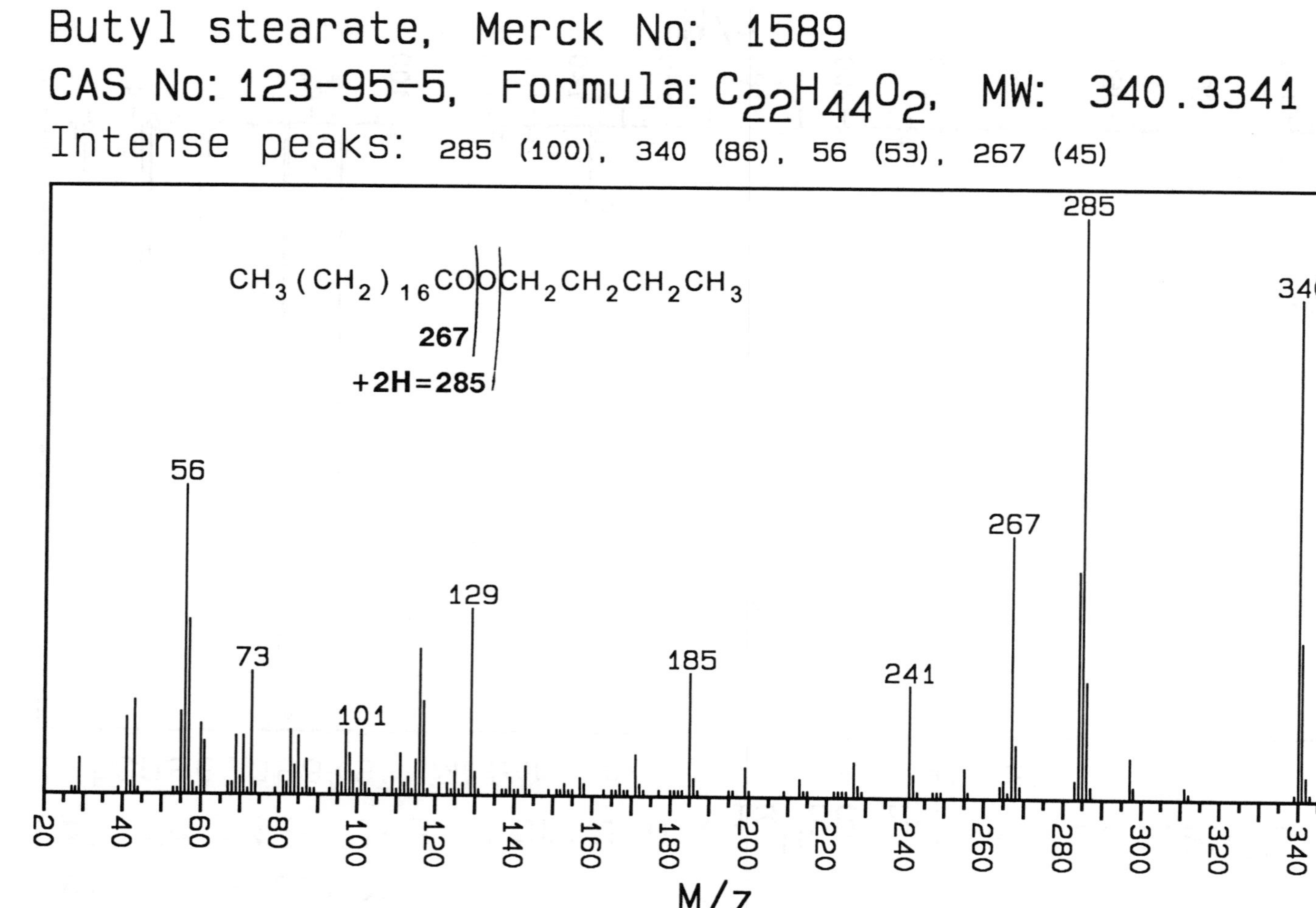

Butyl stearate, Merck No: 1589
CAS No: 123-95-5, Formula: $C_{22}H_{44}O_2$, MW: 340.3341
Intense peaks: 285 (100), 340 (86), 56 (53), 267 (45)
$CH_3(CH_2)_{16}COOCH_2CH_2CH_2CH_3$
267
+2H=285
56
73
101
129
185
241
267
285
340
20
40
60
80
100
120
140
160
180
200
220
240
260
280
300
320
340
360
M/Z

Methoprene, Merck No: 5906

CAS No: 40596-69-8, Formula: $C_{19}H_{34}O_3$, MW: 310.2508

Intense peaks: 73 (100), 110 (33), 69 (26), 81 (25)

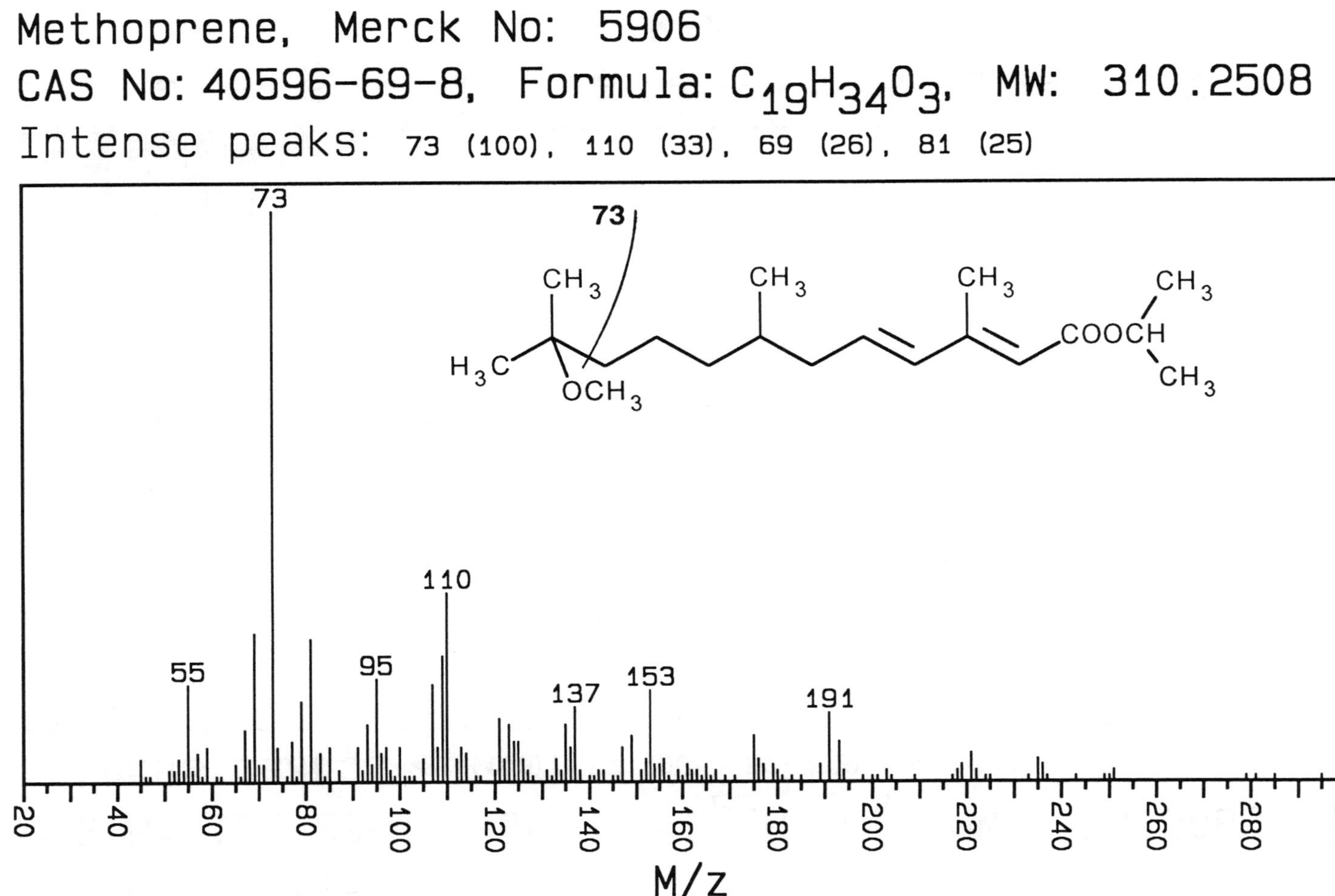

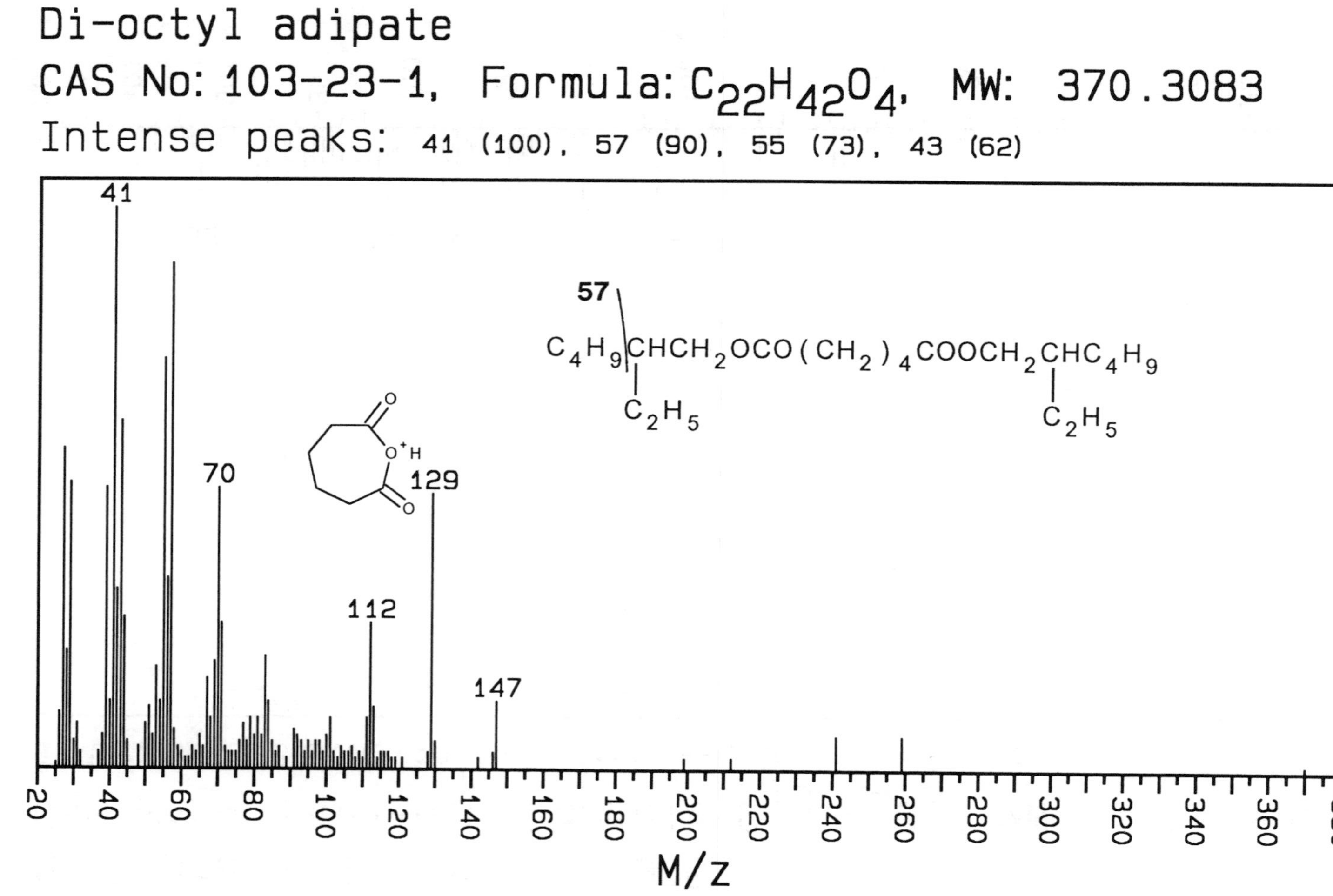
Di-octyl adipate
CAS No: 103-23-1, Formula: $C_{22}H_{42}O_4$, MW: 370.3083
Intense peaks: 41 (100), 57 (90), 55 (73), 43 (62)
57
$C_4H_9CHCH_2OCO(CH_2)_4COOCH_2CHC_4H_9$
$C_2H_5$
$C_2H_5$
41
70
129
112
147
20
40
60
80
100
120
140
160
180
200
220
240
260
280
300
320
340
360
380
M/Z

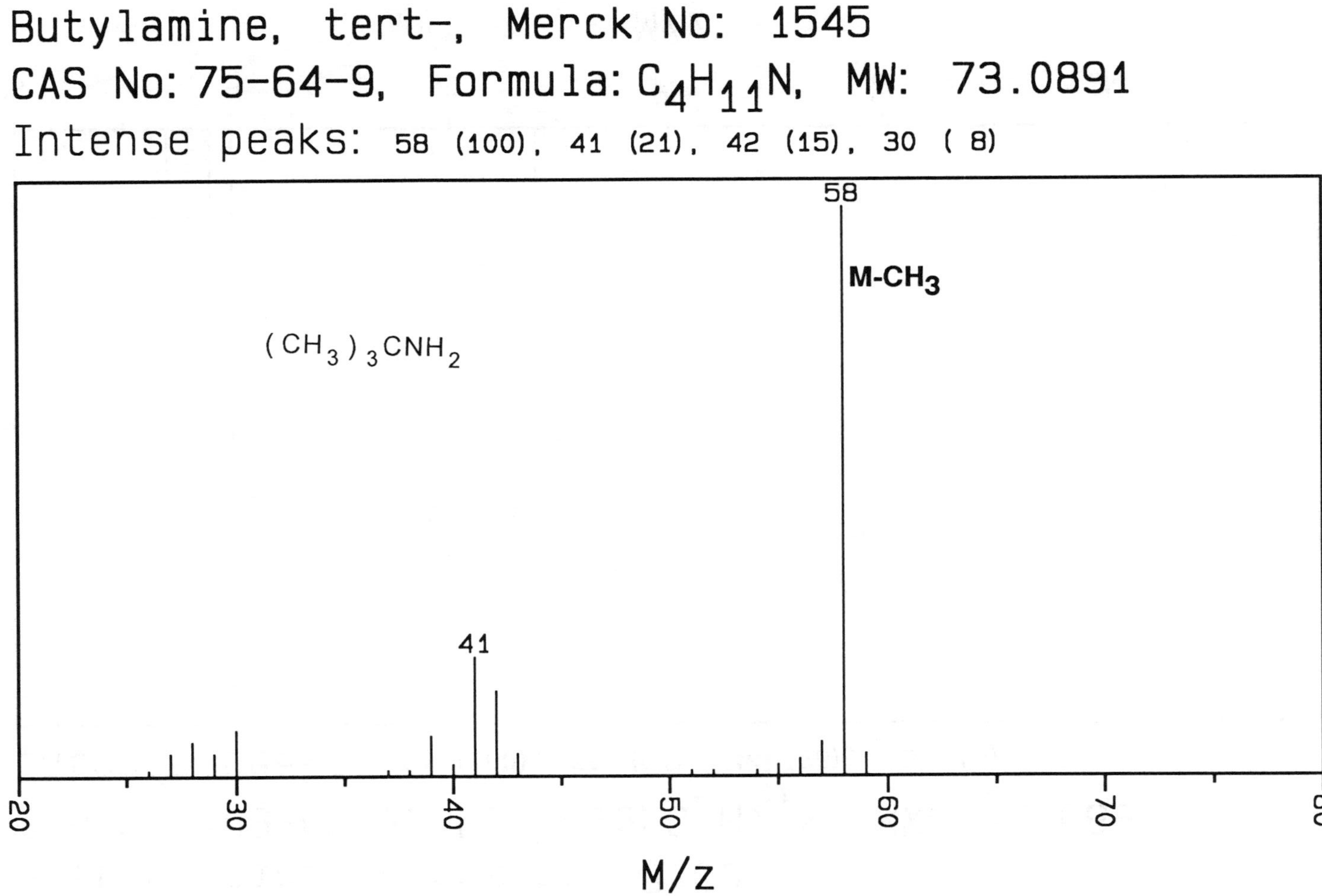

Butylamine, tert-, Merck No: 1545
CAS No: 75-64-9, Formula: $C_4H_{11}N$, MW: 73.0891
Intense peaks: 58 (100), 41 (21), 42 (15), 30 ( 8)
58
M-$CH_3$
$(CH_3)_3CNH_2$
41
20
30
40
50
60
70
80
M/Z

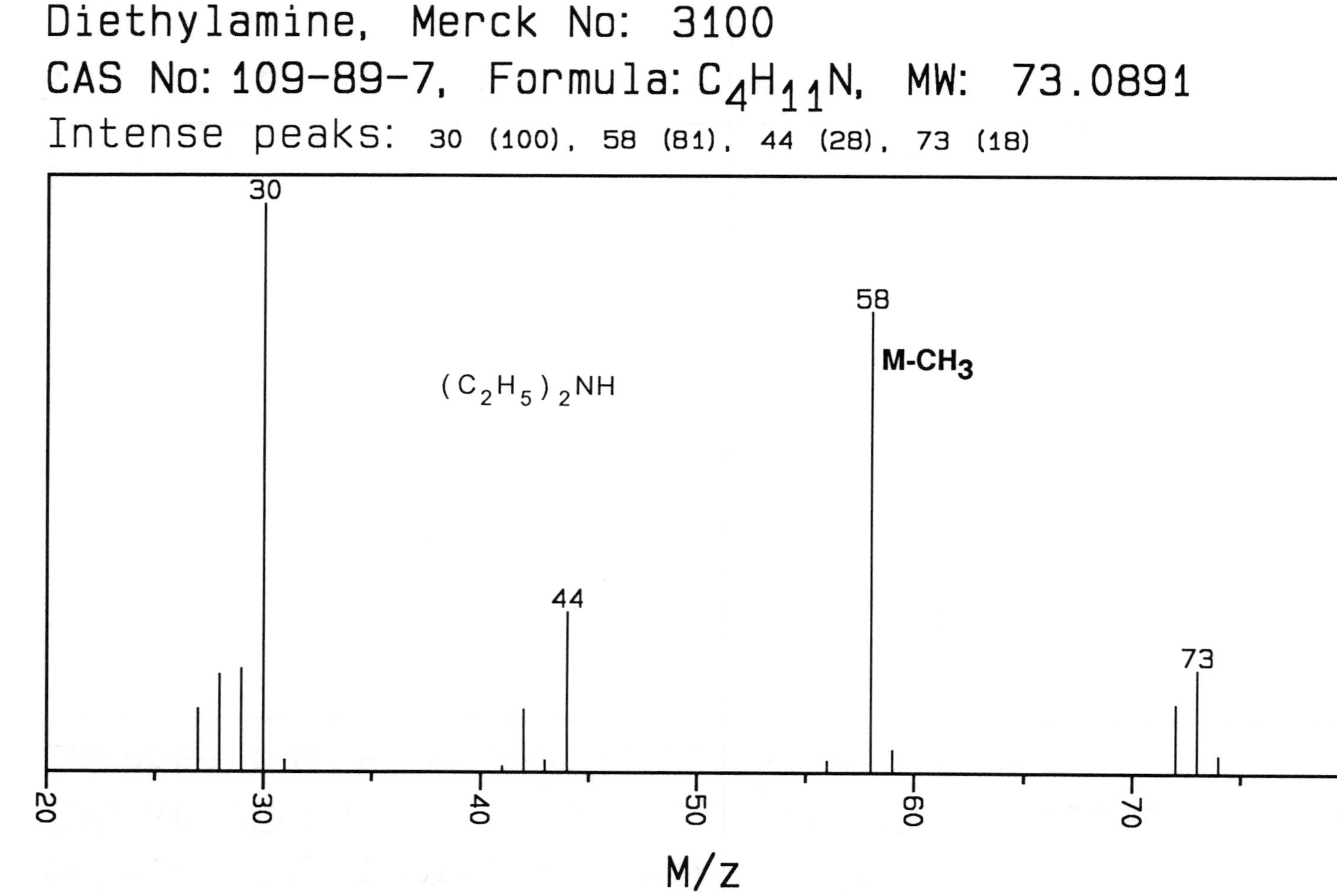
Diethylamine, Merck No: 3100
CAS No: 109-89-7, Formula: $C_4H_{11}N$, MW: 73.0891
Intense peaks: 30 (100), 58 (81), 44 (28), 73 (18)
30
$(C_2H_5)_2NH$
58
M-$CH_3$
44
73
20
30
40
50
60
70
80
M/Z

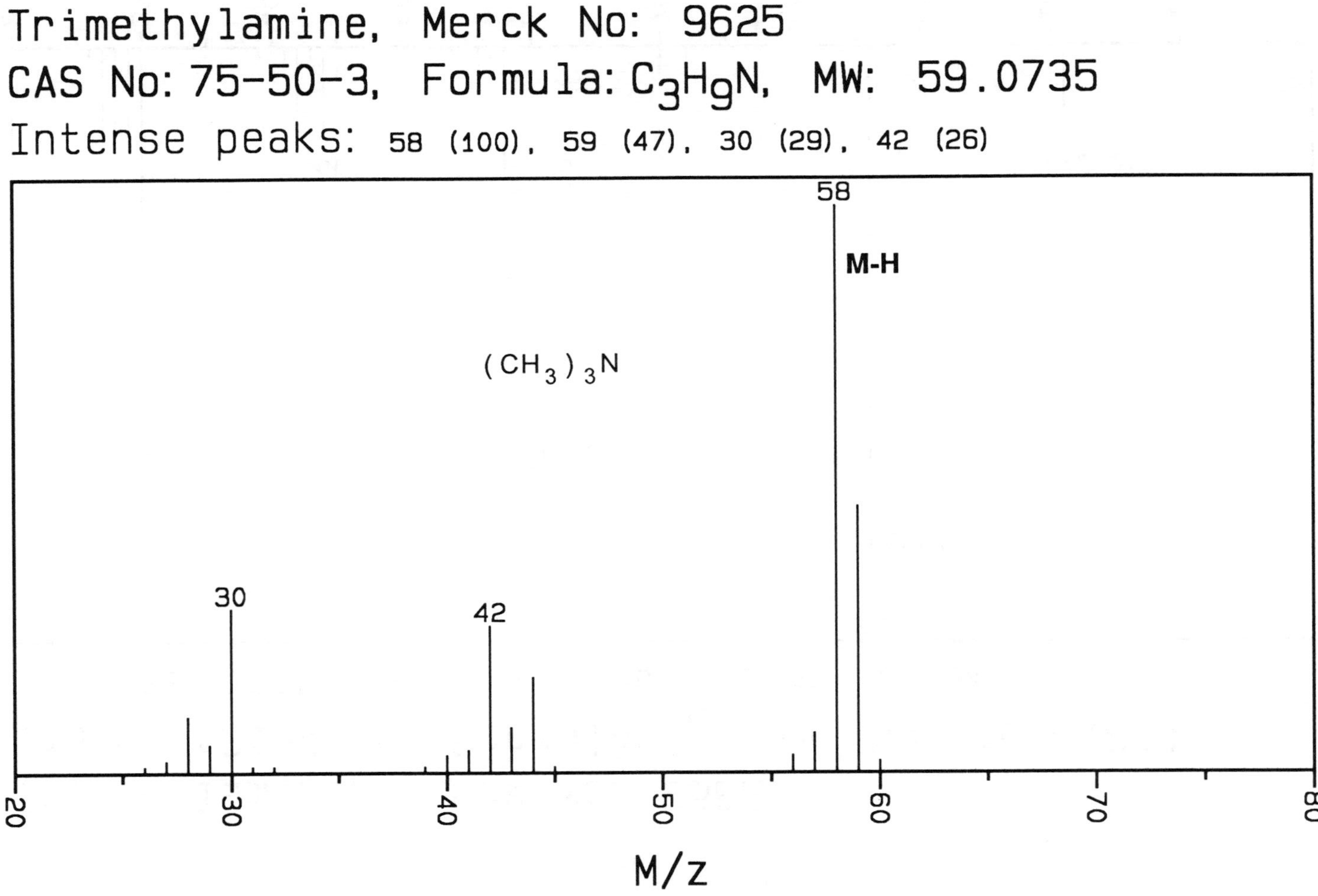

Trimethylamine, Merck No: 9625
CAS No: 75-50-3, Formula: $C_3H_9N$, MW: 59.0735
Intense peaks: 58 (100), 59 (47), 30 (29), 42 (26)
58
M-H
$(CH_3)_3N$
30
42
20
30
40
50
60
70
80
M/Z

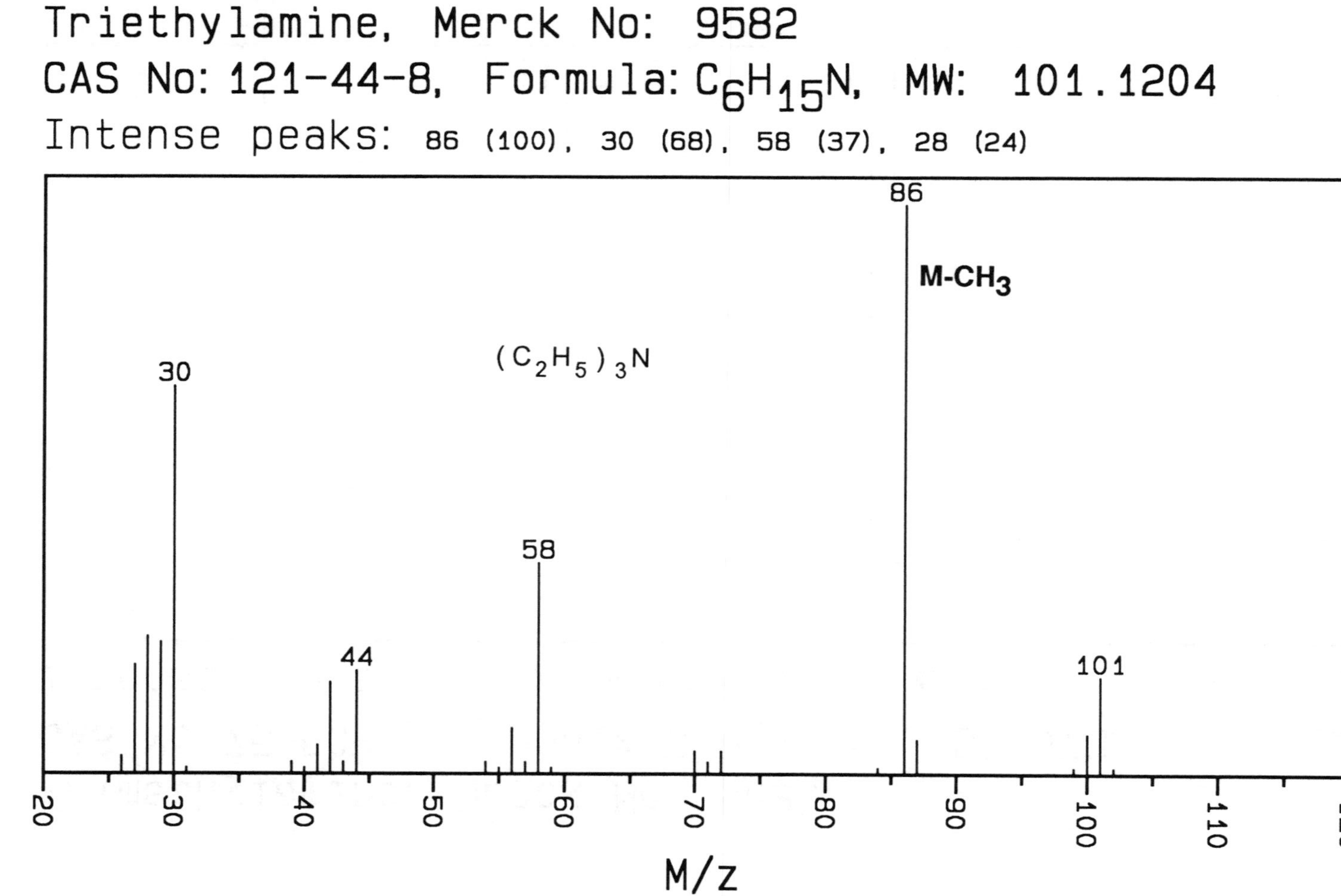
Triethylamine, Merck No: 9582
CAS No: 121-44-8, Formula: $C_6H_{15}N$, MW: 101.1204
Intense peaks: 86 (100), 30 (68), 58 (37), 28 (24)
86
M-$CH_3$
$(C_2H_5)_3N$
30
58
44
101
20
30
40
50
60
70
80
90
100
110
120
M/Z

Diethanolamine, Merck No: 3097

CAS No: 111-42-2, Formula: $C_4H_{11}NO_2$, MW: 105.0791

Intense peaks: 30 (100), 74 (82), 28 (77), 56 (69)

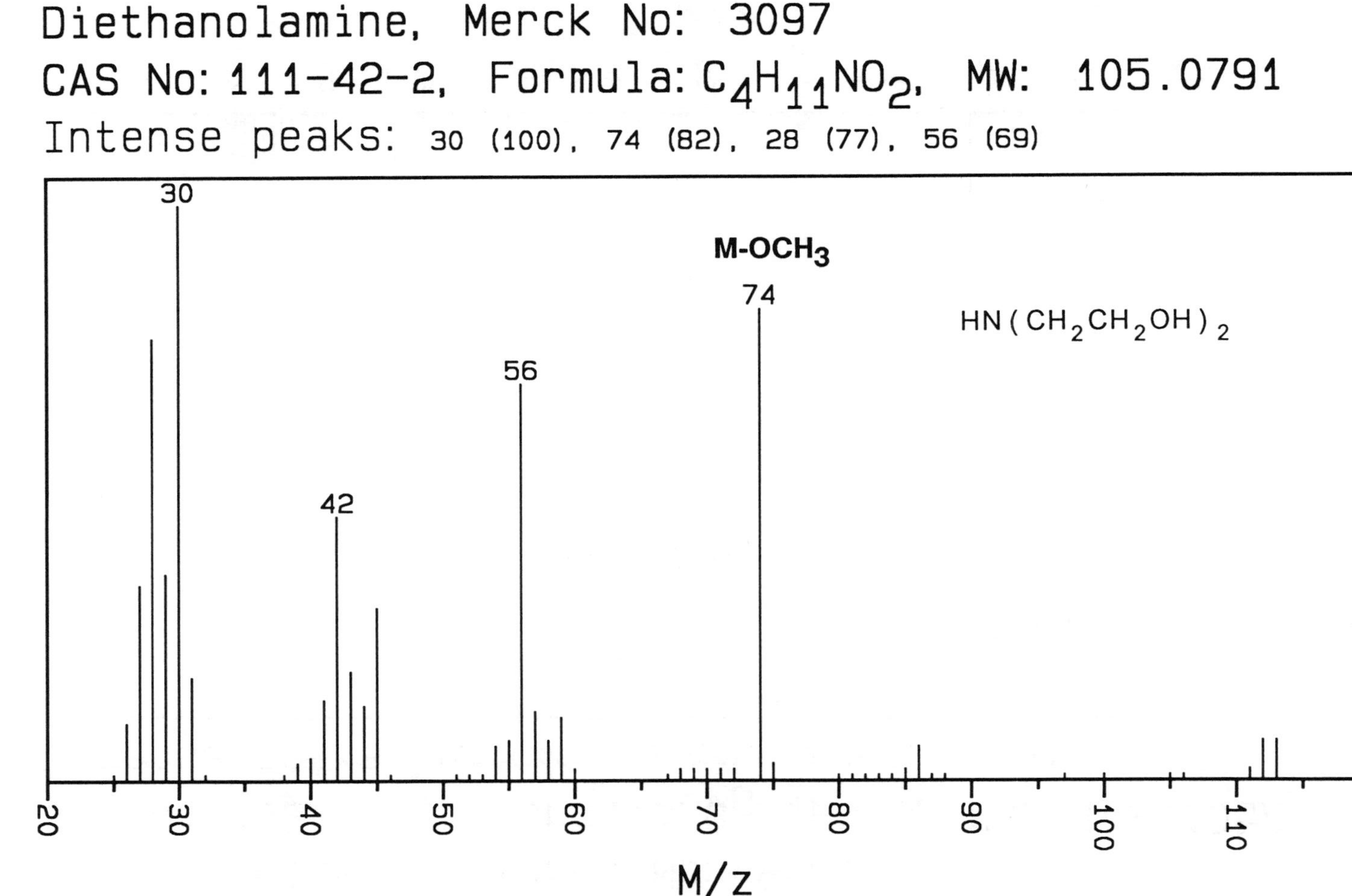

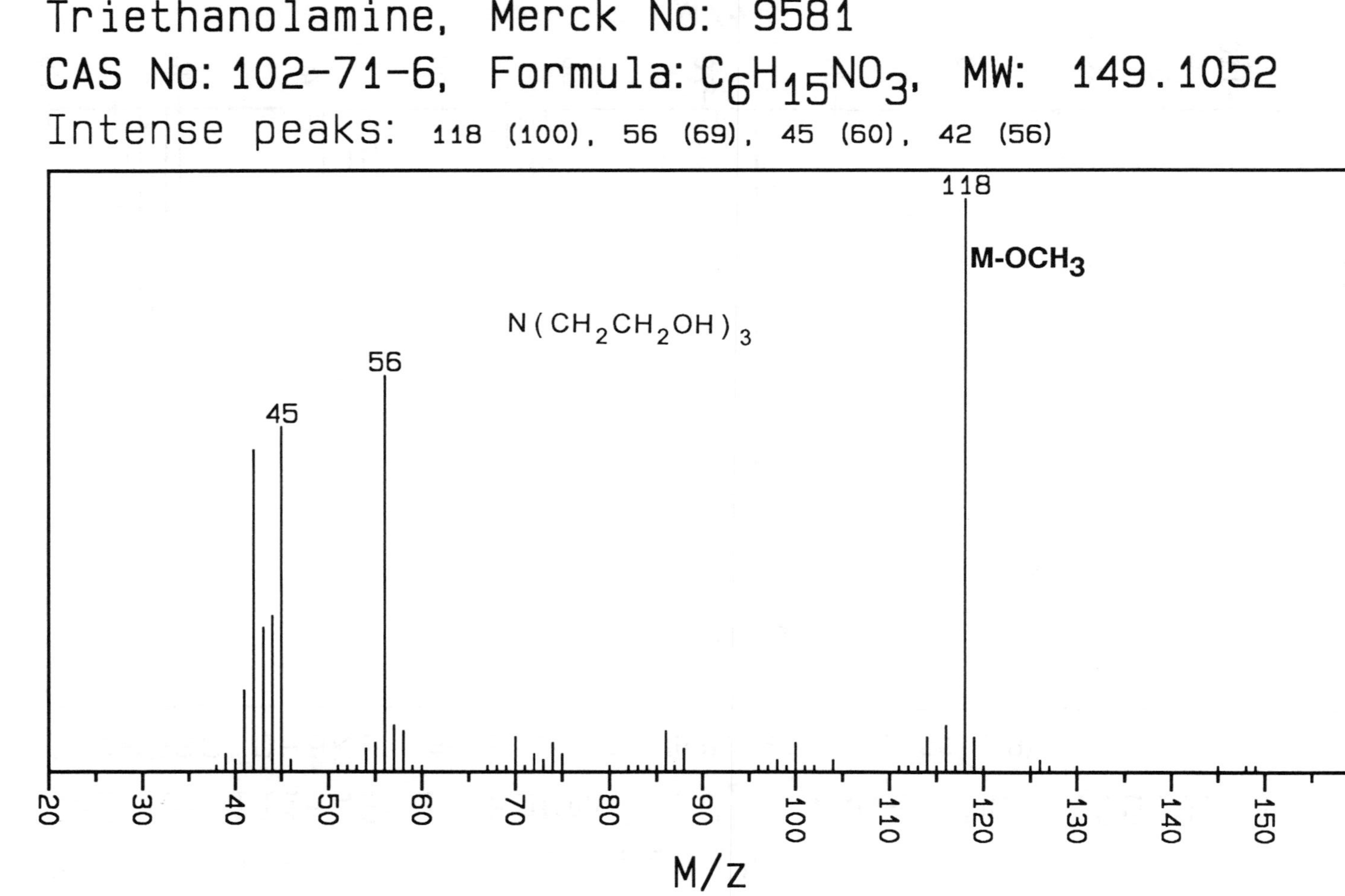
Triethanolamine, Merck No: 9581
CAS No: 102-71-6, Formula: $C_6H_{15}NO_3$, MW: 149.1052
Intense peaks: 118 (100), 56 (69), 45 (60), 42 (56)
118
M-$OCH_3$
$N(CH_2CH_2OH)_3$
56
45
20
30
40
50
60
70
80
90
100
110
120
130
140
150
160
M/Z

N-Nitrosodimethylamine, Merck No: 6558

CAS No: 62-75-9, Formula: $C_2H_6N_2O$, MW: 74.0481

Intense peaks: 74 (100), 42 (60), 43 (35), 44 (18)

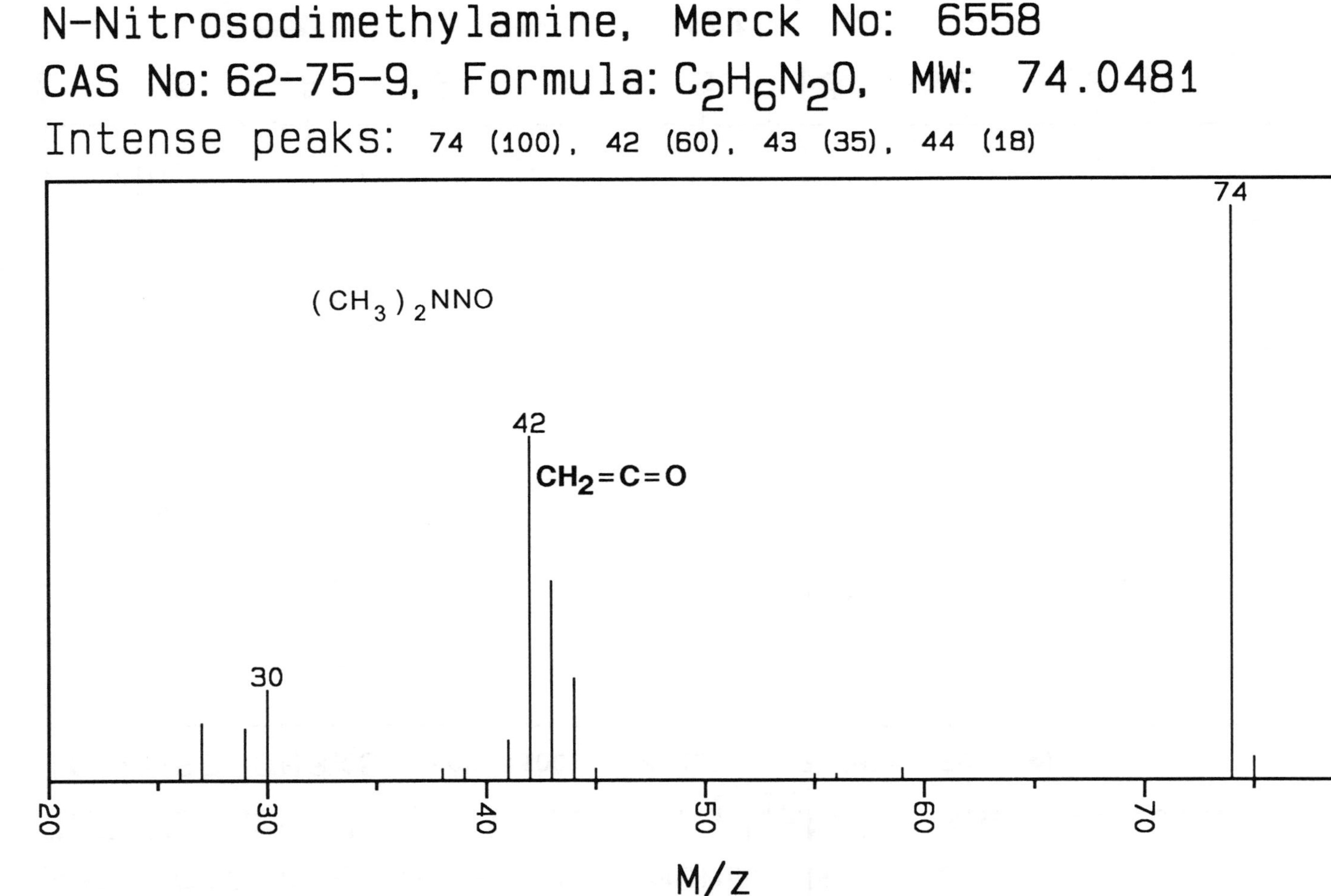

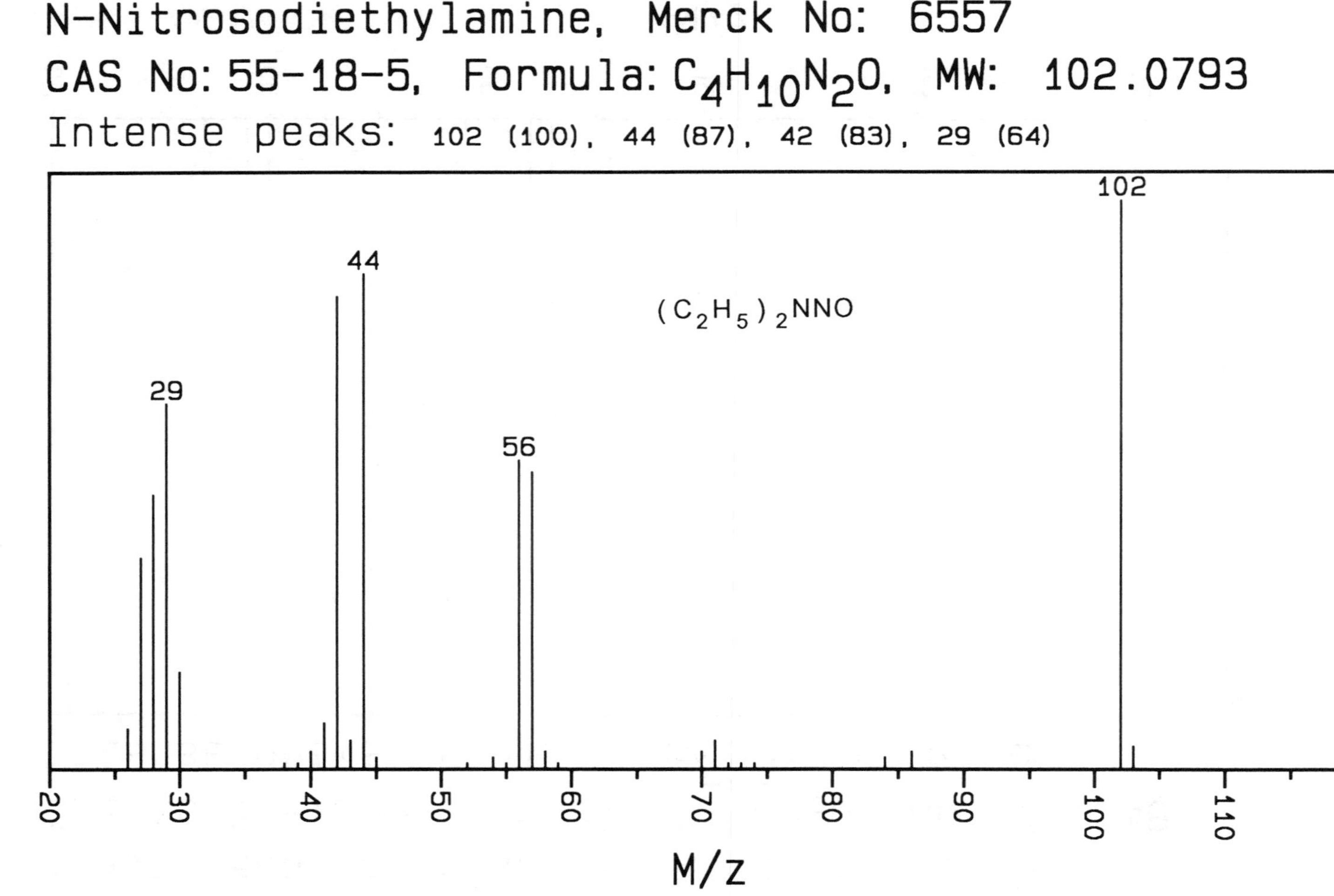
N-Nitrosodiethylamine, Merck No: 6557
CAS No: 55-18-5, Formula: $C_4H_{10}N_2O$, MW: 102.0793
Intense peaks: 102 (100), 44 (87), 42 (83), 29 (64)
102
44
$(C_2H_5)_2NNO$
29
56
20
30
40
50
60
70
80
90
100
110
120
M/Z

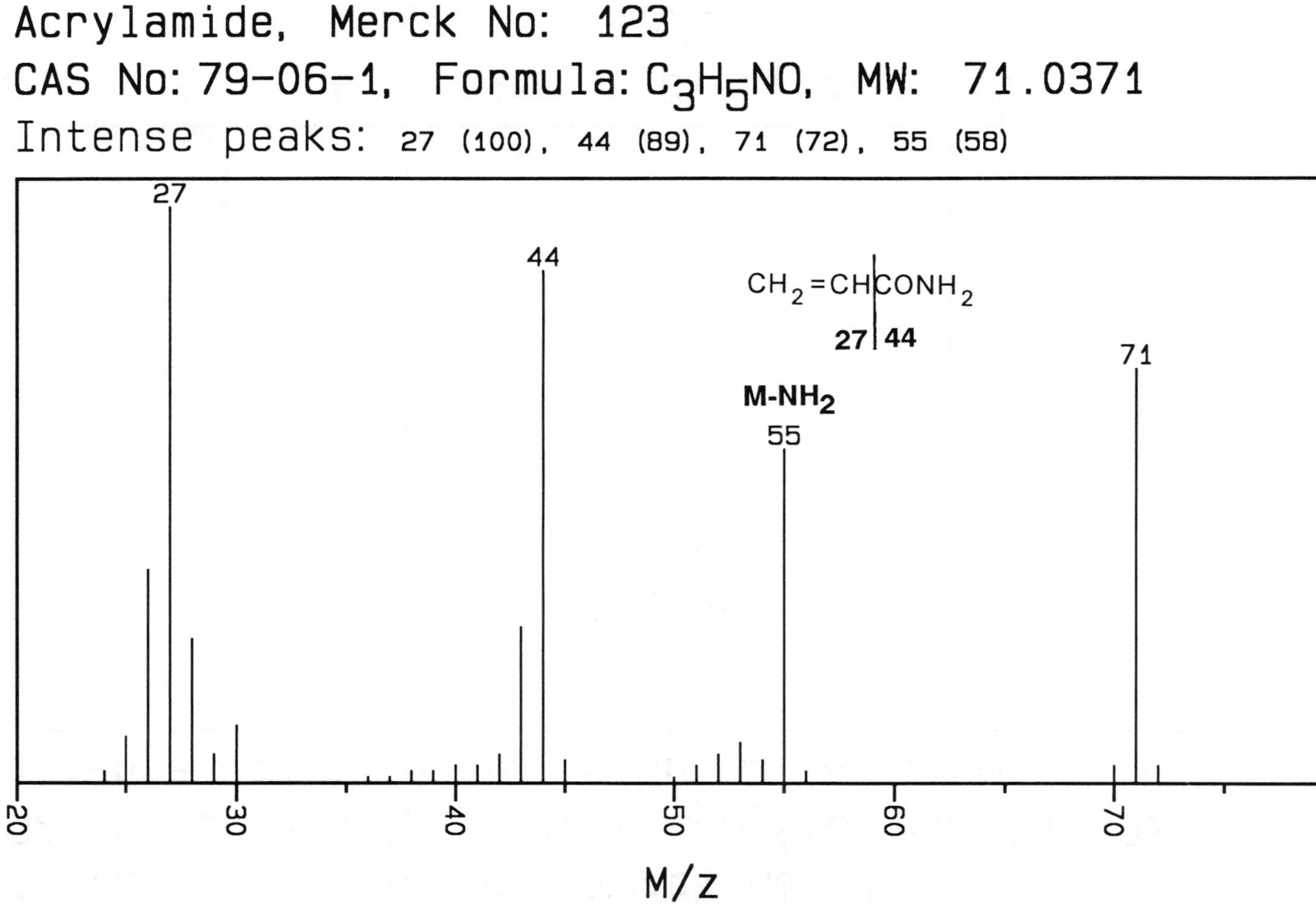
Acrylamide, Merck No: 123
CAS No: 79-06-1, Formula: $C_3H_5NO$, MW: 71.0371
Intense peaks: 27 (100), 44 (89), 71 (72), 55 (58)
27
44
$CH_2=CH|CONH_2$
27 | 44
71
M-$NH_2$
55
20
30
40
50
60
70
80
M/Z

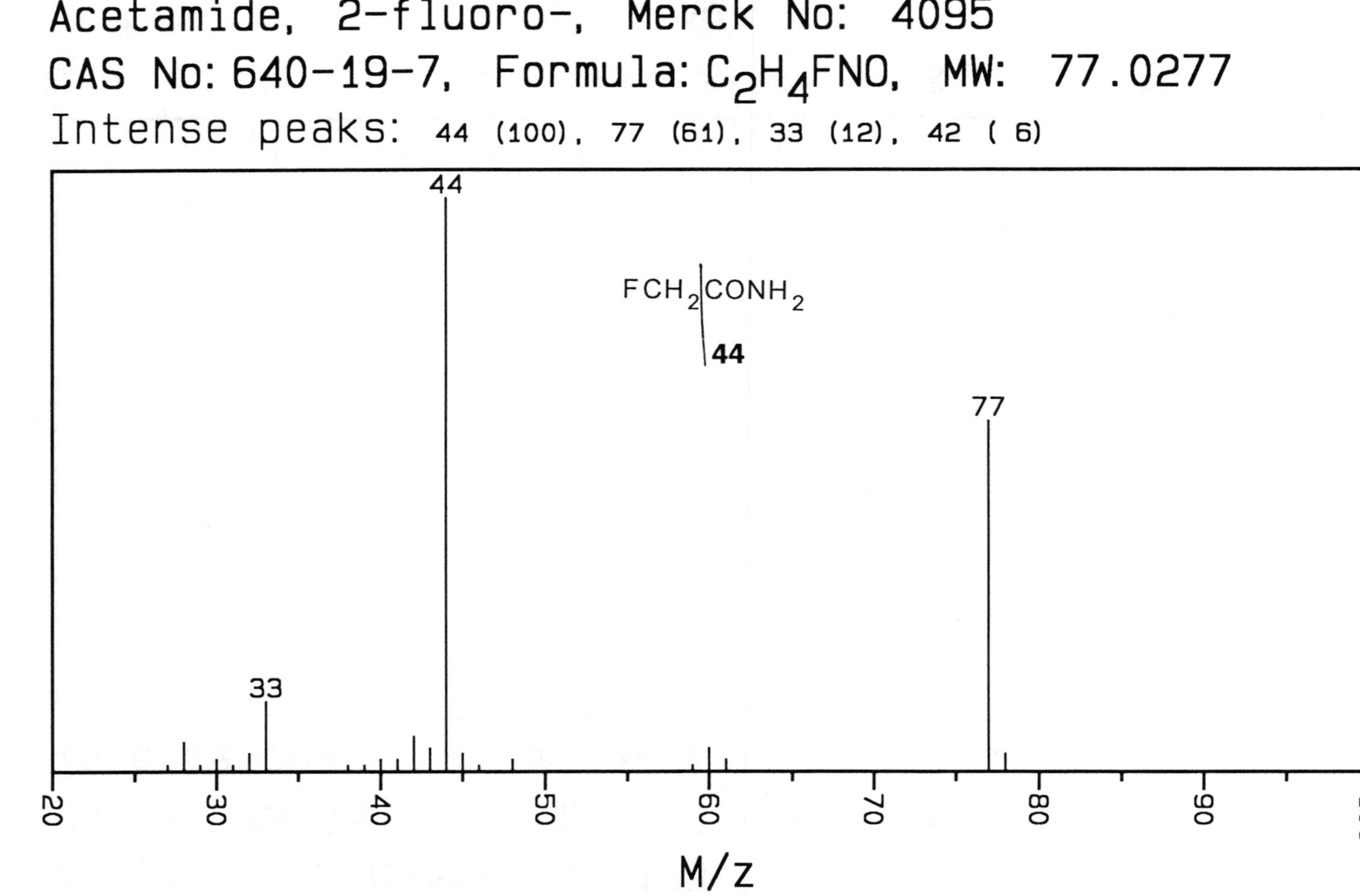
Acetamide, 2-fluoro-, Merck No: 4095
CAS No: 640-19-7, Formula: $C_2H_4FNO$, MW: 77.0277
Intense peaks: 44 (100), 77 (61), 33 (12), 42 ( 6)
44
$FCH_2$|$CONH_2$
44
77
33
20
30
40
50
60
70
80
90
100
M/Z

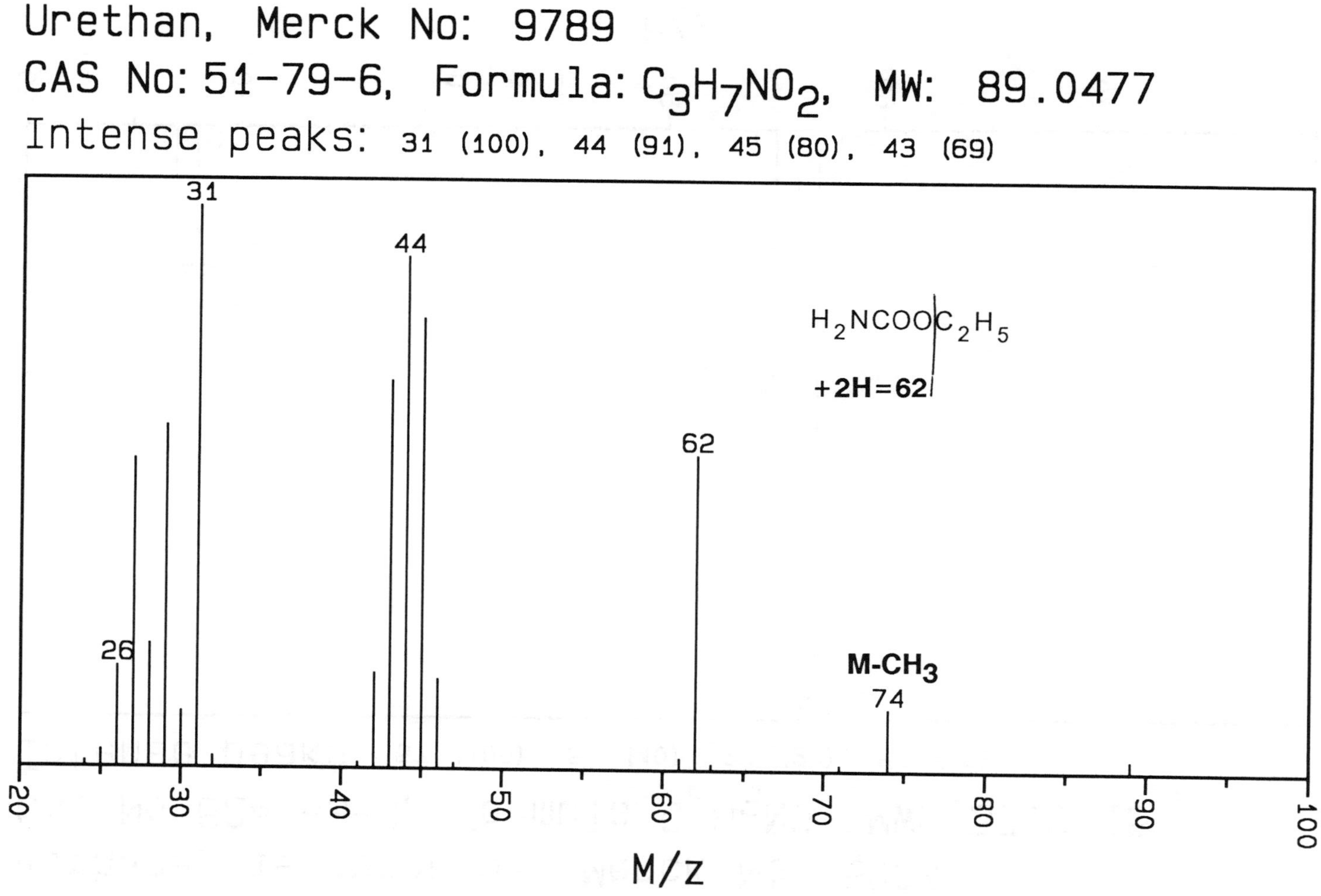

Urethan, Merck No: 9789
CAS No: 51-79-6, Formula: $C_3H_7NO_2$, MW: 89.0477
Intense peaks: 31 (100), 44 (91), 45 (80), 43 (69)
31
44
$H_2NCOOC_2H_5$
+2H=62
62
26
M-$CH_3$
74
20
30
40
50
60
70
80
90
100
M/Z

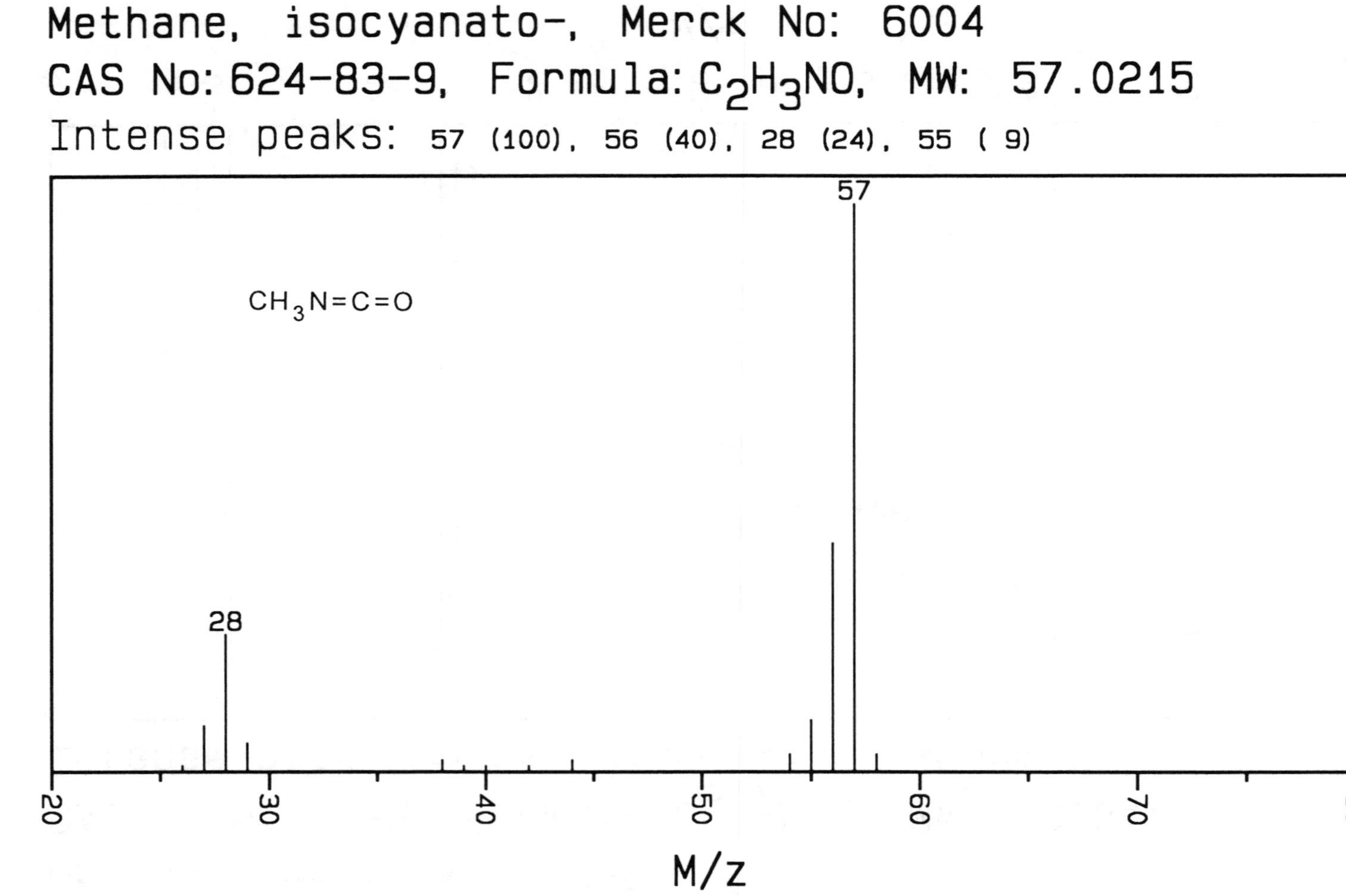
Methane, isocyanato-, Merck No: 6004
CAS No: 624-83-9, Formula: $C_2H_3NO$, MW: 57.0215
Intense peaks: 57 (100), 56 (40), 28 (24), 55 ( 9)
$CH_3N=C=O$
57
28
20
30
40
50
60
70
80
M/Z

Methane, isothiocyanato-, Merck No: 6005

CAS No: 556-61-6, Formula: $C_2H_3NS$, MW: 72.9986

Intense peaks: 73 (100), 72 (48), 45 (28), 44 (12)

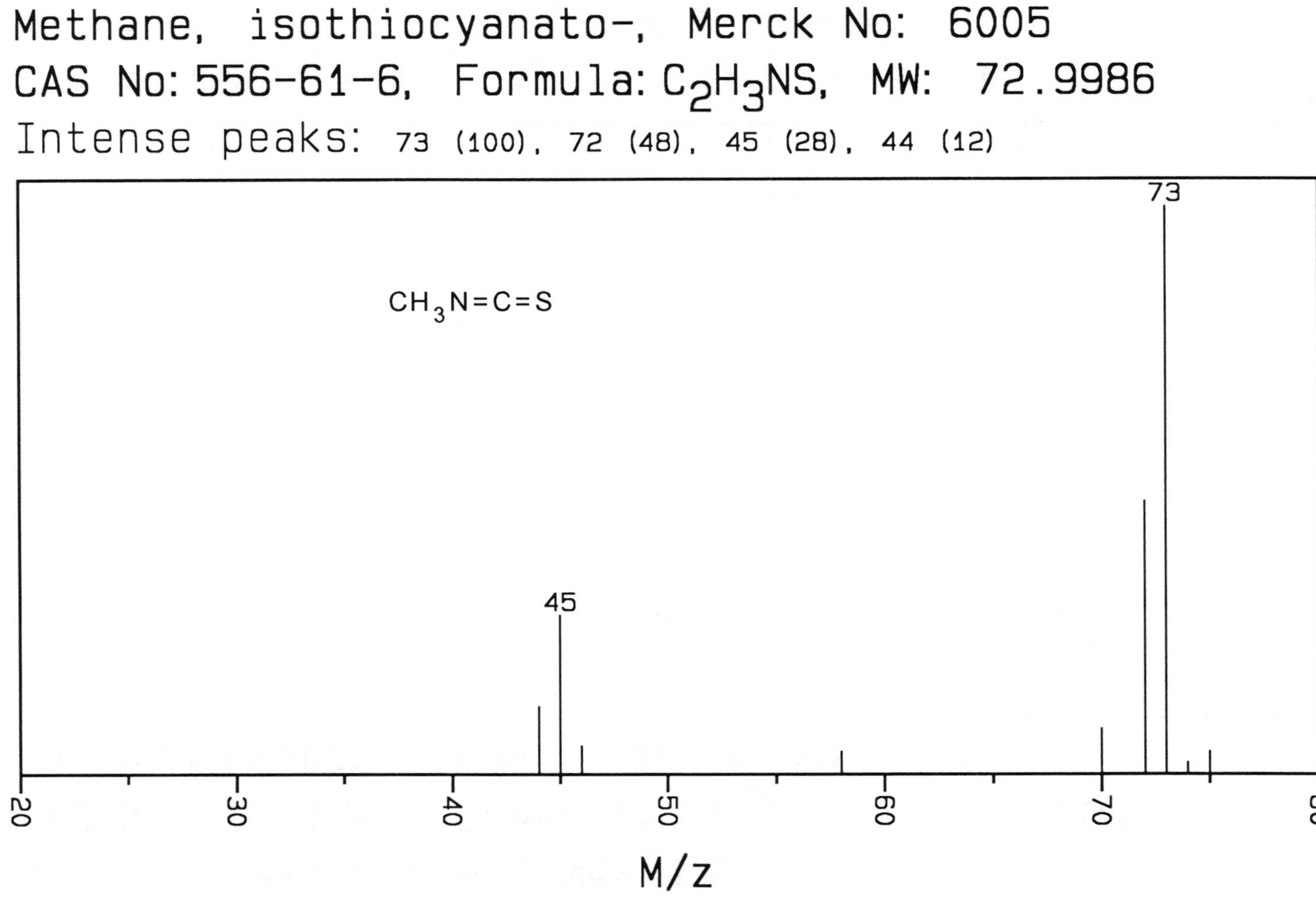

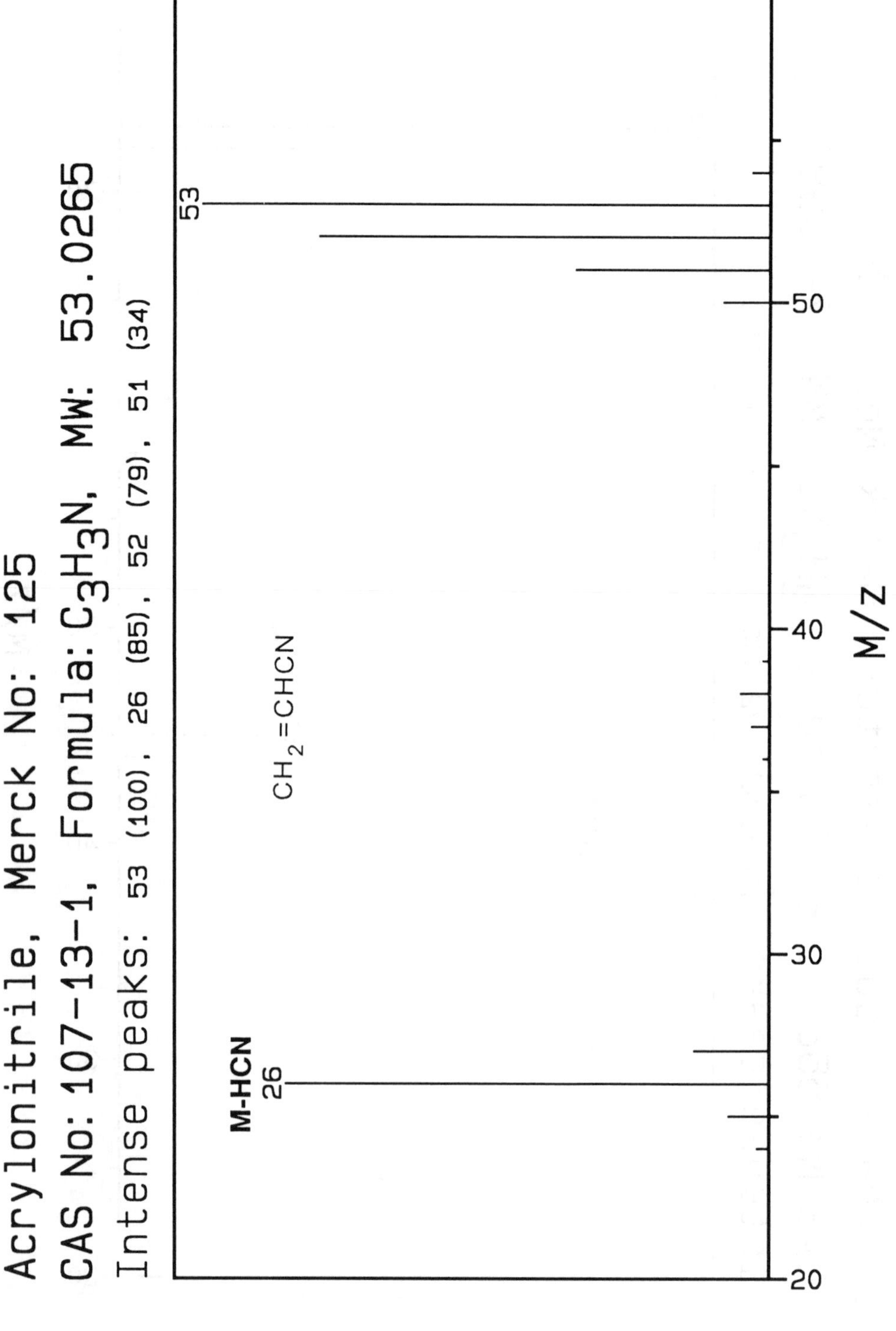
Acrylonitrile, Merck No: 125
CAS No: 107-13-1, Formula: $C_3H_3N$, MW: 53.0265
Intense peaks: 53 (100), 26 (85), 52 (79), 51 (34)
53
$CH_2$=CHCN
M-HCN
26
20
30
40
50
60
M/Z

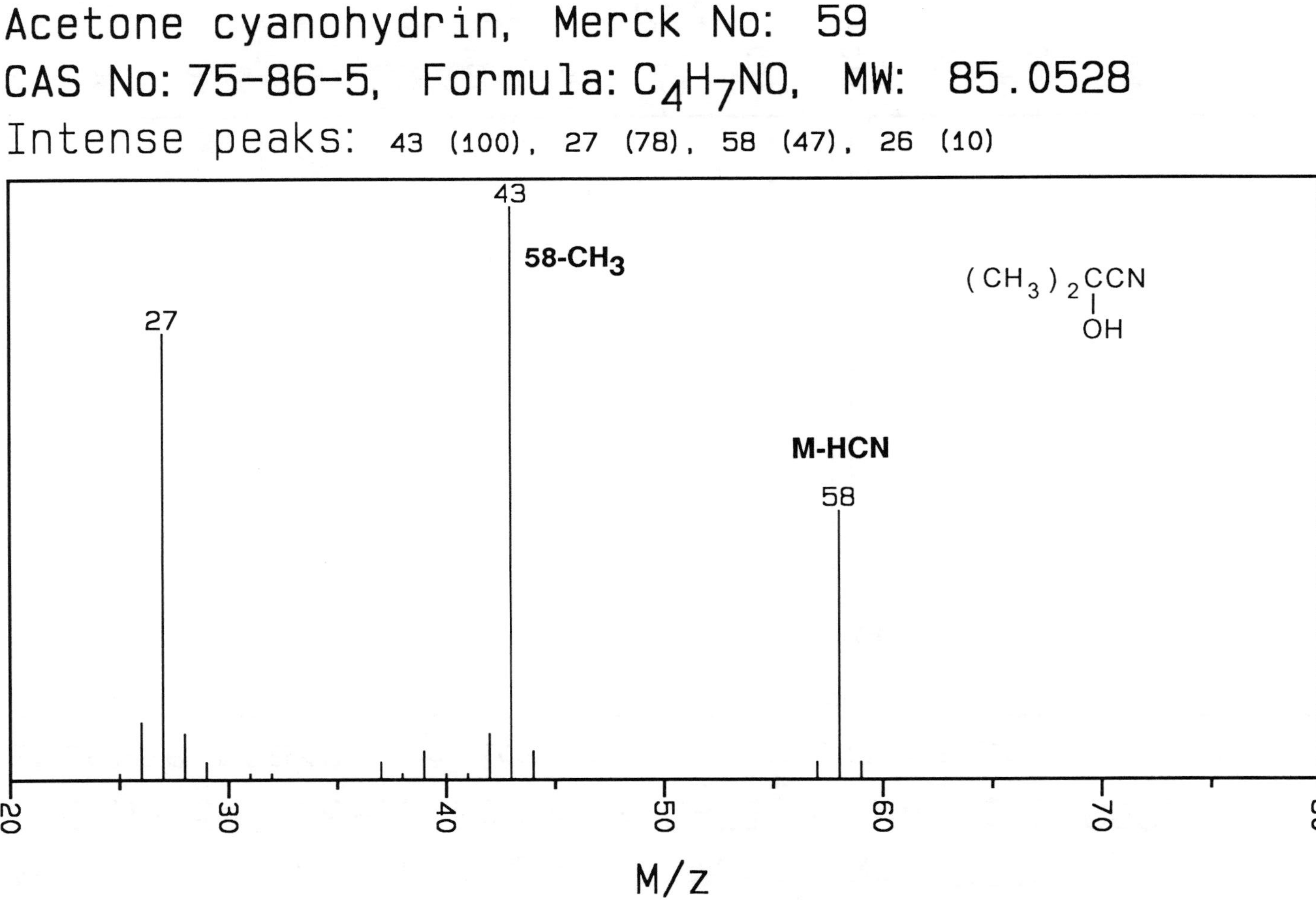
Acetone cyanohydrin, Merck No: 59
CAS No: 75-86-5, Formula: $C_4H_7NO$, MW: 85.0528
Intense peaks: 43 (100), 27 (78), 58 (47), 26 (10)
43
$58\text{-}CH_3$
27
$(CH_3)_2C(OH)CN$
M-HCN
58
20
30
40
50
60
70
80
M/Z

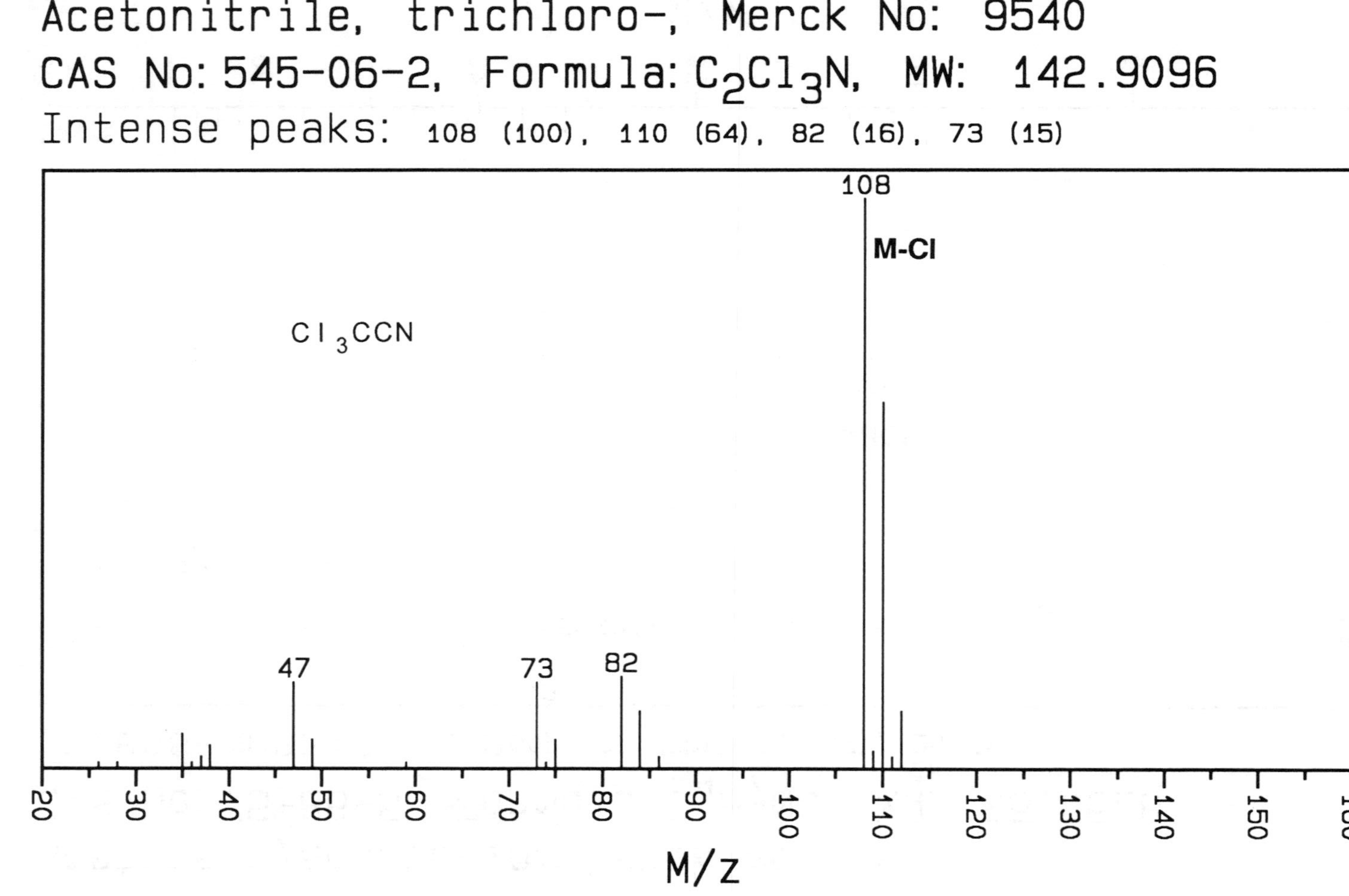
Acetonitrile, trichloro-, Merck No: 9540
CAS No: 545-06-2, Formula: C2Cl3N, MW: 142.9096
Intense peaks: 108 (100), 110 (64), 82 (16), 73 (15)
108
M-Cl
Cl3CCN
47
73
82
20
30
40
50
60
70
80
90
100
110
120
130
140
150
160
M/Z

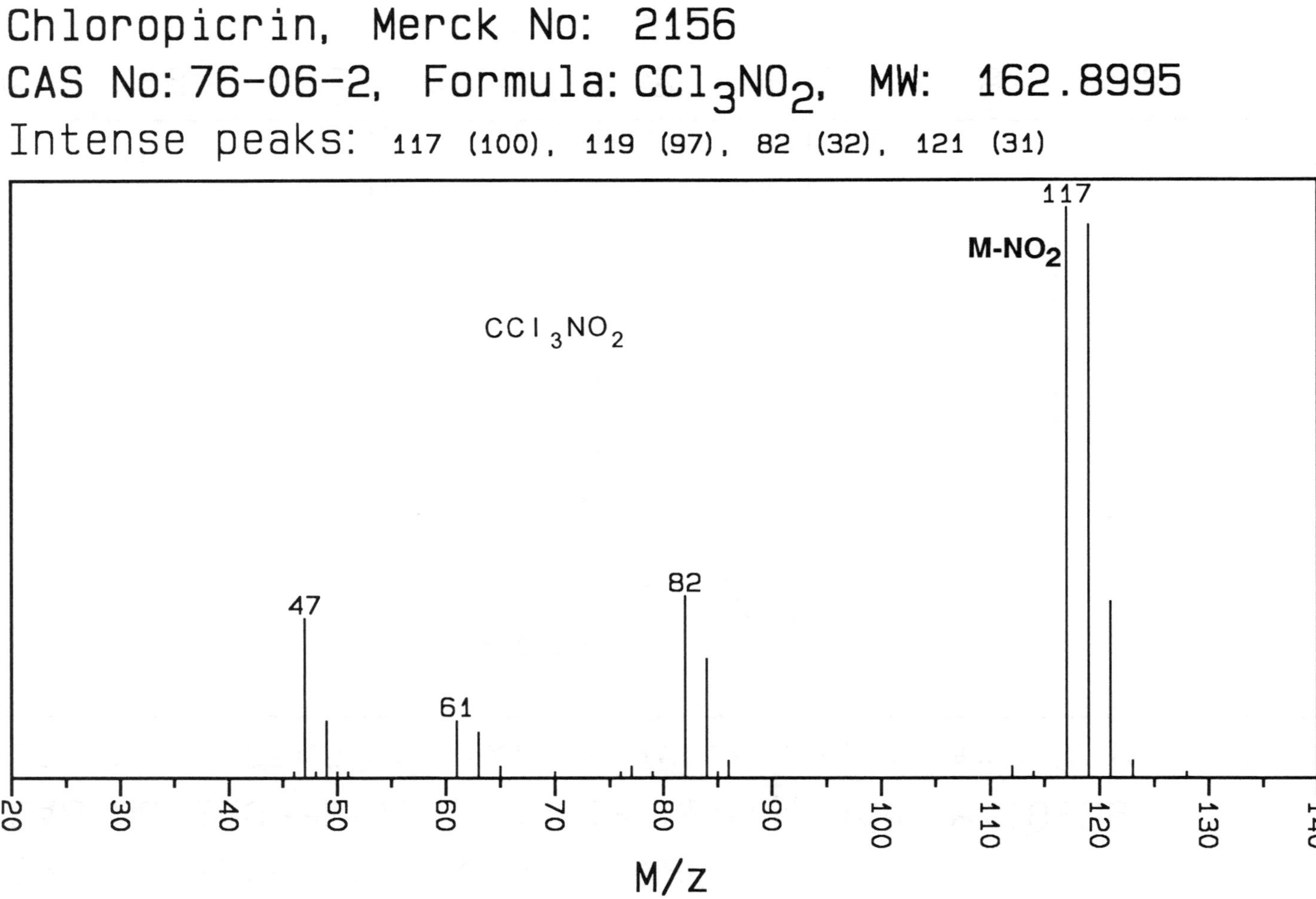
Chloropicrin, Merck No: 2156
CAS No: 76-06-2, Formula: $CCl_3NO_2$, MW: 162.8995
Intense peaks: 117 (100), 119 (97), 82 (32), 121 (31)
117
M-$NO_2$
$CCl_3NO_2$
47
61
82
20
30
40
50
60
70
80
90
100
110
120
130
140
M/Z

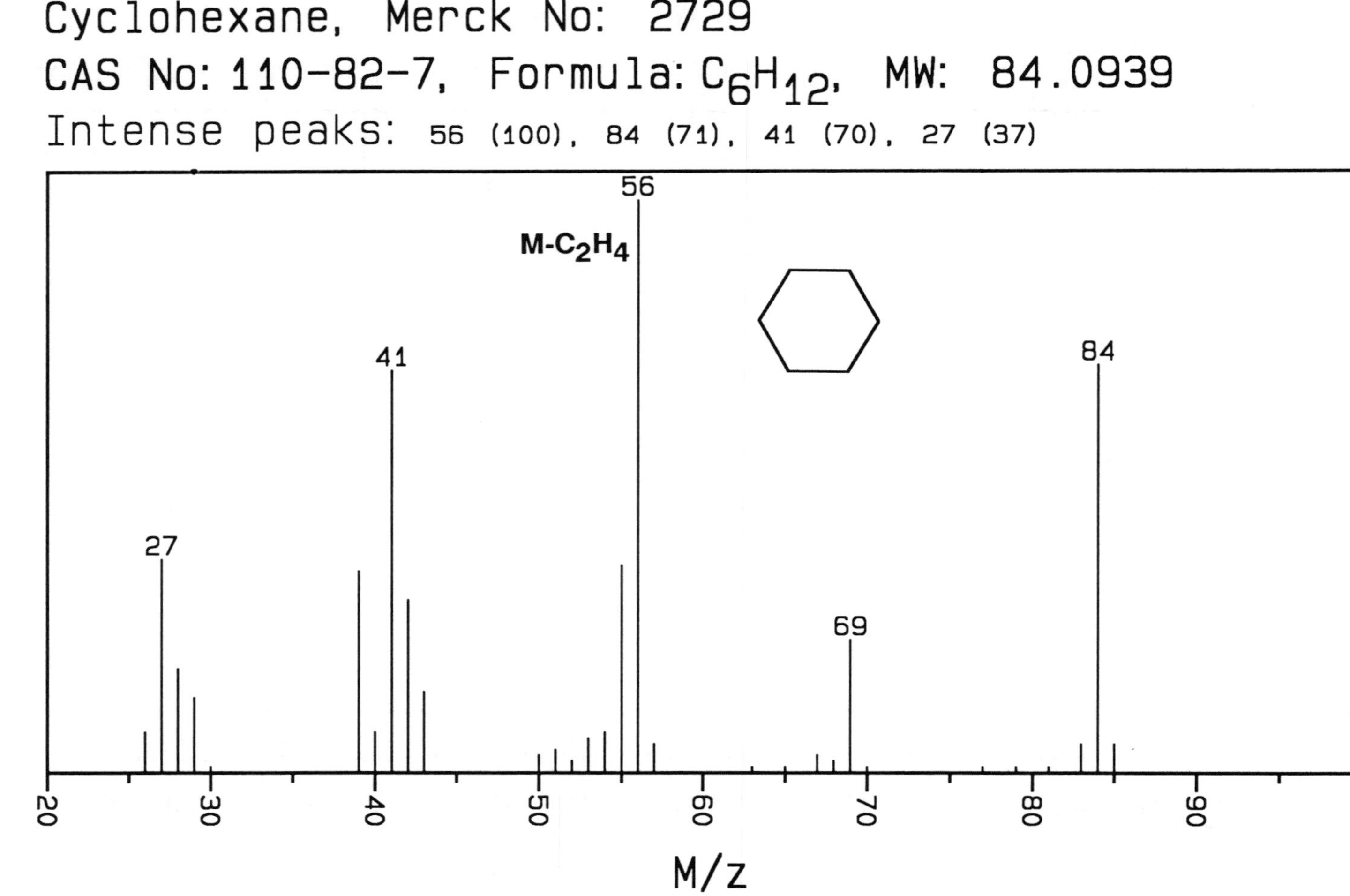
Cyclohexane, Merck No: 2729
CAS No: 110-82-7, Formula: $C_6H_{12}$, MW: 84.0939
Intense peaks: 56 (100), 84 (71), 41 (70), 27 (37)
56
M-$C_2H_4$
41
84
27
69
20
30
40
50
60
70
80
90
100
M/z

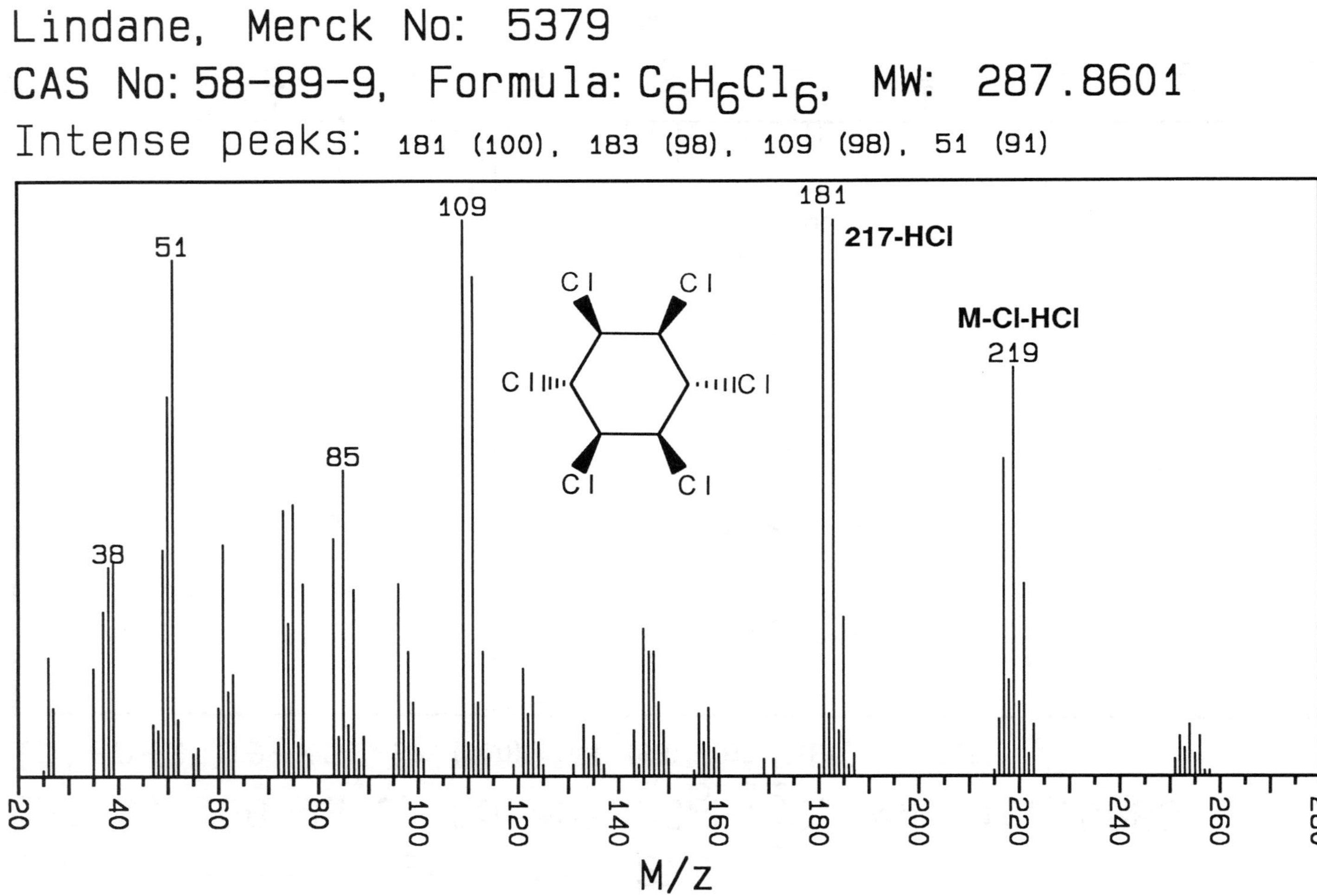

Lindane, Merck No: 5379
CAS No: 58-89-9, Formula: $C_6H_6Cl_6$, MW: 287.8601
Intense peaks: 181 (100), 183 (98), 109 (98), 51 (91)
38
51
85
109
181
217-HCl
M-Cl-HCl
219
Cl
20
40
60
80
100
120
140
160
180
200
220
240
260
280
M/Z

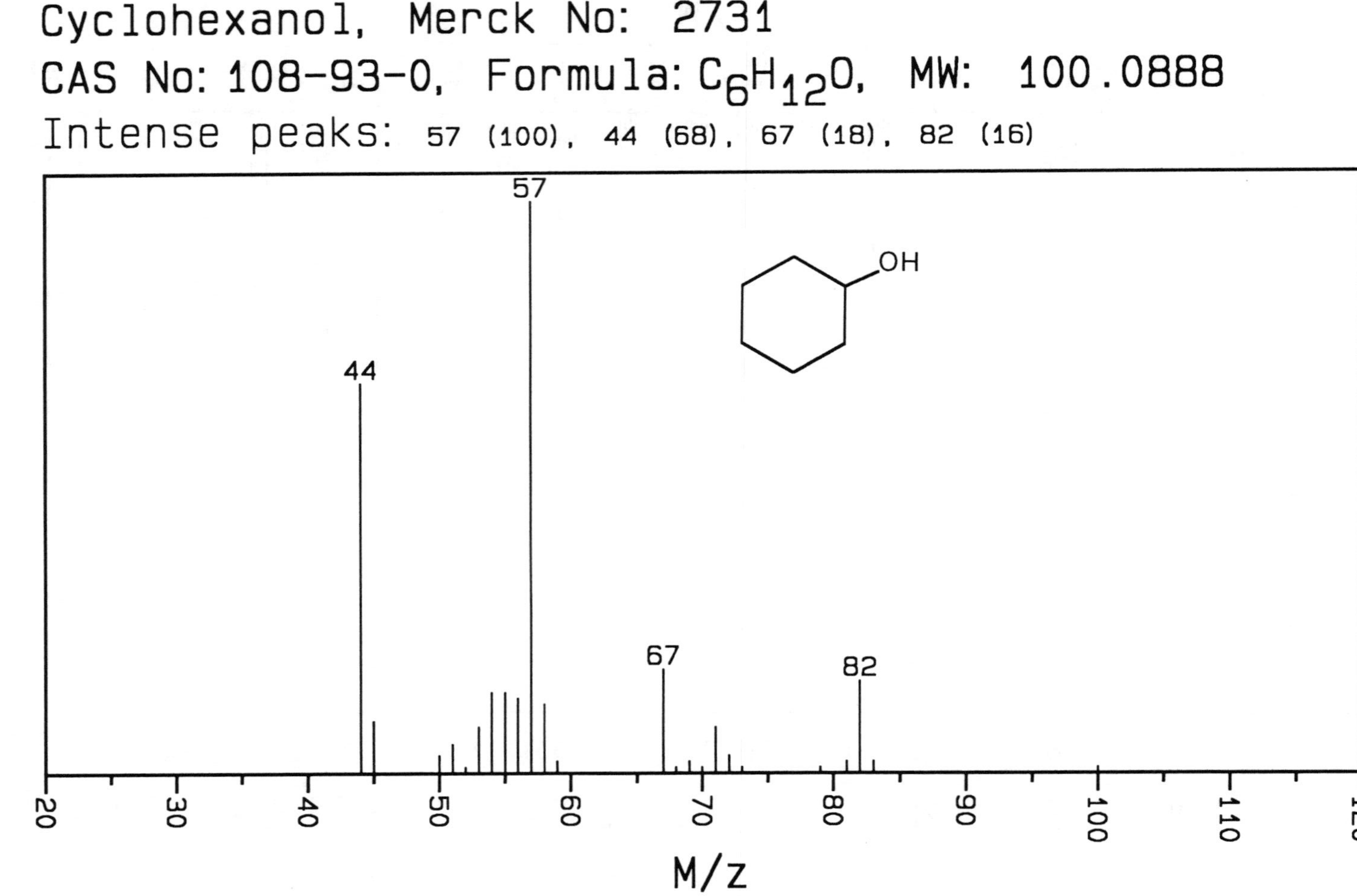
Cyclohexanol, Merck No: 2731
CAS No: 108-93-0, Formula: $C_6H_{12}O$, MW: 100.0888
Intense peaks: 57 (100), 44 (68), 67 (18), 82 (16)
57
OH
44
67
82
20
30
40
50
60
70
80
90
100
110
120
M/Z

Cyclohexanecarboxylic acid, Merck No: 2730
CAS No: 98-89-5, Formula: $C_7H_{12}O_2$, MW: 128.0837
Intense peaks: 55 (100), 73 (88), 83 (66), 41 (54)

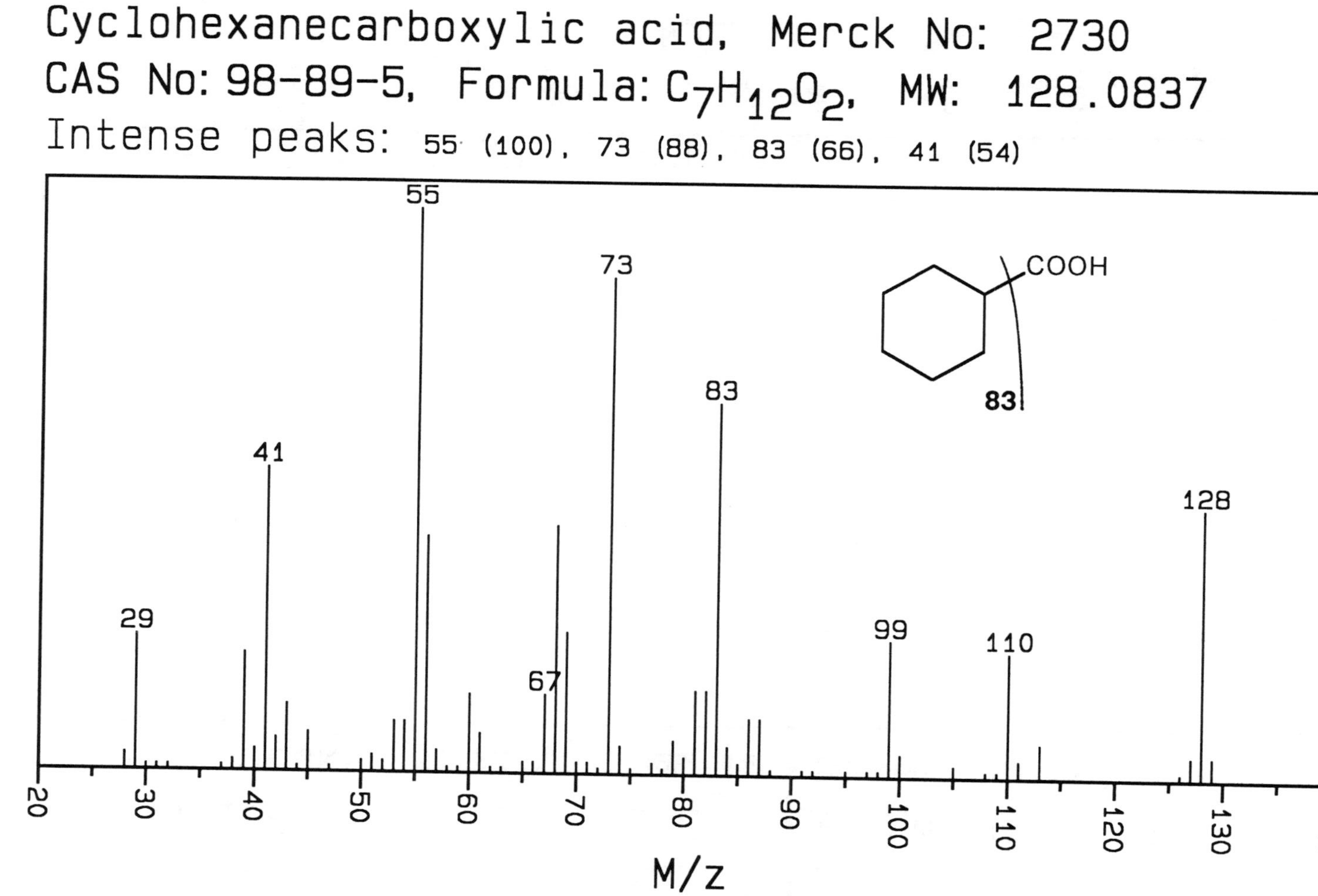

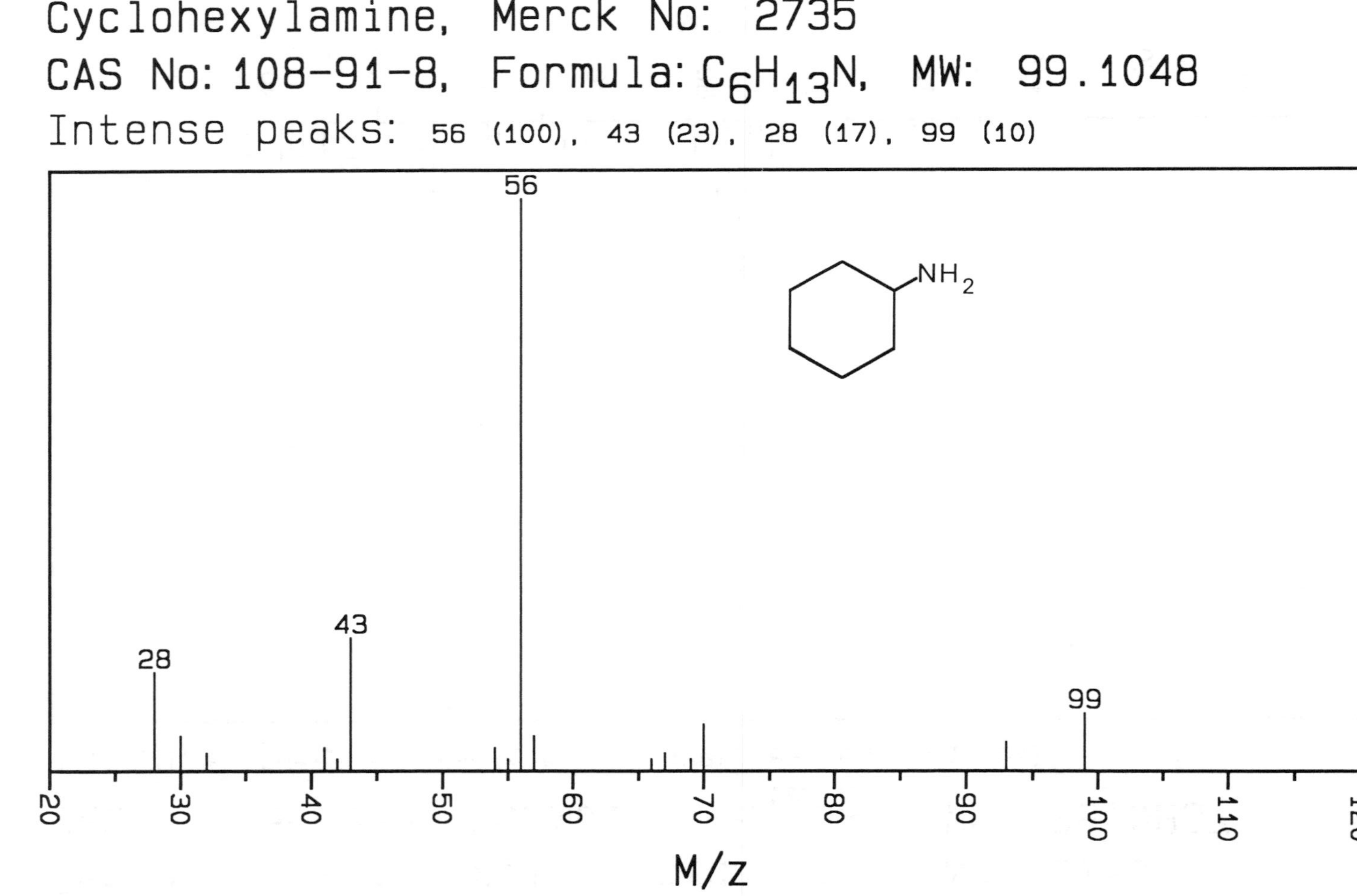
Cyclohexylamine, Merck No: 2735
CAS No: 108-91-8, Formula: C6H13N, MW: 99.1048
Intense peaks: 56 (100), 43 (23), 28 (17), 99 (10)
56
NH2
28
43
99
20
30
40
50
60
70
80
90
100
110
120
M/Z

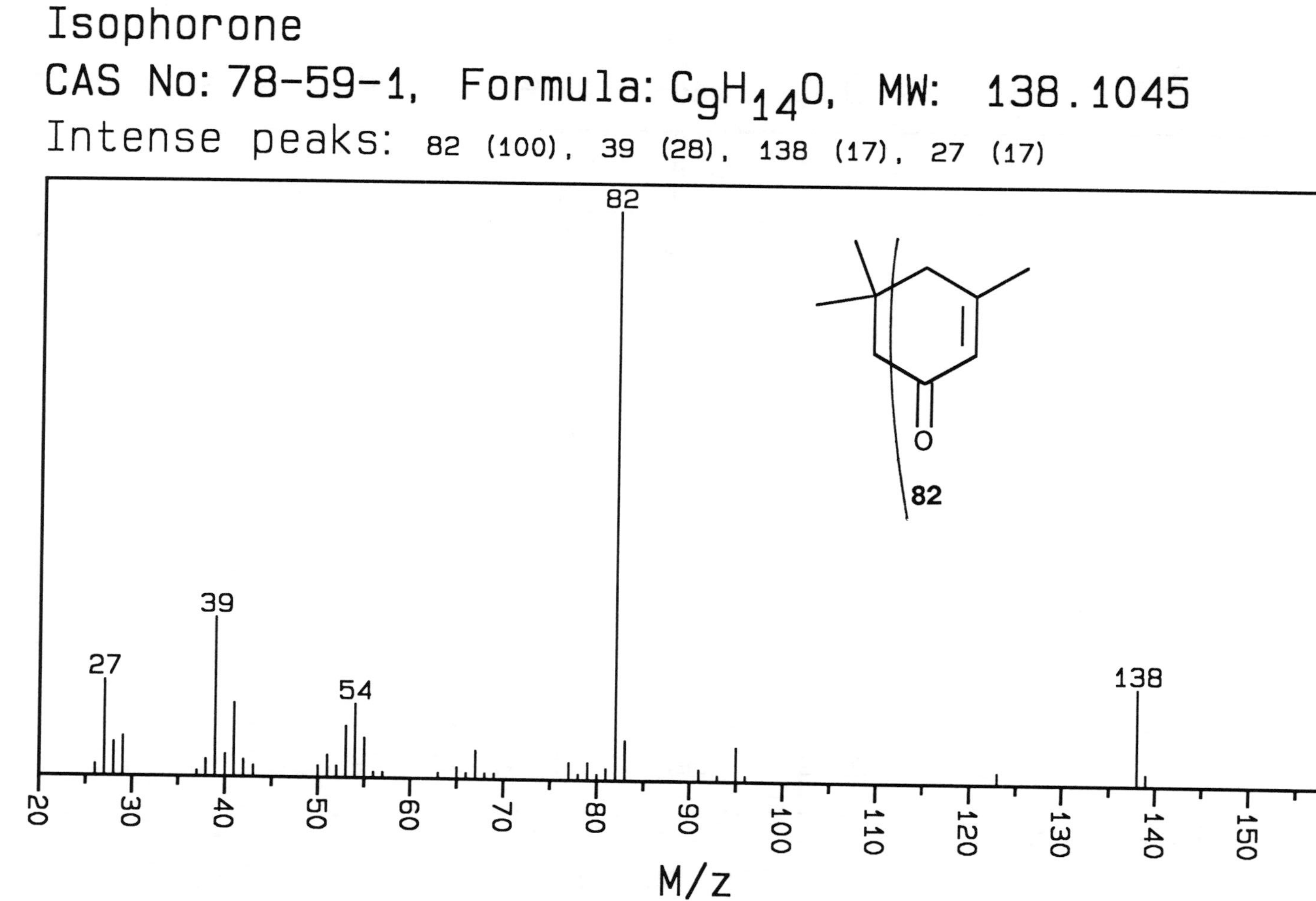
Isophorone
CAS No: 78-59-1, Formula: C9H14O, MW: 138.1045
Intense peaks: 82 (100), 39 (28), 138 (17), 27 (17)
82
39
27
54
138
O
82
20
30
40
50
60
70
80
90
100
110
120
130
140
150
160
M/Z

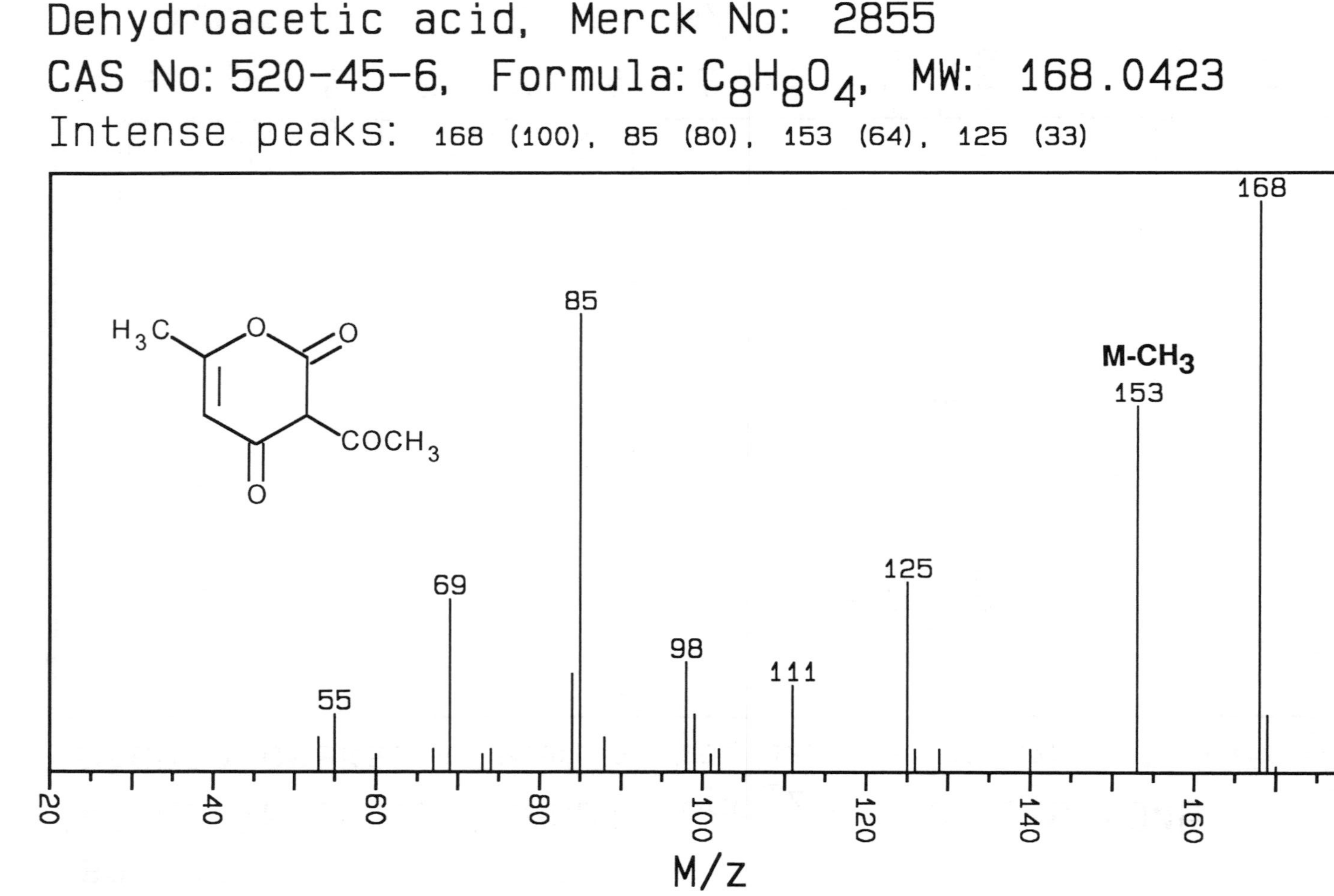
Dehydroacetic acid, Merck No: 2855
CAS No: 520-45-6, Formula: $C_8H_8O_4$, MW: 168.0423
Intense peaks: 168 (100), 85 (80), 153 (64), 125 (33)
168
85
M-$CH_3$
153
$H_3C$
O
O
$COCH_3$
O
69
125
98
111
55
20
40
60
80
100
120
140
160
180
M/Z

Pyracarbolid, Merck No: 7966

CAS No: 24691-76-7, Formula: $C_{13}H_{15}NO_2$, MW: 217.1103

Intense peaks: 125 (100), 43 (90), 55 (44), 97 (44)

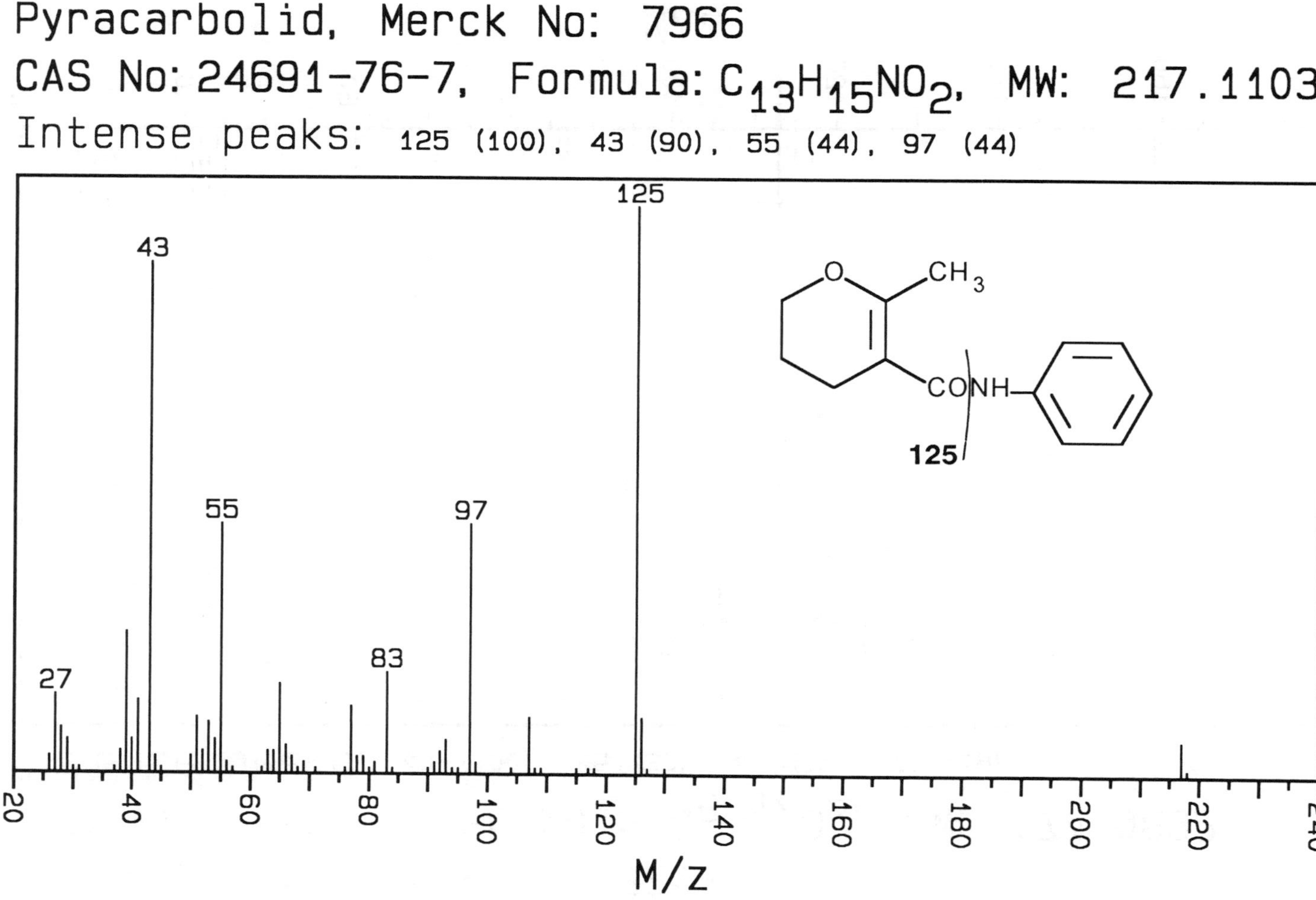

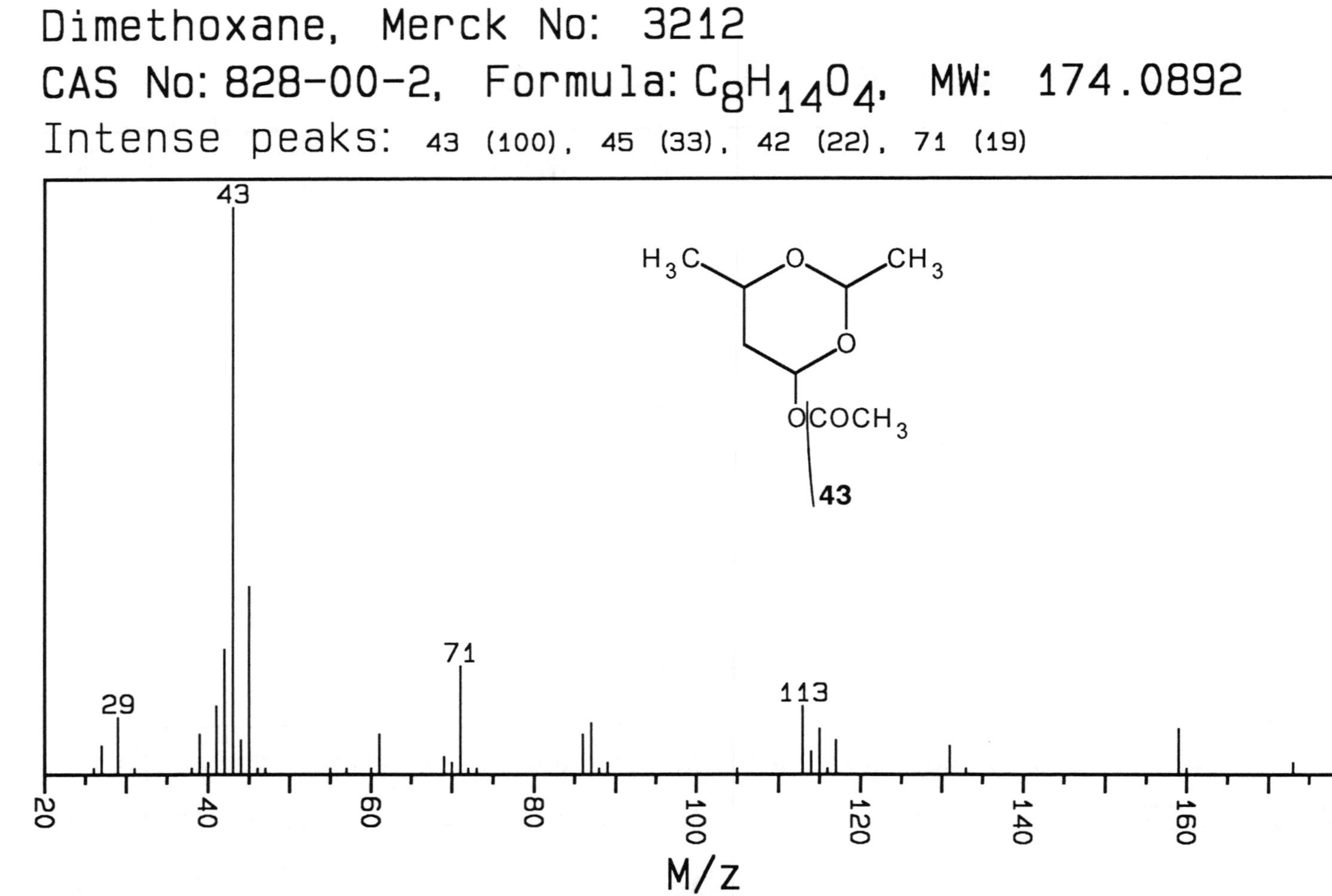
Dimethoxane, Merck No: 3212
CAS No: 828-00-2, Formula: C8H14O4, MW: 174.0892
Intense peaks: 43 (100), 45 (33), 42 (22), 71 (19)
H3C
CH3
O
O
OCOCH3
43
43
29
71
113
20
40
60
80
100
120
140
160
180
M/Z

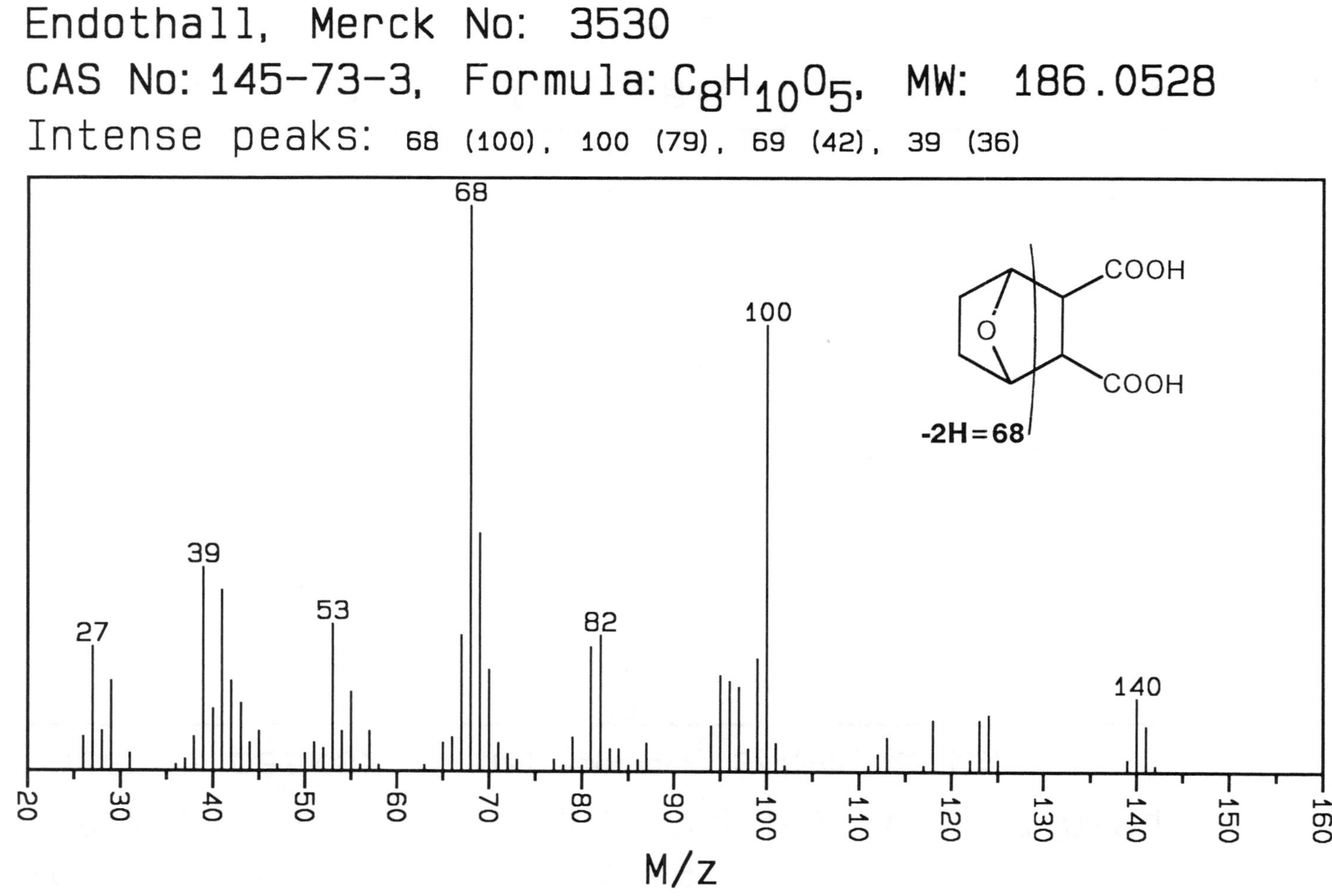

Endothall, Merck No: 3530
CAS No: 145-73-3, Formula: $C_8H_{10}O_5$, MW: 186.0528
Intense peaks: 68 (100), 100 (79), 69 (42), 39 (36)
68
100
39
53
27
82
140
COOH
COOH
O
-2H=68
20
30
40
50
60
70
80
90
100
110
120
130
140
150
160
M/Z

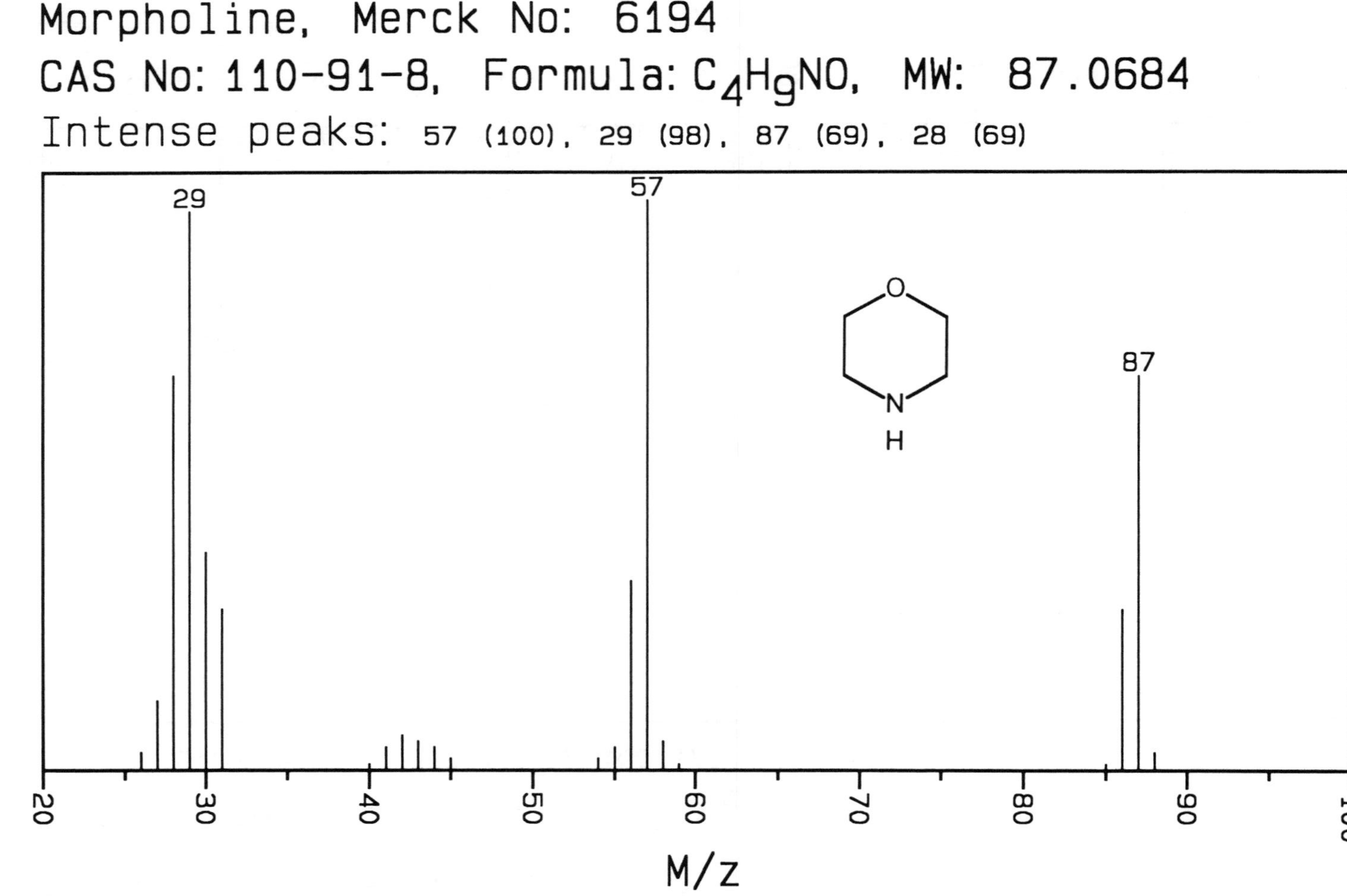
Morpholine, Merck No: 6194
CAS No: 110-91-8, Formula: C4H9NO, MW: 87.0684
Intense peaks: 57 (100), 29 (98), 87 (69), 28 (69)
29
57
87
O
N
H
20
30
40
50
60
70
80
90
100
M/Z

Tridemorph, Merck No: 9576

CAS No: 24602-86-6, Formula: $C_{19}H_{39}NO$, MW: 297.3032

Intense peaks: 128 (100), 129 ( 9), 115 ( 8), 202 ( 6)

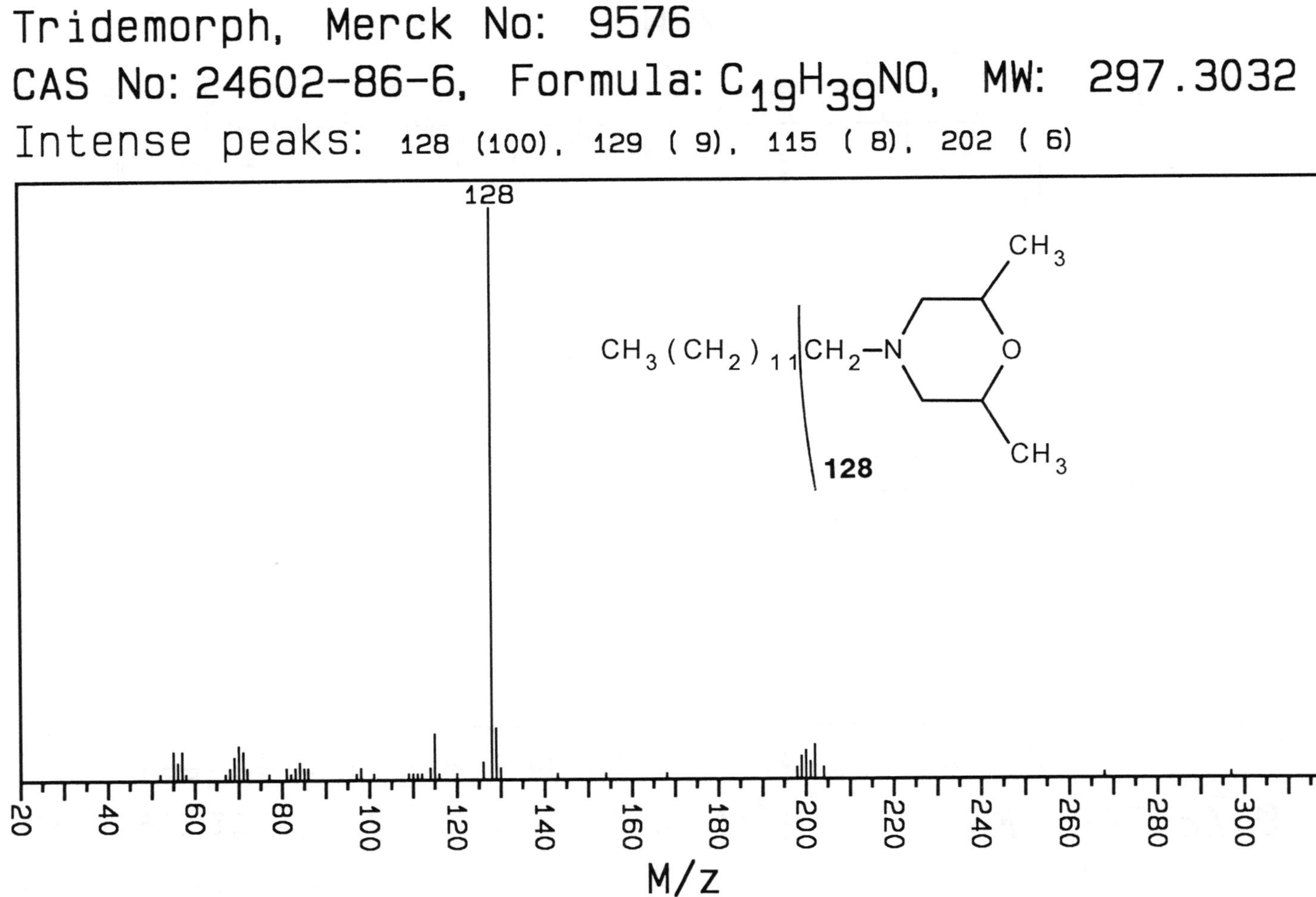

Dodemorph, Merck No: 3405

CAS No: 1593-77-7, Formula: $C_{18}H_{35}NO$, MW: 281.2718

Intense peaks: 154 (100), 55 (34), 41 (24), 141 (19)

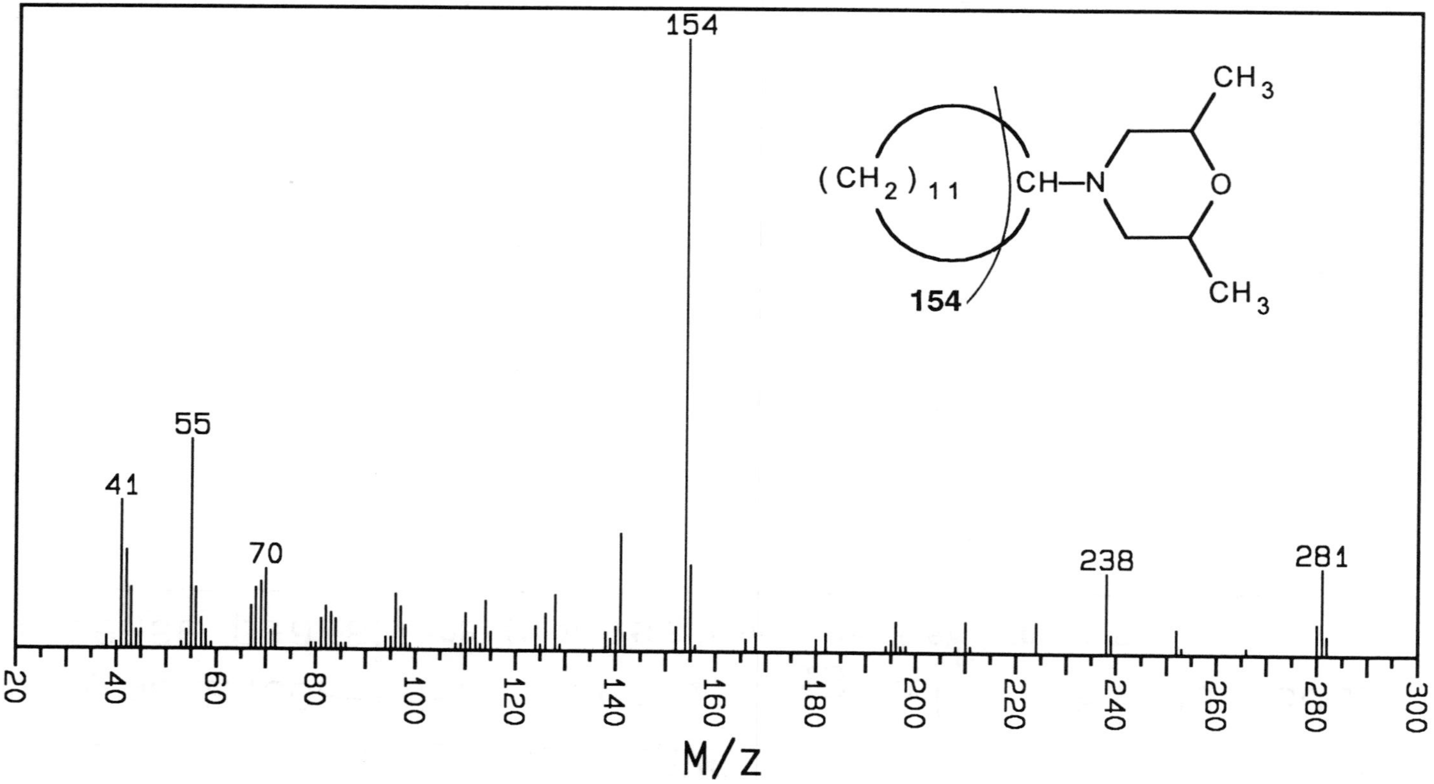

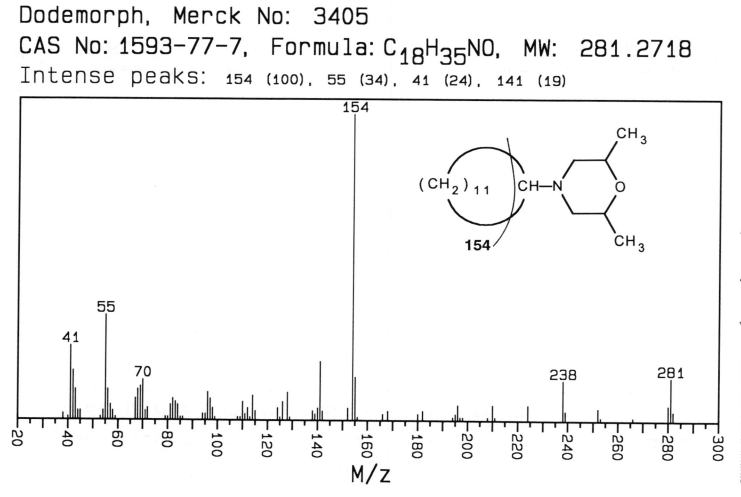

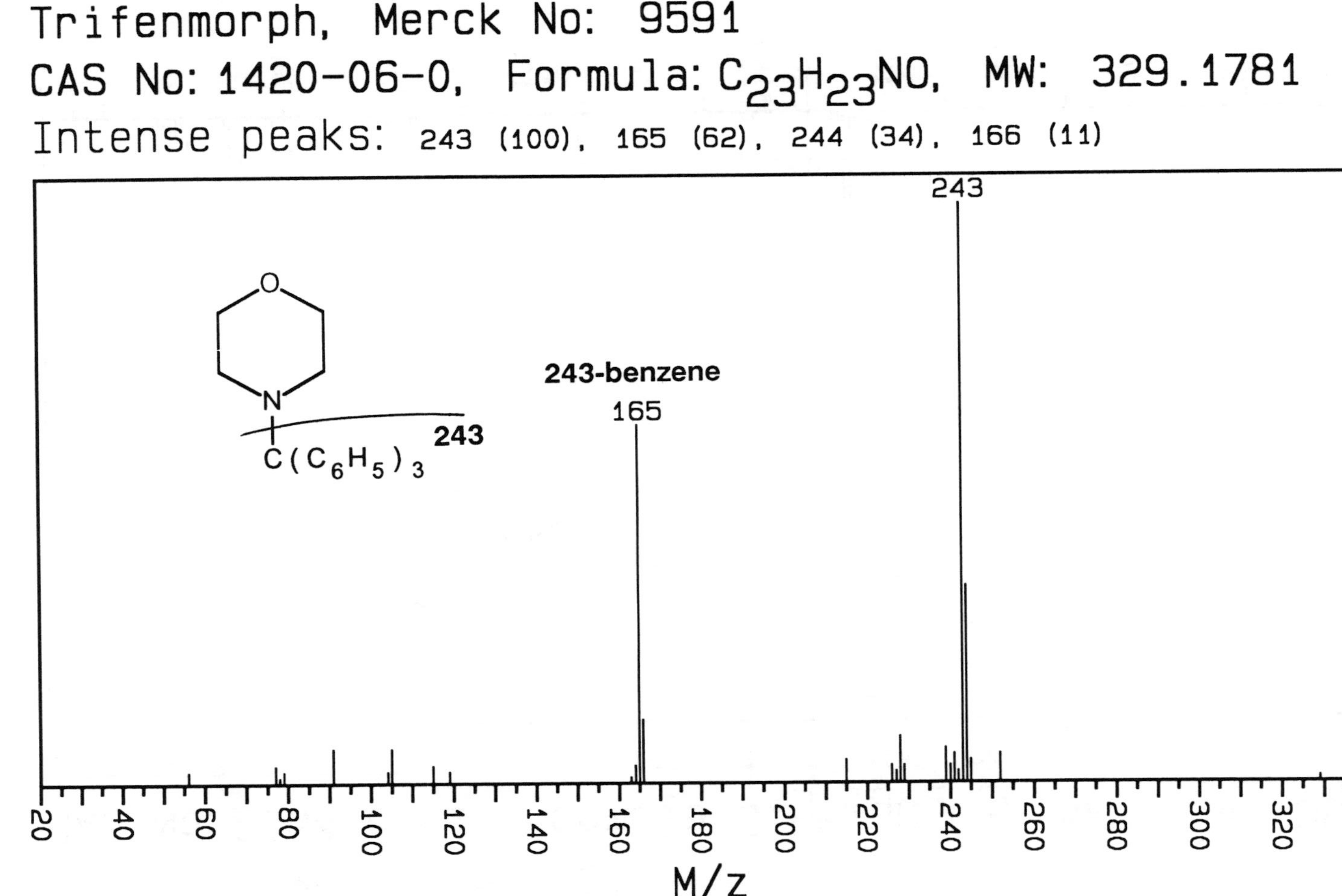

Trifenmorph, Merck No: 9591
CAS No: 1420-06-0, Formula: $C_{23}H_{23}NO$, MW: 329.1781
Intense peaks: 243 (100), 165 (62), 244 (34), 166 (11)
243
O
N
C(C6H5)3
243
243-benzene
165
20
40
60
80
100
120
140
160
180
200
220
240
260
280
300
320
340
M/Z

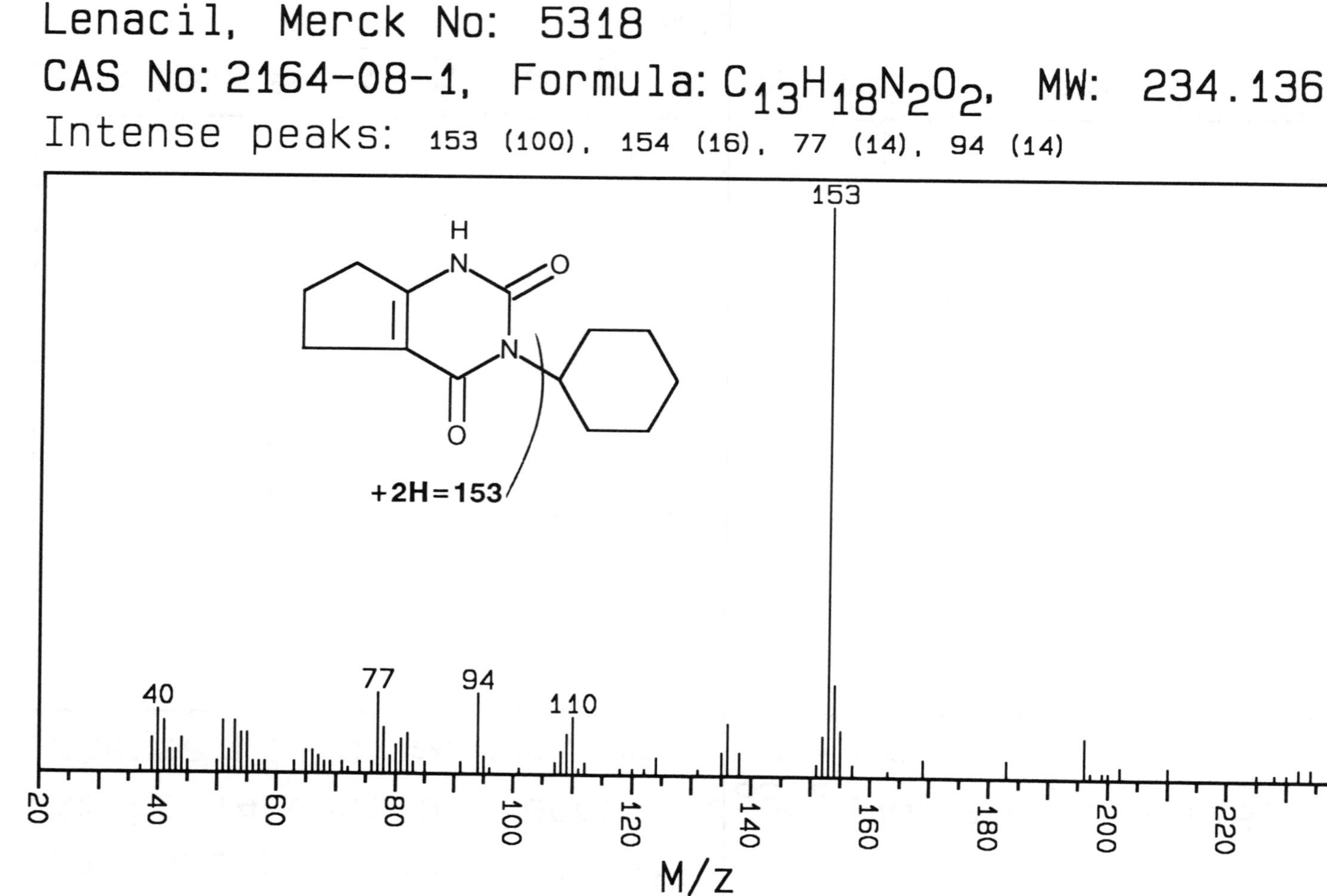

Lenacil, Merck No: 5318
CAS No: 2164-08-1, Formula: C13H18N2O2, MW: 234.1368
Intense peaks: 153 (100), 154 (16), 77 (14), 94 (14)
153
+2H=153
40
77
94
110
20
40
60
80
100
120
140
160
180
200
220
240
M/Z

Dazomet, Merck No: 2827

CAS No: 533-74-4, Formula: $C_5H_{10}N_2S_2$, MW: 162.0285

Intense peaks: 42 (100), 44 (42), 43 (35), 57 (28)

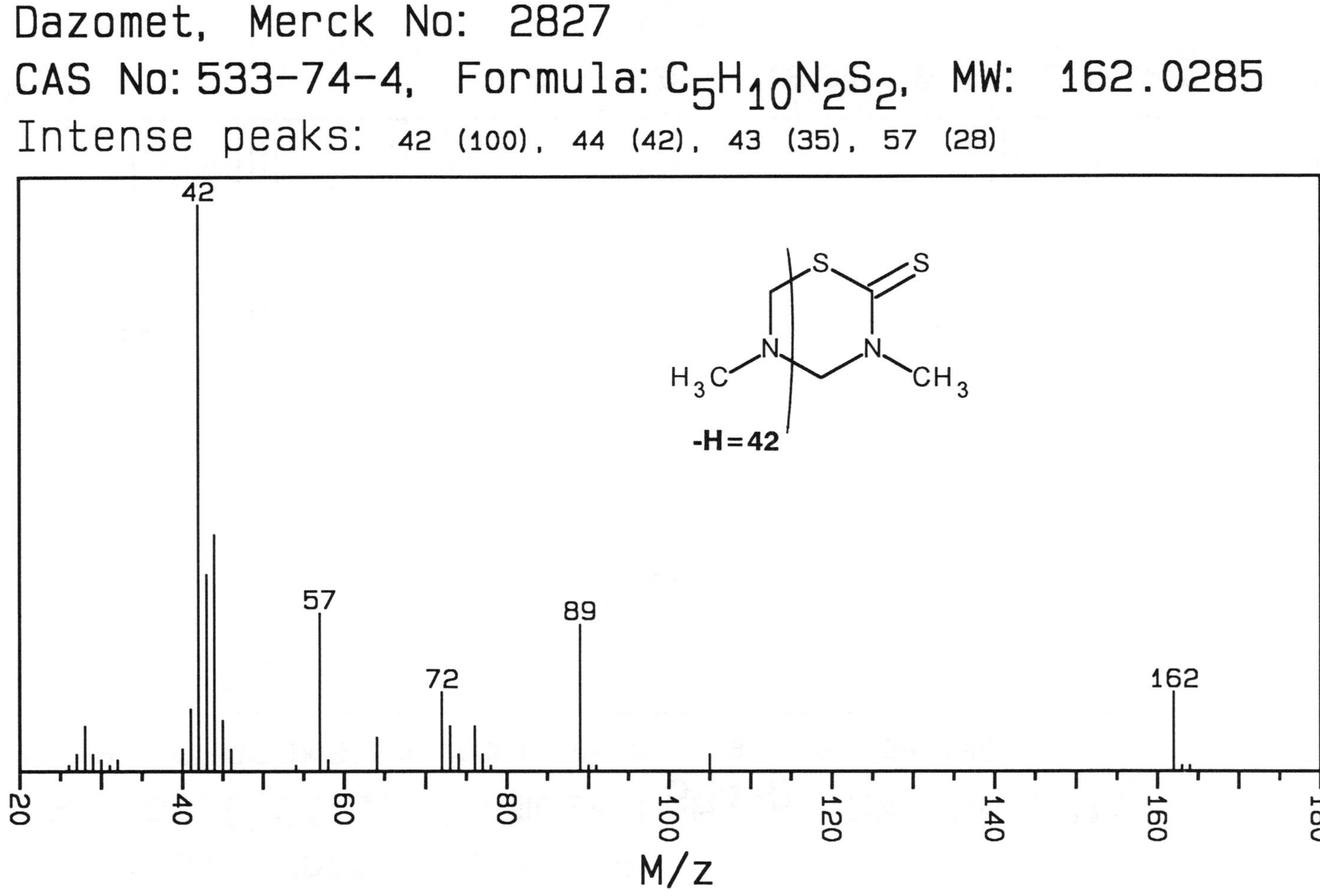

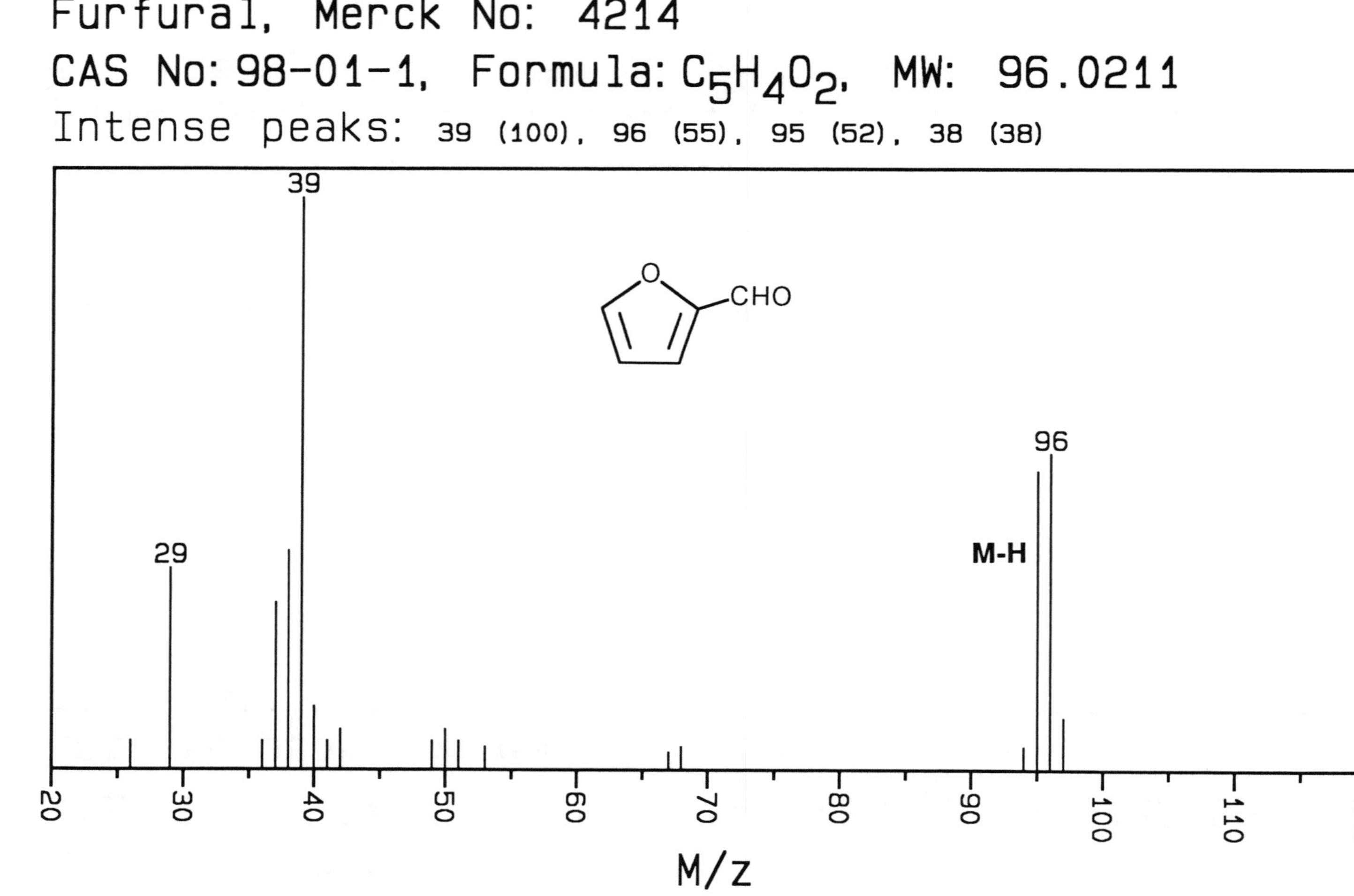
Furfural, Merck No: 4214
CAS No: 98-01-1, Formula: C5H4O2, MW: 96.0211
Intense peaks: 39 (100), 96 (55), 95 (52), 38 (38)
39
O
CHO
96
29
M-H
20
30
40
50
60
70
80
90
100
110
120
M/Z

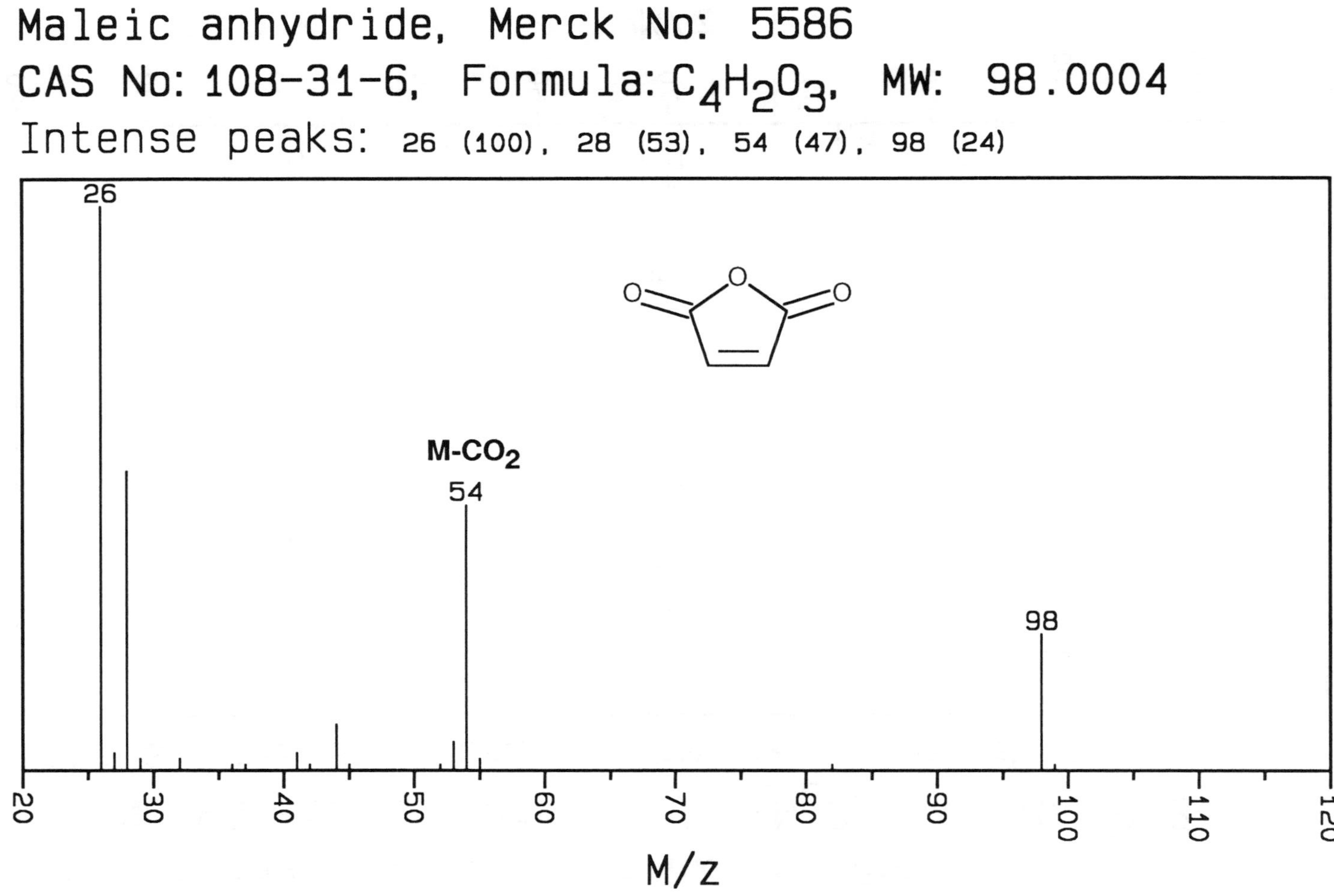
Maleic anhydride, Merck No: 5586
CAS No: 108-31-6, Formula: $C_4H_2O_3$, MW: 98.0004
Intense peaks: 26 (100), 28 (53), 54 (47), 98 (24)
26
O
O
O
M-$CO_2$
54
98
20
30
40
50
60
70
80
90
100
110
120
M/Z

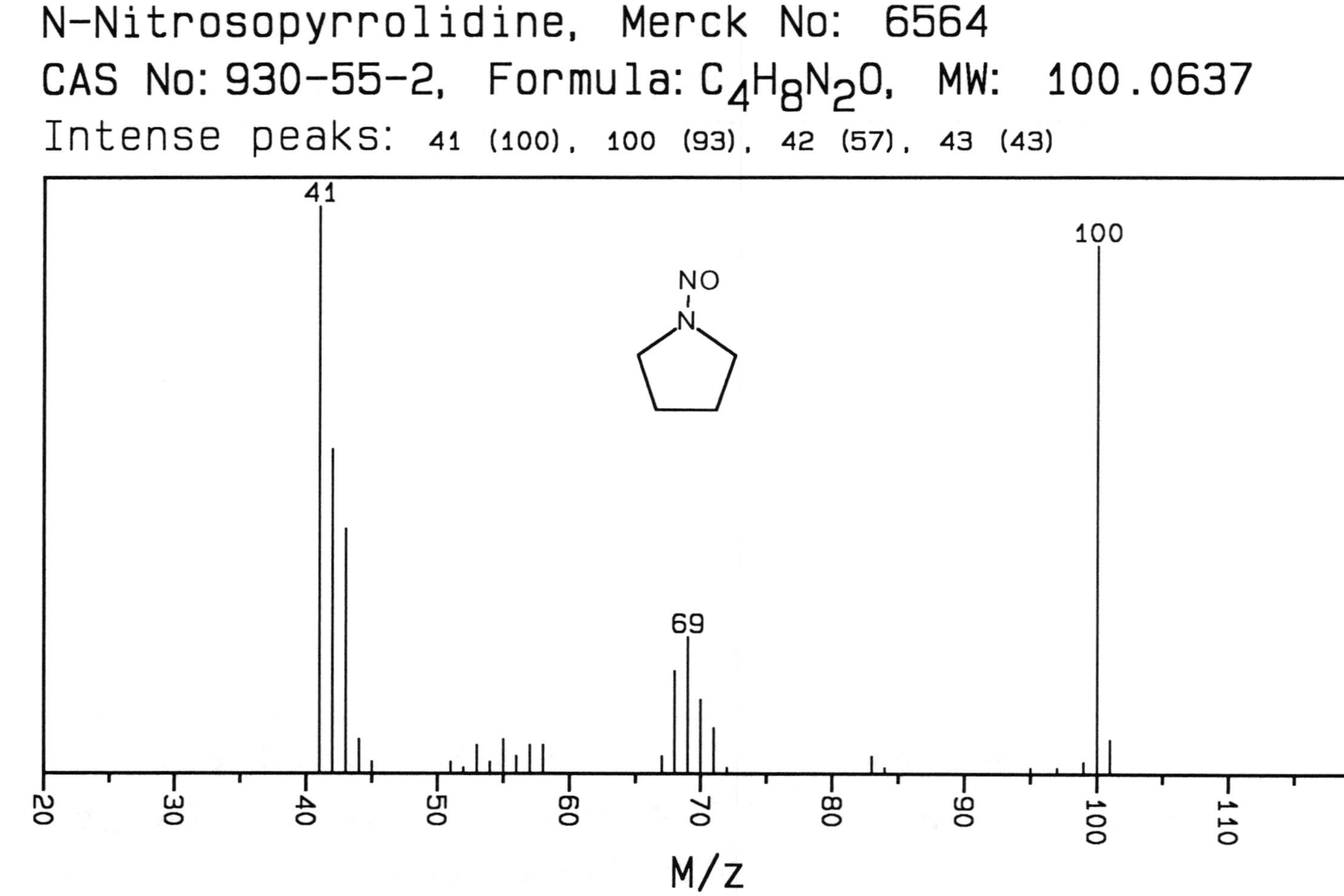
N-Nitrosopyrrolidine, Merck No: 6564
CAS No: 930-55-2, Formula: $C_4H_8N_2O$, MW: 100.0637
Intense peaks: 41 (100), 100 (93), 42 (57), 43 (43)
41
100
NO
N
69
20
30
40
50
60
70
80
90
100
110
120
M/Z

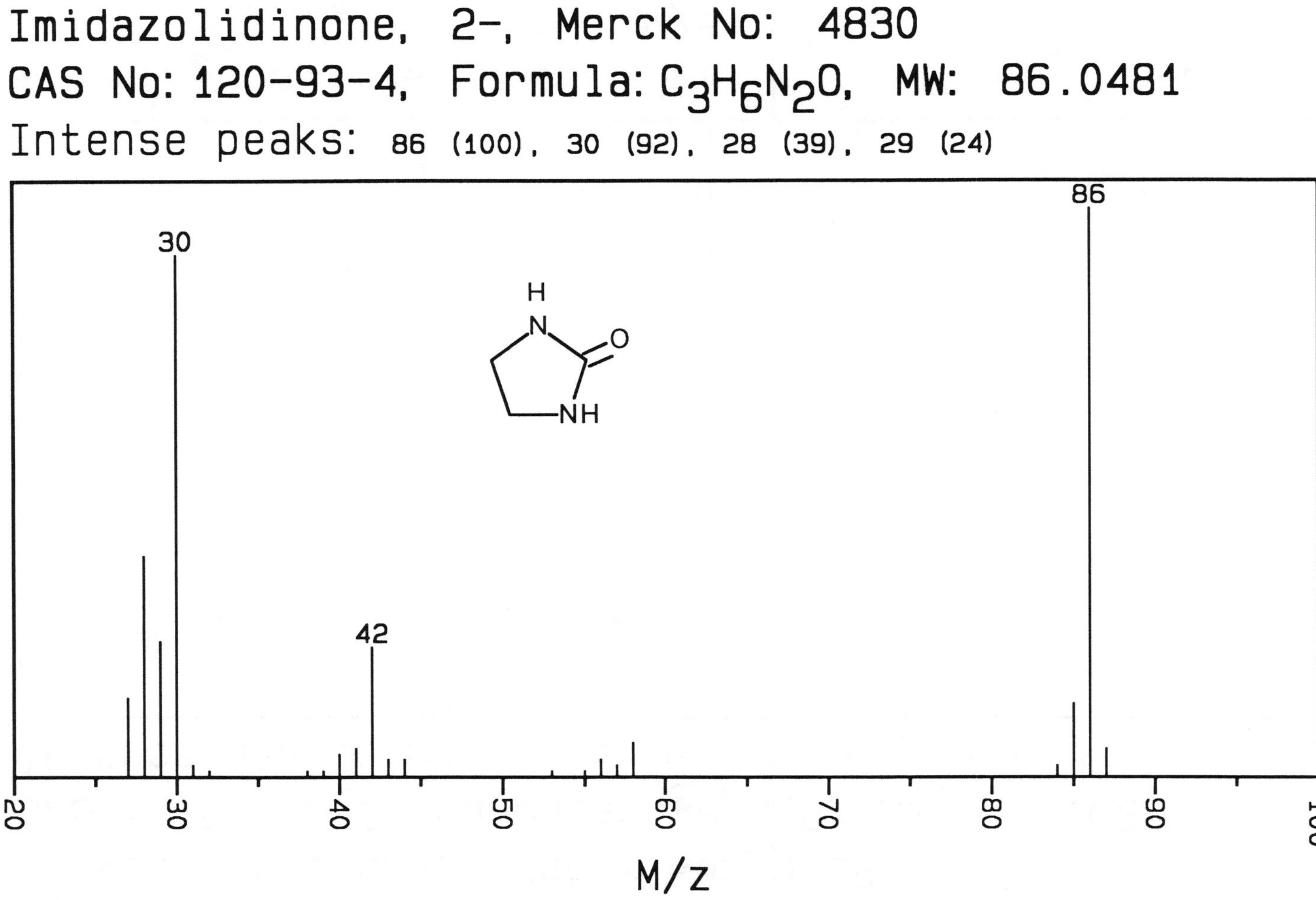
Imidazolidinone, 2-, Merck No: 4830
CAS No: 120-93-4, Formula: $C_3H_6N_2O$, MW: 86.0481
Intense peaks: 86 (100), 30 (92), 28 (39), 29 (24)
86
30
42
H
N
O
NH
20
30
40
50
60
70
80
90
100
M/Z

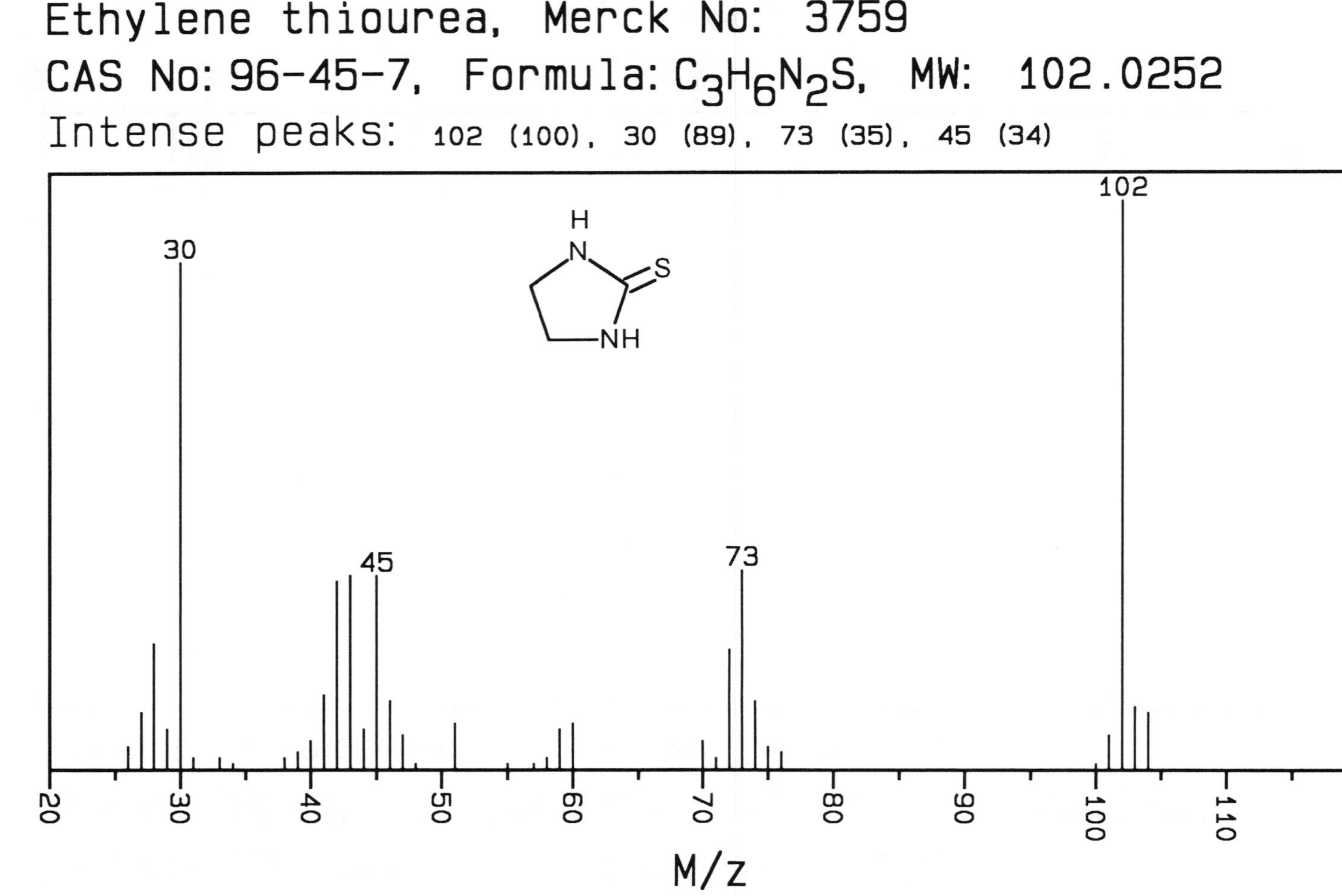
Ethylene thiourea, Merck No: 3759
CAS No: 96-45-7, Formula: $C_3H_6N_2S$, MW: 102.0252
Intense peaks: 102 (100), 30 (89), 73 (35), 45 (34)
H
N
S
NH
30
45
73
102
20
30
40
50
60
70
80
90
100
110
120
M/Z

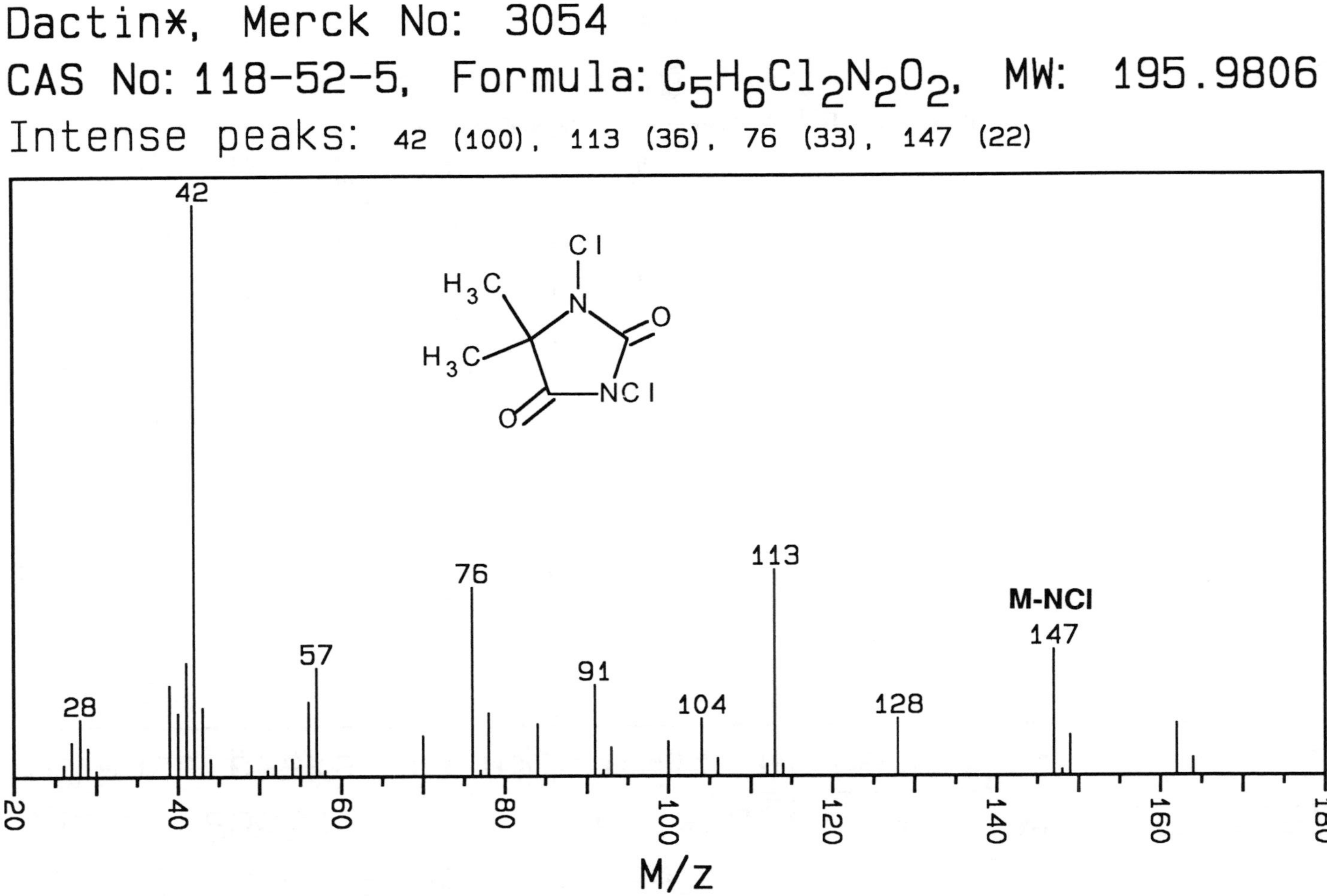
Dactin*, Merck No: 3054
CAS No: 118-52-5, Formula: $C_5H_6Cl_2N_2O_2$, MW: 195.9806
Intense peaks: 42 (100), 113 (36), 76 (33), 147 (22)
Cl
N
NCl
O
O
$H_3C$
$H_3C$
42
28
57
76
91
104
113
128
M-NCl
147
20
40
60
80
100
120
140
160
180
M/Z

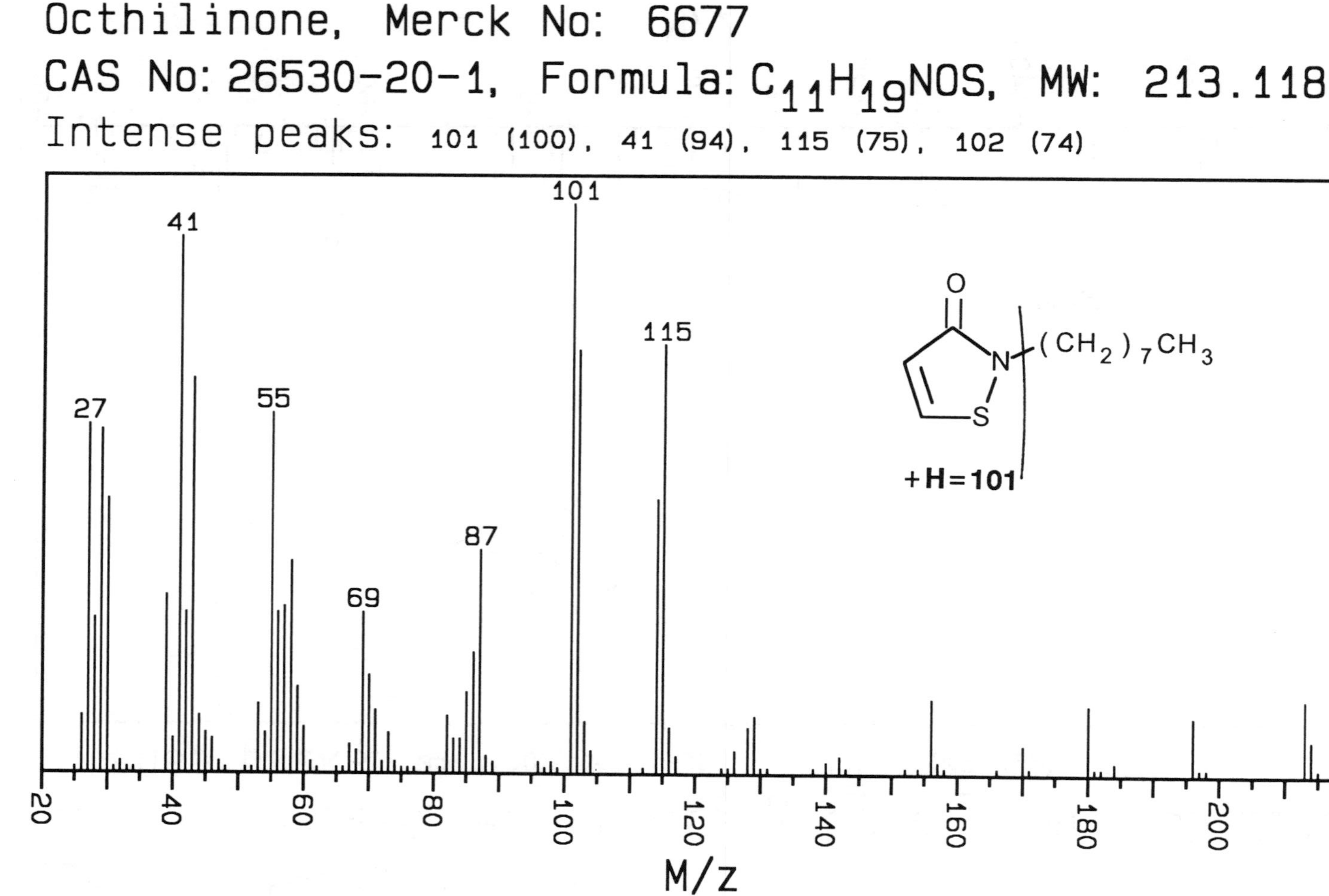
Octhilinone, Merck No: 6677
CAS No: 26530-20-1, Formula: C11H19NOS, MW: 213.1187
Intense peaks: 101 (100), 41 (94), 115 (75), 102 (74)
27
41
55
69
87
101
115
(CH2)7CH3
+H=101
20
40
60
80
100
120
140
160
180
200
220
M/Z

Tebuthiuron, Merck No: 9053

CAS No: 34014-18-1, Formula: $C_9H_{16}N_4OS$, MW: 228.1045

Intense peaks: 57 (100), 98 (85), 171 (70), 41 (55)

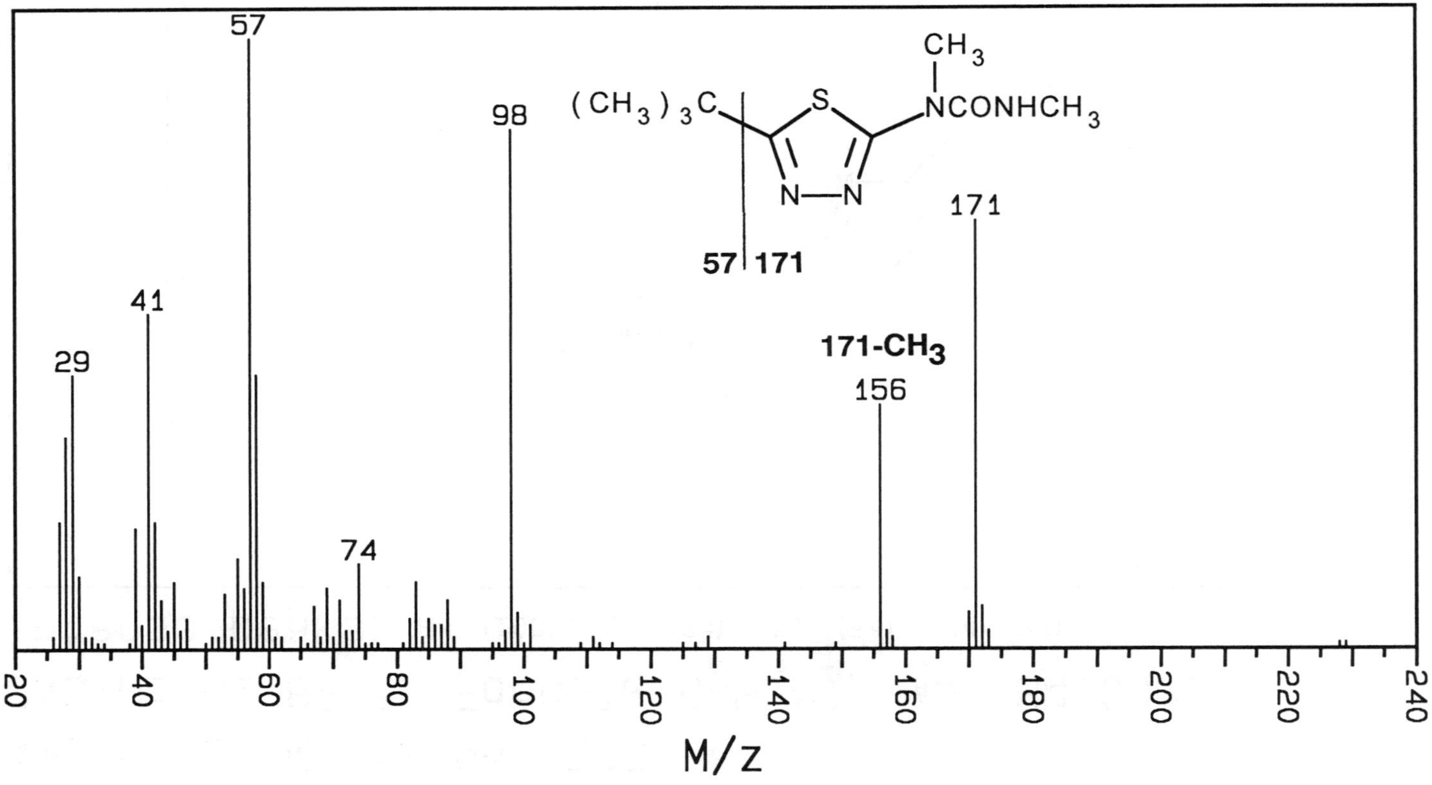

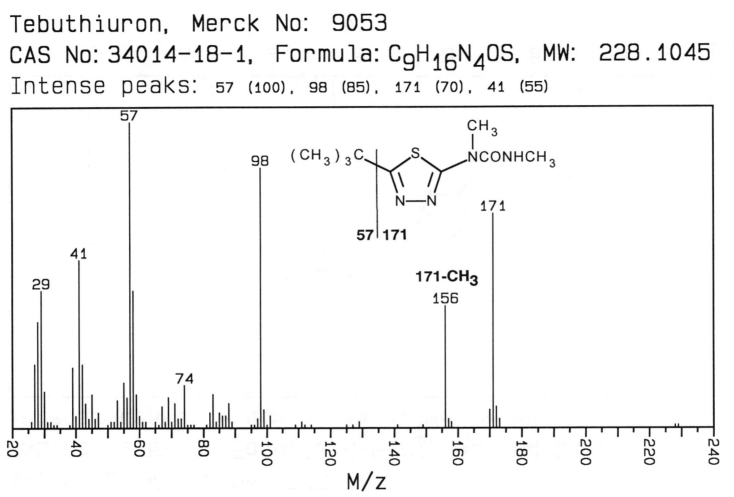

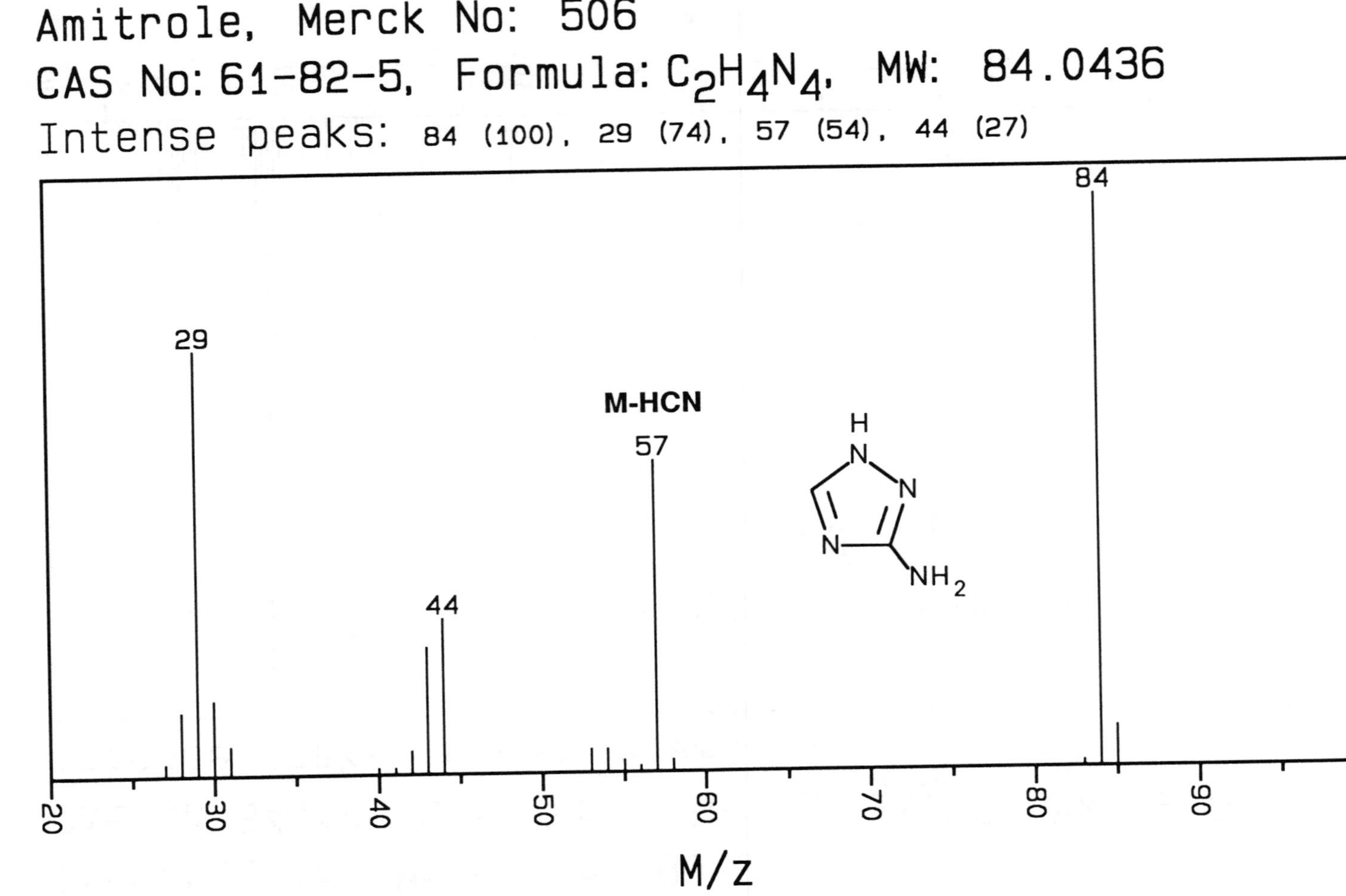

Amitrole, Merck No: 506
CAS No: 61-82-5, Formula: $C_2H_4N_4$, MW: 84.0436
Intense peaks: 84 (100), 29 (74), 57 (54), 44 (27)
84
29
M-HCN
57
44
H
N
N
N
$NH_2$
20
30
40
50
60
70
80
90
100
M/Z

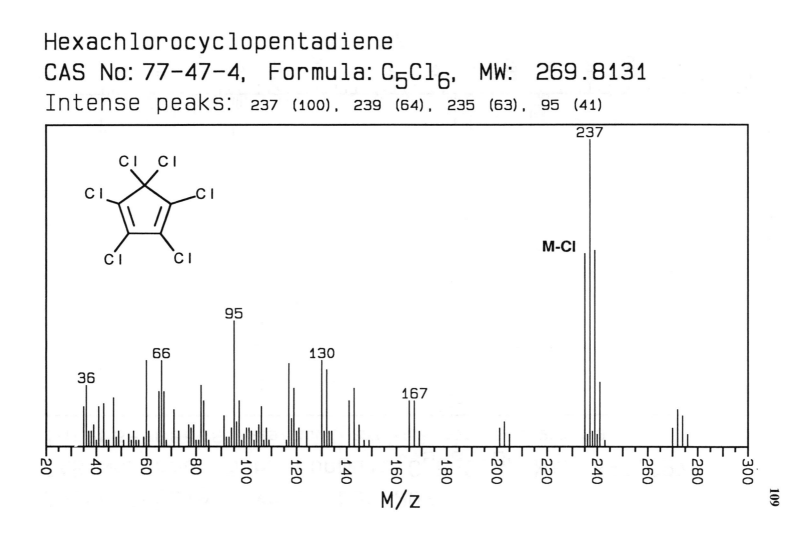

Hexachlorocyclopentadiene

CAS No: 77-47-4, Formula: $C_5Cl_6$, MW: 269.8131

Intense peaks: 237 (100), 239 (64), 235 (63), 95 (41)

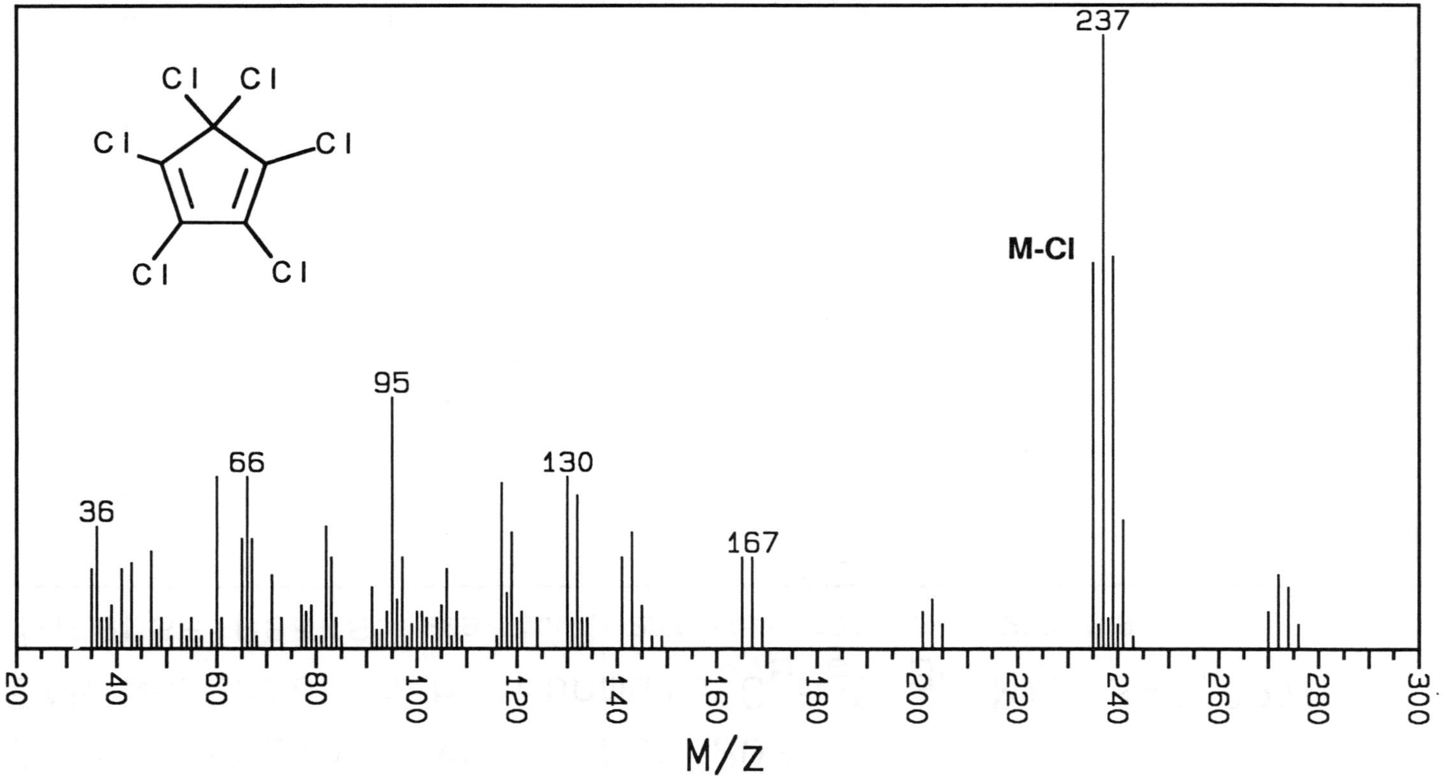

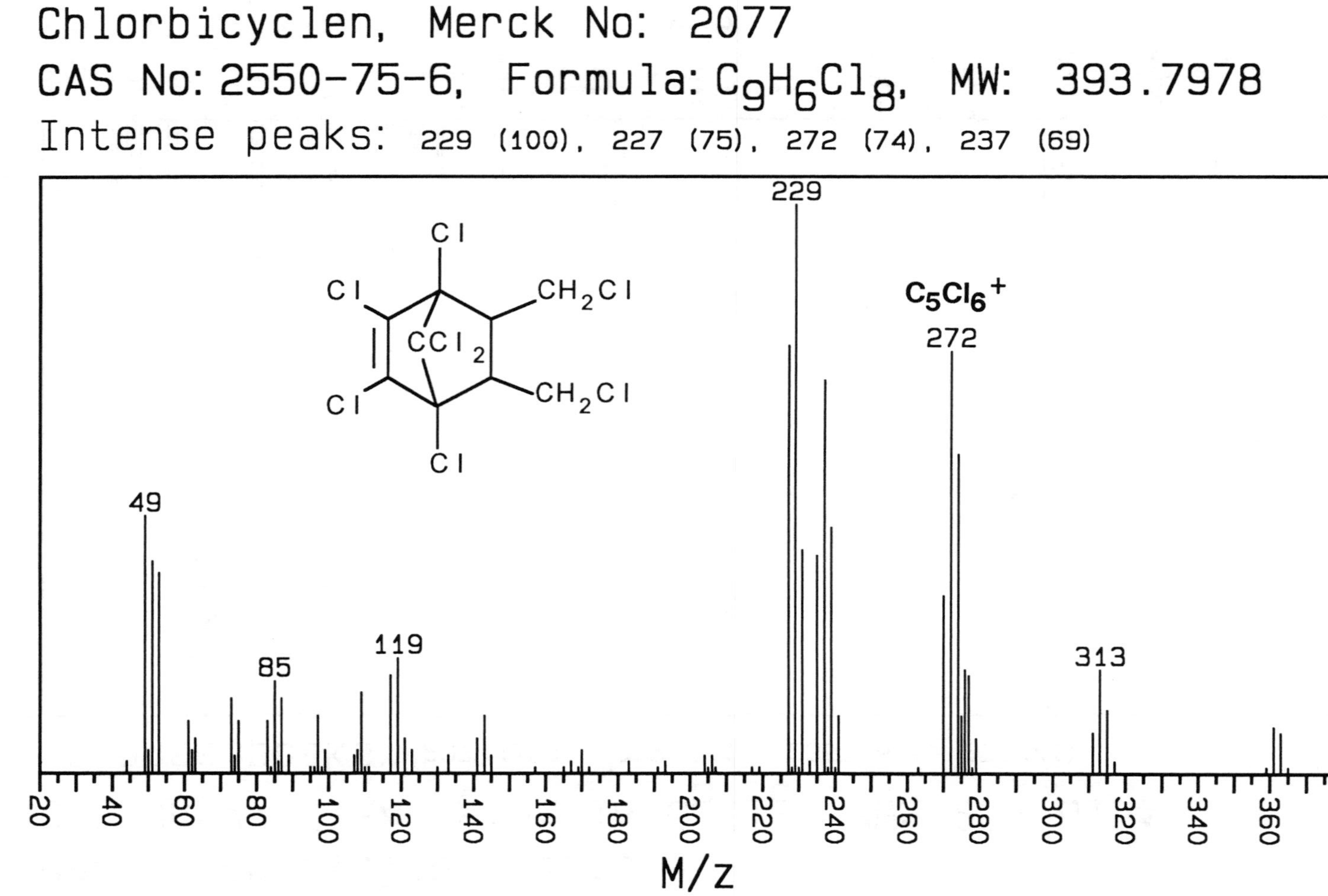
Chlorbicyclen, Merck No: 2077
CAS No: 2550-75-6, Formula: C9H6Cl8, MW: 393.7978
Intense peaks: 229 (100), 227 (75), 272 (74), 237 (69)
Cl
Cl
CH2Cl
CCl2
Cl
CH2Cl
Cl
229
C5Cl6+
272
49
85
119
313
20
40
60
80
100
120
140
160
180
200
220
240
260
280
300
320
340
360
380
M/Z

Isobenzan, Merck No: 5010

CAS No: 297-78-9, Formula: $C_9H_4Cl_8O$, MW: 407.7769

Intense peaks: 103 (100), 311 (51), 313 (45), 105 (33)

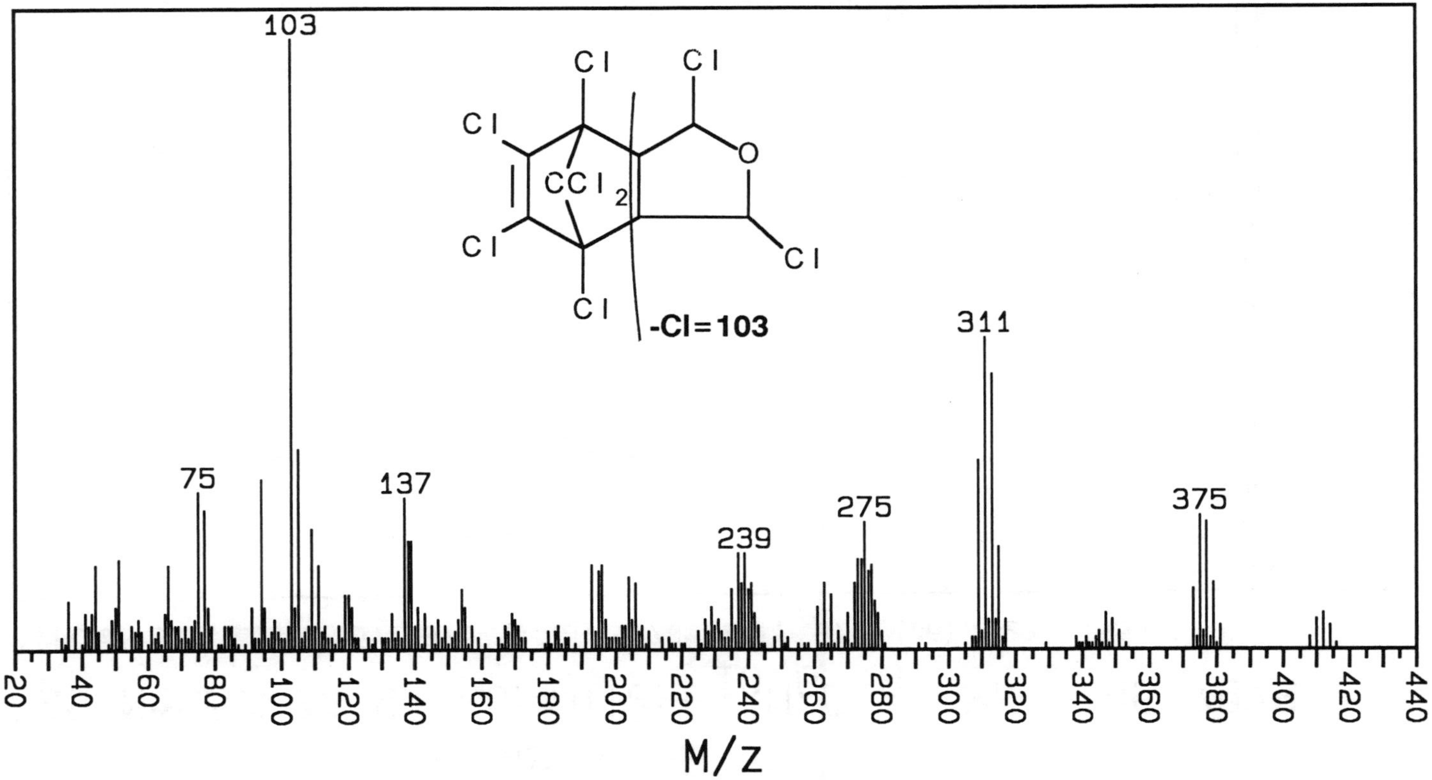

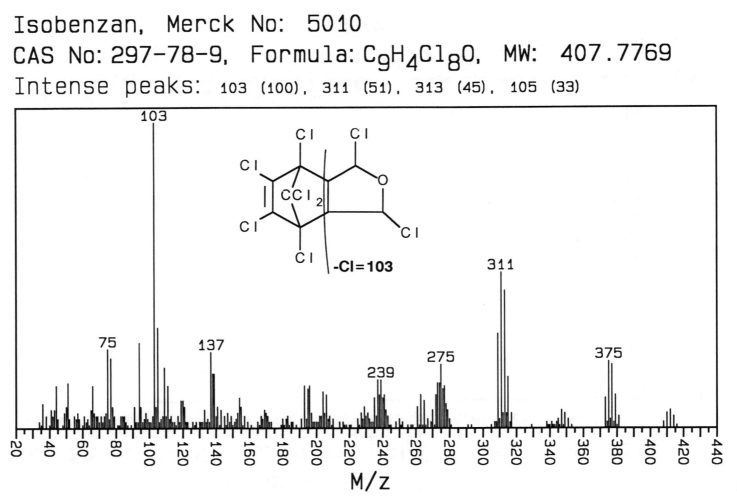

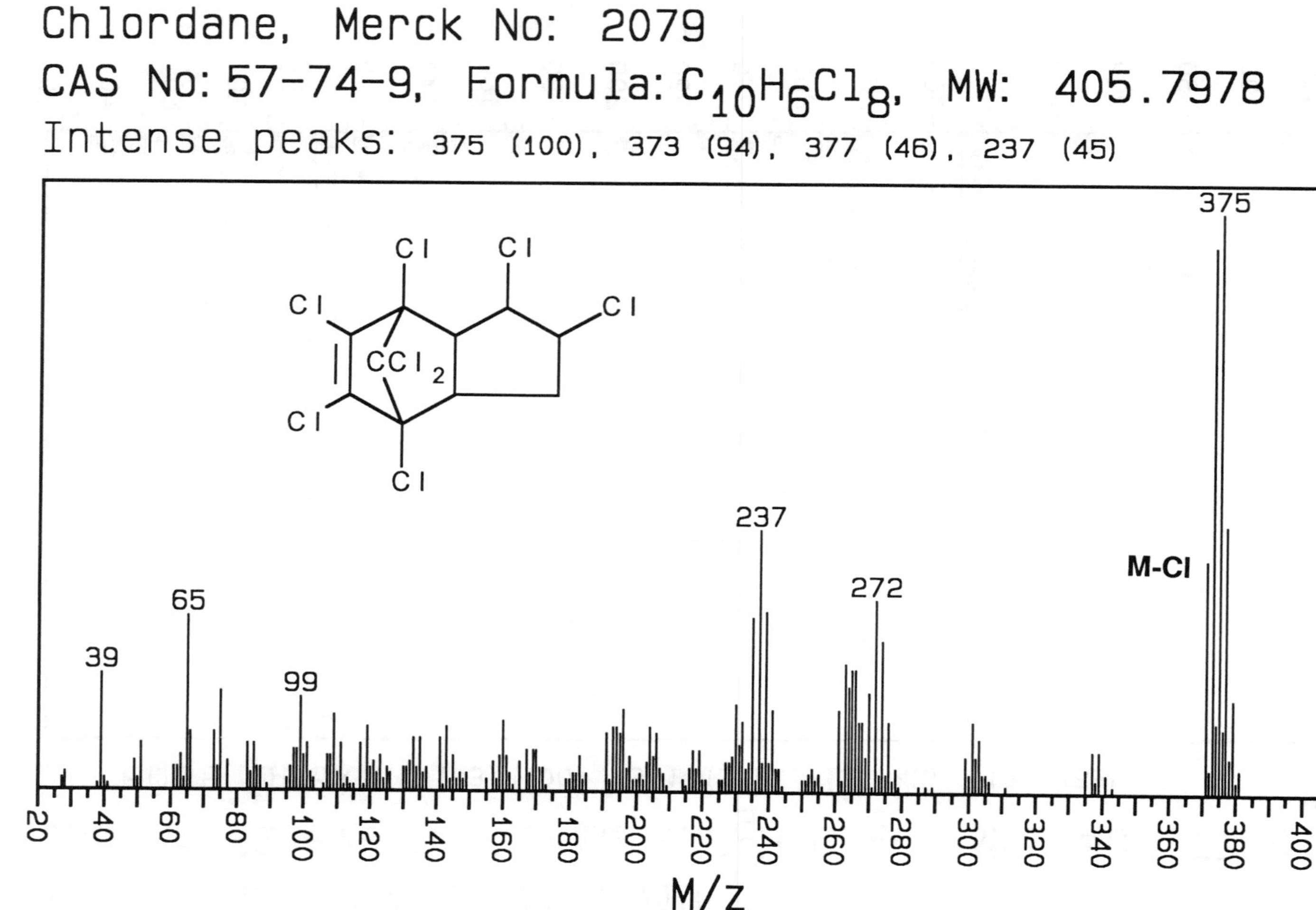
Chlordane, Merck No: 2079
CAS No: 57-74-9, Formula: $C_{10}H_6Cl_8$, MW: 405.7978
Intense peaks: 375 (100), 373 (94), 377 (46), 237 (45)
Cl
Cl
Cl
Cl
CCl2
Cl
Cl
Cl
375
M-Cl
272
237
99
65
39
20
40
60
80
100
120
140
160
180
200
220
240
260
280
300
320
340
360
380
400
420
M/Z

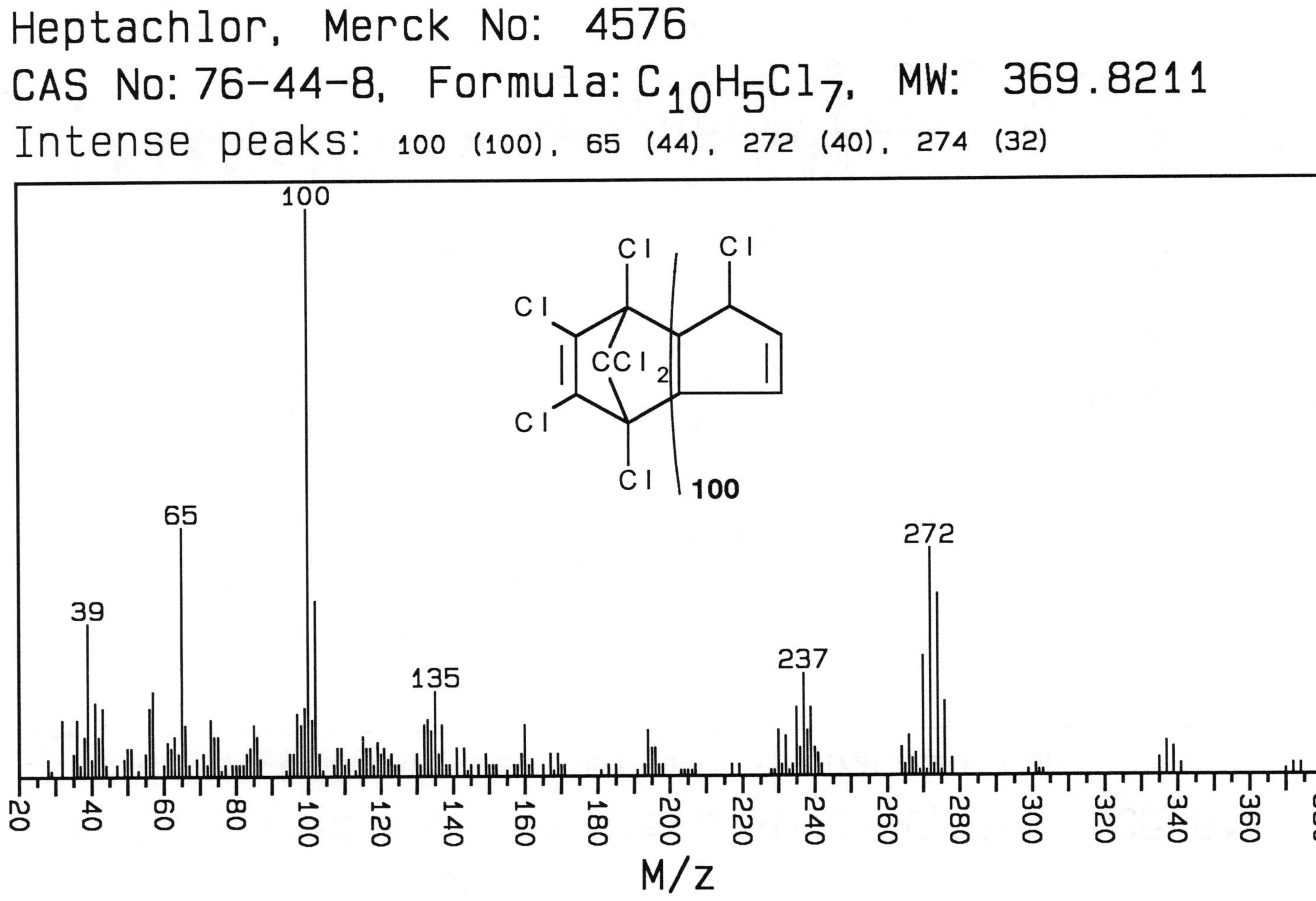
Heptachlor, Merck No: 4576
CAS No: 76-44-8, Formula: $C_{10}H_5Cl_7$, MW: 369.8211
Intense peaks: 100 (100), 65 (44), 272 (40), 274 (32)
100
65
39
135
237
272
Cl
Cl
Cl
CCl2
Cl
Cl
Cl
100
20
40
60
80
100
120
140
160
180
200
220
240
260
280
300
320
340
360
380
M/Z

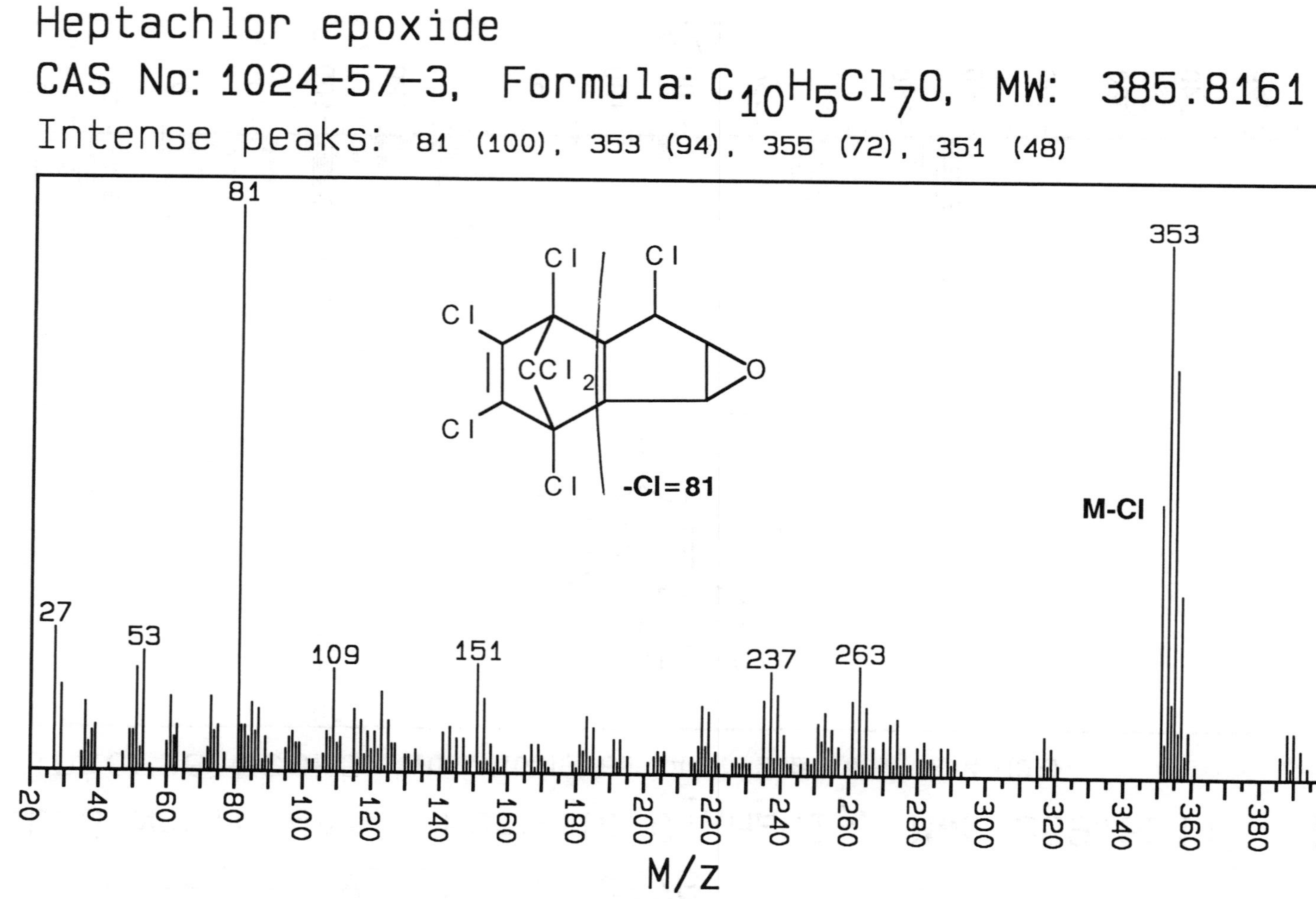
Heptachlor epoxide
CAS No: 1024-57-3, Formula: C10H5Cl7O, MW: 385.8161
Intense peaks: 81 (100), 353 (94), 355 (72), 351 (48)
81
353
Cl
CCl2
O
-Cl=81
M-Cl
27
53
109
151
237
263
20
40
60
80
100
120
140
160
180
200
220
240
260
280
300
320
340
360
380
400
M/Z

Endosulfan, Merck No: 3529

CAS No: 115-29-7, Formula: $C_9H_6Cl_6O_3S$, MW: 403.8169

Intense peaks: 195 (100), 241 (80), 197 (77), 237 (70)

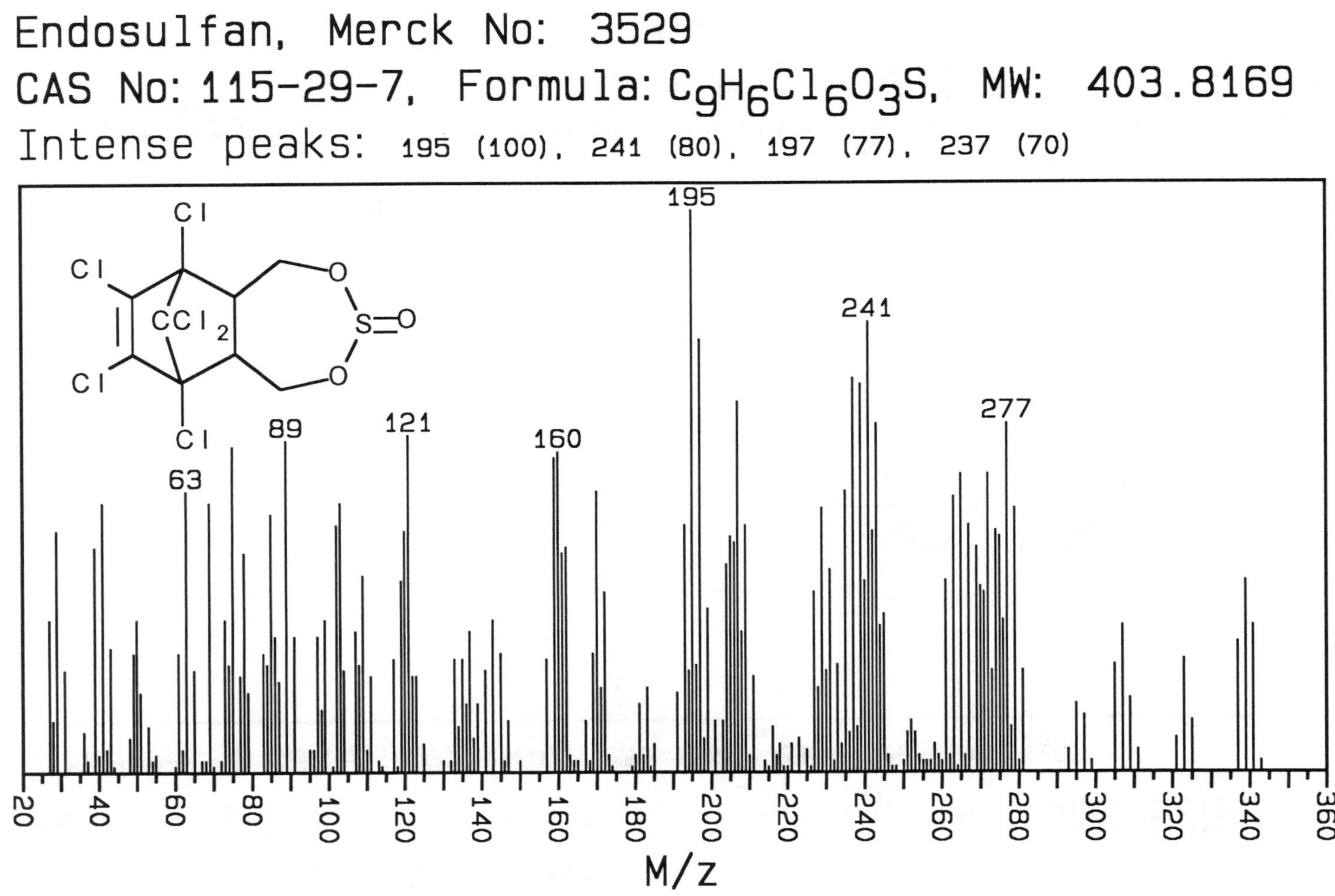

Pentac, Merck No: 3095

CAS No: 2227-17-0, Formula: $C_{10}Cl_{10}$, MW: 469.6886

Intense peaks: 237 (100), 239 (64), 235 (62), 241 (20)

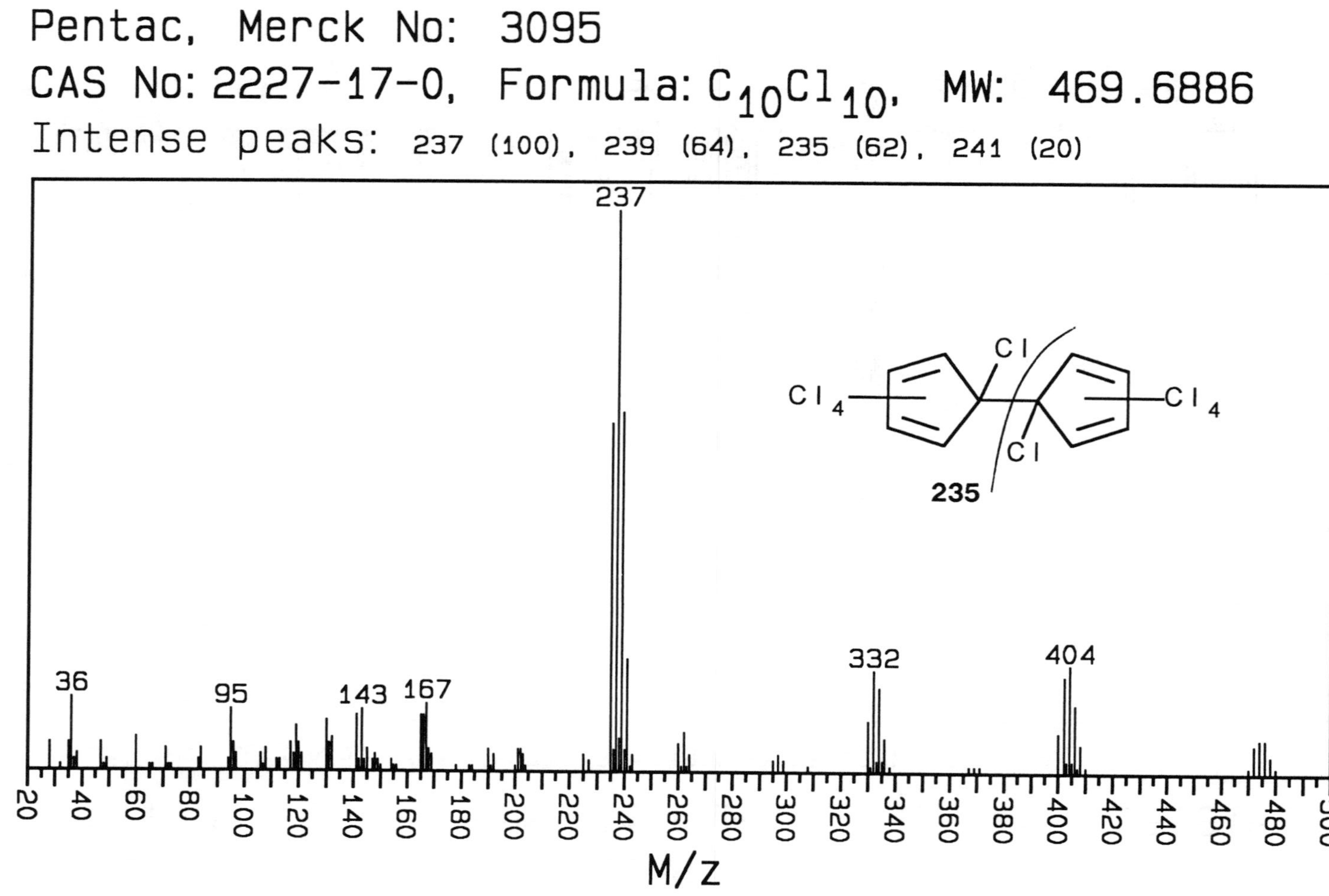

Kepone*, Merck No: 2081

CAS No: 143-50-0, Formula: $C_{10}Cl_{10}O$, MW: 485.6835

Intense peaks: 272 (100), 274 (79), 270 (51), 237 (45)

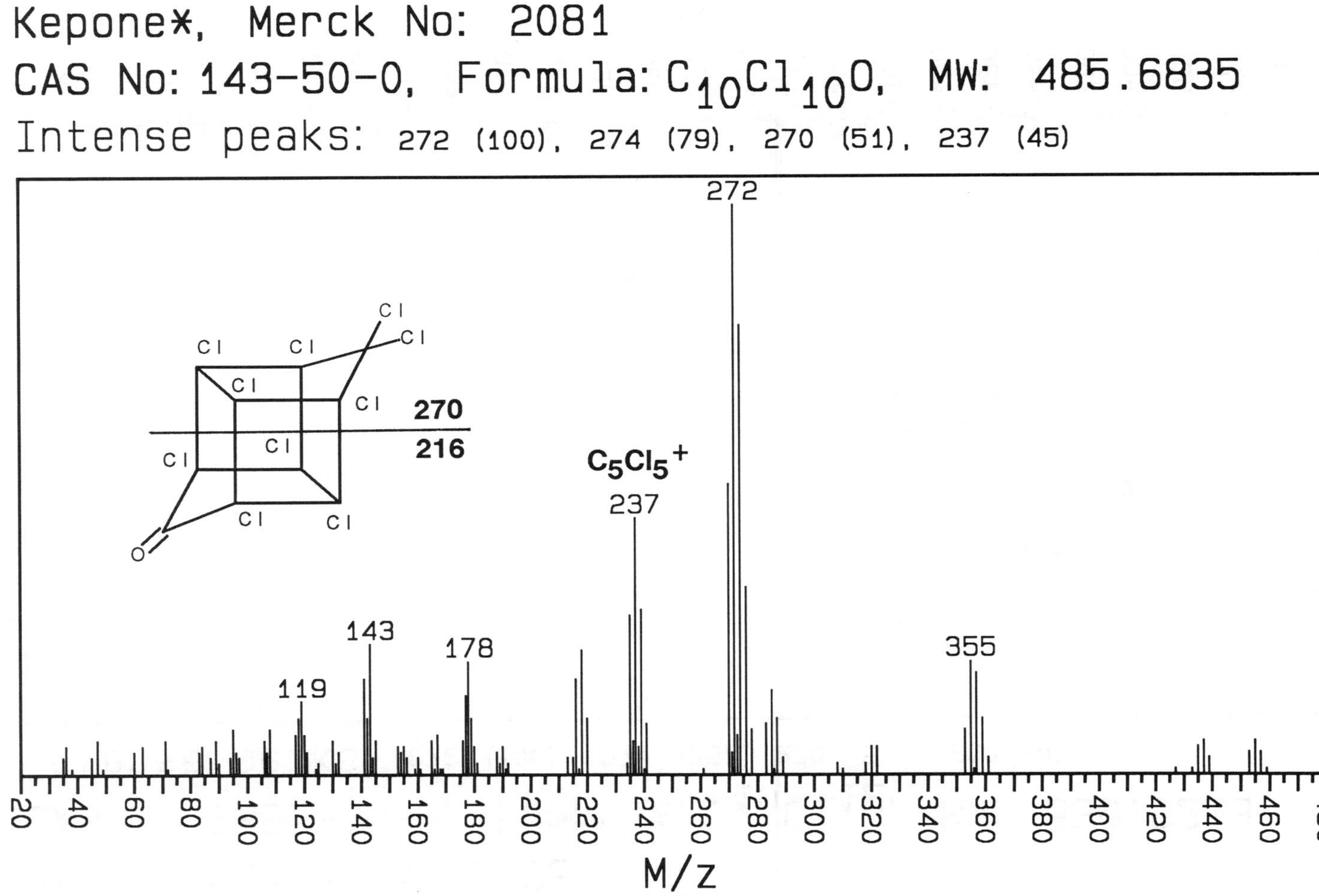

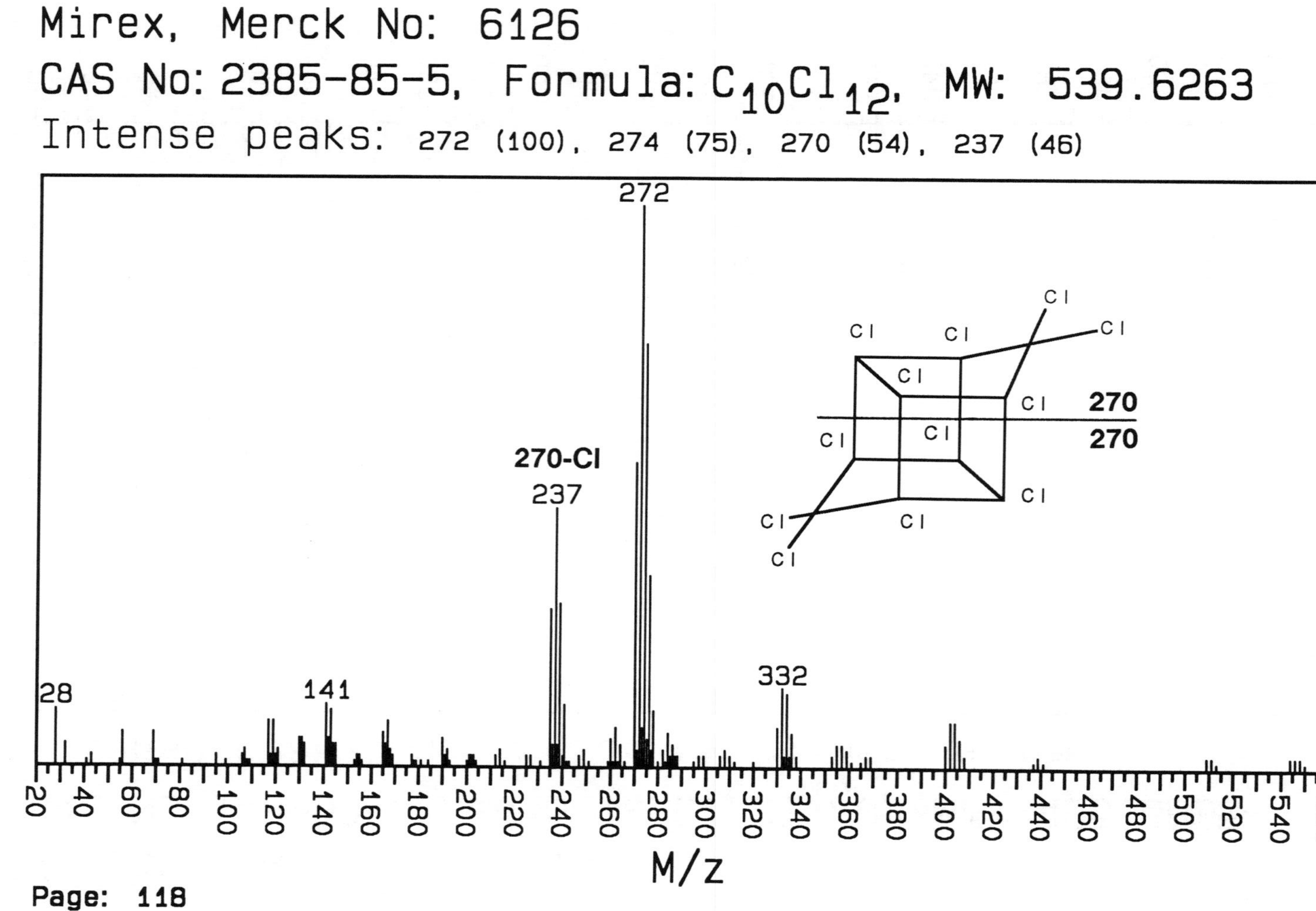
Mirex, Merck No: 6126
CAS No: 2385-85-5, Formula: $C_{10}Cl_{12}$, MW: 539.6263
Intense peaks: 272 (100), 274 (75), 270 (54), 237 (46)
272
270-Cl
237
28
141
332
Cl
270
270
20
40
60
80
100
120
140
160
180
200
220
240
260
280
300
320
340
360
380
400
420
440
460
480
500
520
540
560
M/Z
Page: 118

Aldrin, Merck No: 219

CAS No: 309-00-2, Formula: $C_{12}H_8Cl_6$, MW: 361.8757

Intense peaks: 66 (100), 79 (43), 91 (34), 263 (32)

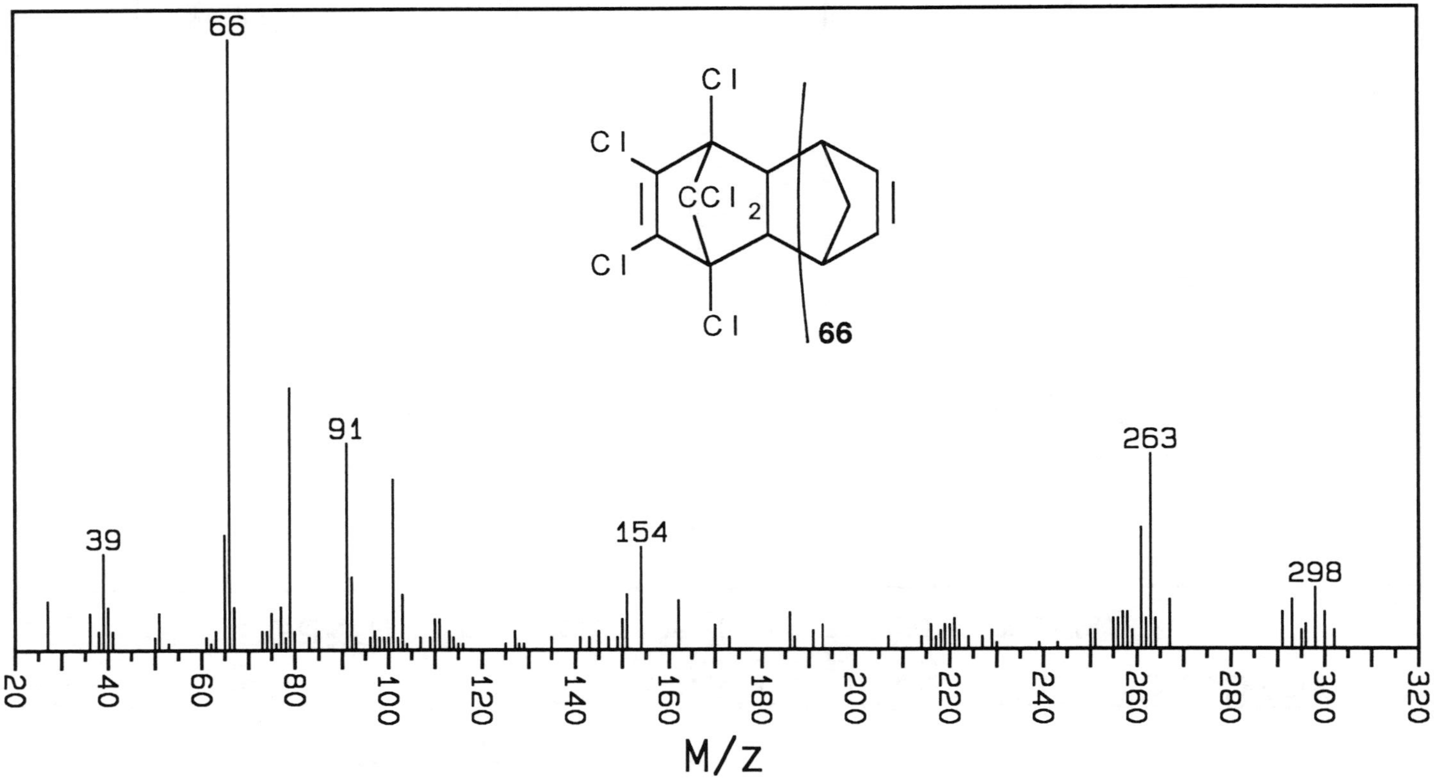

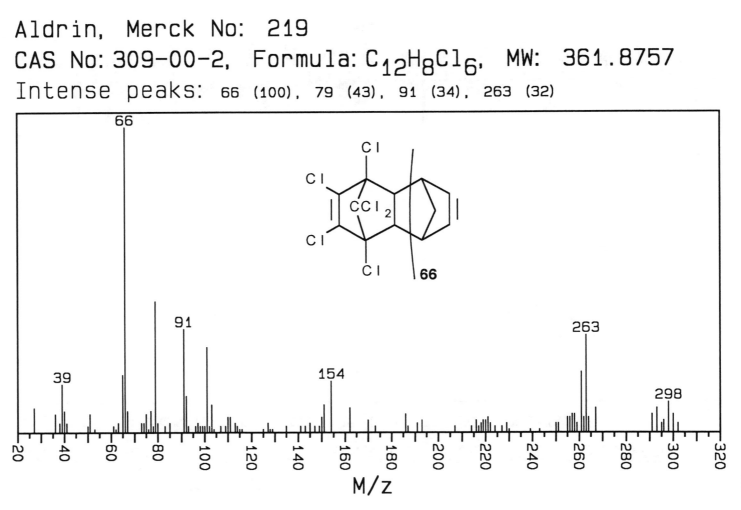

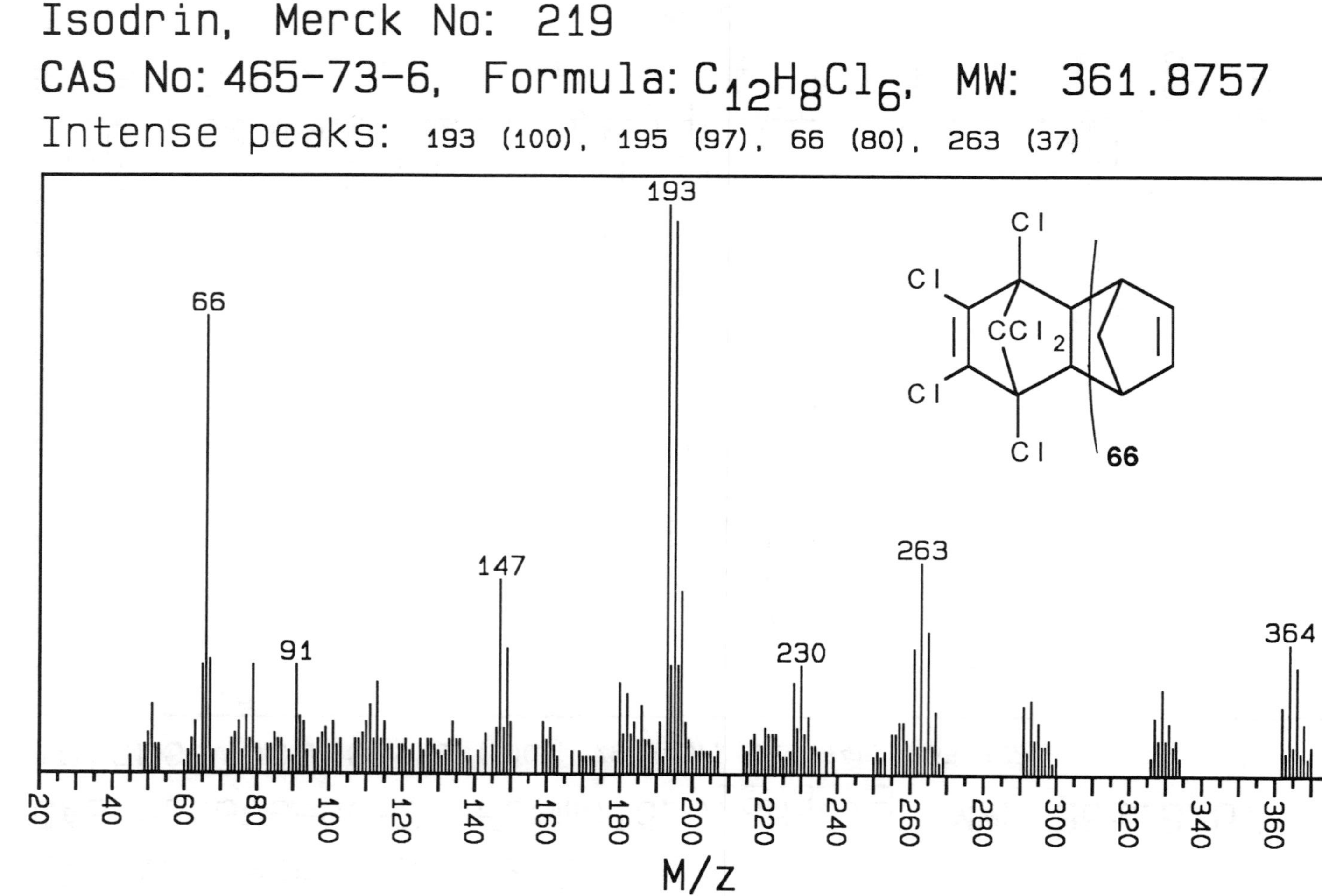
Isodrin, Merck No: 219
CAS No: 465-73-6, Formula: $C_{12}H_8Cl_6$, MW: 361.8757
Intense peaks: 193 (100), 195 (97), 66 (80), 263 (37)
193
66
147
91
230
263
364
Cl
Cl
Cl
Cl
$CCl_2$
66
M/Z

Dieldrin, Merck No: 3093

CAS No: 60-57-1, Formula: $C_{12}H_8Cl_6O$, MW: 377.8706

Intense peaks: 79 (100), 82 (42), 81 (35), 108 (21)

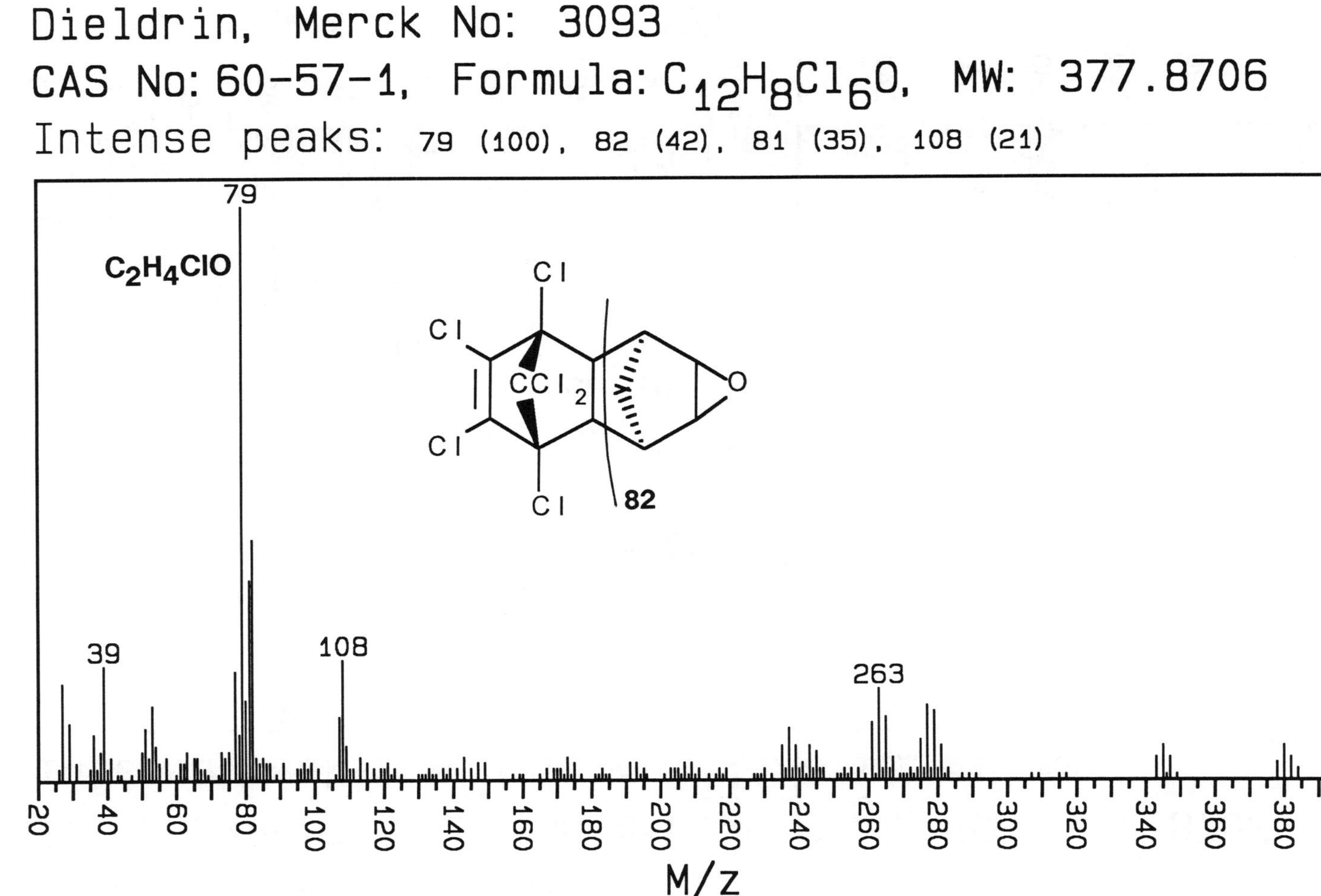

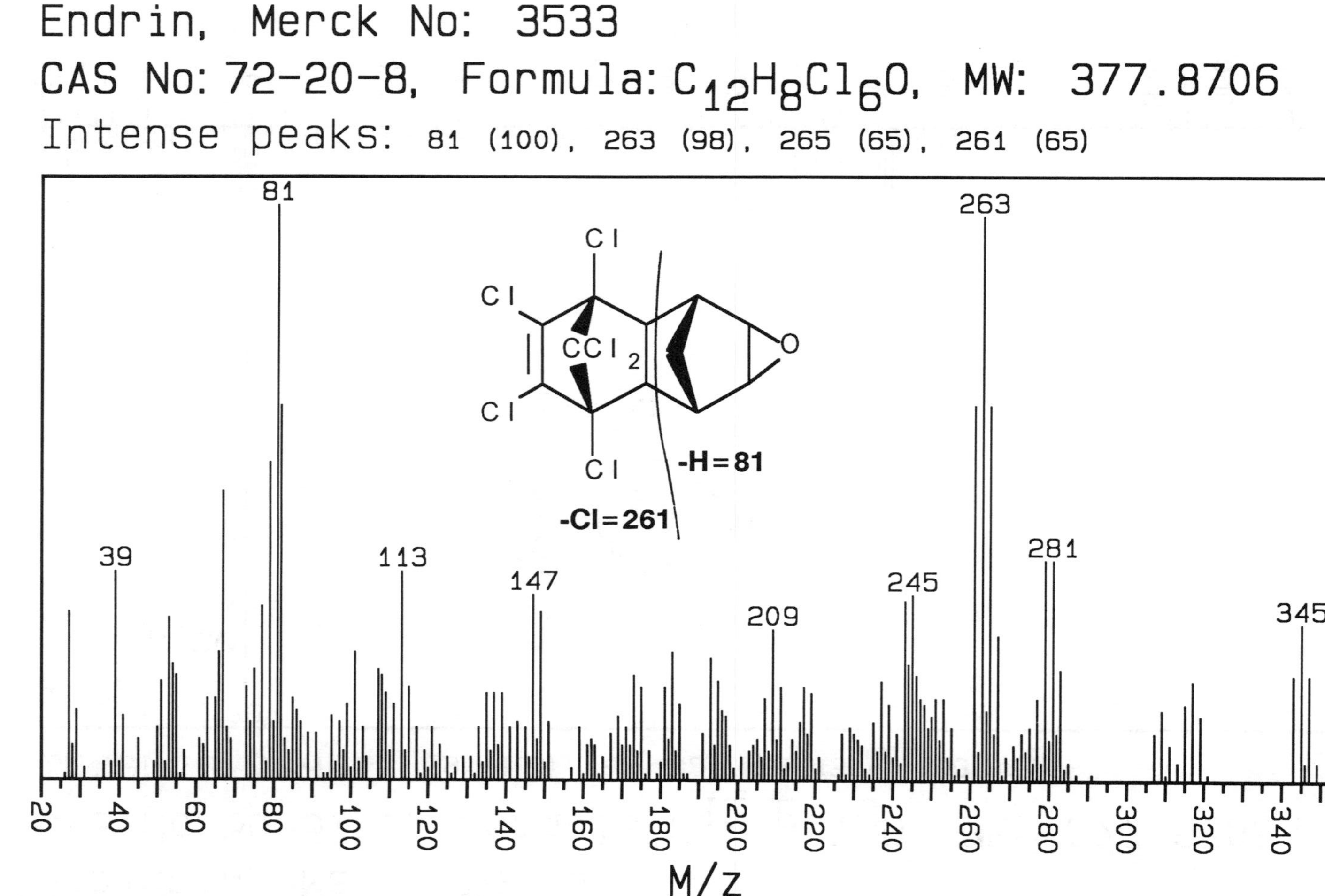
Endrin, Merck No: 3533
CAS No: 72-20-8, Formula: $C_{12}H_8Cl_6O$, MW: 377.8706
Intense peaks: 81 (100), 263 (98), 265 (65), 261 (65)
81
263
Cl
Cl
$CCl_2$
O
Cl
Cl
-H=81
-Cl=261
39
113
147
209
245
281
345
20
40
60
80
100
120
140
160
180
200
220
240
260
280
300
320
340
360
M/Z

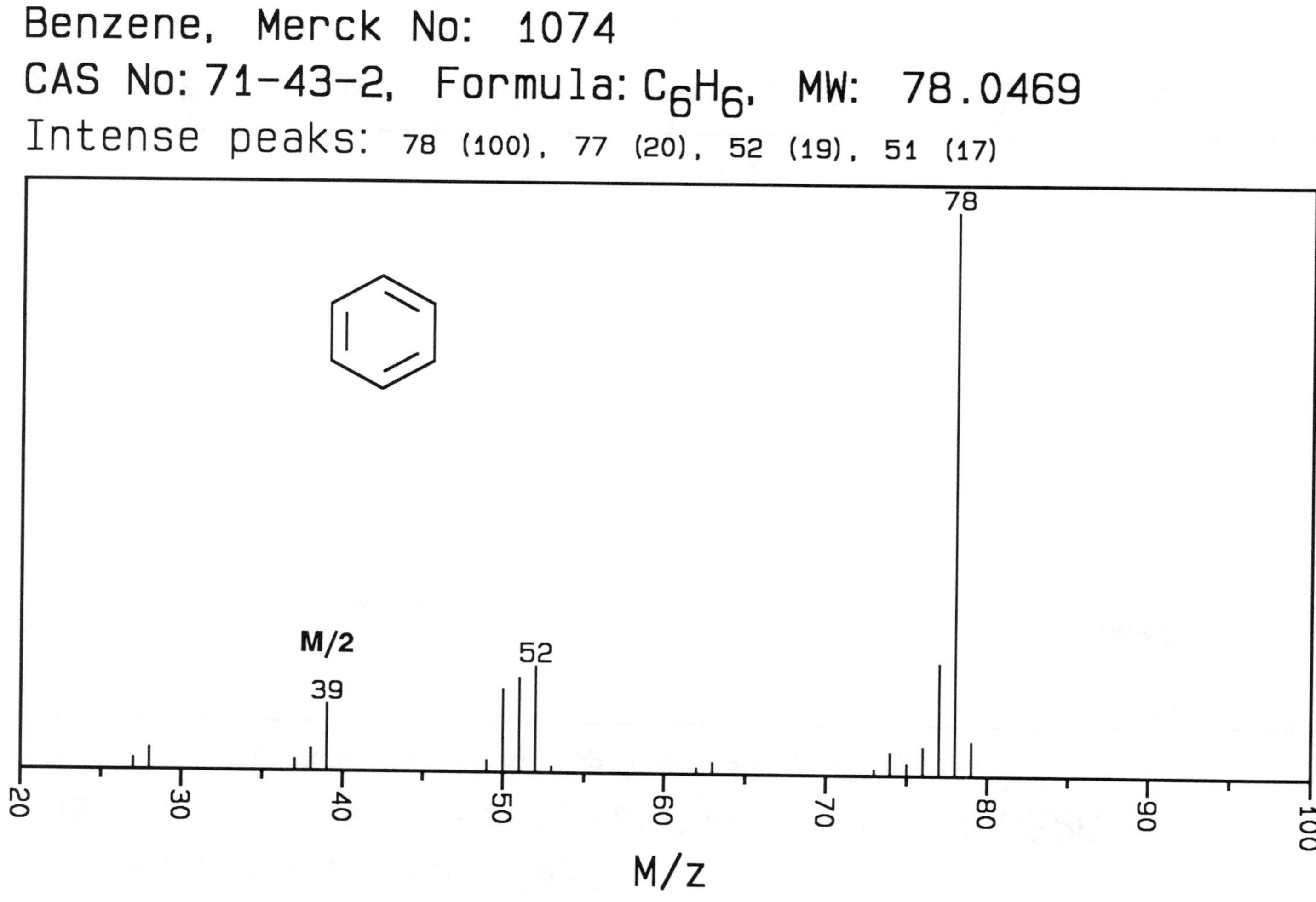
Benzene, Merck No: 1074
CAS No: 71-43-2, Formula: $C_6H_6$, MW: 78.0469
Intense peaks: 78 (100), 77 (20), 52 (19), 51 (17)
78
M/2
39
52
20
30
40
50
60
70
80
90
100
M/z

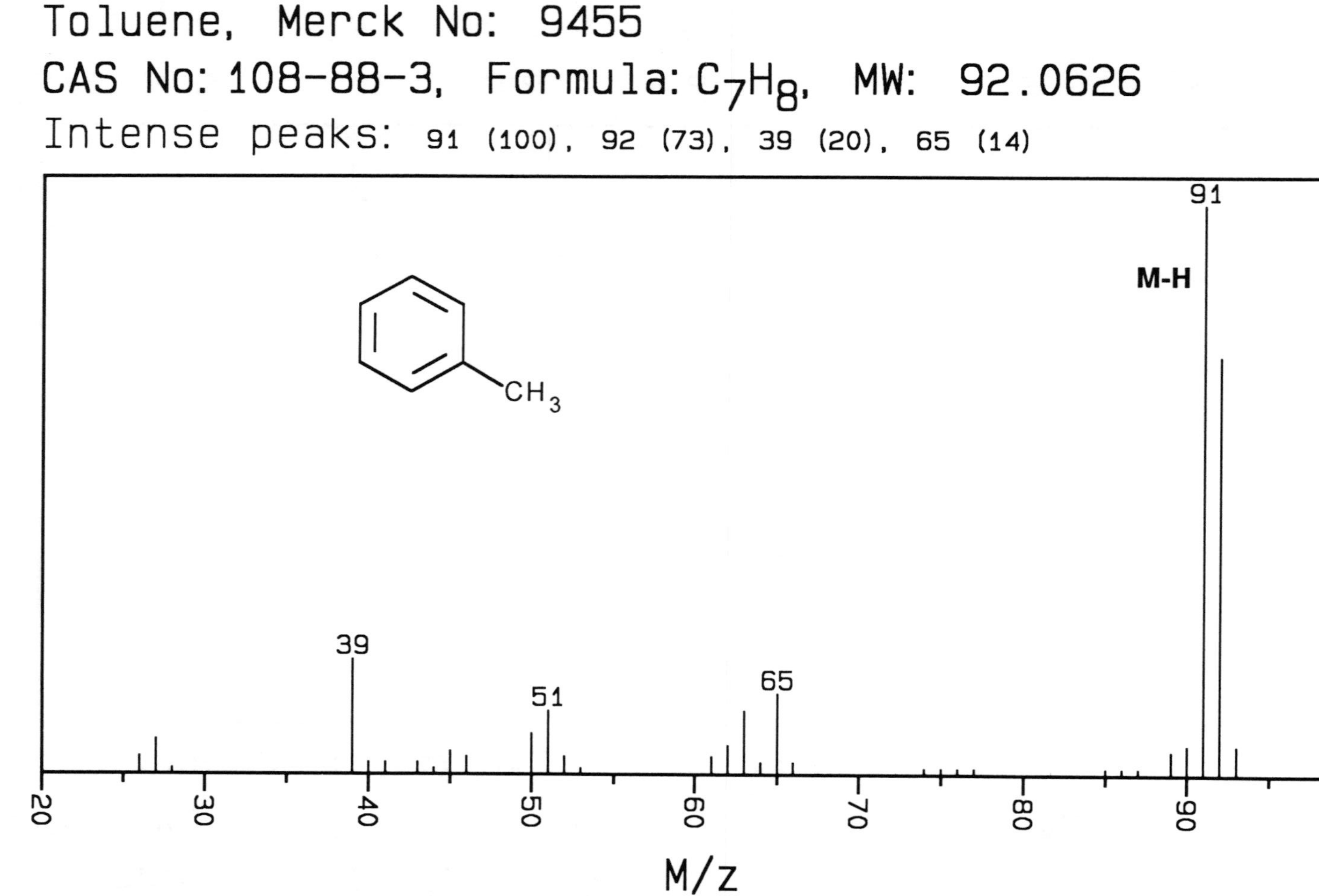
Toluene, Merck No: 9455
CAS No: 108-88-3, Formula: $C_7H_8$, MW: 92.0626
Intense peaks: 91 (100), 92 (73), 39 (20), 65 (14)
$CH_3$
91
M-H
39
51
65
20
30
40
50
60
70
80
90
100
M/Z

Pyridine, 4-methyl-, Merck No: 7374

CAS No: 108-89-4, Formula: $C_6H_7N$, MW: 93.0578

Intense peaks: 93 (100), 39 (44), 66 (42), 65 (26)

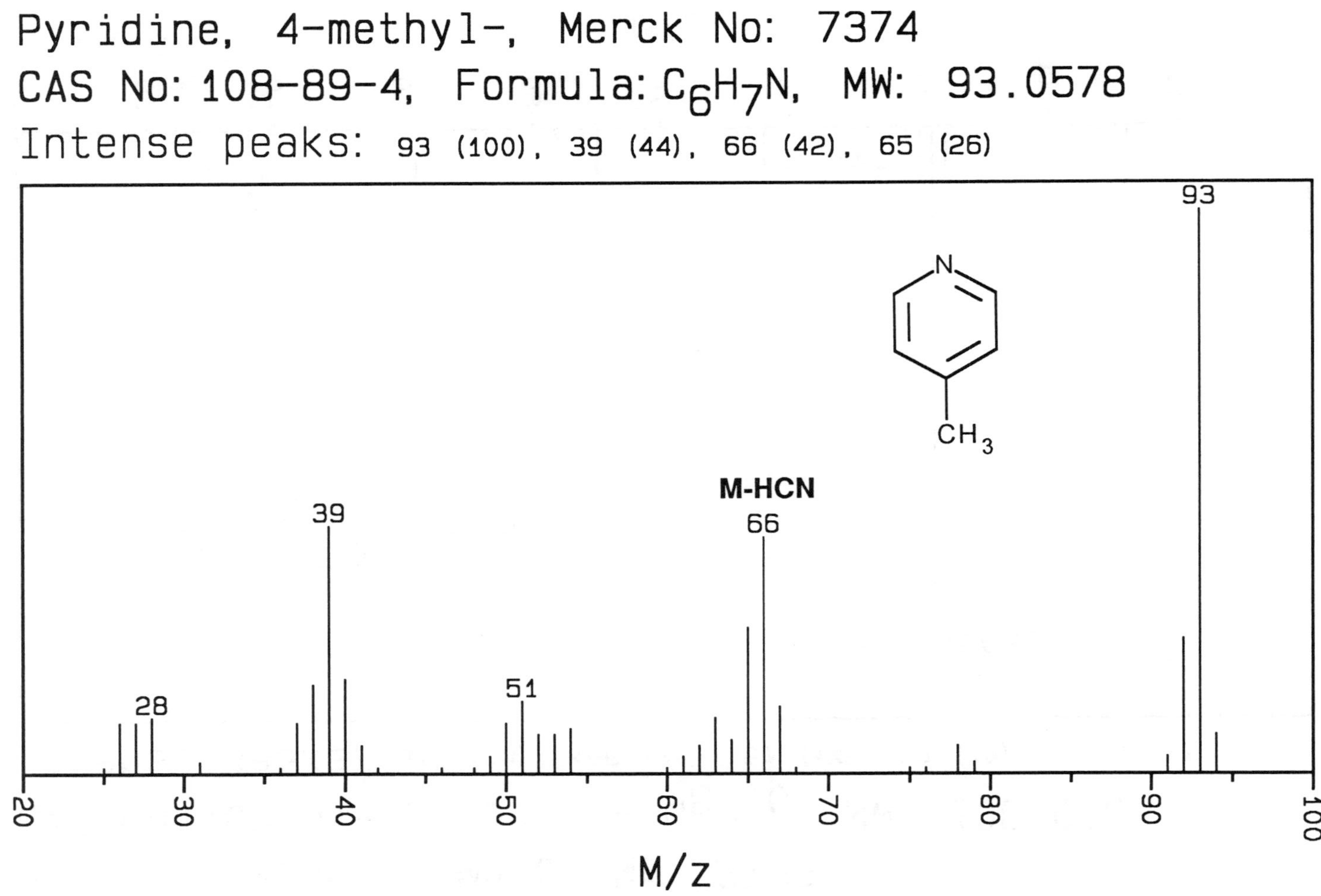

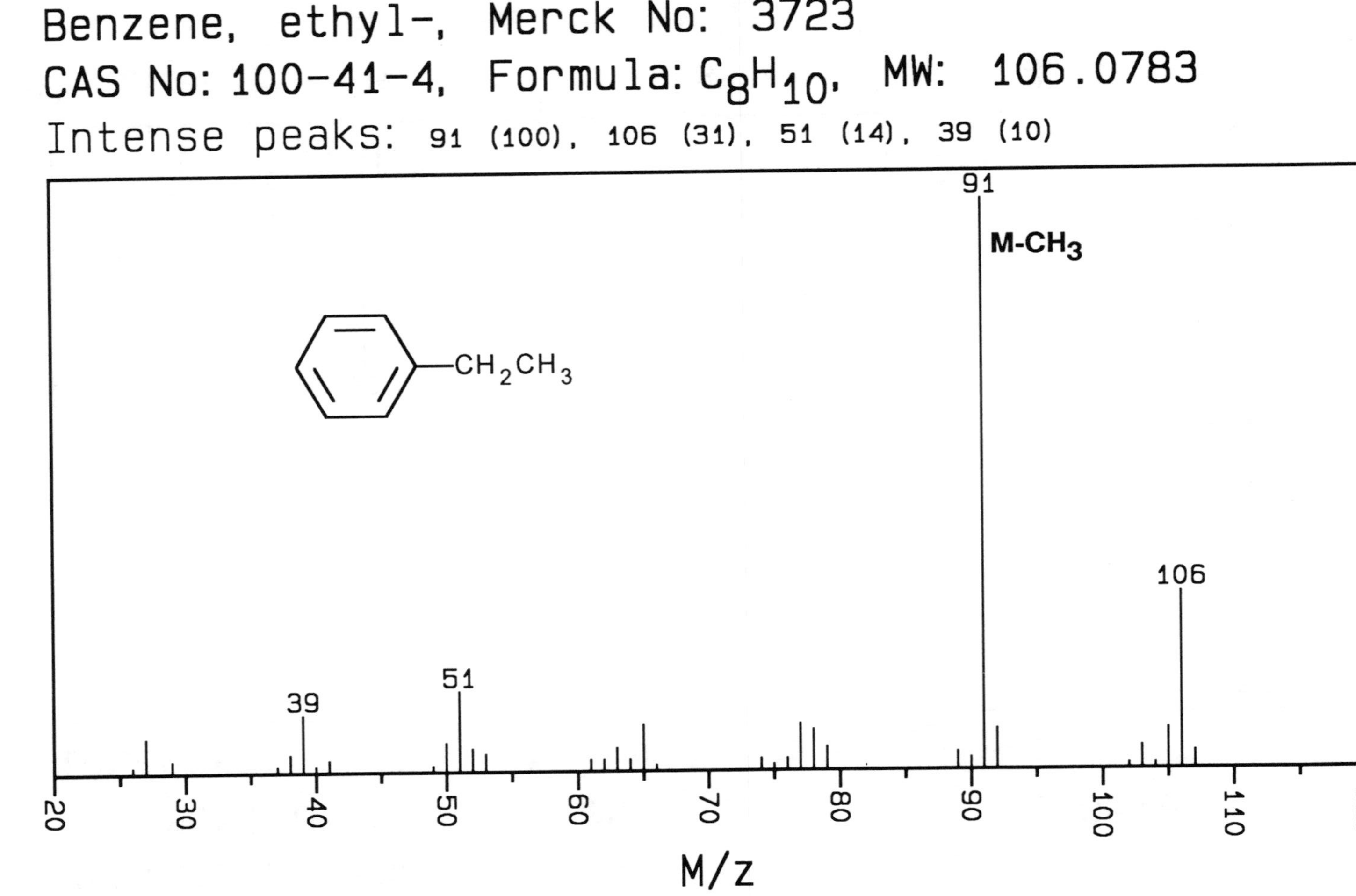
Benzene, ethyl-, Merck No: 3723
CAS No: 100-41-4, Formula: $C_8H_{10}$, MW: 106.0783
Intense peaks: 91 (100), 106 (31), 51 (14), 39 (10)
91
M-$CH_3$
$CH_2CH_3$
106
39
51
20
30
40
50
60
70
80
90
100
110
120
M/Z

Xylene, Merck No: 9988

CAS No: 1330-20-7, Formula: $C_8H_{10}$, MW: 106.0782

Intense peaks: 91 (100), 106 (57), 105 (25), 77 (13)

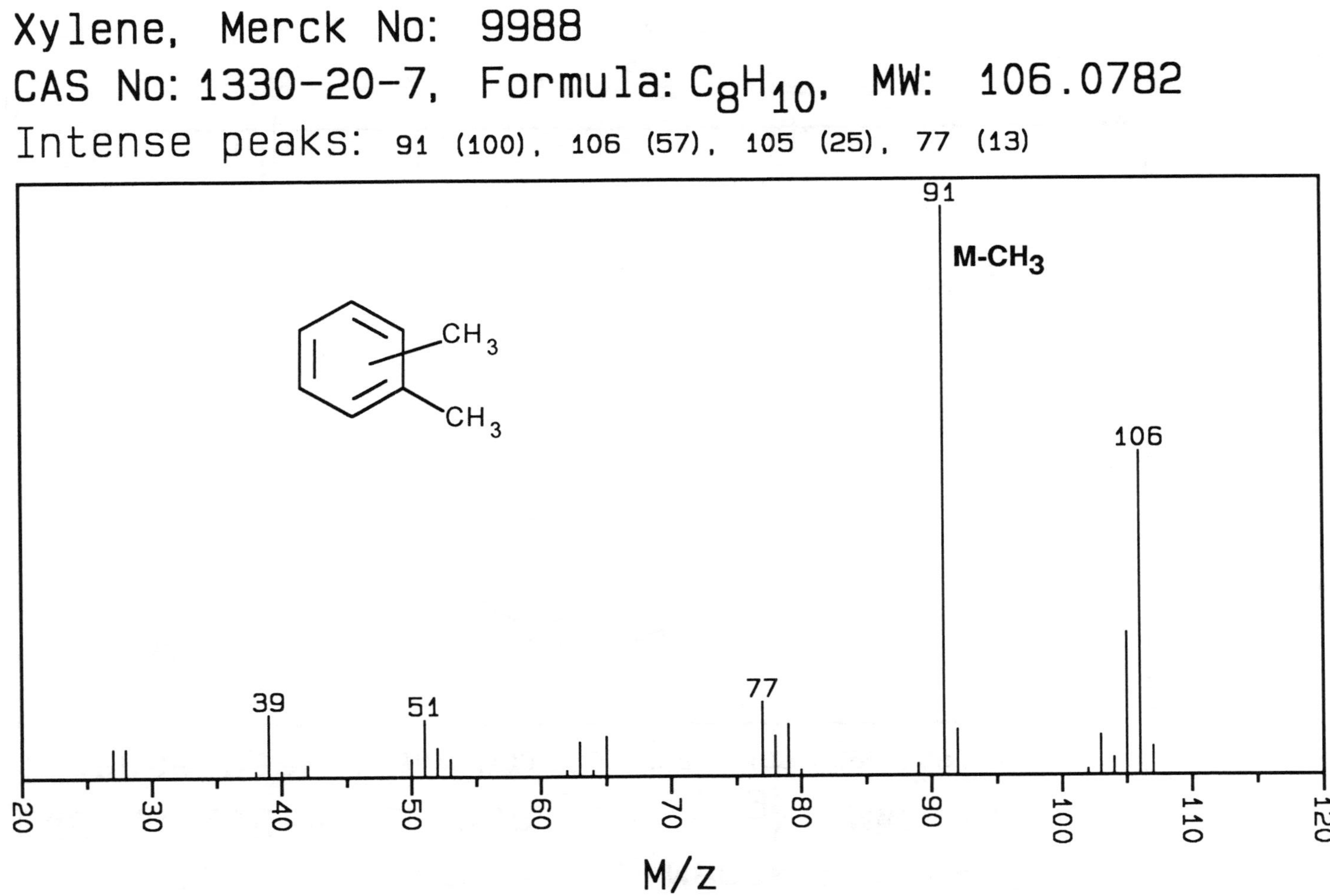

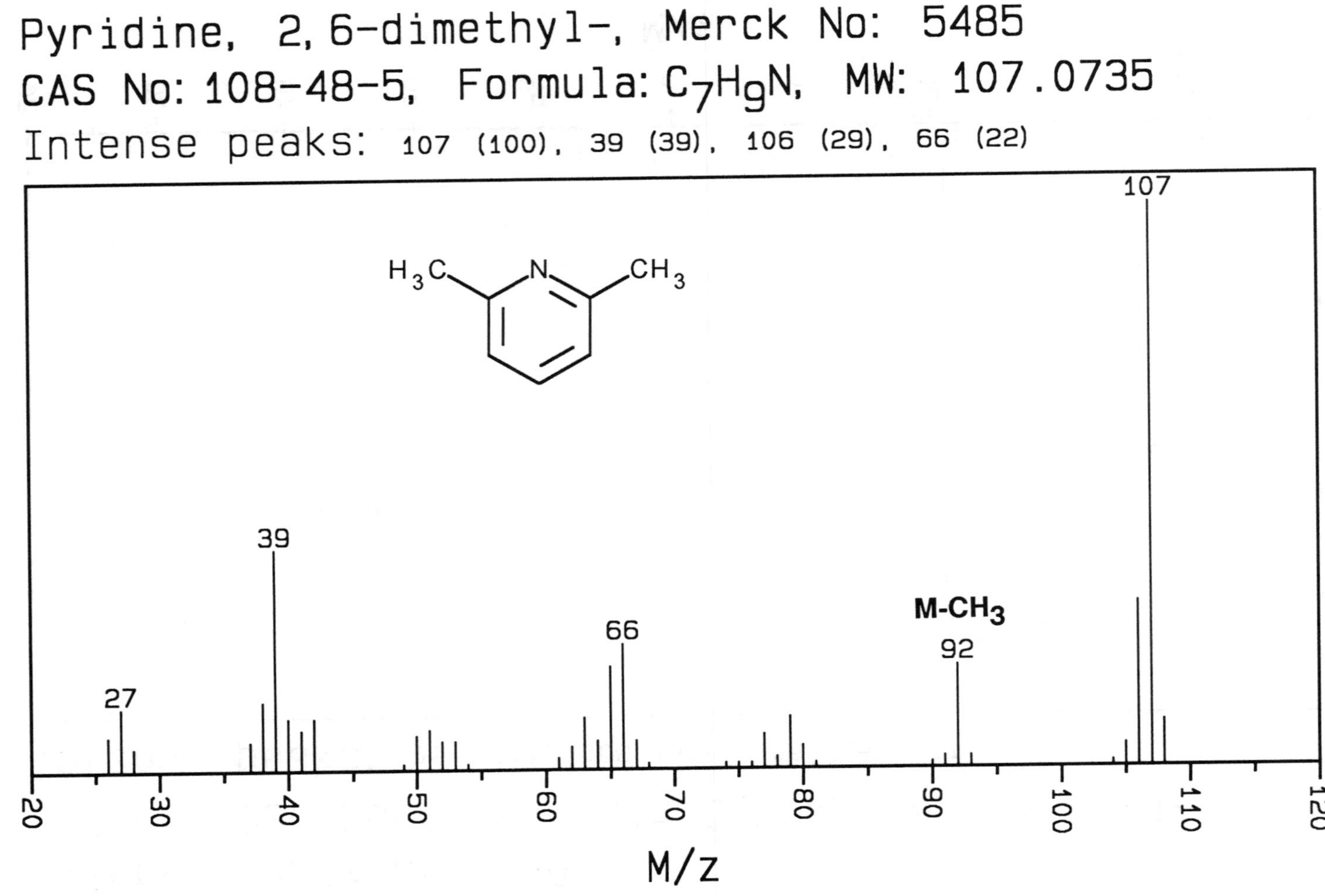
Pyridine, 2,6-dimethyl-, Merck No: 5485
CAS No: 108-48-5, Formula: C7H9N, MW: 107.0735
Intense peaks: 107 (100), 39 (39), 106 (29), 66 (22)
H3C
N
CH3
107
39
66
M-CH3
92
27
20
30
40
50
60
70
80
90
100
110
120
M/Z

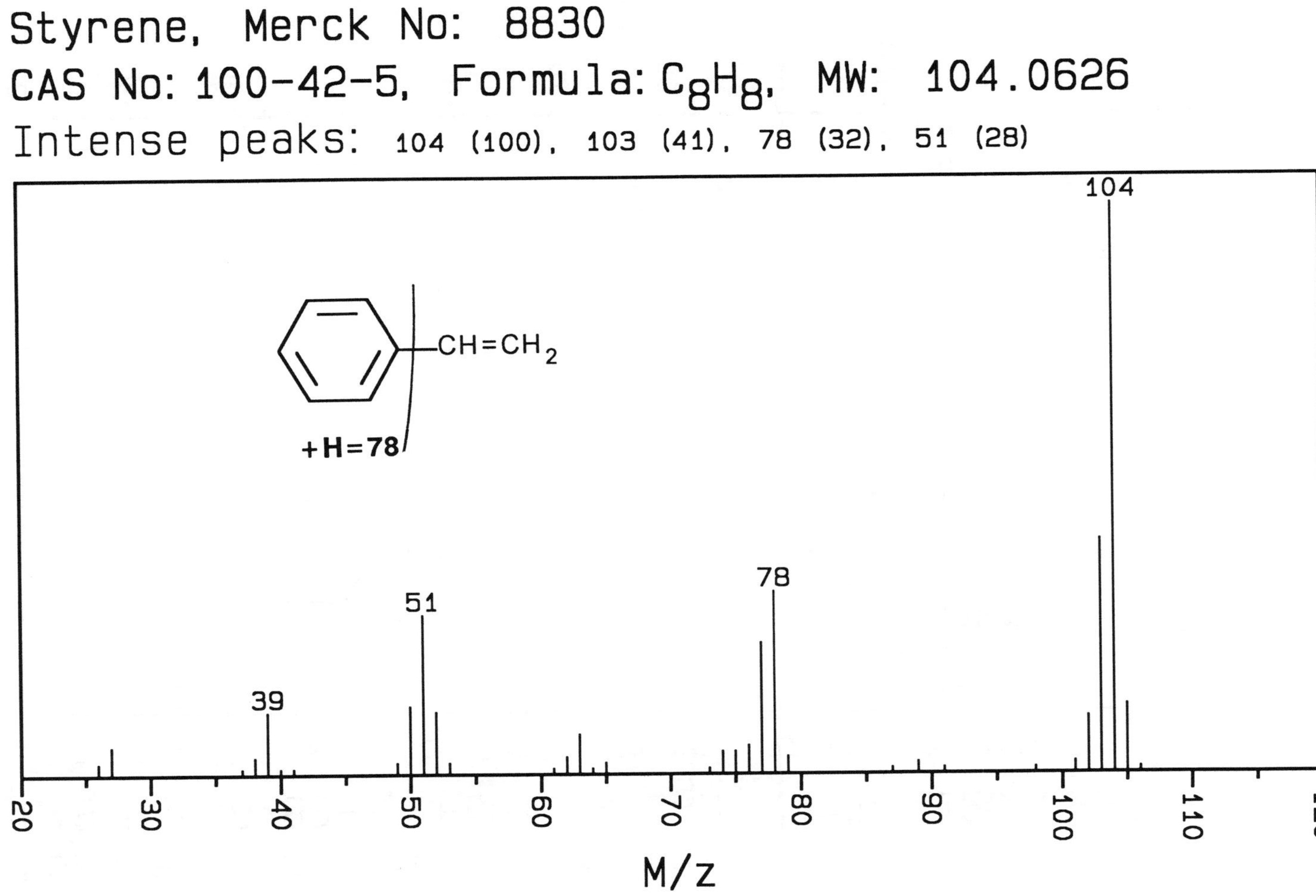
Styrene, Merck No: 8830
CAS No: 100-42-5, Formula: $C_8H_8$, MW: 104.0626
Intense peaks: 104 (100), 103 (41), 78 (32), 51 (28)
104
CH=CH$_2$
+H=78
78
51
39
20
30
40
50
60
70
80
90
100
110
120
M/Z

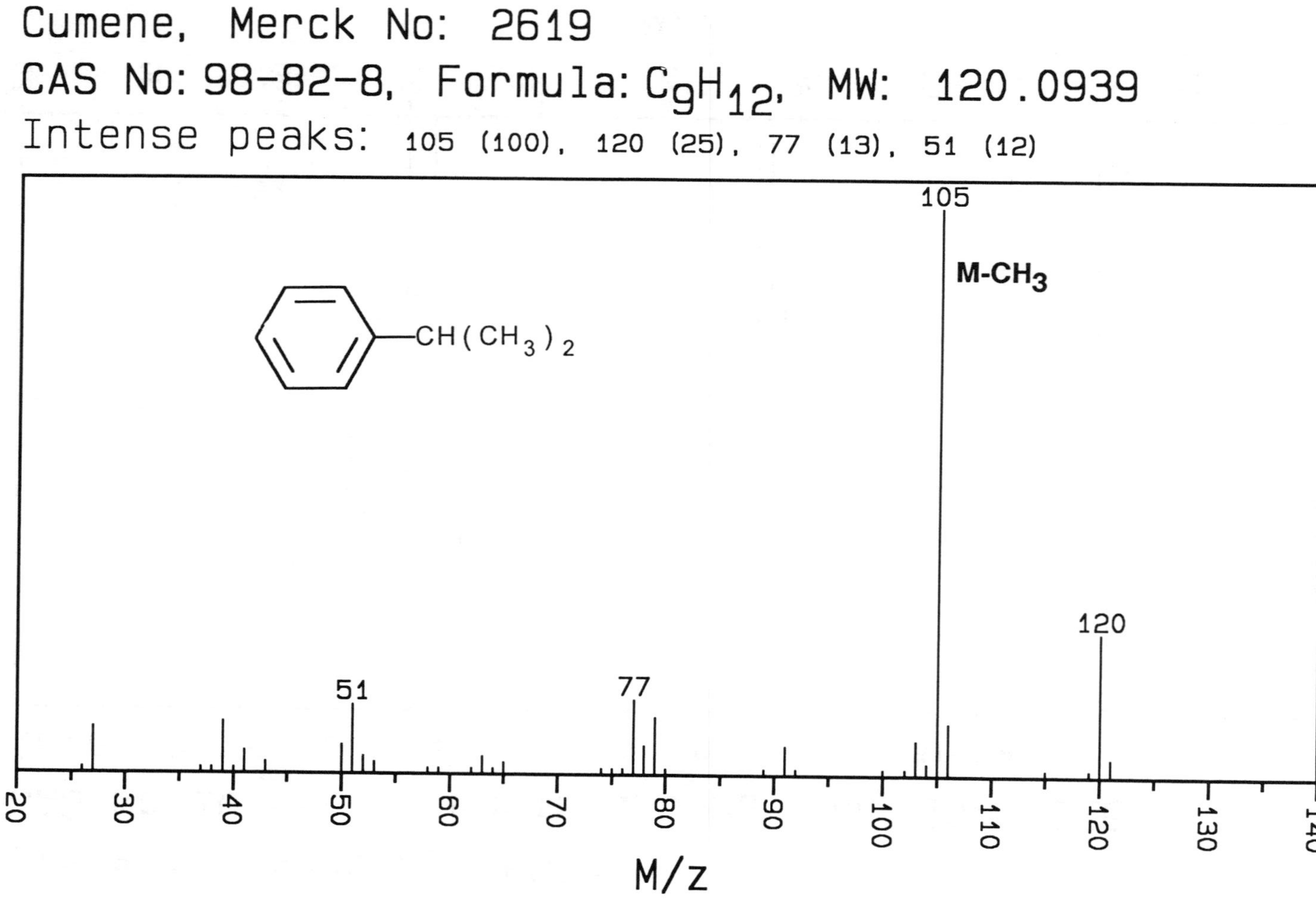
Cumene, Merck No: 2619
CAS No: 98-82-8, Formula: $C_9H_{12}$, MW: 120.0939
Intense peaks: 105 (100), 120 (25), 77 (13), 51 (12)
CH(CH$_3$)$_2$
105
M-CH$_3$
120
51
77
20
30
40
50
60
70
80
90
100
110
120
130
140
M/Z

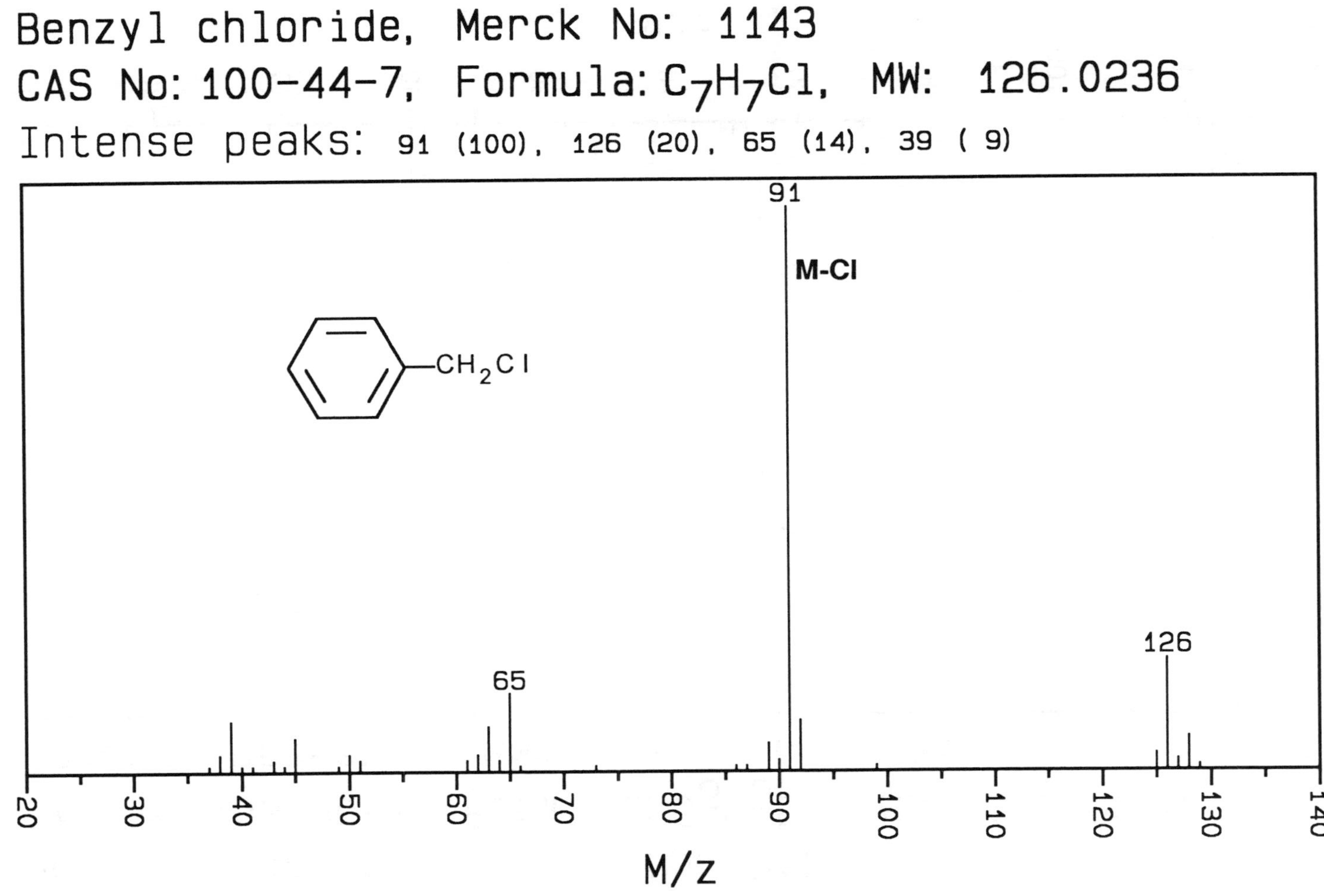
Benzyl chloride, Merck No: 1143
CAS No: 100-44-7, Formula: C7H7Cl, MW: 126.0236
Intense peaks: 91 (100), 126 (20), 65 (14), 39 ( 9)
91
M-Cl
CH2Cl
65
126
20
30
40
50
60
70
80
90
100
110
120
130
140
M/z

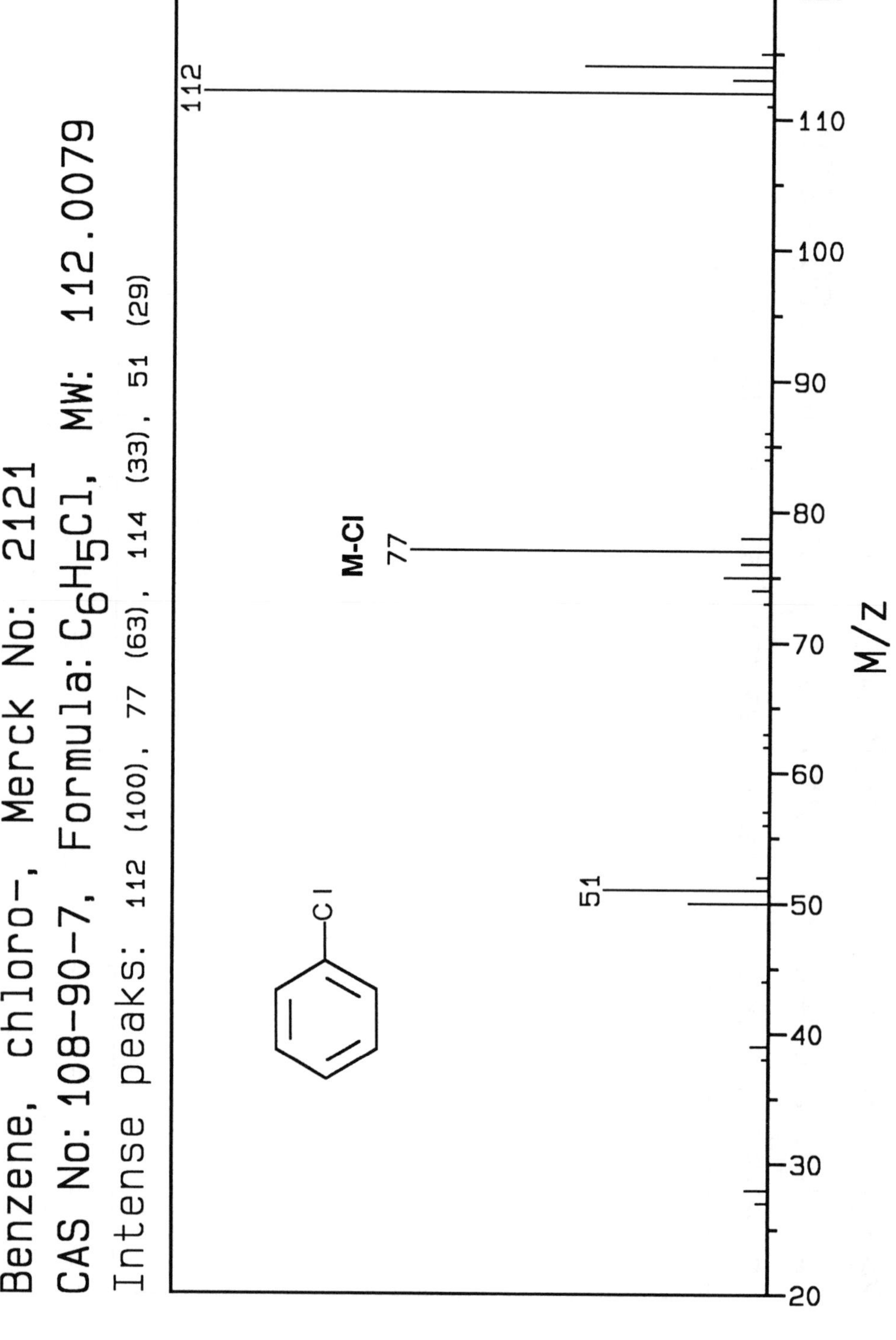
Benzene, chloro-, Merck No: 2121
CAS No: 108-90-7, Formula: C6H5Cl, MW: 112.0079
Intense peaks: 112 (100), 77 (63), 114 (33), 51 (29)
Cl
112
M-Cl
77
51
20
30
40
50
60
70
80
90
100
110
120
M/Z

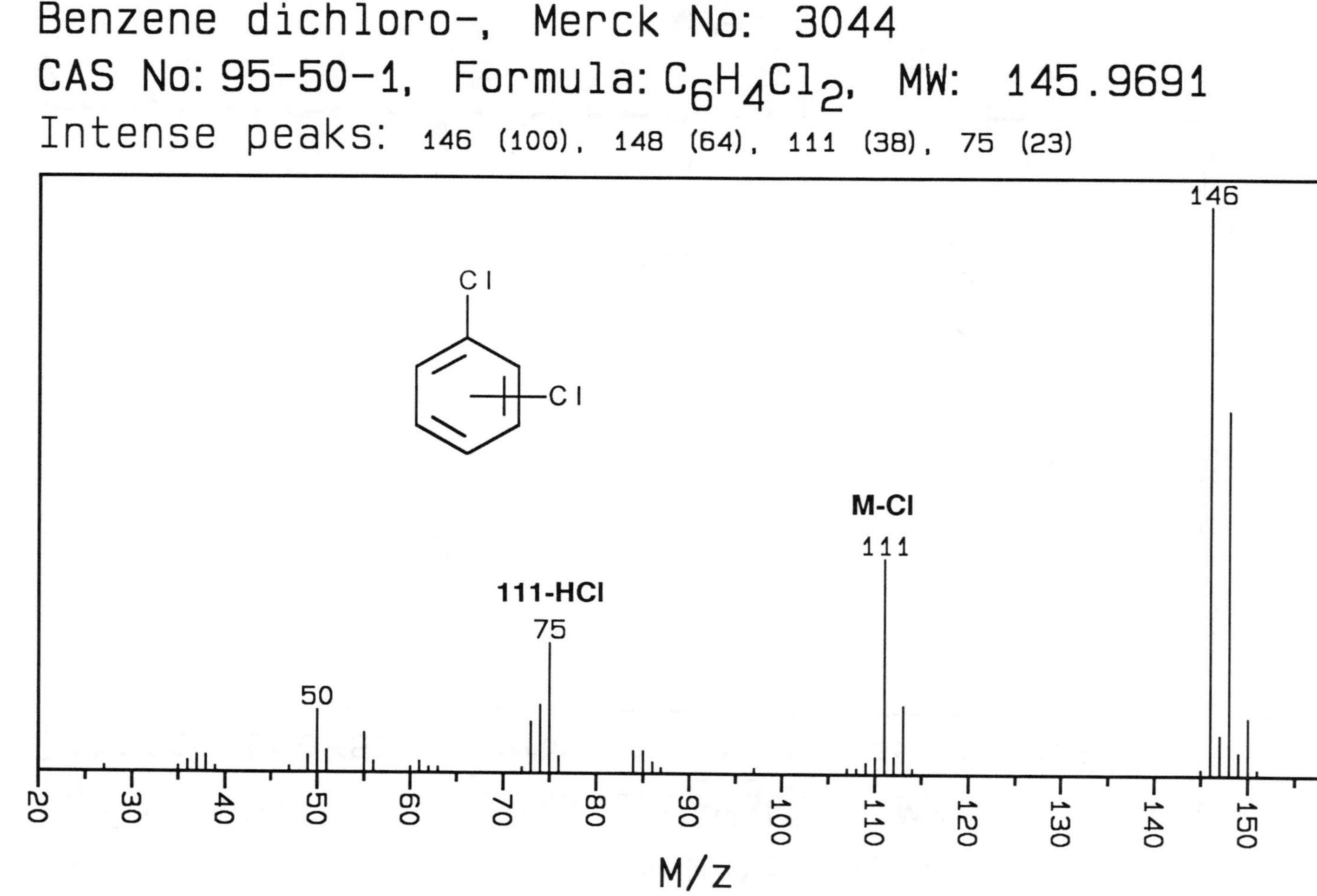
Benzene dichloro-, Merck No: 3044
CAS No: 95-50-1, Formula: $C_6H_4Cl_2$, MW: 145.9691
Intense peaks: 146 (100), 148 (64), 111 (38), 75 (23)
Cl
Cl
146
M-Cl
111
111-HCl
75
50
20
30
40
50
60
70
80
90
100
110
120
130
140
150
160
M/Z

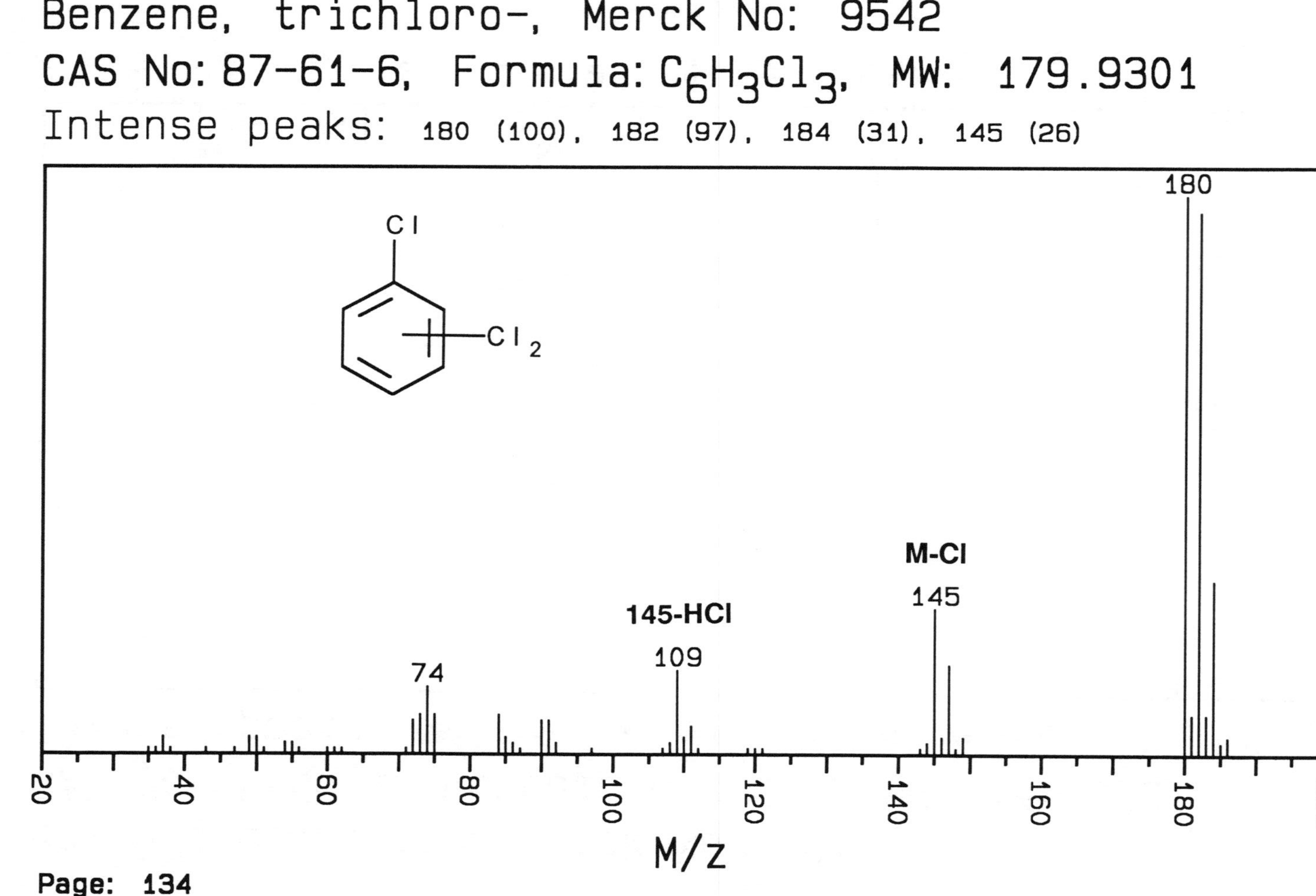
Benzene, trichloro-, Merck No: 9542
CAS No: 87-61-6, Formula: $C_6H_3Cl_3$, MW: 179.9301
Intense peaks: 180 (100), 182 (97), 184 (31), 145 (26)
Cl
Cl 2
180
M-Cl
145
145-HCl
109
74
20
40
60
80
100
120
140
160
180
200
M/Z
Page: 134

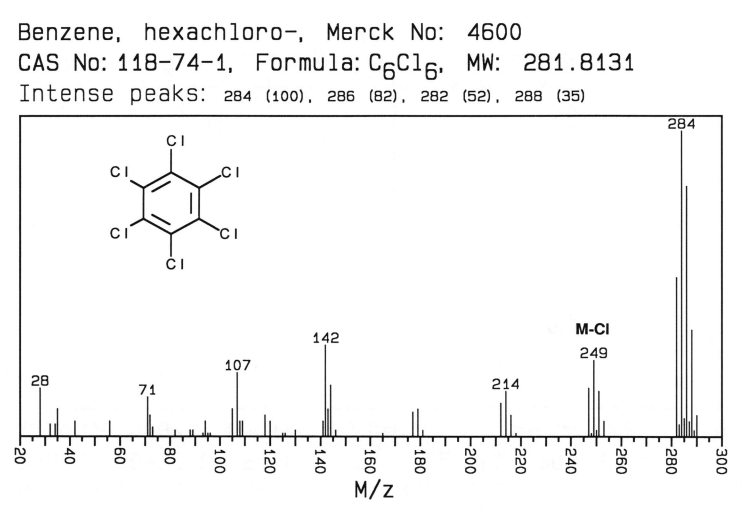

Benzene, hexachloro-, Merck No: 4600

CAS No: 118-74-1, Formula: $C_6Cl_6$, MW: 281.8131

Intense peaks: 284 (100), 286 (82), 282 (52), 288 (35)

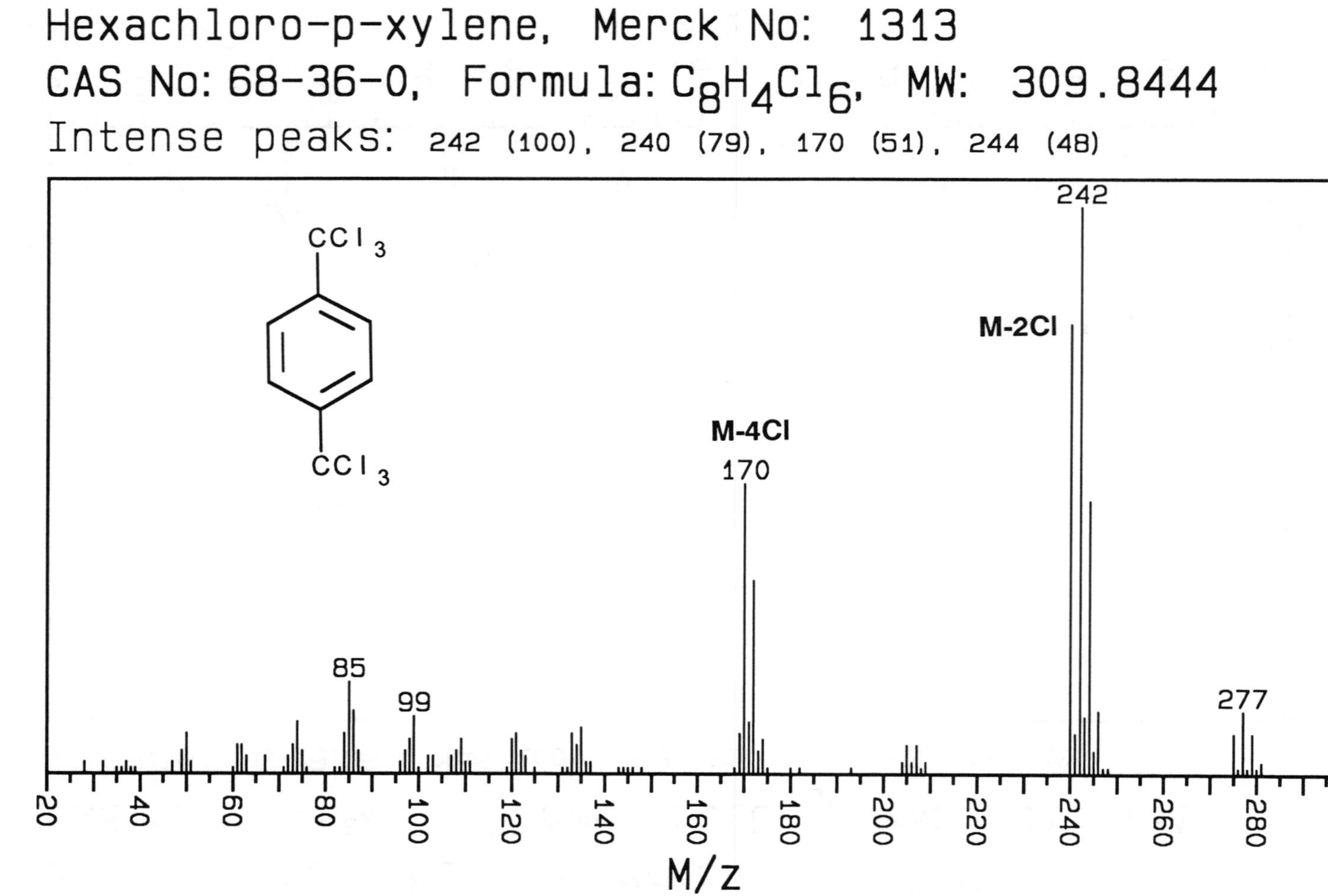
Hexachloro-p-xylene, Merck No: 1313
CAS No: 68-36-0, Formula: $C_8H_4Cl_6$, MW: 309.8444
Intense peaks: 242 (100), 240 (79), 170 (51), 244 (48)
$CCl_3$
$CCl_3$
242
M-2Cl
M-4Cl
170
85
99
277
20
40
60
80
100
120
140
160
180
200
220
240
260
280
300
M/Z

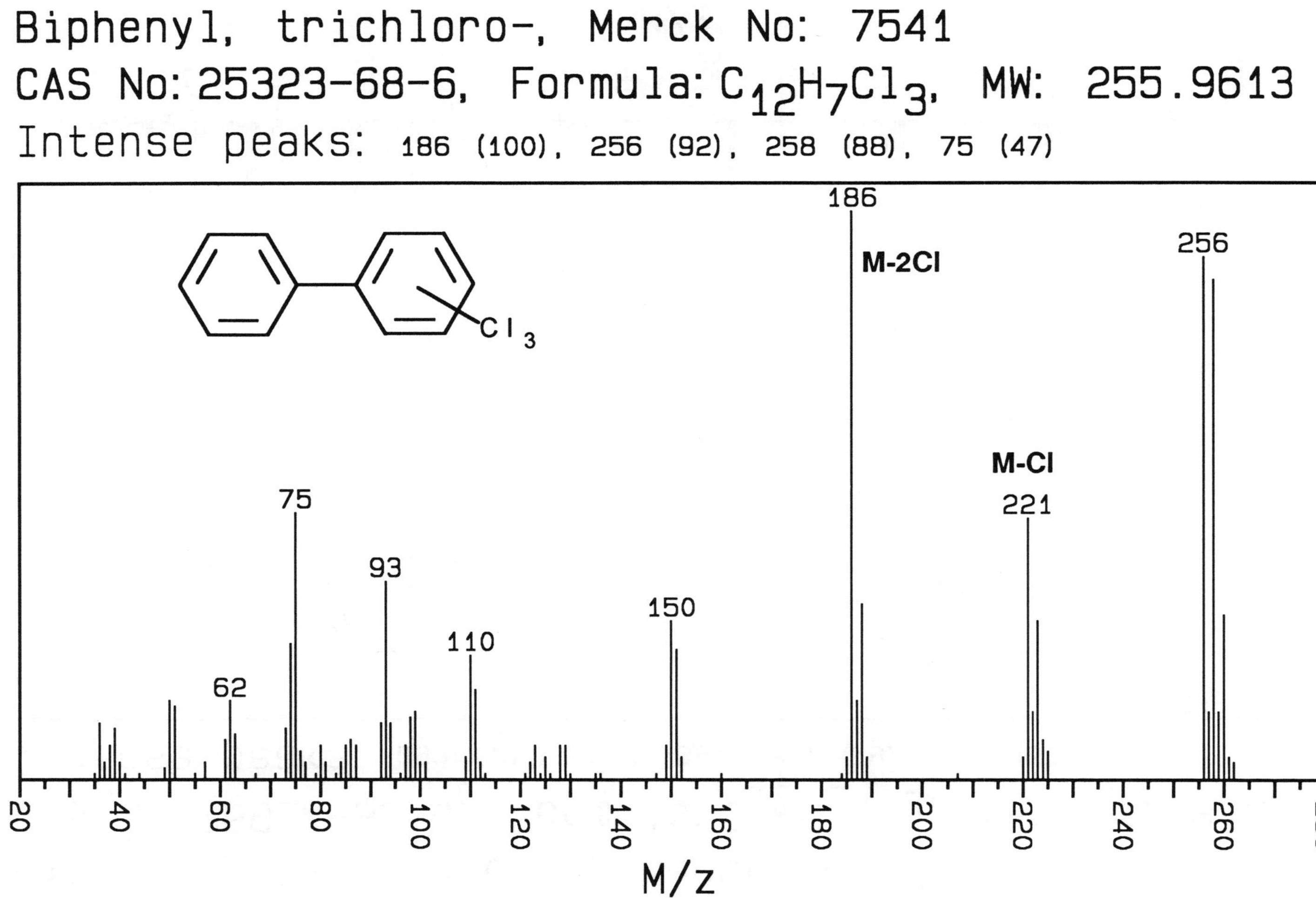

Biphenyl, trichloro-, Merck No: 7541
CAS No: 25323-68-6, Formula: C12H7Cl3, MW: 255.9613
Intense peaks: 186 (100), 256 (92), 258 (88), 75 (47)
Cl3
186
M-2Cl
256
M-Cl
221
75
93
110
150
62
20
40
60
80
100
120
140
160
180
200
220
240
260
280
M/Z

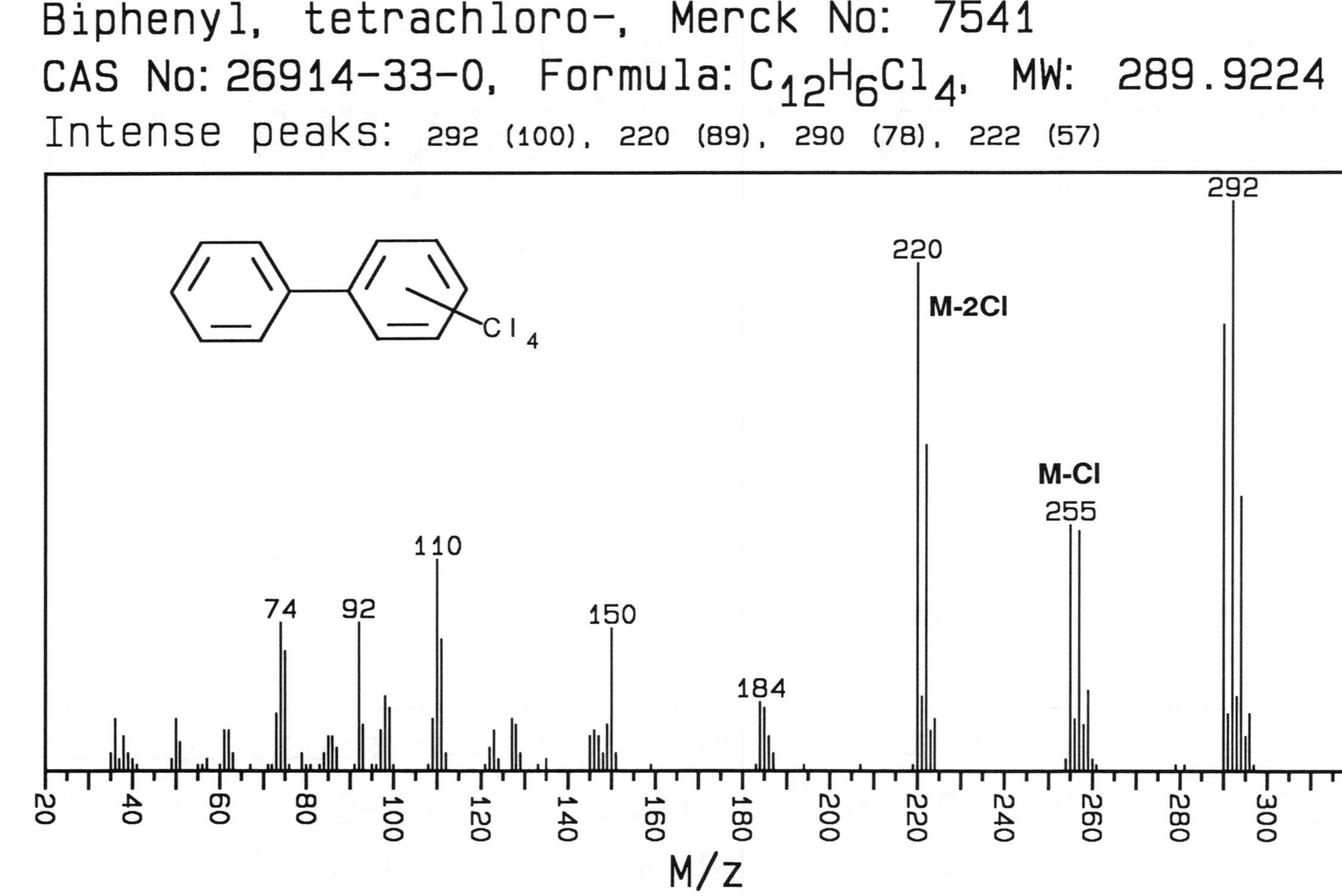
Biphenyl, tetrachloro-, Merck No: 7541
CAS No: 26914-33-0, Formula: $C_{12}H_6Cl_4$, MW: 289.9224
Intense peaks: 292 (100), 220 (89), 290 (78), 222 (57)
$Cl_4$
292
220
M-2Cl
M-Cl
255
110
74
92
150
184
20
40
60
80
100
120
140
160
180
200
220
240
260
280
300
320
M/Z

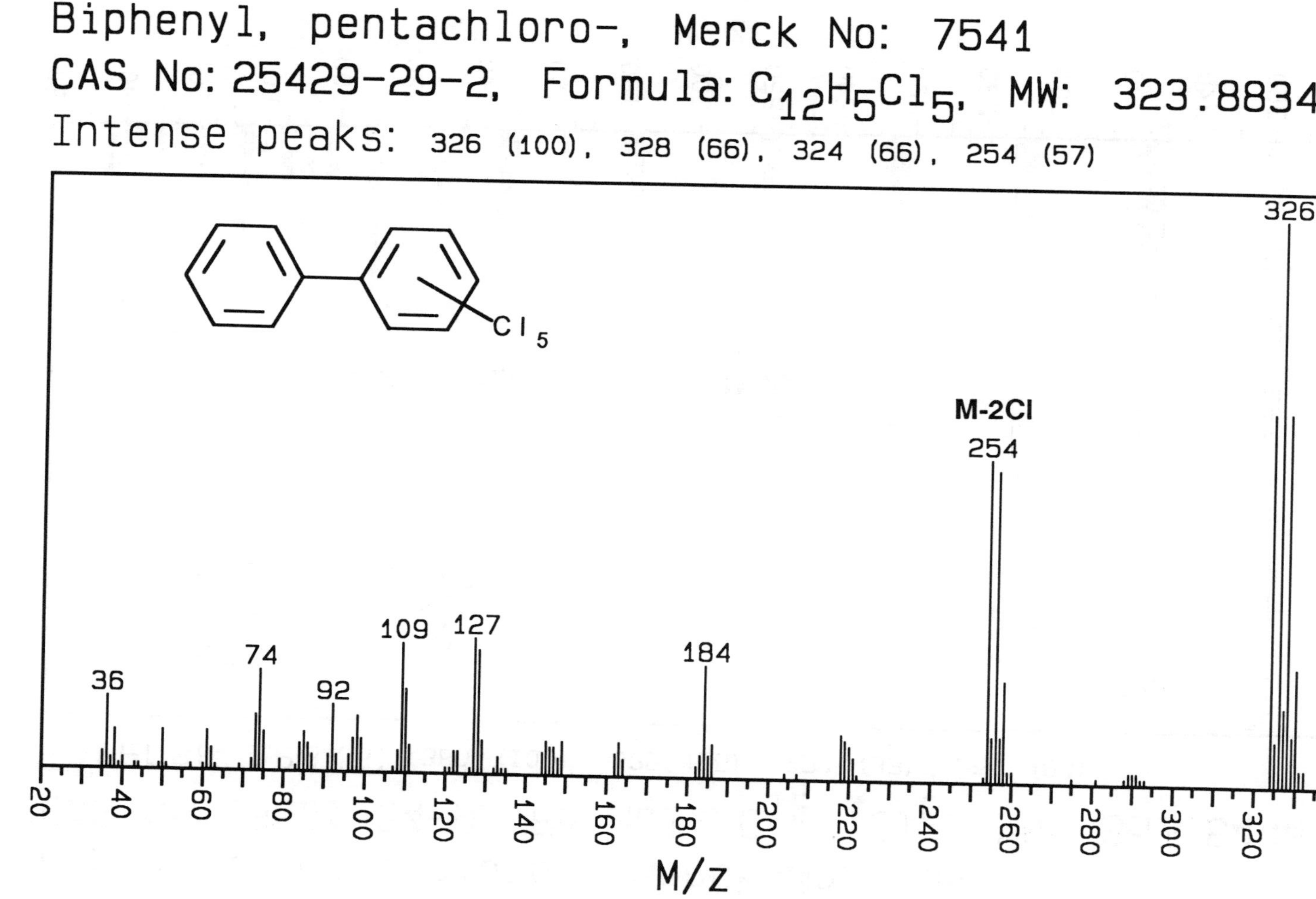

Biphenyl, pentachloro-, Merck No: 7541
CAS No: 25429-29-2, Formula: $C_{12}H_5Cl_5$, MW: 323.8834
Intense peaks: 326 (100), 328 (66), 324 (66), 254 (57)
$Cl_5$
326
M-2Cl
254
36
74
92
109
127
184
20
40
60
80
100
120
140
160
180
200
220
240
260
280
300
320
340
M/Z

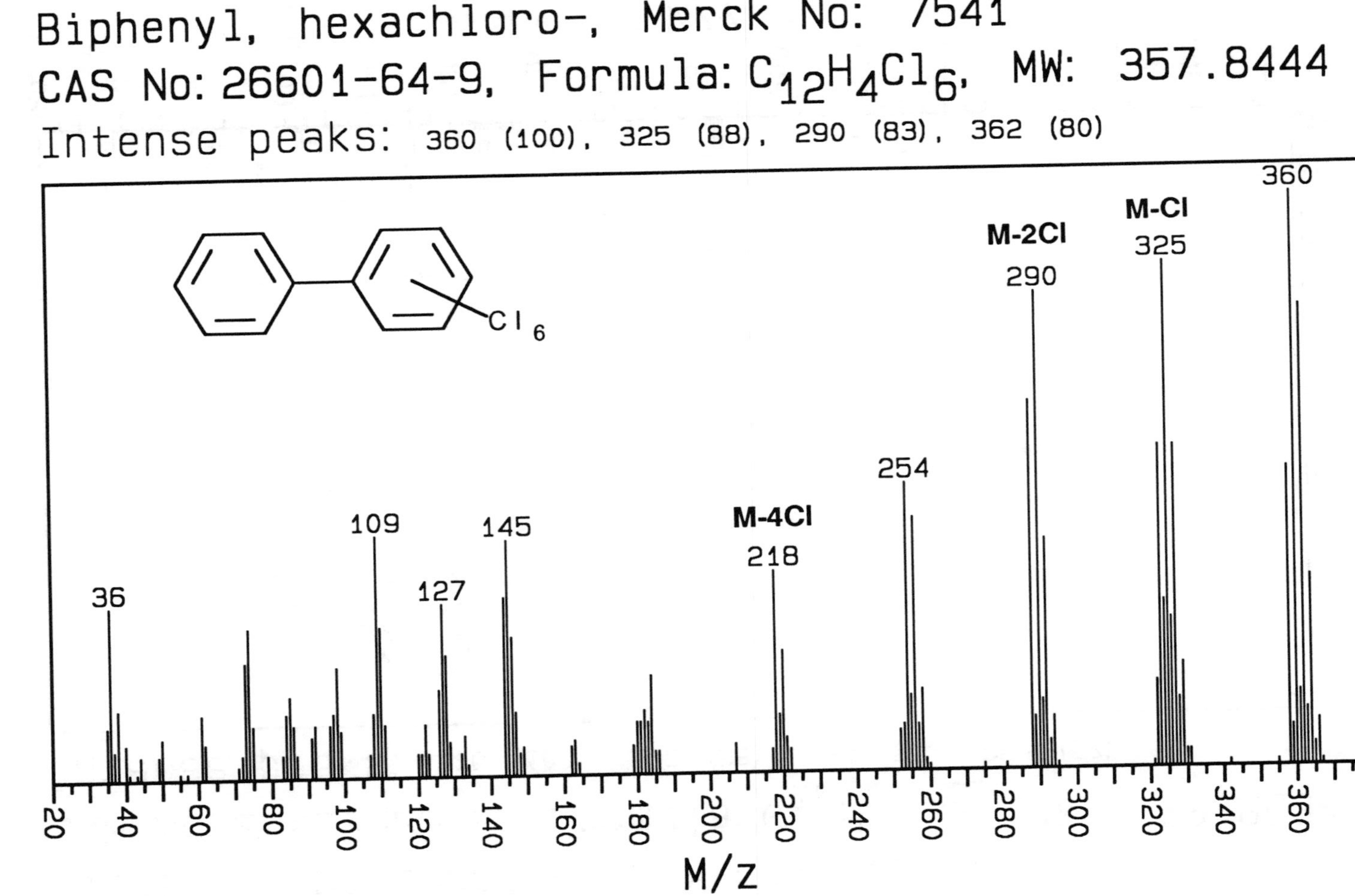
Biphenyl, hexachloro-, Merck No: 7541
CAS No: 26601-64-9, Formula: $C_{12}H_4Cl_6$, MW: 357.8444
Intense peaks: 360 (100), 325 (88), 290 (83), 362 (80)
Cl 6
360
M-Cl
325
M-2Cl
290
254
M-4Cl
218
109
145
127
36
20
40
60
80
100
120
140
160
180
200
220
240
260
280
300
320
340
360
380
M/Z

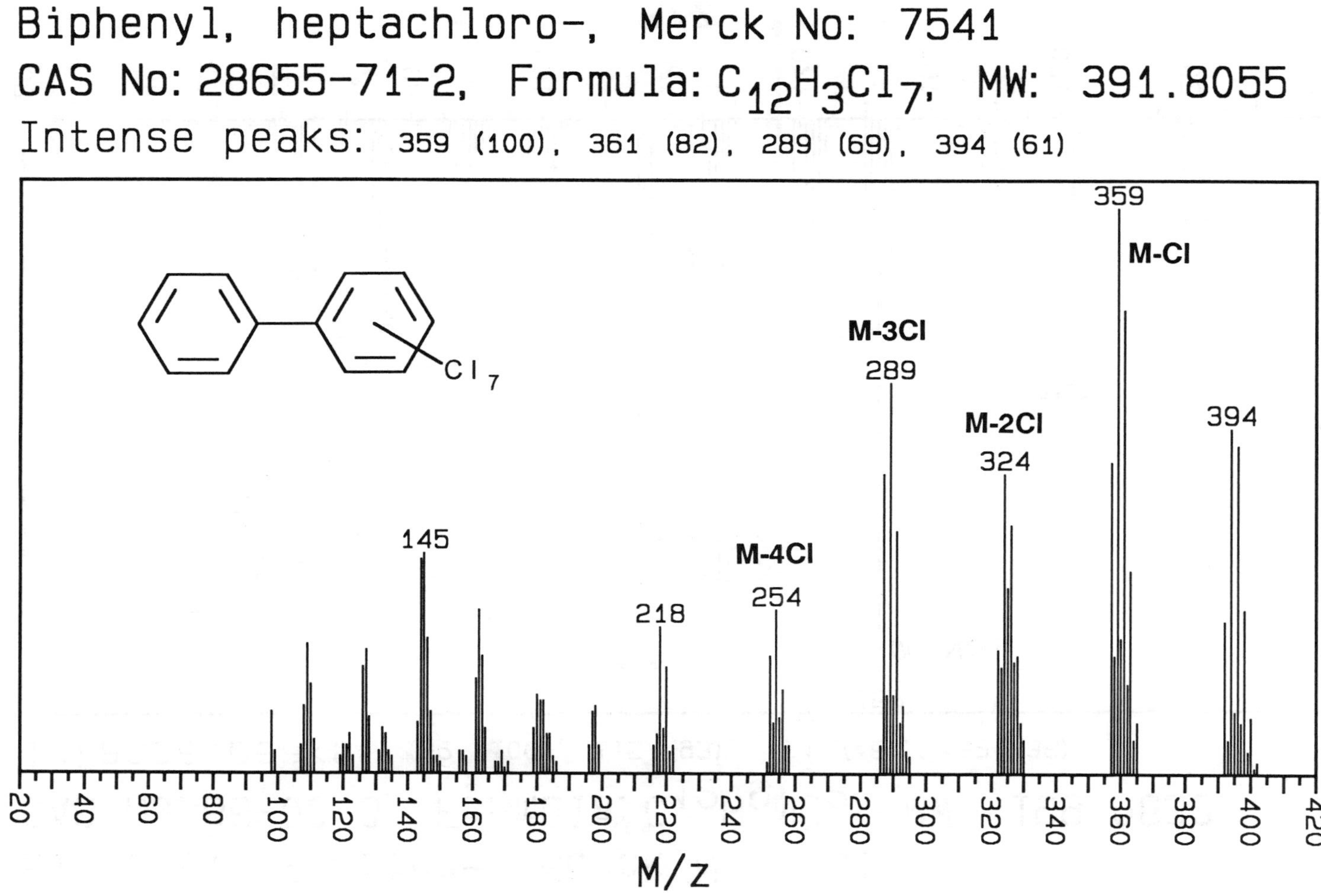
Biphenyl, heptachloro-, Merck No: 7541
CAS No: 28655-71-2, Formula: $C_{12}H_3Cl_7$, MW: 391.8055
Intense peaks: 359 (100), 361 (82), 289 (69), 394 (61)
$Cl_7$
359
M-Cl
M-3Cl
289
M-2Cl
324
394
145
M-4Cl
254
218
20
40
60
80
100
120
140
160
180
200
220
240
260
280
300
320
340
360
380
400
420
M/Z

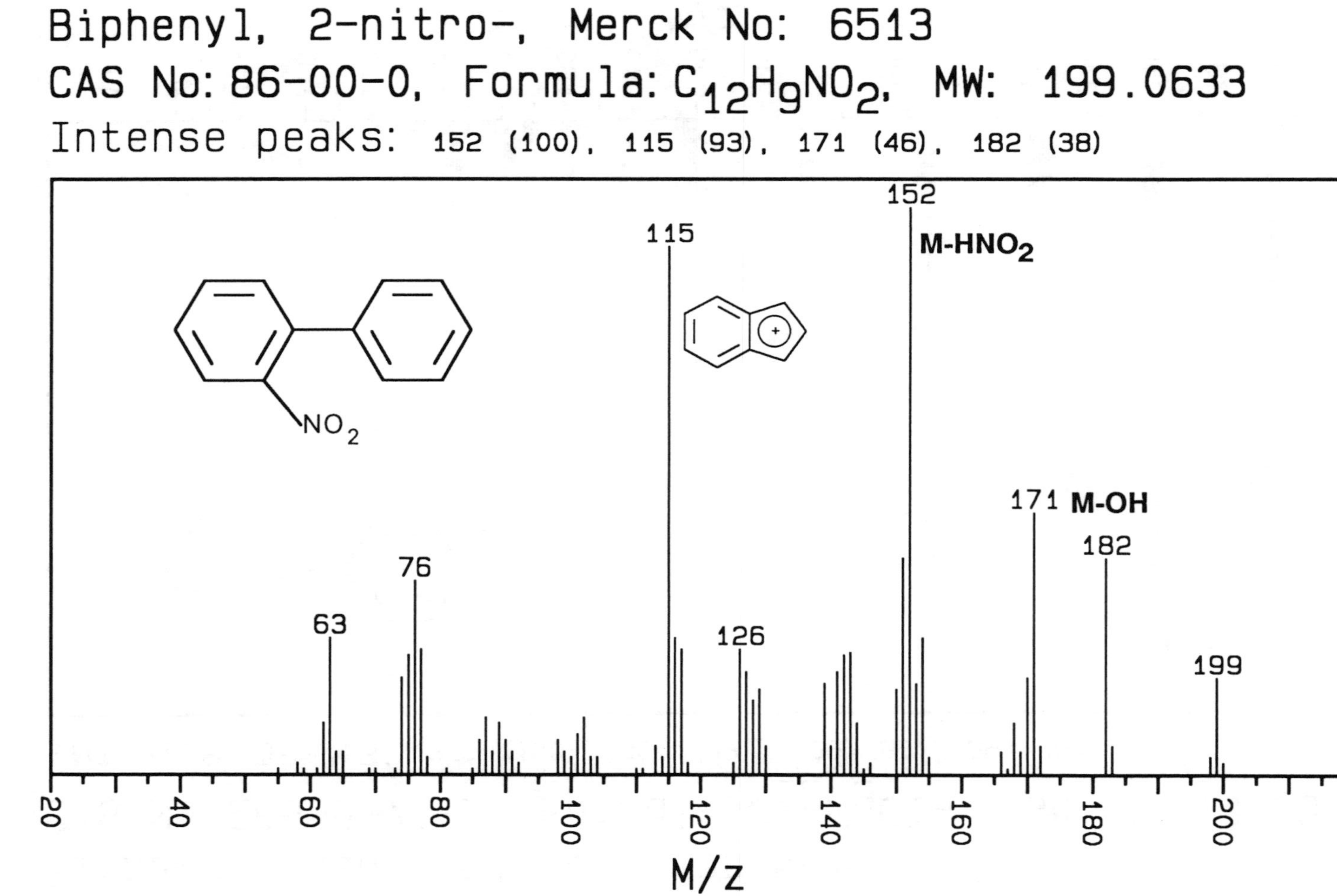
Biphenyl, 2-nitro-, Merck No: 6513
CAS No: 86-00-0, Formula: $C_{12}H_9NO_2$, MW: 199.0633
Intense peaks: 152 (100), 115 (93), 171 (46), 182 (38)
$NO_2$
115
152
M-$HNO_2$
171 M-OH
182
199
126
76
63
M/Z

Phenol, 2-phenyl-6-chloro-, Merck No: 7251

CAS No: 85-97-2, Formula: $C_{12}H_9ClO$, MW: 204.0342

Intense peaks: 204 (100), 139 (55), 141 (42), 69 (38)

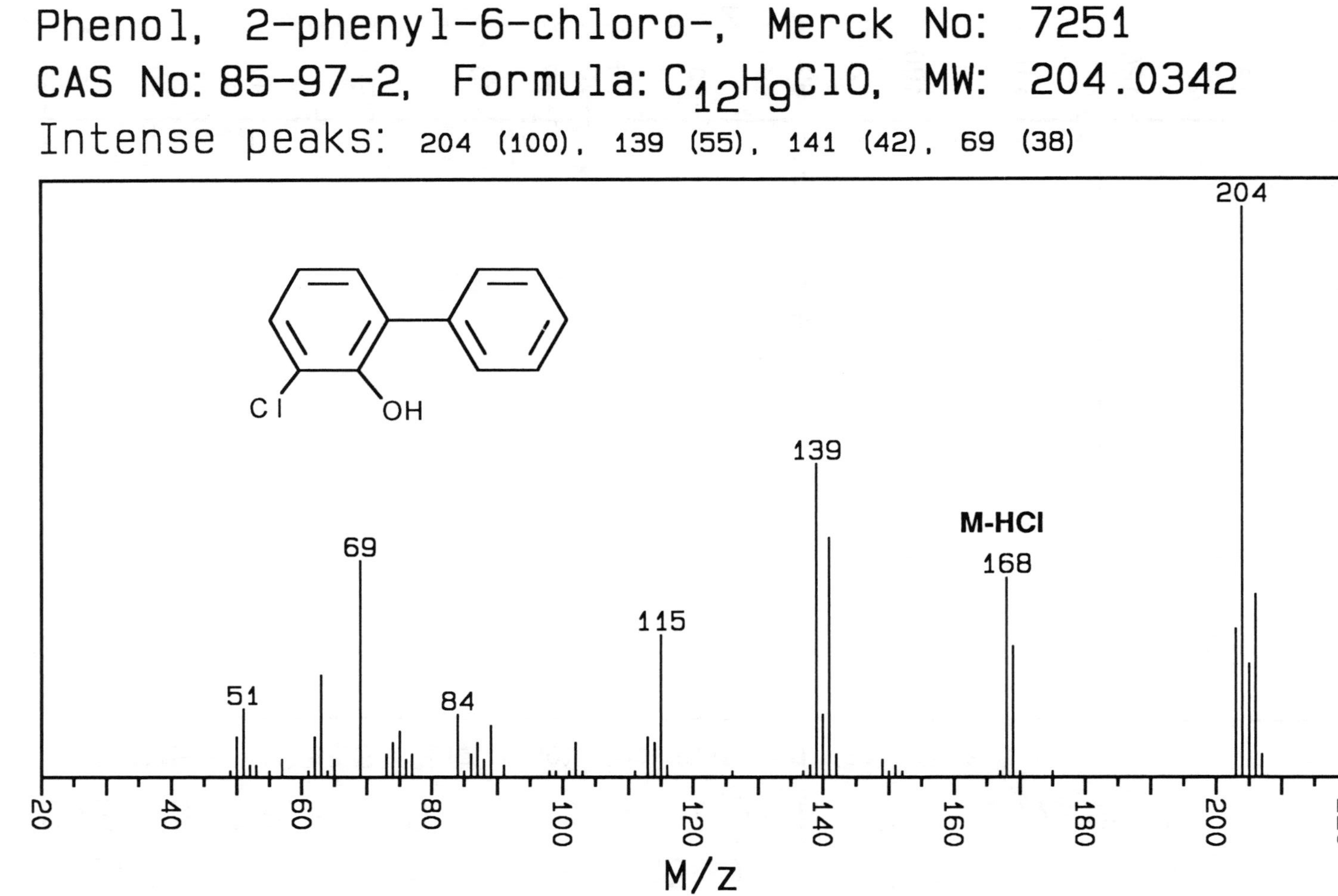

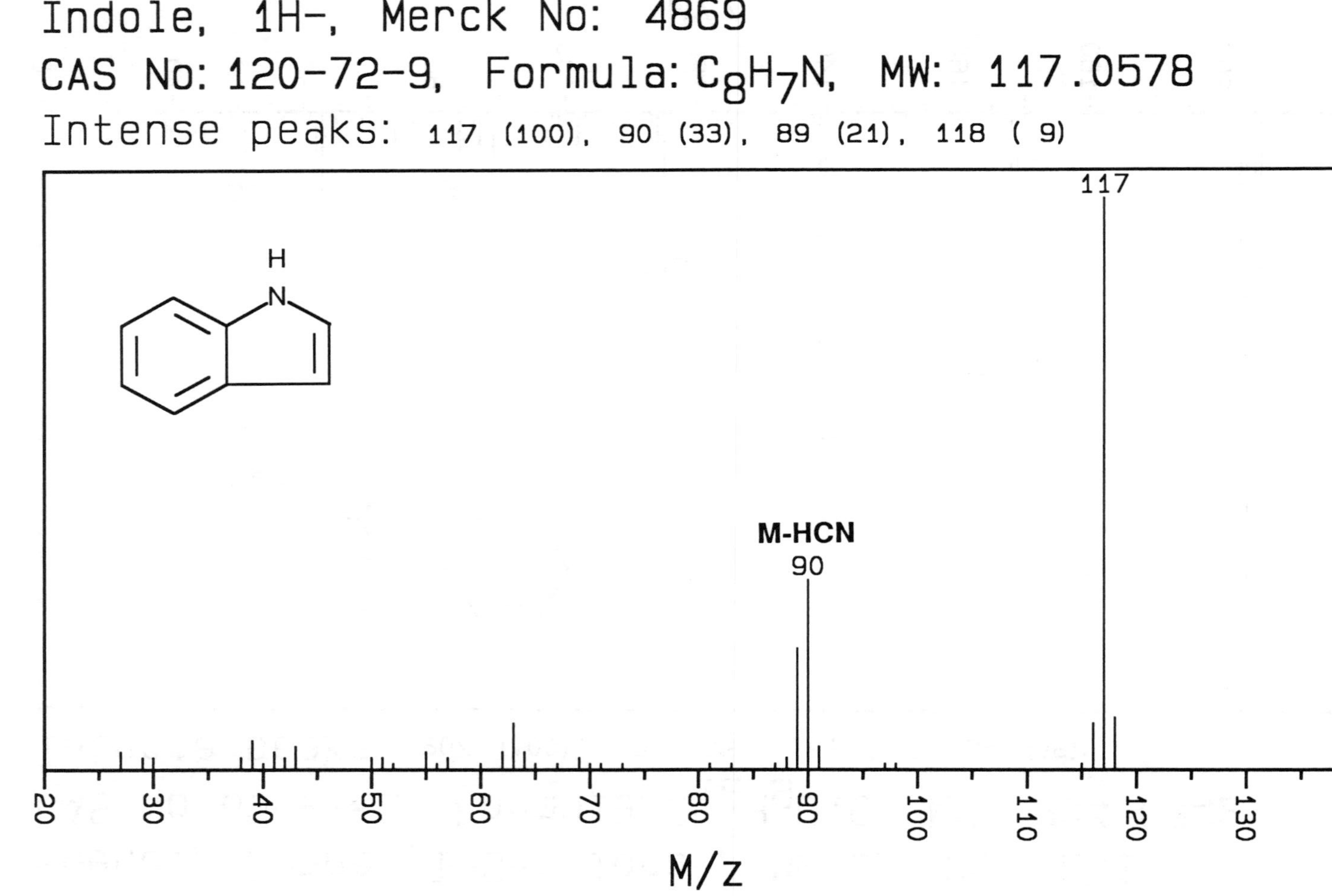
Indole, 1H-, Merck No: 4869
CAS No: 120-72-9, Formula: $C_8H_7N$, MW: 117.0578
Intense peaks: 117 (100), 90 (33), 89 (21), 118 ( 9)
117
M-HCN
90
H
N
20
30
40
50
60
70
80
90
100
110
120
130
140
M/Z

Azulene, Merck No: 939

CAS No: 275-51-4, Formula: $C_{10}H_8$, MW: 128.0626

Intense peaks: 128 (100), 51 (16), 102 (14), 127 (12)

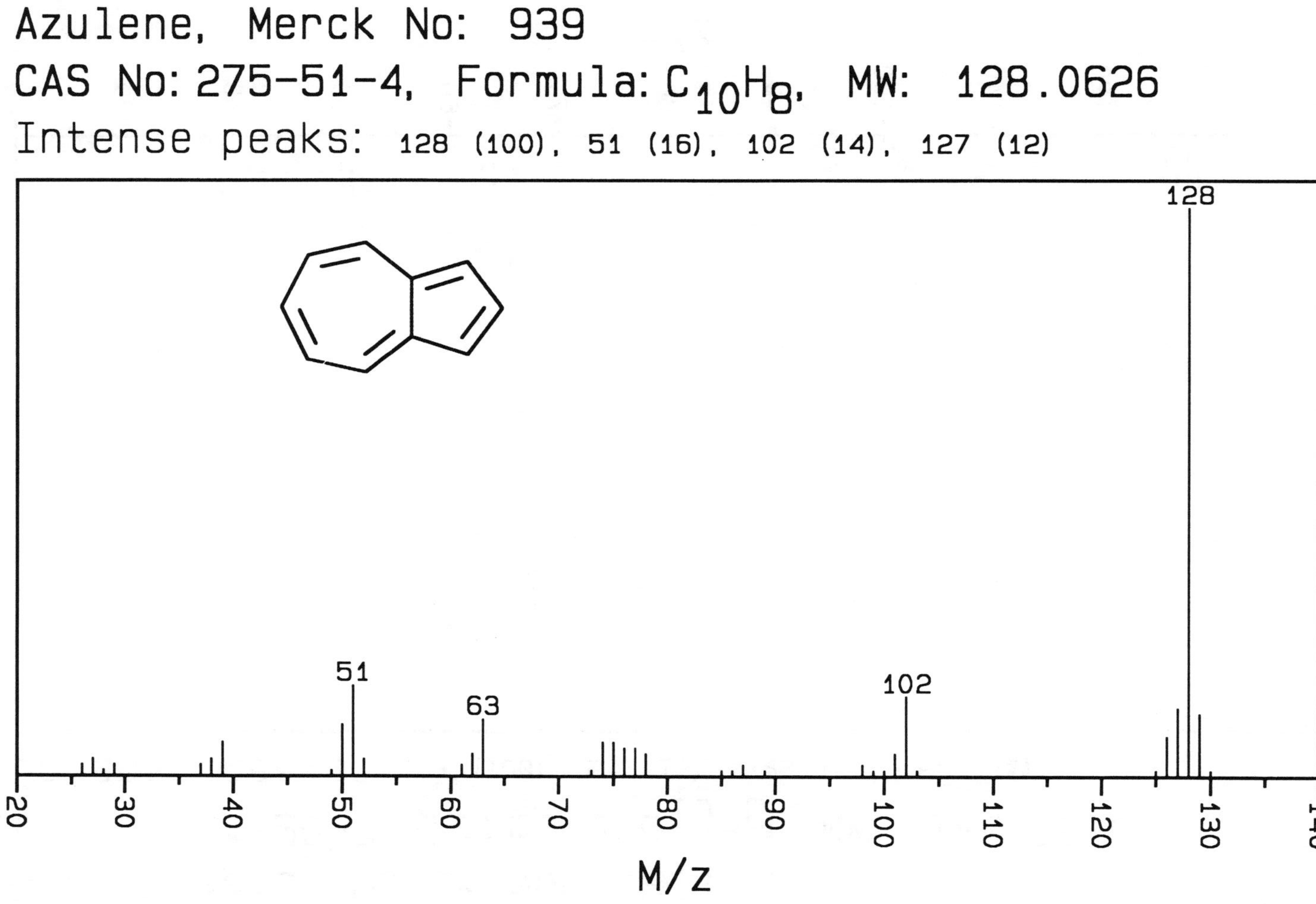

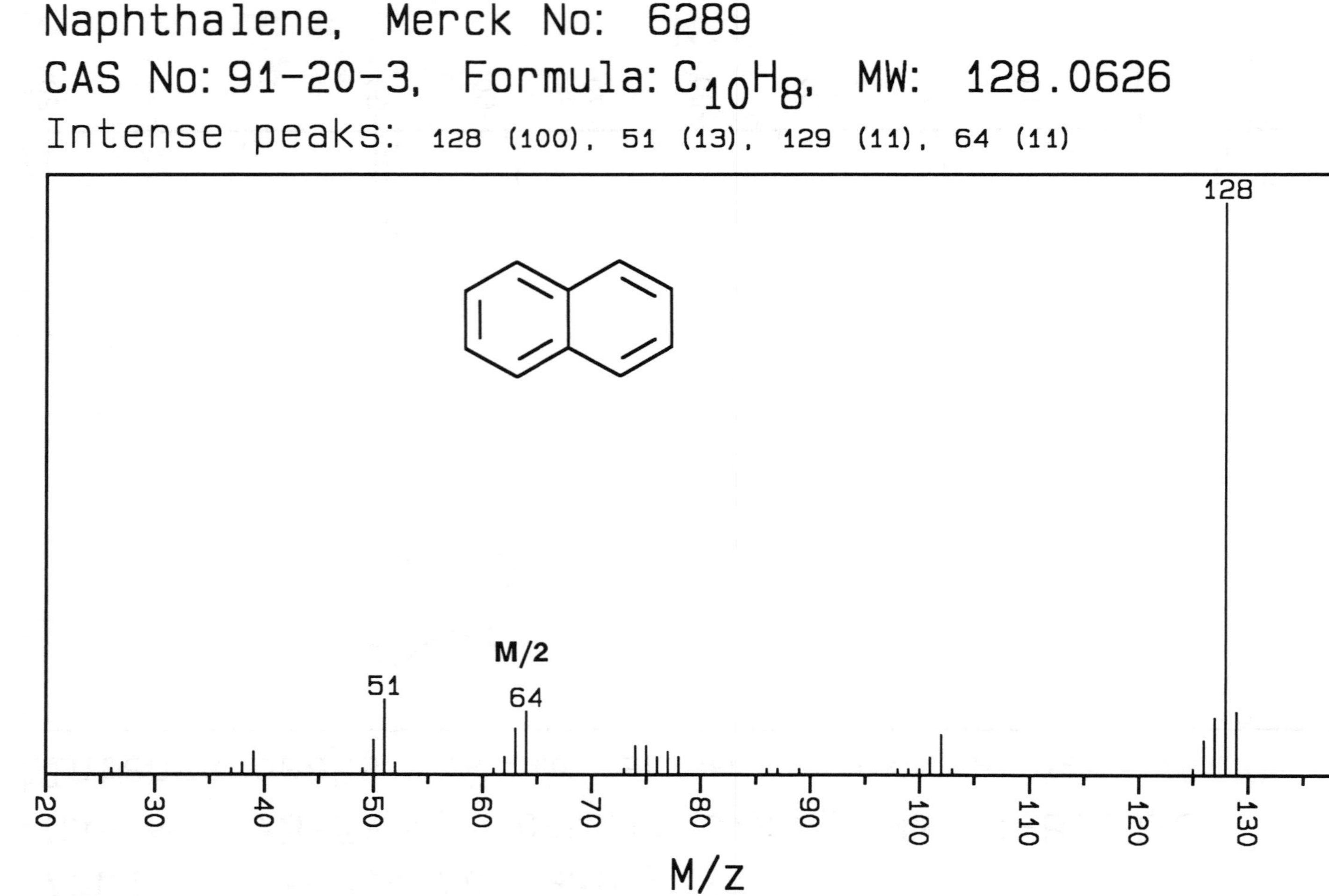

Naphthalene, Merck No: 6289
CAS No: 91-20-3, Formula: C10H8, MW: 128.0626
Intense peaks: 128 (100), 51 (13), 129 (11), 64 (11)
128
M/2
51
64
20
30
40
50
60
70
80
90
100
110
120
130
140
M/Z

Naphthalene, chloro-, Merck No: 2149

CAS No: 90-13-1, Formula: $C_{10}H_7Cl$, MW: 162.0236

Intense peaks: 162 (100), 164 (33), 127 (30), 126 (16)

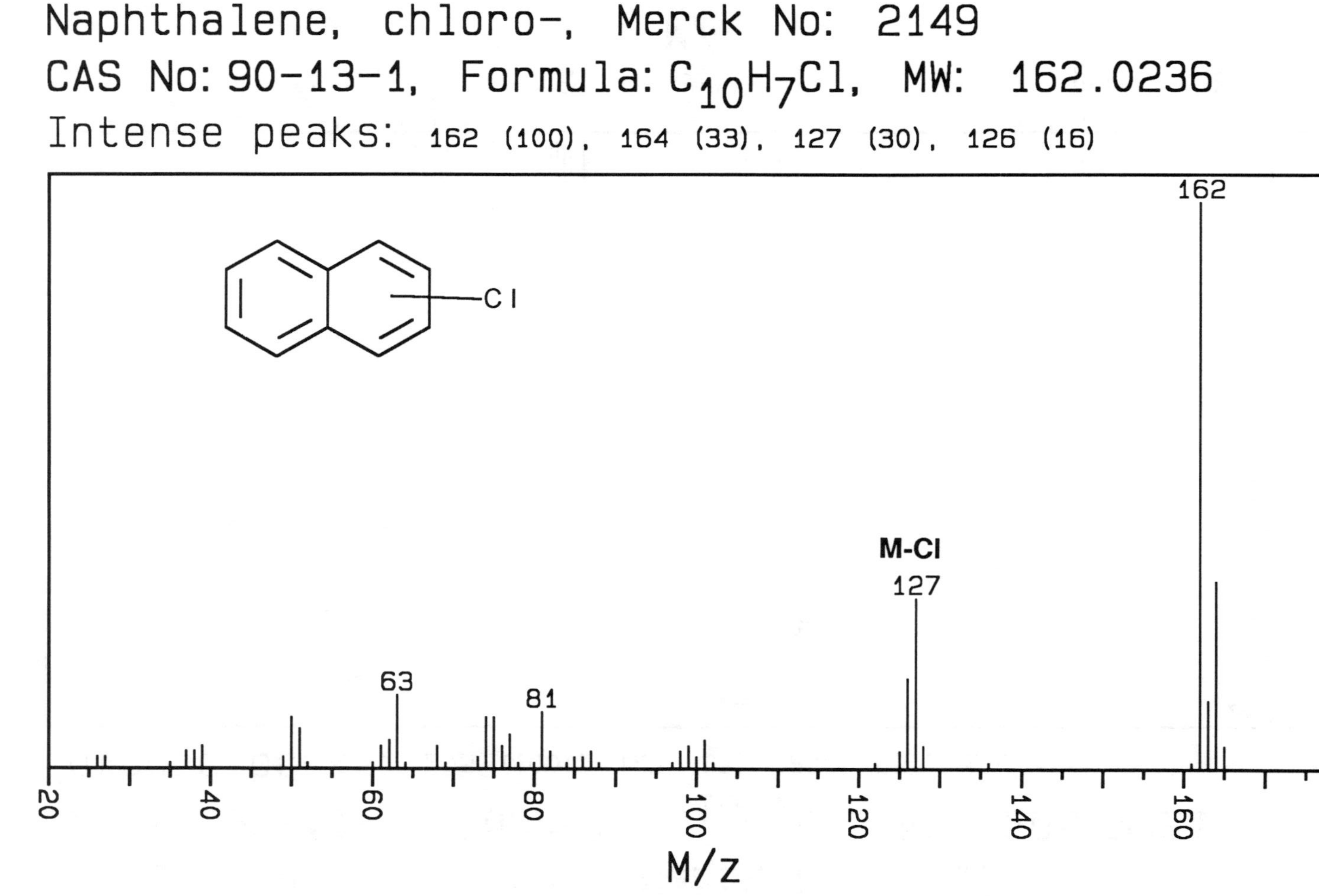

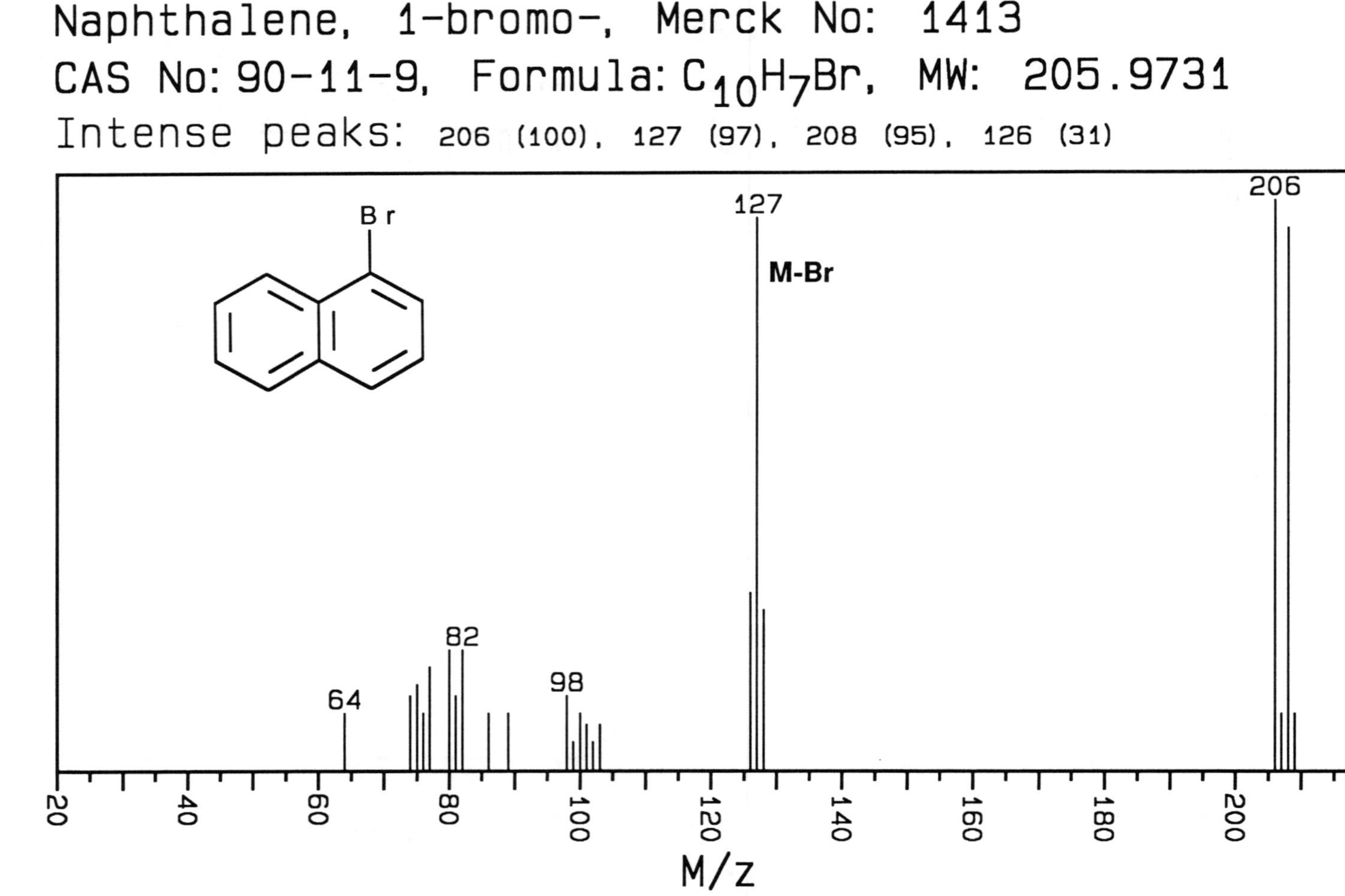
Naphthalene, 1-bromo-, Merck No: 1413
CAS No: 90-11-9, Formula: $C_{10}H_7Br$, MW: 205.9731
Intense peaks: 206 (100), 127 (97), 208 (95), 126 (31)
Br
127
M-Br
206
64
82
98
20
40
60
80
100
120
140
160
180
200
220
M/Z

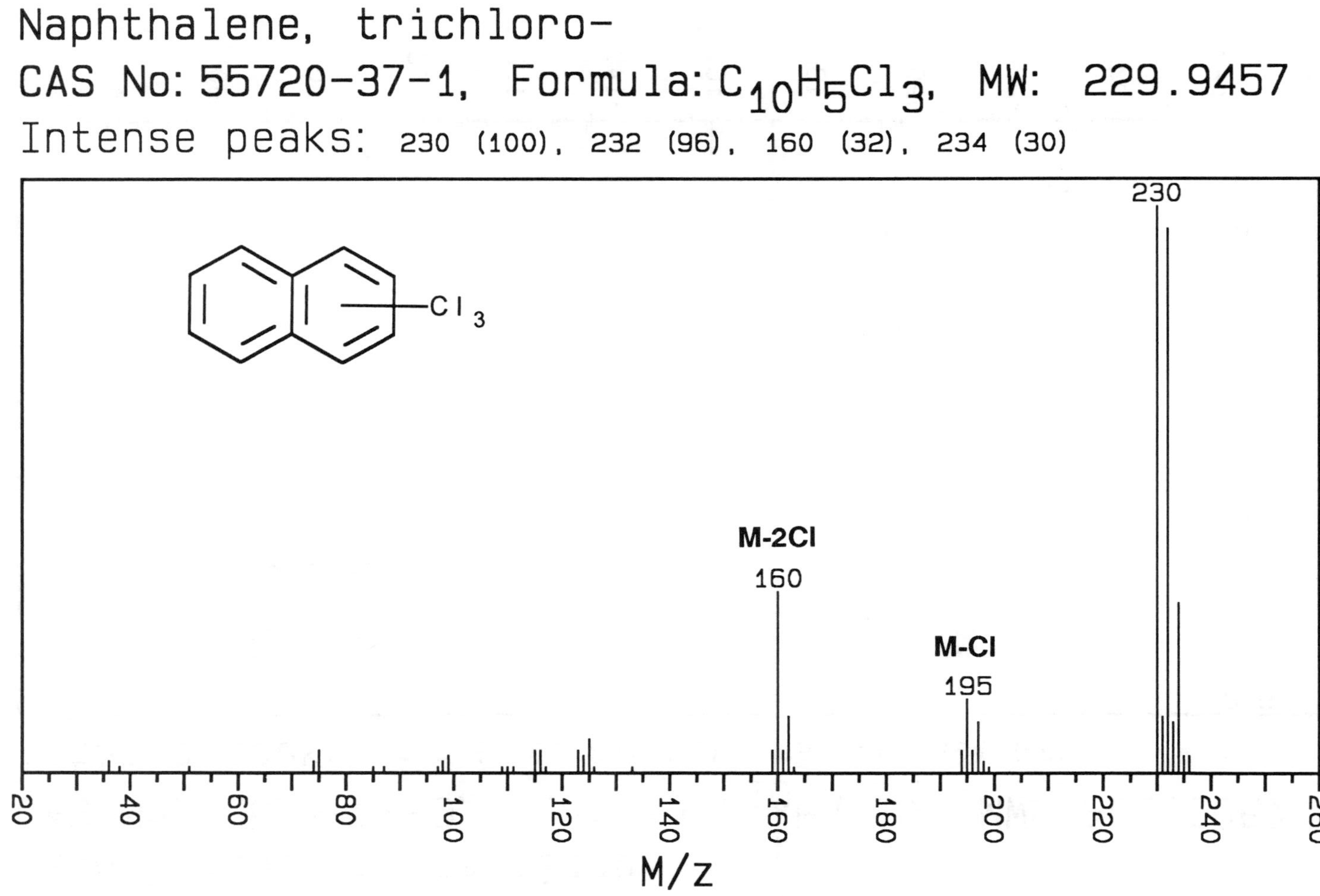
Naphthalene, trichloro-
CAS No: 55720-37-1, Formula: C10H5Cl3, MW: 229.9457
Intense peaks: 230 (100), 232 (96), 160 (32), 234 (30)
Cl3
230
M-2Cl
160
M-Cl
195
20
40
60
80
100
120
140
160
180
200
220
240
260
M/Z

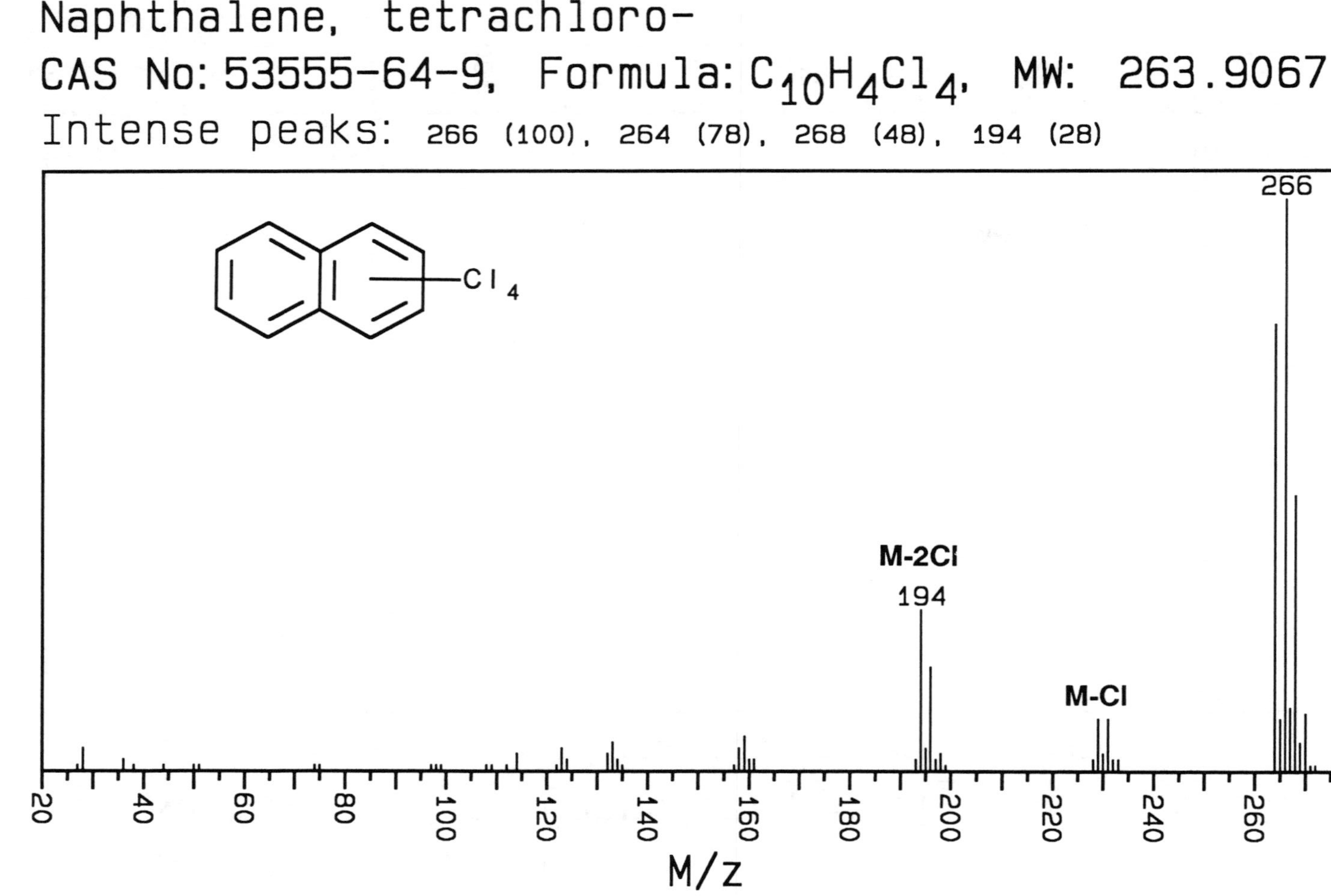

Naphthalene, tetrachloro-
CAS No: 53555-64-9, Formula: $C_{10}H_4Cl_4$, MW: 263.9067
Intense peaks: 266 (100), 264 (78), 268 (48), 194 (28)
$Cl_4$
266
M-2Cl
194
M-Cl
20
40
60
80
100
120
140
160
180
200
220
240
260
280
M/Z

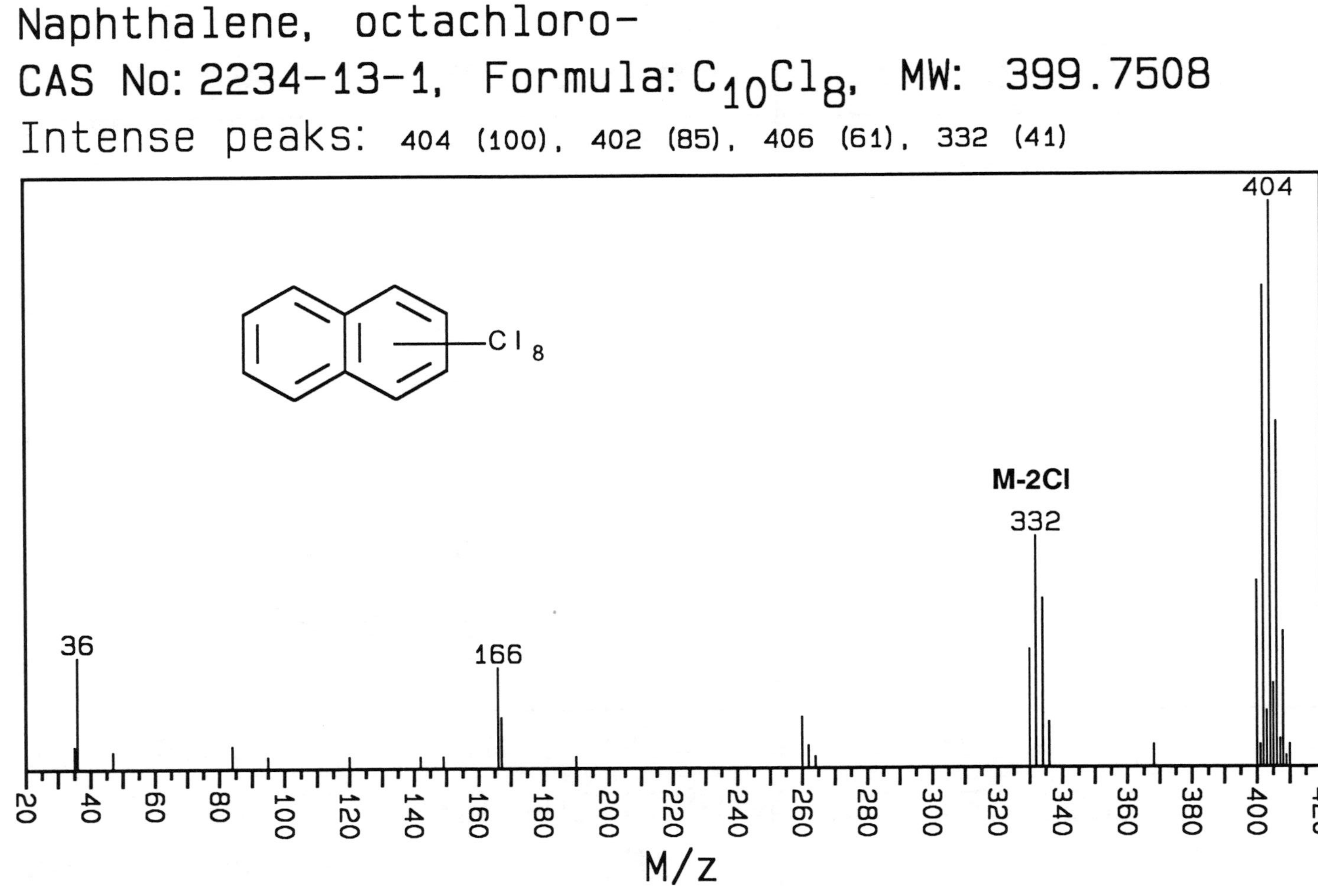
Naphthalene, octachloro-
CAS No: 2234-13-1, Formula: $C_{10}Cl_8$, MW: 399.7508
Intense peaks: 404 (100), 402 (85), 406 (61), 332 (41)
Cl 8
404
M-2Cl
332
36
166
20
40
60
80
100
120
140
160
180
200
220
240
260
280
300
320
340
360
380
400
420
M/Z

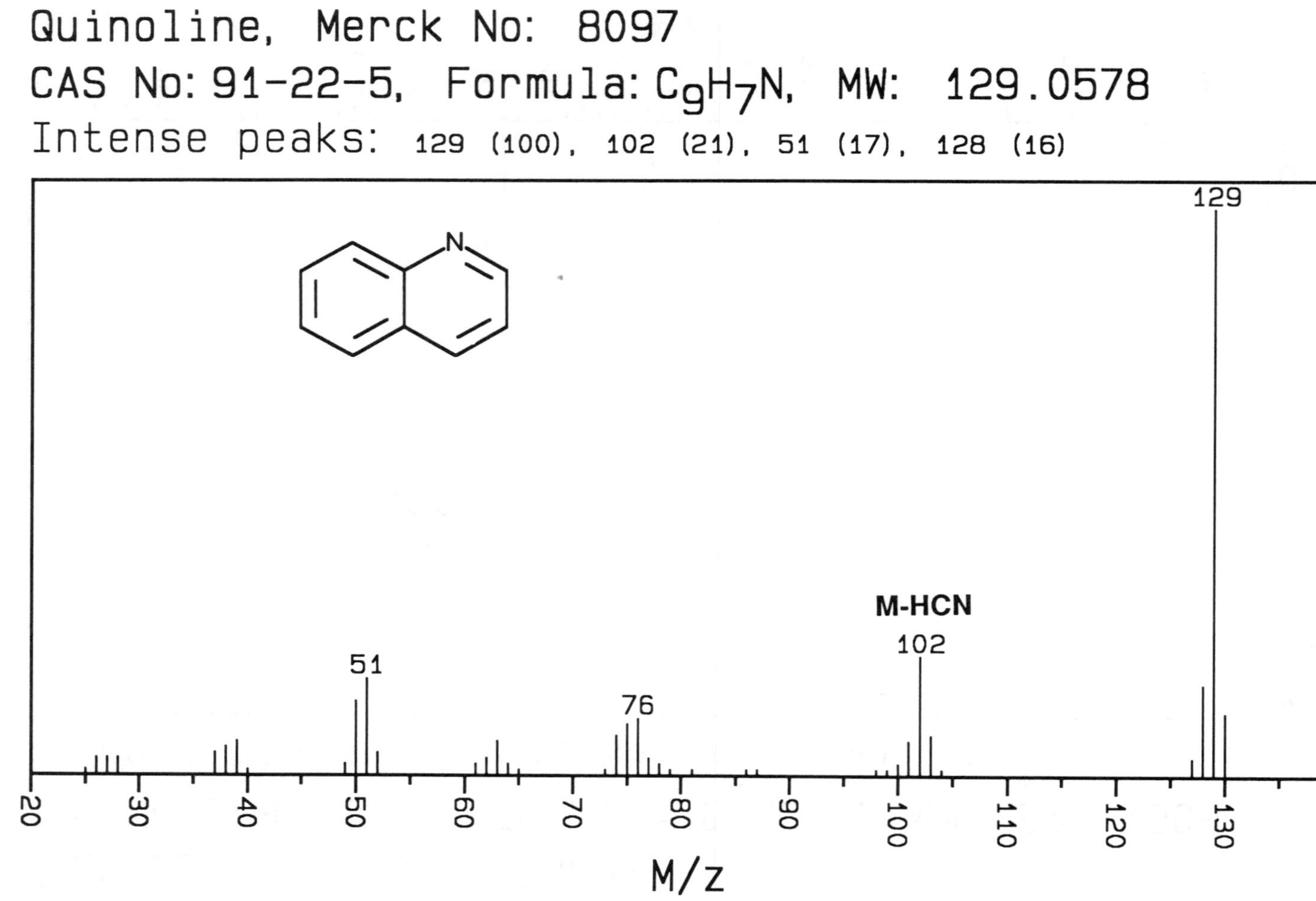
Quinoline, Merck No: 8097
CAS No: 91-22-5, Formula: $C_9H_7N$, MW: 129.0578
Intense peaks: 129 (100), 102 (21), 51 (17), 128 (16)
N
129
M-HCN
102
76
51
20
30
40
50
60
70
80
90
100
110
120
130
140
M/Z

Quinoline, 2-methyl-, Merck No: 8055

CAS No: 91-63-4, Formula: $C_{10}H_9N$, MW: 143.0735

Intense peaks: 143 (100), 128 (21), 115 (21), 142 (16)

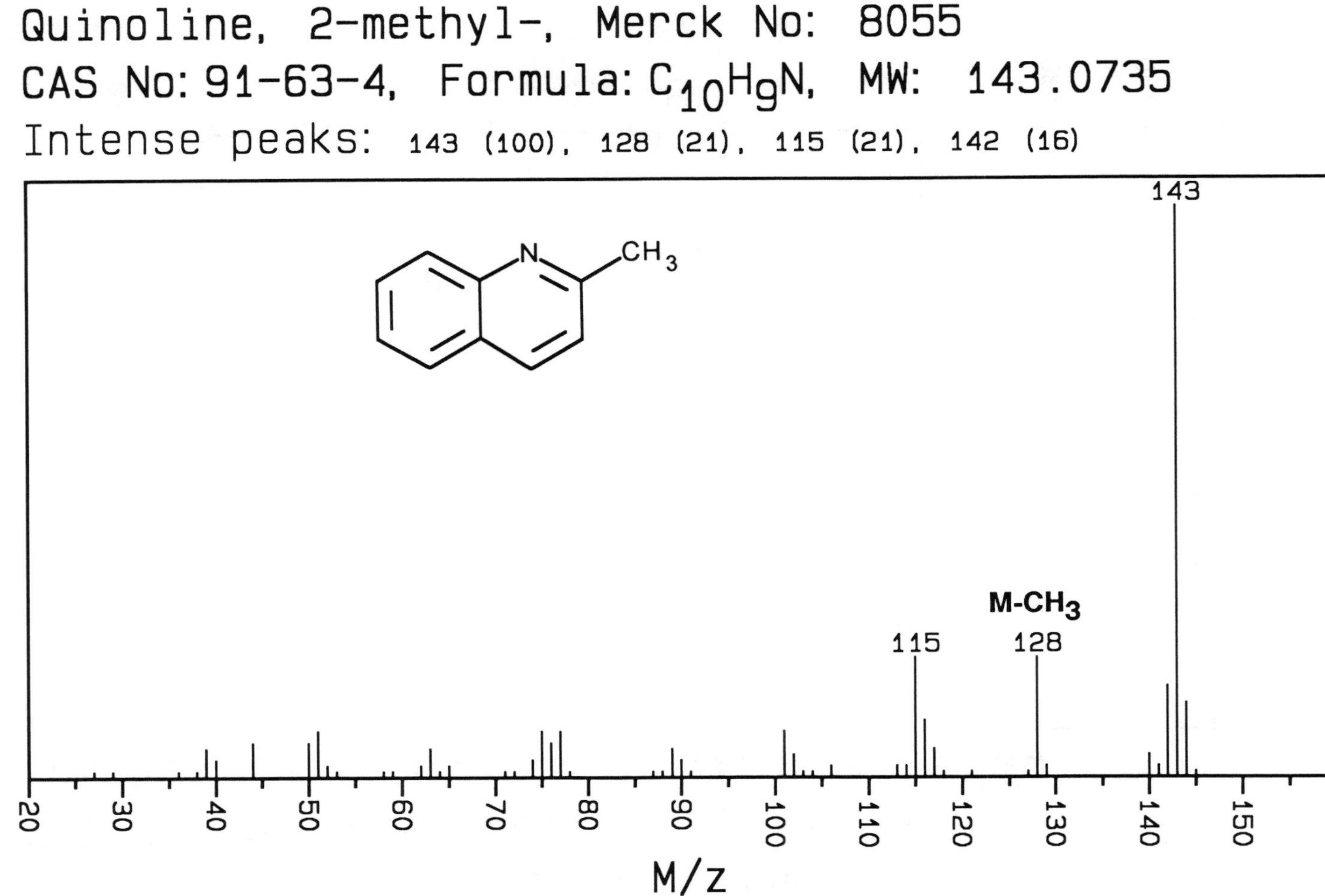

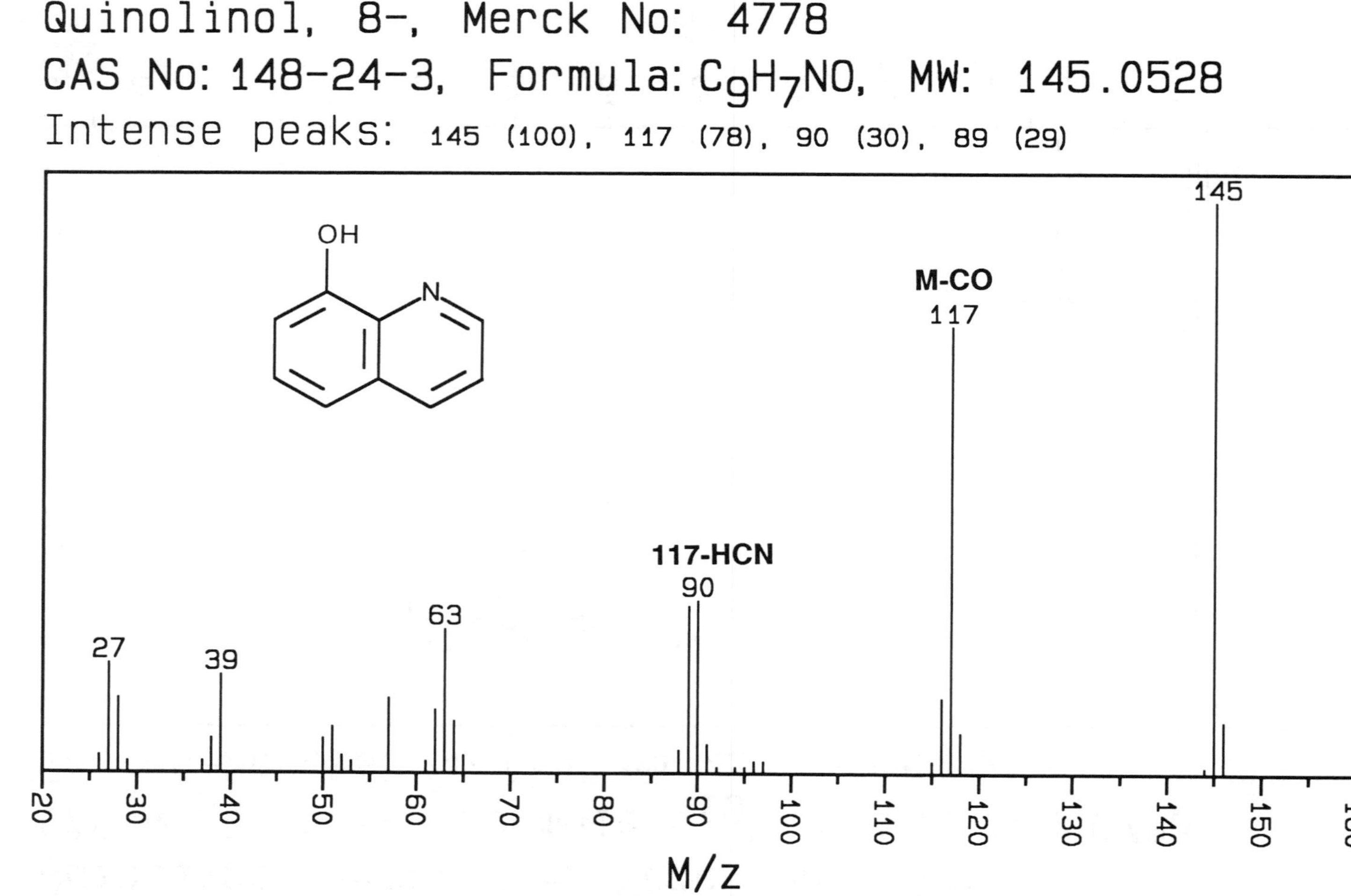
Quinolinol, 8-, Merck No: 4778
CAS No: 148-24-3, Formula: C9H7NO, MW: 145.0528
Intense peaks: 145 (100), 117 (78), 90 (30), 89 (29)
OH
N
145
M-CO
117
117-HCN
90
63
39
27
20
30
40
50
60
70
80
90
100
110
120
130
140
150
160
M/Z

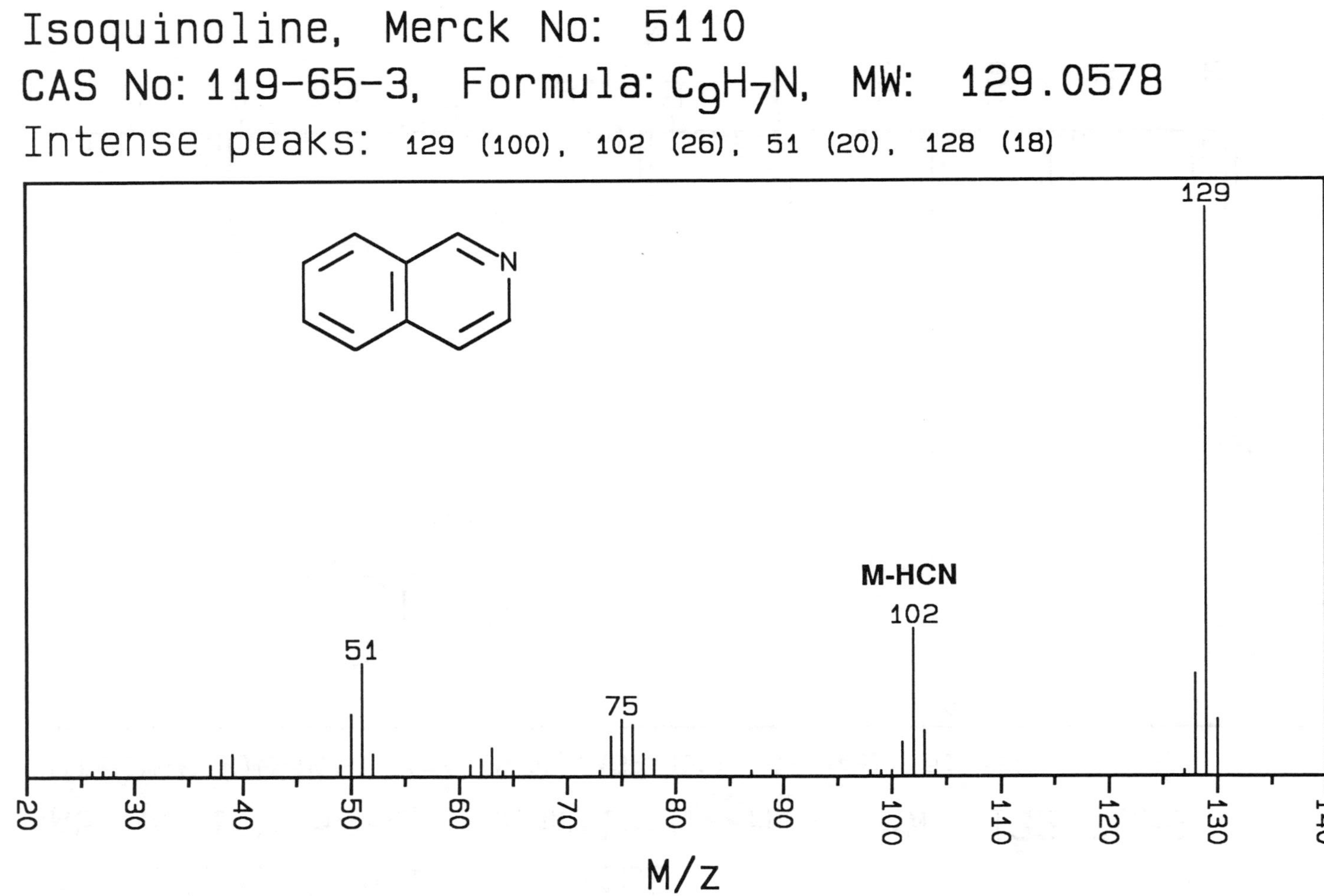
Isoquinoline, Merck No: 5110
CAS No: 119-65-3, Formula: $C_9H_7N$, MW: 129.0578
Intense peaks: 129 (100), 102 (26), 51 (20), 128 (18)
N
129
M-HCN
102
51
75
20
30
40
50
60
70
80
90
100
110
120
130
140
M/z

Tetralin*, Merck No: 9152

CAS No: 119-64-2, Formula: $C_{10}H_{12}$, MW: 132.0939

Intense peaks: 104 (100), 132 (53), 91 (43), 51 (17)

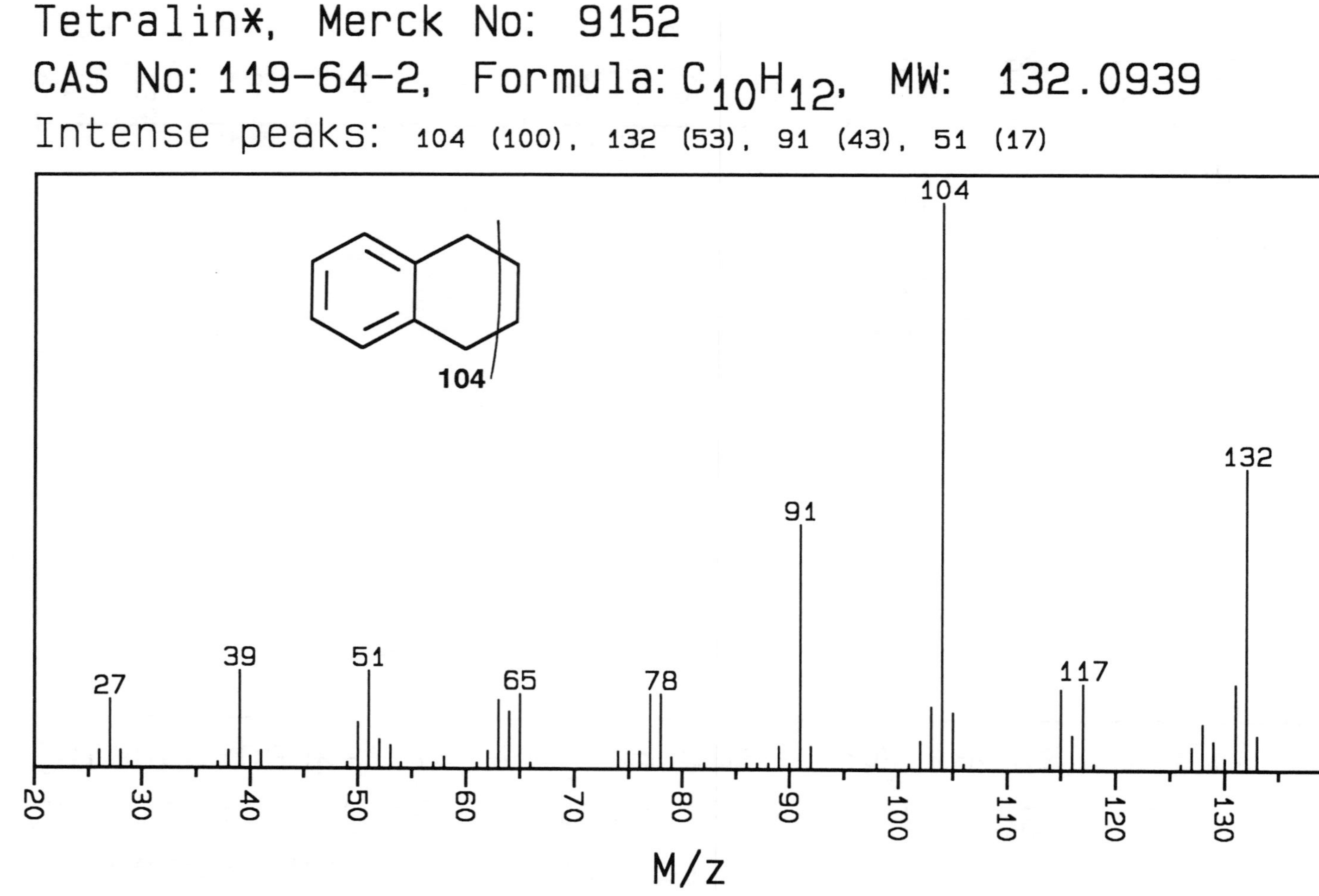

Acenaphthylene

CAS No: 208-96-8, Formula: $C_{12}H_8$, MW: 152.0626

Intense peaks: 152 (100), 151 (20), 76 (17), 153 (14)

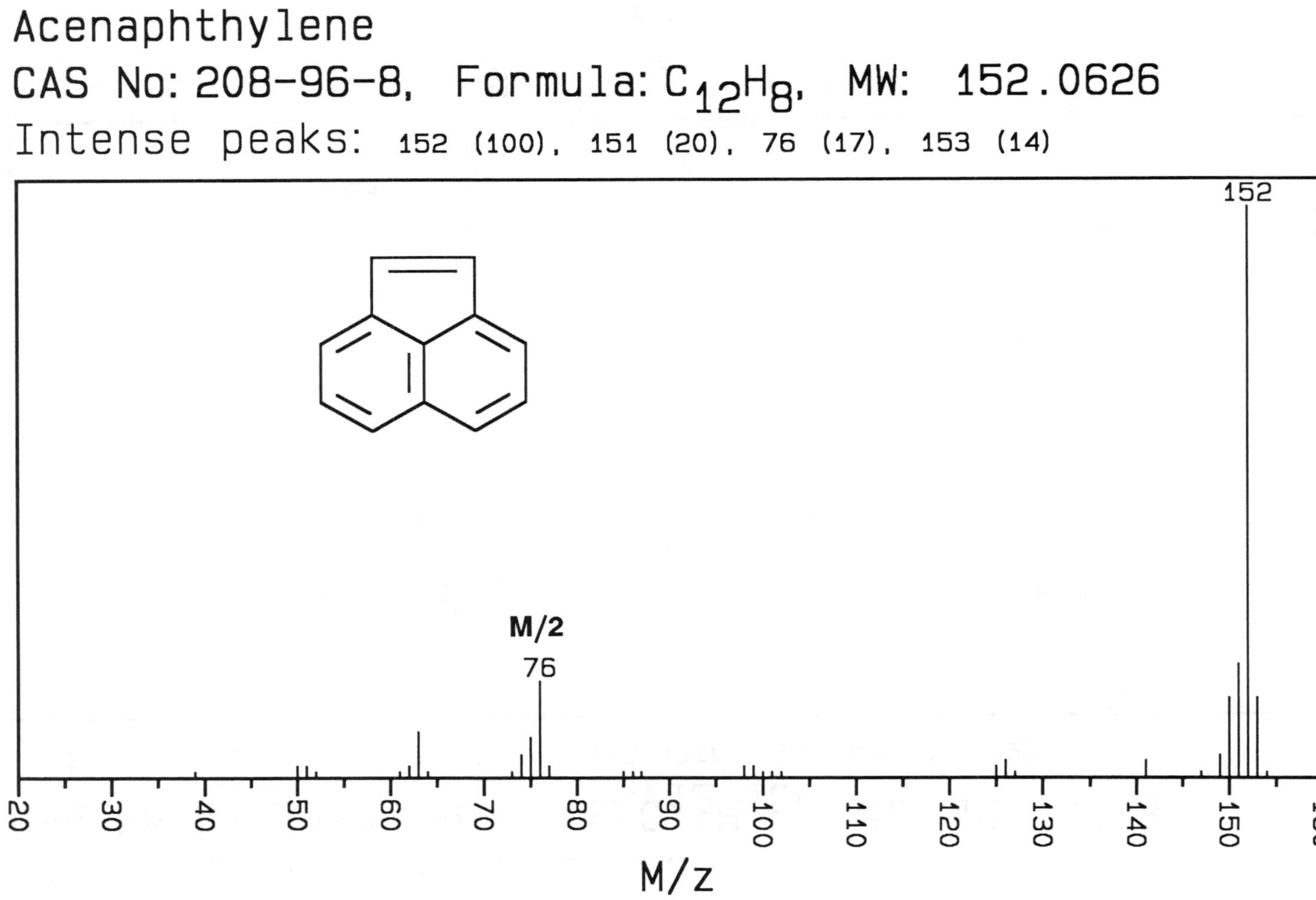

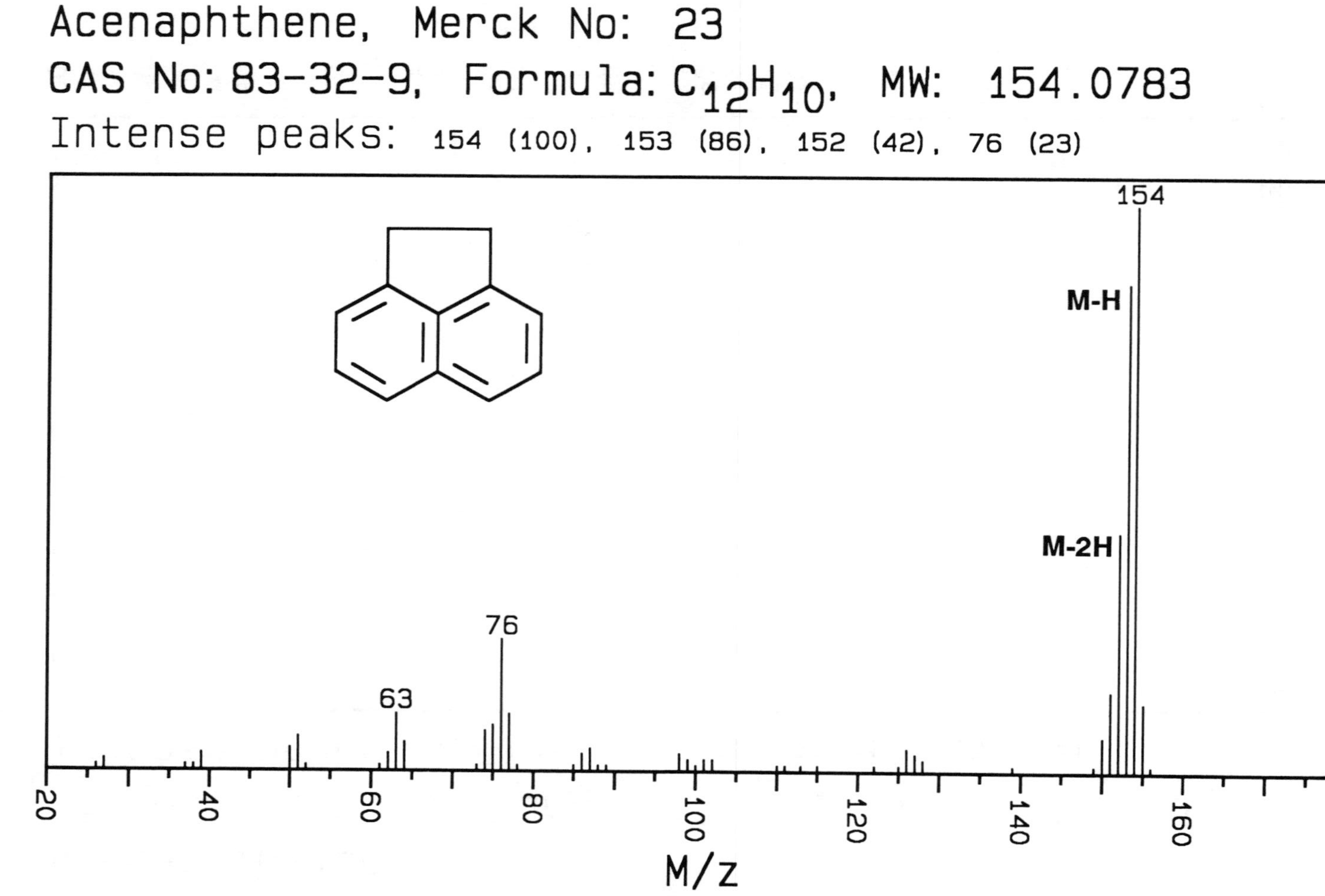
Acenaphthene, Merck No: 23
CAS No: 83-32-9, Formula: C12H10, MW: 154.0783
Intense peaks: 154 (100), 153 (86), 152 (42), 76 (23)
154
M-H
M-2H
76
63
20
40
60
80
100
120
140
160
180
M/Z

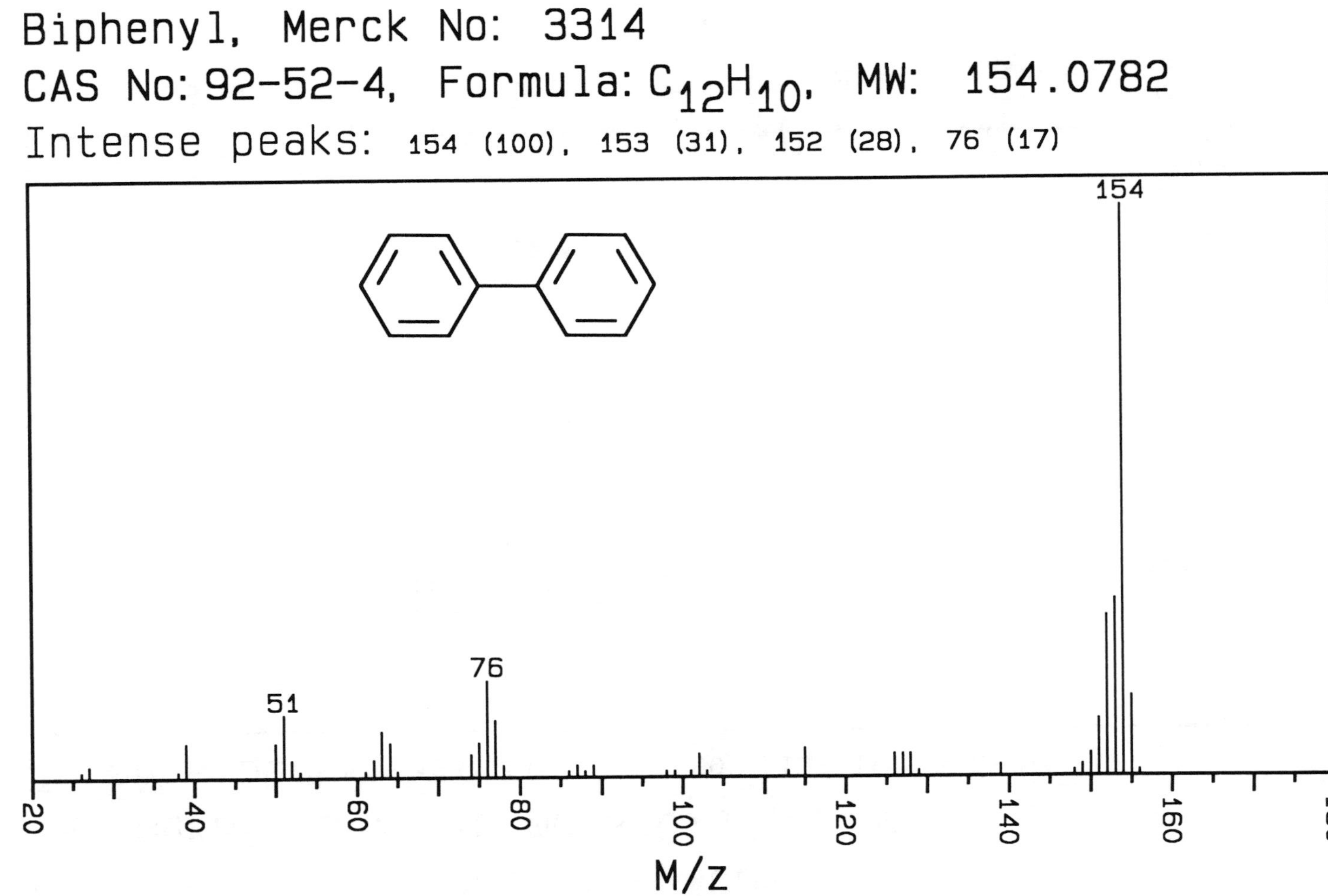
Biphenyl, Merck No: 3314
CAS No: 92-52-4, Formula: C12H10, MW: 154.0782
Intense peaks: 154 (100), 153 (31), 152 (28), 76 (17)
154
76
51
20
40
60
80
100
120
140
160
180
M/Z

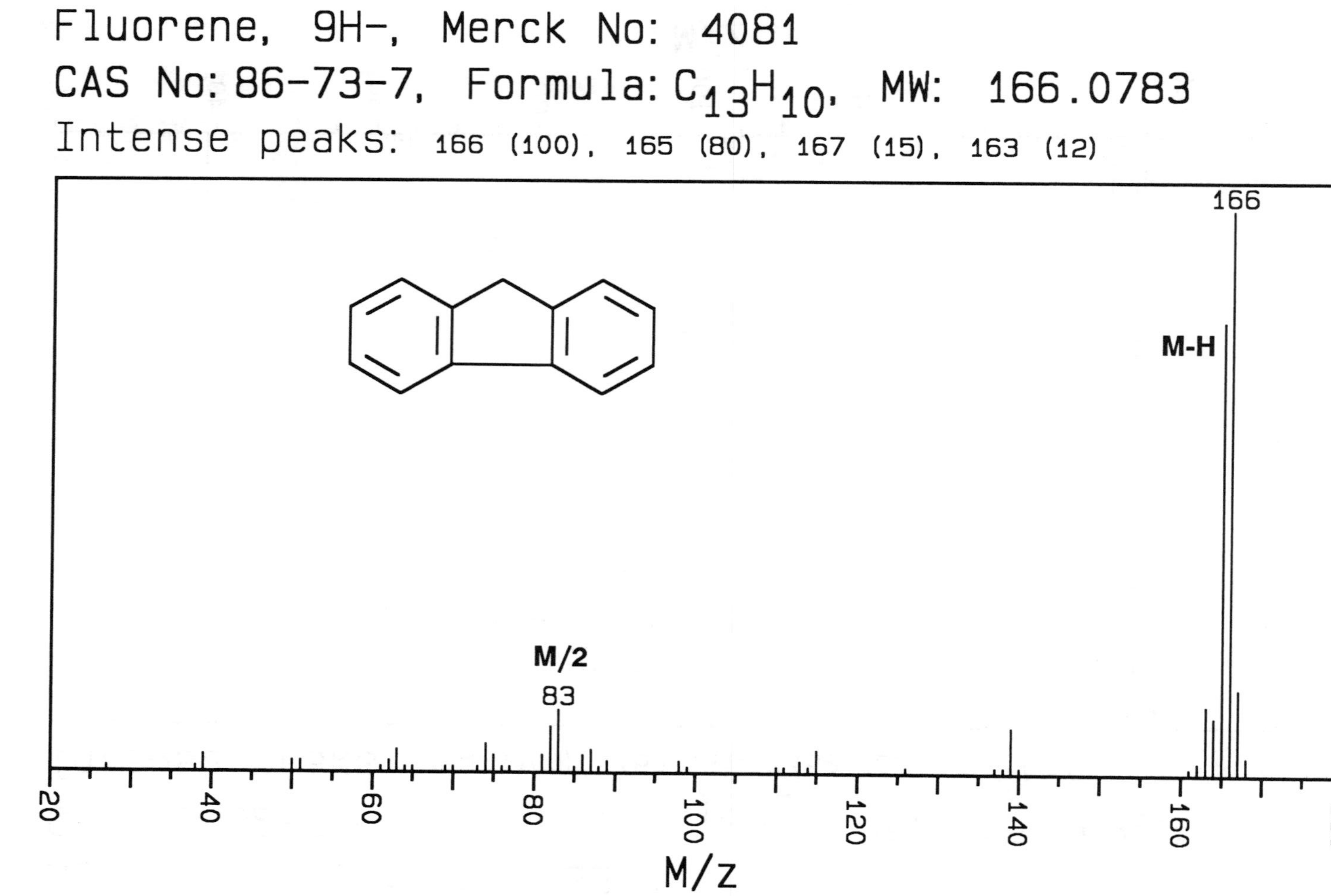
Fluorene, 9H-, Merck No: 4081
CAS No: 86-73-7, Formula: $C_{13}H_{10}$, MW: 166.0783
Intense peaks: 166 (100), 165 (80), 167 (15), 163 (12)
166
M-H
M/2
83
20
40
60
80
100
120
140
160
180
M/Z

Carbazole, 9H-, Merck No: 1792

CAS No: 86-74-8, Formula: $C_{12}H_9N$, MW: 167.0735

Intense peaks: 167 (100), 166 (14), 44 (13), 168 (12)

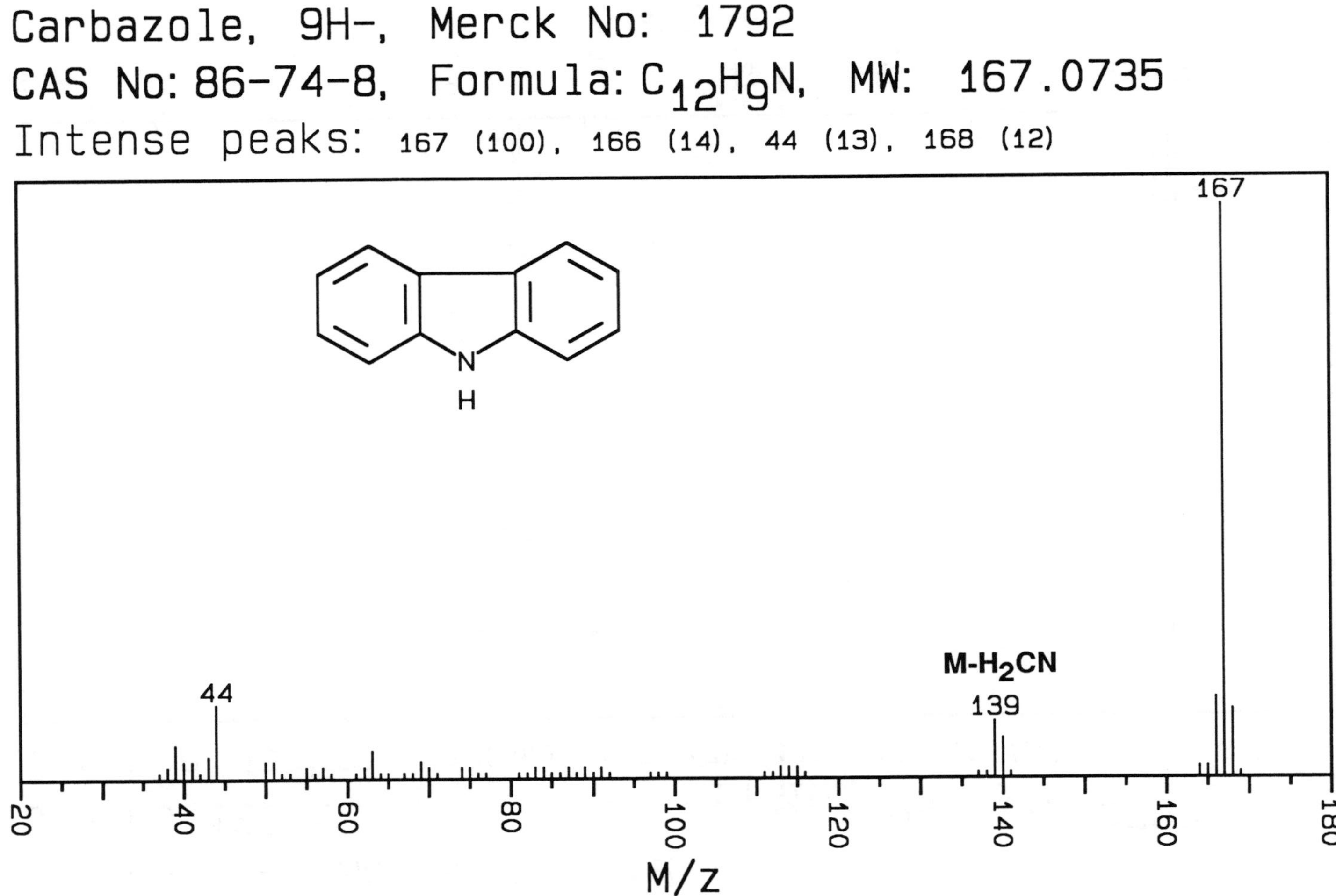

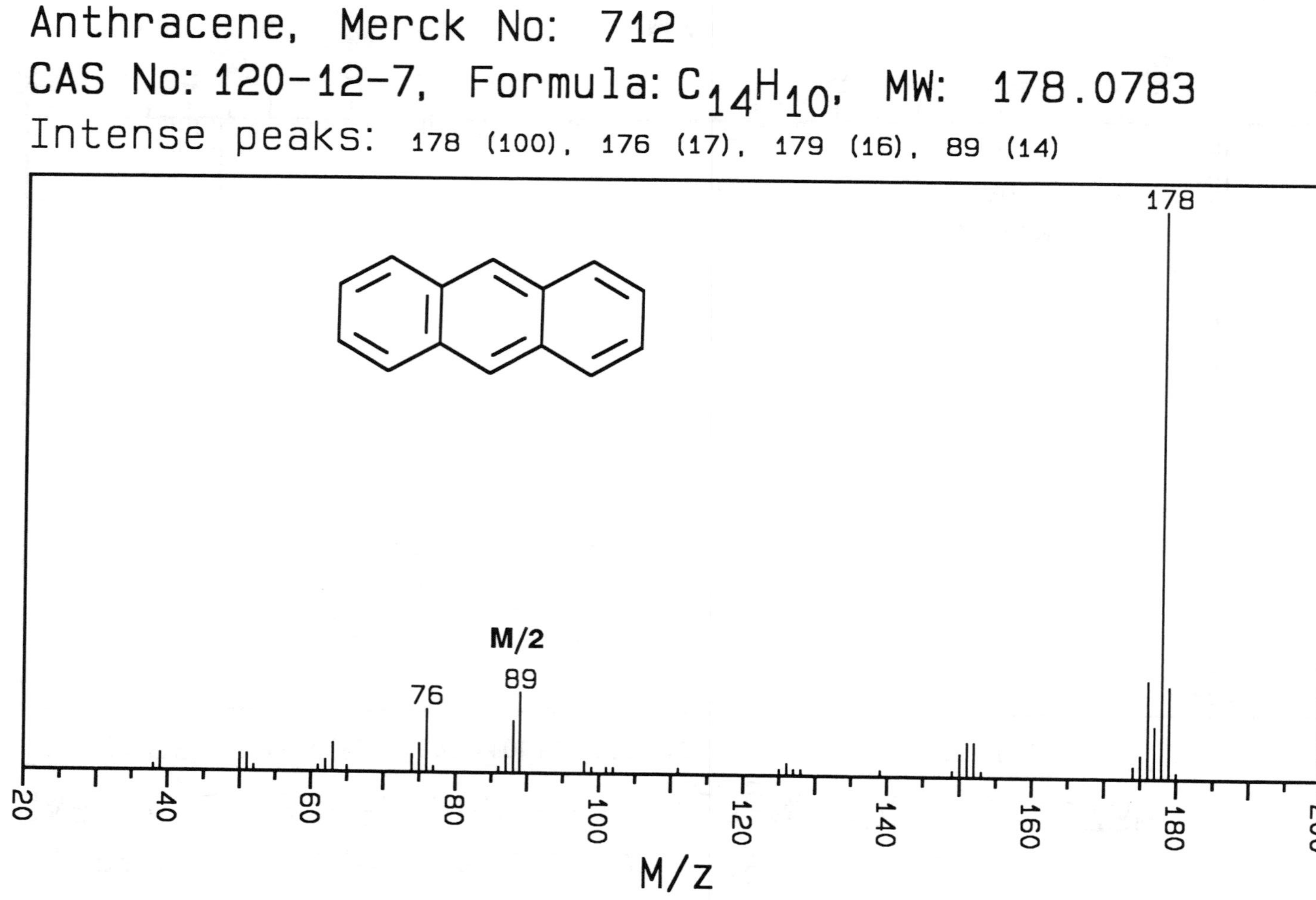
Anthracene, Merck No: 712
CAS No: 120-12-7, Formula: $C_{14}H_{10}$, MW: 178.0783
Intense peaks: 178 (100), 176 (17), 179 (16), 89 (14)
178
M/2
89
76
20
40
60
80
100
120
140
160
180
200
M/Z

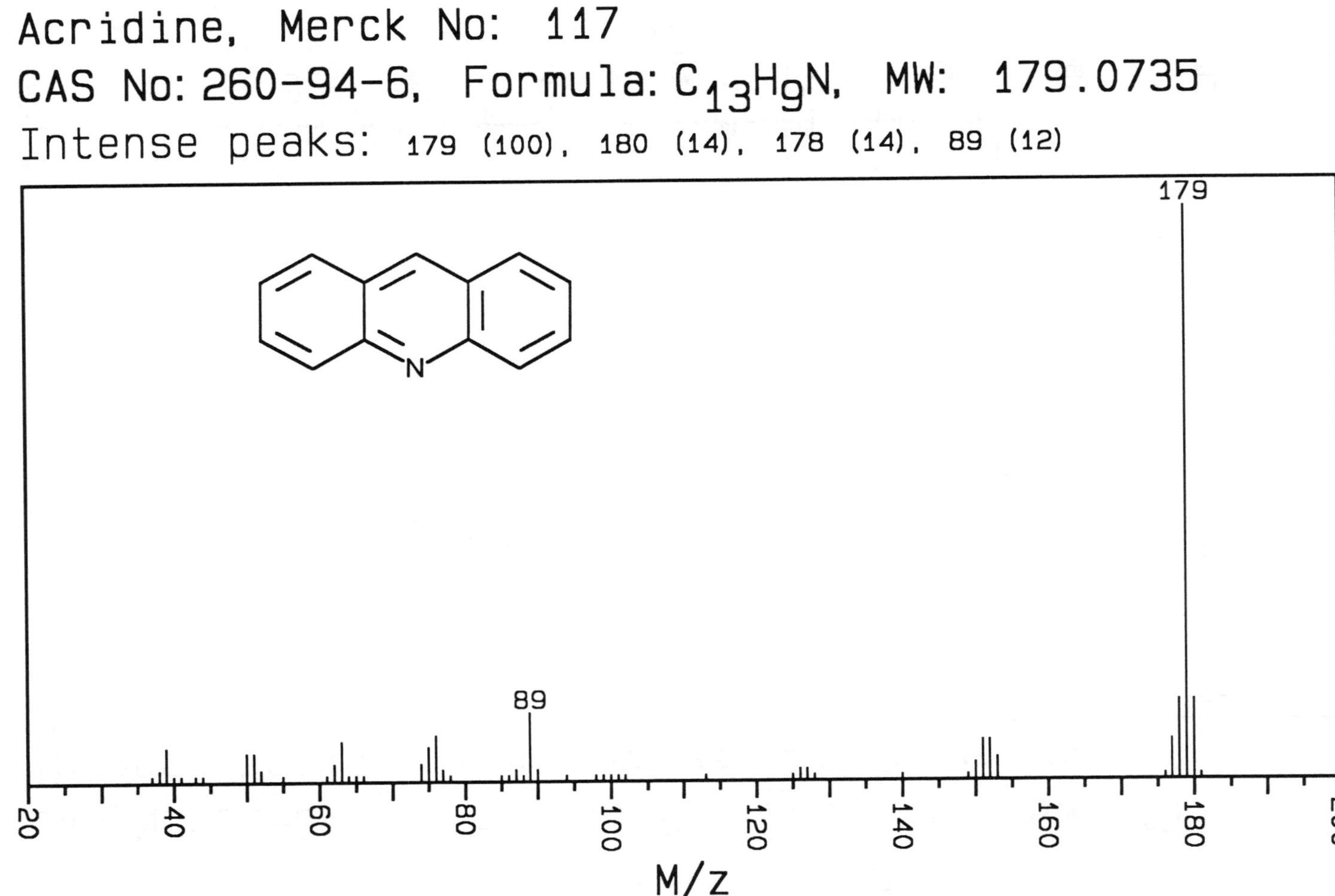

Acridine, Merck No: 117
CAS No: 260-94-6, Formula: $C_{13}H_9N$, MW: 179.0735
Intense peaks: 179 (100), 180 (14), 178 (14), 89 (12)
179
89
N
20
40
60
80
100
120
140
160
180
200
M/Z

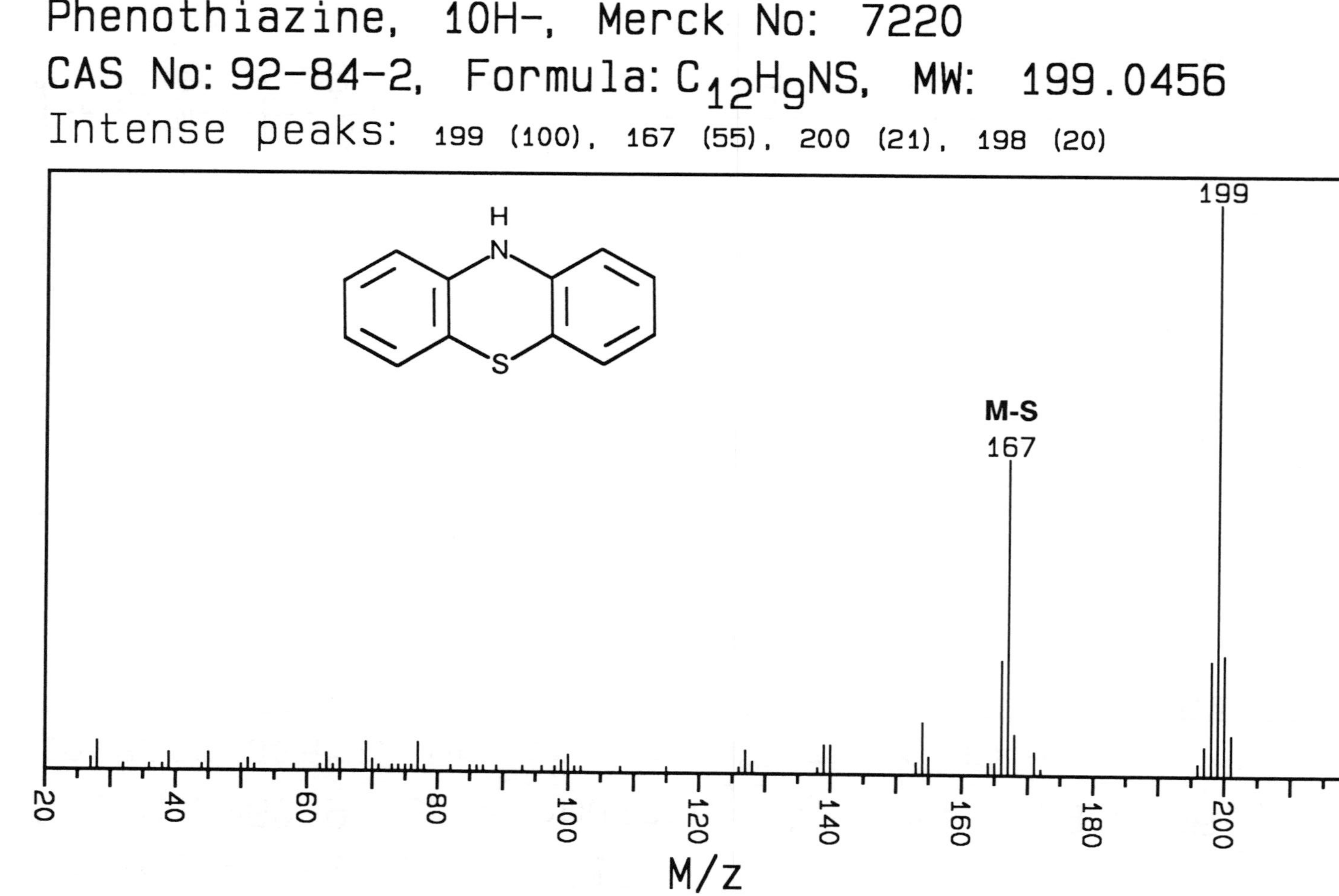
Phenothiazine, 10H-, Merck No: 7220
CAS No: 92-84-2, Formula: C12H9NS, MW: 199.0456
Intense peaks: 199 (100), 167 (55), 200 (21), 198 (20)
H
N
S
199
M-S
167
20
40
60
80
100
120
140
160
180
200
220
M/Z

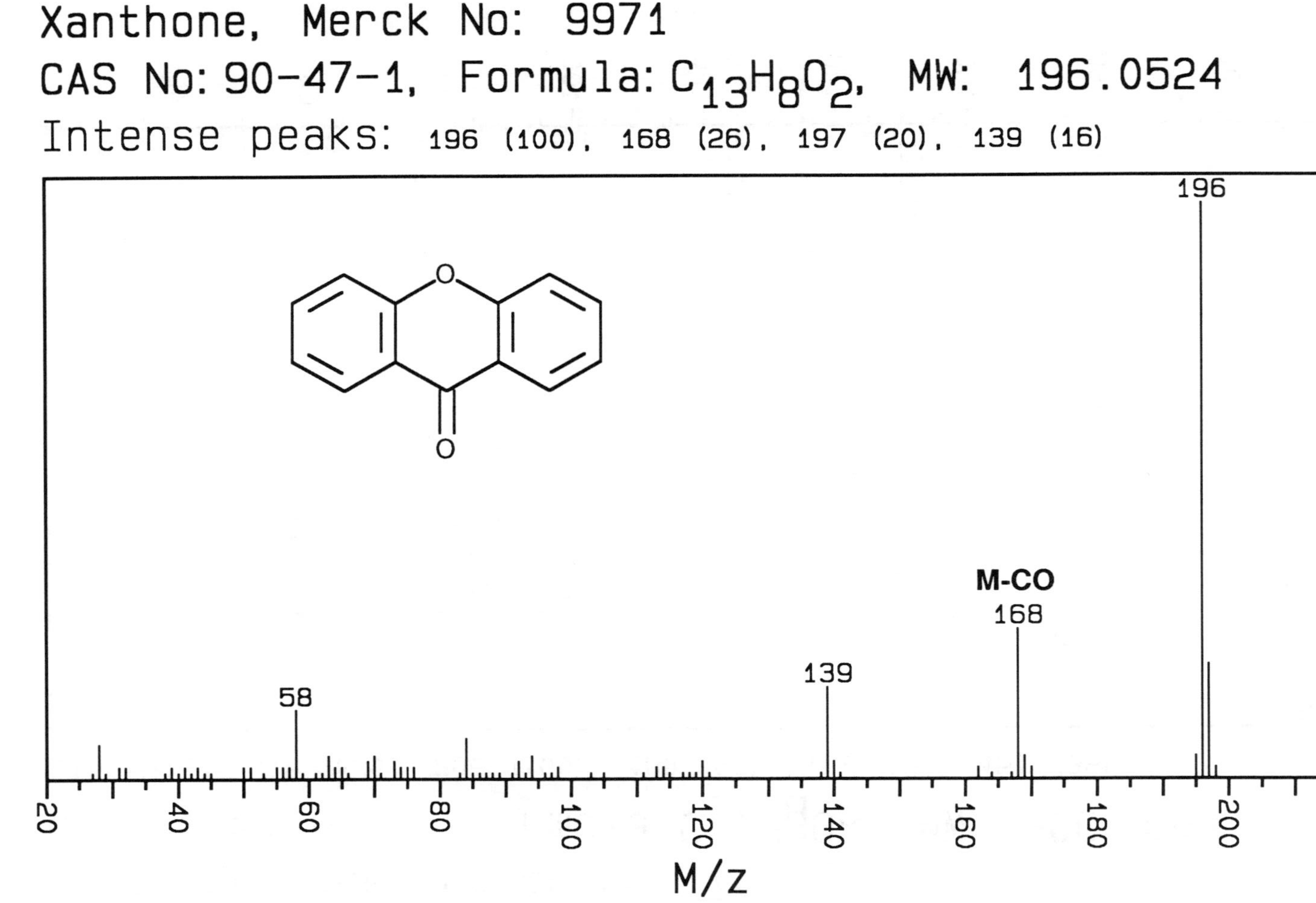
Xanthone, Merck No: 9971
CAS No: 90-47-1, Formula: $C_{13}H_8O_2$, MW: 196.0524
Intense peaks: 196 (100), 168 (26), 197 (20), 139 (16)
196
M-CO
168
139
58
O
O
20
40
60
80
100
120
140
160
180
200
220
M/Z

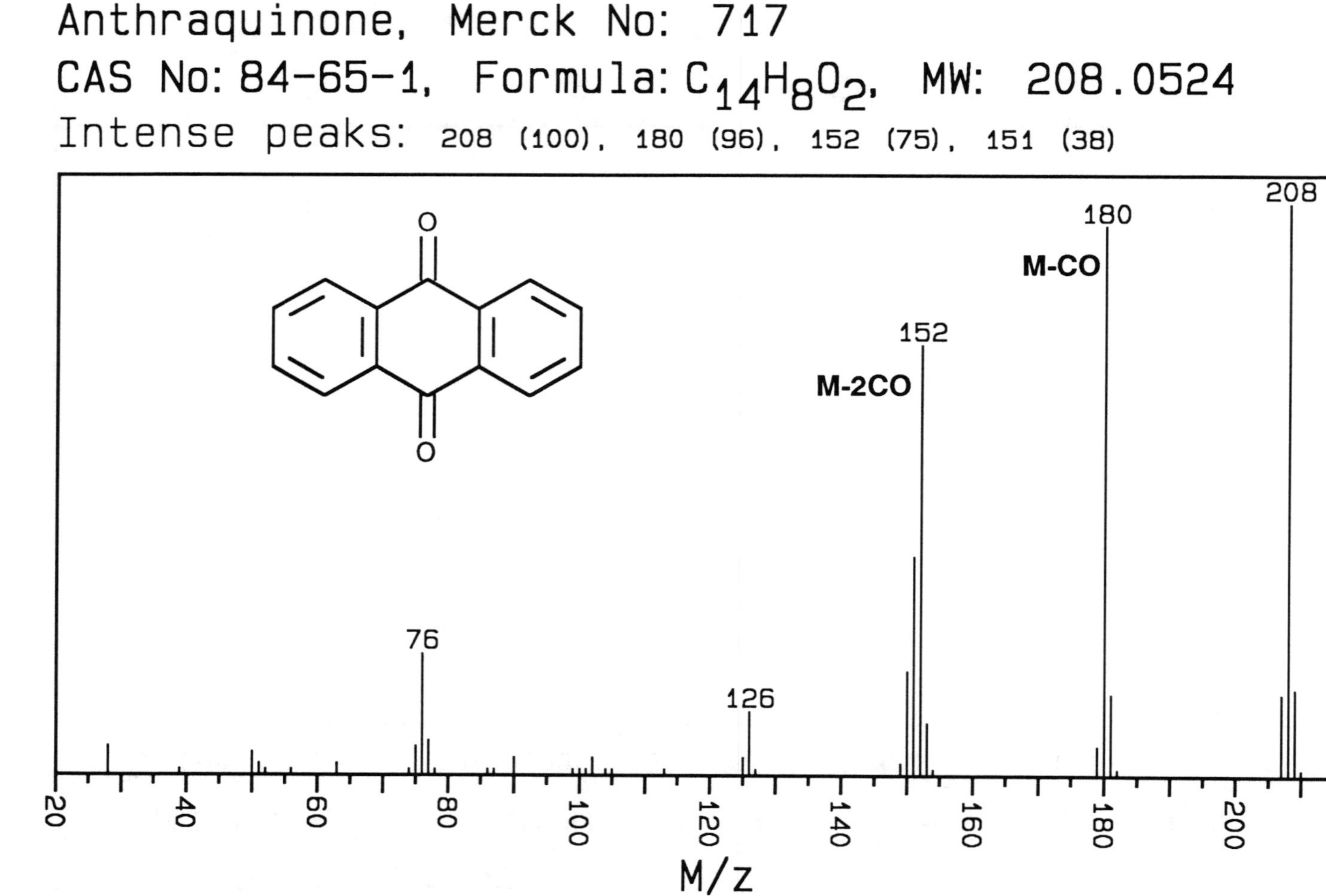
Anthraquinone, Merck No: 717
CAS No: 84-65-1, Formula: $C_{14}H_8O_2$, MW: 208.0524
Intense peaks: 208 (100), 180 (96), 152 (75), 151 (38)
O
O
208
180
M-CO
152
M-2CO
126
76
20
40
60
80
100
120
140
160
180
200
220
M/Z

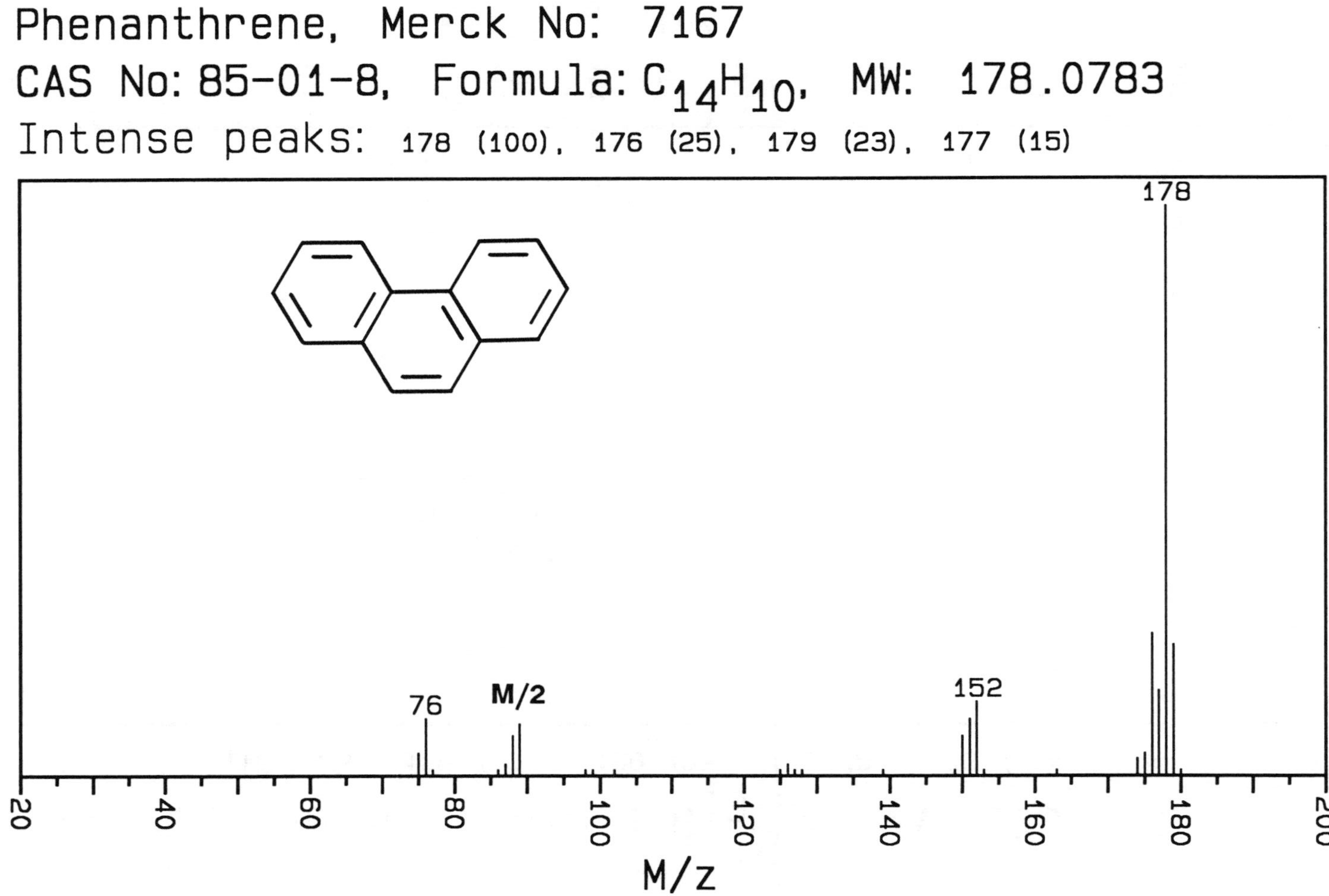
Phenanthrene, Merck No: 7167
CAS No: 85-01-8, Formula: C14H10, MW: 178.0783
Intense peaks: 178 (100), 176 (25), 179 (23), 177 (15)
178
152
76
M/2
20
40
60
80
100
120
140
160
180
200
M/z

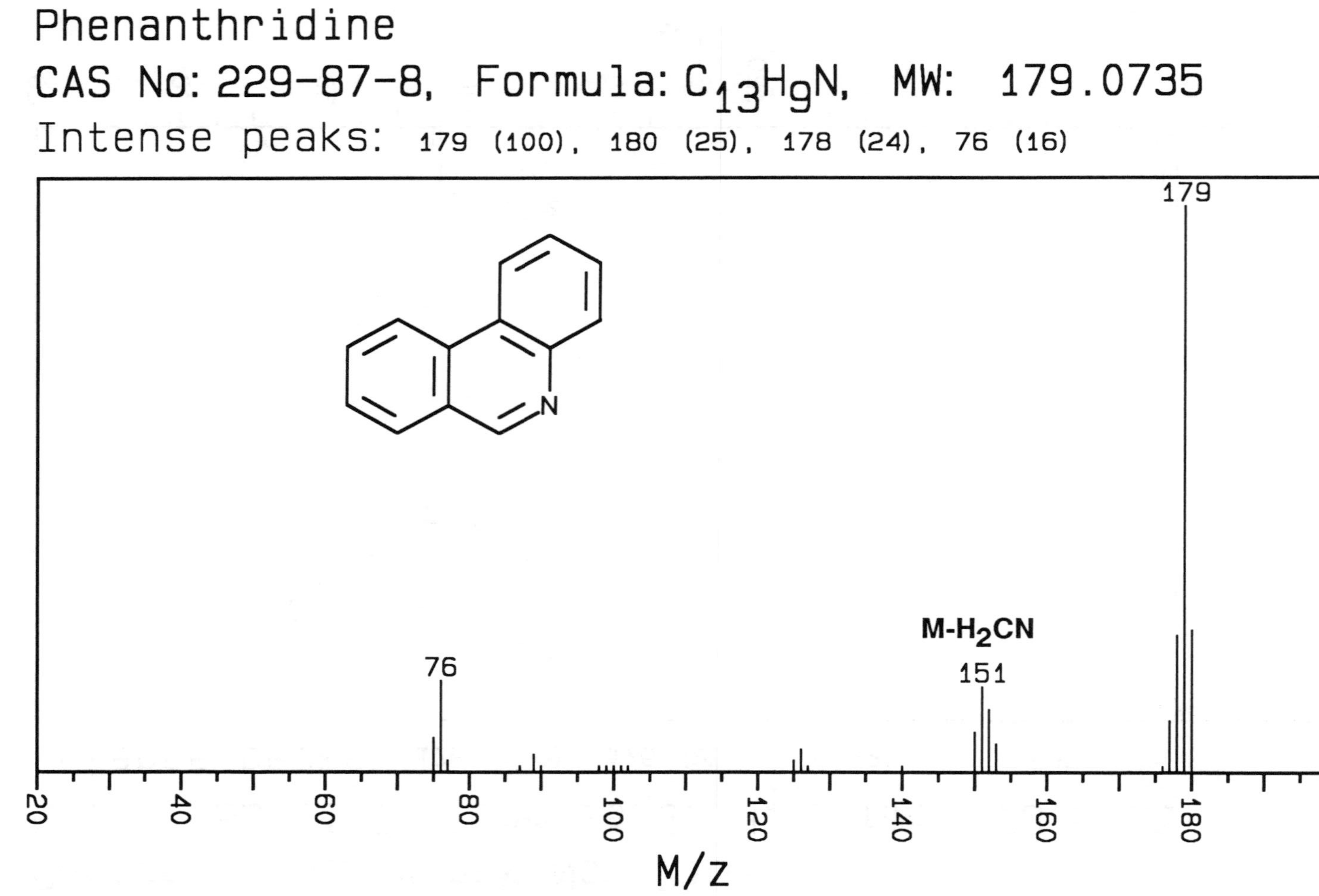

Phenanthridine
CAS No: 229-87-8, Formula: $C_{13}H_9N$, MW: 179.0735
Intense peaks: 179 (100), 180 (25), 178 (24), 76 (16)
179
M-$H_2$CN
151
76
N
M/Z
20
40
60
80
100
120
140
160
180
200

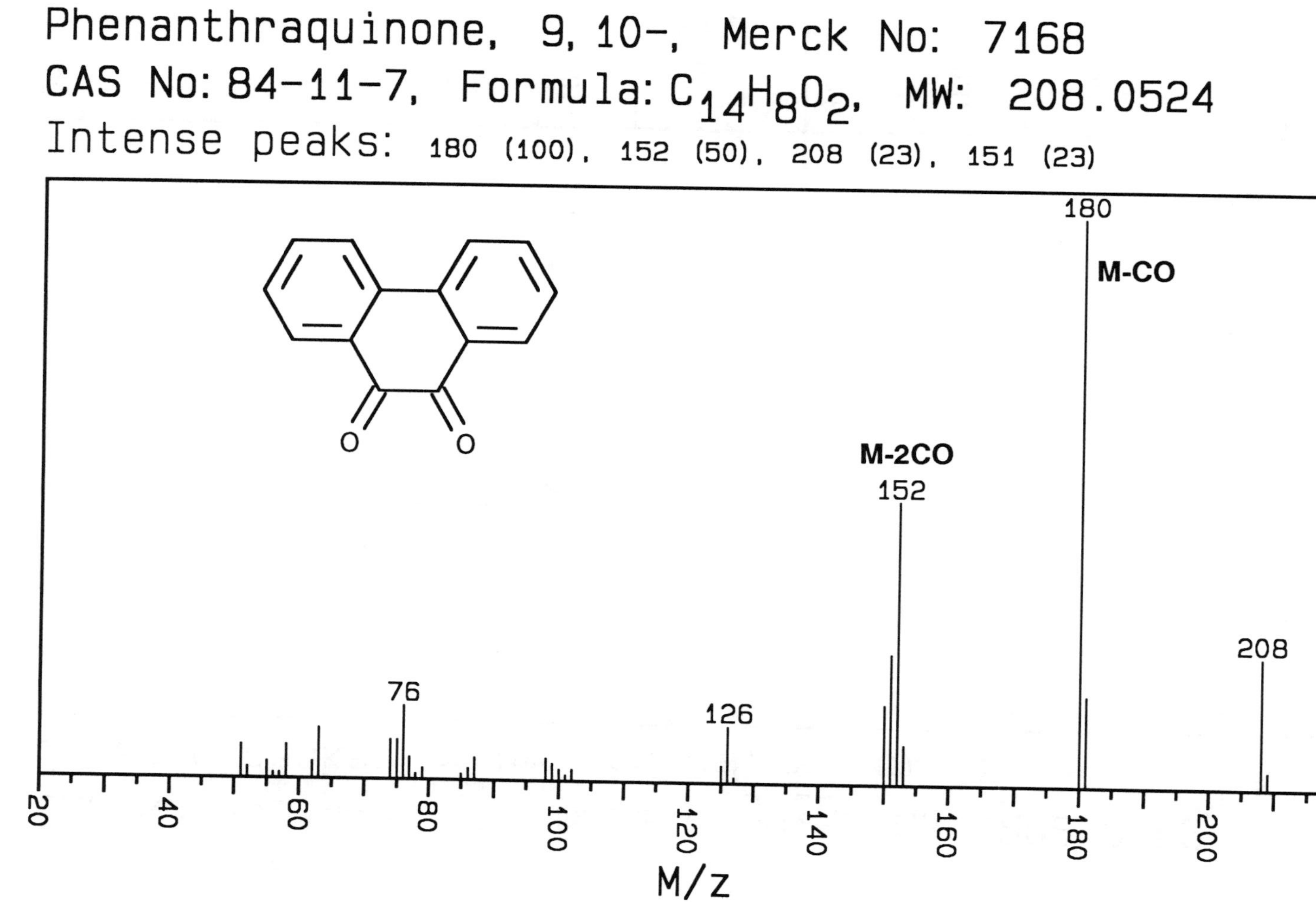

Phenanthraquinone, 9,10-, Merck No: 7168
CAS No: 84-11-7, Formula: $C_{14}H_8O_2$, MW: 208.0524
Intense peaks: 180 (100), 152 (50), 208 (23), 151 (23)
180
M-CO
O
O
M-2CO
152
208
76
126
20
40
60
80
100
120
140
160
180
200
220
M/Z

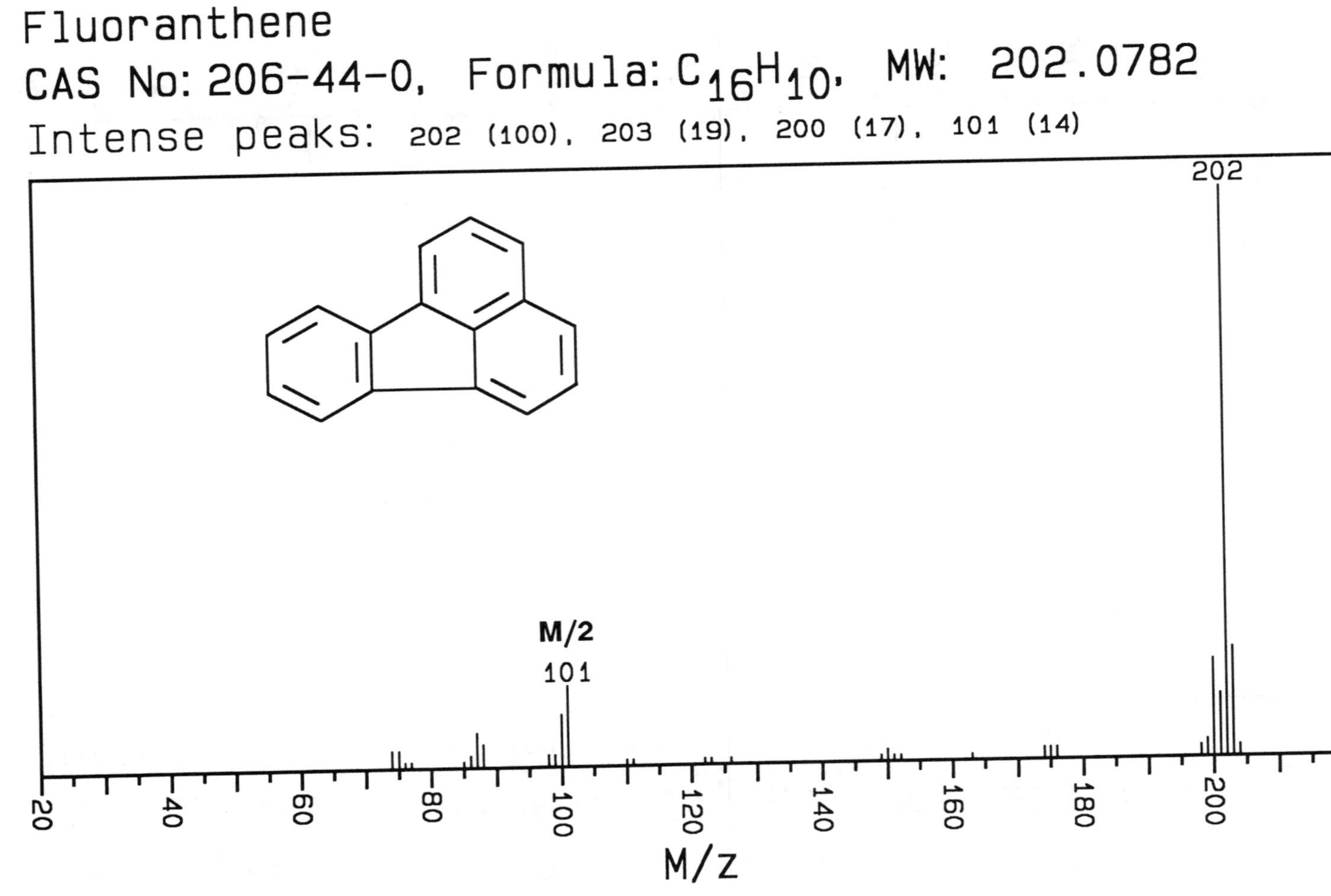
Fluoranthene
CAS No: 206-44-0, Formula: $C_{16}H_{10}$, MW: 202.0782
Intense peaks: 202 (100), 203 (19), 200 (17), 101 (14)
202
M/2
101
20
40
60
80
100
120
140
160
180
200
220
M/Z

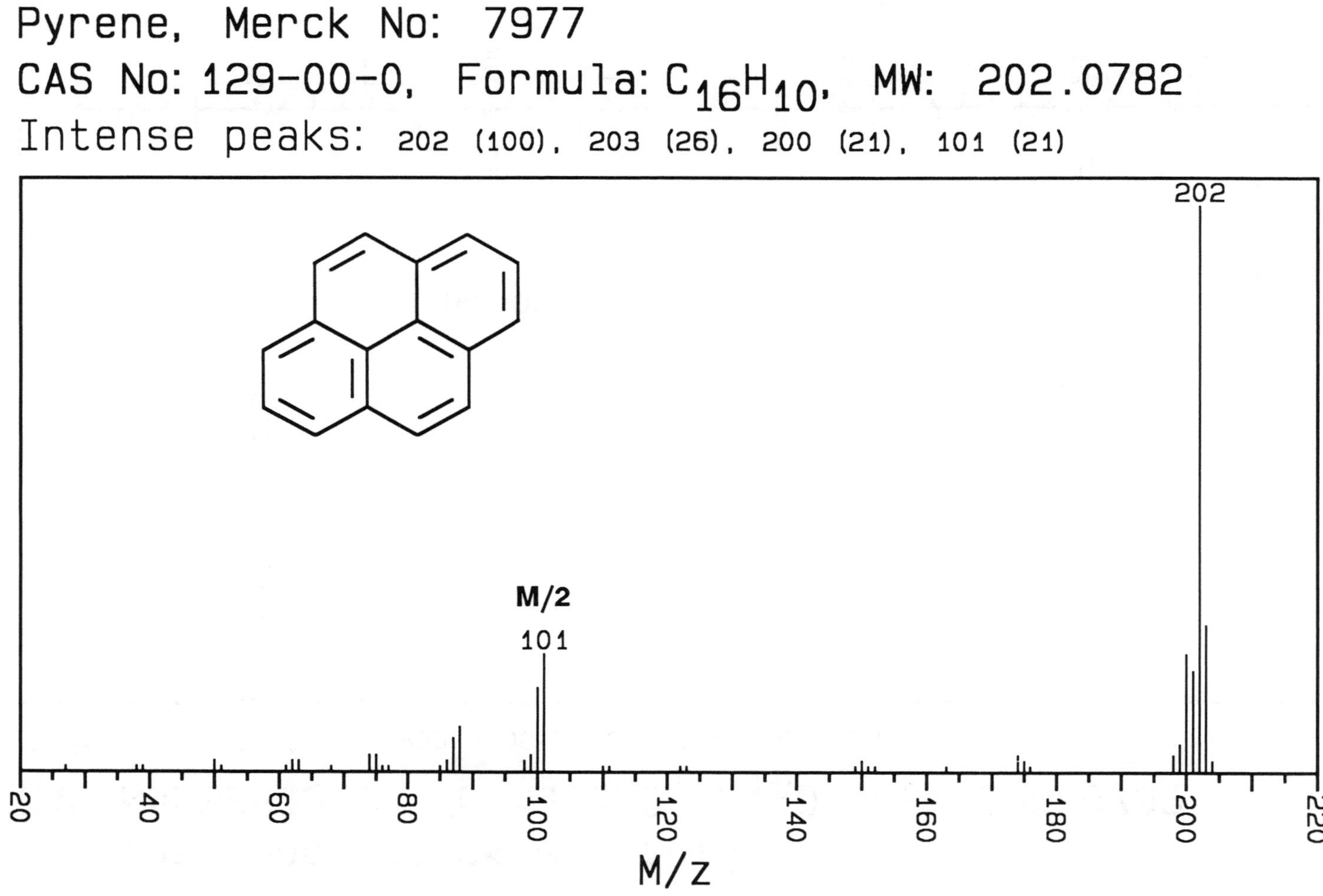
Pyrene, Merck No: 7977
CAS No: 129-00-0, Formula: $C_{16}H_{10}$, MW: 202.0782
Intense peaks: 202 (100), 203 (26), 200 (21), 101 (21)
202
M/2
101
20
40
60
80
100
120
140
160
180
200
220
M/Z

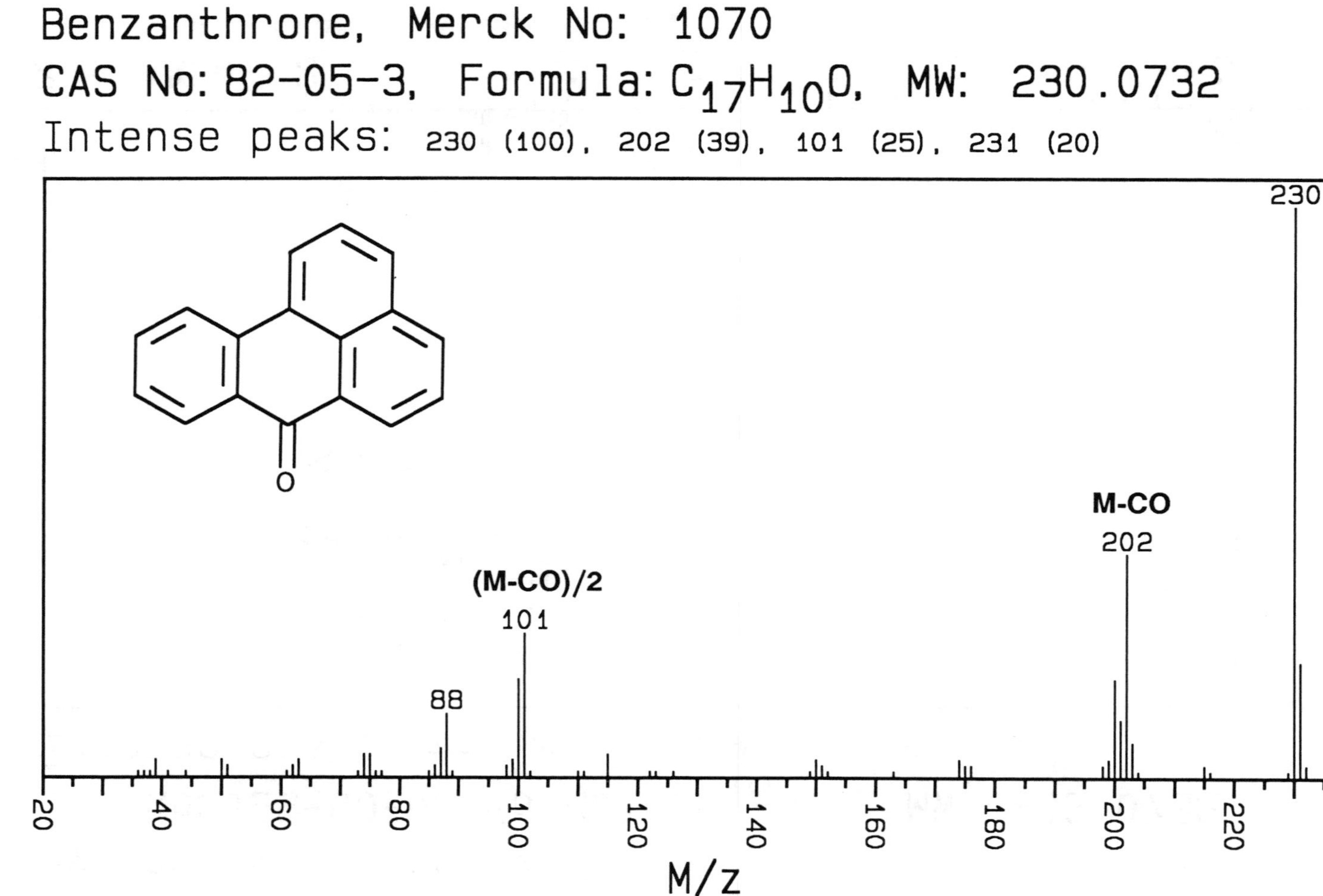
Benzanthrone, Merck No: 1070
CAS No: 82-05-3, Formula: $C_{17}H_{10}O$, MW: 230.0732
Intense peaks: 230 (100), 202 (39), 101 (25), 231 (20)
230
M-CO
202
(M-CO)/2
101
88
O
20
40
60
80
100
120
140
160
180
200
220
240
M/Z

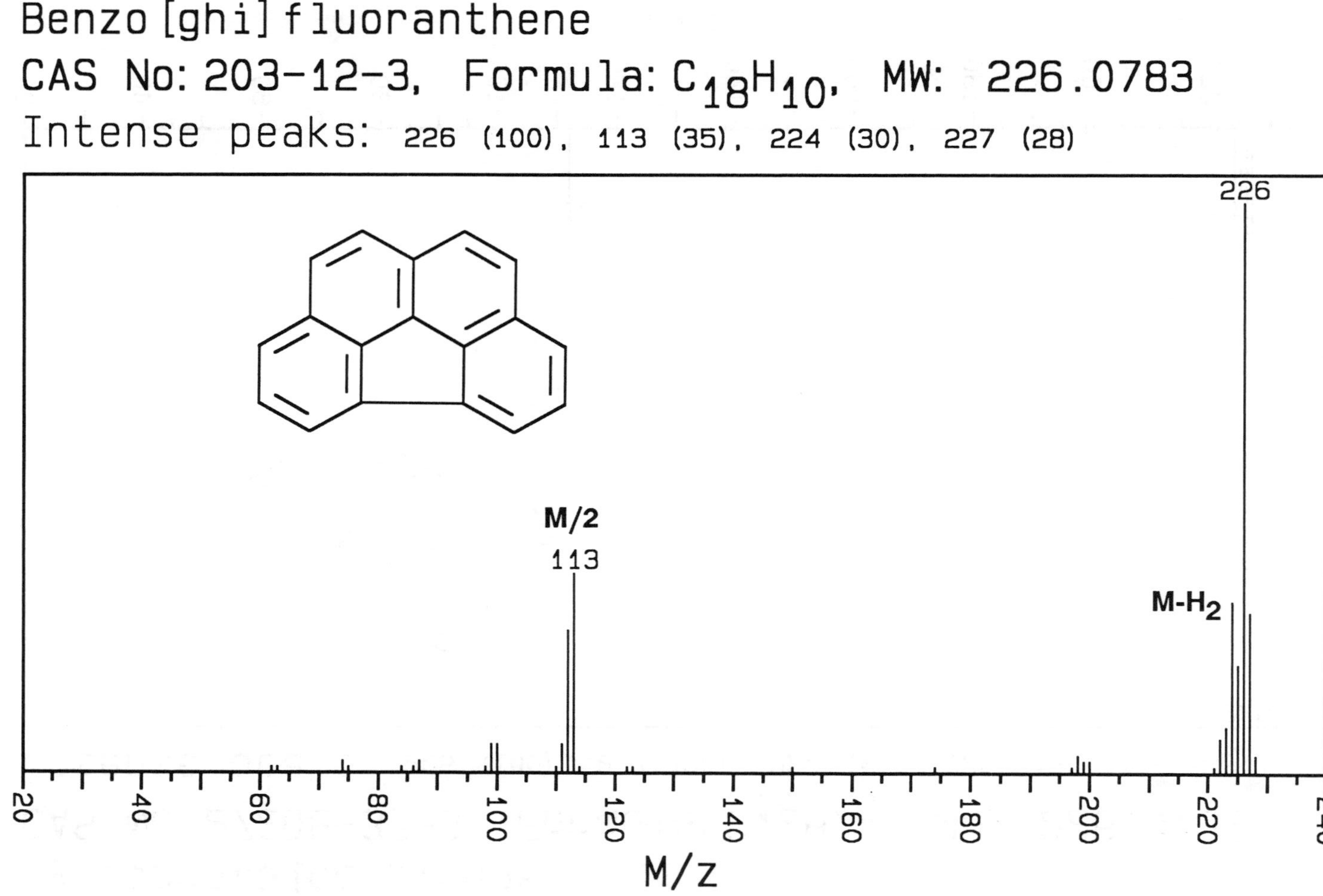
Benzo[ghi]fluoranthene
CAS No: 203-12-3, Formula: $C_{18}H_{10}$, MW: 226.0783
Intense peaks: 226 (100), 113 (35), 224 (30), 227 (28)
226
M/2
113
M-$H_2$
20
40
60
80
100
120
140
160
180
200
220
240
M/Z

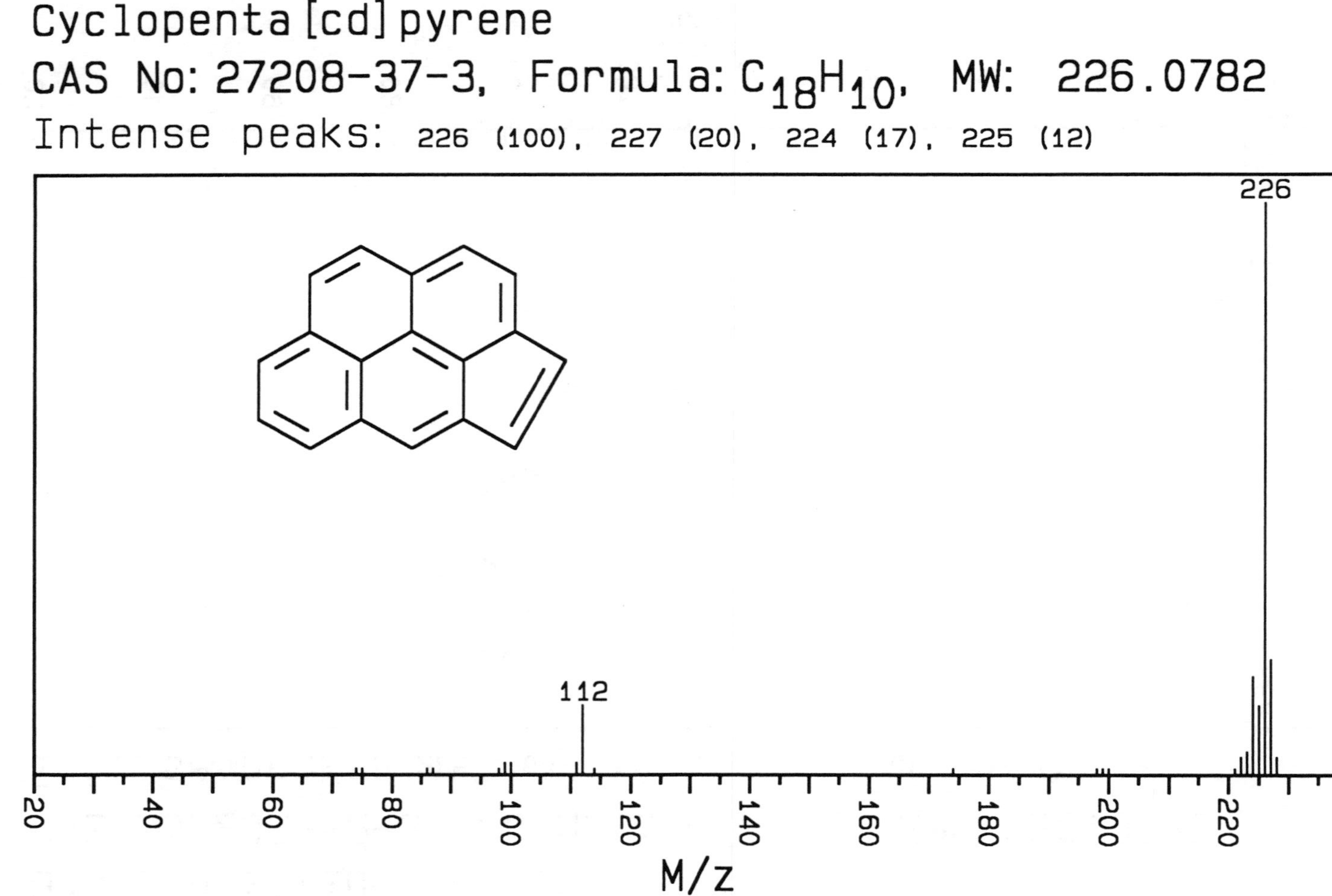

Cyclopenta[cd]pyrene
CAS No: 27208-37-3, Formula: C18H10, MW: 226.0782
Intense peaks: 226 (100), 227 (20), 224 (17), 225 (12)
226
112
20
40
60
80
100
120
140
160
180
200
220
240
M/Z

Benz[a]anthracene, Merck No: 1069

CAS No: 56-55-3, Formula: $C_{18}H_{12}$, MW: 228.0939

Intense peaks: 228 (100), 229 (20), 226 (20), 43 (18)

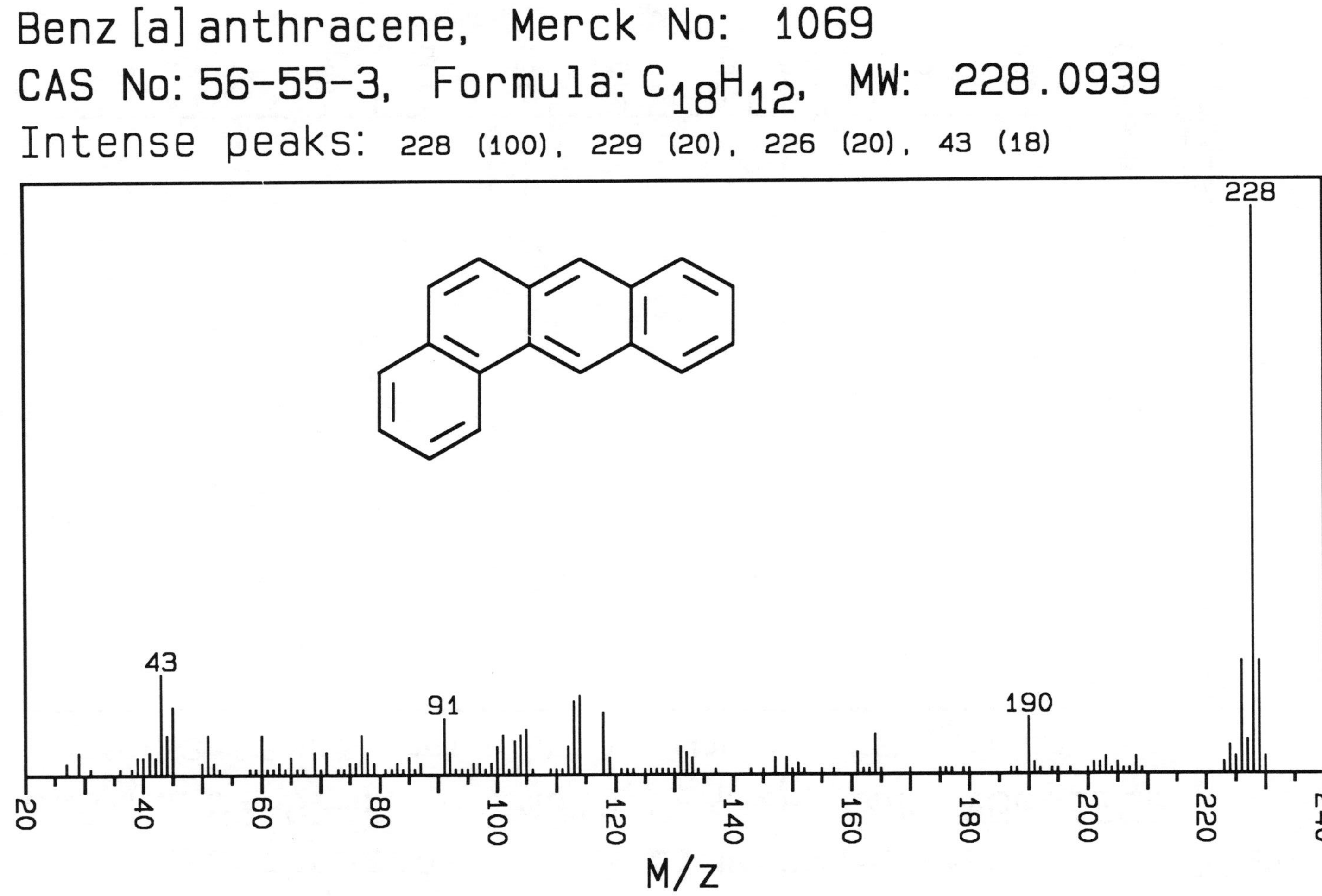

Benz[a]anthracene, 7,12-dimethyl-, Merck No: 3224

CAS No: 57-97-6, Formula: $C_{20}H_{16}$, MW: 256.1252

Intense peaks: 256 (100), 241 (40), 239 (37), 240 (24)

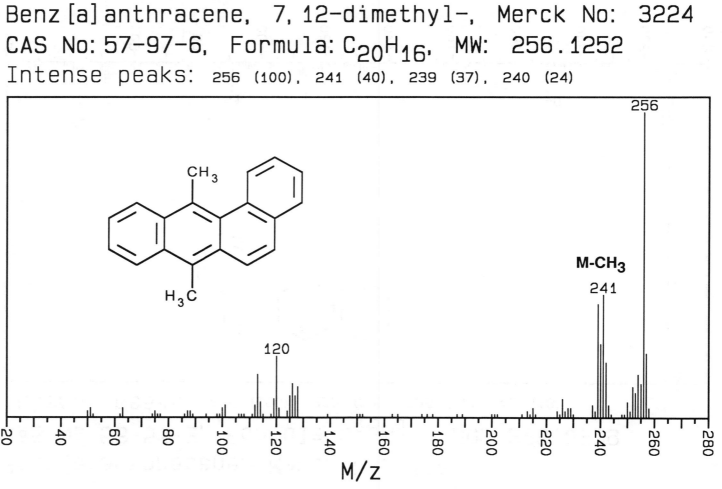

$CH_3$
$H_3C$
256
M-$CH_3$
241
120
20 40 60 80 100 120 140 160 180 200 220 240 260 280
M/Z

Chrysene, Merck No: 2259

CAS No: 218-01-9, Formula: $C_{18}H_{12}$, MW: 228.0939

Intense peaks: 228 (100), 226 (21), 229 (20), 114 (15)

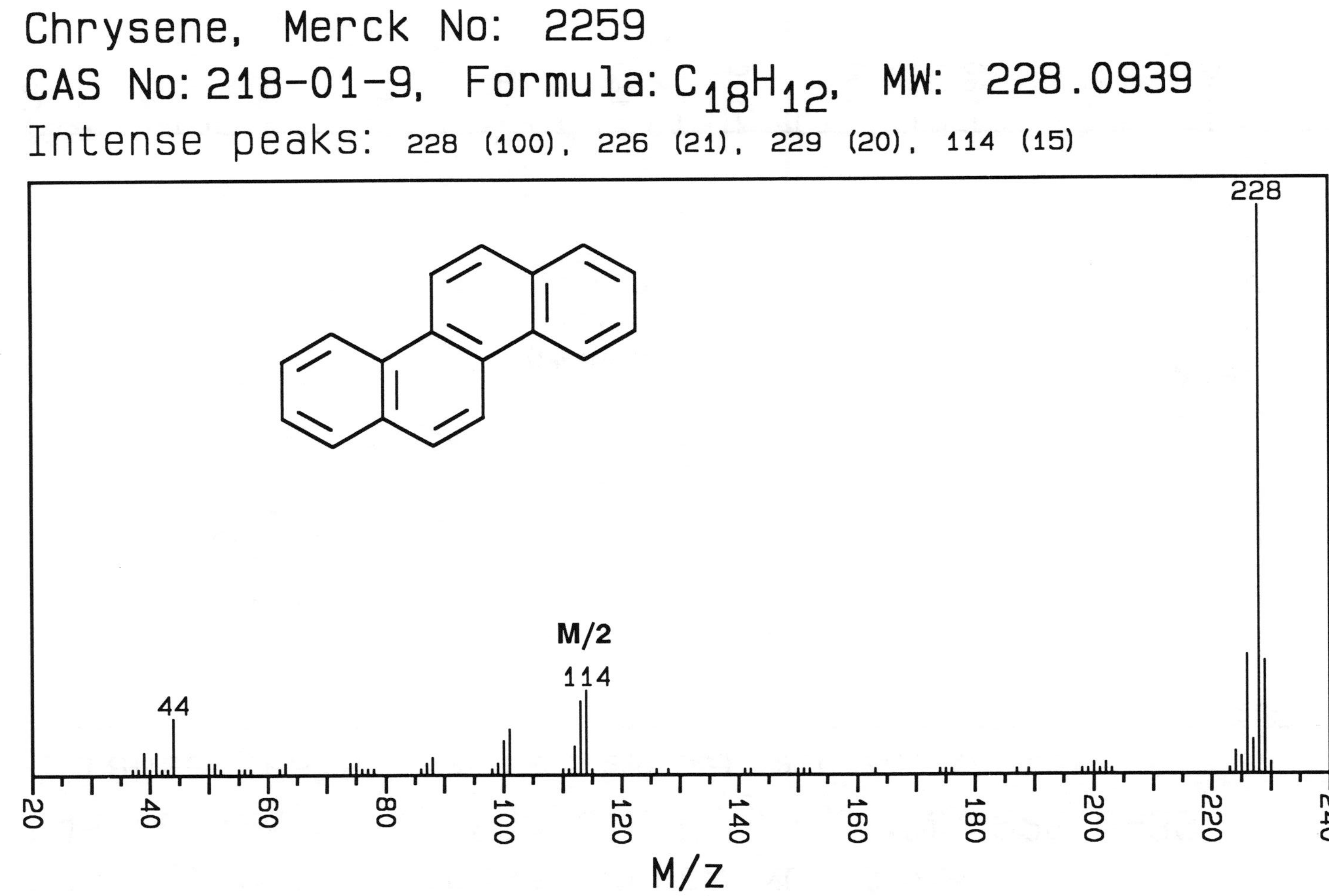

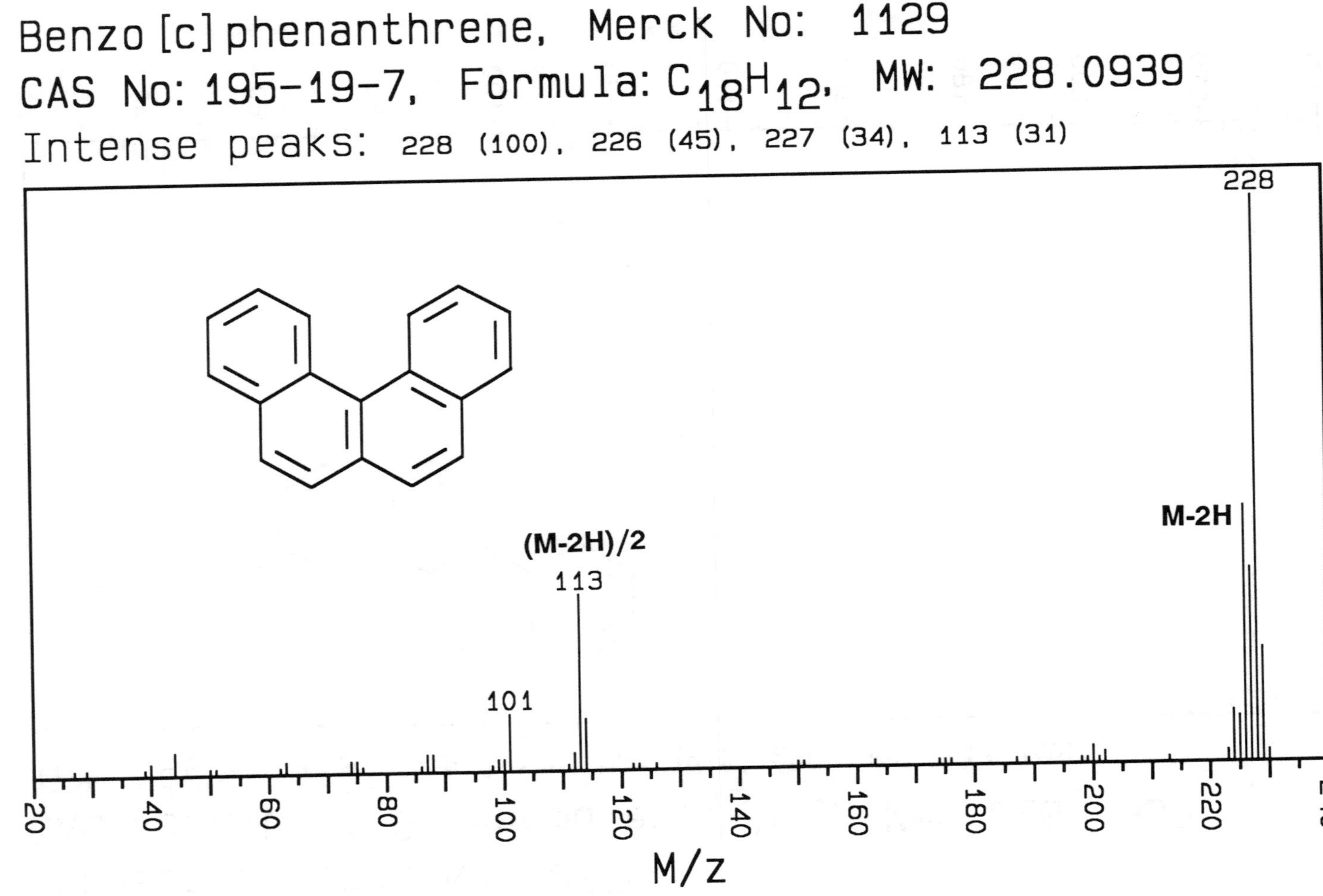

Benzo[c]phenanthrene, Merck No: 1129
CAS No: 195-19-7, Formula: C18H12, MW: 228.0939
Intense peaks: 228 (100), 226 (45), 227 (34), 113 (31)
228
M-2H
(M-2H)/2
113
101
20
40
60
80
100
120
140
160
180
200
220
240
M/Z

Benzo[j]fluoranthene

CAS No: 205-82-3, Formula: $C_{20}H_{12}$, MW: 252.0939

Intense peaks: 252 (100), 253 (23), 126 (22), 250 (21)

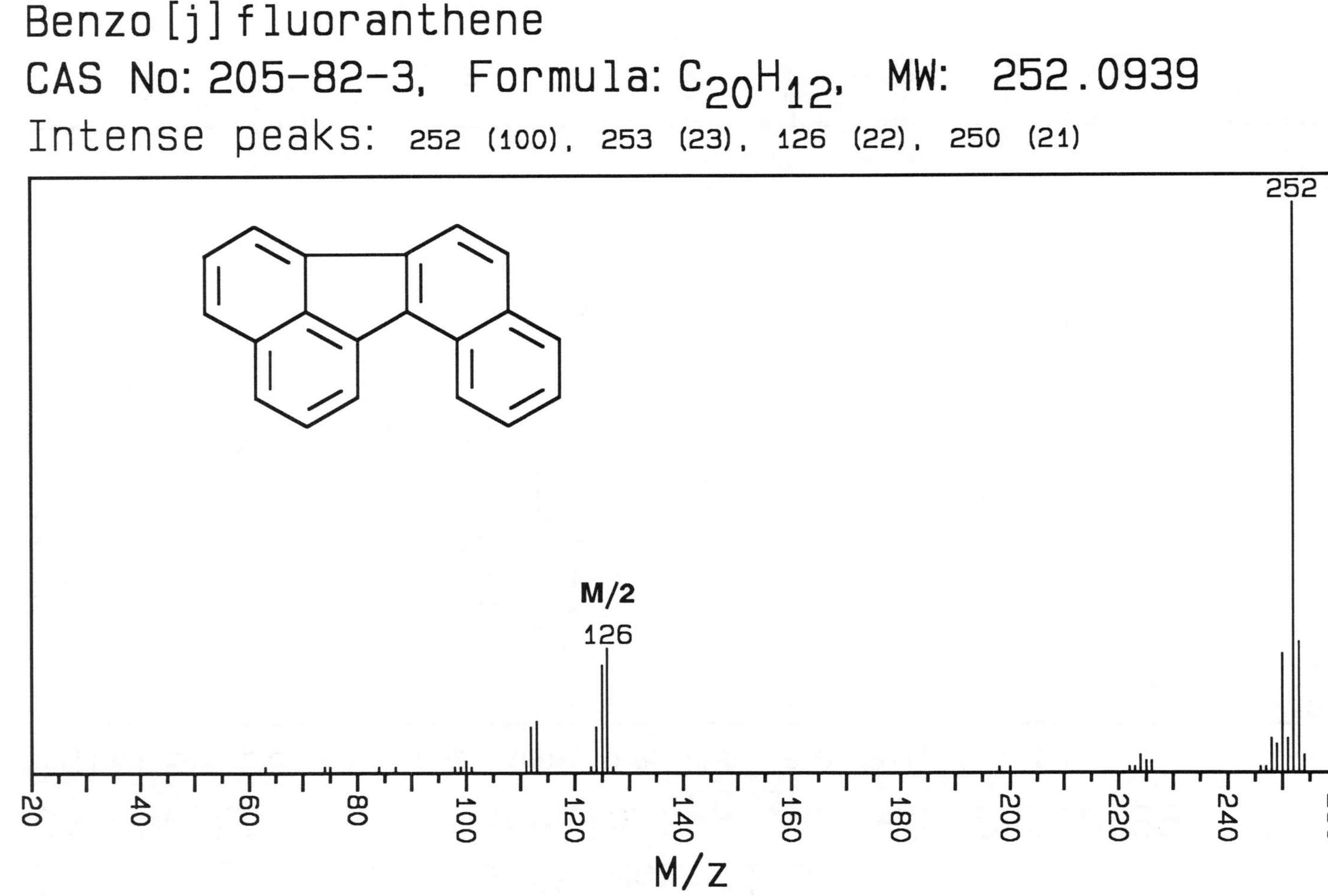

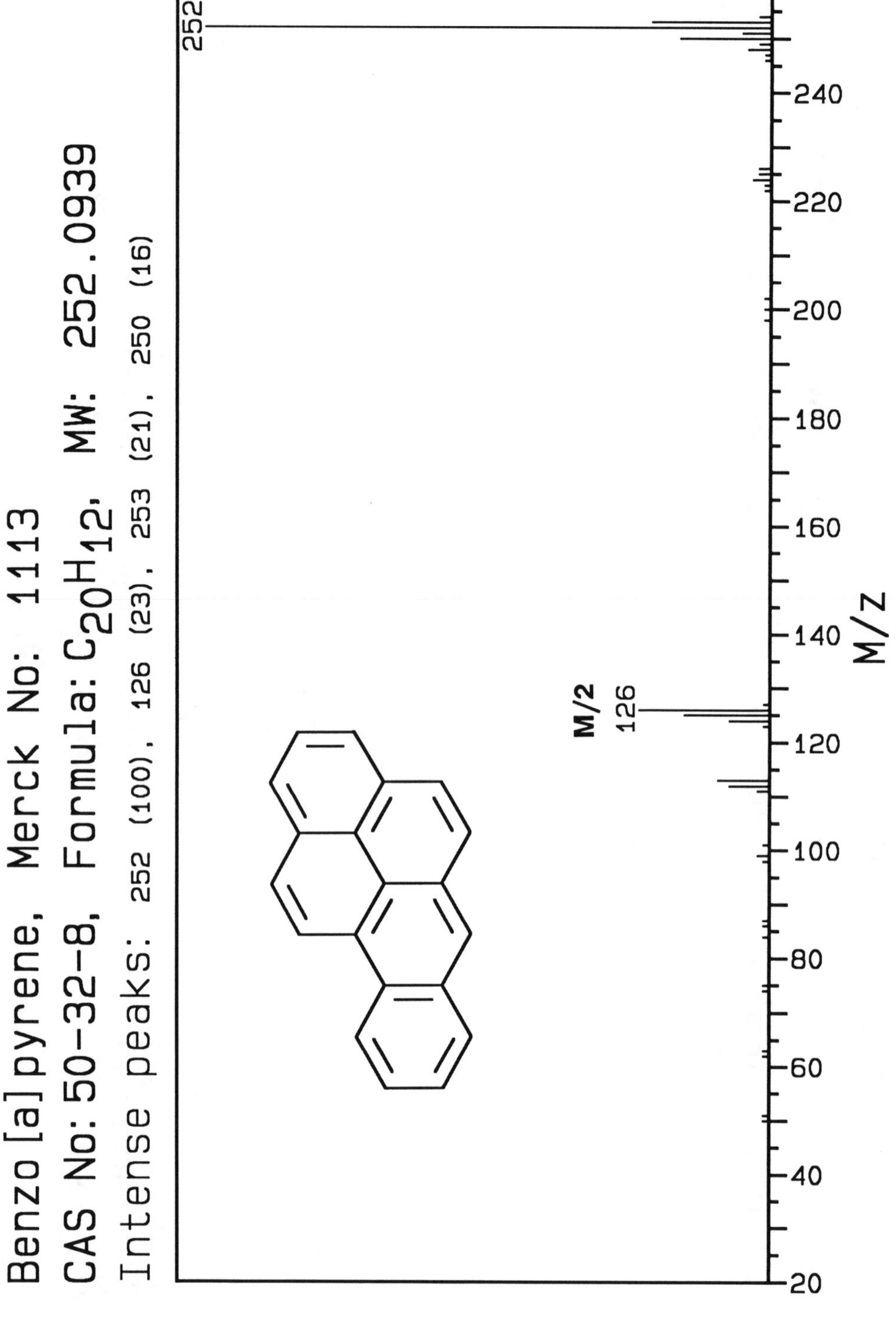
Benzo[a]pyrene, Merck No: 1113
CAS No: 50-32-8, Formula: $C_{20}H_{12}$, MW: 252.0939
Intense peaks: 252 (100), 126 (23), 253 (21), 250 (16)
252
M/2
126
20
40
60
80
100
120
140
160
180
200
220
240
260
M/Z

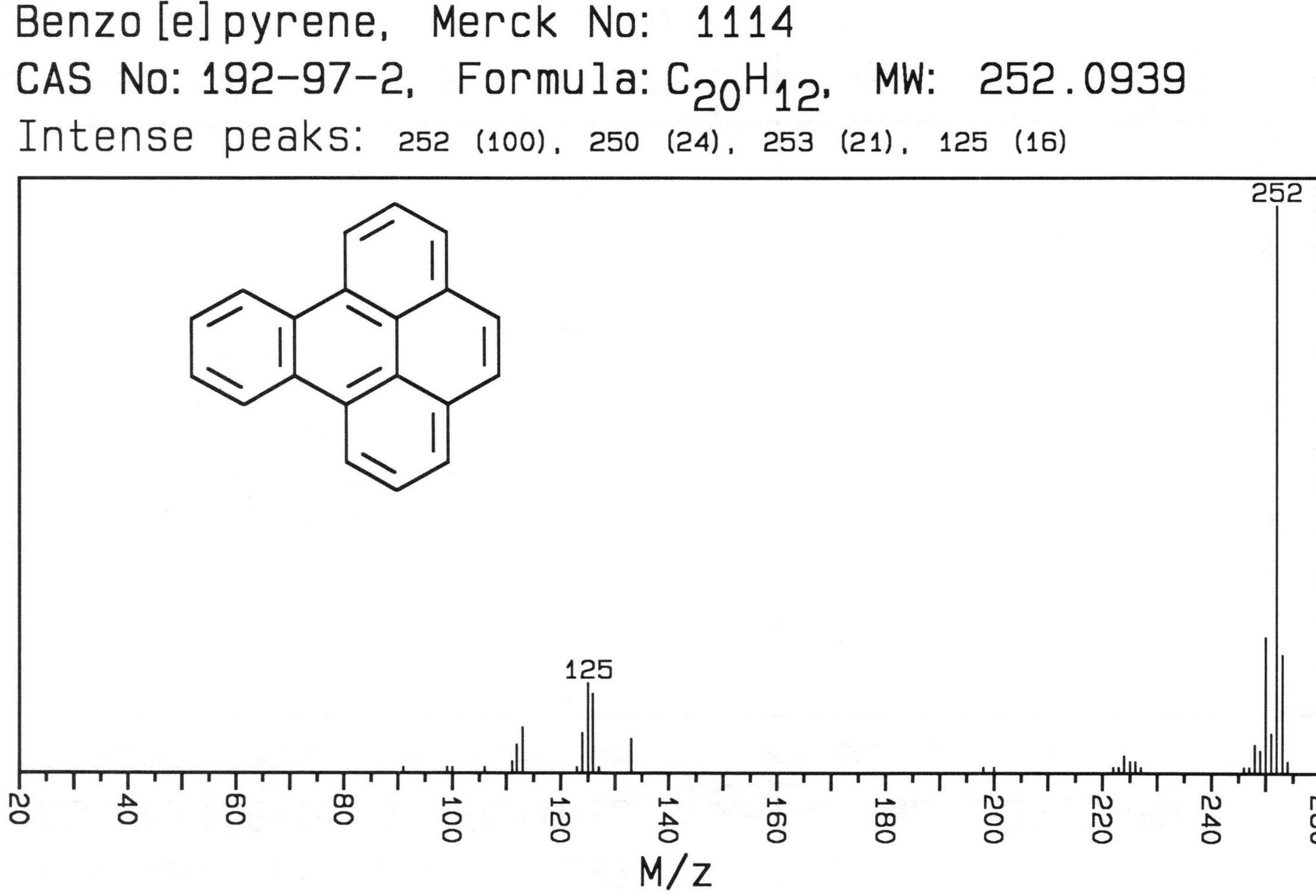
Benzo[e]pyrene, Merck No: 1114
CAS No: 192-97-2, Formula: $C_{20}H_{12}$, MW: 252.0939
Intense peaks: 252 (100), 250 (24), 253 (21), 125 (16)
252
125
20
40
60
80
100
120
140
160
180
200
220
240
260
M/Z

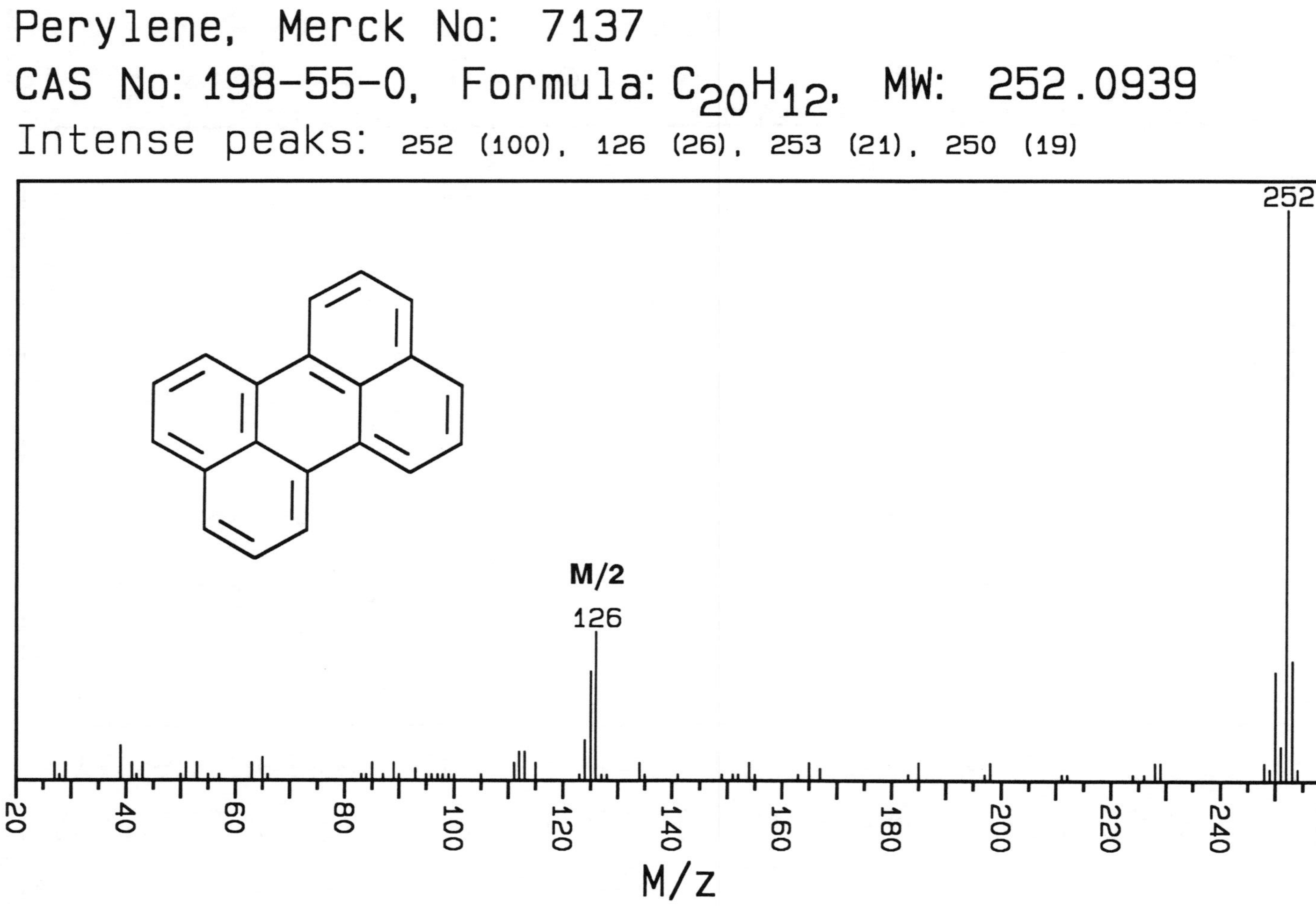

Perylene, Merck No: 7137
CAS No: 198-55-0, Formula: $C_{20}H_{12}$, MW: 252.0939
Intense peaks: 252 (100), 126 (26), 253 (21), 250 (19)
252
M/2
126
20
40
60
80
100
120
140
160
180
200
220
240
260
M/Z

Anthanthrene

CAS No: 191-26-4, Formula: $C_{22}H_{12}$, MW: 276.0939

Intense peaks: 276 (100), 138 (24), 277 (23), 274 (16)

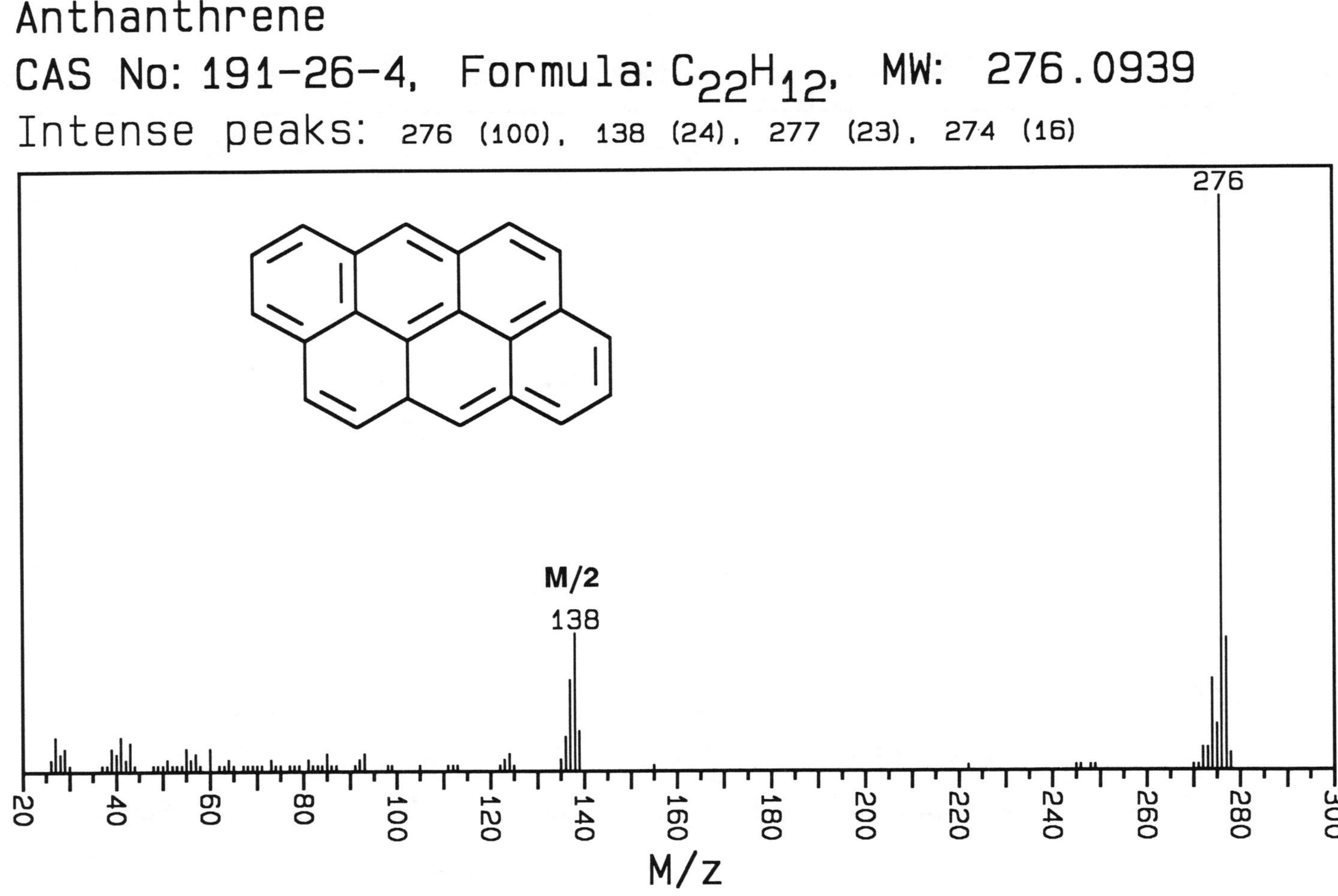

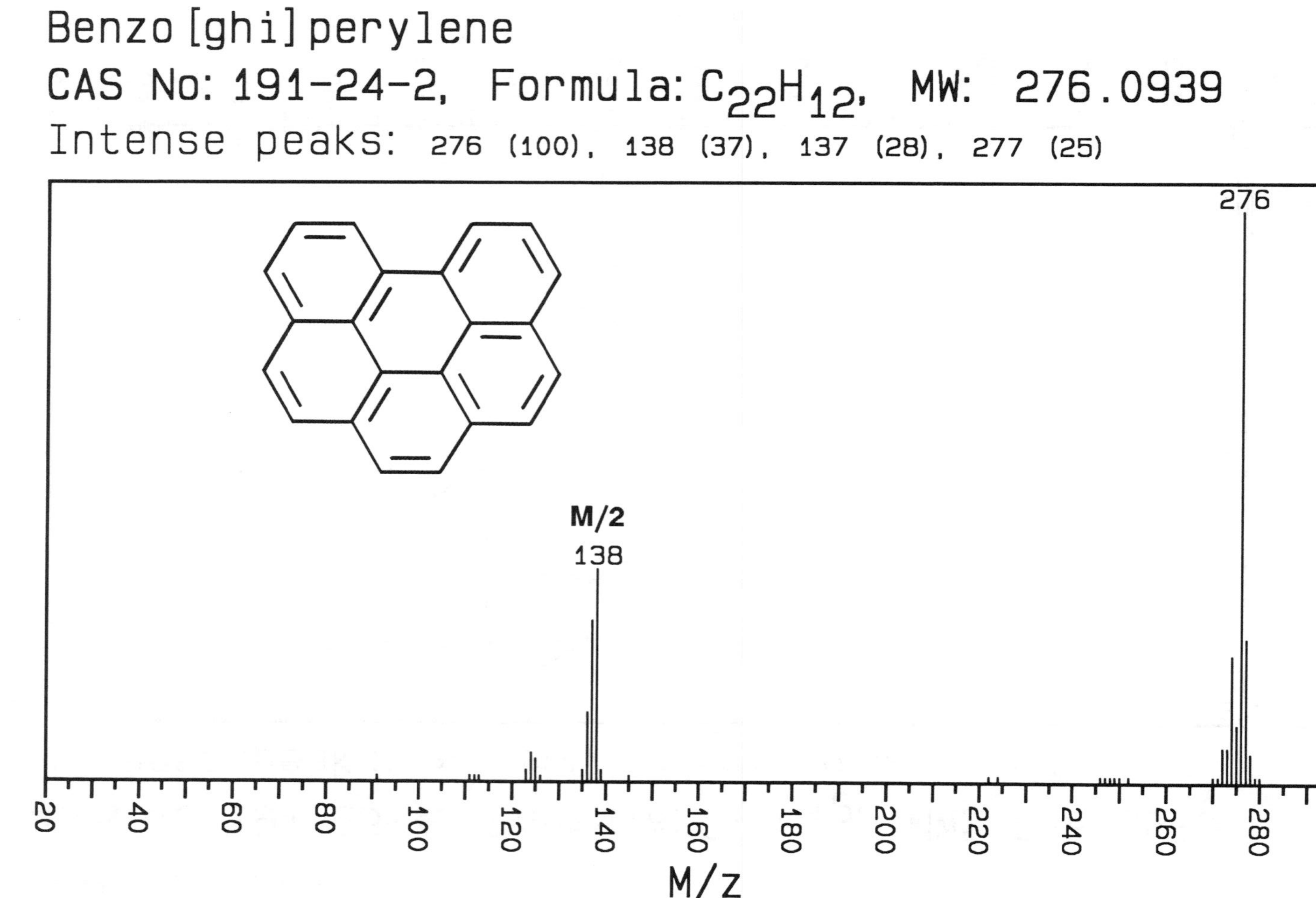
Benzo[ghi]perylene
CAS No: 191-24-2, Formula: $C_{22}H_{12}$, MW: 276.0939
Intense peaks: 276 (100), 138 (37), 137 (28), 277 (25)
276
M/2
138
20
40
60
80
100
120
140
160
180
200
220
240
260
280
300
M/Z

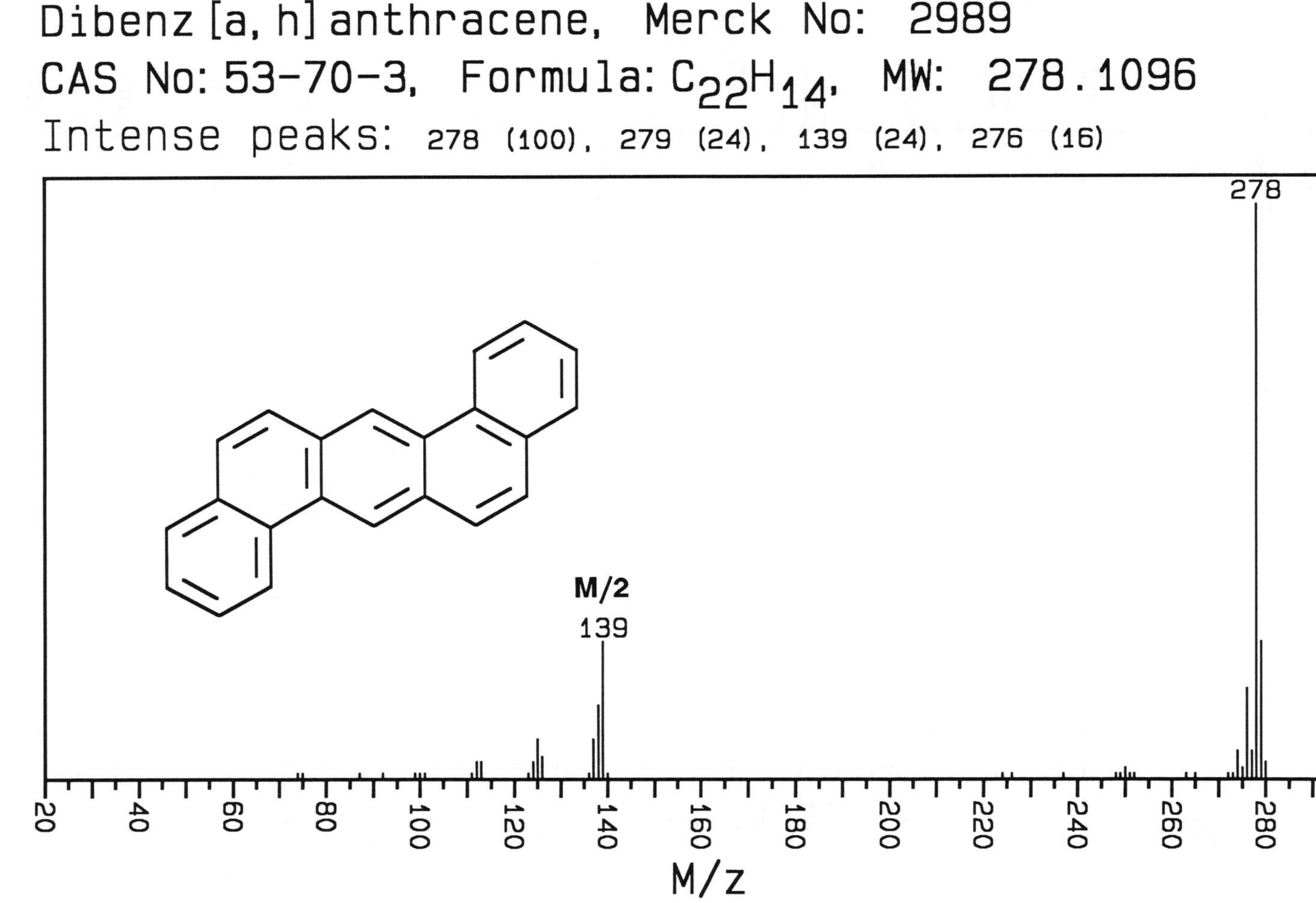
Dibenz[a,h]anthracene, Merck No: 2989
CAS No: 53-70-3, Formula: $C_{22}H_{14}$, MW: 278.1096
Intense peaks: 278 (100), 279 (24), 139 (24), 276 (16)
278
M/2
139
20
40
60
80
100
120
140
160
180
200
220
240
260
280
300
M/z

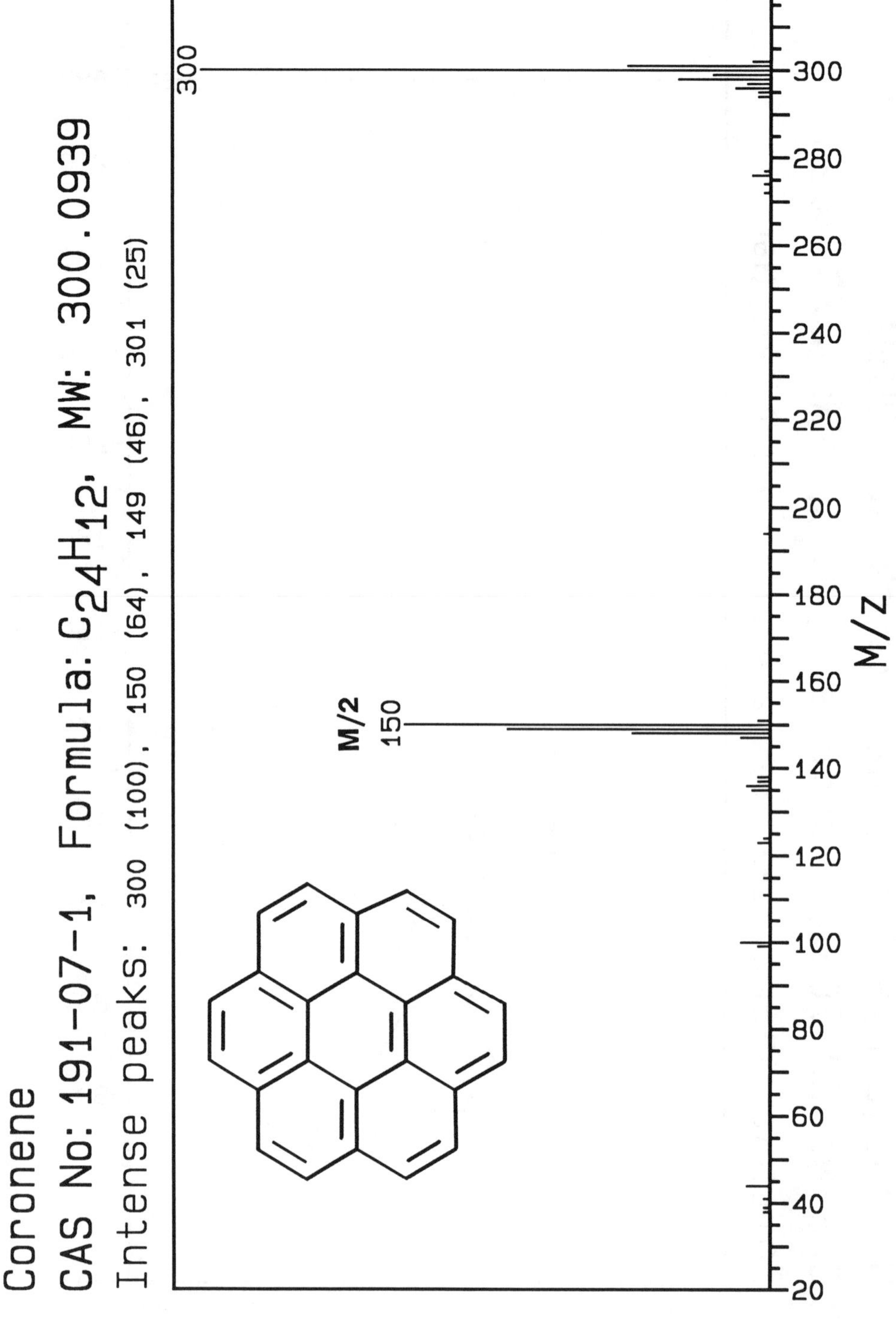
Coronene
CAS No: 191-07-1, Formula: $C_{24}H_{12}$, MW: 300.0939
Intense peaks: 300 (100), 150 (64), 149 (46), 301 (25)
300
M/2
150
M/Z
20
40
60
80
100
120
140
160
180
200
220
240
260
280
300
320

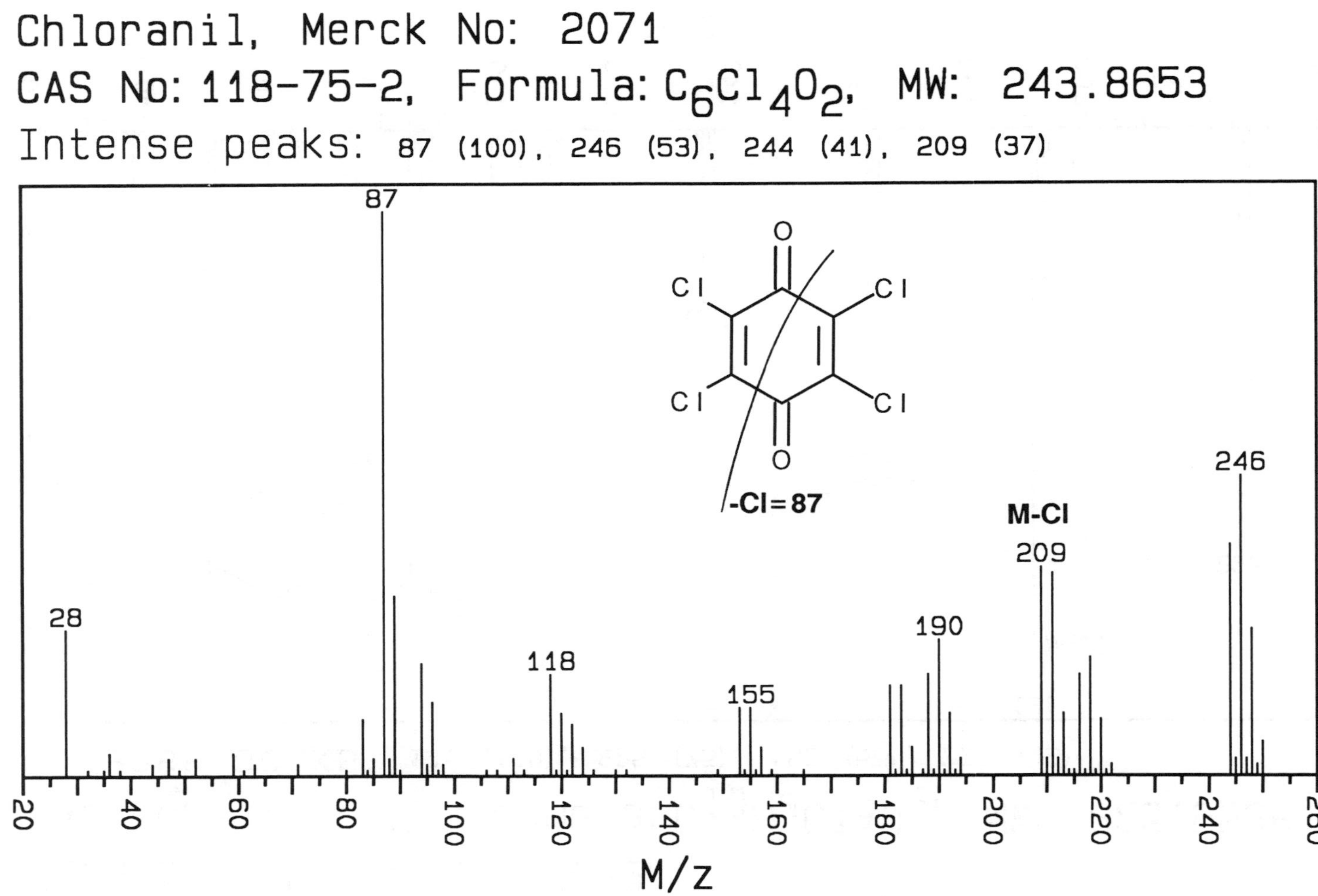
Chloranil, Merck No: 2071
CAS No: 118-75-2, Formula: $C_6Cl_4O_2$, MW: 243.8653
Intense peaks: 87 (100), 246 (53), 244 (41), 209 (37)
87
28
118
155
190
M-Cl
209
246
-Cl=87
O
Cl
Cl
Cl
Cl
O
20
40
60
80
100
120
140
160
180
200
220
240
260
M/Z

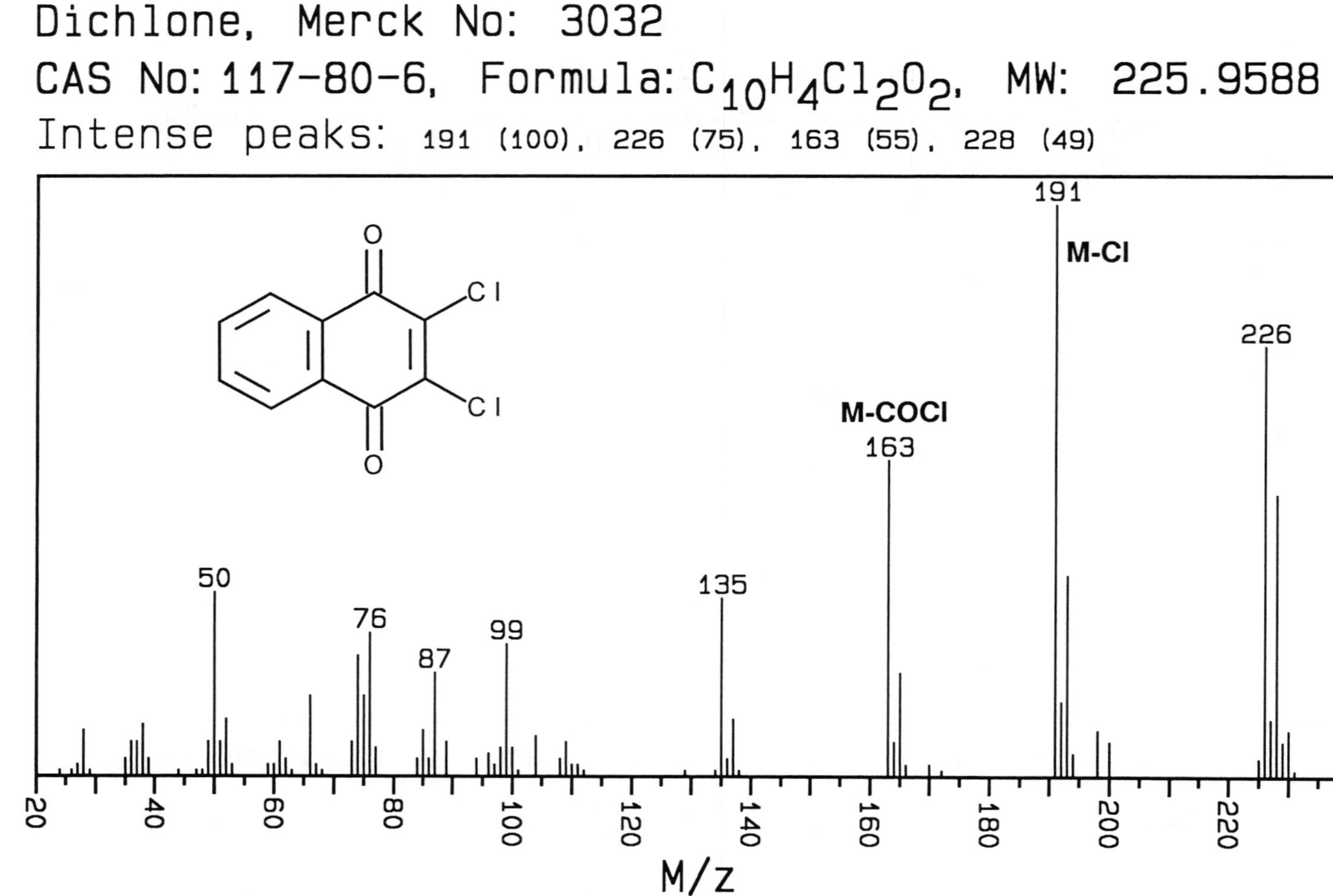
Dichlone, Merck No: 3032
CAS No: 117-80-6, Formula: $C_{10}H_4Cl_2O_2$, MW: 225.9588
Intense peaks: 191 (100), 226 (75), 163 (55), 228 (49)
O
Cl
Cl
O
191
M-Cl
226
M-COCl
163
50
135
76
99
87
20
40
60
80
100
120
140
160
180
200
220
240
M/Z

Juglone, Merck No: 5150

CAS No: 481-39-0, Formula: $C_{10}H_6O_3$, MW: 174.0317

Intense peaks: 174 (100), 118 (46), 120 (33), 92 (32)

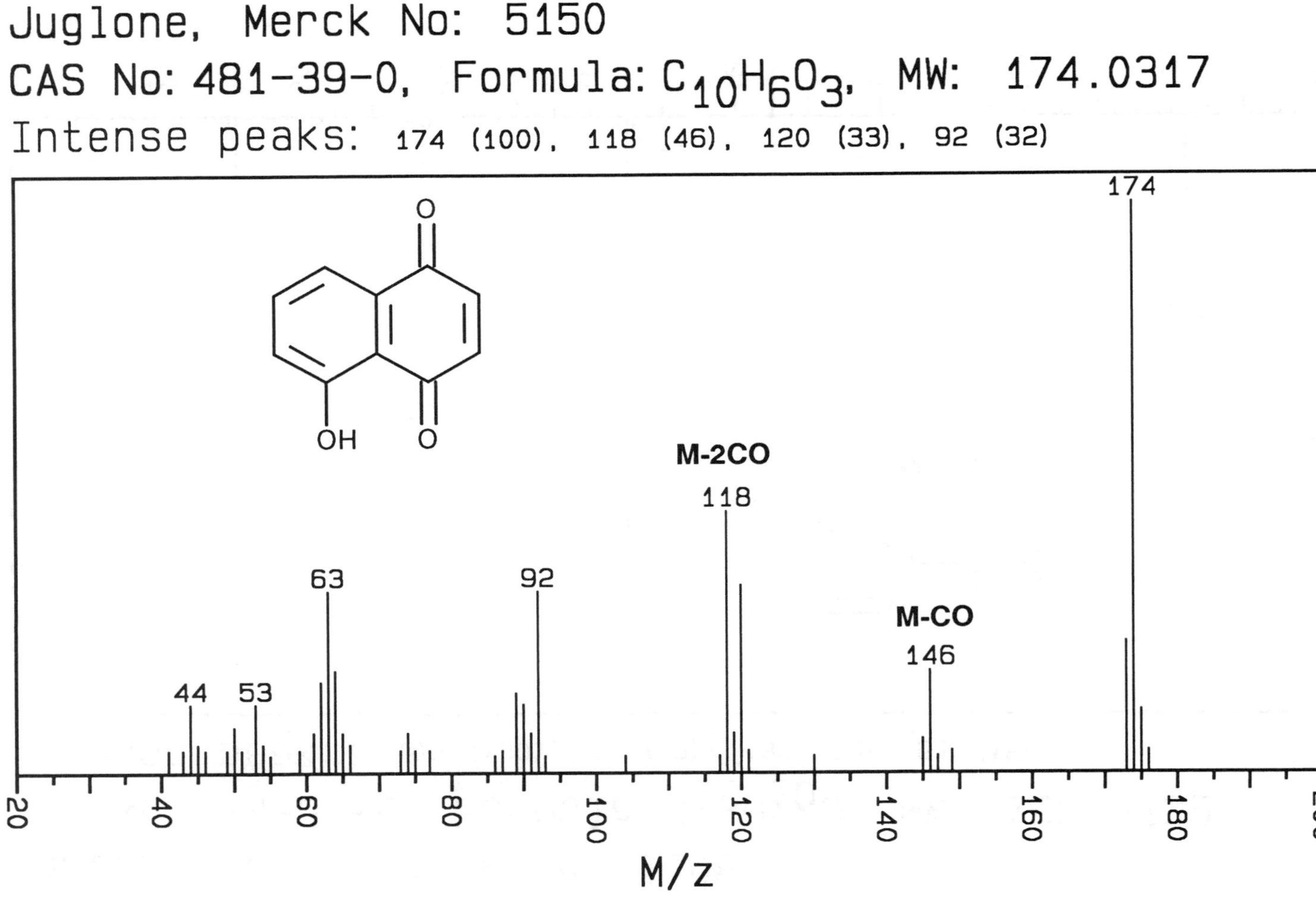

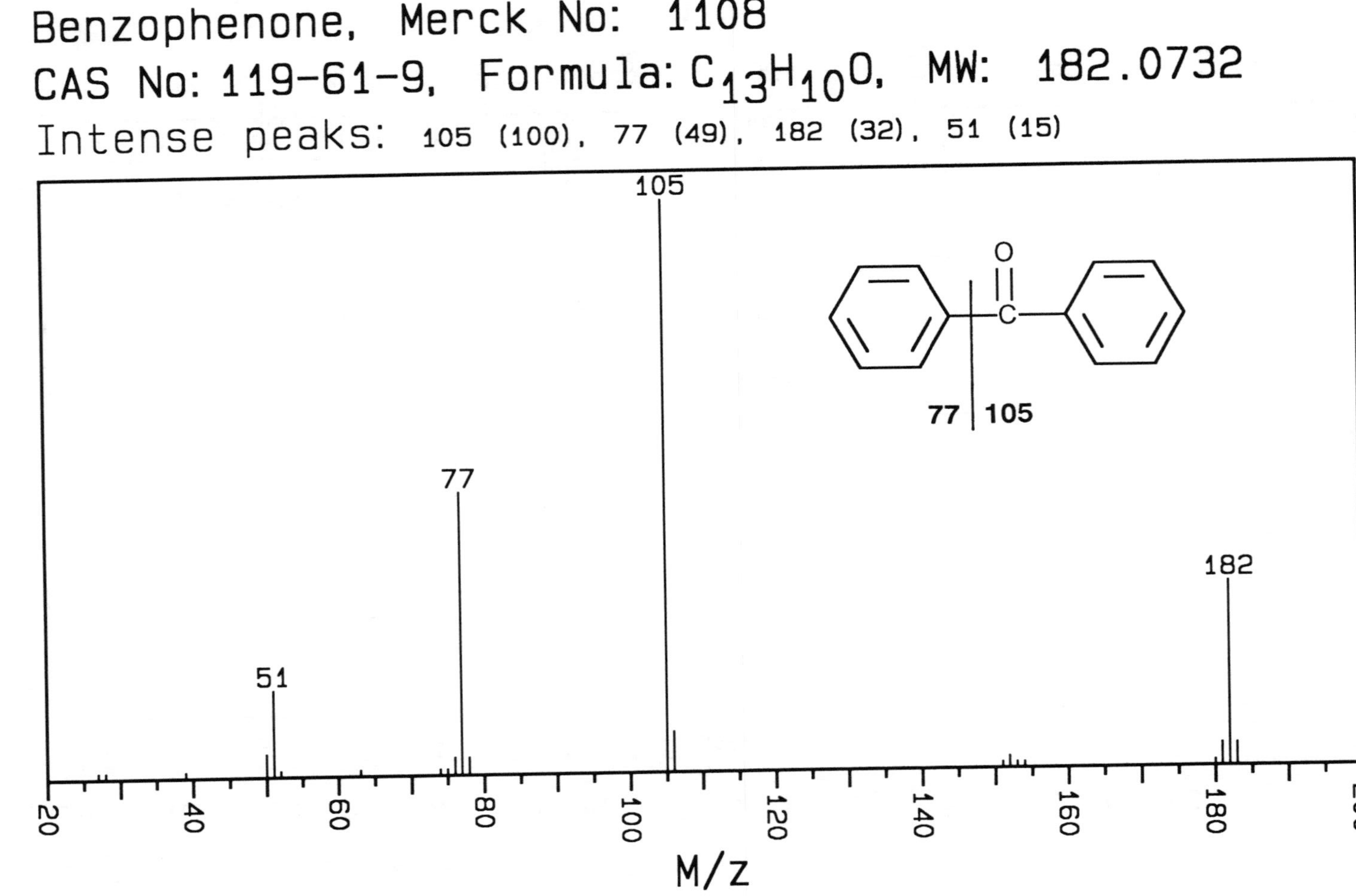
Benzophenone, Merck No: 1108
CAS No: 119-61-9, Formula: $C_{13}H_{10}O$, MW: 182.0732
Intense peaks: 105 (100), 77 (49), 182 (32), 51 (15)
105
O
C
77
105
77
182
51
20
40
60
80
100
120
140
160
180
200
M/Z

Diphenylacetic acid, Merck No: 3316

CAS No: 117-34-0, Formula: $C_{14}H_{12}O_2$, MW: 212.0837

Intense peaks: 167 (100), 165 (52), 152 (29), 168 (18)

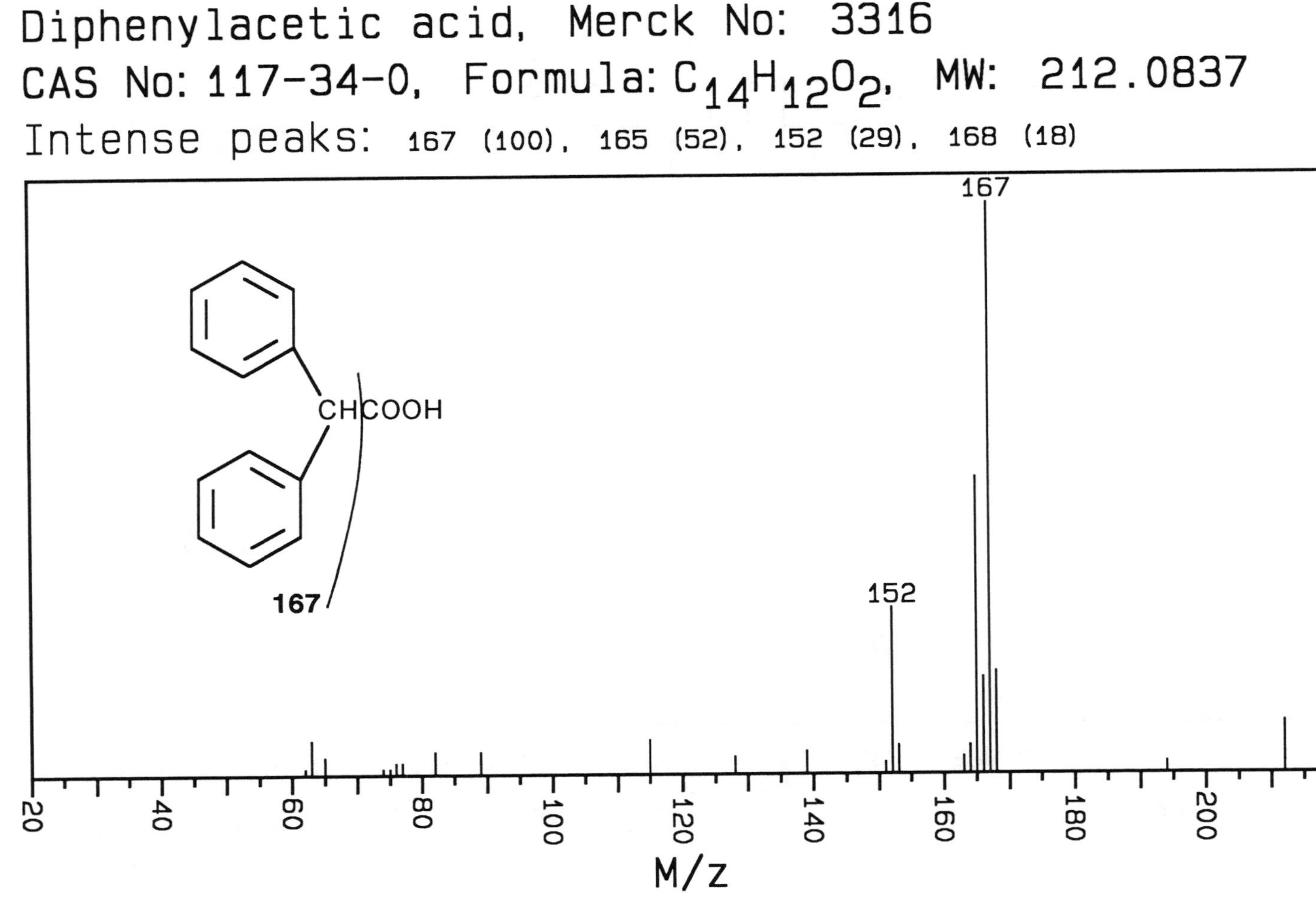

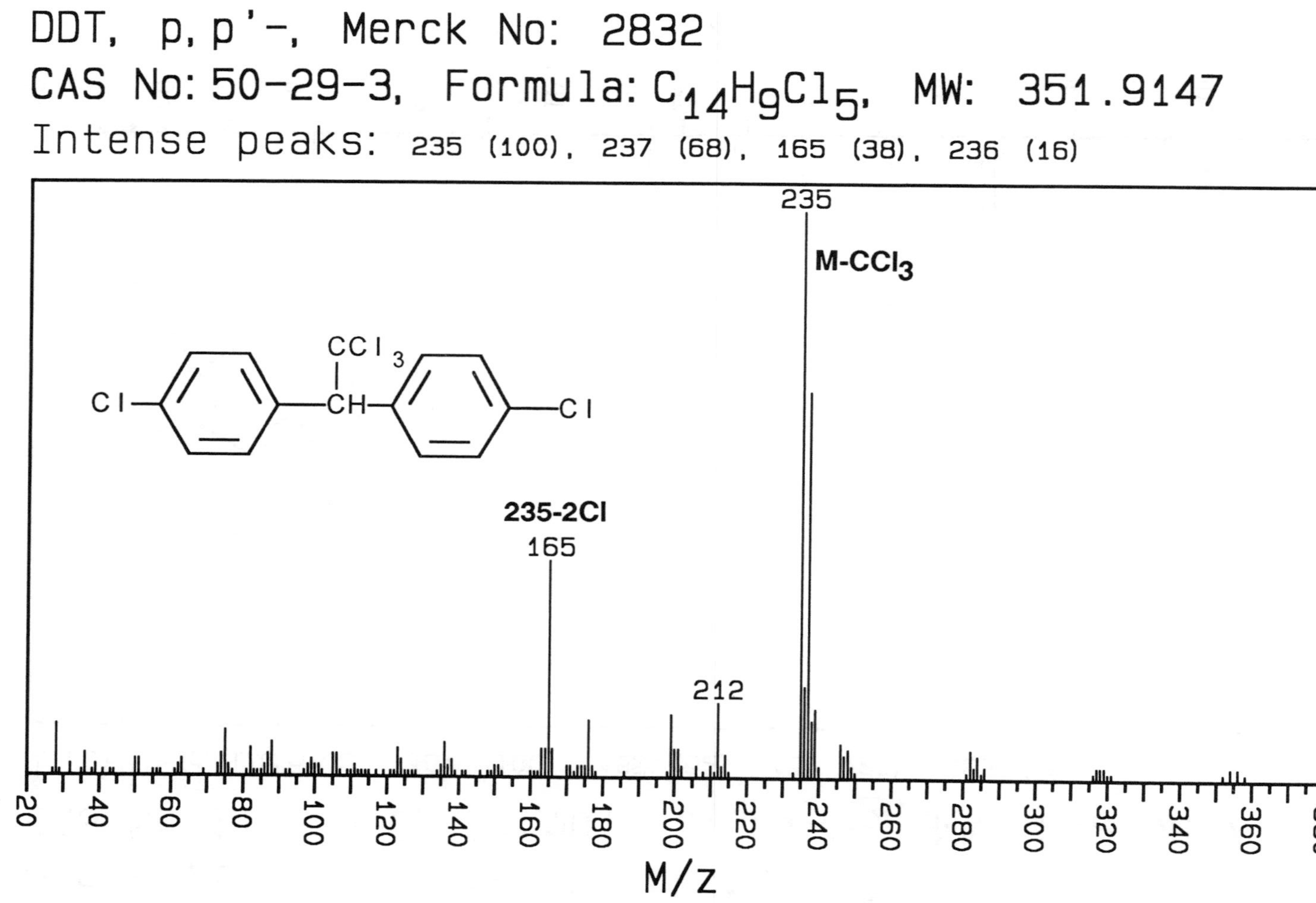
DDT, p,p'-, Merck No: 2832
CAS No: 50-29-3, Formula: C14H9Cl5, MW: 351.9147
Intense peaks: 235 (100), 237 (68), 165 (38), 236 (16)
235
M-CCl3
CCl3
Cl
CH
Cl
235-2Cl
165
212
20
40
60
80
100
120
140
160
180
200
220
240
260
280
300
320
340
360
380
M/Z

DDE, p,p'-

CAS No: 72-55-9, Formula: $C_{14}H_8Cl_4$, MW: 315.9381

Intense peaks: 246 (100), 318 (79), 248 (66), 316 (61)

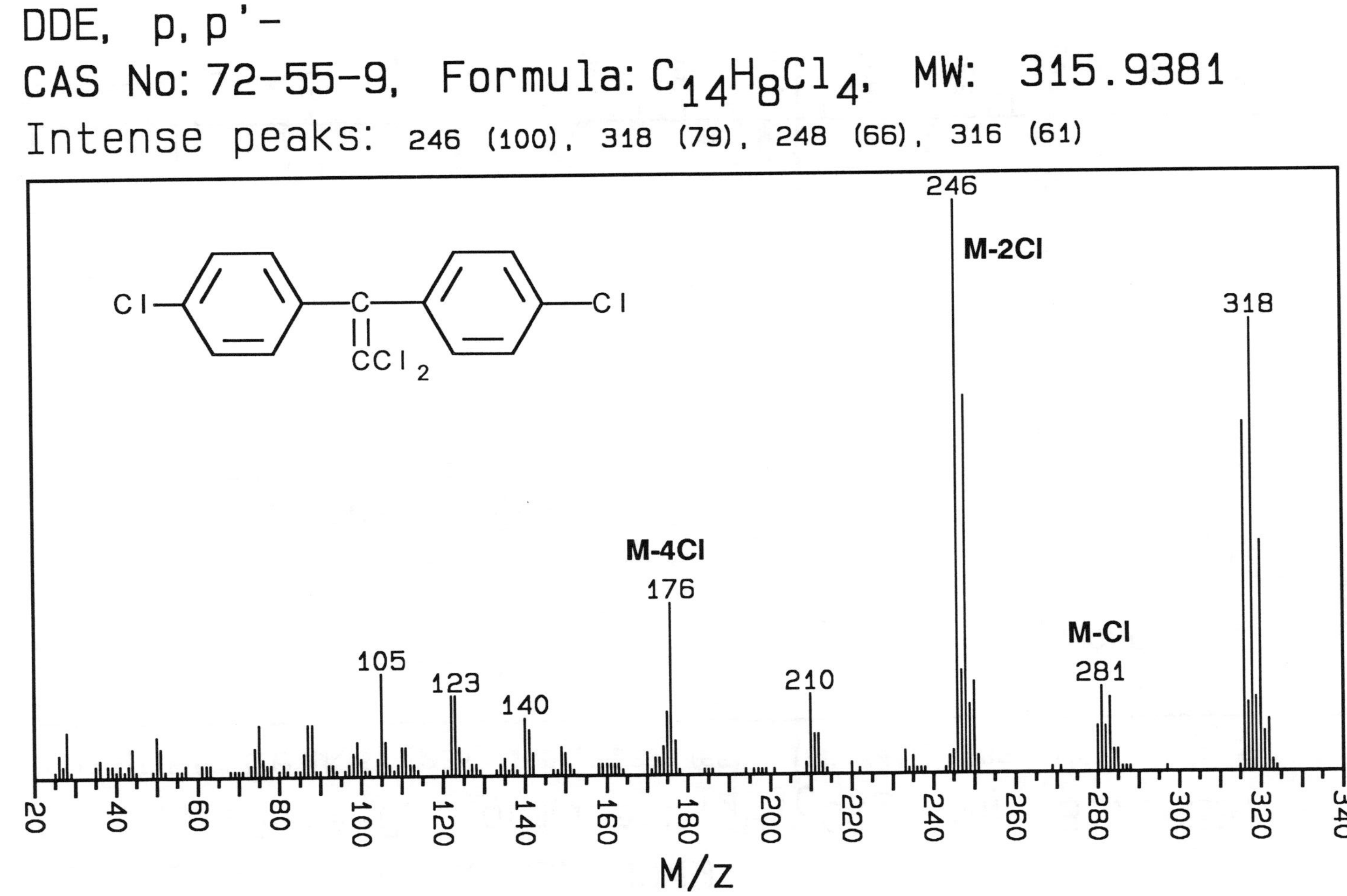

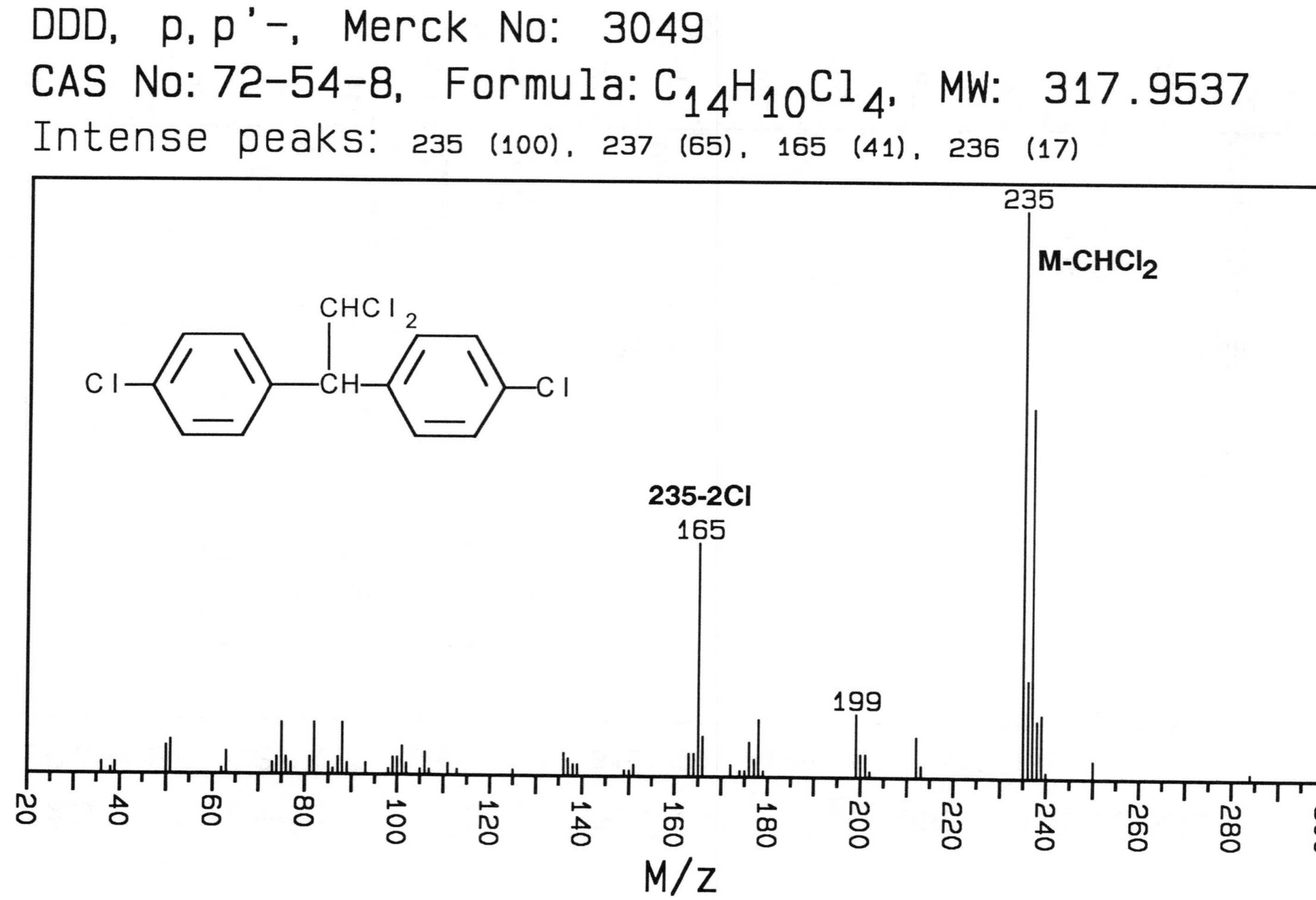
DDD, p,p'-, Merck No: 3049
CAS No: 72-54-8, Formula: $C_{14}H_{10}Cl_4$, MW: 317.9537
Intense peaks: 235 (100), 237 (65), 165 (41), 236 (17)
235
M-CHCl2
CHCl2
Cl
CH
Cl
235-2Cl
165
199
20
40
60
80
100
120
140
160
180
200
220
240
260
280
300
M/Z

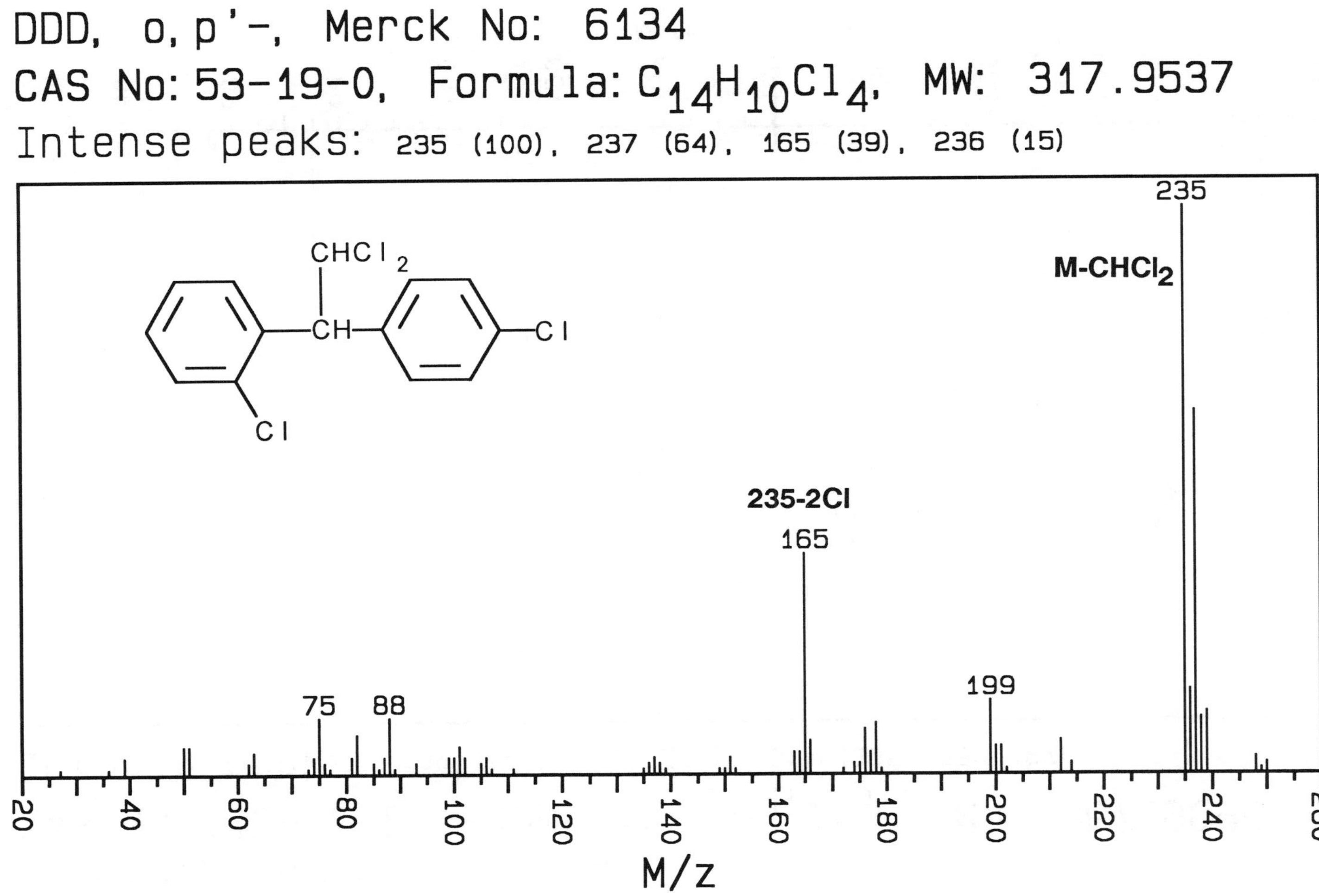
DDD, o,p'-, Merck No: 6134
CAS No: 53-19-0, Formula: C14H10Cl4, MW: 317.9537
Intense peaks: 235 (100), 237 (64), 165 (39), 236 (15)
CHCl2
CH
Cl
Cl
235
M-CHCl2
235-2Cl
165
199
75
88
20
40
60
80
100
120
140
160
180
200
220
240
260
M/Z

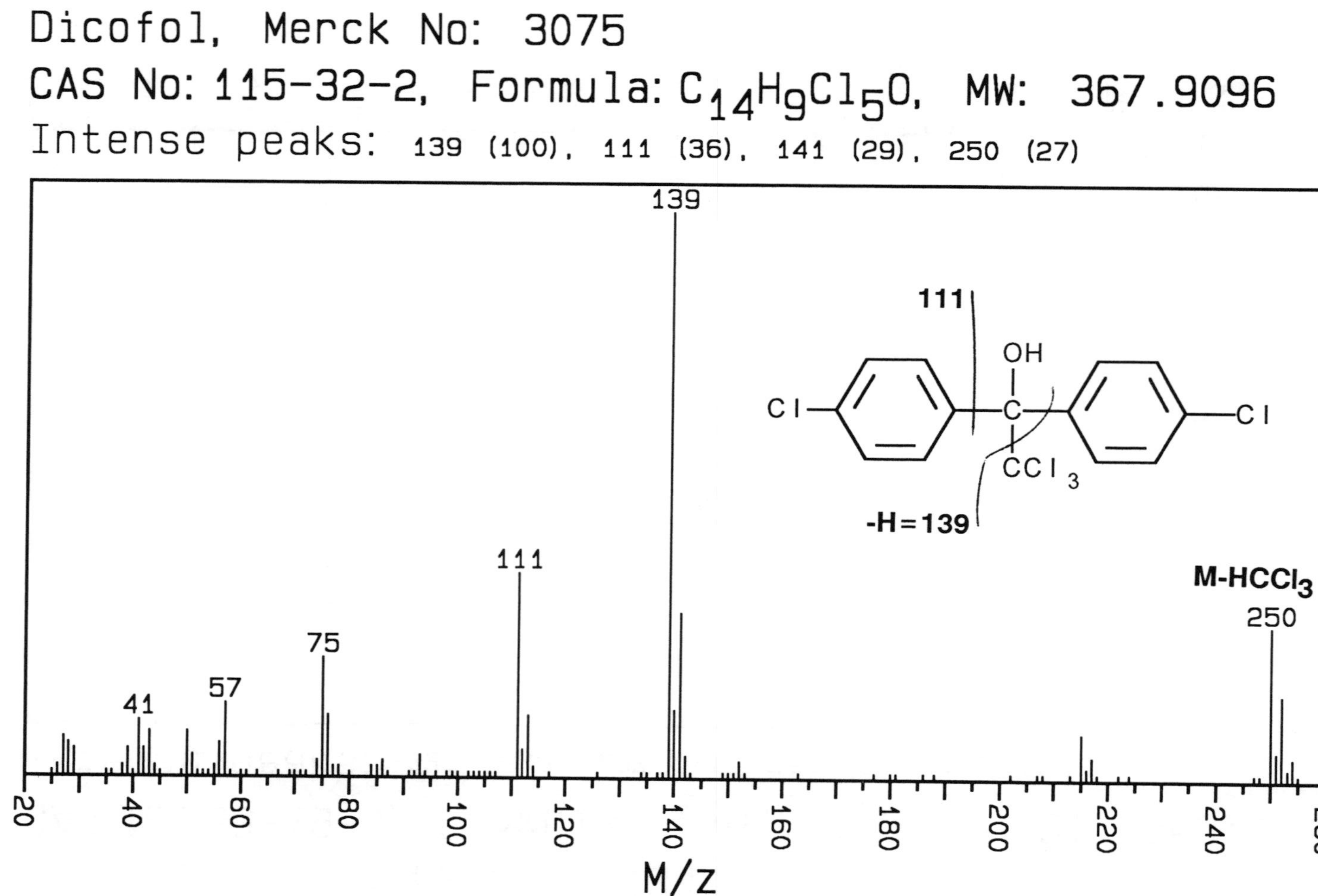

Dicofol, Merck No: 3075
CAS No: 115-32-2, Formula: C14H9Cl5O, MW: 367.9096
Intense peaks: 139 (100), 111 (36), 141 (29), 250 (27)
139
111
OH
Cl
C
Cl
CCl3
-H=139
111
75
57
41
M-HCCl3
250
20
40
60
80
100
120
140
160
180
200
220
240
260
M/Z

Chlorfenethol, Merck No: 2086

CAS No: 80-06-8, Formula: $C_{14}H_{12}Cl_2O$, MW: 266.0265

Intense peaks: 251 (100), 139 (88), 43 (86), 253 (62)

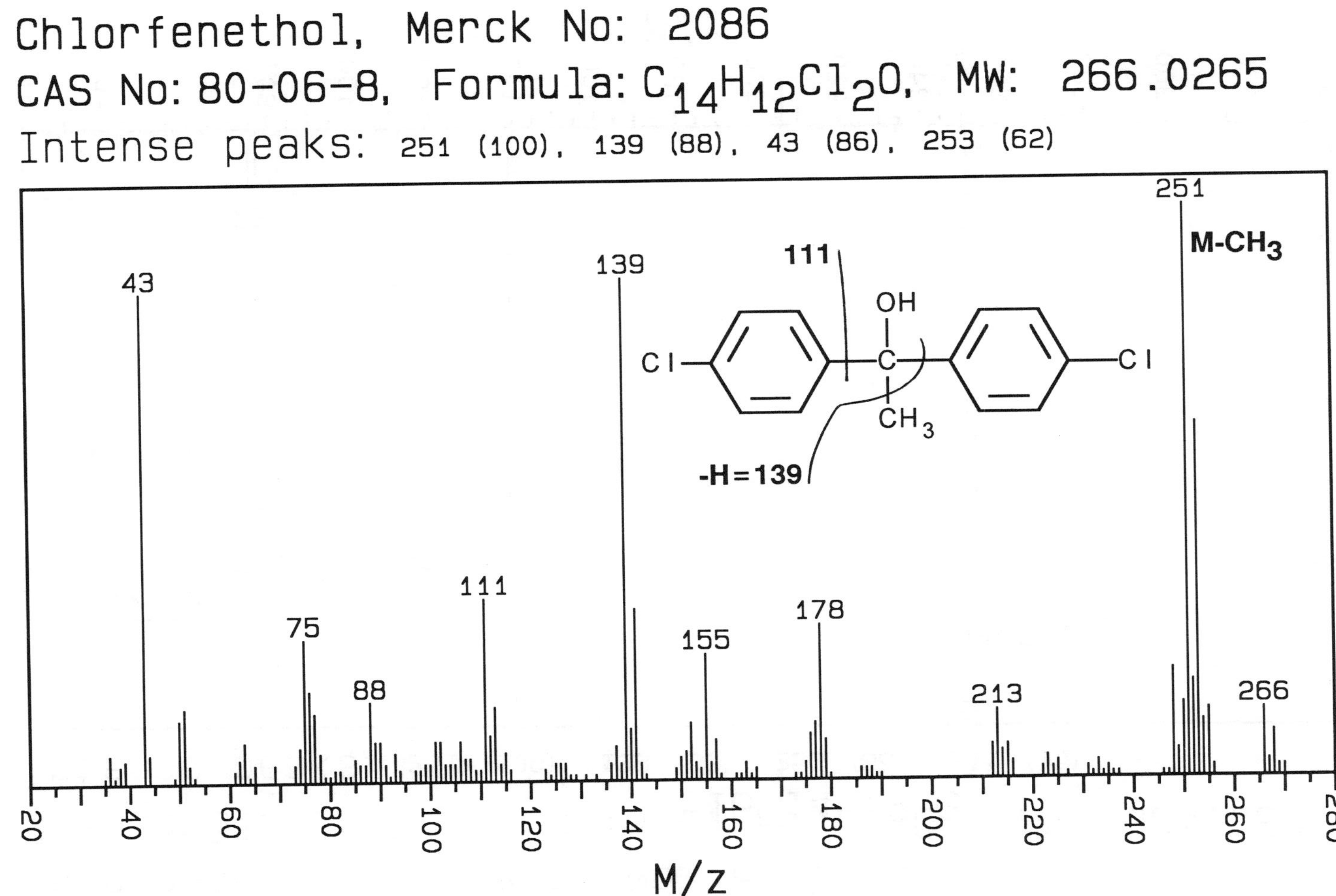

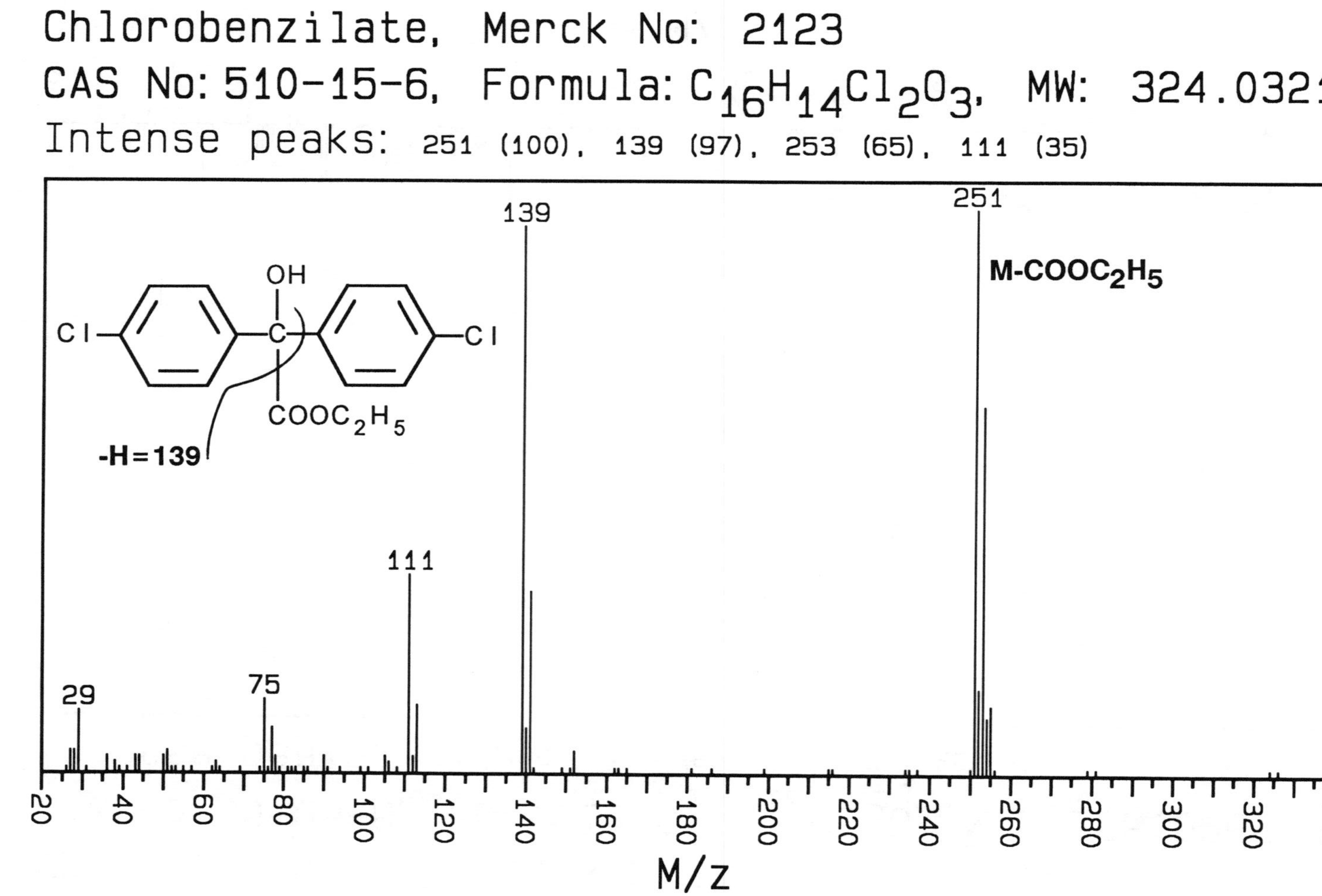
Chlorobenzilate, Merck No: 2123
CAS No: 510-15-6, Formula: $C_{16}H_{14}Cl_2O_3$, MW: 324.0321
Intense peaks: 251 (100), 139 (97), 253 (65), 111 (35)
139
251
M-$COOC_2H_5$
OH
Cl
C
Cl
$COOC_2H_5$
-H=139
111
29
75
20
40
60
80
100
120
140
160
180
200
220
240
260
280
300
320
340
M/Z

Bromopropylate, Merck No: 1422

CAS No: 18181-80-1, Formula: $C_{17}H_{16}Br_2O_3$, MW: 425.9466

Intense peaks: 43 (100), 75 (98), 41 (59), 27 (58)

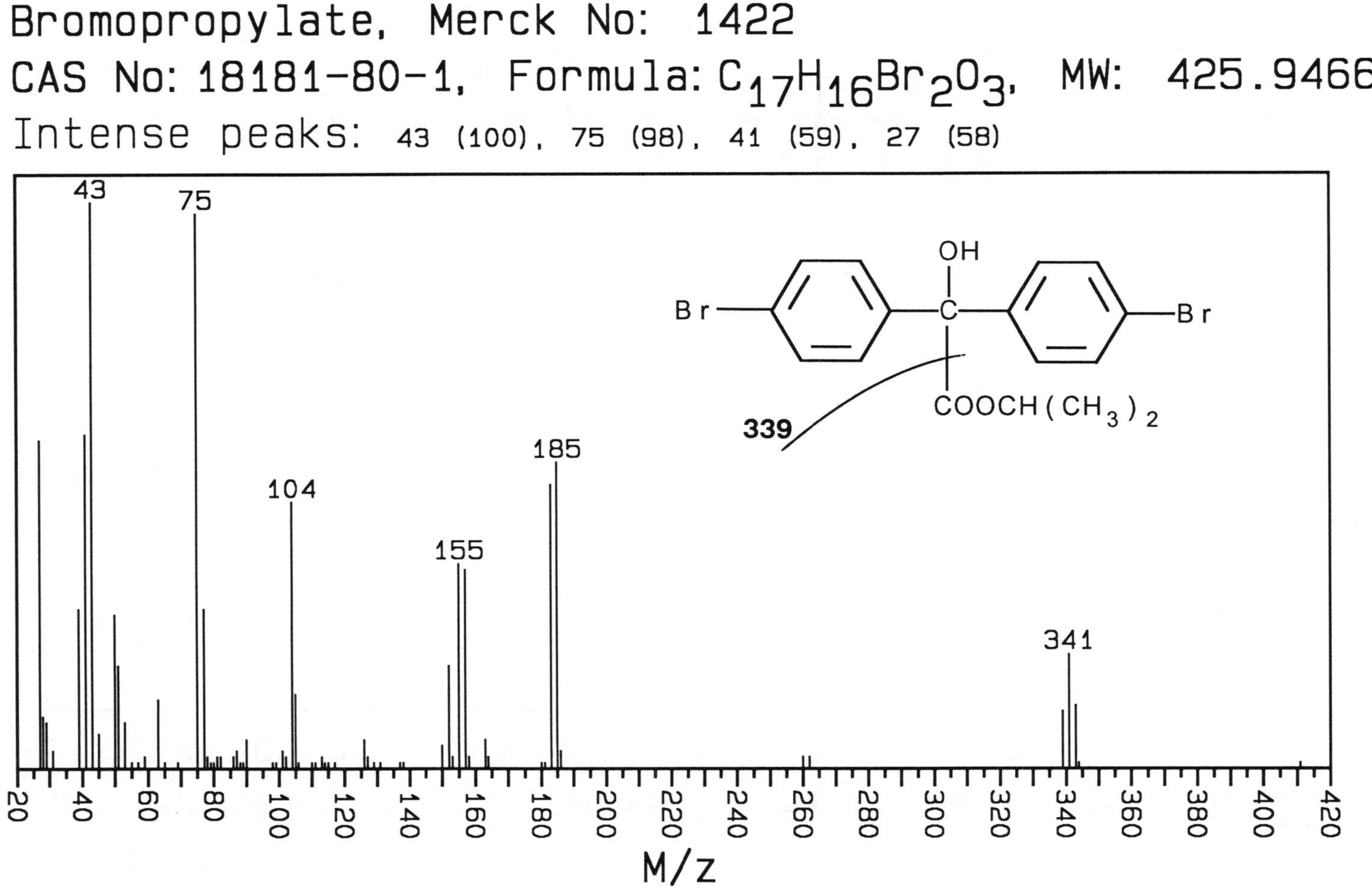

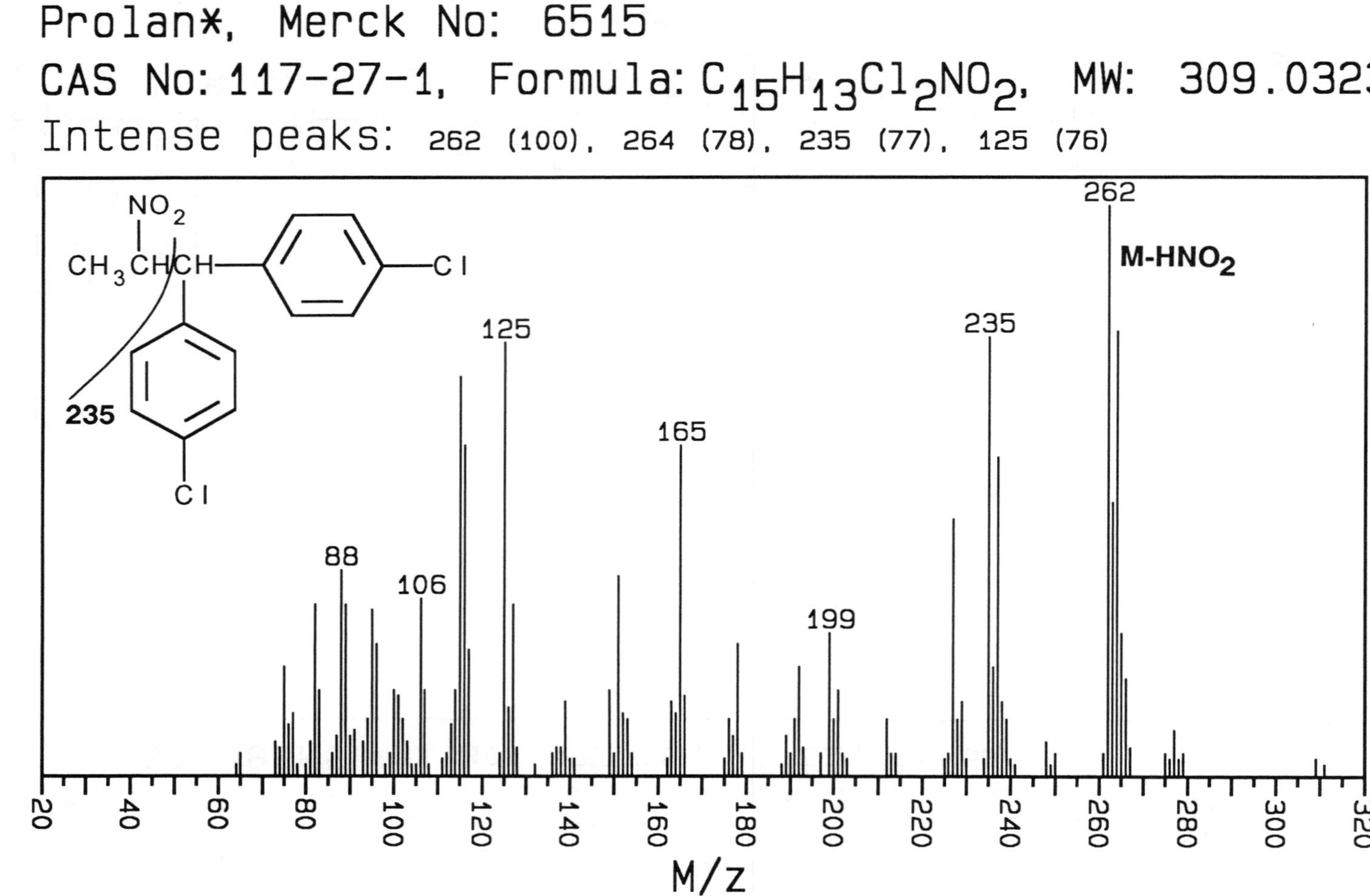

Prolan*, Merck No: 6515
CAS No: 117-27-1, Formula: $C_{15}H_{13}Cl_2NO_2$, MW: 309.0323
Intense peaks: 262 (100), 264 (78), 235 (77), 125 (76)
$NO_2$
$CH_3CHCH$
Cl
Cl
235
262
M-$HNO_2$
235
125
165
199
88
106
20
40
60
80
100
120
140
160
180
200
220
240
260
280
300
320
M/Z

Bulan*, Merck No: 1470

CAS No: 117-26-0, Formula: $C_{16}H_{15}Cl_2NO_2$, MW: 323.0481

Intense peaks: 125 (100), 235 (80), 241 (59), 276 (56)

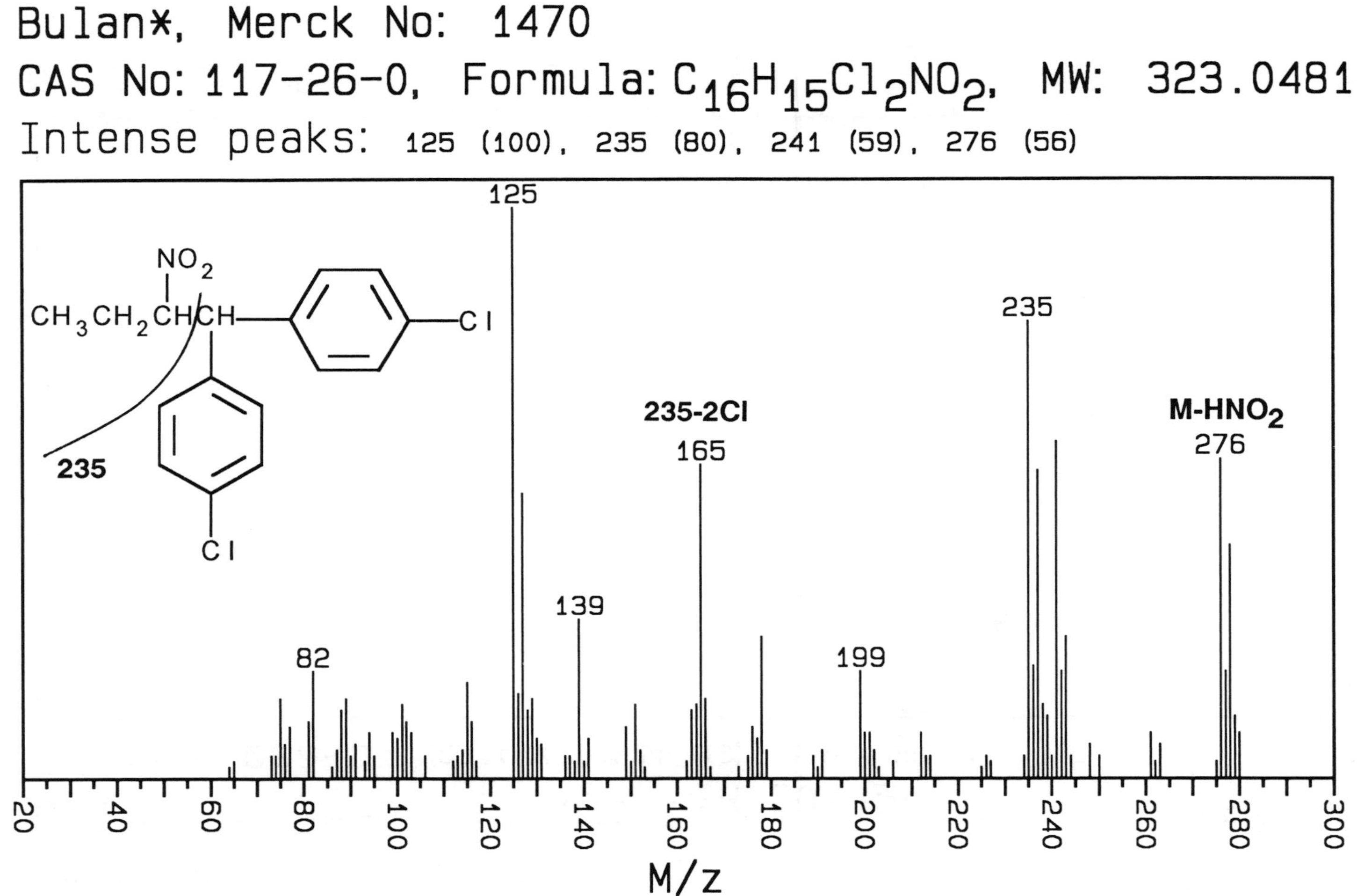

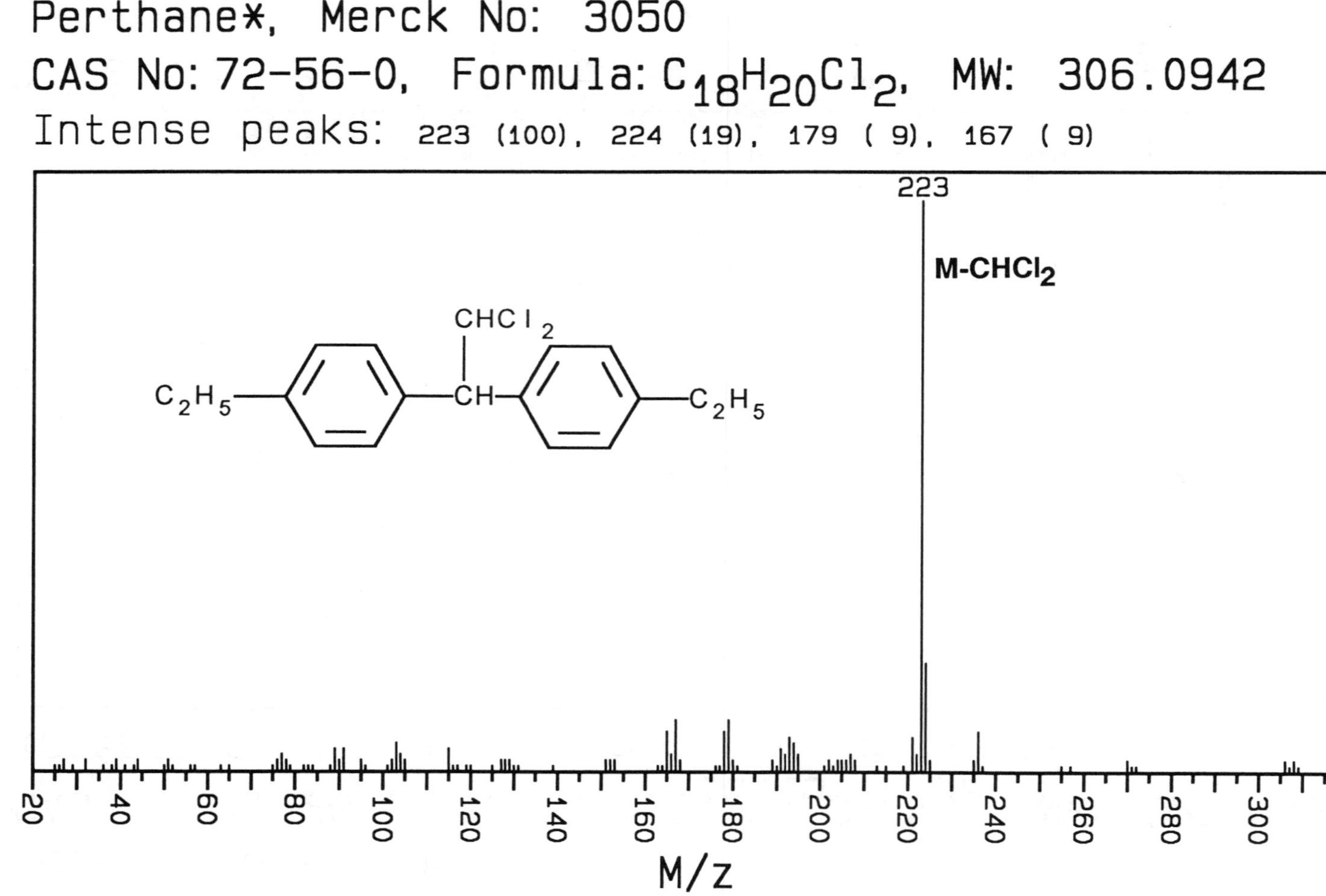
Perthane*, Merck No: 3050
CAS No: 72-56-0, Formula: C18H20Cl2, MW: 306.0942
Intense peaks: 223 (100), 224 (19), 179 ( 9), 167 ( 9)
223
M-CHCl2
CHCl2
C2H5
CH
C2H5
20
40
60
80
100
120
140
160
180
200
220
240
260
280
300
320
M/Z

Methoxychlor, Merck No: 5913

CAS No: 72-43-5, Formula: $C_{16}H_{15}Cl_3O_2$, MW: 344.0138

Intense peaks: 227 (100), 228 (23), 152 (10), 138 (10)

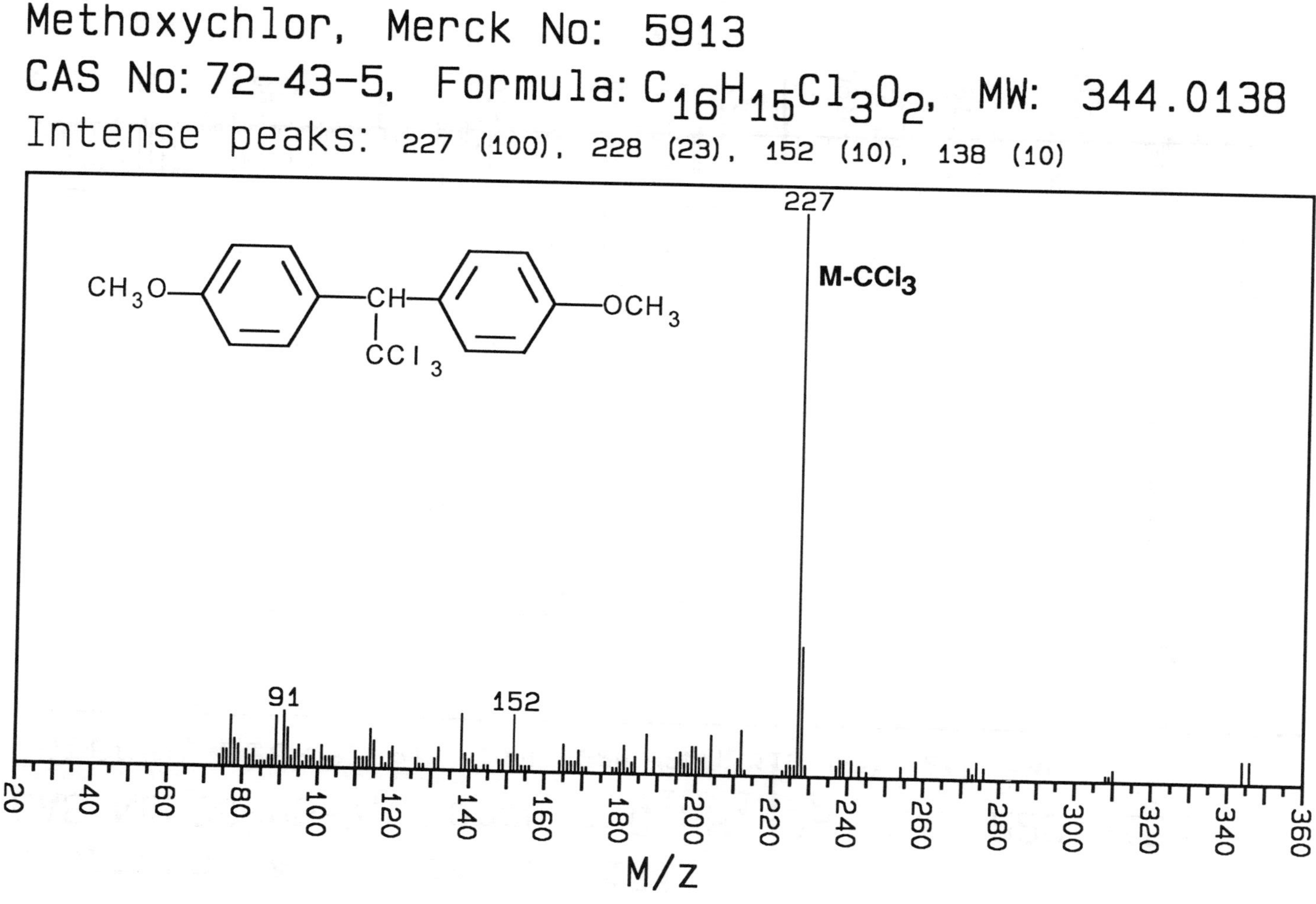

Bisphenol A, Merck No: 1311

CAS No: 80-05-7, Formula: $C_{15}H_{16}O_2$, MW: 228.1151

Intense peaks: 213 (100), 228 (26), 119 (25), 214 (14)

213

M-$CH_3$

228

119

91

57

29

M/Z

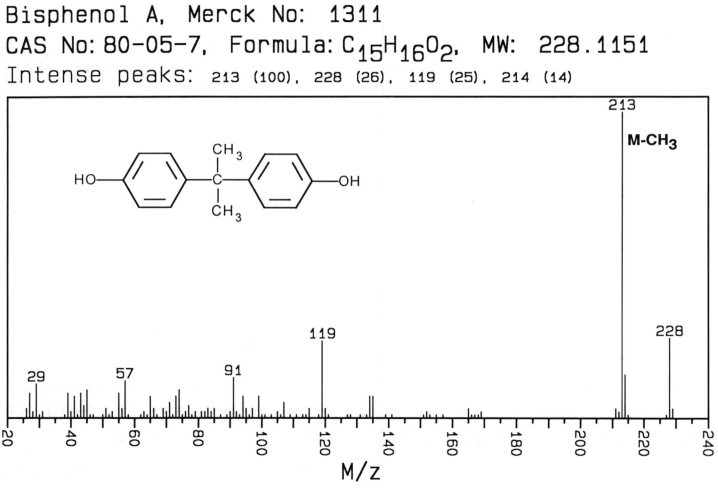

Dichlorophene, Merck No: 3059

CAS No: 97-23-4, Formula: $C_{13}H_{10}Cl_2O_2$, MW: 268.0058

Intense peaks: 128 (100), 141 (62), 268 (32), 152 (32)

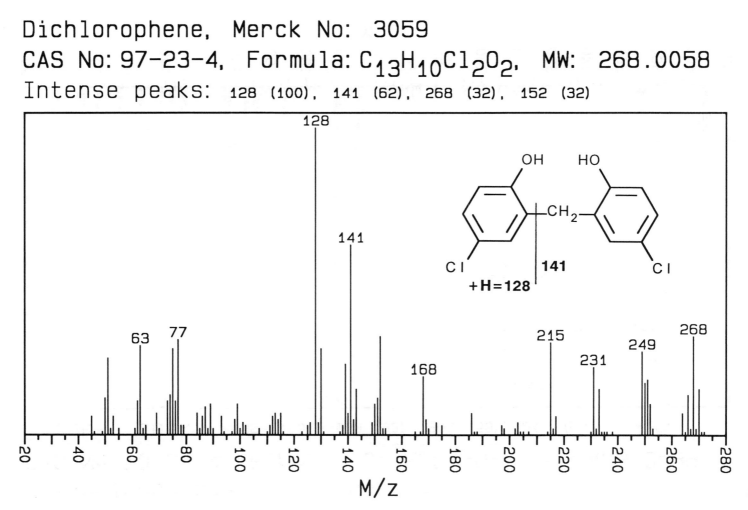

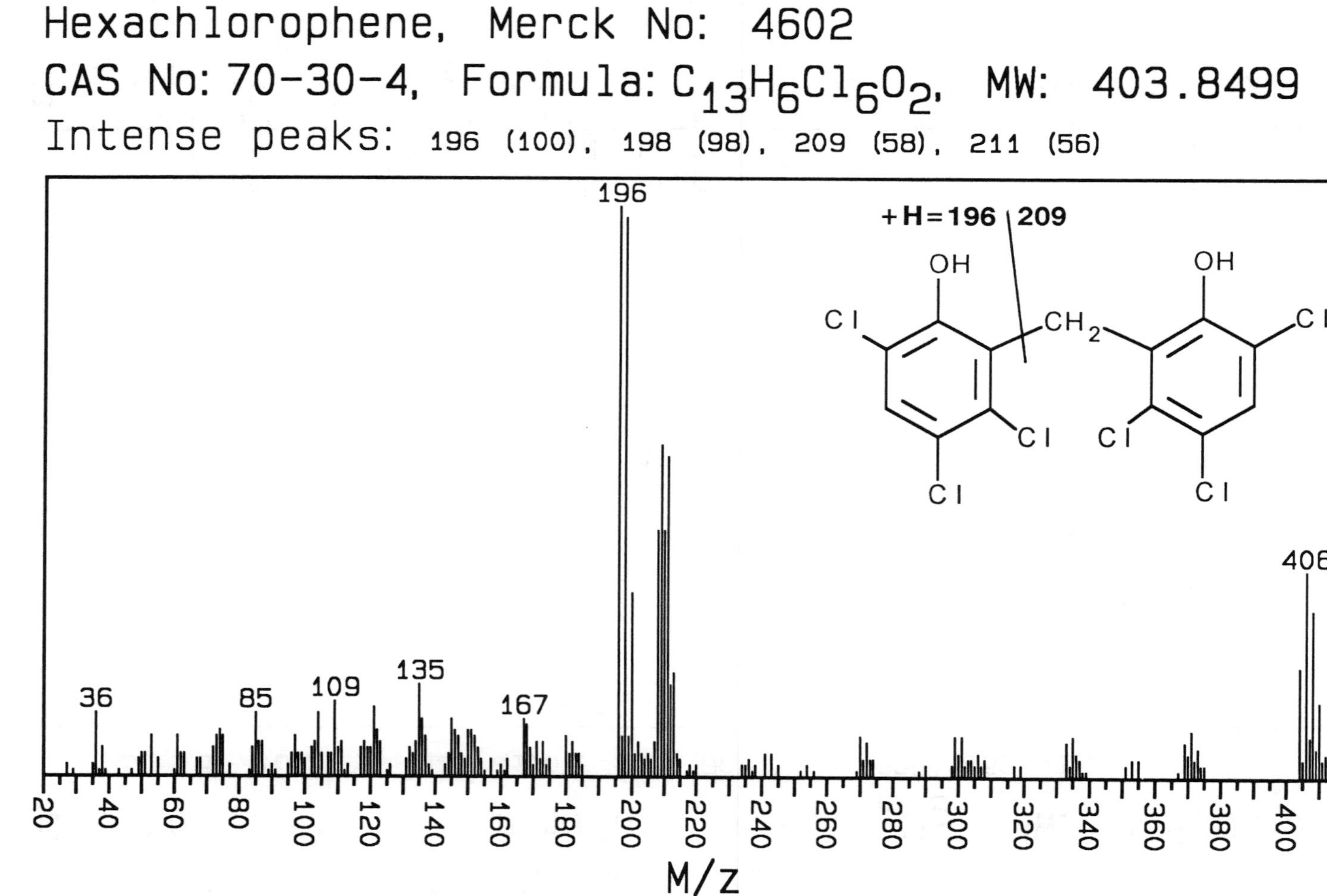

Hexachlorophene, Merck No: 4602
CAS No: 70-30-4, Formula: C13H6Cl6O2, MW: 403.8499
Intense peaks: 196 (100), 198 (98), 209 (58), 211 (56)
+H=196 209
OH
OH
CH2
Cl
Cl
Cl
Cl
Cl
Cl
406
196
167
135
109
85
36
20
40
60
80
100
120
140
160
180
200
220
240
260
280
300
320
340
360
380
400
420
M/Z

Ancymidol, Merck No: 665

CAS No: 12771-68-5, Formula: $C_{15}H_{16}N_2O_2$, MW: 256.1212

Intense peaks: 107 (100), 39 (73), 228 (71), 121 (64)

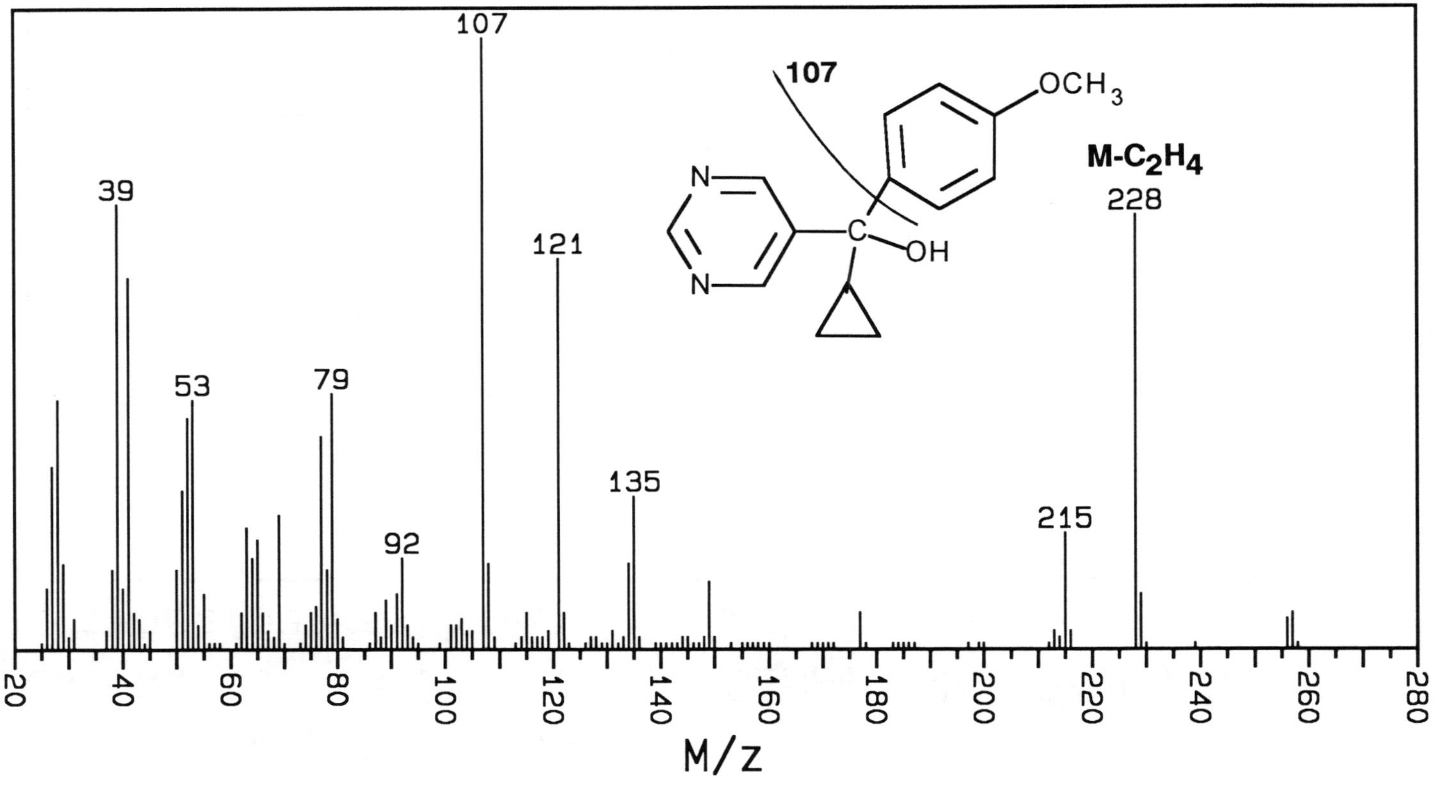

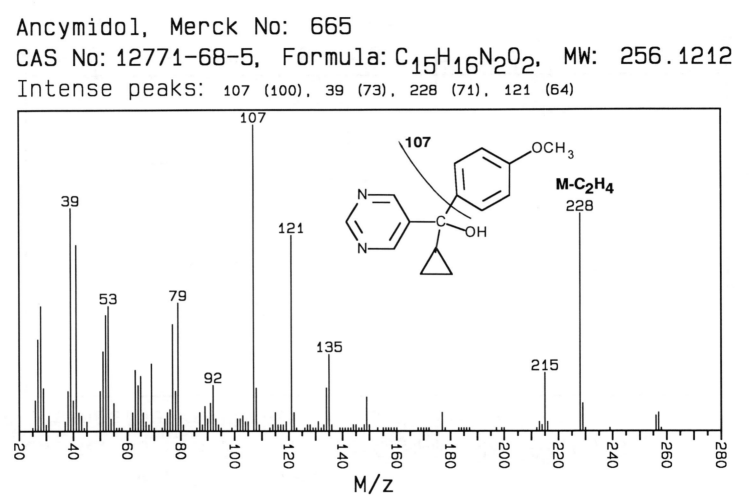

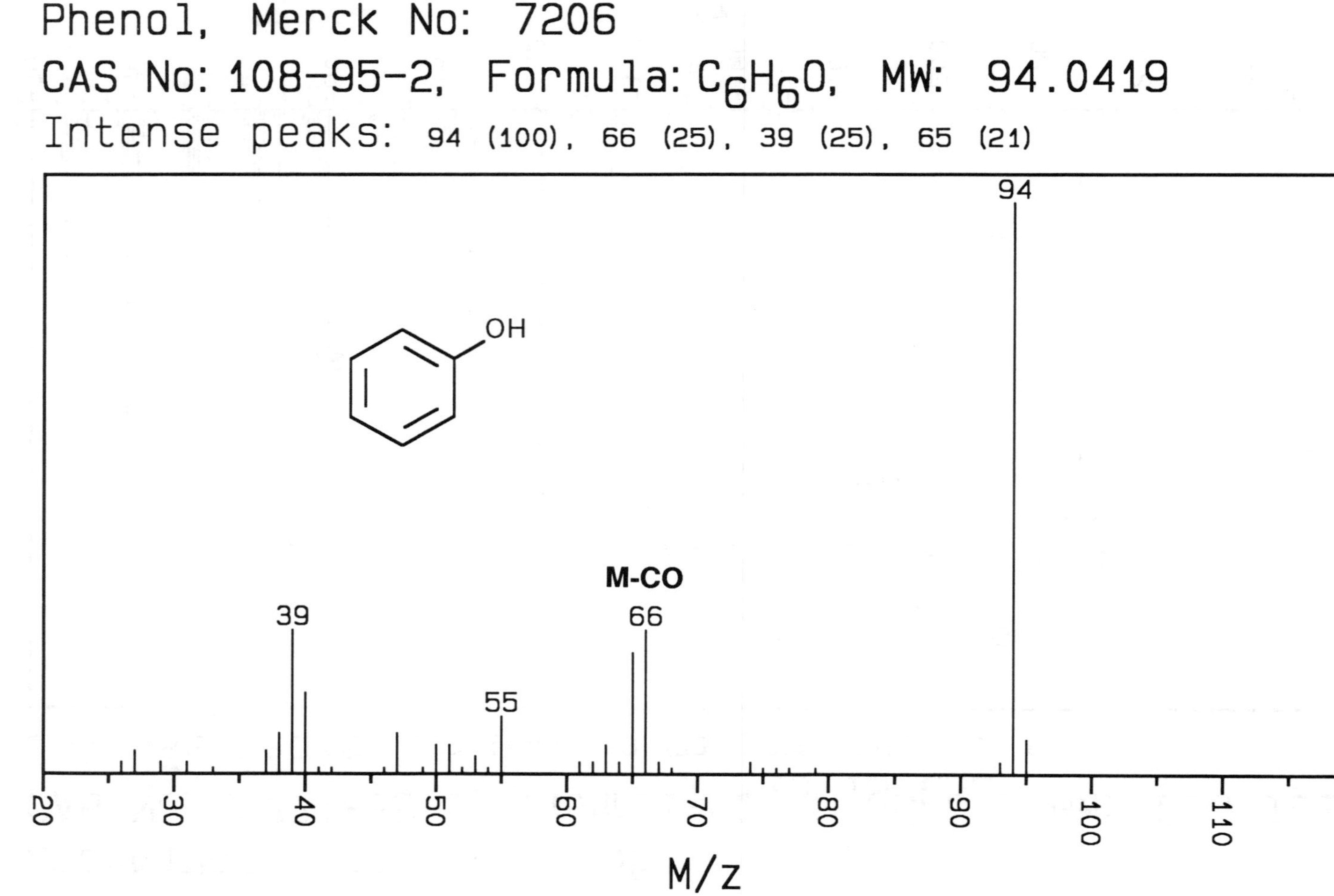
Phenol, Merck No: 7206
CAS No: 108-95-2, Formula: C6H6O, MW: 94.0419
Intense peaks: 94 (100), 66 (25), 39 (25), 65 (21)
OH
94
M-CO
66
39
55
20
30
40
50
60
70
80
90
100
110
120
M/Z

Resorcinol, Merck No: 8158

CAS No: 108-46-3, Formula: $C_6H_6O_2$, MW: 110.0368

Intense peaks: 110 (100), 82 (12), 81 (11), 69 ( 9)

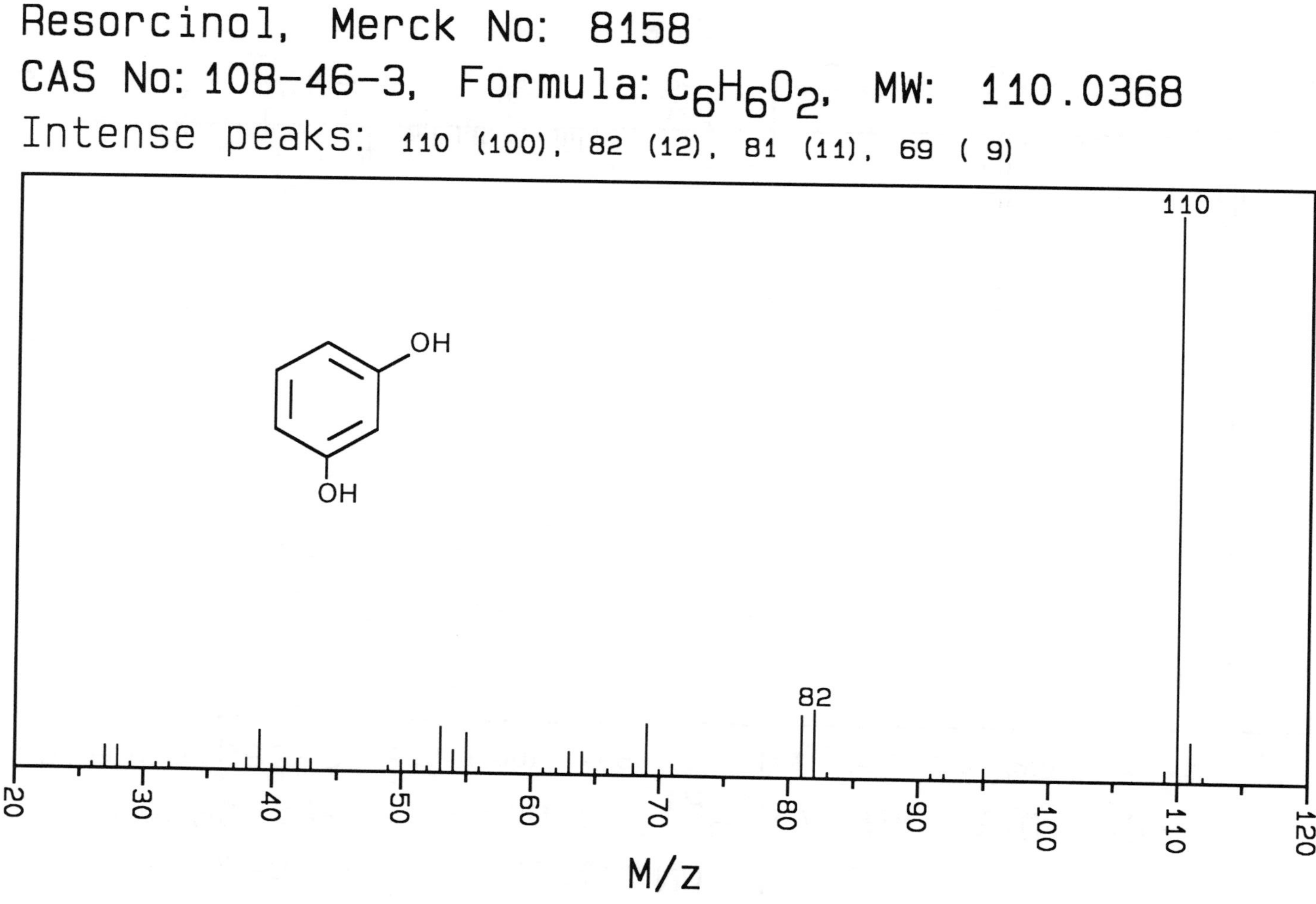

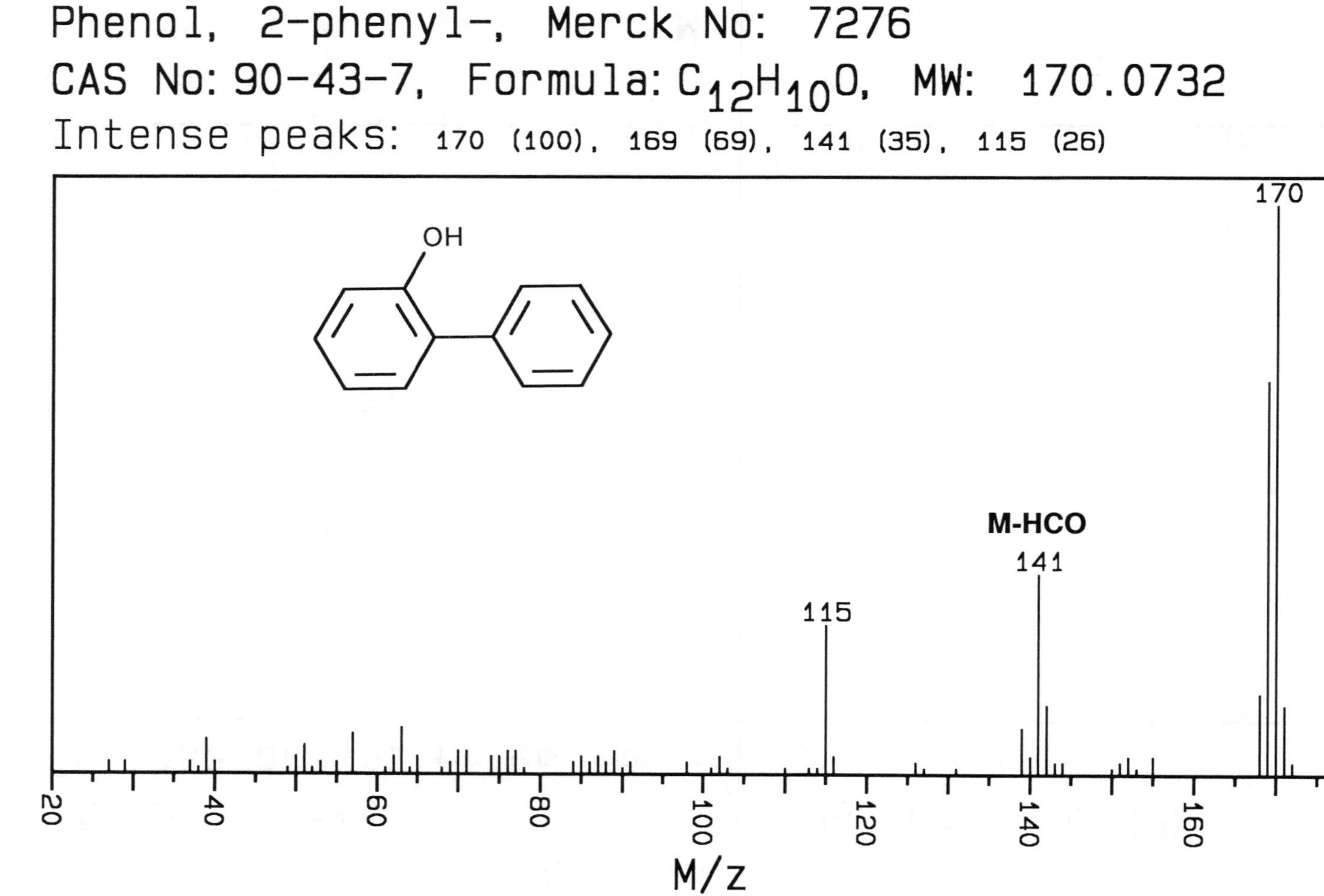
Phenol, 2-phenyl-, Merck No: 7276
CAS No: 90-43-7, Formula: $C_{12}H_{10}O$, MW: 170.0732
Intense peaks: 170 (100), 169 (69), 141 (35), 115 (26)
OH
170
M-HCO
141
115
20
40
60
80
100
120
140
160
180
M/Z

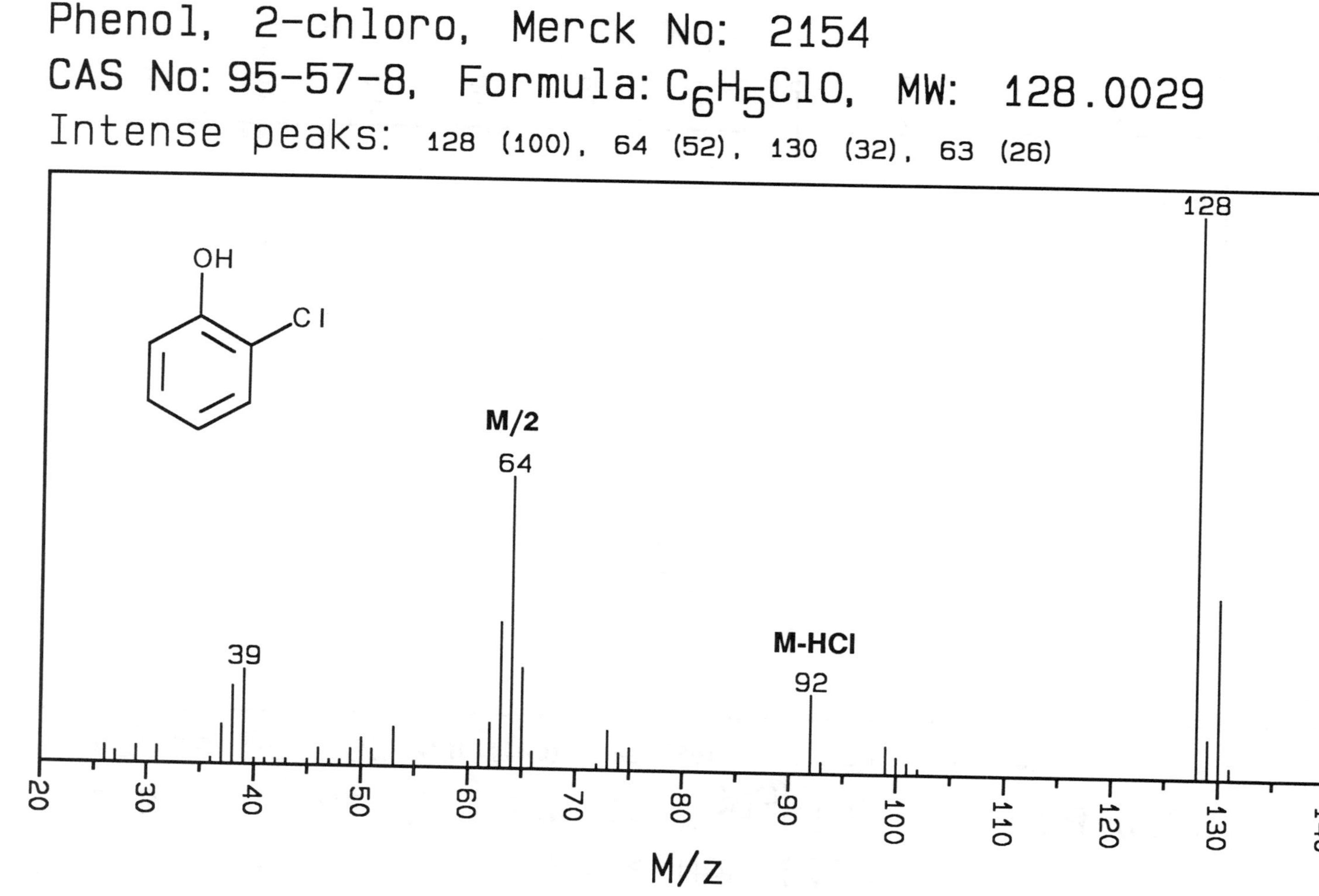
Phenol, 2-chloro, Merck No: 2154
CAS No: 95-57-8, Formula: $C_6H_5ClO$, MW: 128.0029
Intense peaks: 128 (100), 64 (52), 130 (32), 63 (26)
OH
Cl
128
M/2
64
39
M-HCl
92
20
30
40
50
60
70
80
90
100
110
120
130
140
M/Z

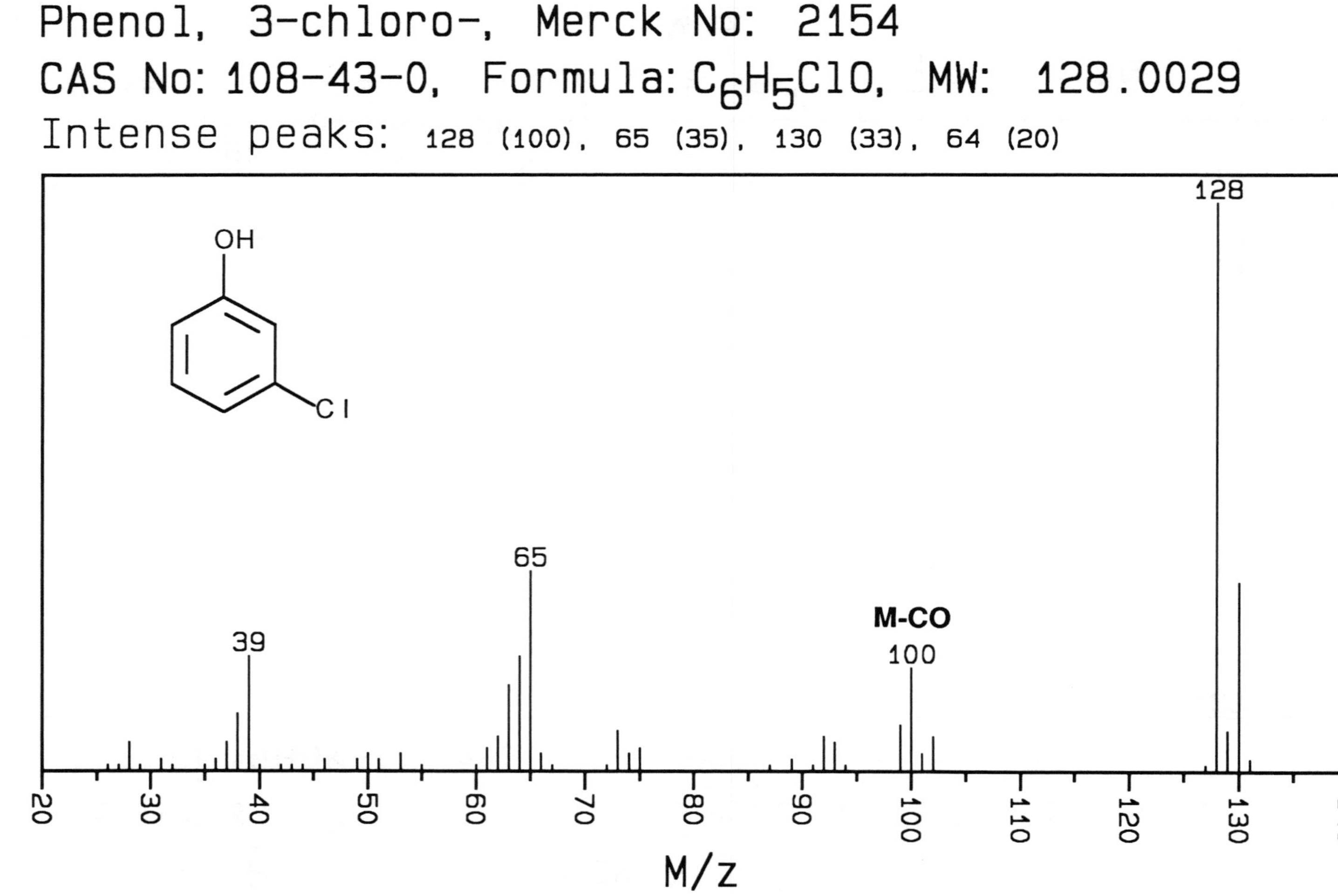
Phenol, 3-chloro-, Merck No: 2154
CAS No: 108-43-0, Formula: $C_6H_5ClO$, MW: 128.0029
Intense peaks: 128 (100), 65 (35), 130 (33), 64 (20)
OH
Cl
128
65
39
M-CO
100
20
30
40
50
60
70
80
90
100
110
120
130
140
M/Z

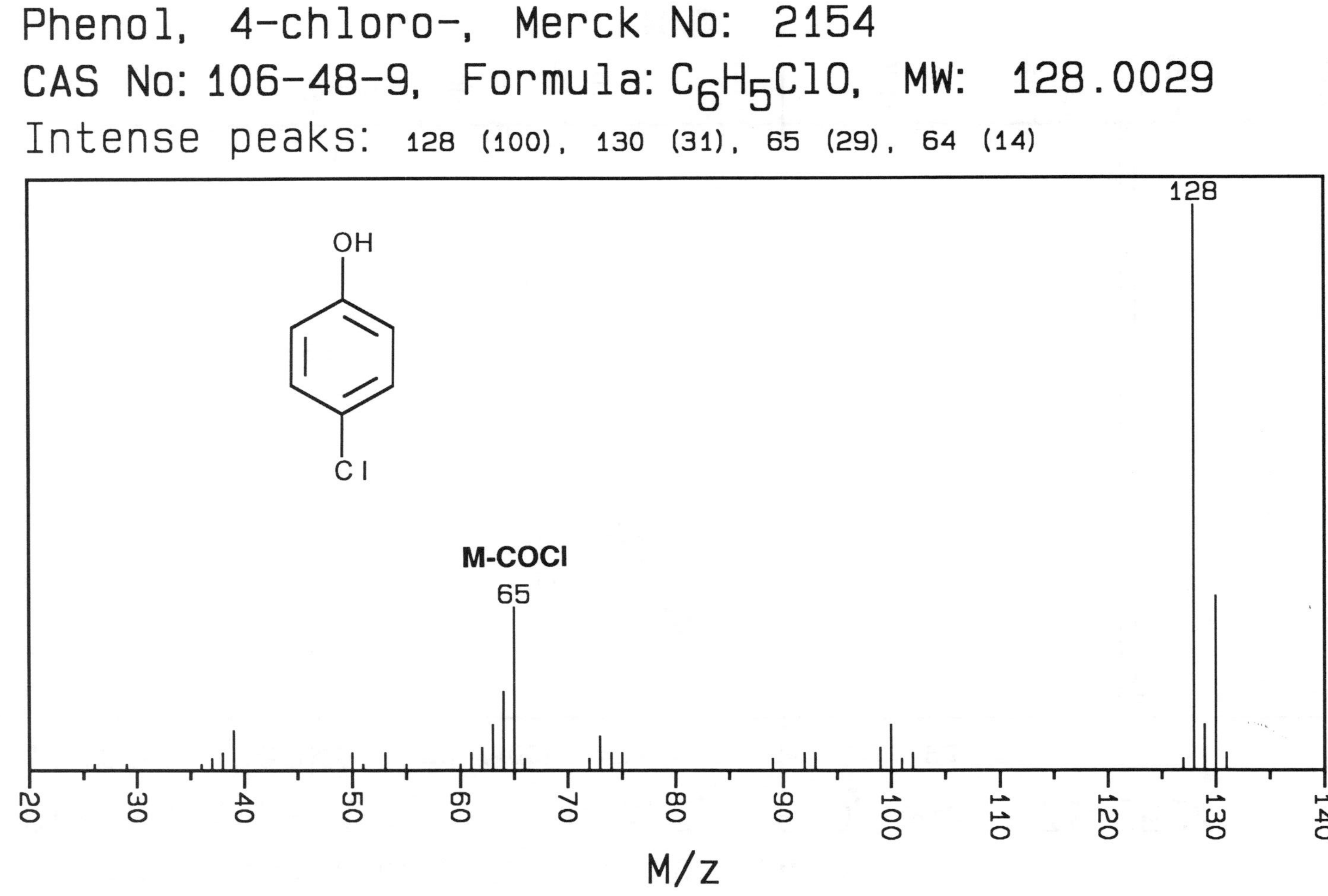
Phenol, 4-chloro-, Merck No: 2154
CAS No: 106-48-9, Formula: $C_6H_5ClO$, MW: 128.0029
Intense peaks: 128 (100), 130 (31), 65 (29), 64 (14)
OH
Cl
M-COCl
65
128
20
30
40
50
60
70
80
90
100
110
120
130
140
M/Z

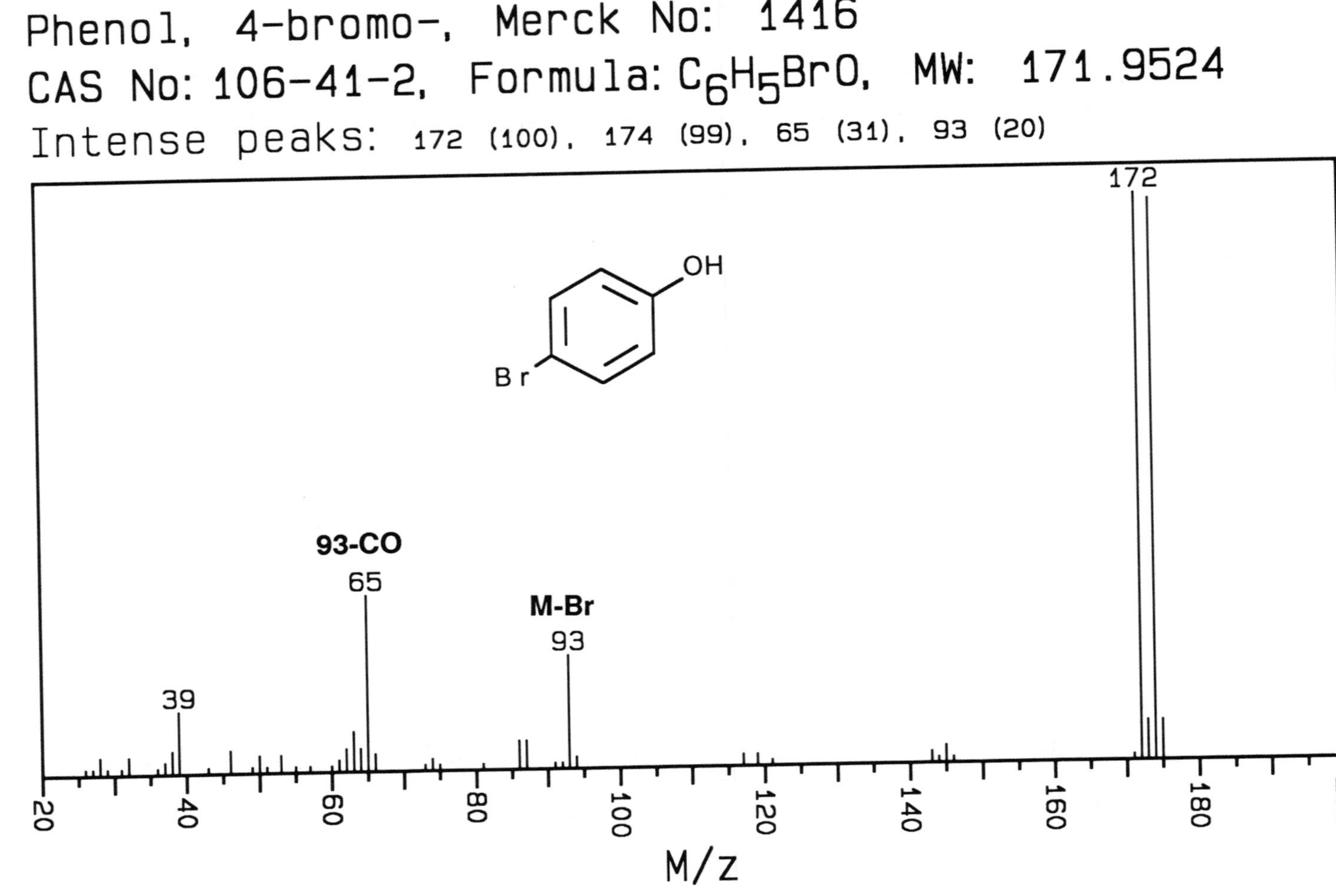

Phenol, 4-bromo-, Merck No: 1416
CAS No: 106-41-2, Formula: $C_6H_5BrO$, MW: 171.9524
Intense peaks: 172 (100), 174 (99), 65 (31), 93 (20)
172
OH
Br
93-CO
65
M-Br
93
39
20
40
60
80
100
120
140
160
180
200
M/Z

Phenol, 2,6-dichloro-, Merck No: 3062
CAS No: 87-65-0, Formula: $C_6H_4Cl_2O$, MW: 161.9639
Intense peaks: 162 (100), 164 (64), 63 (21), 98 (15)

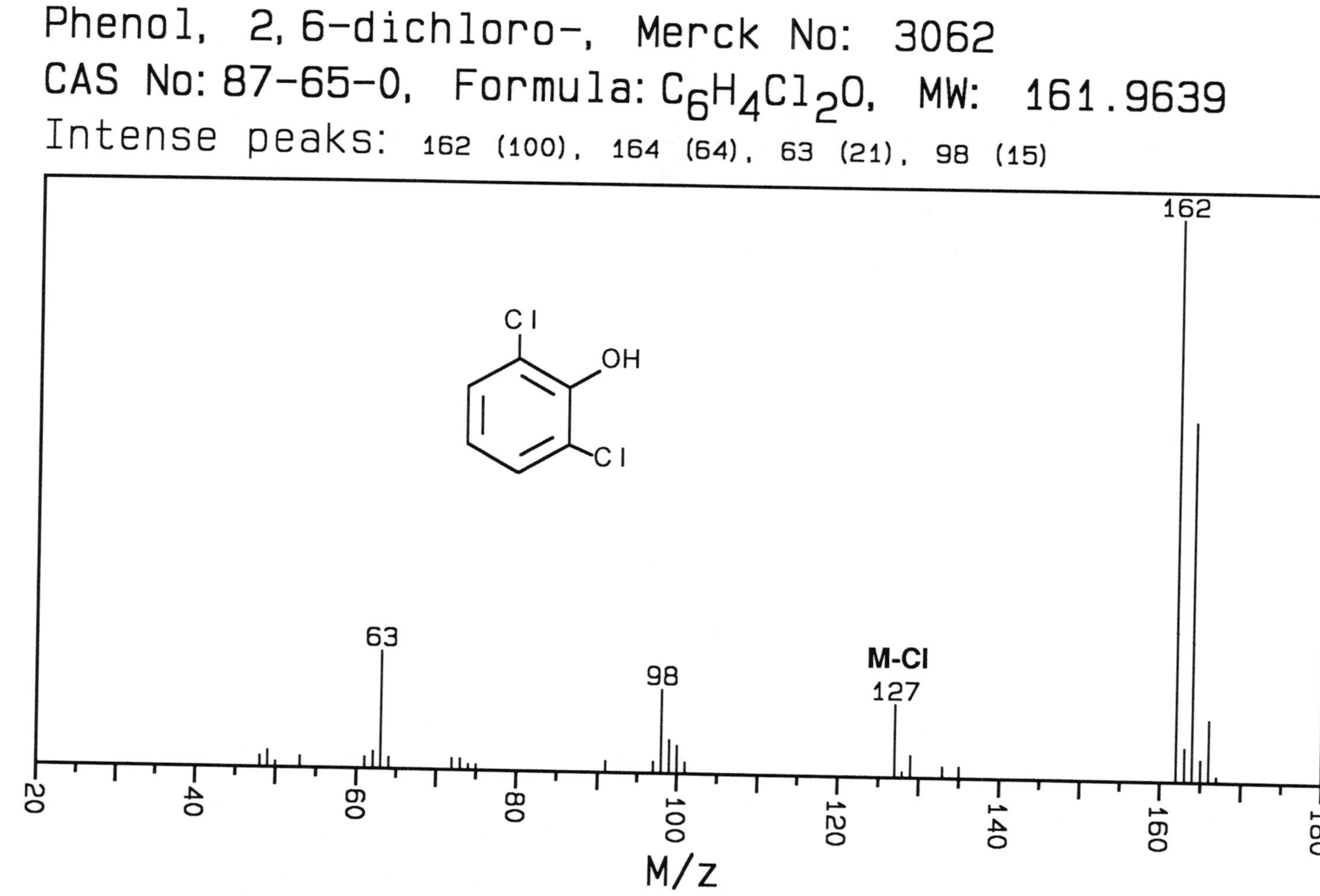

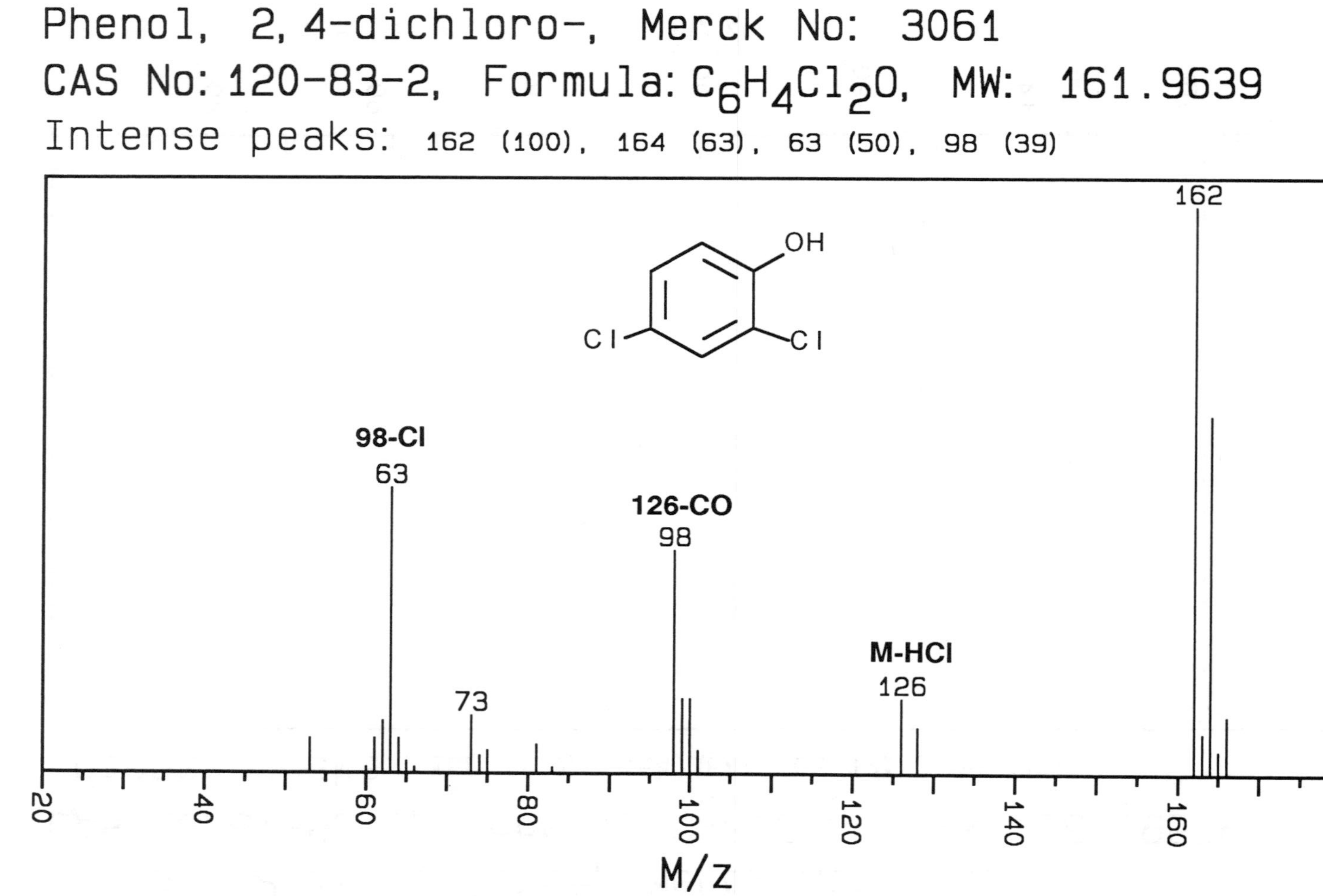
Phenol, 2,4-dichloro-, Merck No: 3061
CAS No: 120-83-2, Formula: $C_6H_4Cl_2O$, MW: 161.9639
Intense peaks: 162 (100), 164 (63), 63 (50), 98 (39)
OH
Cl
Cl
162
98-Cl
63
126-CO
98
M-HCl
126
73
20
40
60
80
100
120
140
160
180
M/Z

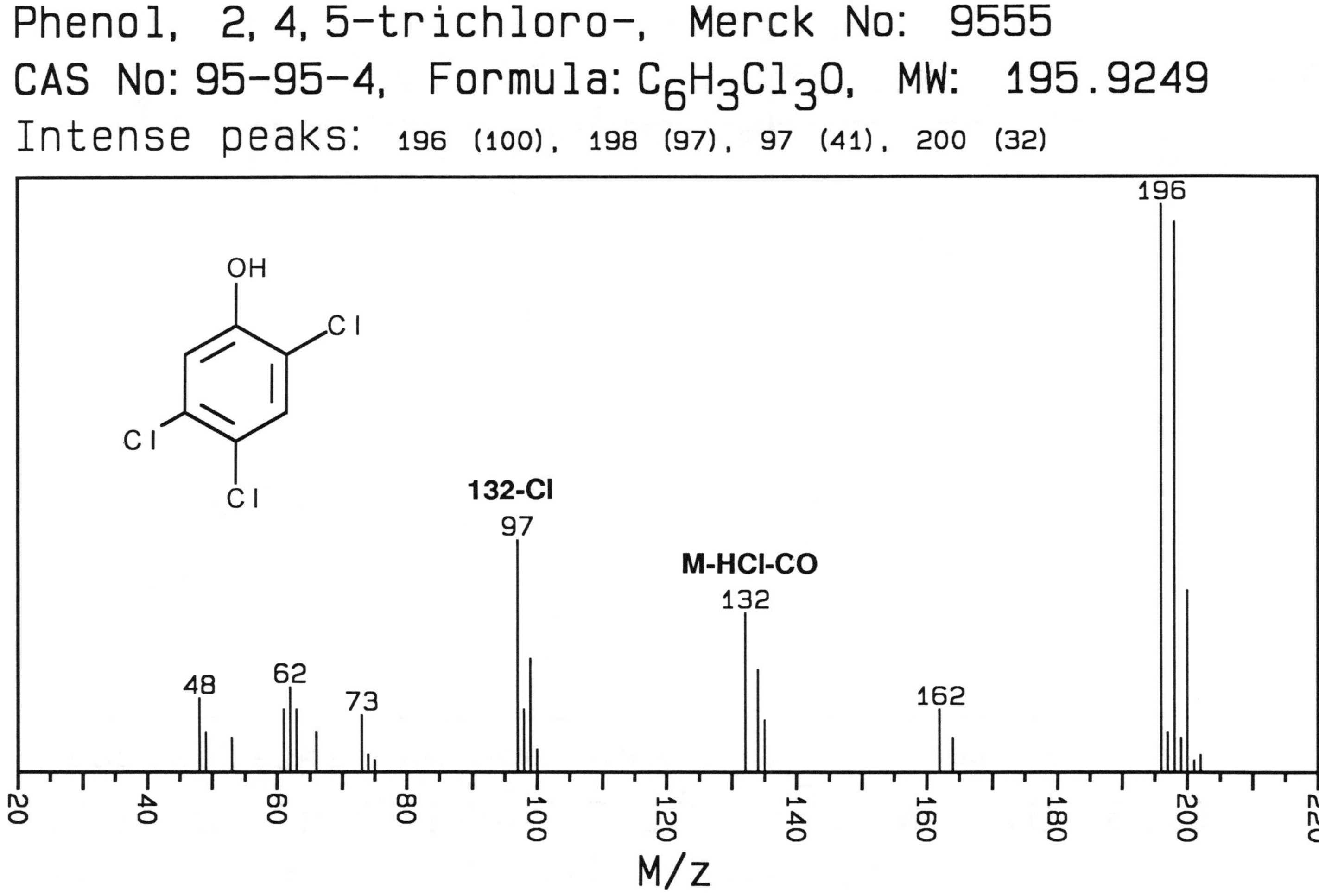
Phenol, 2,4,5-trichloro-, Merck No: 9555
CAS No: 95-95-4, Formula: C6H3Cl3O, MW: 195.9249
Intense peaks: 196 (100), 198 (97), 97 (41), 200 (32)
OH
Cl
Cl
Cl
132-Cl
97
M-HCl-CO
132
48
62
73
162
196
20
40
60
80
100
120
140
160
180
200
220
M/Z

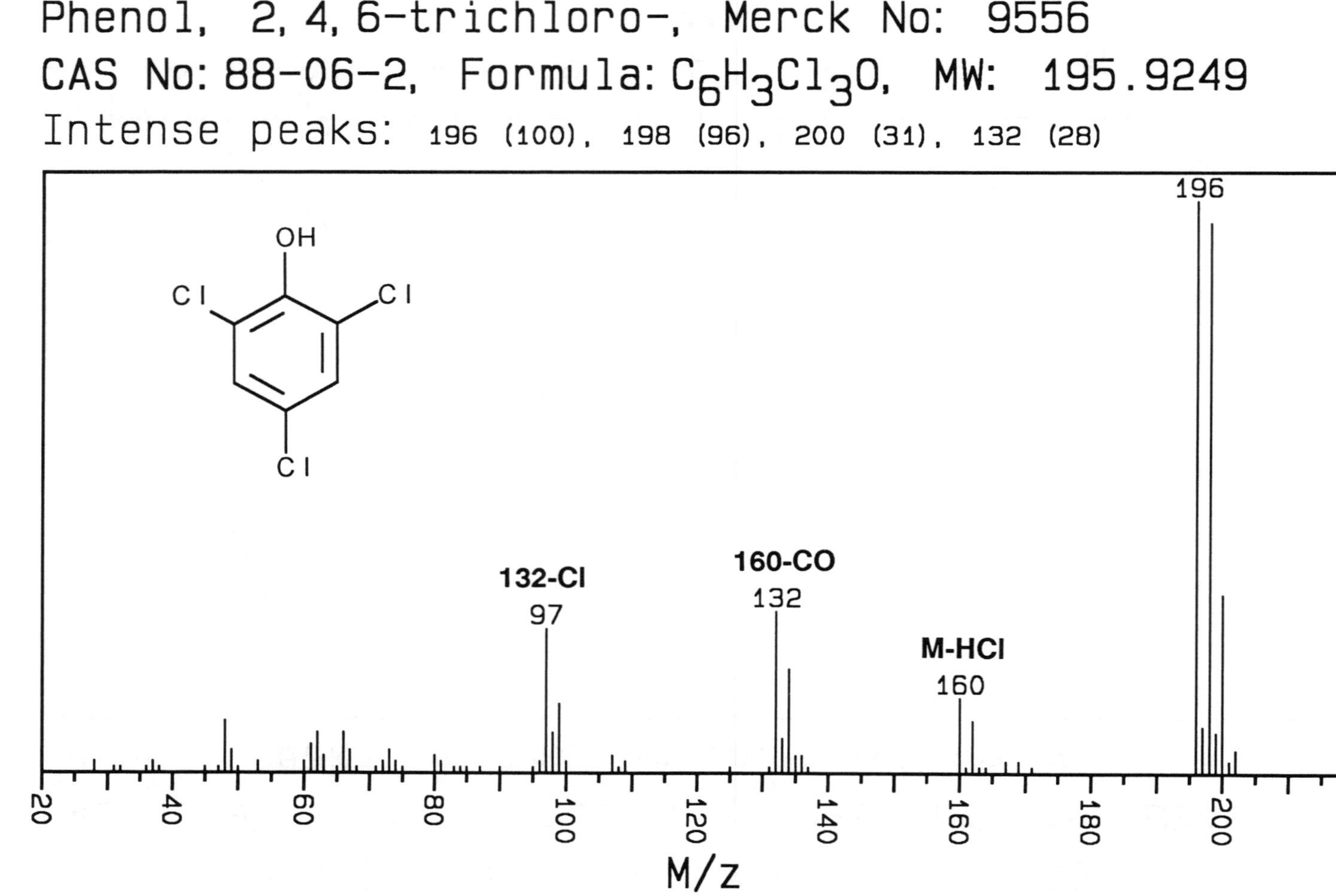
Phenol, 2,4,6-trichloro-, Merck No: 9556
CAS No: 88-06-2, Formula: $C_6H_3Cl_3O$, MW: 195.9249
Intense peaks: 196 (100), 198 (96), 200 (31), 132 (28)
OH
Cl
Cl
Cl
196
132-Cl
97
160-CO
132
M-HCl
160
20
40
60
80
100
120
140
160
180
200
220
M/Z

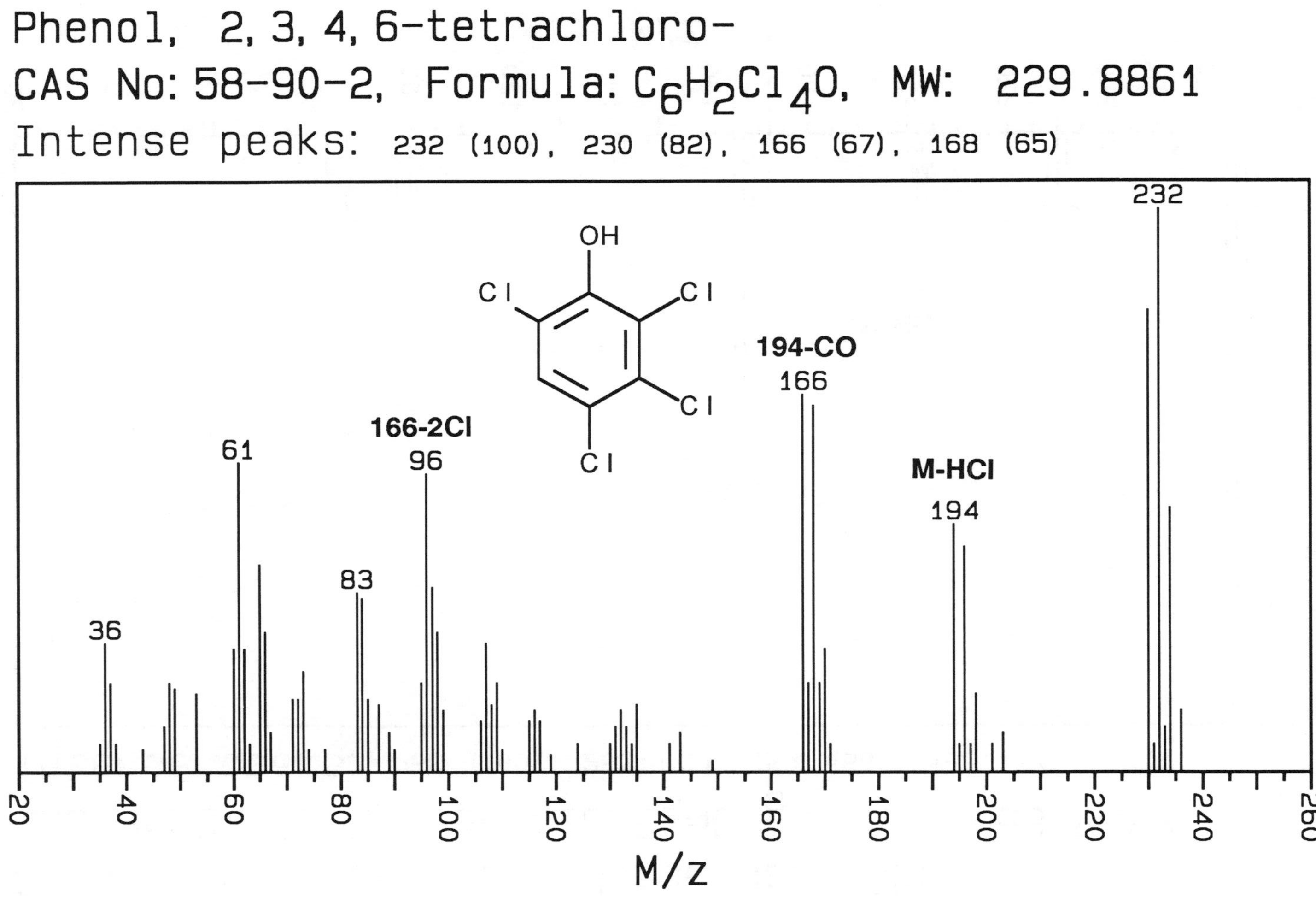
Phenol, 2,3,4,6-tetrachloro-
CAS No: 58-90-2, Formula: $C_6H_2Cl_4O$, MW: 229.8861
Intense peaks: 232 (100), 230 (82), 166 (67), 168 (65)
232
OH
Cl
Cl
Cl
Cl
194-CO
166
166-2Cl
96
61
M-HCl
194
83
36
20
40
60
80
100
120
140
160
180
200
220
240
260
M/Z

Phenol, pentachloro-, Merck No: 7059
CAS No: 87-86-5, Formula: $C_6HCl_5O$, MW: 263.8469
Intense peaks: 266 (100), 268 (70), 264 (68), 165 (54)

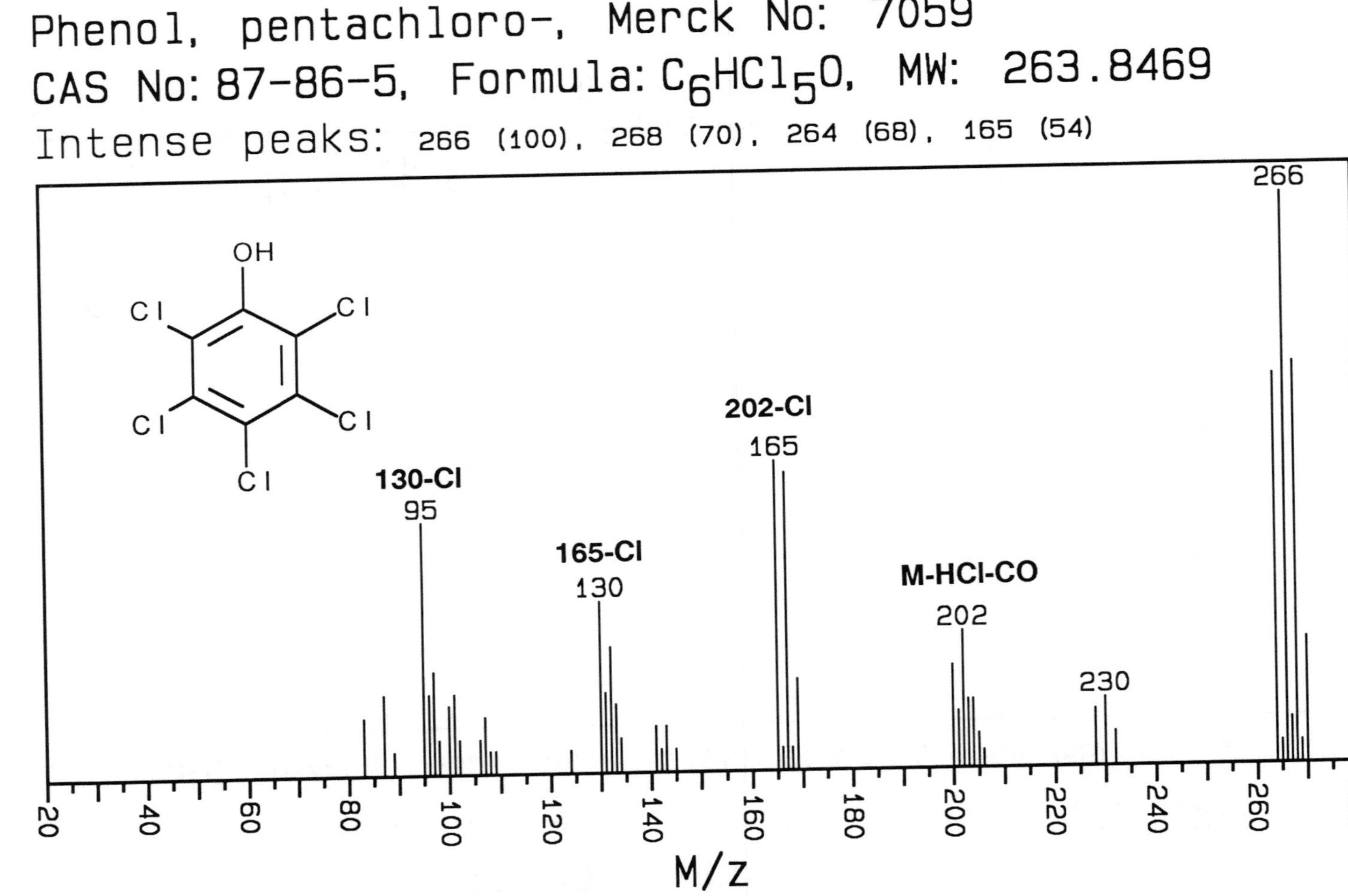

Dioxin, 1, 2, 3, 4-tetrachloro-

CAS No: 30746-58-8, Formula: $C_{12}H_4Cl_4O_2$, MW: 319.8965

Intense peaks: 322 (100), 320 (81), 324 (51), 257 (23)

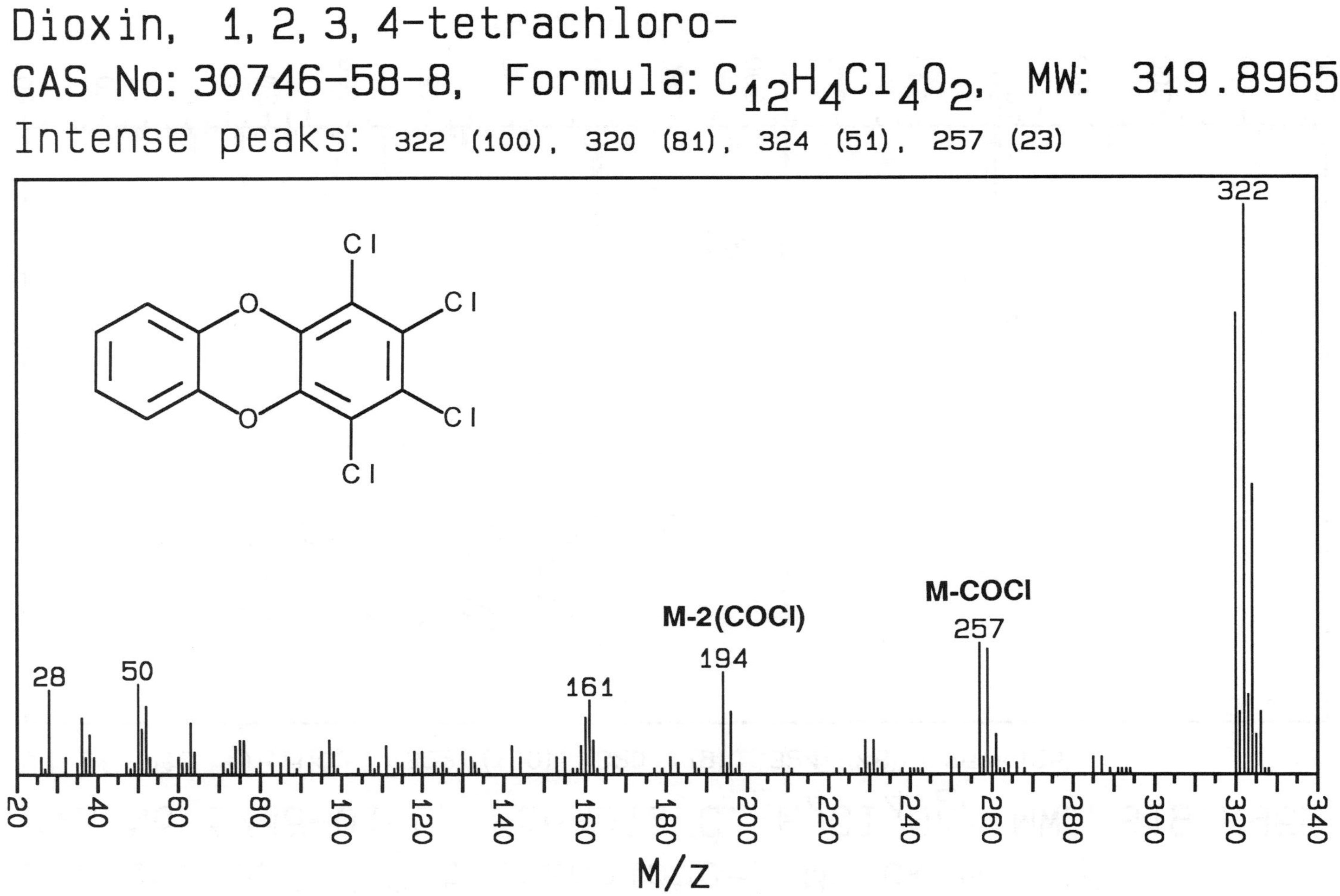

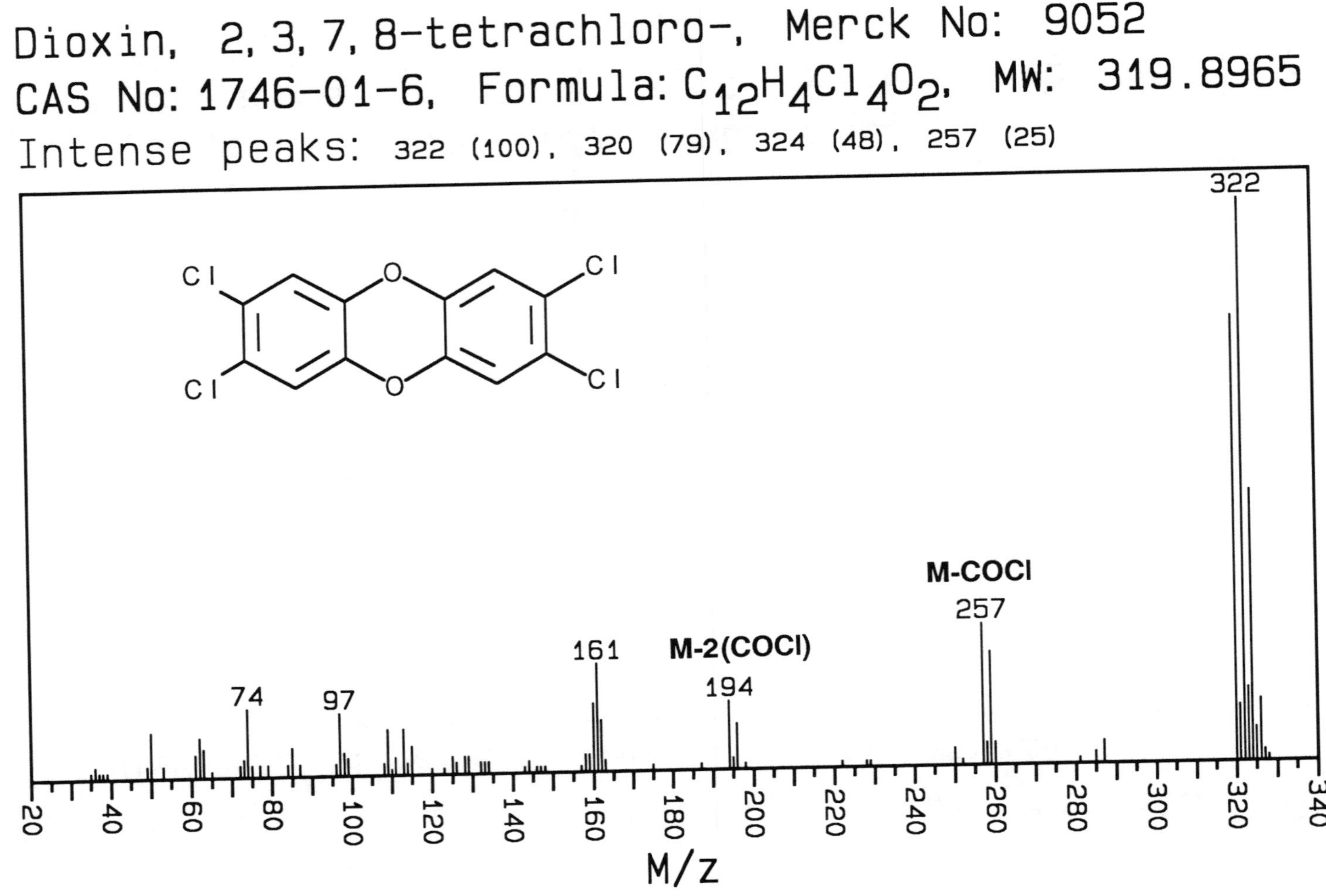
Dioxin, 2,3,7,8-tetrachloro-, Merck No: 9052
CAS No: 1746-01-6, Formula: $C_{12}H_4Cl_4O_2$, MW: 319.8965
Intense peaks: 322 (100), 320 (79), 324 (48), 257 (25)
Cl
Cl
Cl
Cl
O
O
322
M-COCl
257
161
M-2(COCl)
194
74
97
20
40
60
80
100
120
140
160
180
200
220
240
260
280
300
320
340
M/Z

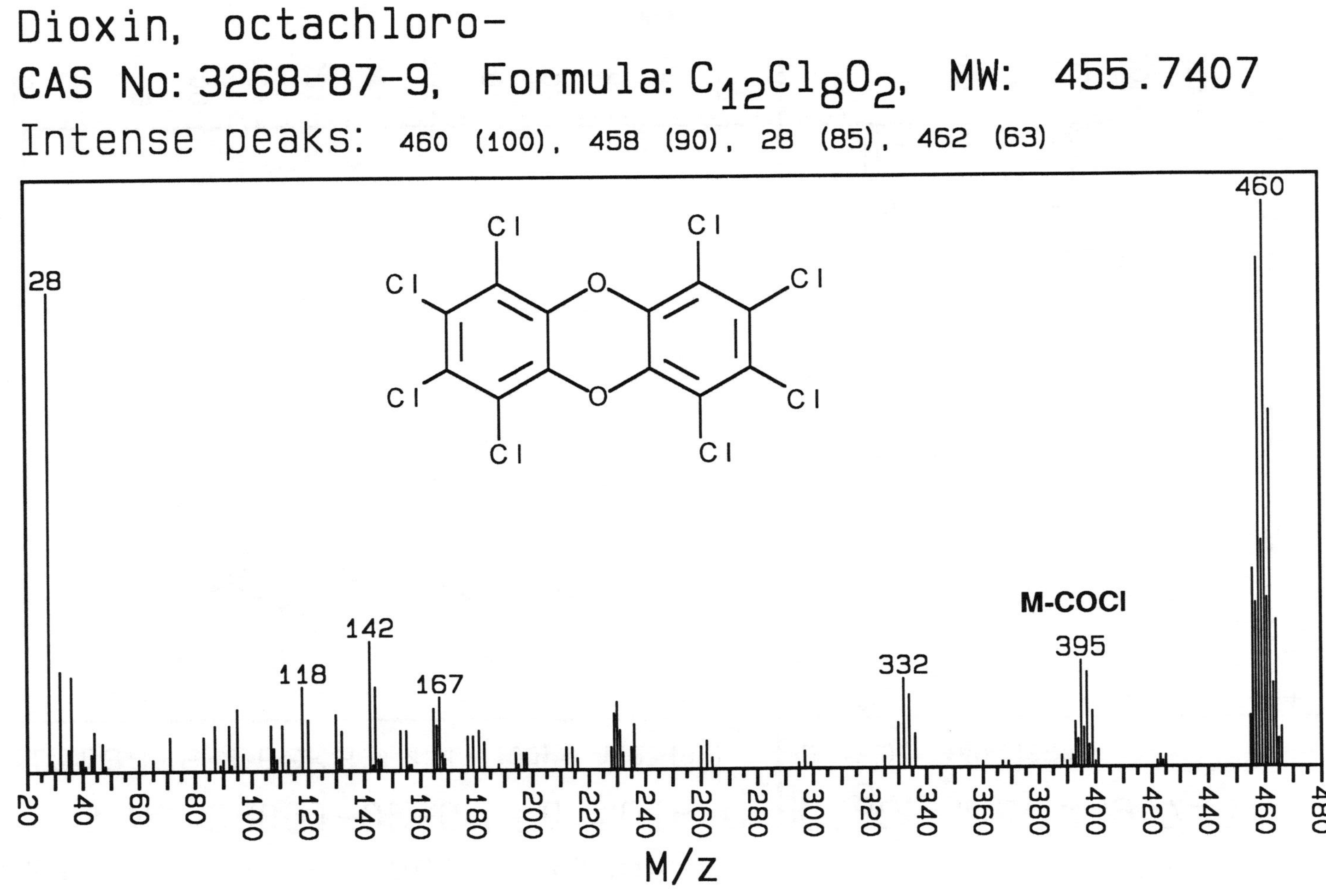
Dioxin, octachloro-
CAS No: 3268-87-9, Formula: $C_{12}Cl_8O_2$, MW: 455.7407
Intense peaks: 460 (100), 458 (90), 28 (85), 462 (63)
Cl
O
460
28
142
118
167
332
M-COCl
395
20
40
60
80
100
120
140
160
180
200
220
240
260
280
300
320
340
360
380
400
420
440
460
480
M/Z

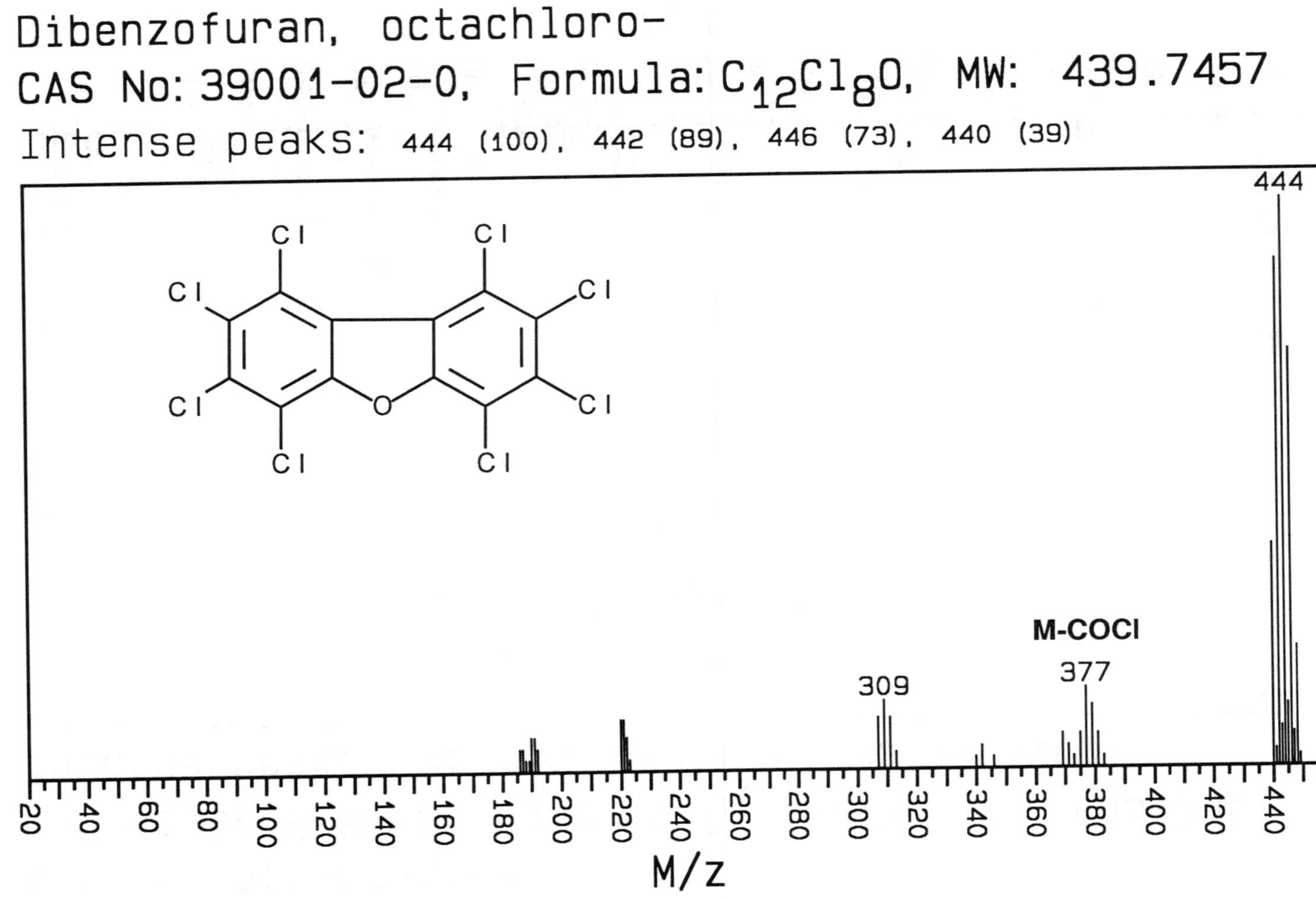

Dibenzofuran, octachloro-
CAS No: 39001-02-0, Formula: $C_{12}Cl_8O$, MW: 439.7457
Intense peaks: 444 (100), 442 (89), 446 (73), 440 (39)
444
M-COCl
377
309
Cl
O
M/Z
20
40
60
80
100
120
140
160
180
200
220
240
260
280
300
320
340
360
380
400
420
440
460

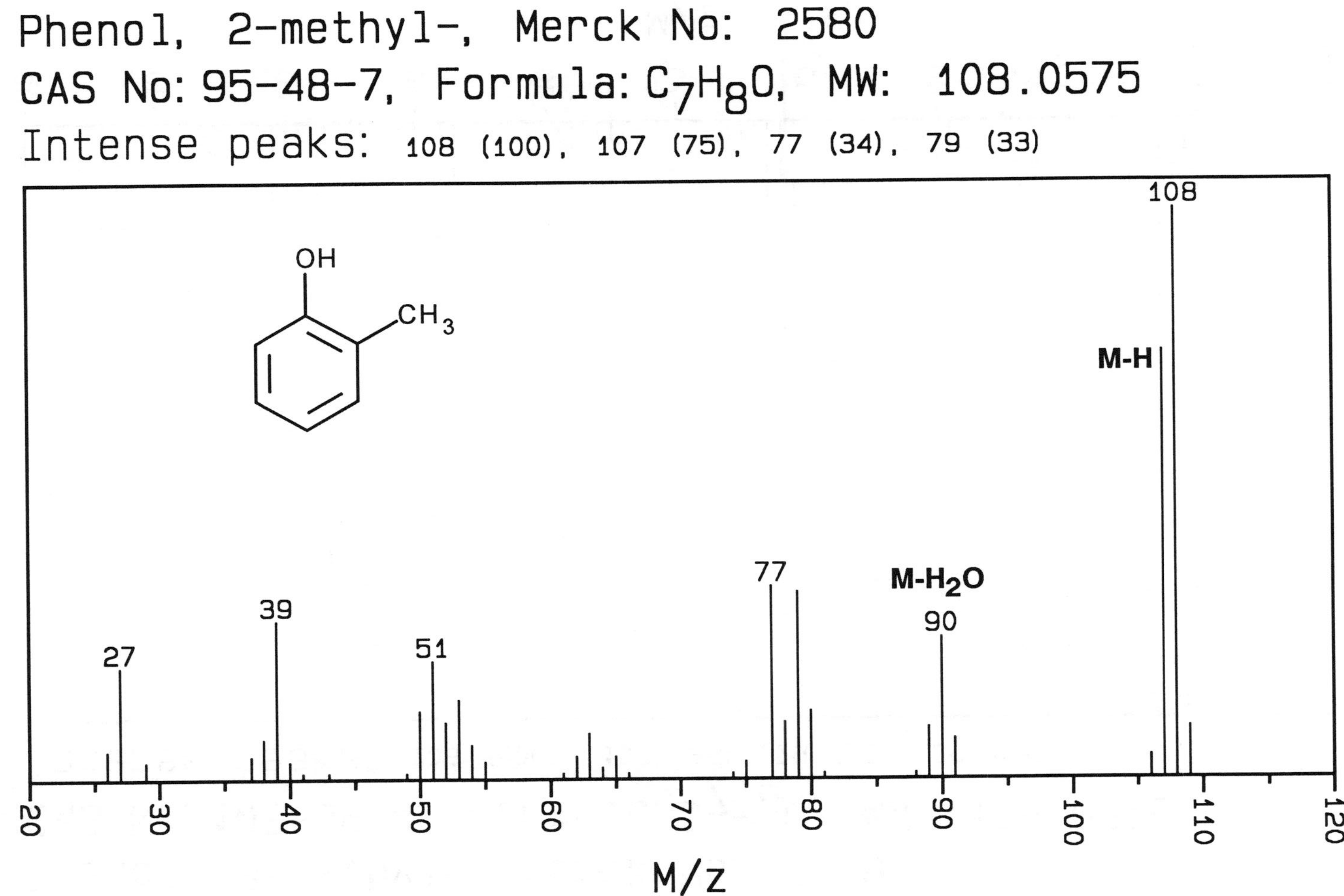

Phenol, 2-methyl-, Merck No: 2580
CAS No: 95-48-7, Formula: $C_7H_8O$, MW: 108.0575
Intense peaks: 108 (100), 107 (75), 77 (34), 79 (33)
OH
$CH_3$
108
M-H
77
M-$H_2$O
90
39
51
27
20
30
40
50
60
70
80
90
100
110
120
M/Z

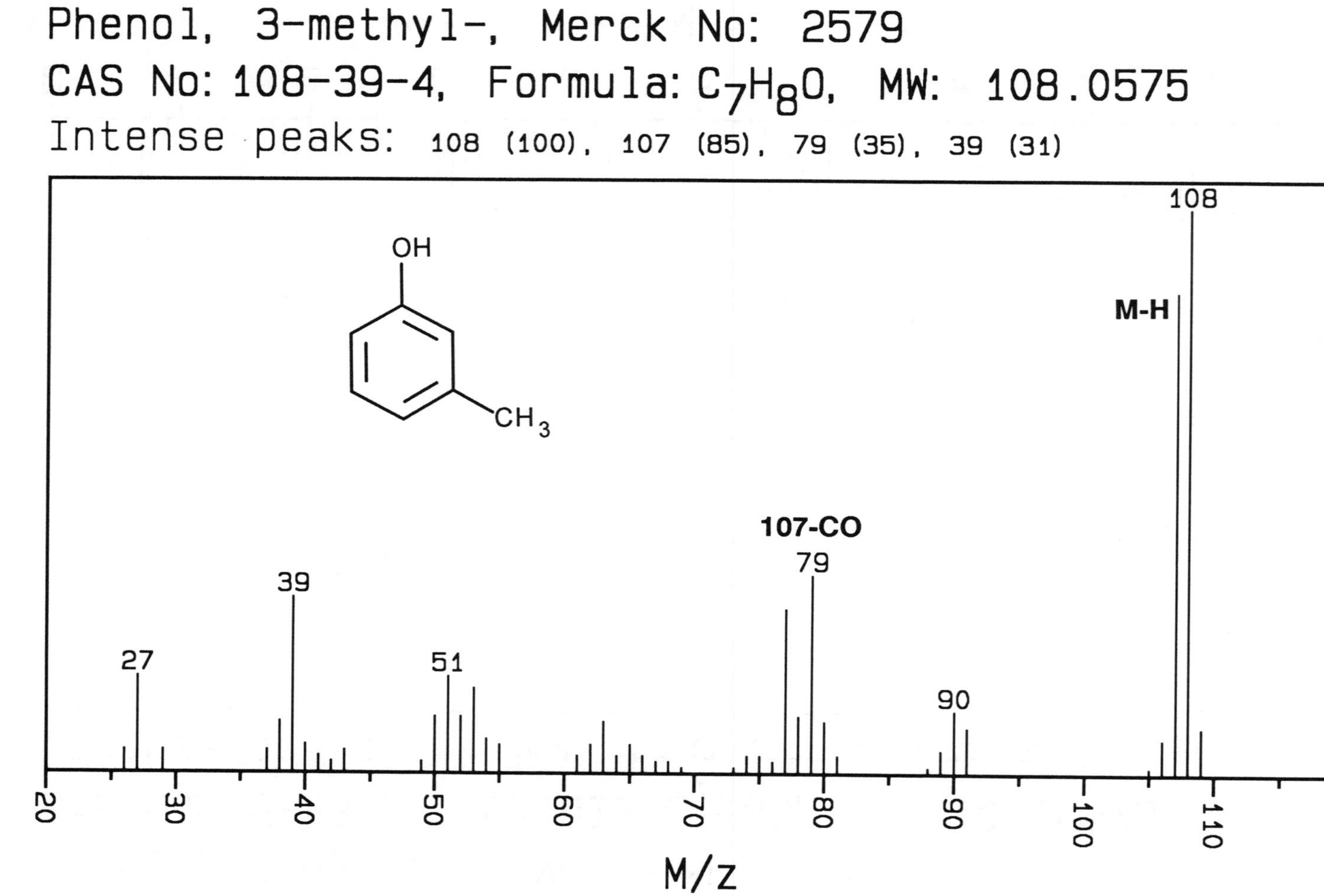
Phenol, 3-methyl-, Merck No: 2579
CAS No: 108-39-4, Formula: $C_7H_8O$, MW: 108.0575
Intense peaks: 108 (100), 107 (85), 79 (35), 39 (31)
OH
CH3
108
M-H
107-CO
79
39
27
51
90
20
30
40
50
60
70
80
90
100
110
120
M/Z

Phenol, 4-chloro-3-methyl, Merck No: 2133

CAS No: 59-50-7, Formula: $C_7H_7ClO$, MW: 142.0185

Intense peaks: 142 (100), 107 (80), 144 (32), 77 (24)

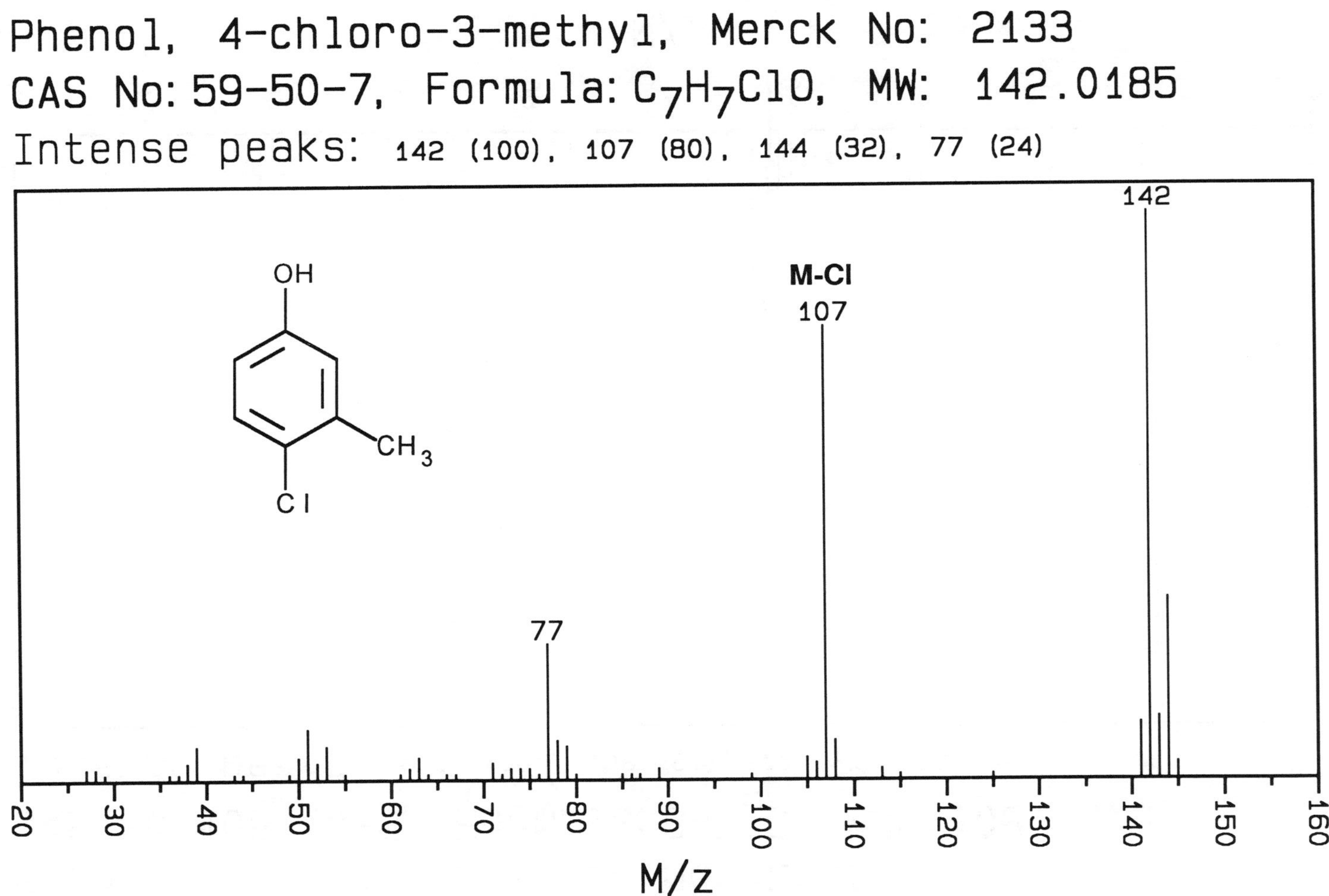

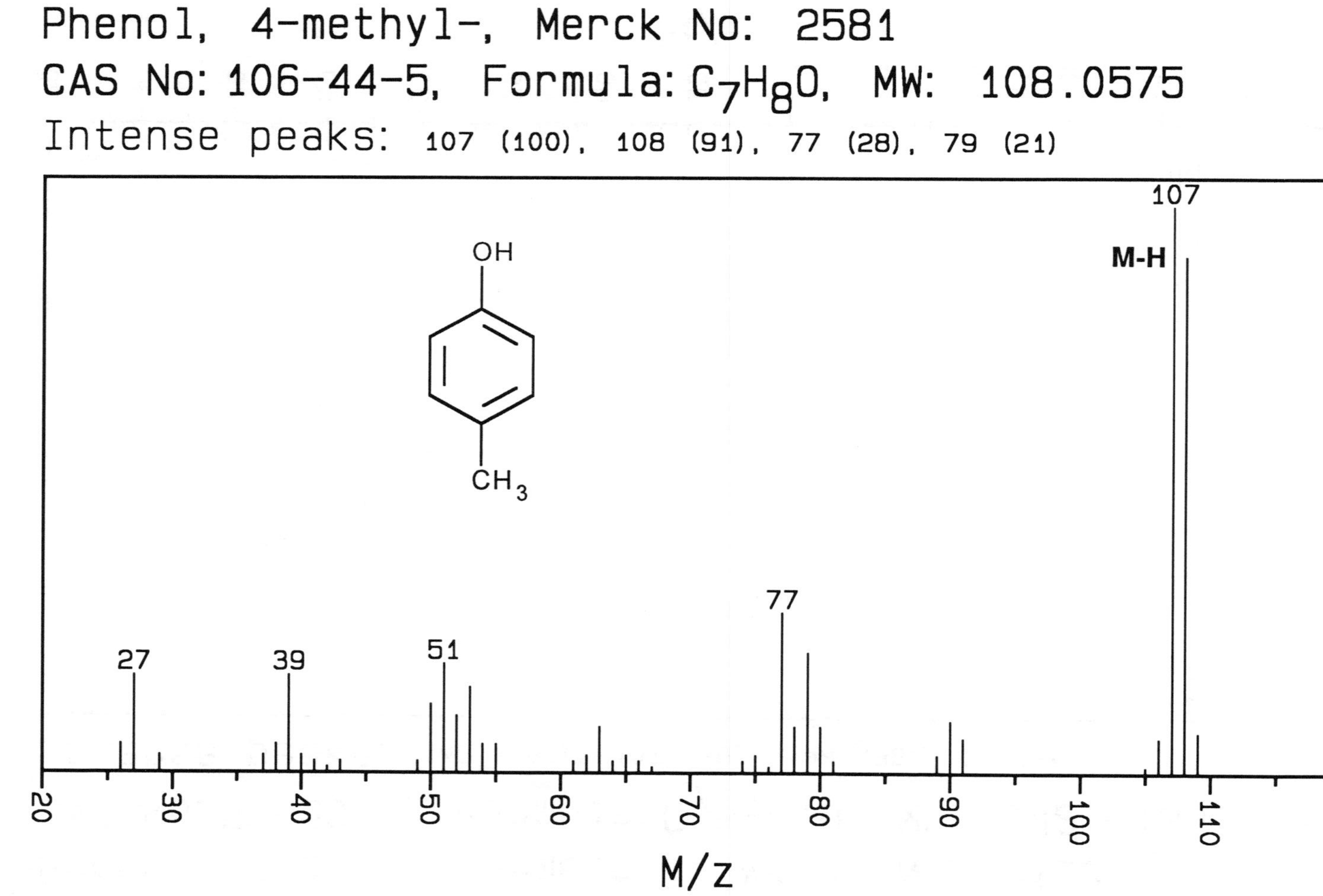
Phenol, 4-methyl-, Merck No: 2581
CAS No: 106-44-5, Formula: C7H8O, MW: 108.0575
Intense peaks: 107 (100), 108 (91), 77 (28), 79 (21)
OH
CH3
107
M-H
77
51
39
27
20
30
40
50
60
70
80
90
100
110
120
M/Z

Phlorol, Merck No: 7302

CAS No: 90-00-6, Formula: $C_8H_{10}O$, MW: 122.0732

Intense peaks: 107 (100), 122 (36), 77 (27), 39 (17)

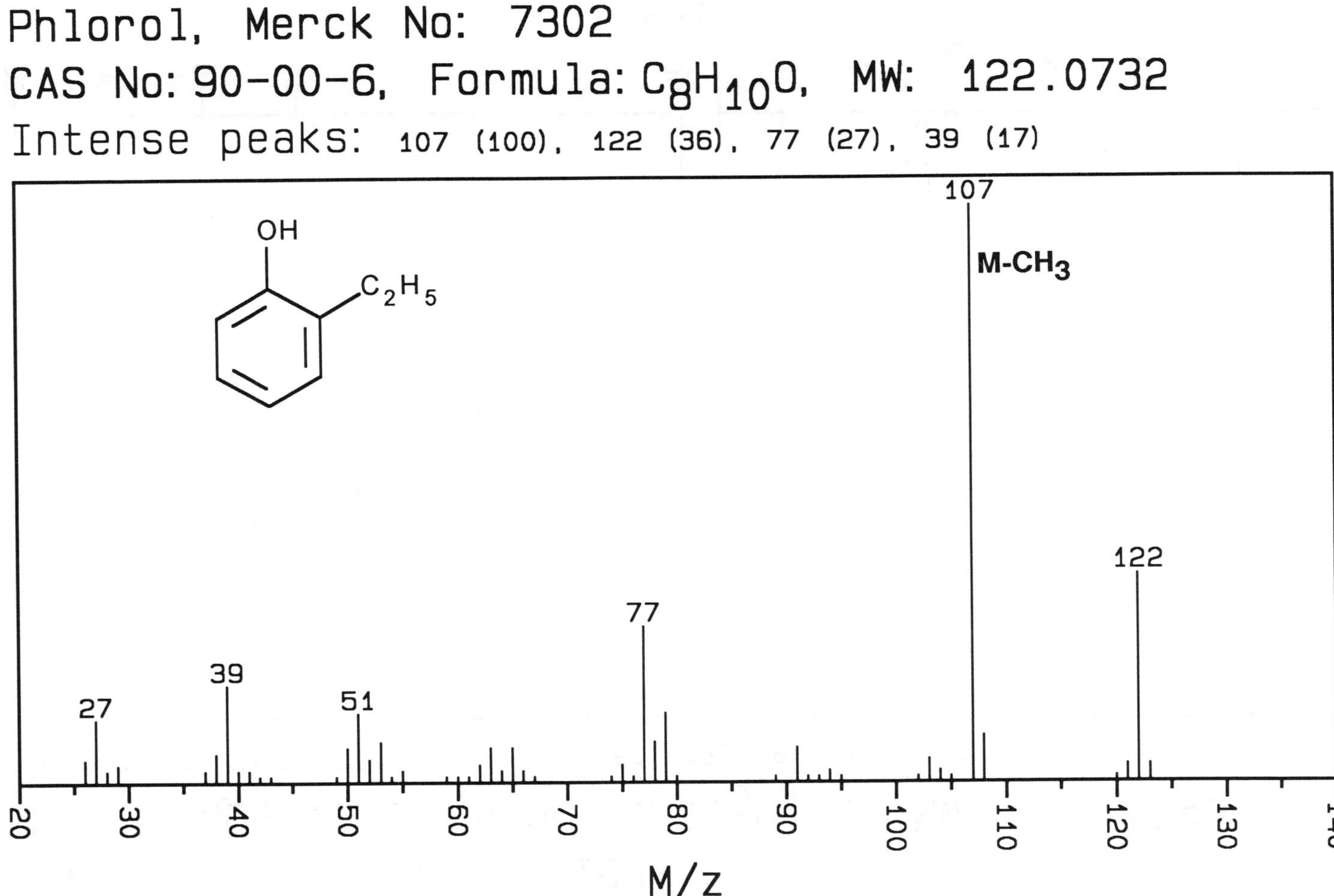

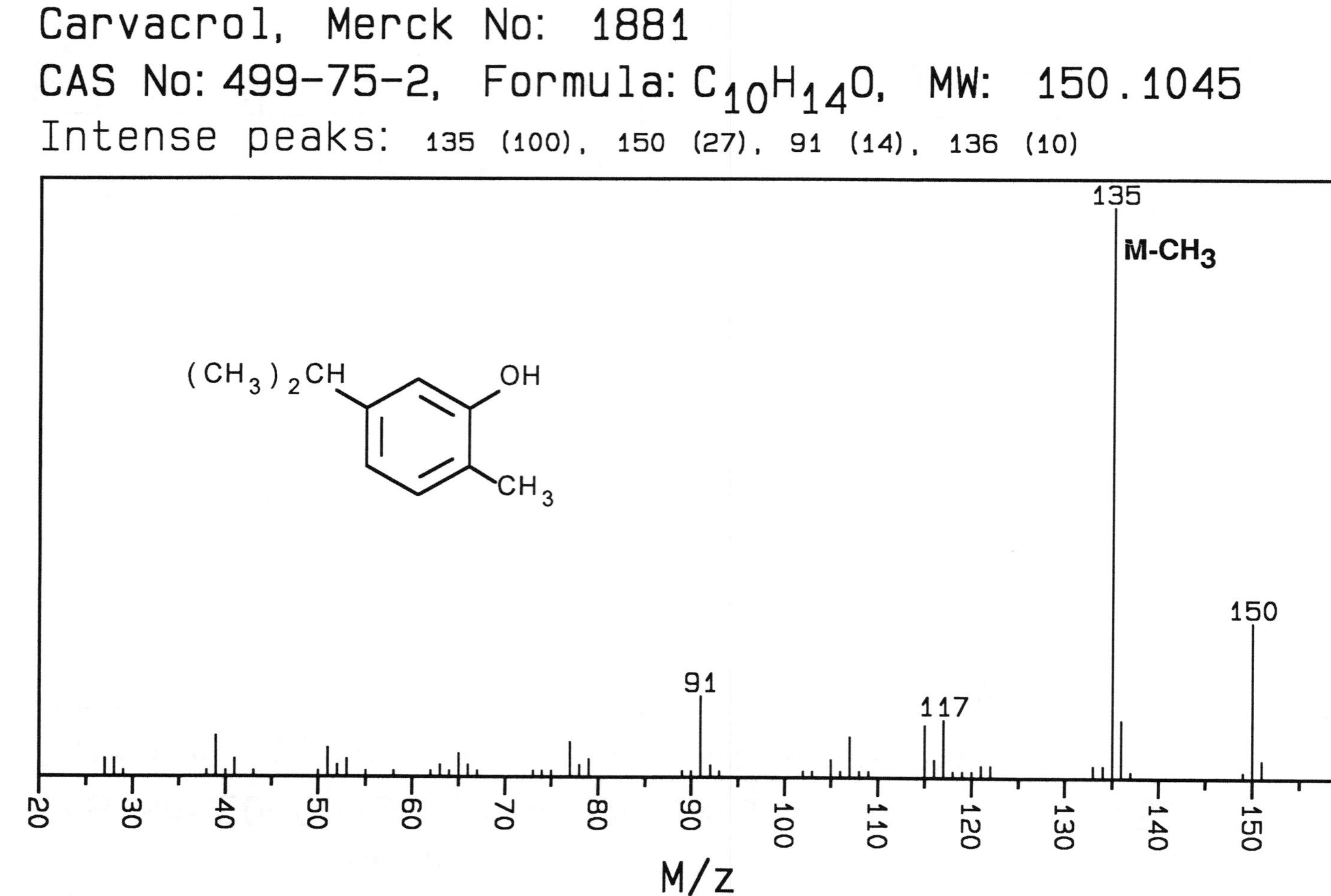
Carvacrol, Merck No: 1881
CAS No: 499-75-2, Formula: $C_{10}H_{14}O$, MW: 150.1045
Intense peaks: 135 (100), 150 (27), 91 (14), 136 (10)
$(CH_3)_2CH$
OH
$CH_3$
135
M-$CH_3$
150
91
117
20
30
40
50
60
70
80
90
100
110
120
130
140
150
160
M/Z

Thymol, Merck No: 9333

CAS No: 89-83-8, Formula: $C_{10}H_{14}O$, MW: 150.1045

Intense peaks: 135 (100), 150 (38), 91 (15), 136 (10)

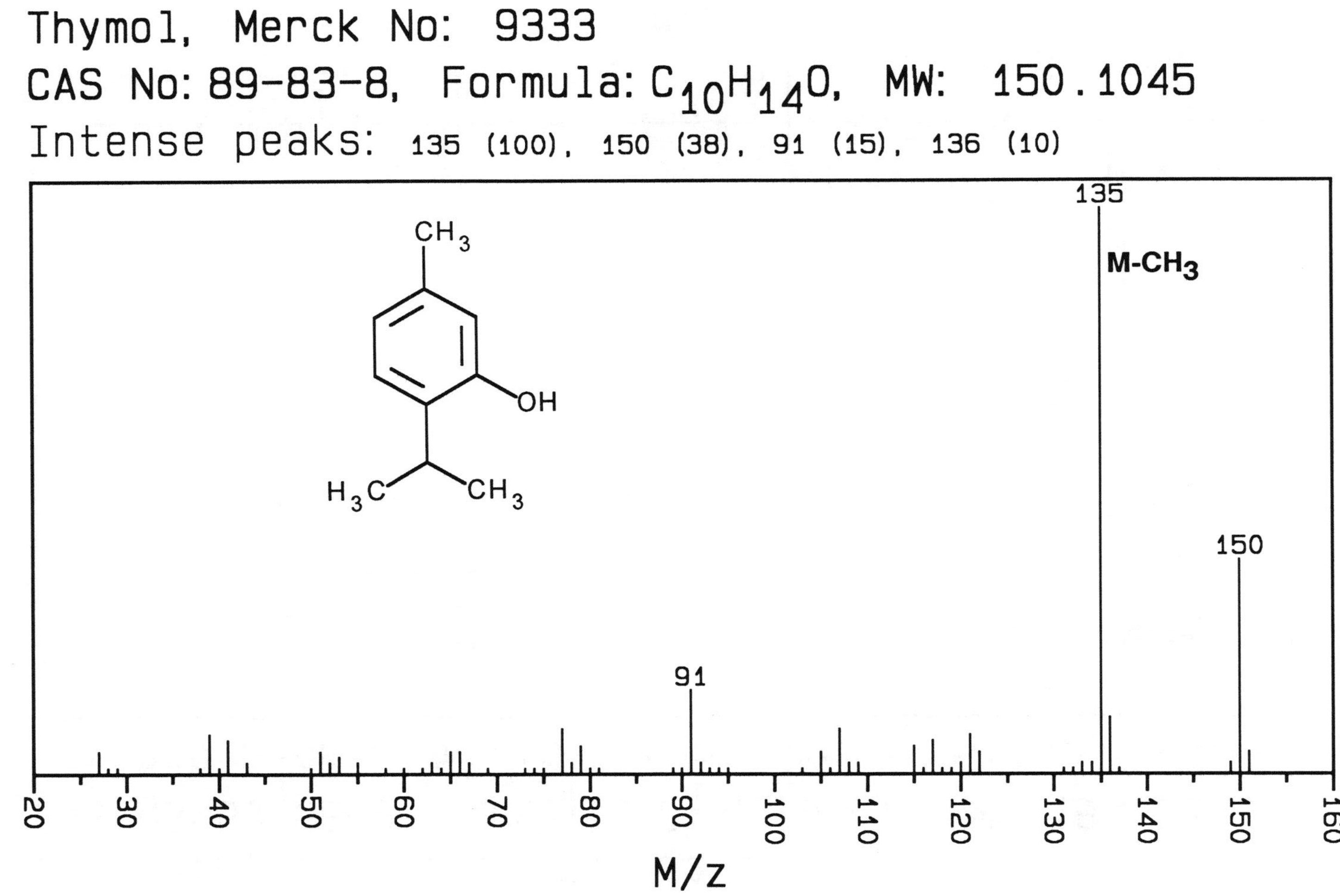

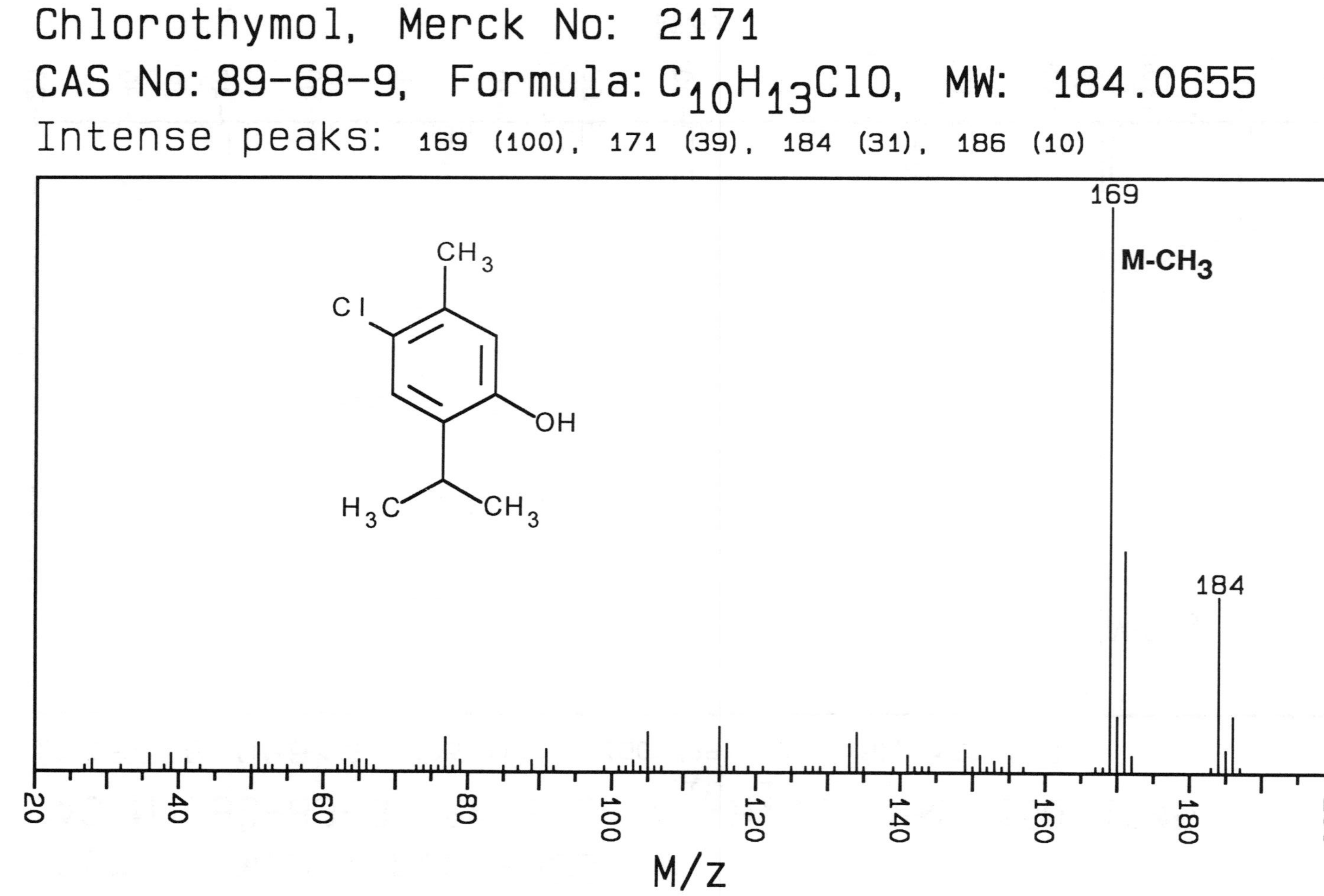
Chlorothymol, Merck No: 2171
CAS No: 89-68-9, Formula: C10H13ClO, MW: 184.0655
Intense peaks: 169 (100), 171 (39), 184 (31), 186 (10)
169
M-CH3
184
CH3
Cl
OH
H3C
CH3
20
40
60
80
100
120
140
160
180
200
M/Z

Phenol, 4-tert-pentyl-, Merck No: 7098

CAS No: 80-46-6, Formula: $C_{11}H_{16}O$, MW: 164.1201

Intense peaks: 107 (100), 164 (44), 135 (38), 95 (29)

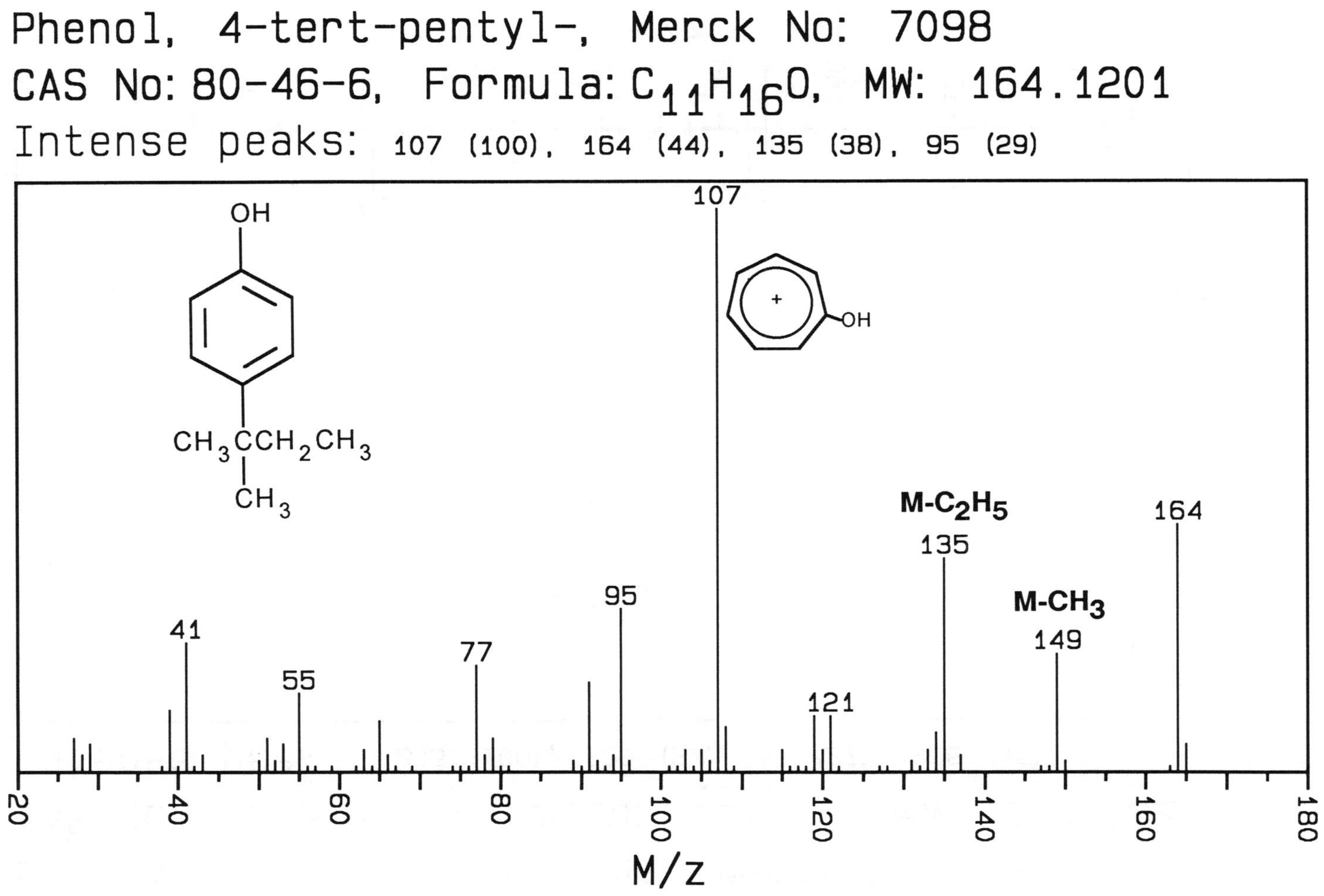

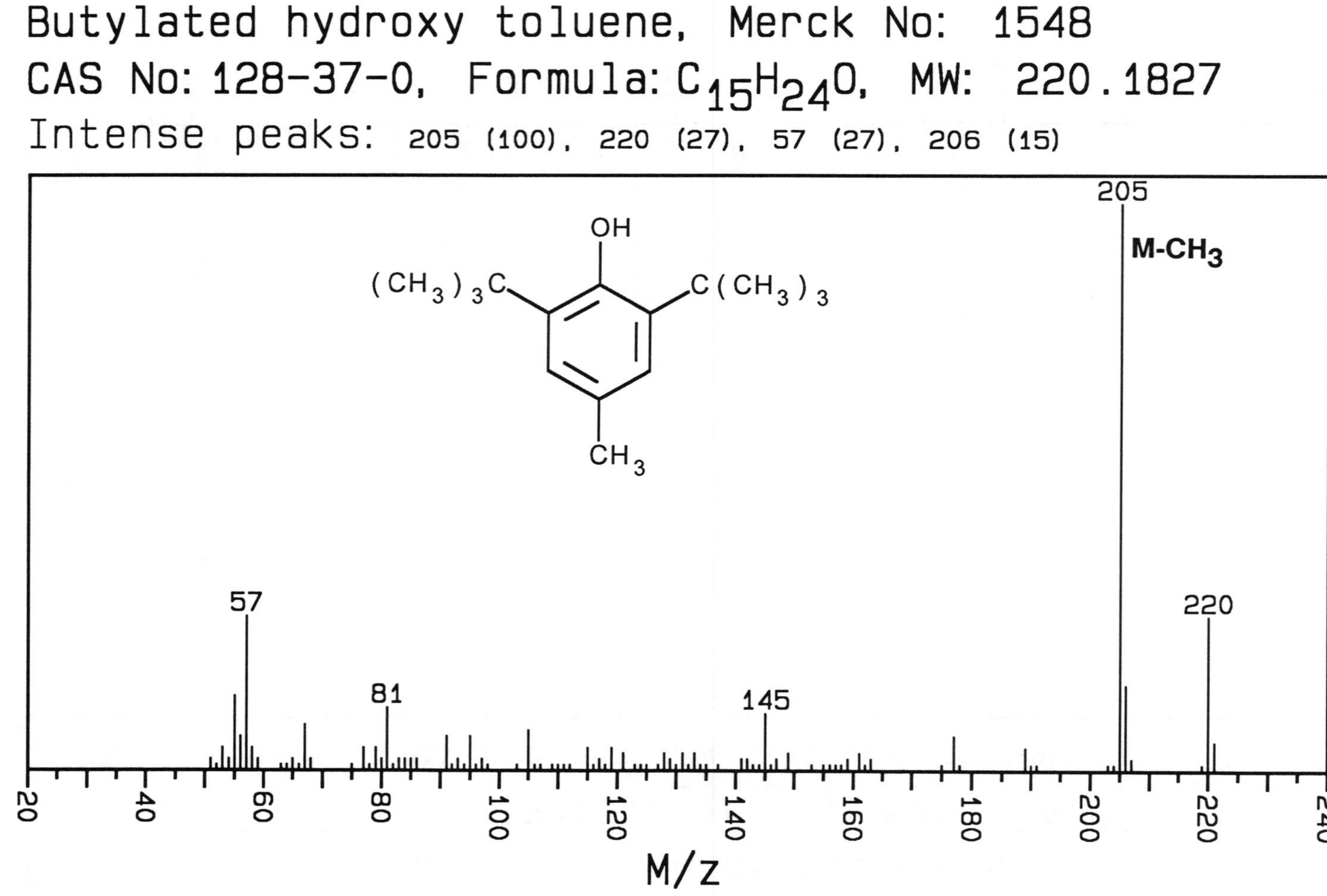
Butylated hydroxy toluene, Merck No: 1548
CAS No: 128-37-0, Formula: $C_{15}H_{24}O$, MW: 220.1827
Intense peaks: 205 (100), 220 (27), 57 (27), 206 (15)
OH
$(CH_3)_3C$
$C(CH_3)_3$
$CH_3$
205
M-$CH_3$
220
57
81
145
20
40
60
80
100
120
140
160
180
200
220
240
M/Z

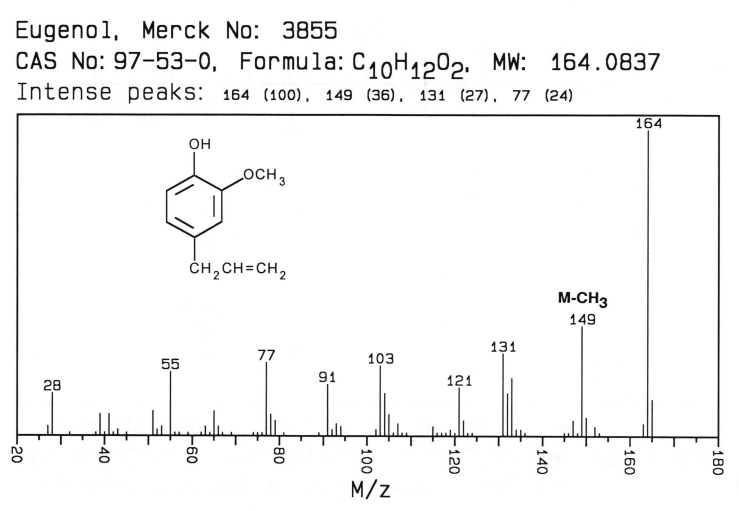

Eugenol, Merck No: 3855

CAS No: 97-53-0, Formula: $C_{10}H_{12}O_2$, MW: 164.0837

Intense peaks: 164 (100), 149 (36), 131 (27), 77 (24)

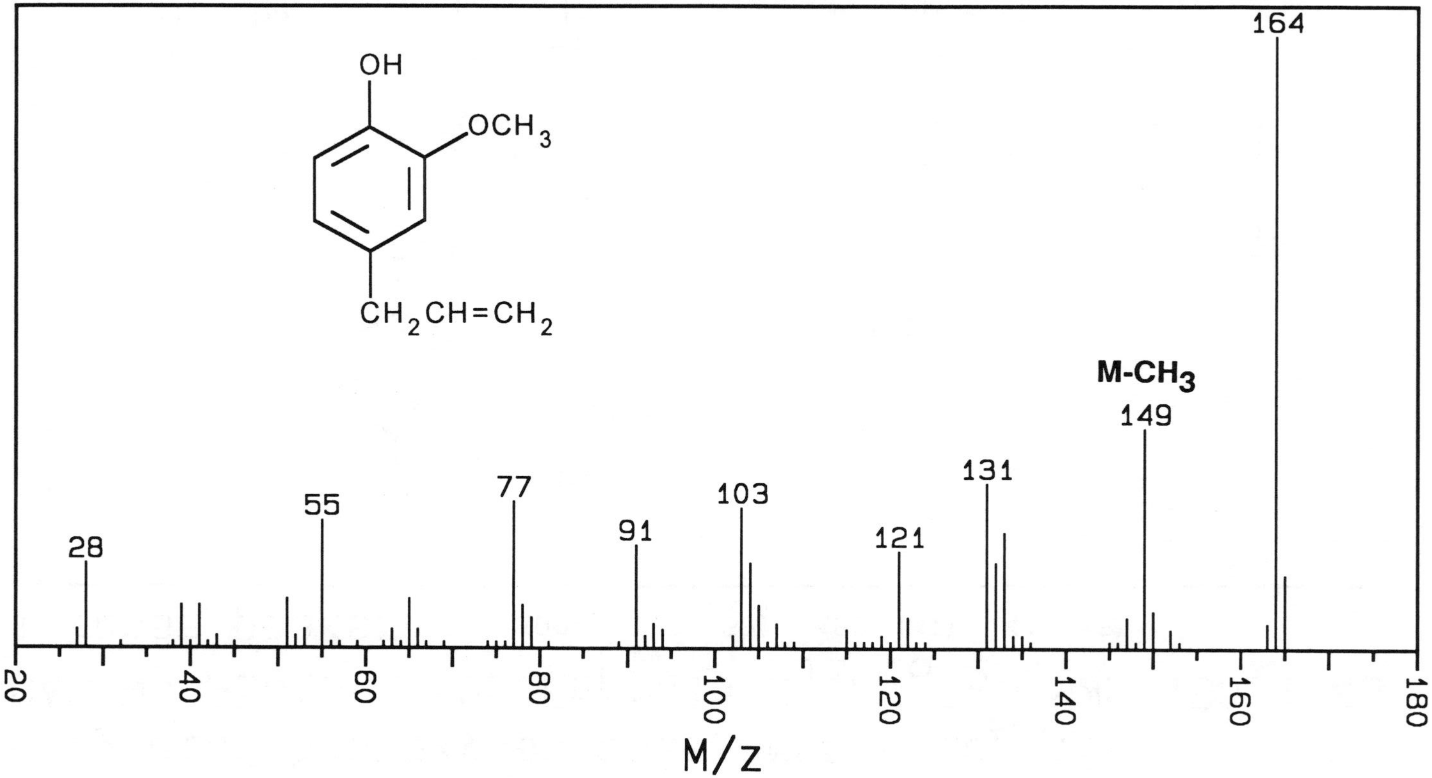

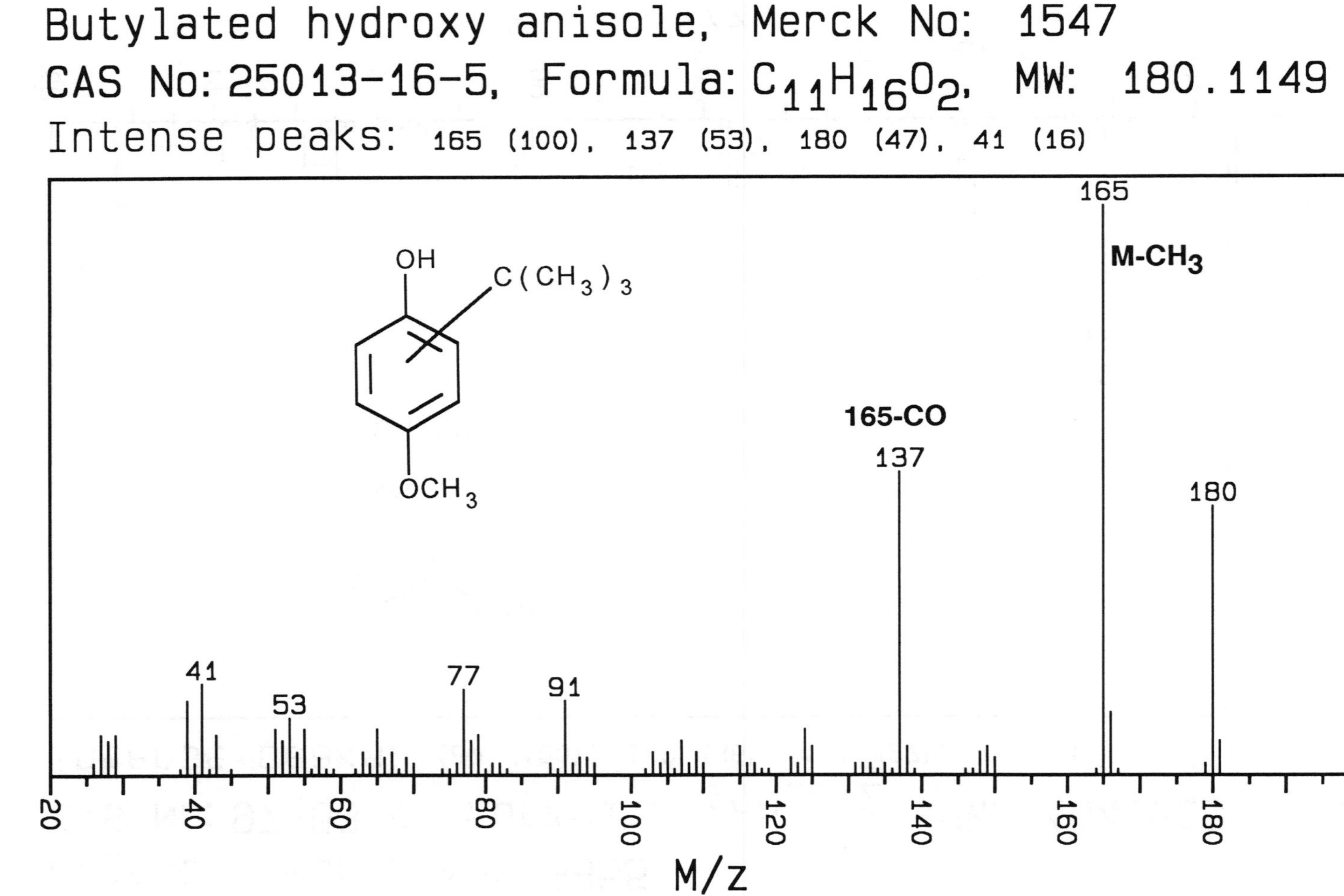
Butylated hydroxy anisole, Merck No: 1547
CAS No: 25013-16-5, Formula: C11H16O2, MW: 180.1149
Intense peaks: 165 (100), 137 (53), 180 (47), 41 (16)
OH
C(CH3)3
OCH3
165
M-CH3
180
165-CO
137
91
77
53
41
20
40
60
80
100
120
140
160
180
200
M/Z

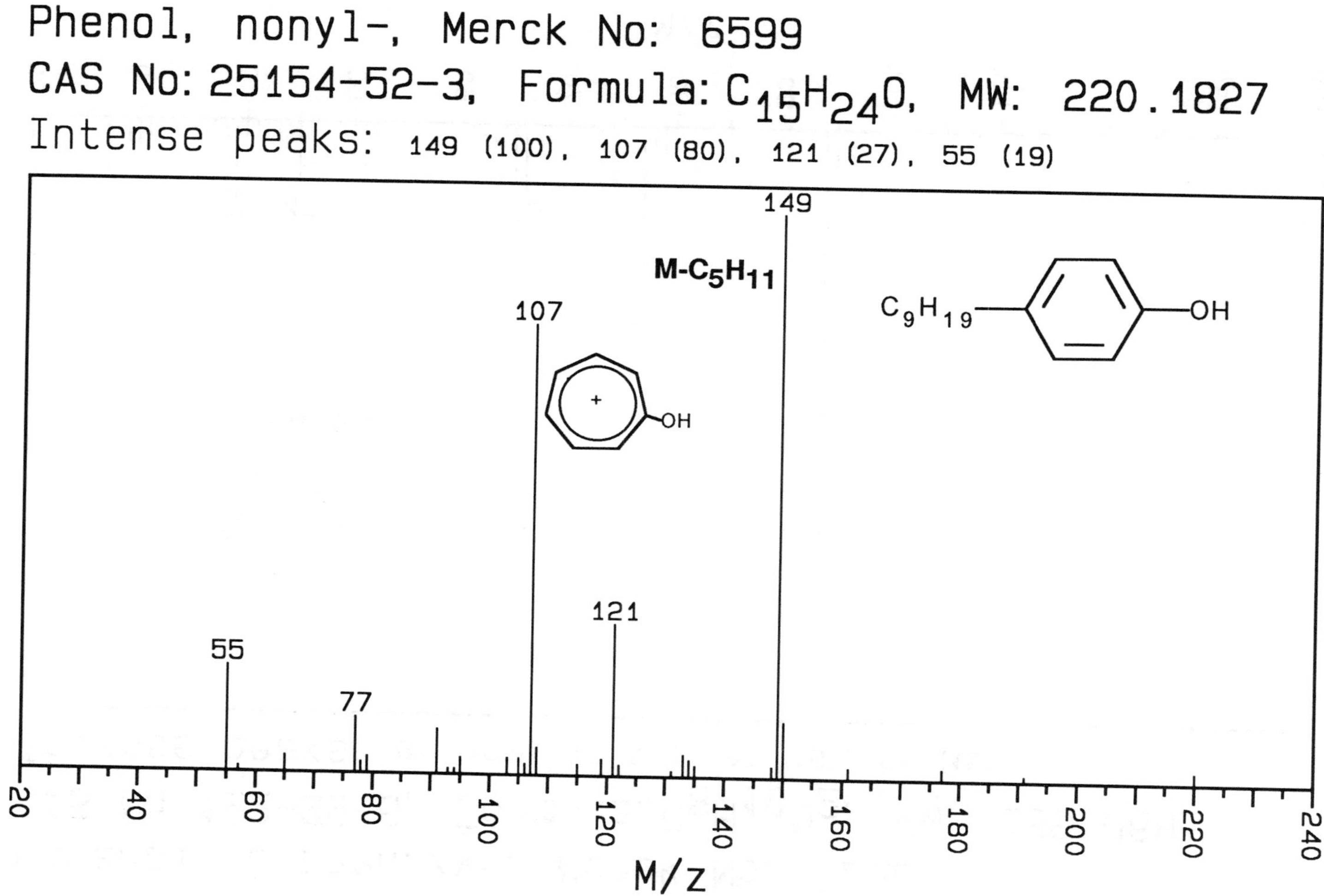
Phenol, nonyl-, Merck No: 6599
CAS No: 25154-52-3, Formula: C15H24O, MW: 220.1827
Intense peaks: 149 (100), 107 (80), 121 (27), 55 (19)
149
M-C5H11
107
C9H19
OH
OH
+
121
55
77
20
40
60
80
100
120
140
160
180
200
220
240
M/Z

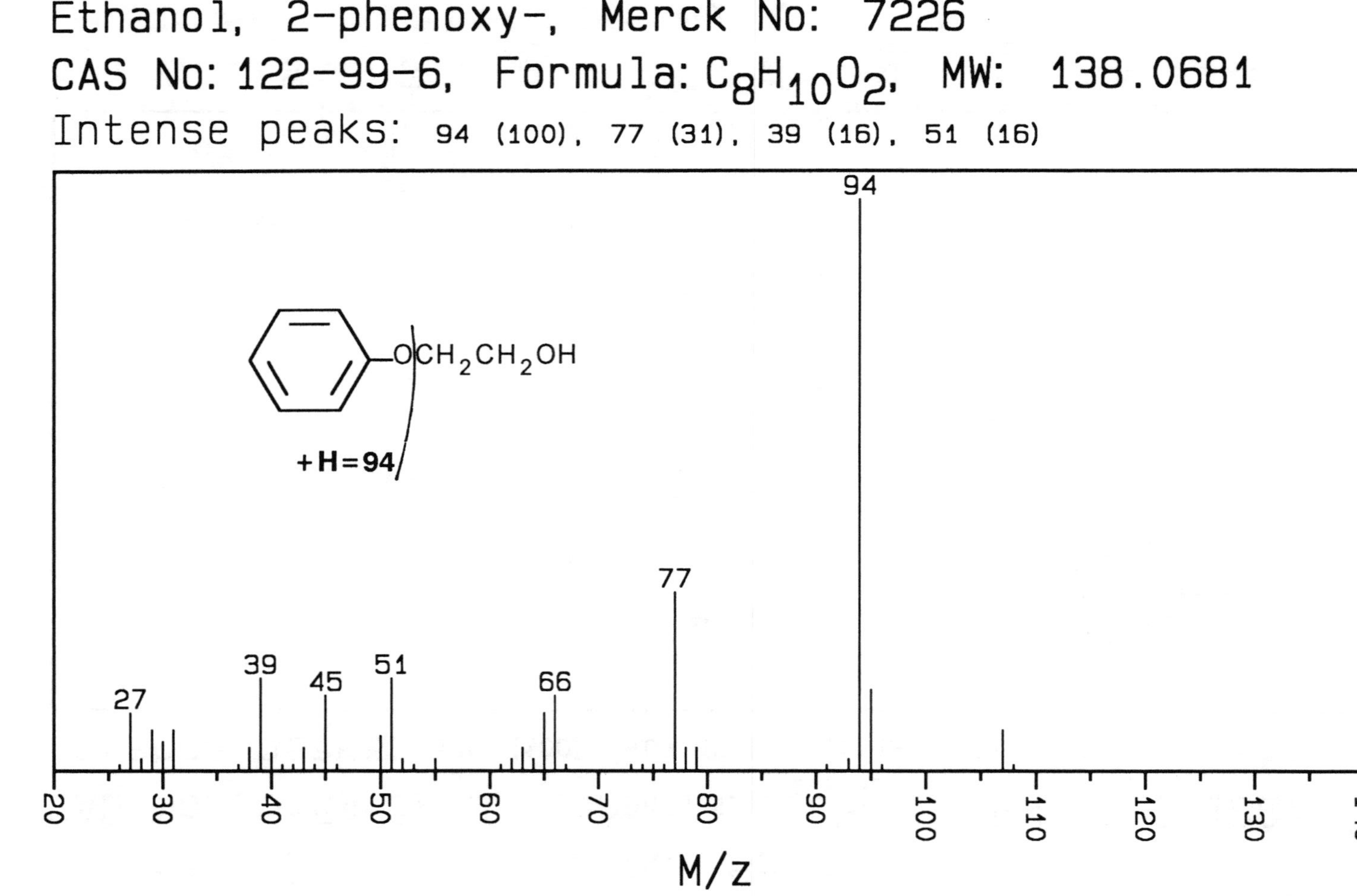
Ethanol, 2-phenoxy-, Merck No: 7226
CAS No: 122-99-6, Formula: $C_8H_{10}O_2$, MW: 138.0681
Intense peaks: 94 (100), 77 (31), 39 (16), 51 (16)
94
OCH2CH2OH
+H=94
77
27
39
45
51
66
20
30
40
50
60
70
80
90
100
110
120
130
140
M/Z

Meclofenoxate, Merck No: 5660

CAS No: 51-68-3, Formula: $C_{12}H_{16}ClNO_3$, MW: 257.0819

Intense peaks: 58 (100), 42 ( 8), 71 ( 6), 111 ( 5)

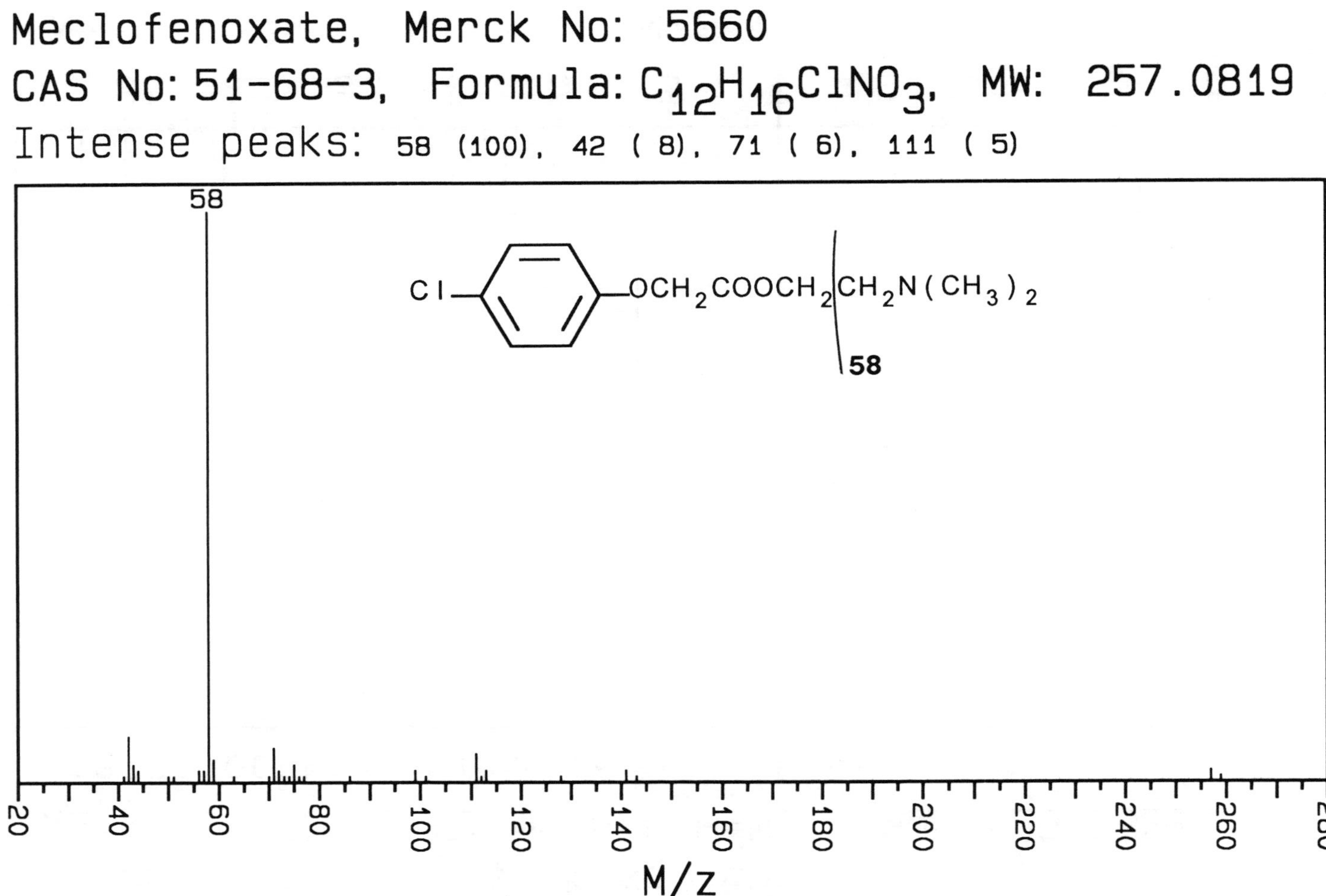

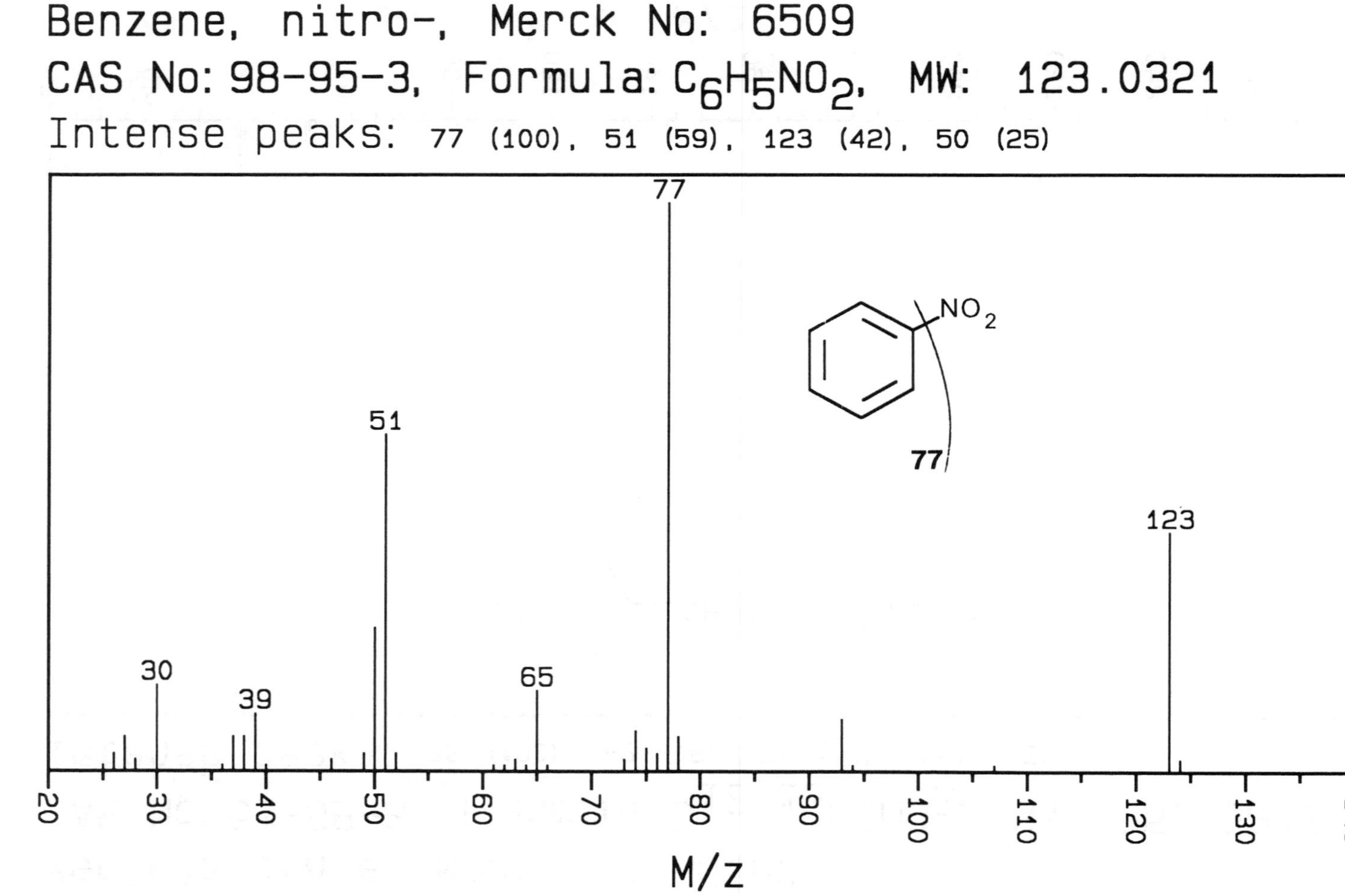
Benzene, nitro-, Merck No: 6509
CAS No: 98-95-3, Formula: C6H5NO2, MW: 123.0321
Intense peaks: 77 (100), 51 (59), 123 (42), 50 (25)
77
NO2
77
51
123
30
39
65
20
30
40
50
60
70
80
90
100
110
120
130
140
M/Z

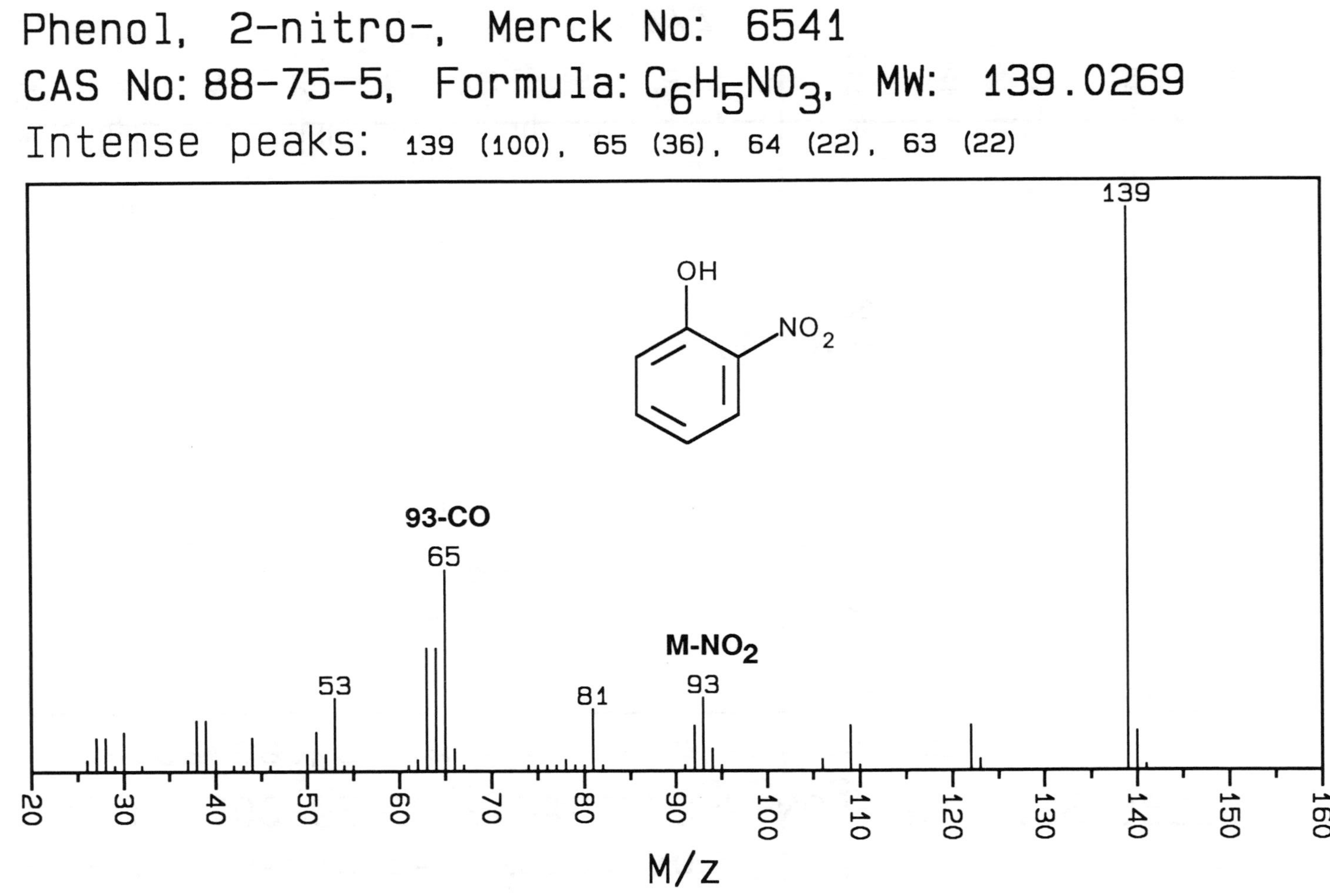

Phenol, 2-nitro-, Merck No: 6541
CAS No: 88-75-5, Formula: $C_6H_5NO_3$, MW: 139.0269
Intense peaks: 139 (100), 65 (36), 64 (22), 63 (22)
139
OH
$NO_2$
93-CO
65
M-$NO_2$
53
81
93
20
30
40
50
60
70
80
90
100
110
120
130
140
150
160
M/Z

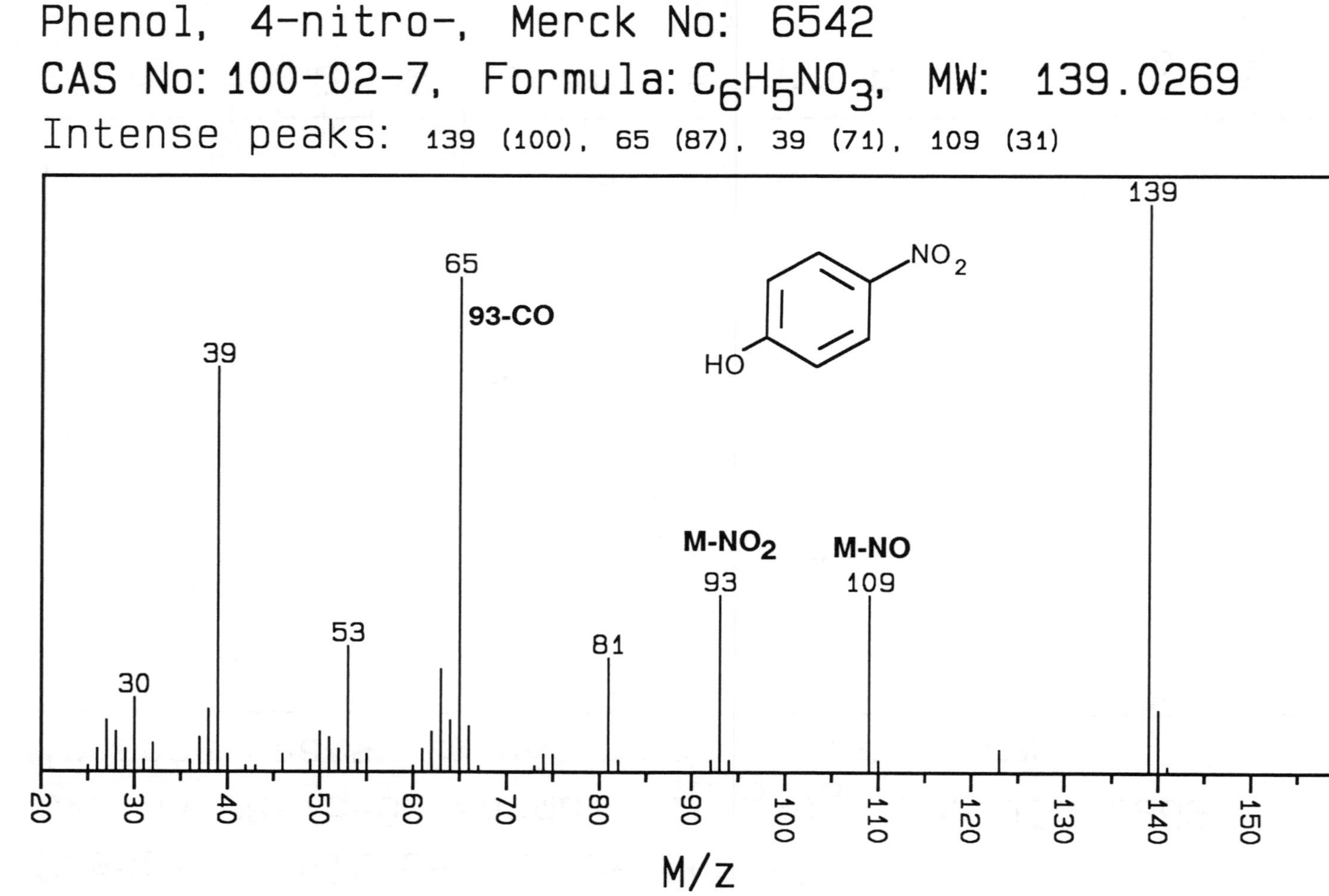
Phenol, 4-nitro-, Merck No: 6542
CAS No: 100-02-7, Formula: C6H5NO3, MW: 139.0269
Intense peaks: 139 (100), 65 (87), 39 (71), 109 (31)
139
65
93-CO
NO2
HO
39
M-NO2
M-NO
93
109
53
81
30
20
30
40
50
60
70
80
90
100
110
120
130
140
150
160
M/Z

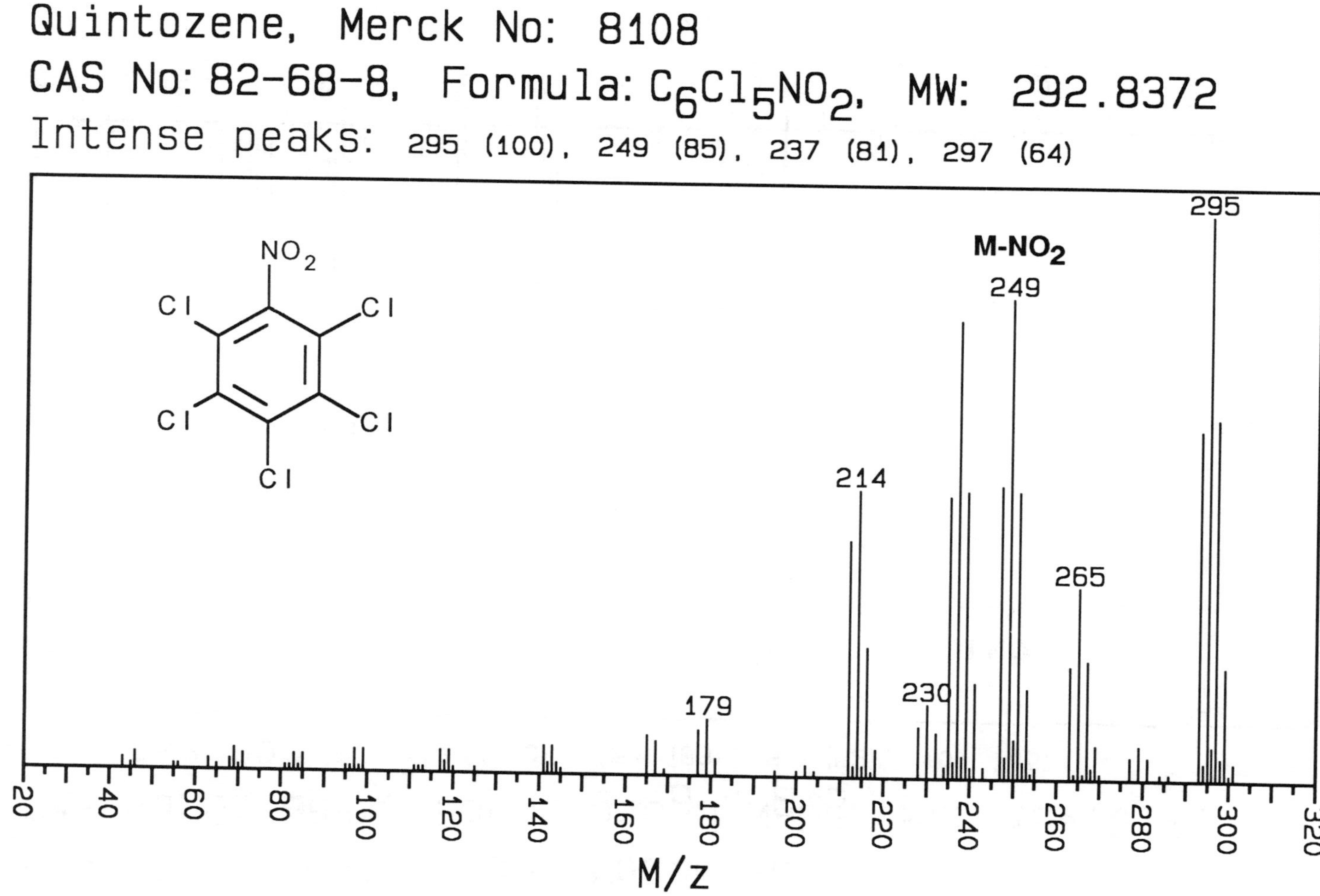
Quintozene, Merck No: 8108
CAS No: 82-68-8, Formula: $C_6Cl_5NO_2$, MW: 292.8372
Intense peaks: 295 (100), 249 (85), 237 (81), 297 (64)
$NO_2$
Cl
Cl
Cl
Cl
Cl
295
M-$NO_2$
249
214
265
230
179
20
40
60
80
100
120
140
160
180
200
220
240
260
280
300
320
M/Z

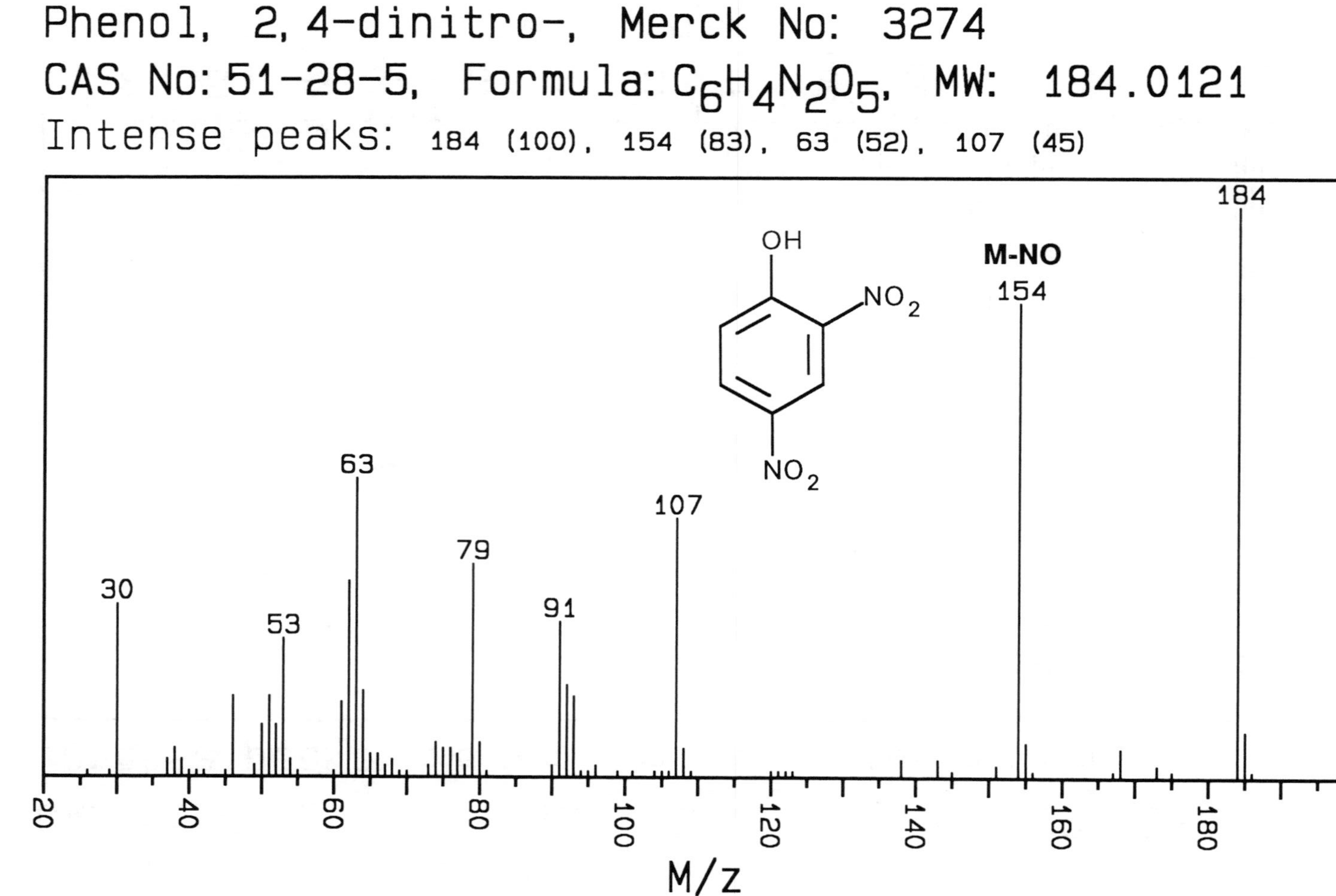
Phenol, 2,4-dinitro-, Merck No: 3274
CAS No: 51-28-5, Formula: $C_6H_4N_2O_5$, MW: 184.0121
Intense peaks: 184 (100), 154 (83), 63 (52), 107 (45)
184
M-NO
154
OH
$NO_2$
$NO_2$
107
91
79
63
53
30
20
40
60
80
100
120
140
160
180
200
M/Z

Phenol, 2-methyl-4,6-dinitro-, Merck No: 3272
CAS No: 534-52-1, Formula: $C_7H_6N_2O_5$, MW: 198.0277
Intense peaks: 121 (100), 53 (96), 39 (93), 52 (79)

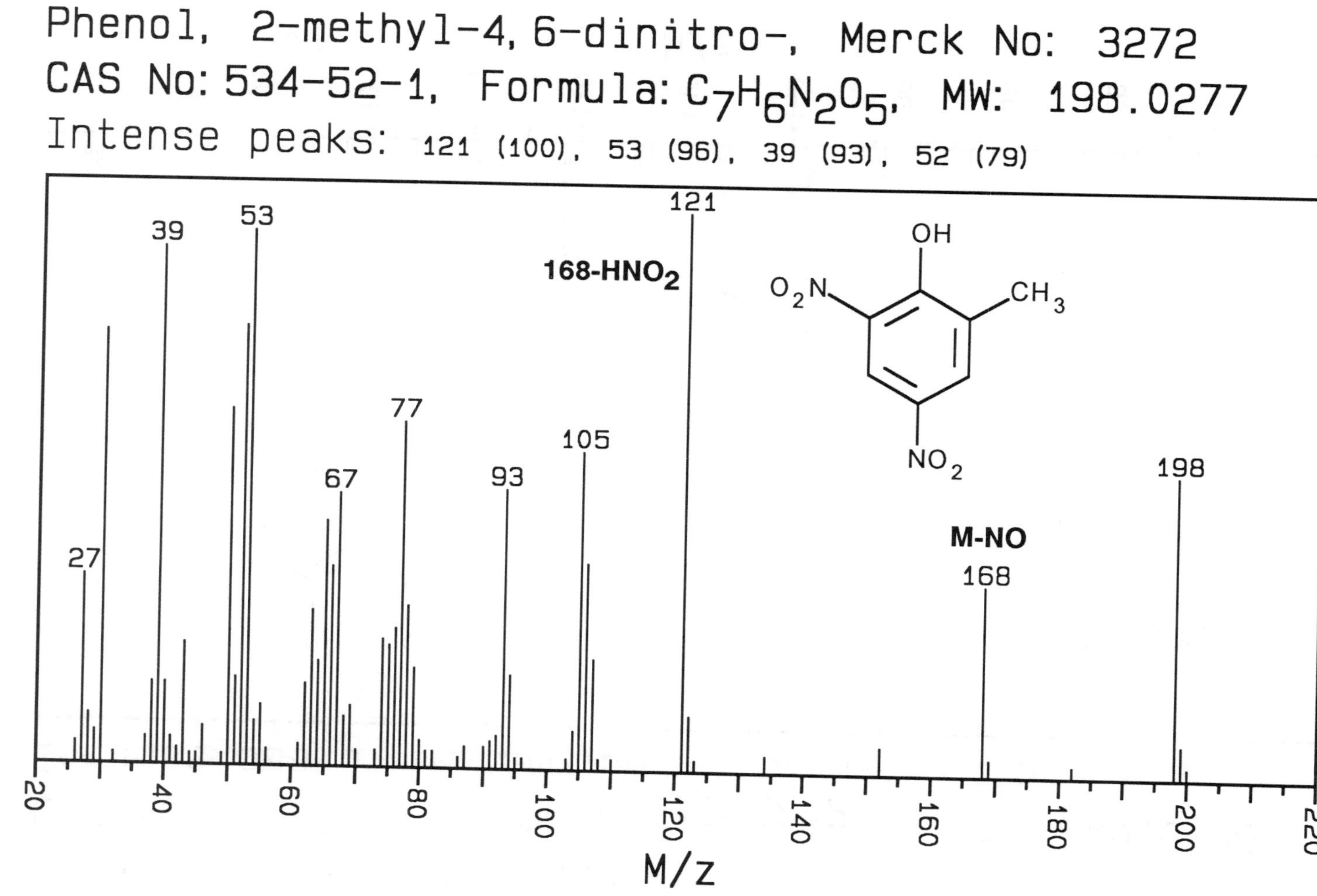

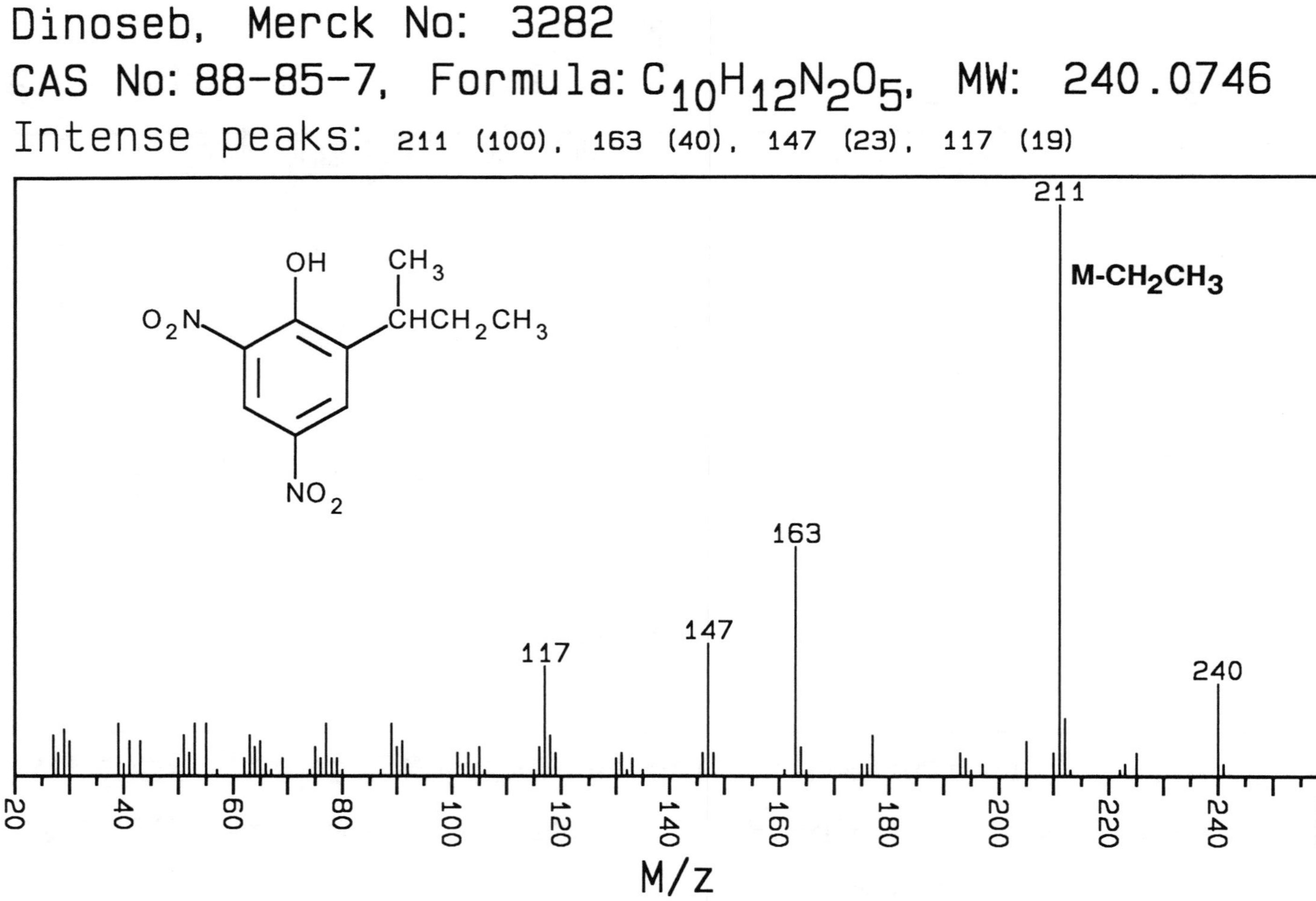
Dinoseb, Merck No: 3282
CAS No: 88-85-7, Formula: $C_{10}H_{12}N_2O_5$, MW: 240.0746
Intense peaks: 211 (100), 163 (40), 147 (23), 117 (19)
OH
$CH_3$
$CHCH_2CH_3$
$O_2N$
$NO_2$
211
M-$CH_2CH_3$
163
147
117
240
20
40
60
80
100
120
140
160
180
200
220
240
260
M/Z

Phenol, 2-cyclohexyl-4,6-dinitro-, Merck No: 2739
CAS No: 131-89-5, Formula: $C_{12}H_{14}N_2O_5$, MW: 266.0903
Intense peaks: 266 (100), 231 (97), 55 (81), 41 (79)

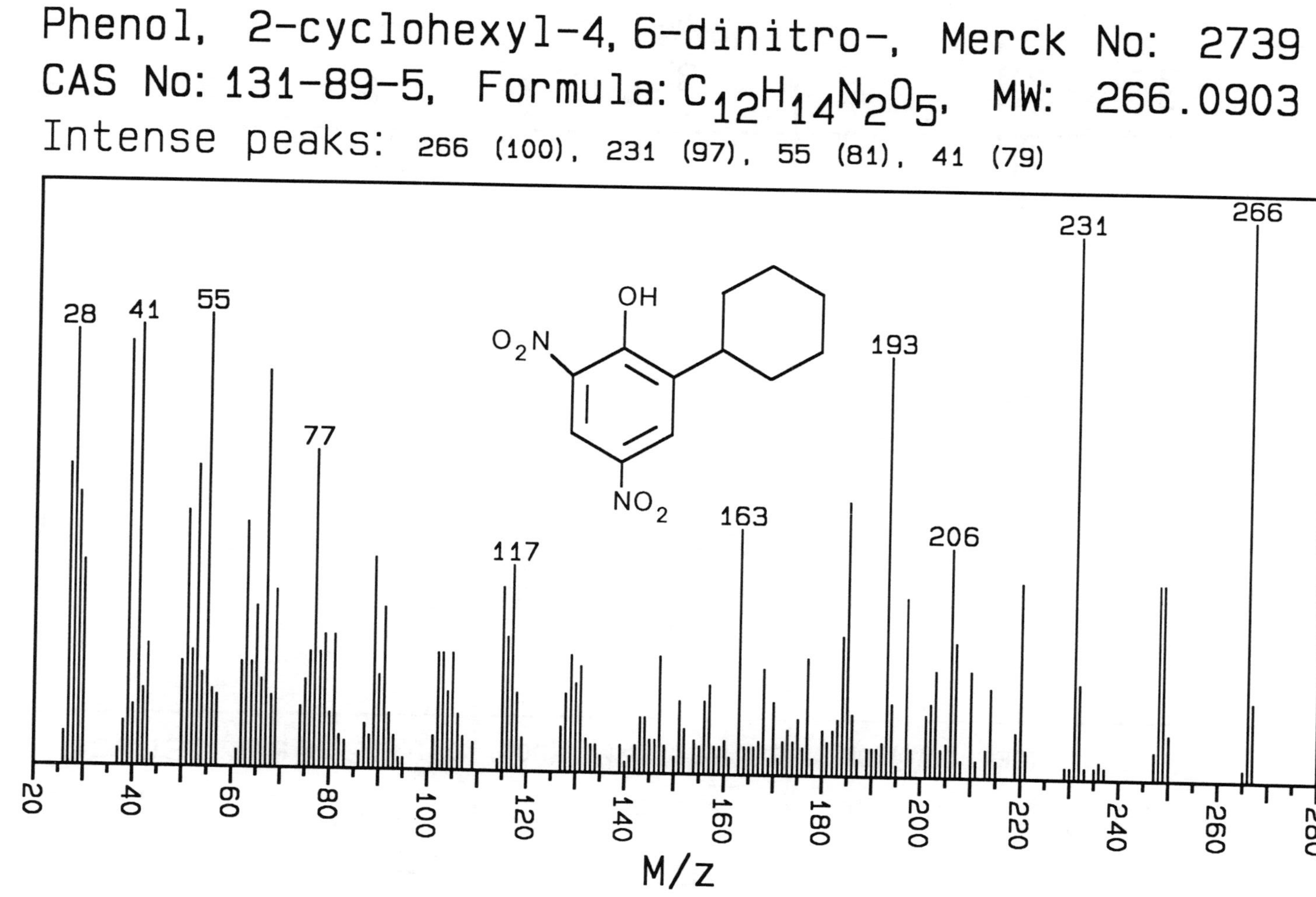

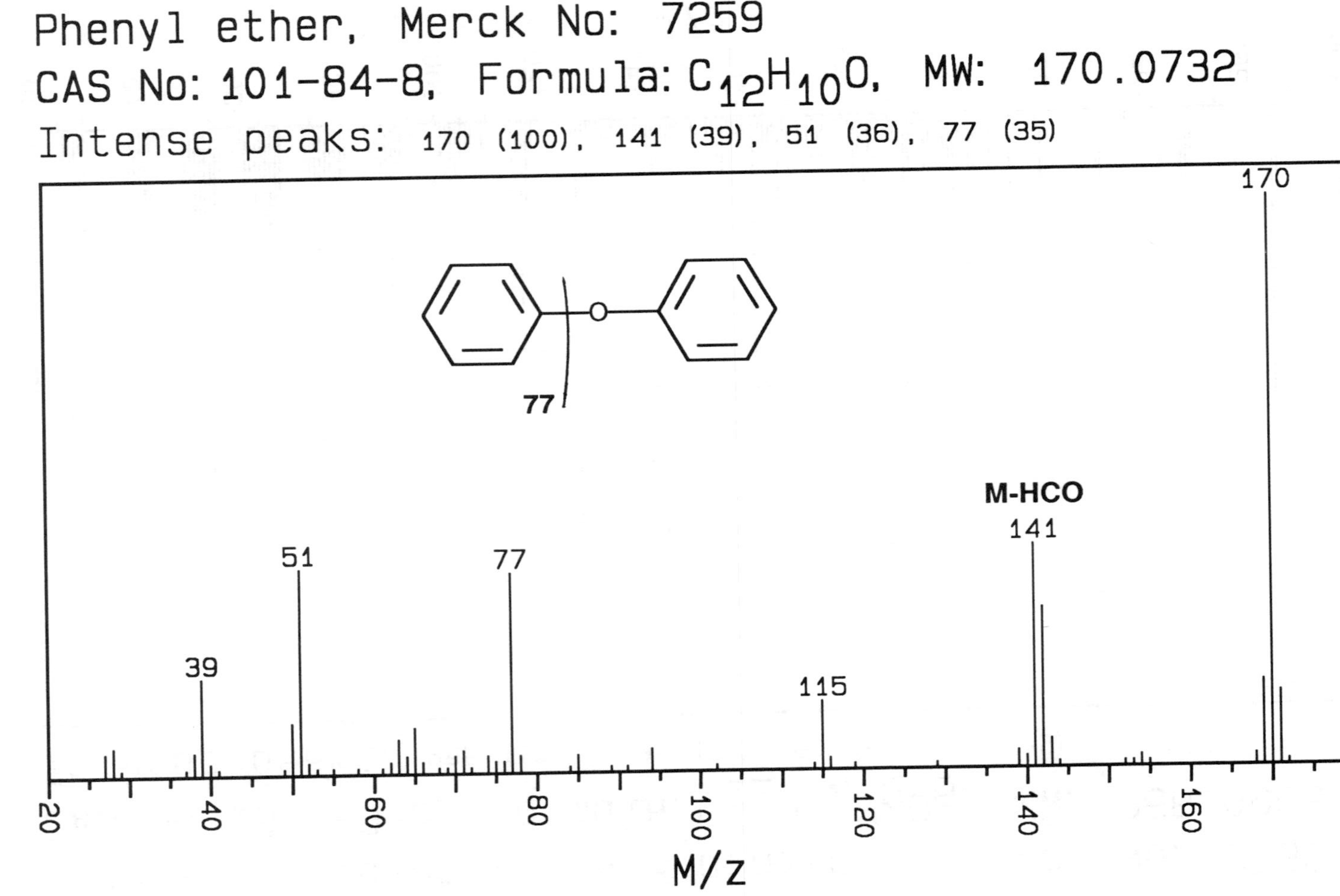
Phenyl ether, Merck No: 7259
CAS No: 101-84-8, Formula: C12H10O, MW: 170.0732
Intense peaks: 170 (100), 141 (39), 51 (36), 77 (35)
170
77
M-HCO
141
51
77
39
115
20
40
60
80
100
120
140
160
180
M/Z

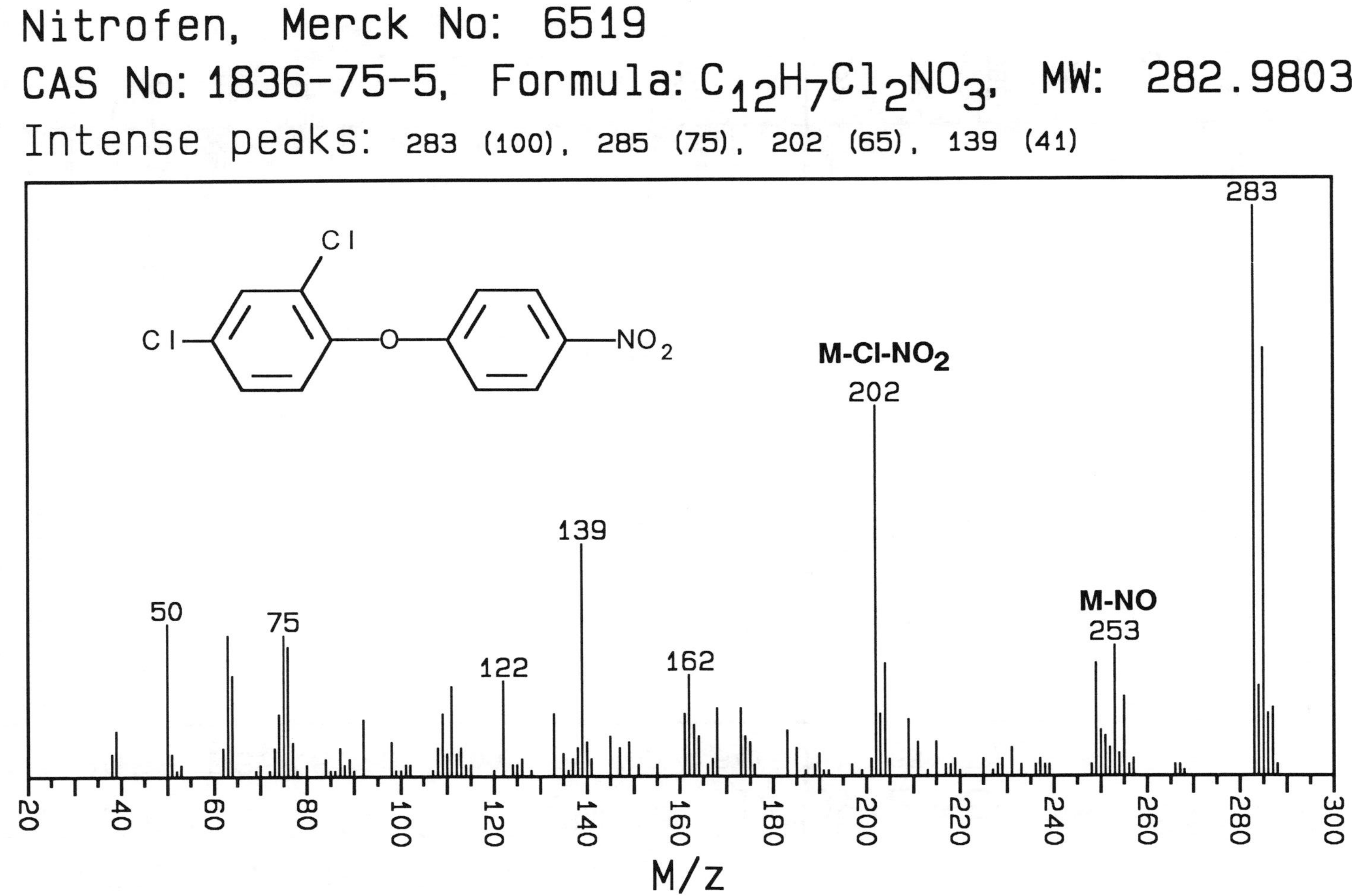
Nitrofen, Merck No: 6519
CAS No: 1836-75-5, Formula: C12H7Cl2NO3, MW: 282.9803
Intense peaks: 283 (100), 285 (75), 202 (65), 139 (41)
Cl
Cl
O
NO2
283
M-Cl-NO2
202
139
50
75
122
162
M-NO
253
20
40
60
80
100
120
140
160
180
200
220
240
260
280
300
M/Z

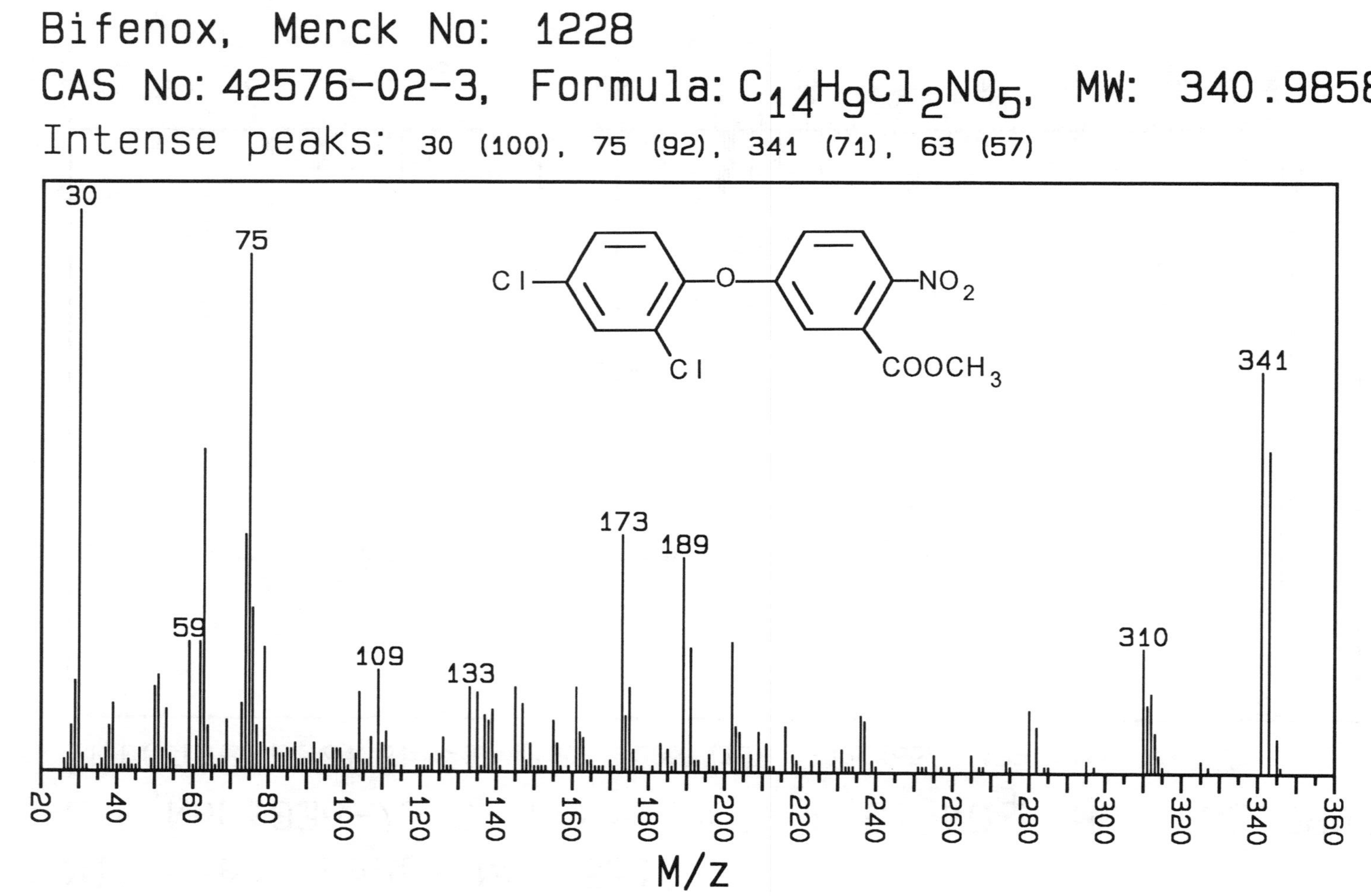
Bifenox, Merck No: 1228
CAS No: 42576-02-3, Formula: $C_{14}H_9Cl_2NO_5$, MW: 340.9858
Intense peaks: 30 (100), 75 (92), 341 (71), 63 (57)
Cl
O
NO2
Cl
COOCH3
30
75
59
109
133
173
189
310
341
20
40
60
80
100
120
140
160
180
200
220
240
260
280
300
320
340
360
M/Z

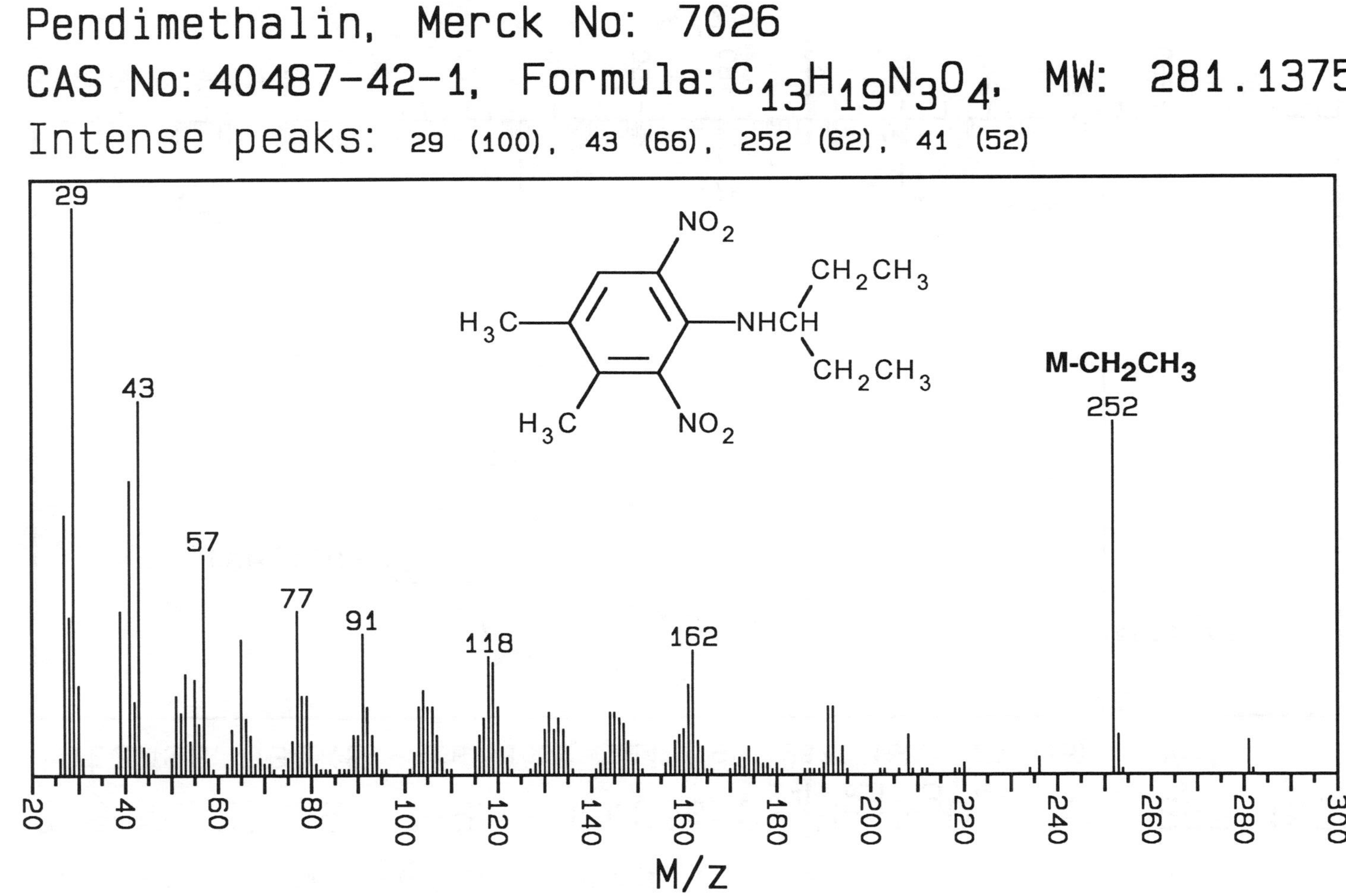
Pendimethalin, Merck No: 7026
CAS No: 40487-42-1, Formula: $C_{13}H_{19}N_3O_4$, MW: 281.1375
Intense peaks: 29 (100), 43 (66), 252 (62), 41 (52)
29
43
57
77
91
118
162
252
M-$CH_2CH_3$
$NO_2$
$NO_2$
$H_3C$
$H_3C$
NHCH
$CH_2CH_3$
$CH_2CH_3$
20
40
60
80
100
120
140
160
180
200
220
240
260
280
300
M/Z

Butralin, Merck No: 1532

CAS No: 33629-47-9, Formula: $C_{14}H_{21}N_3O_4$, MW: 295.1532

Intense peaks: 266 (100), 224 (19), 267 (16), 220 (15)

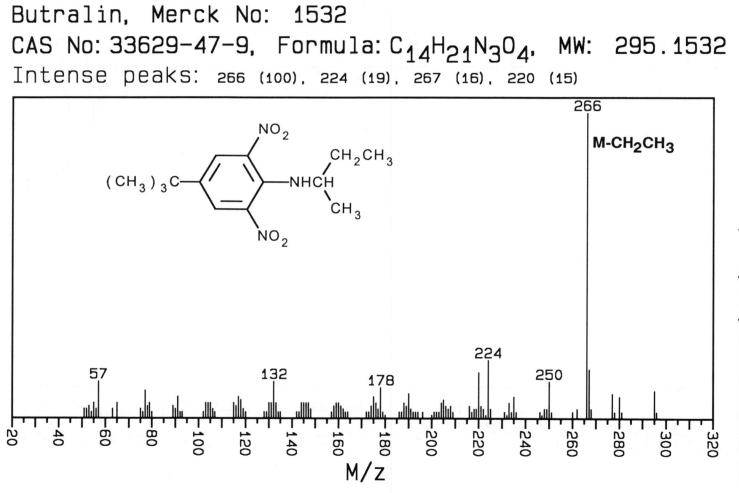

Isopropalin, Merck No: 5089

CAS No: 33820-53-0, Formula: $C_{15}H_{23}N_3O_4$, MW: 309.1688

Intense peaks: 43 (100), 280 (86), 41 (84), 27 (69)

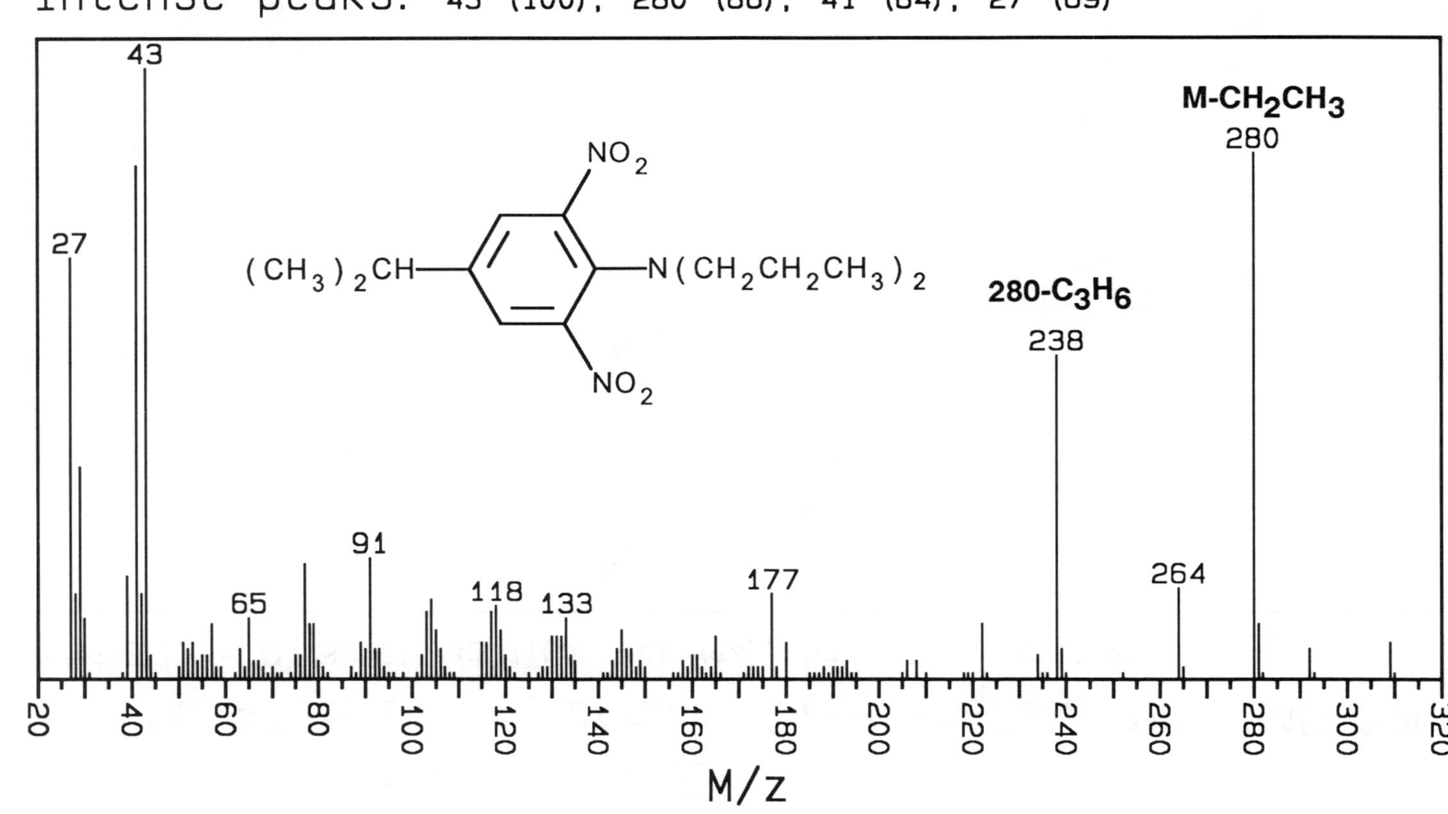

Nitralin, Merck No: 6486

CAS No: 4726-14-1, Formula: $C_{13}H_{19}N_3O_6S$, MW: 345.0995

Intense peaks: 43 (100), 41 (64), 274 (45), 316 (39)

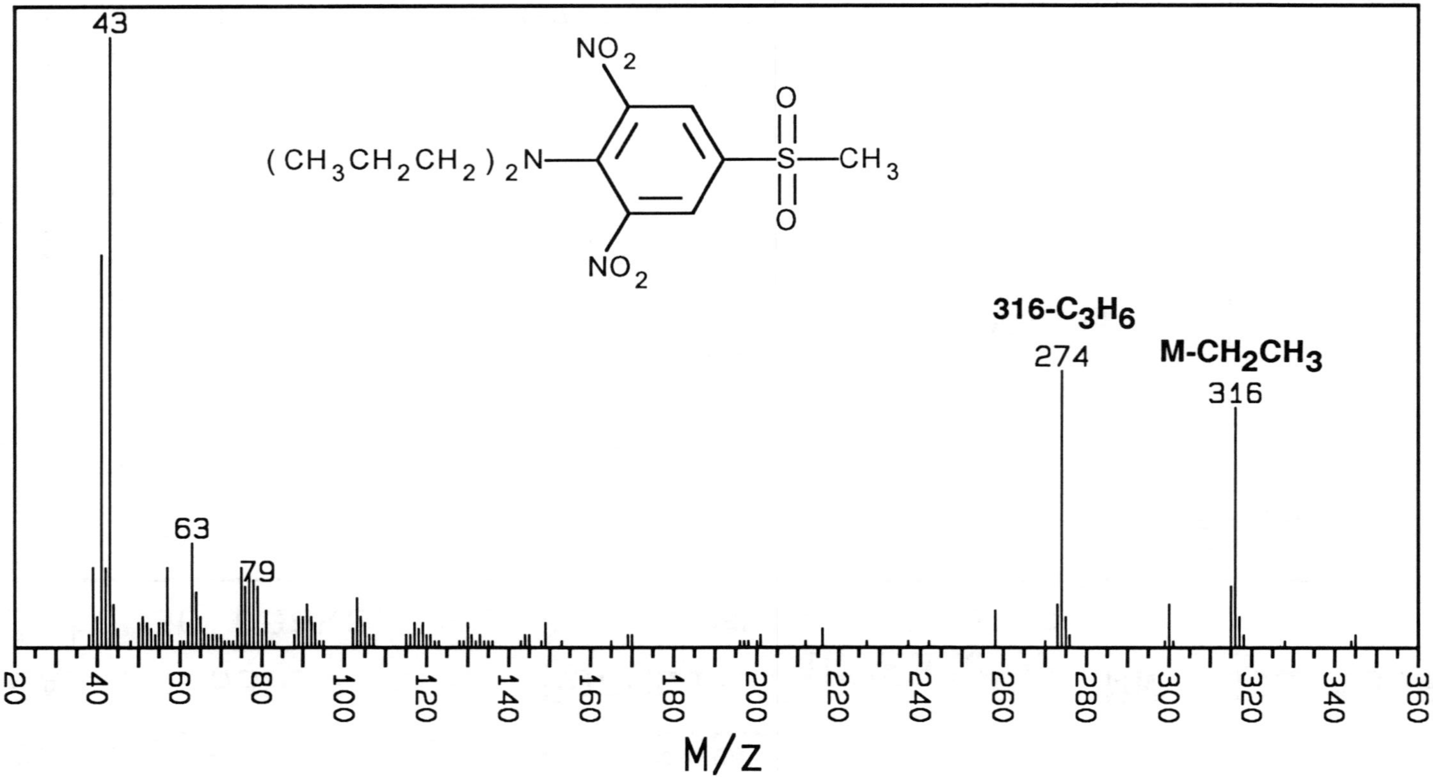

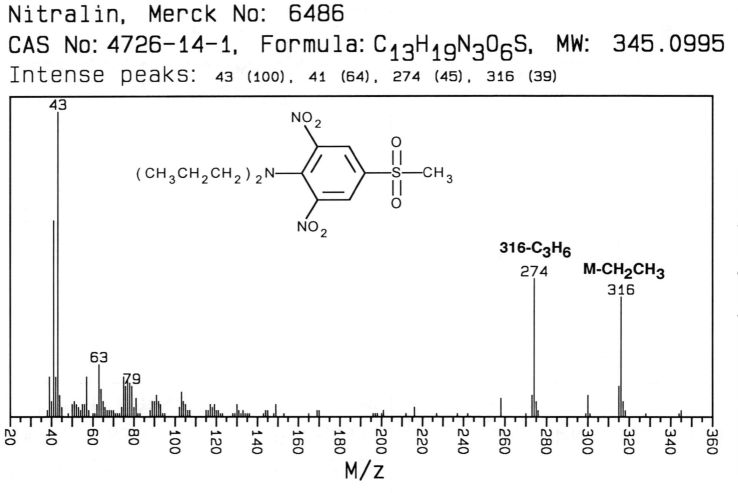

Oryzalin, Merck No: 6840

CAS No: 19044-88-3, Formula: $C_{12}H_{18}N_4O_6S$, MW: 346.0947

Intense peaks: 43 (100), 41 (66), 27 (55), 317 (38)

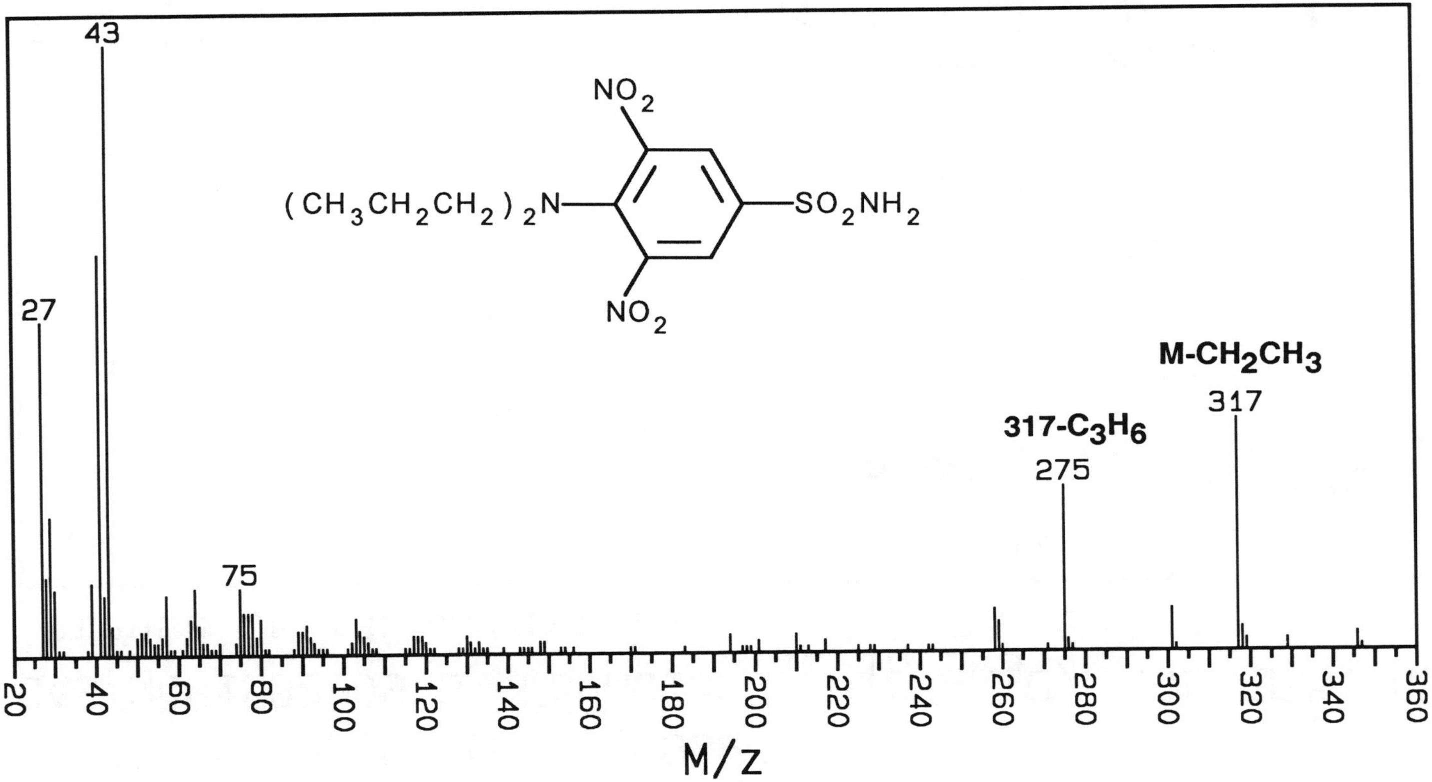

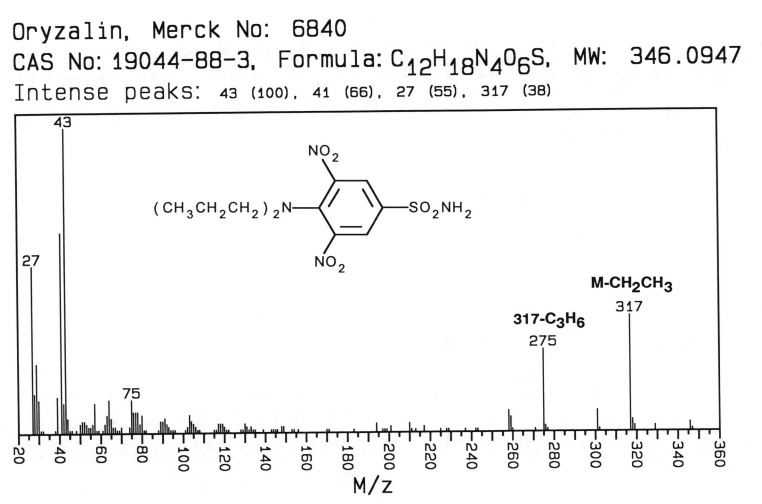

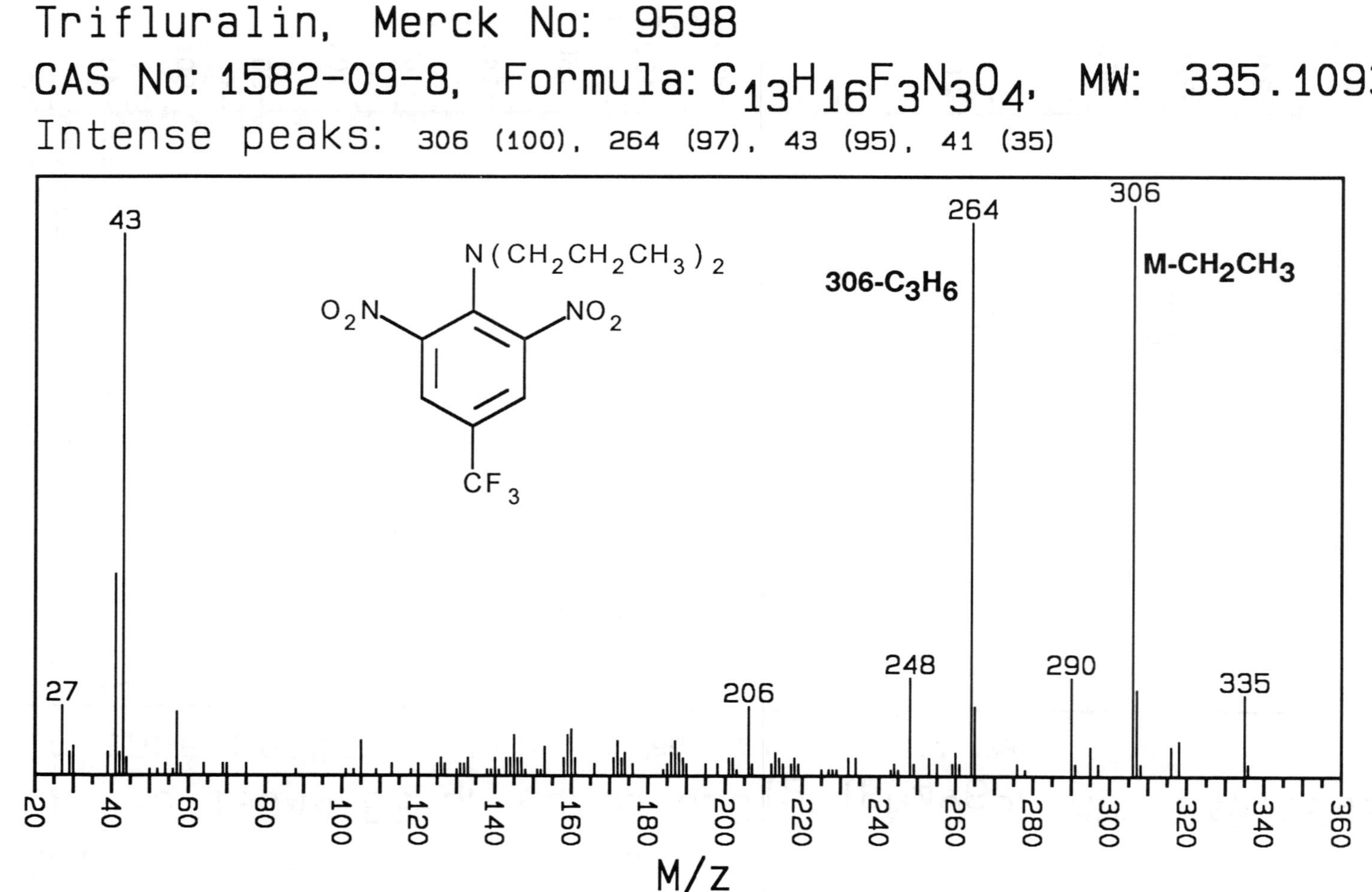
Trifluralin, Merck No: 9598
CAS No: 1582-09-8, Formula: $C_{13}H_{16}F_3N_3O_4$, MW: 335.1093
Intense peaks: 306 (100), 264 (97), 43 (95), 41 (35)
$N(CH_2CH_2CH_3)_2$
$O_2N$
$NO_2$
$CF_3$
306-$C_3H_6$
M-$CH_2CH_3$
43
27
206
248
264
290
306
335
M/Z

Benfluralin, Merck No: 1048

CAS No: 1861-40-1, Formula: $C_{13}H_{16}F_3N_3O_4$, MW: 335.1093

Intense peaks: 292 (100), 41 (39), 43 (38), 264 (29)

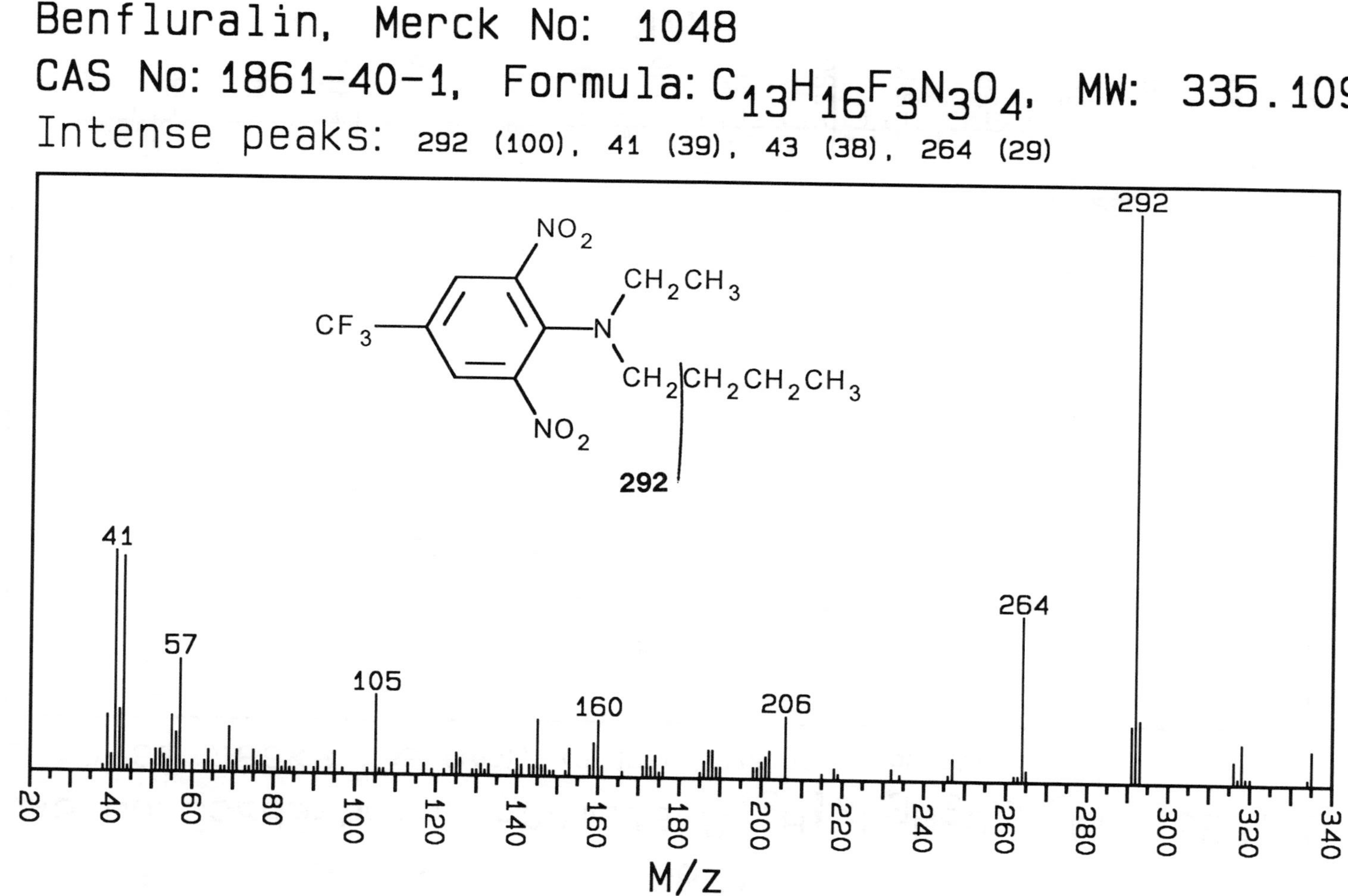

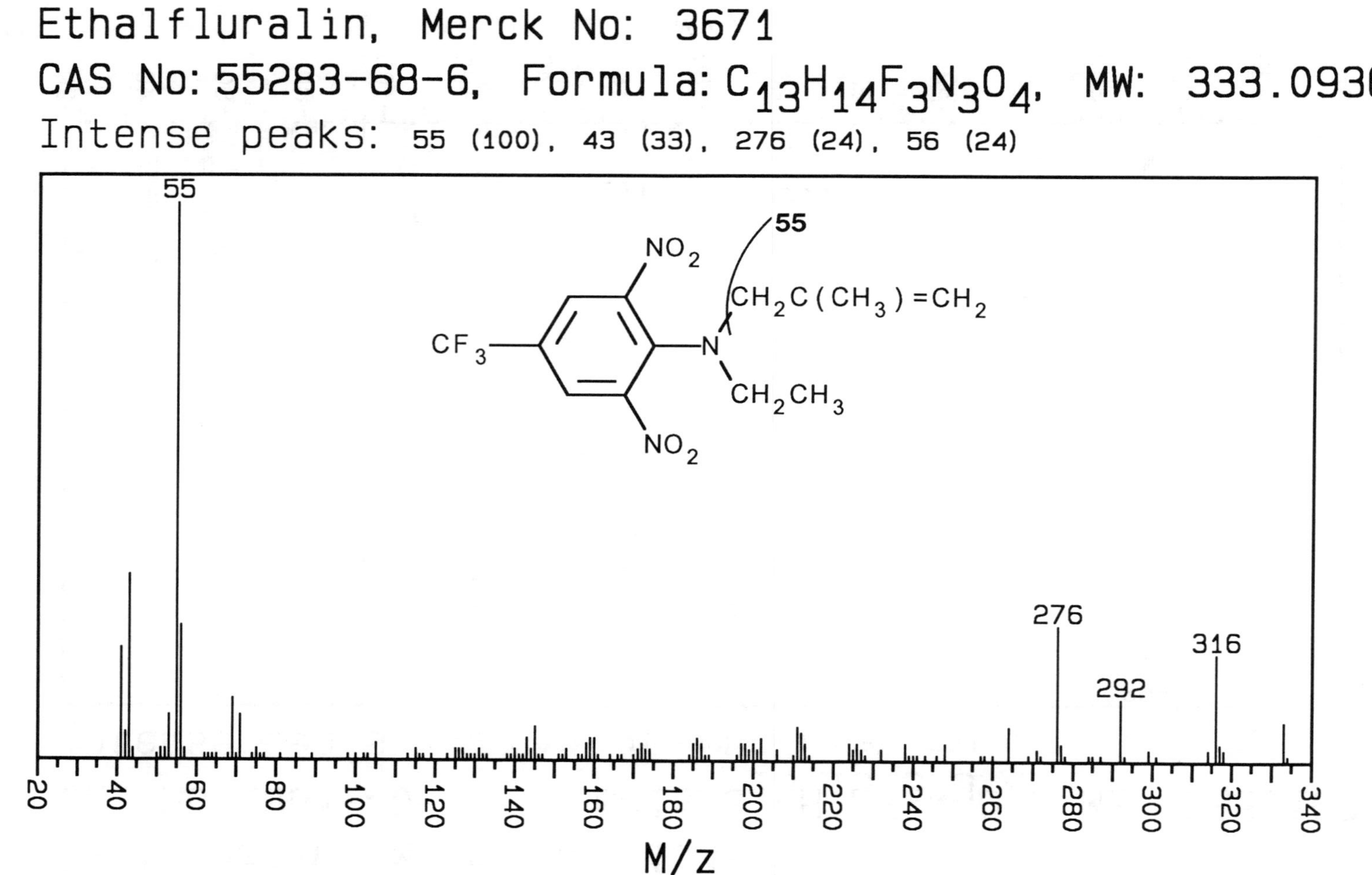
Ethalfluralin, Merck No: 3671
CAS No: 55283-68-6, Formula: C13H14F3N3O4, MW: 333.0936
Intense peaks: 55 (100), 43 (33), 276 (24), 56 (24)
55
NO2
55
CH2C(CH3)=CH2
CF3
N
CH2CH3
NO2
276
316
292
20
40
60
80
100
120
140
160
180
200
220
240
260
280
300
320
340
M/Z

Fluchloralin, Merck No: 4052 MW: 355.0547

CAS No: 33245-39-5, Formula: $C_{12}H_{13}ClF_3N_3O_4$

Intense peaks: 63 (100), 43 (92), 27 (81), 306 (52)

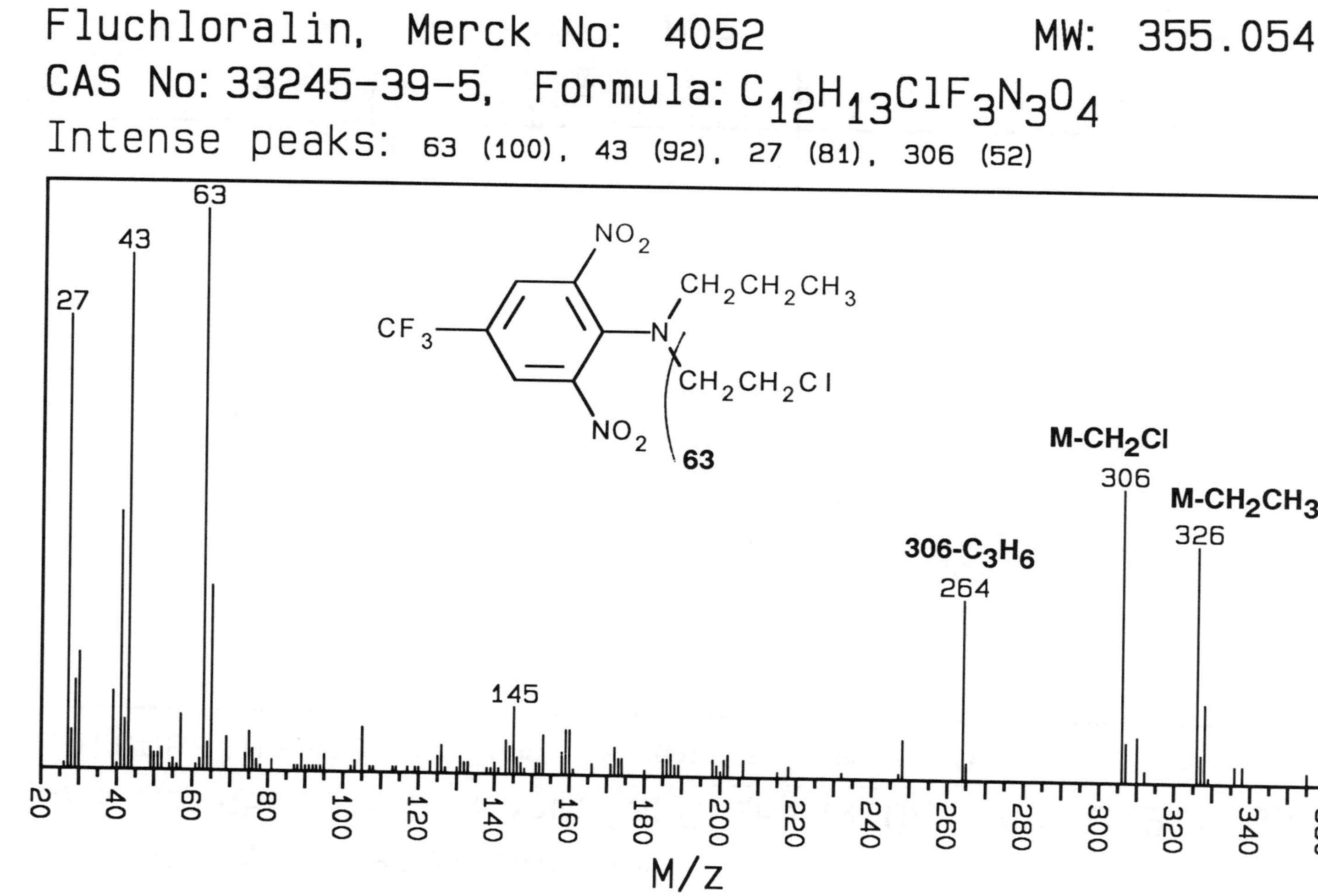

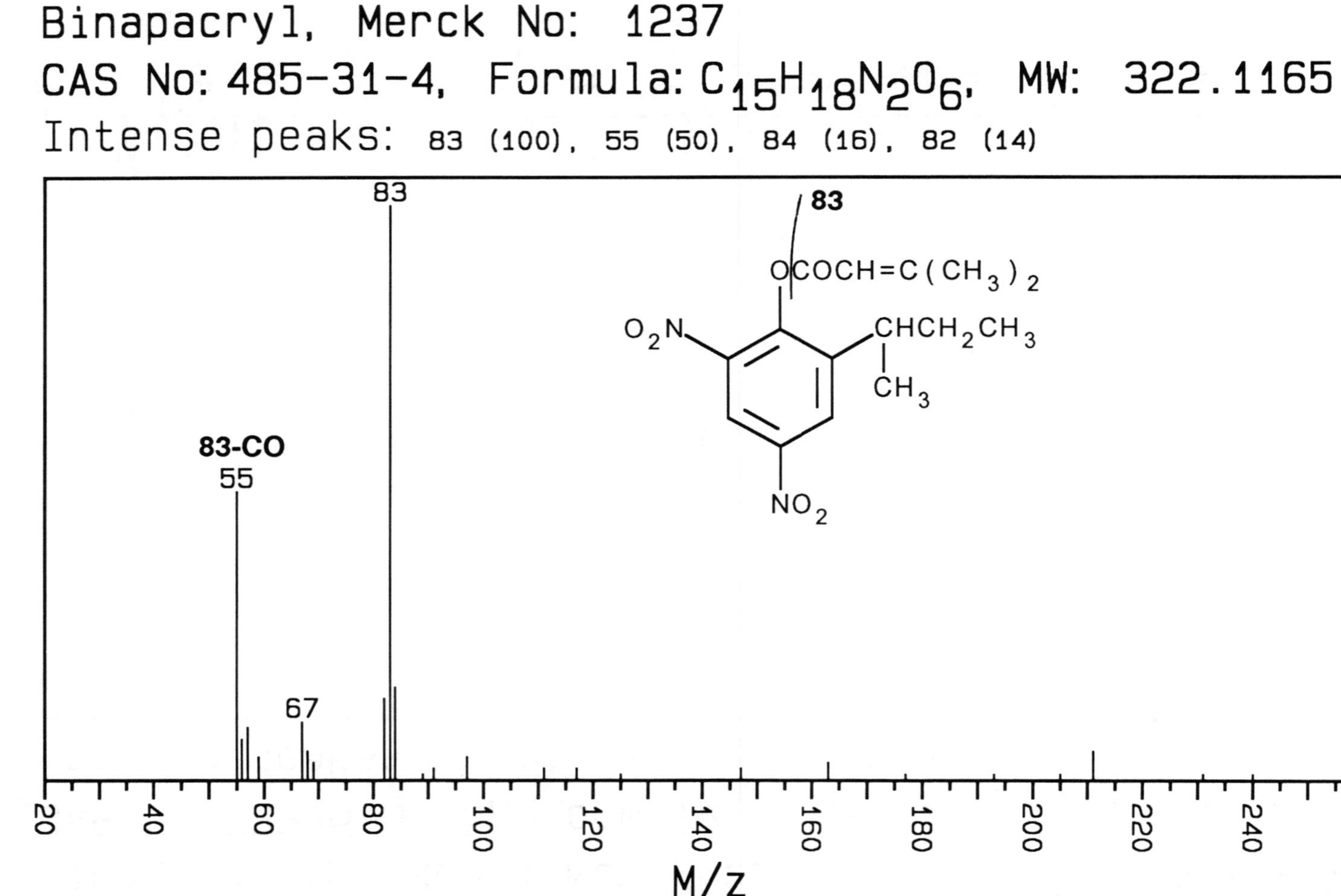

Binapacryl, Merck No: 1237
CAS No: 485-31-4, Formula: $C_{15}H_{18}N_2O_6$, MW: 322.1165
Intense peaks: 83 (100), 55 (50), 84 (16), 82 (14)
83
83
OCOCH=C(CH3)2
O2N
CHCH2CH3
CH3
NO2
83-CO
55
67
20
40
60
80
100
120
140
160
180
200
220
240
260
M/Z

Dinobuton, Merck No: 3280

CAS No: 973-21-7, Formula: $C_{14}H_{18}N_2O_7$, MW: 326.1114

Intense peaks: 43 (100), 211 (45), 41 (15), 163 ( 9)

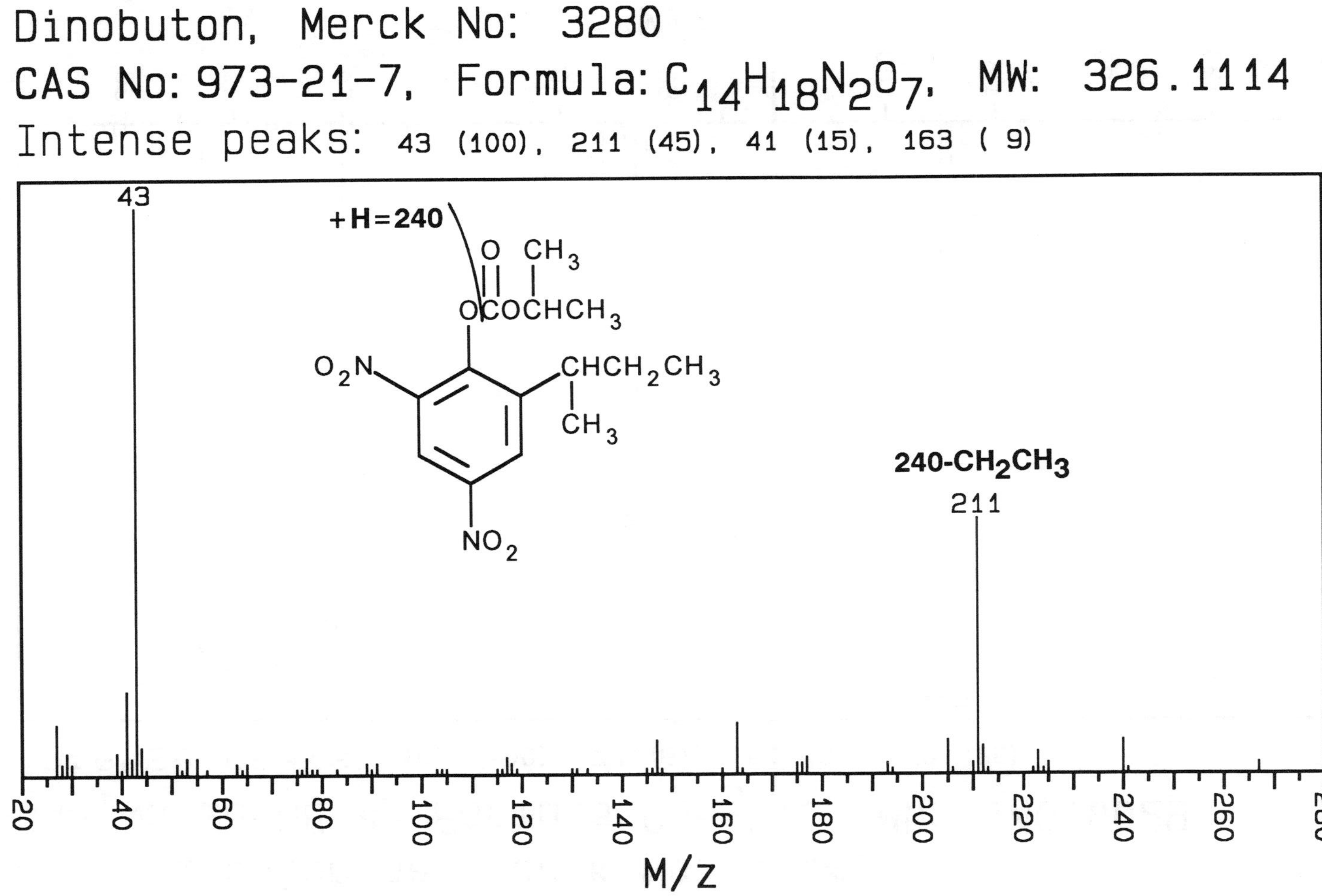

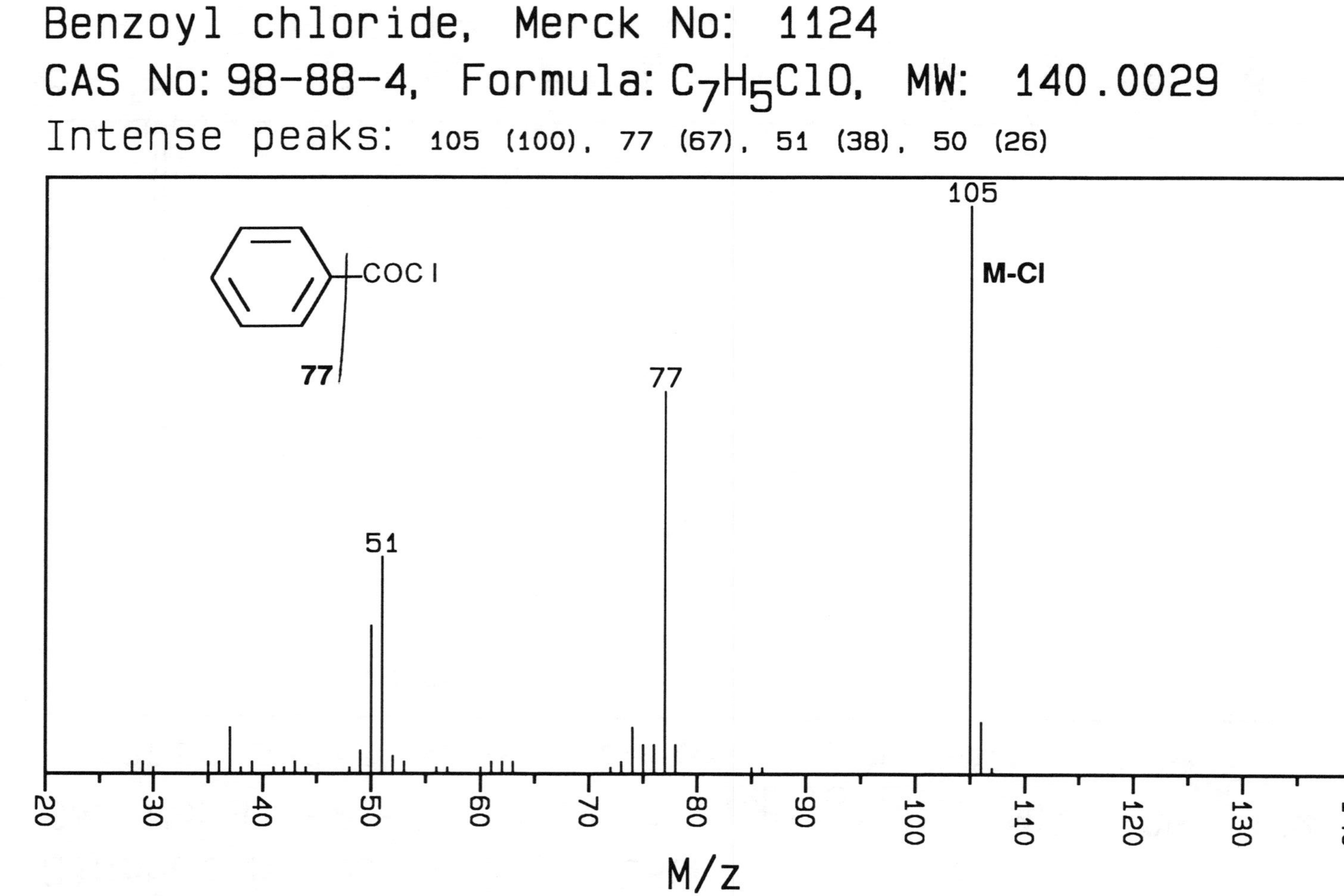
Benzoyl chloride, Merck No: 1124
CAS No: 98-88-4, Formula: $C_7H_5ClO$, MW: 140.0029
Intense peaks: 105 (100), 77 (67), 51 (38), 50 (26)
105
M-Cl
COCl
77
77
51
20
30
40
50
60
70
80
90
100
110
120
130
140
M/z

Benzyl benzoate, Merck No: 1141
CAS No: 120-51-4, Formula: $C_{14}H_{12}O_2$, MW: 212.0837
Intense peaks: 105 (100), 91 (42), 77 (28), 51 (12)

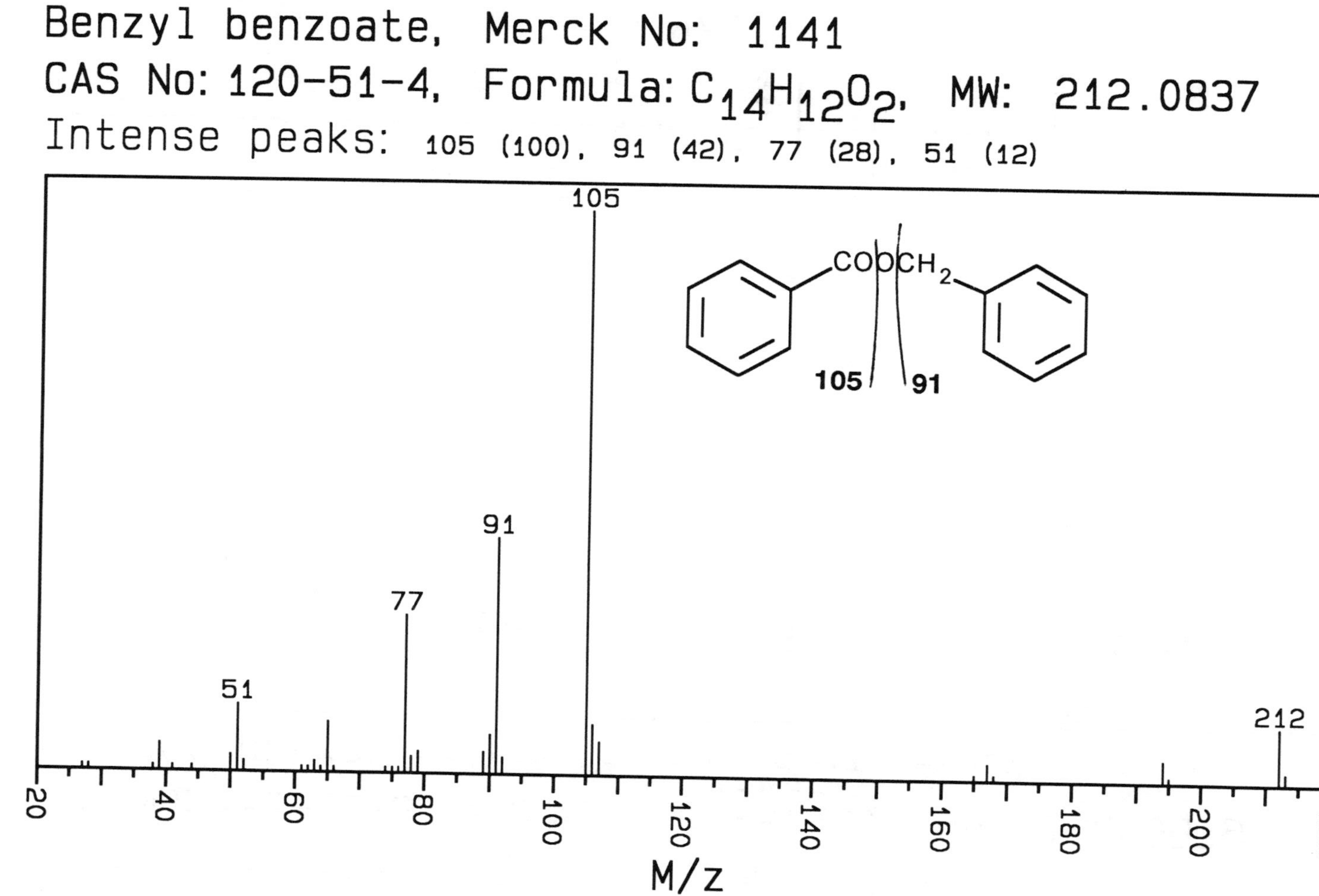

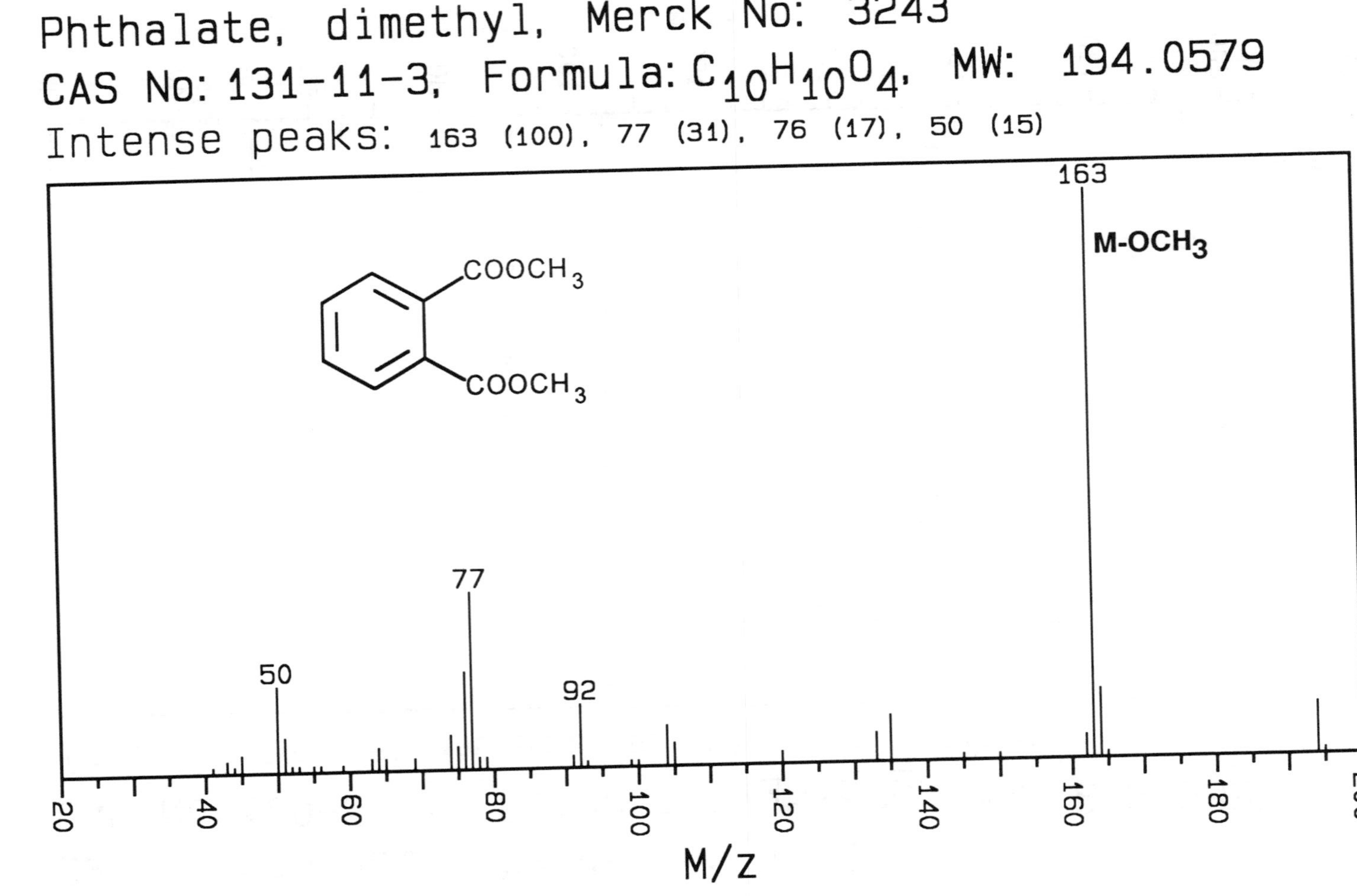
Phthalate, dimethyl, Merck No: 3243
CAS No: 131-11-3, Formula: $C_{10}H_{10}O_4$, MW: 194.0579
Intense peaks: 163 (100), 77 (31), 76 (17), 50 (15)
COOCH3
COOCH3
163
M-OCH3
77
50
92
20
40
60
80
100
120
140
160
180
200
M/Z

Phthalate, diethyl-, Merck No: 7345

CAS No: 84-66-2, Formula: $C_{12}H_{14}O_4$, MW: 222.0892

Intense peaks: 149 (100), 177 (28), 150 (13), 176 ( 9)

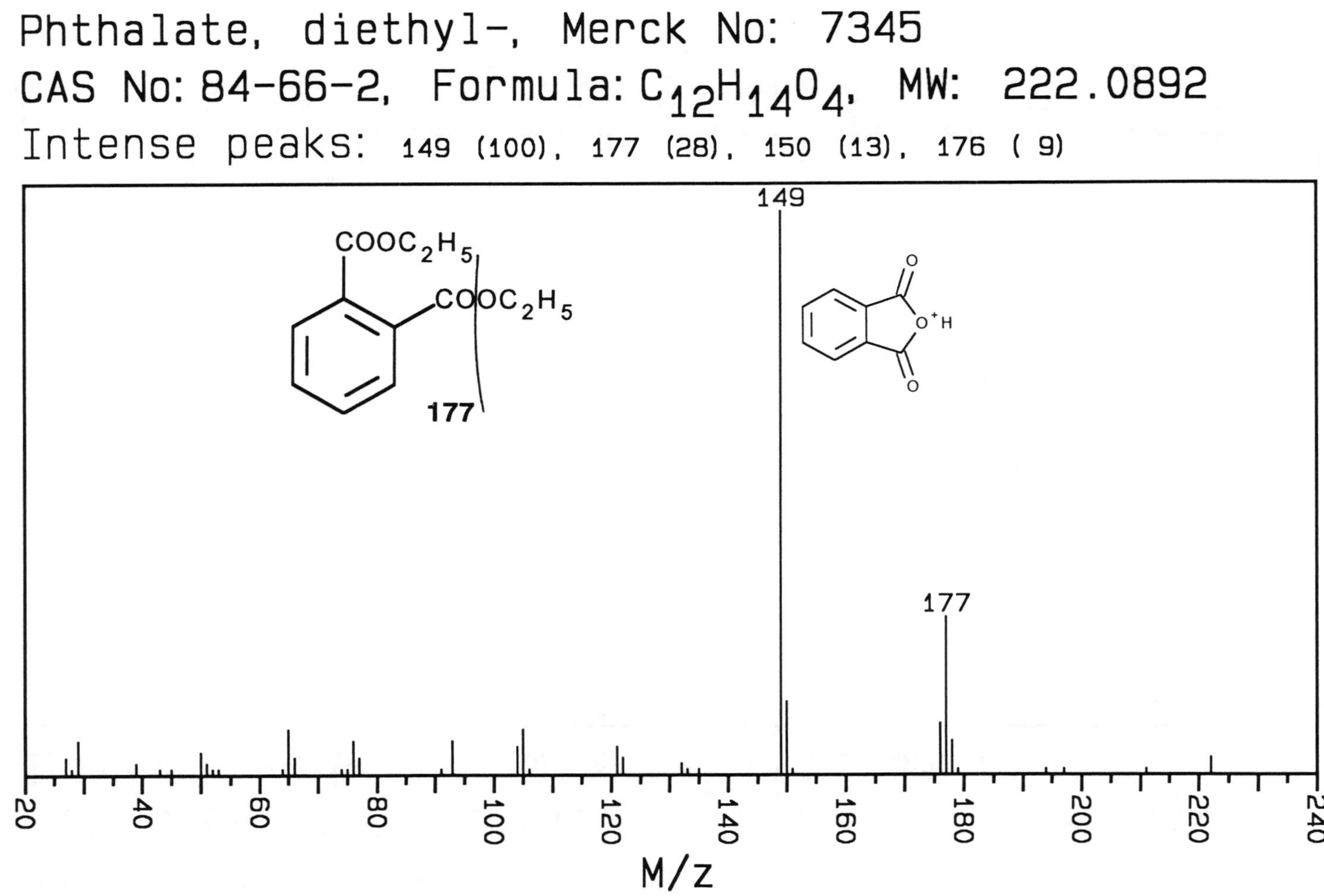

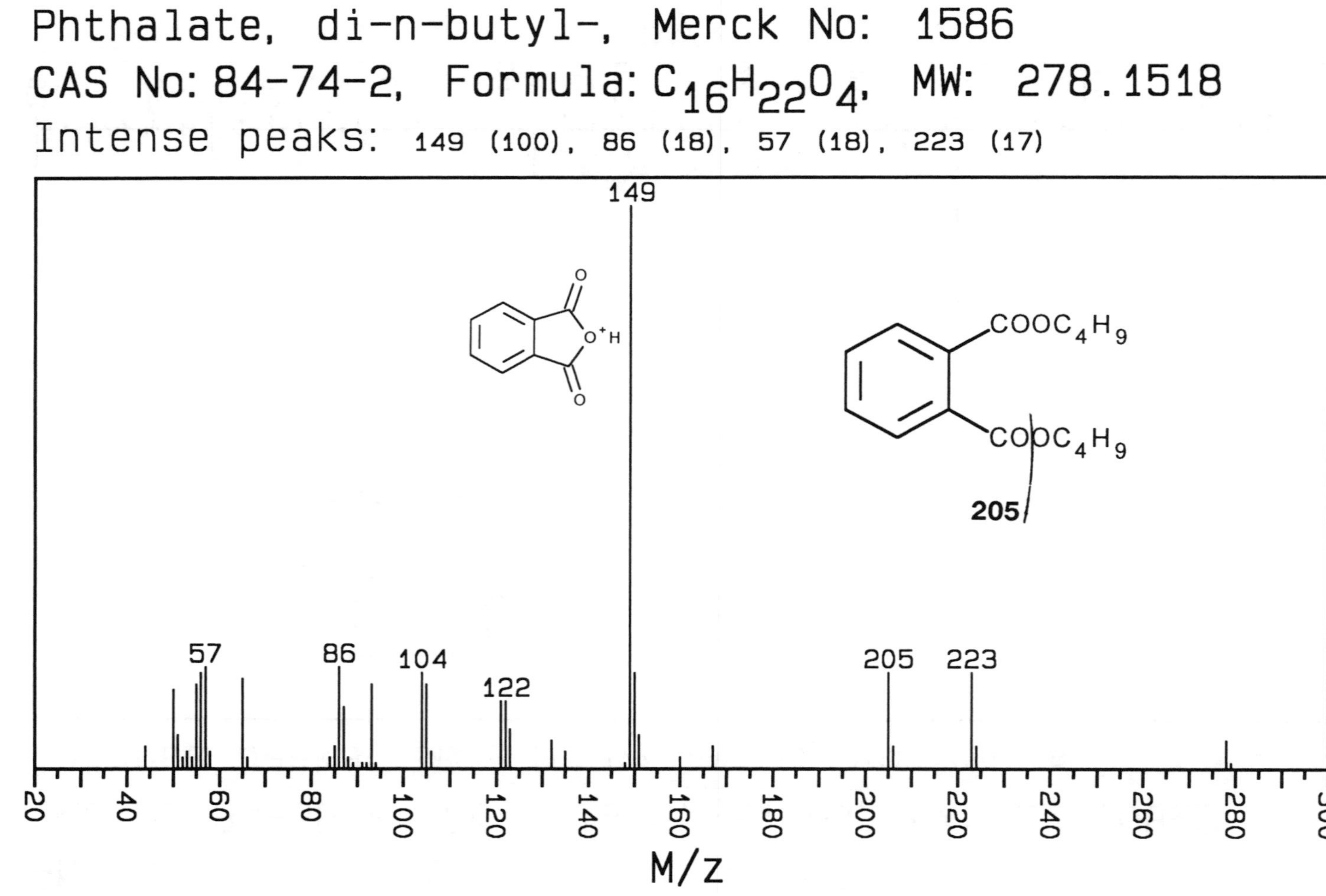
Phthalate, di-n-butyl-, Merck No: 1586
CAS No: 84-74-2, Formula: $C_{16}H_{22}O_4$, MW: 278.1518
Intense peaks: 149 (100), 86 (18), 57 (18), 223 (17)
149
O
O⁺H
O
COOC$_4$H$_9$
COOC$_4$H$_9$
205
57
86
104
122
205
223
20
40
60
80
100
120
140
160
180
200
220
240
260
280
300
M/Z

Phthalate benzyl butyl

CAS No: 85-68-7, Formula: $C_{19}H_{20}O_4$, MW: 312.1362

Intense peaks: 149 (100), 91 (61), 206 (27), 104 (27)

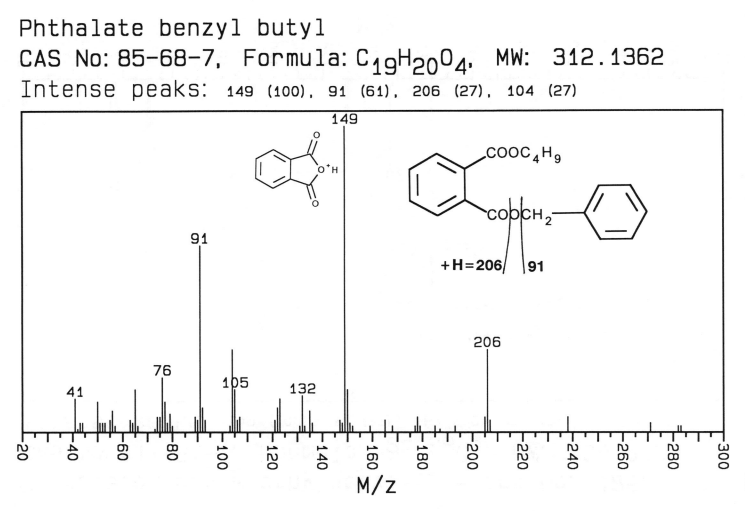

149

91

206

76

105

132

41

$COOC_4H_9$

$COOCH_2$

+H=206

91

M/Z

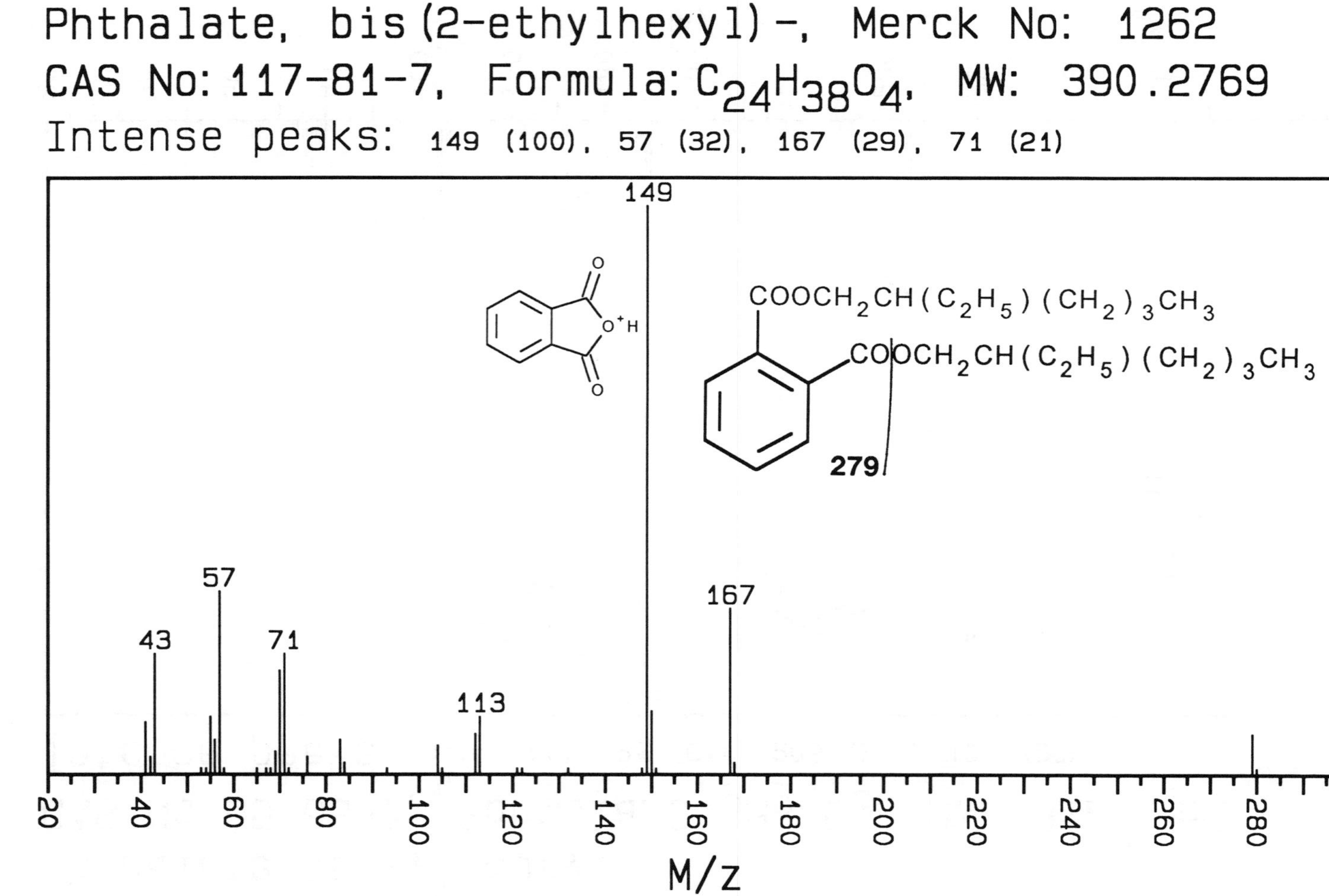
Phthalate, bis(2-ethylhexyl)-, Merck No: 1262
CAS No: 117-81-7, Formula: C24H38O4, MW: 390.2769
Intense peaks: 149 (100), 57 (32), 167 (29), 71 (21)
149
COOCH2CH(C2H5)(CH2)3CH3
COOCH2CH(C2H5)(CH2)3CH3
279
57
167
43
71
113
20
40
60
80
100
120
140
160
180
200
220
240
260
280
300
M/Z

Phthalate, phenyl, Merck No: 7278

CAS No: 84-62-8, Formula: $C_{20}H_{14}O_4$, MW: 318.0892

Intense peaks: 65 (100), 44 (97), 66 (94), 105 (71)

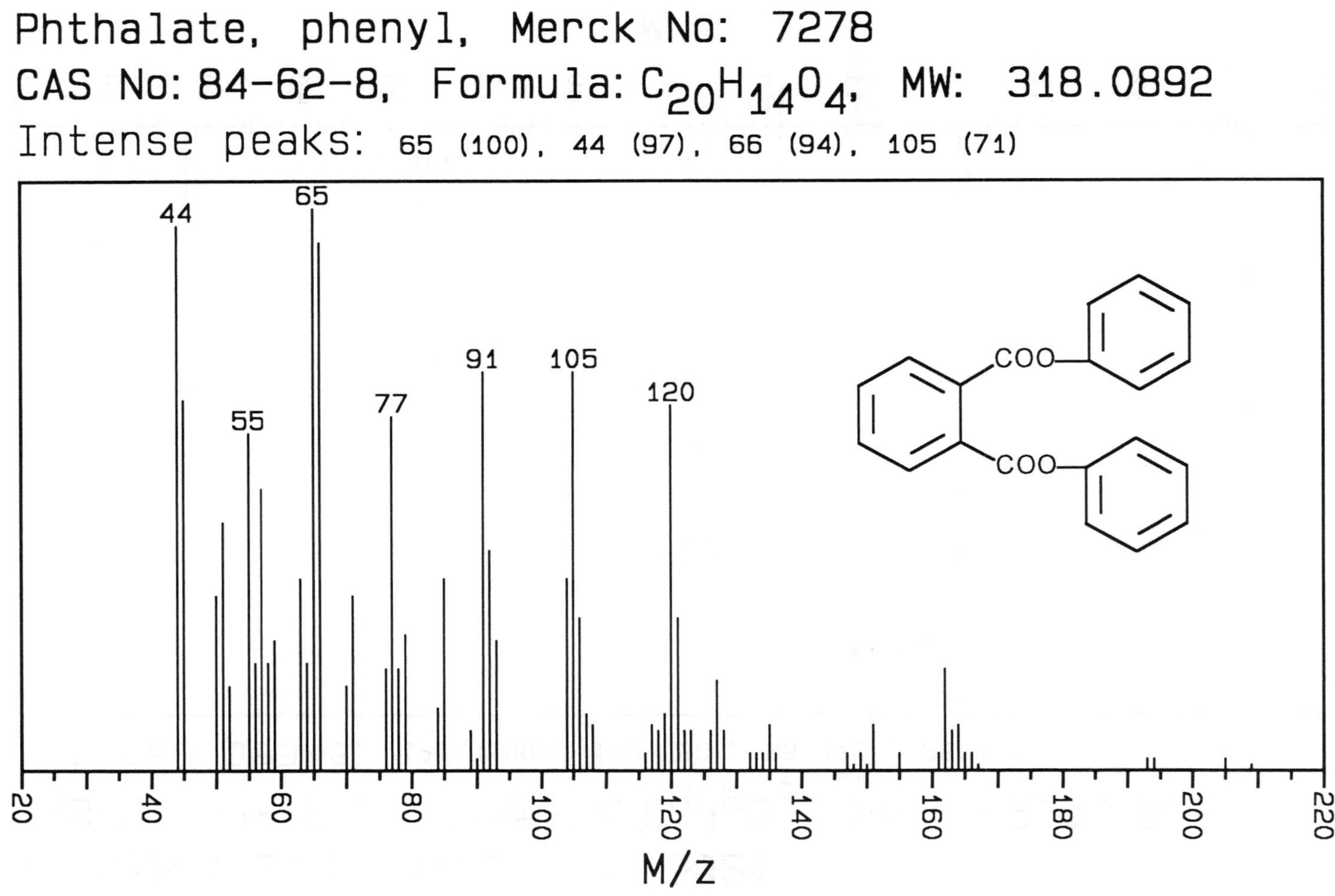

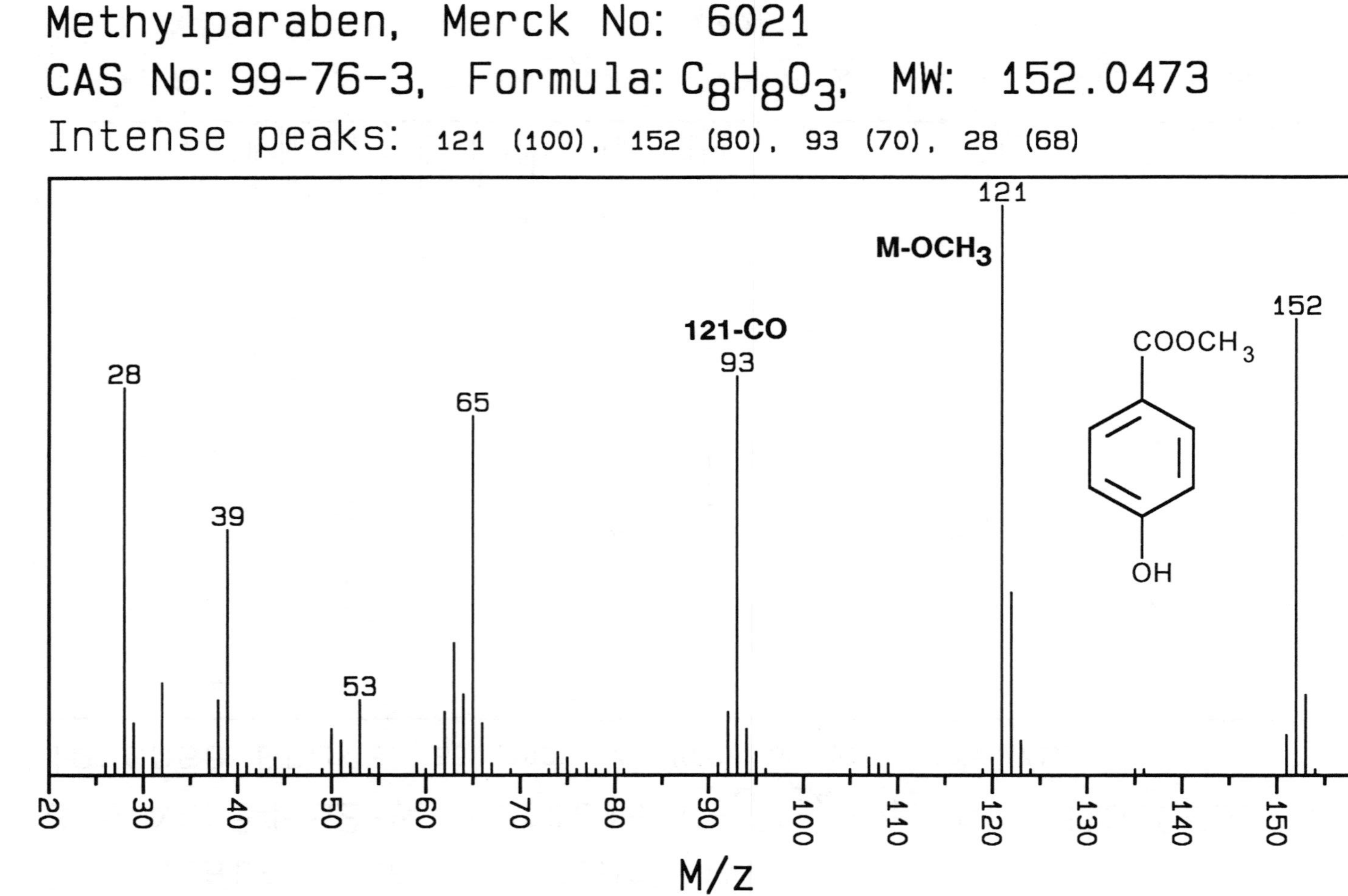
Methylparaben, Merck No: 6021
CAS No: 99-76-3, Formula: $C_8H_8O_3$, MW: 152.0473
Intense peaks: 121 (100), 152 (80), 93 (70), 28 (68)
121
M-$OCH_3$
152
121-CO
93
$COOCH_3$
28
65
39
OH
53
20
30
40
50
60
70
80
90
100
110
120
130
140
150
160
M/Z

DCPA, Merck No: 2830

CAS No: 1861-32-1, Formula: $C_{10}H_6Cl_4O_4$, MW: 329.9021

Intense peaks: 301 (100), 299 (80), 303 (48), 332 (32)

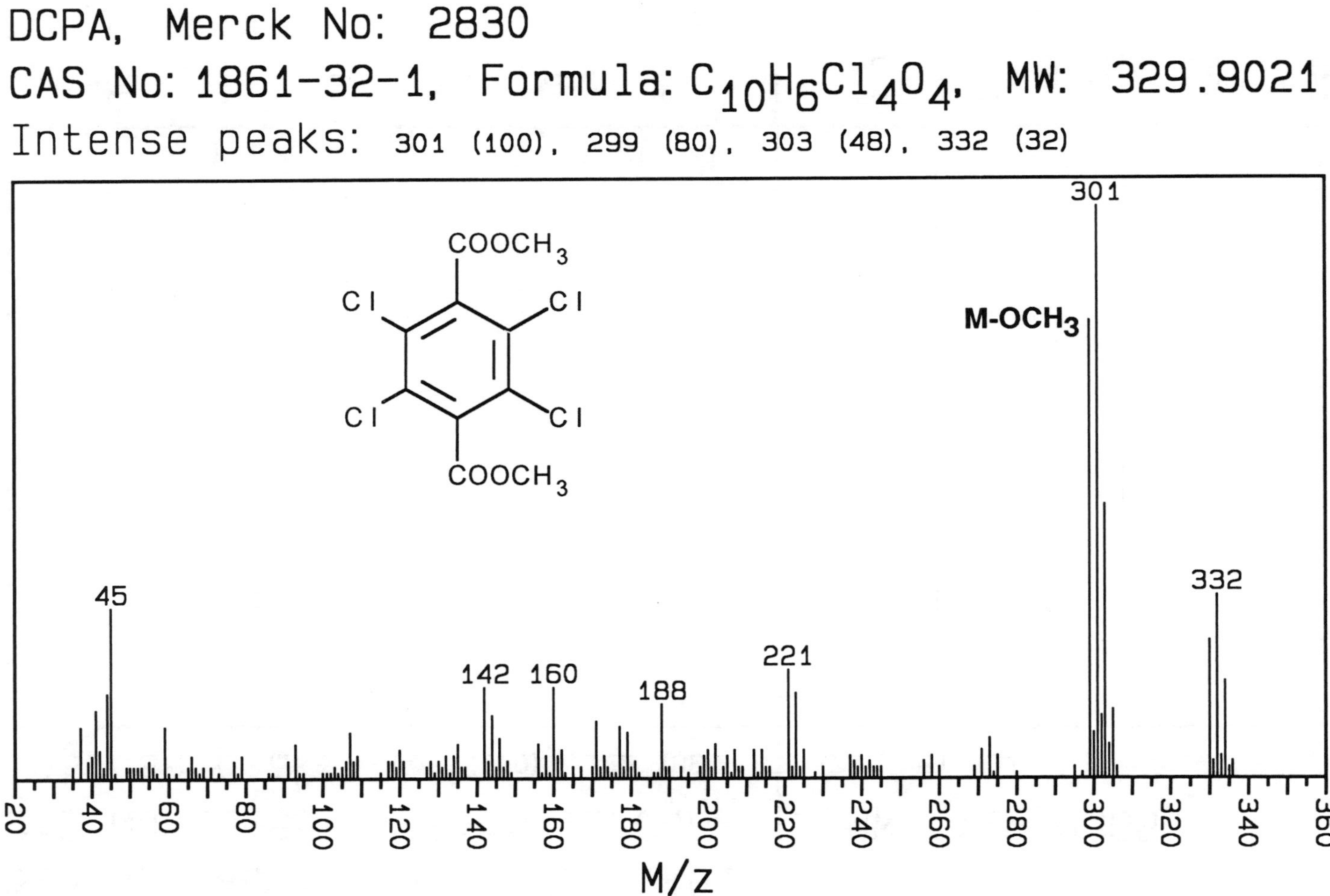

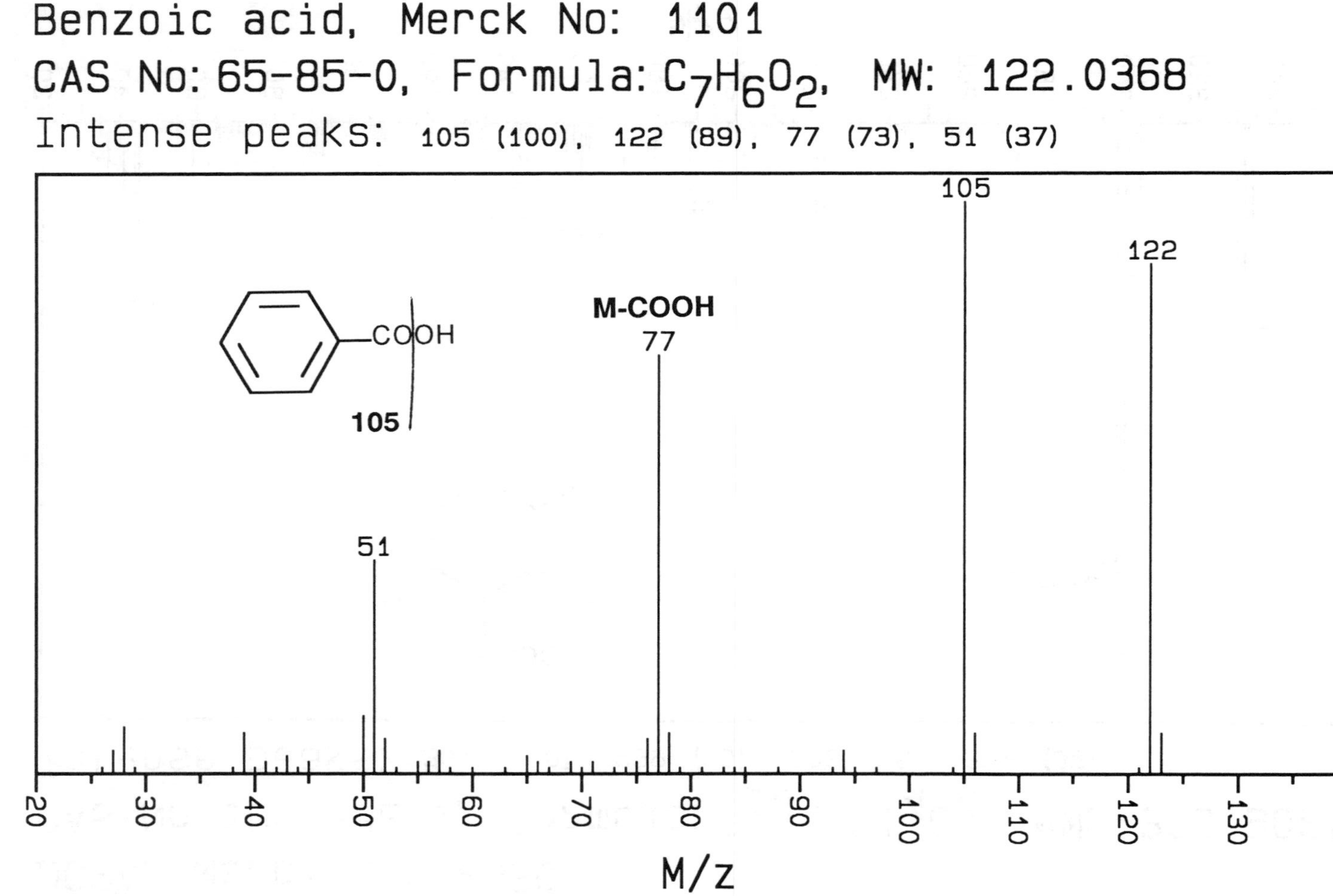
Benzoic acid, Merck No: 1101
CAS No: 65-85-0, Formula: $C_7H_6O_2$, MW: 122.0368
Intense peaks: 105 (100), 122 (89), 77 (73), 51 (37)
COOH
105
M-COOH
77
51
105
122
20 30 40 50 60 70 80 90 100 110 120 130 140
M/Z

Benzoic acid, 2-chloro-, Merck No: 2125

CAS No: 118-91-2, Formula: $C_7H_5ClO_2$, MW: 155.9978

Intense peaks: 139 (100), 156 (68), 50 (51), 75 (47)

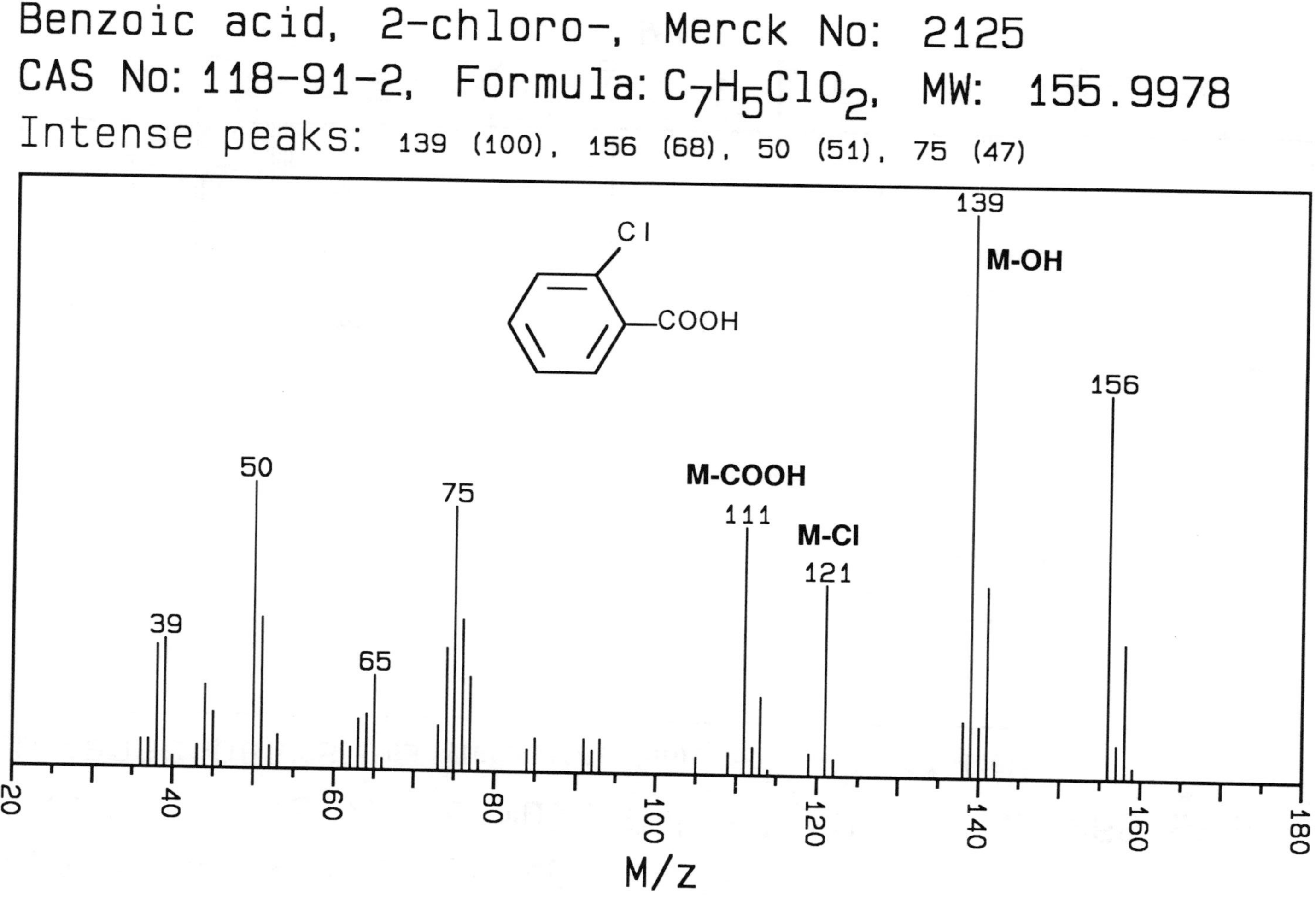

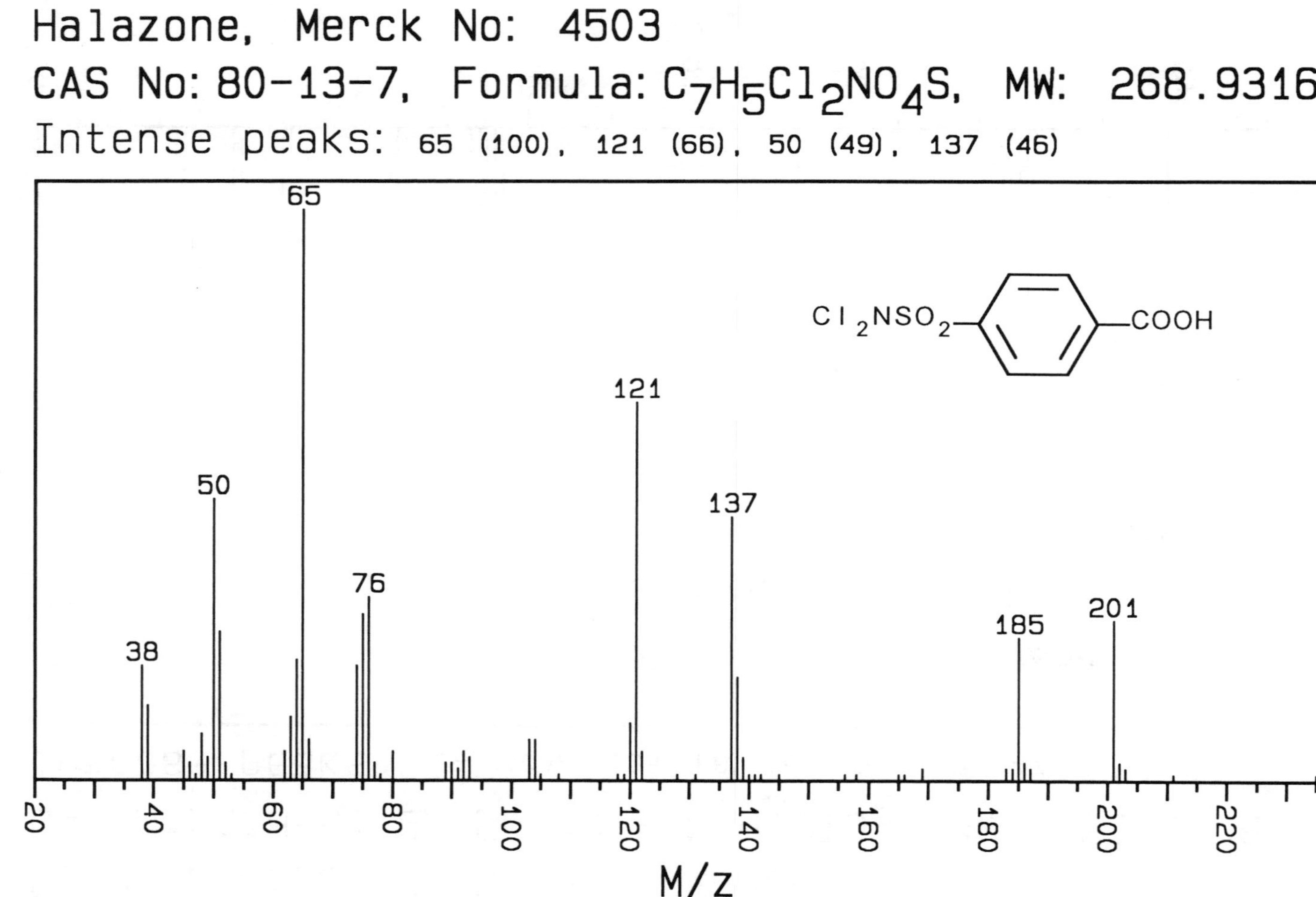
Halazone, Merck No: 4503
CAS No: 80-13-7, Formula: C7H5Cl2NO4S, MW: 268.9316
Intense peaks: 65 (100), 121 (66), 50 (49), 137 (46)
Cl2NSO2
COOH
65
121
50
137
76
38
185
201
20
40
60
80
100
120
140
160
180
200
220
240
M/Z

Chloramben, Merck No: 2063

CAS No: 133-90-4, Formula: $C_7H_5Cl_2NO_2$, MW: 204.9697

Intense peaks: 205 (100), 207 (63), 188 (28), 63 (19)

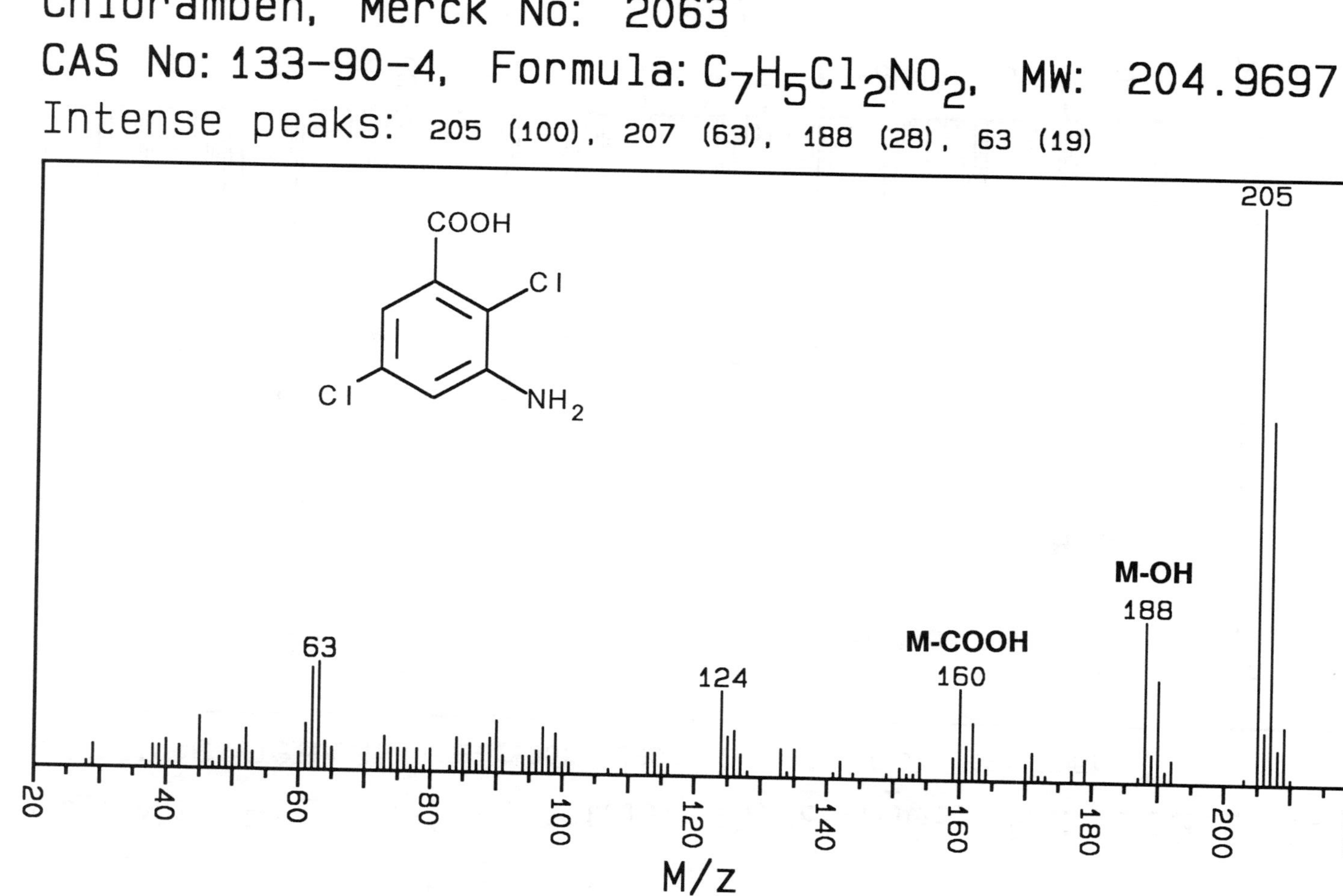

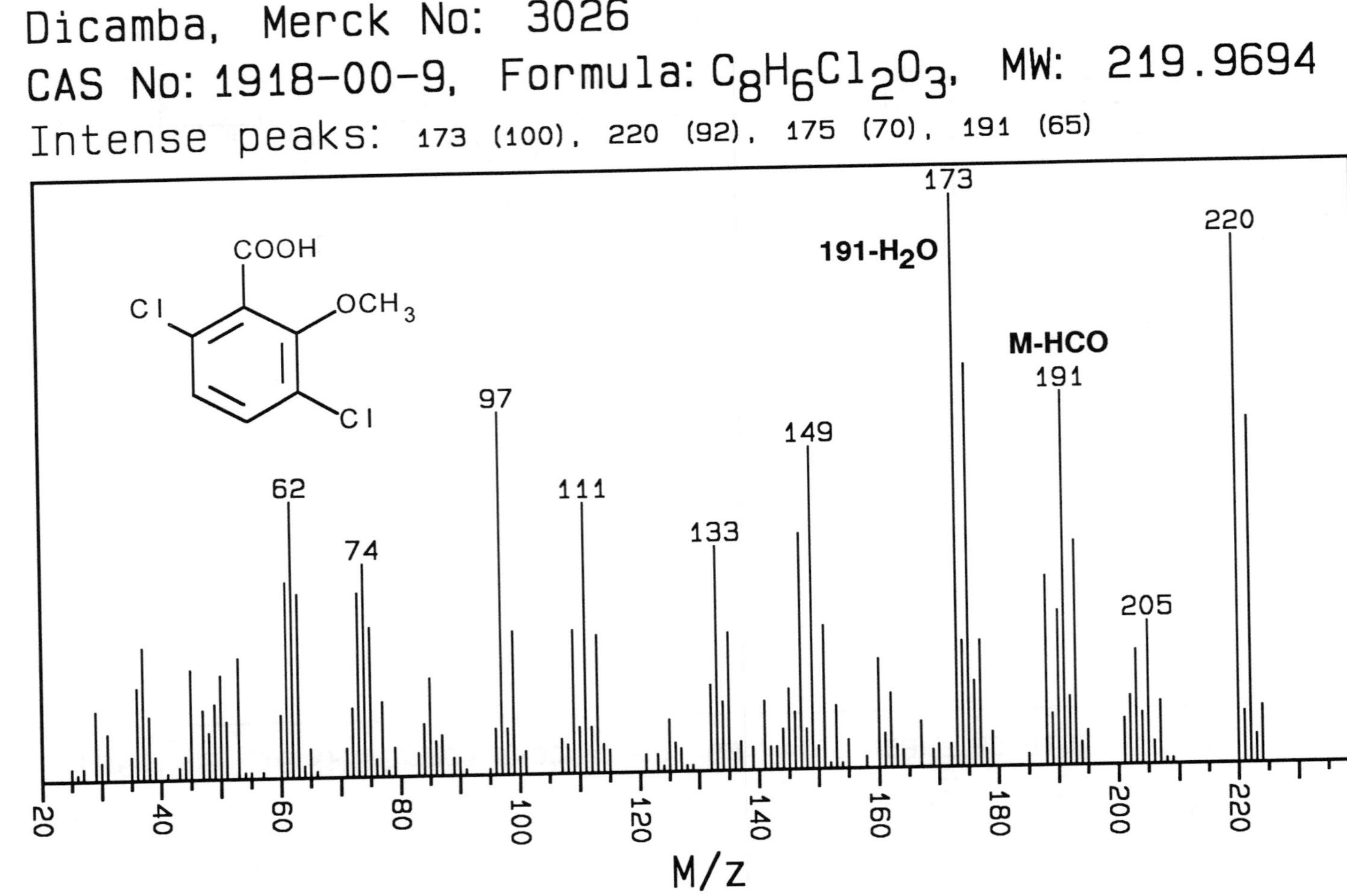
Dicamba, Merck No: 3026
CAS No: 1918-00-9, Formula: $C_8H_6Cl_2O_3$, MW: 219.9694
Intense peaks: 173 (100), 220 (92), 175 (70), 191 (65)
COOH
Cl
OCH3
Cl
62
74
97
111
133
149
191-H2O
173
M-HCO
191
205
220
20
40
60
80
100
120
140
160
180
200
220
240
M/Z

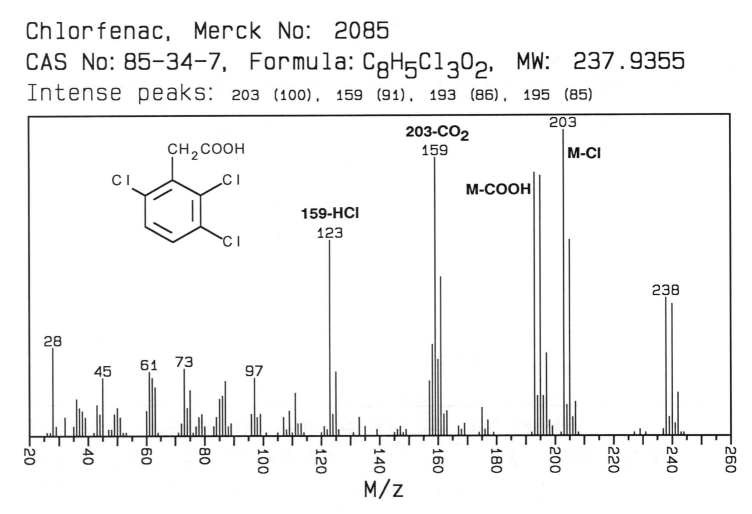

Chlorfenac, Merck No: 2085

CAS No: 85-34-7, Formula: $C_8H_5Cl_3O_2$, MW: 237.9355

Intense peaks: 203 (100), 159 (91), 193 (86), 195 (85)

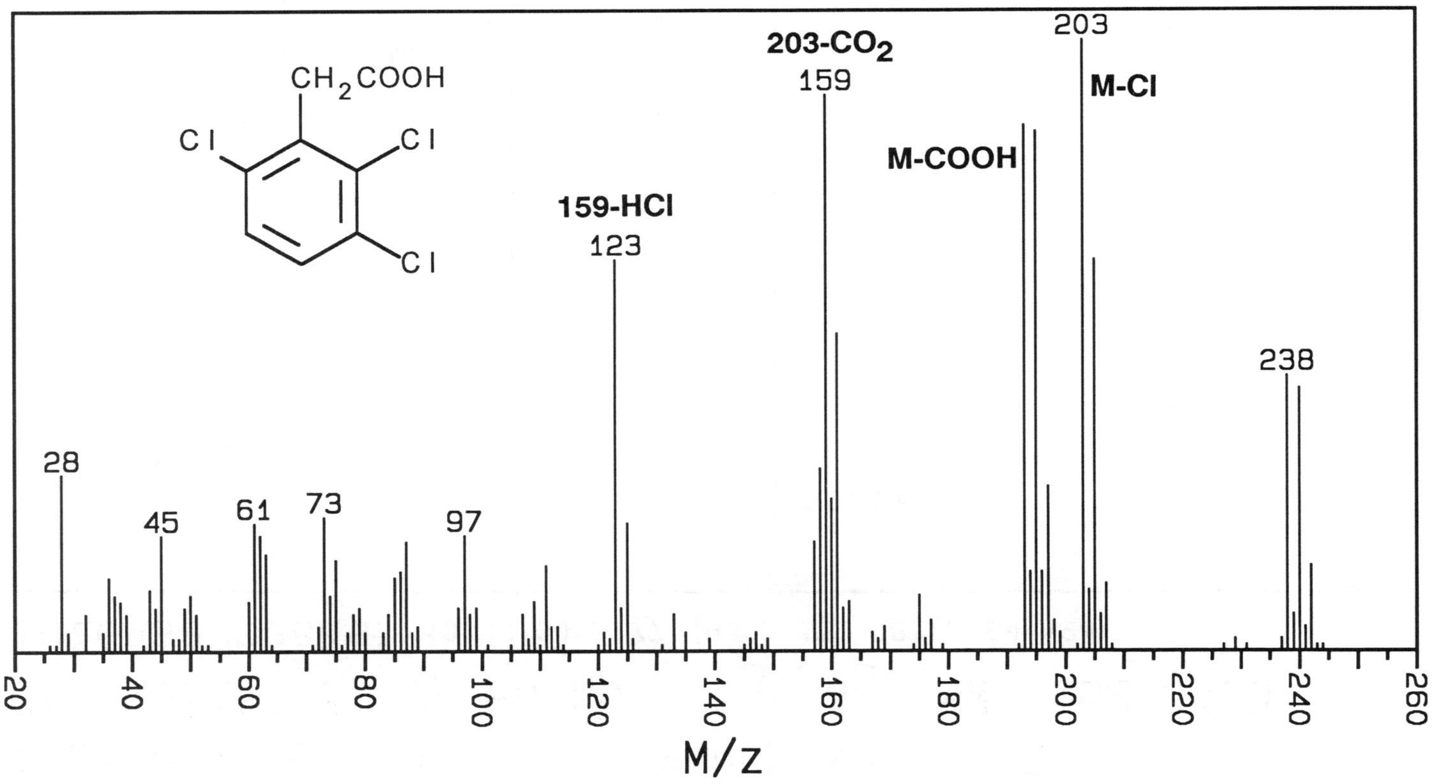

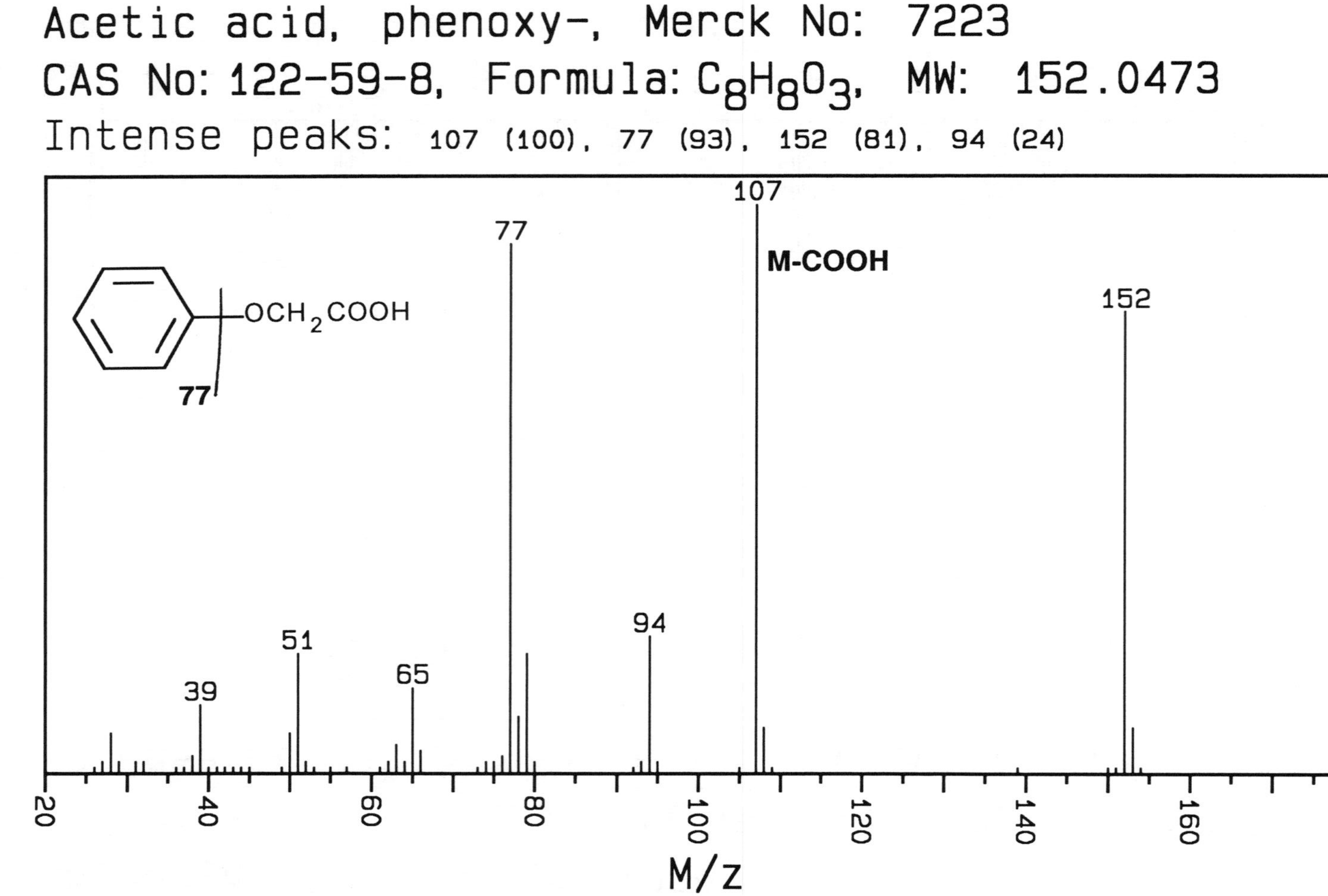

Acetic acid, phenoxy-, Merck No: 7223
CAS No: 122-59-8, Formula: $C_8H_8O_3$, MW: 152.0473
Intense peaks: 107 (100), 77 (93), 152 (81), 94 (24)
$OCH_2COOH$
77
77
107
M-COOH
152
39
51
65
94
20
40
60
80
100
120
140
160
180
M/Z

Methoxone*, Merck No: 5645

CAS No: 94-74-6, Formula: $C_9H_9ClO_3$, MW: 200.0241

Intense peaks: 141 (100), 200 (75), 77 (63), 143 (33)

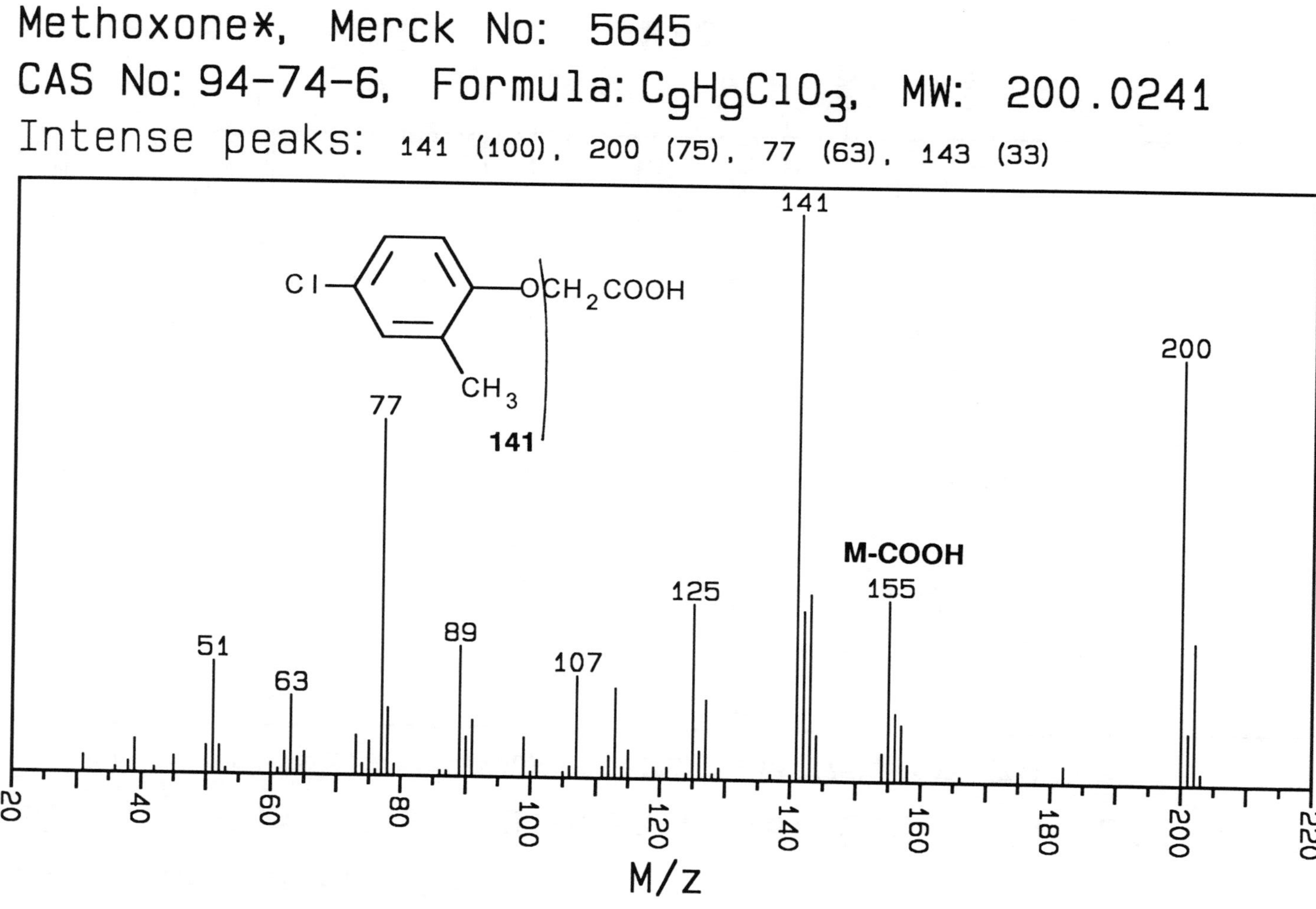

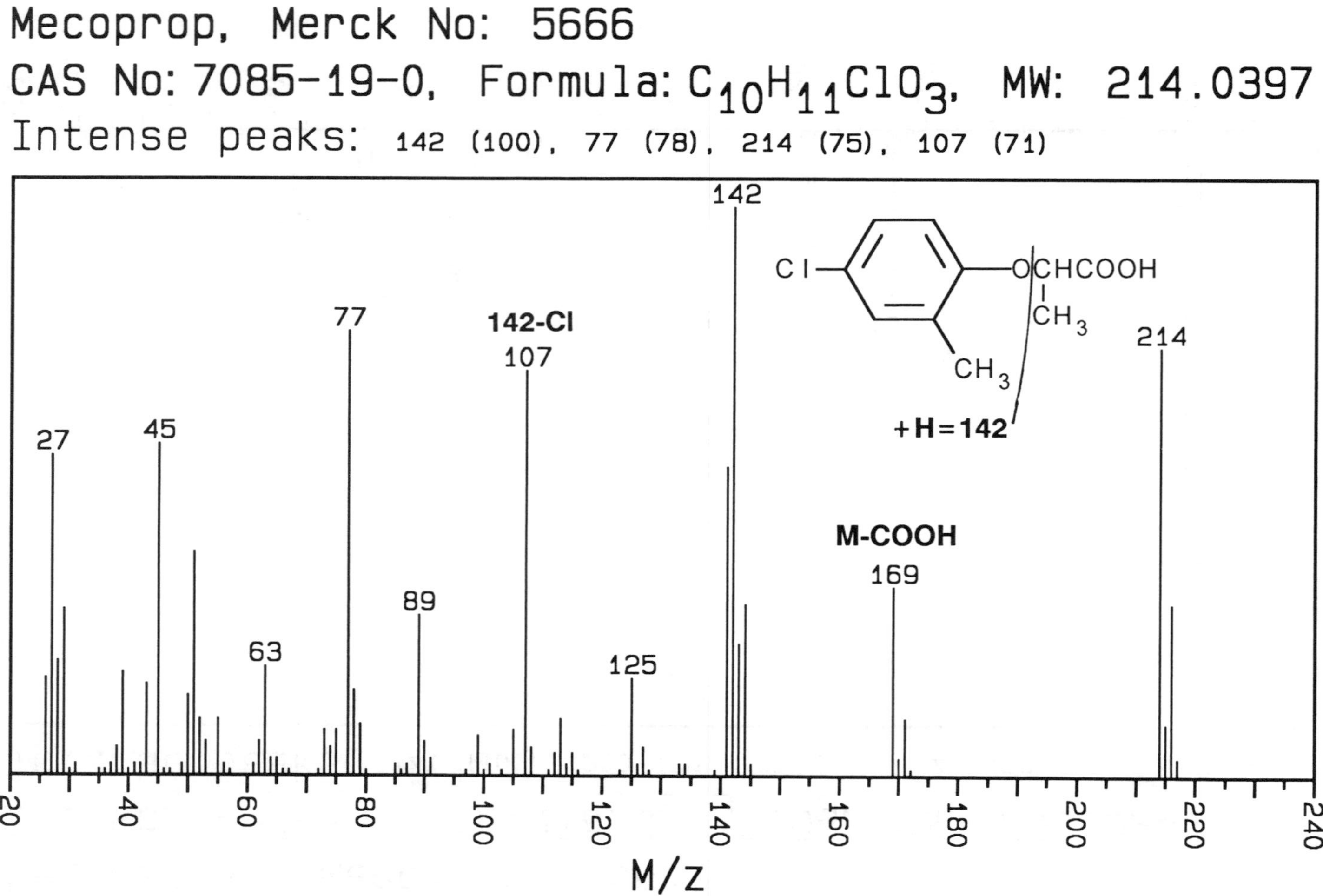
Mecoprop, Merck No: 5666
CAS No: 7085-19-0, Formula: $C_{10}H_{11}ClO_3$, MW: 214.0397
Intense peaks: 142 (100), 77 (78), 214 (75), 107 (71)
Cl
OCHCOOH
CH3
CH3
+H=142
142-Cl
M-COOH
27
45
63
77
89
107
125
142
169
214
20
40
60
80
100
120
140
160
180
200
220
240
M/Z

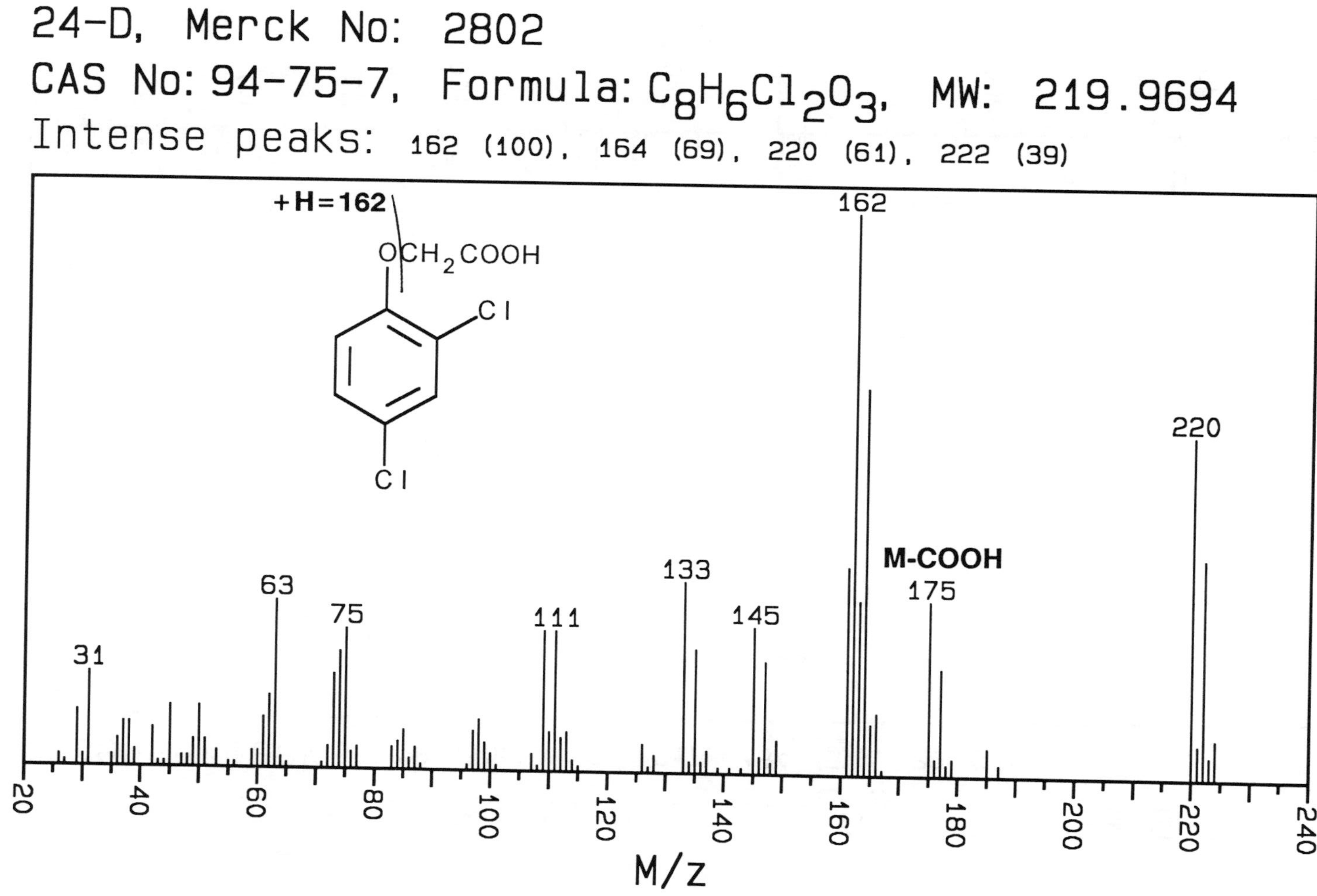
24-D, Merck No: 2802
CAS No: 94-75-7, Formula: $C_8H_6Cl_2O_3$, MW: 219.9694
Intense peaks: 162 (100), 164 (69), 220 (61), 222 (39)
+H=162
$OCH_2COOH$
Cl
Cl
162
220
M-COOH
175
133
145
111
75
63
31
20
40
60
80
100
120
140
160
180
200
220
240
M/Z

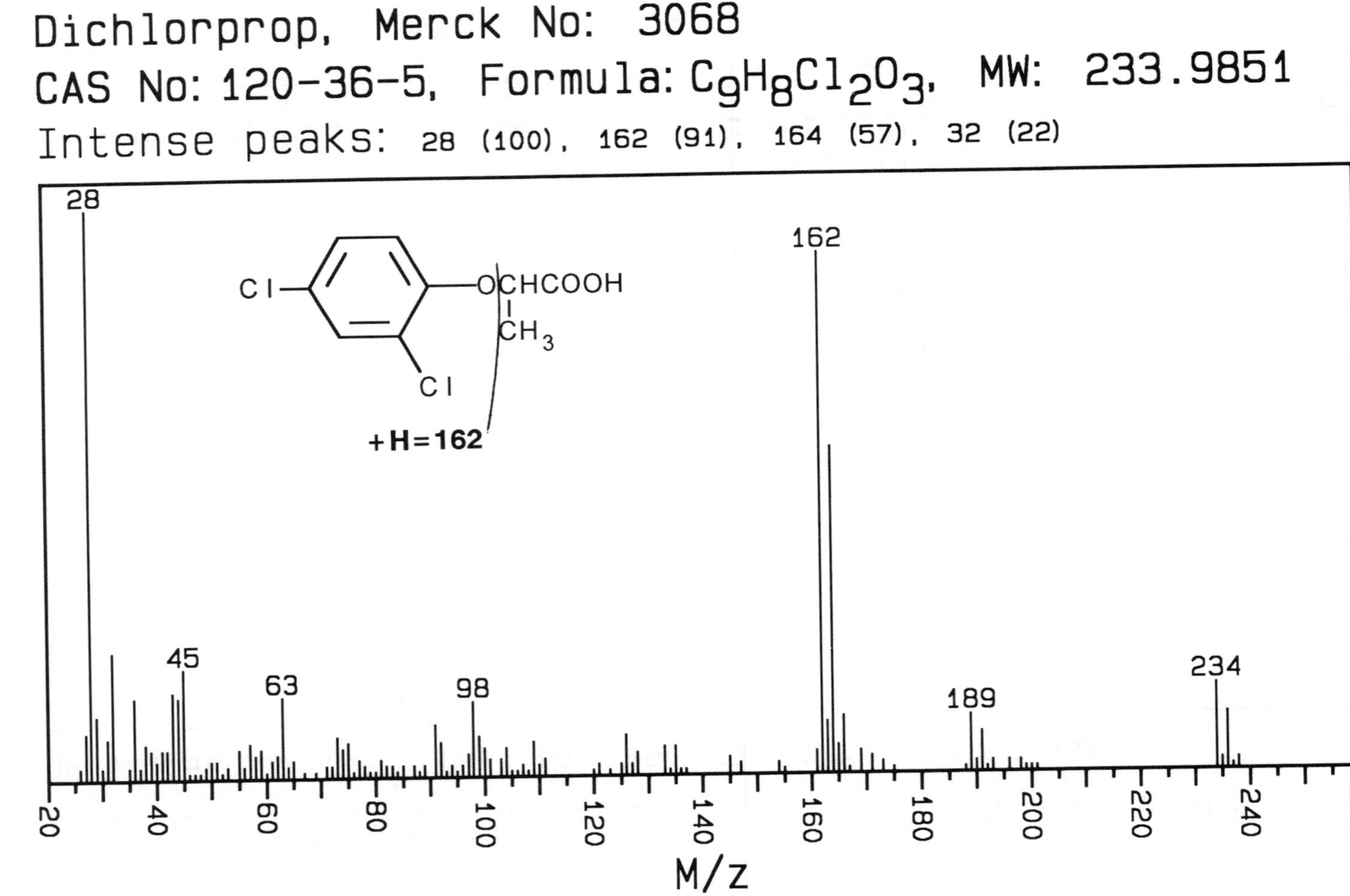
Dichlorprop, Merck No: 3068
CAS No: 120-36-5, Formula: C9H8Cl2O3, MW: 233.9851
Intense peaks: 28 (100), 162 (91), 164 (57), 32 (22)
Cl
OCHCOOH
CH3
Cl
+H=162
28
45
63
98
162
189
234
20
40
60
80
100
120
140
160
180
200
220
240
260
M/Z

24-DB, Merck No: 2828

CAS No: 94-82-6, Formula: $C_{10}H_{10}Cl_2O_3$, MW: 248.0007

Intense peaks: 162 (100), 164 (66), 87 (33), 43 (29)

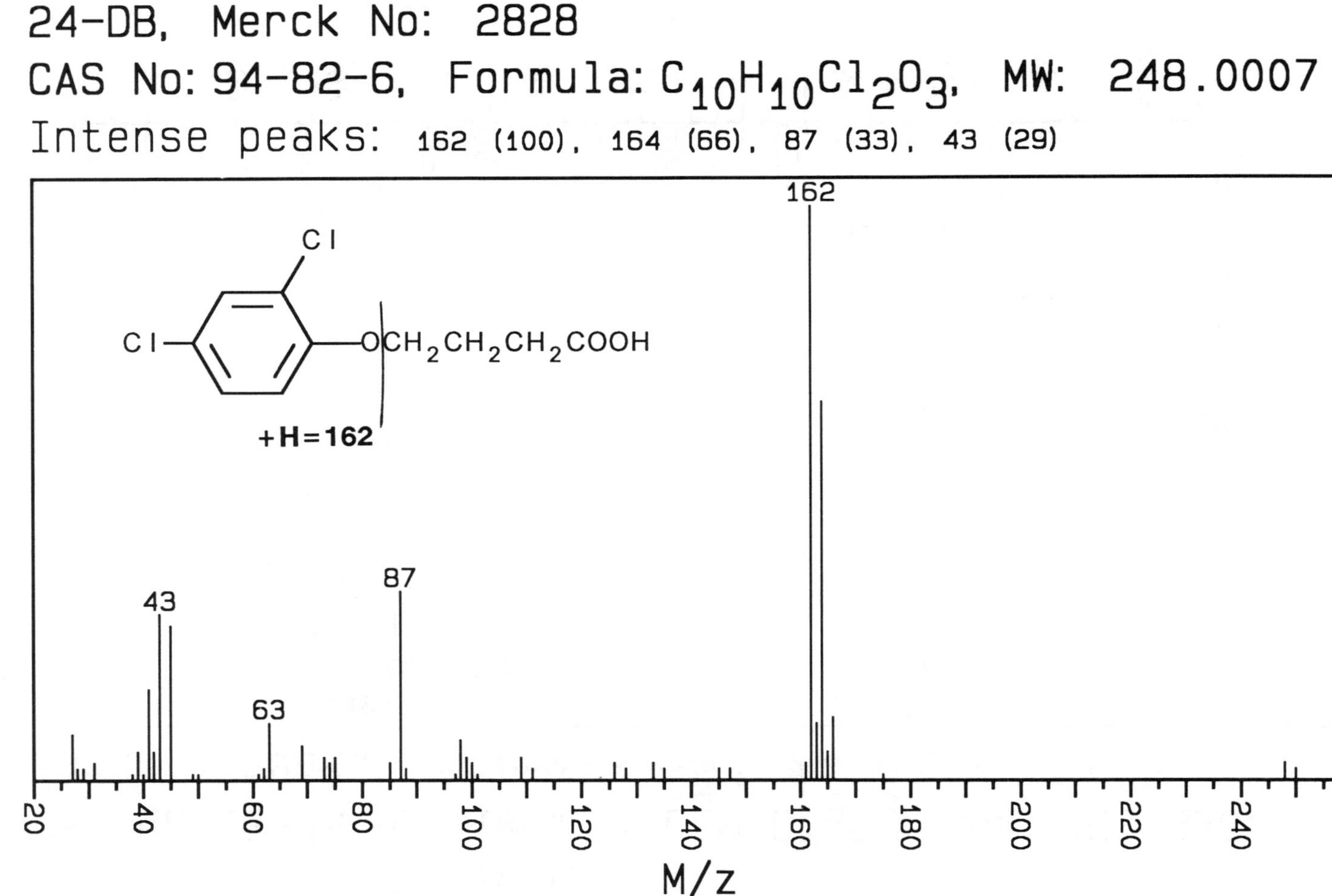

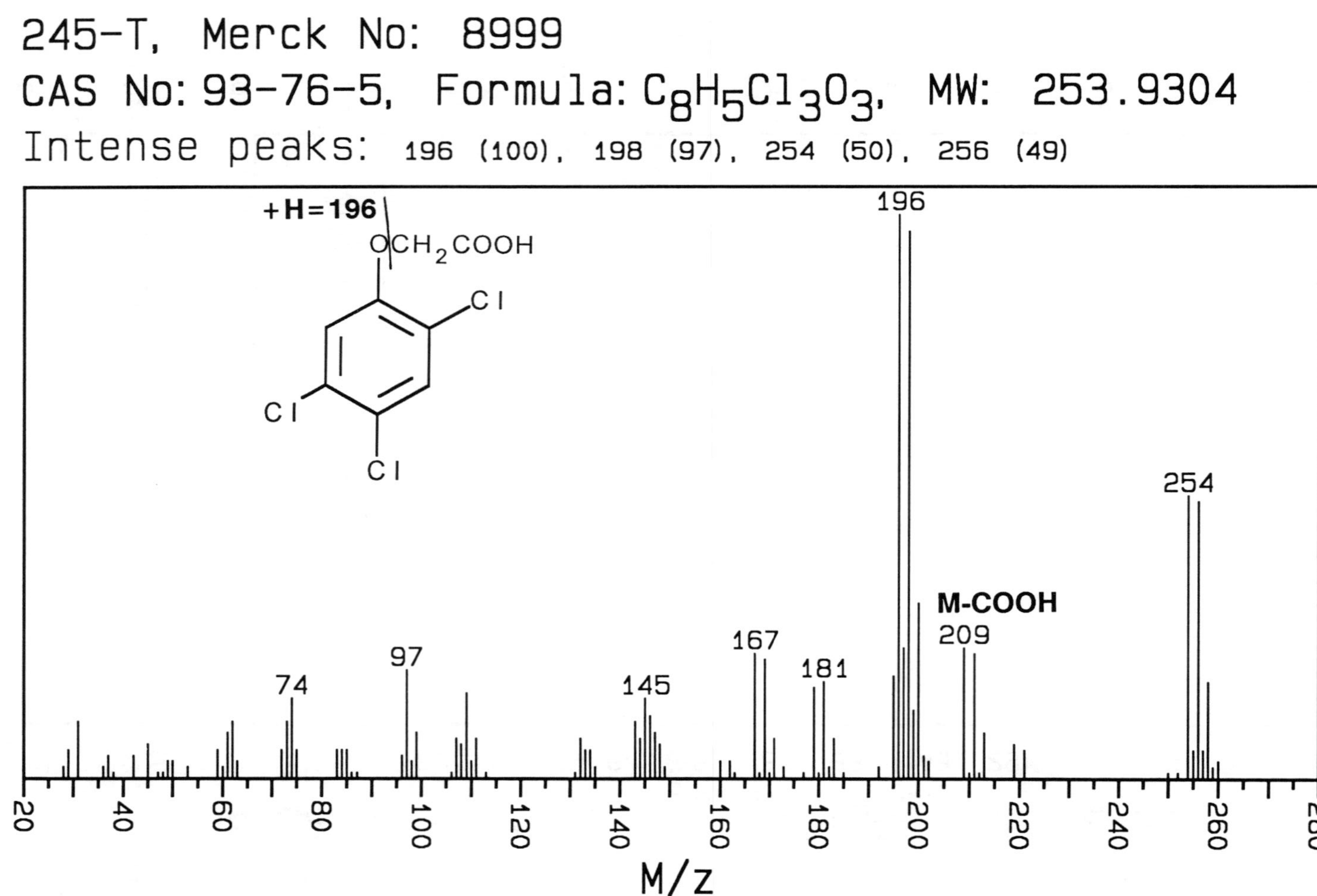
245-T, Merck No: 8999
CAS No: 93-76-5, Formula: $C_8H_5Cl_3O_3$, MW: 253.9304
Intense peaks: 196 (100), 198 (97), 254 (50), 256 (49)
+H=196
$OCH_2COOH$
Cl
Cl
Cl
196
254
M-COOH
209
167
181
145
97
74
20
40
60
80
100
120
140
160
180
200
220
240
260
280
M/Z

Silvex, Merck No: 8483

CAS No: 93-72-1, Formula: $C_9H_7Cl_3O_3$, MW: 267.9461

Intense peaks: 196 (100), 198 (95), 97 (40), 200 (29)

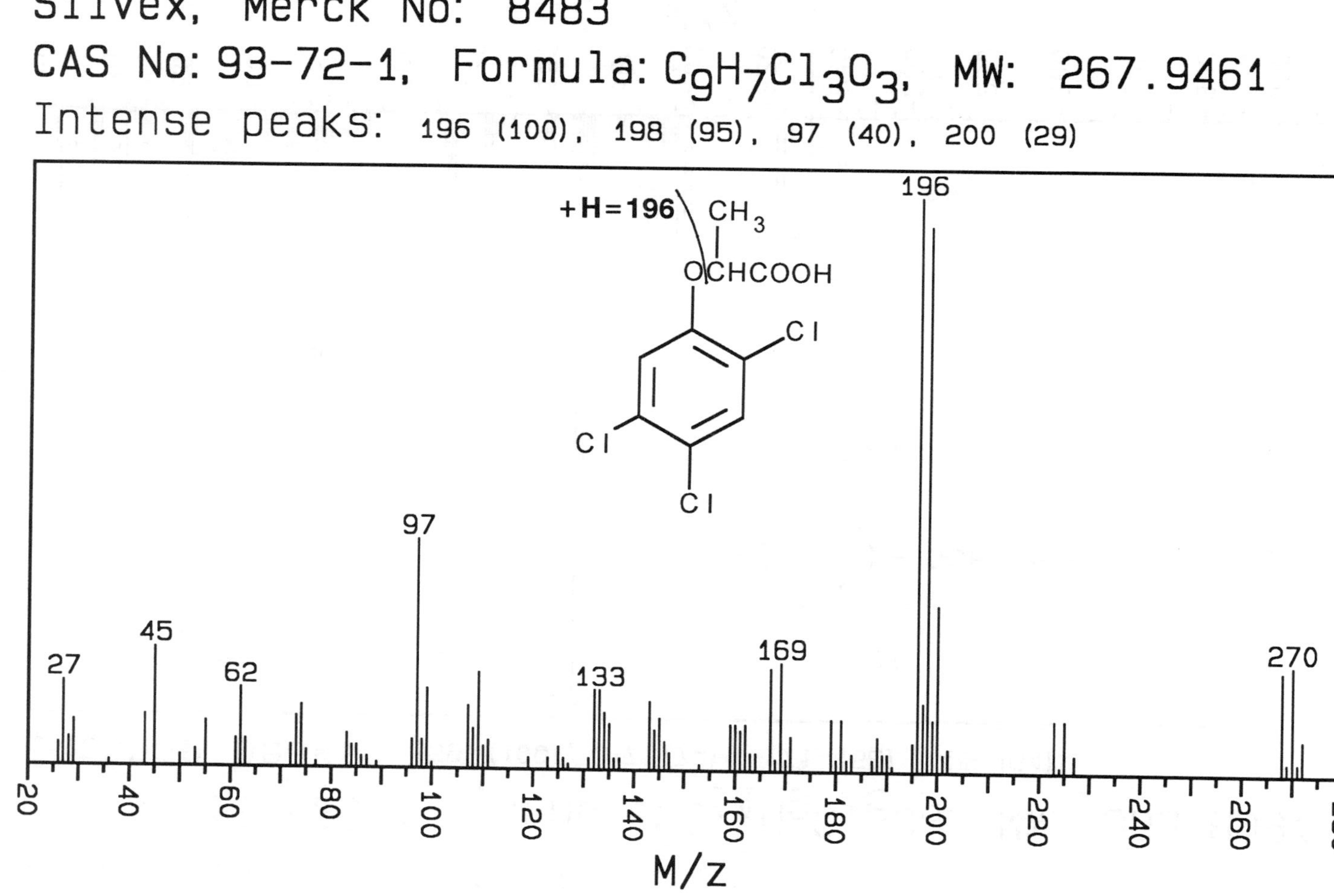

Erbon, Merck No: 3587

CAS No: 136-25-4, Formula: $C_{11}H_9Cl_5O_3$, MW: 363.8994

Intense peaks: 169 (100), 97 (74), 171 (68), 45 (64)

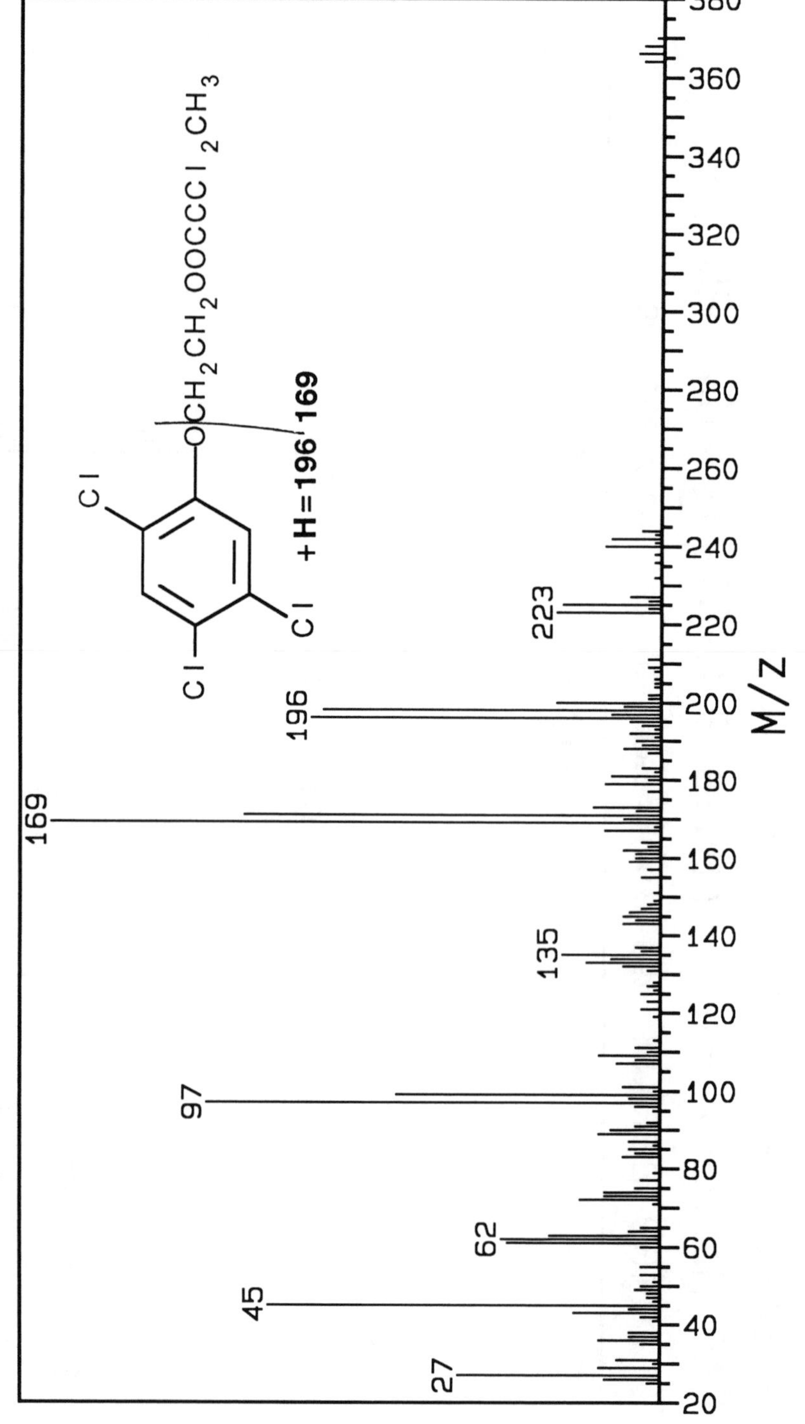

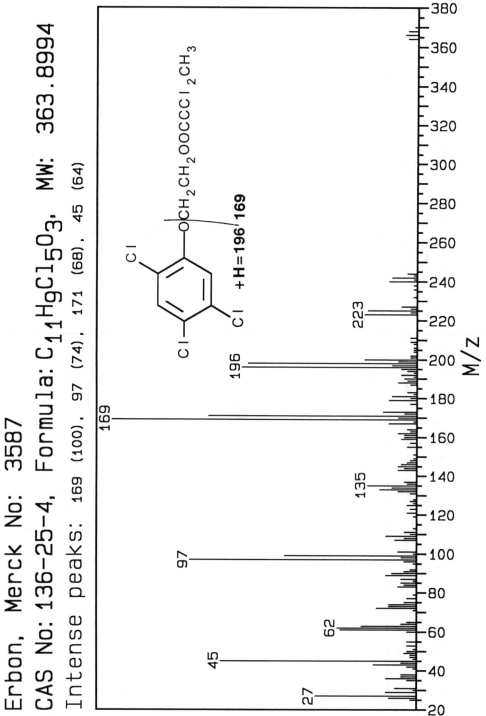

Naphthaleneacetic acid, 1-, Merck No: 6290

CAS No: 86-87-3, Formula: $C_{12}H_{10}O_2$, MW: 186.0681

Intense peaks: 141 (100), 186 (49), 115 (29), 142 (17)

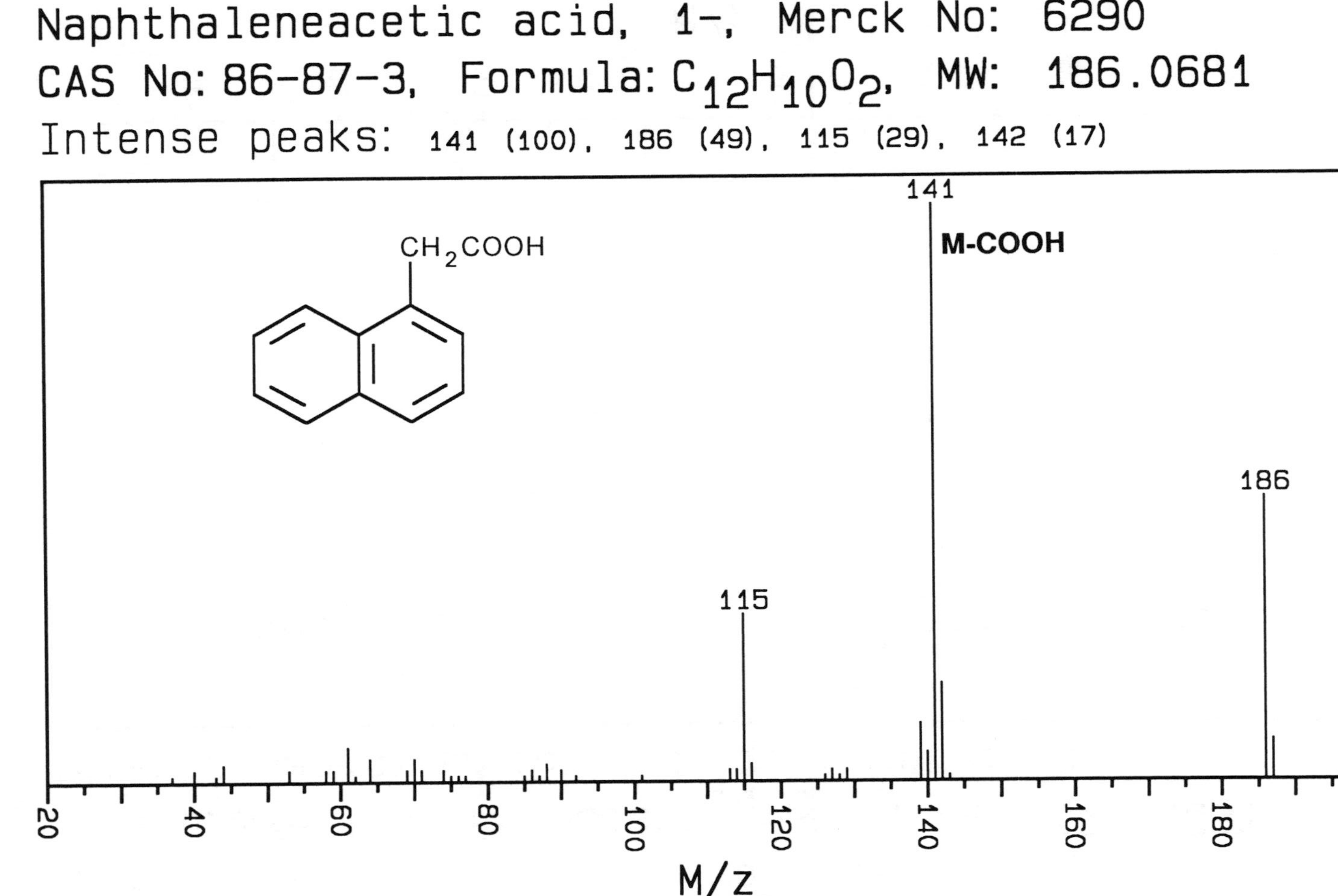

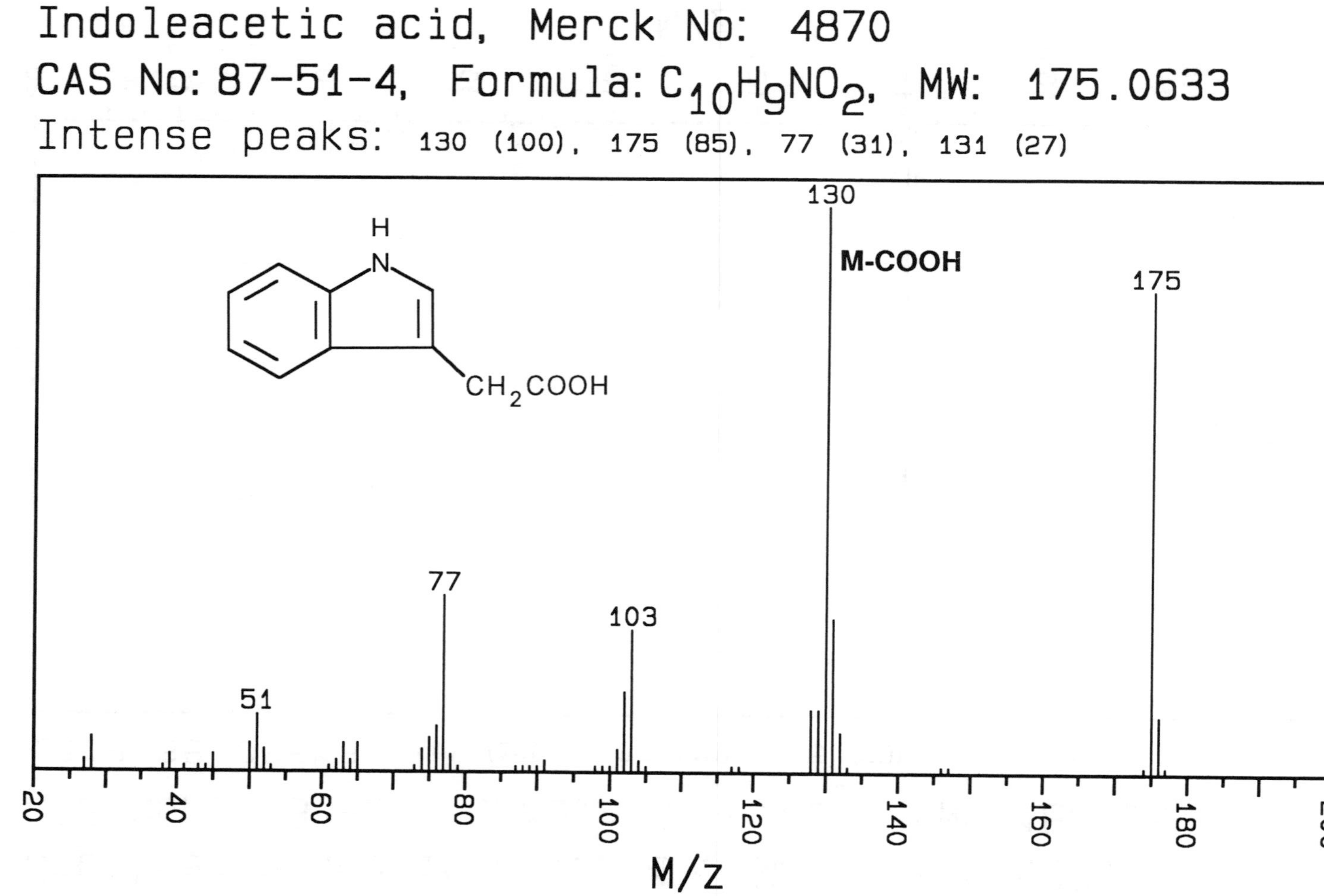
Indoleacetic acid, Merck No: 4870
CAS No: 87-51-4, Formula: $C_{10}H_9NO_2$, MW: 175.0633
Intense peaks: 130 (100), 175 (85), 77 (31), 131 (27)
H
N
$CH_2COOH$
130
M-COOH
175
77
103
51
20
40
60
80
100
120
140
160
180
200
M/Z

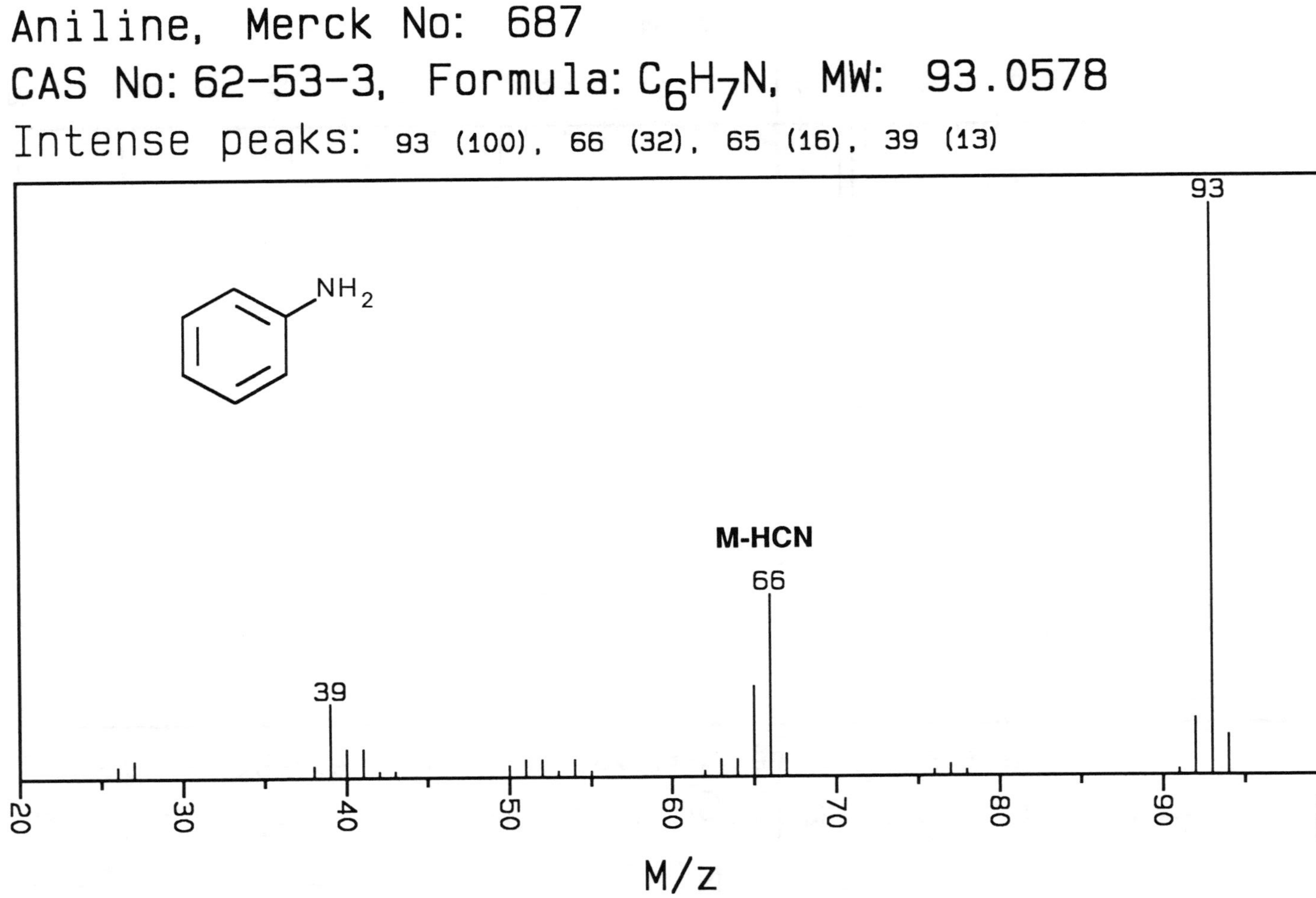

Aniline, Merck No: 687
CAS No: 62-53-3, Formula: $C_6H_7N$, MW: 93.0578
Intense peaks: 93 (100), 66 (32), 65 (16), 39 (13)
$NH_2$
93
M-HCN
66
39
20
30
40
50
60
70
80
90
100
M/z

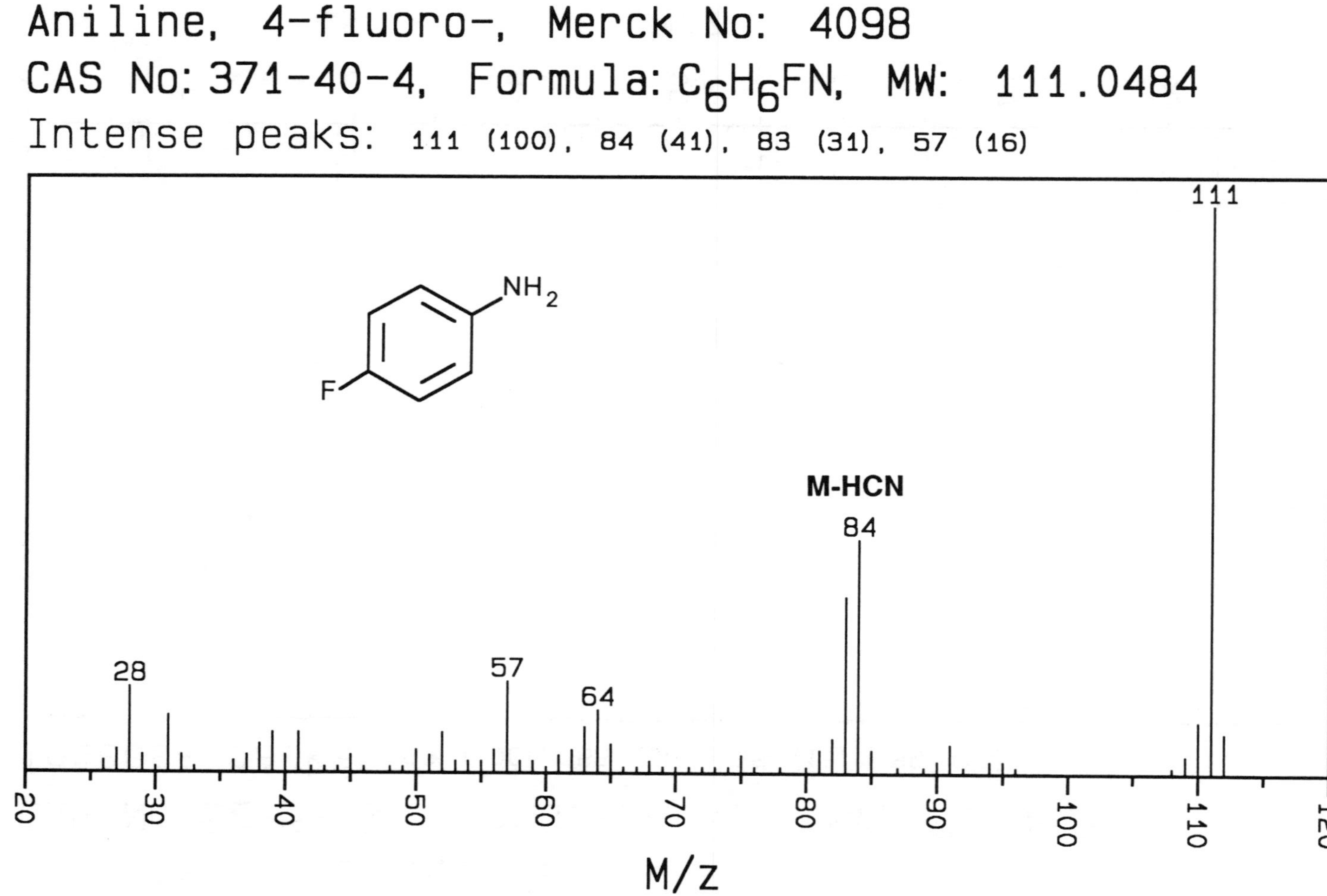
Aniline, 4-fluoro-, Merck No: 4098
CAS No: 371-40-4, Formula: C6H6FN, MW: 111.0484
Intense peaks: 111 (100), 84 (41), 83 (31), 57 (16)
NH2
F
111
M-HCN
84
64
57
28
20
30
40
50
60
70
80
90
100
110
120
M/Z

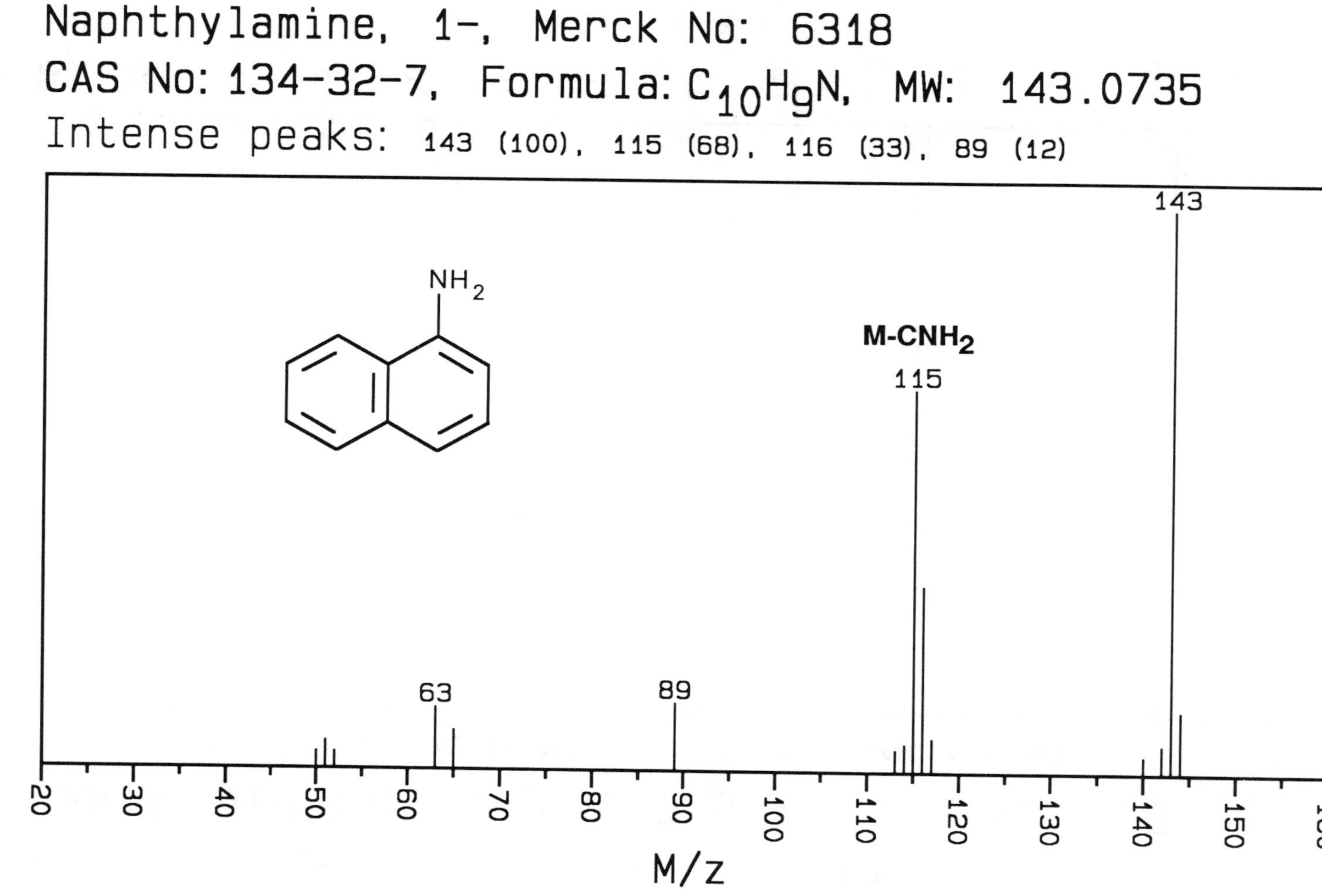
Naphthylamine, 1-, Merck No: 6318
CAS No: 134-32-7, Formula: $C_{10}H_9N$, MW: 143.0735
Intense peaks: 143 (100), 115 (68), 116 (33), 89 (12)
$NH_2$
143
M-$CNH_2$
115
63
89
20
30
40
50
60
70
80
90
100
110
120
130
140
150
160
M/Z

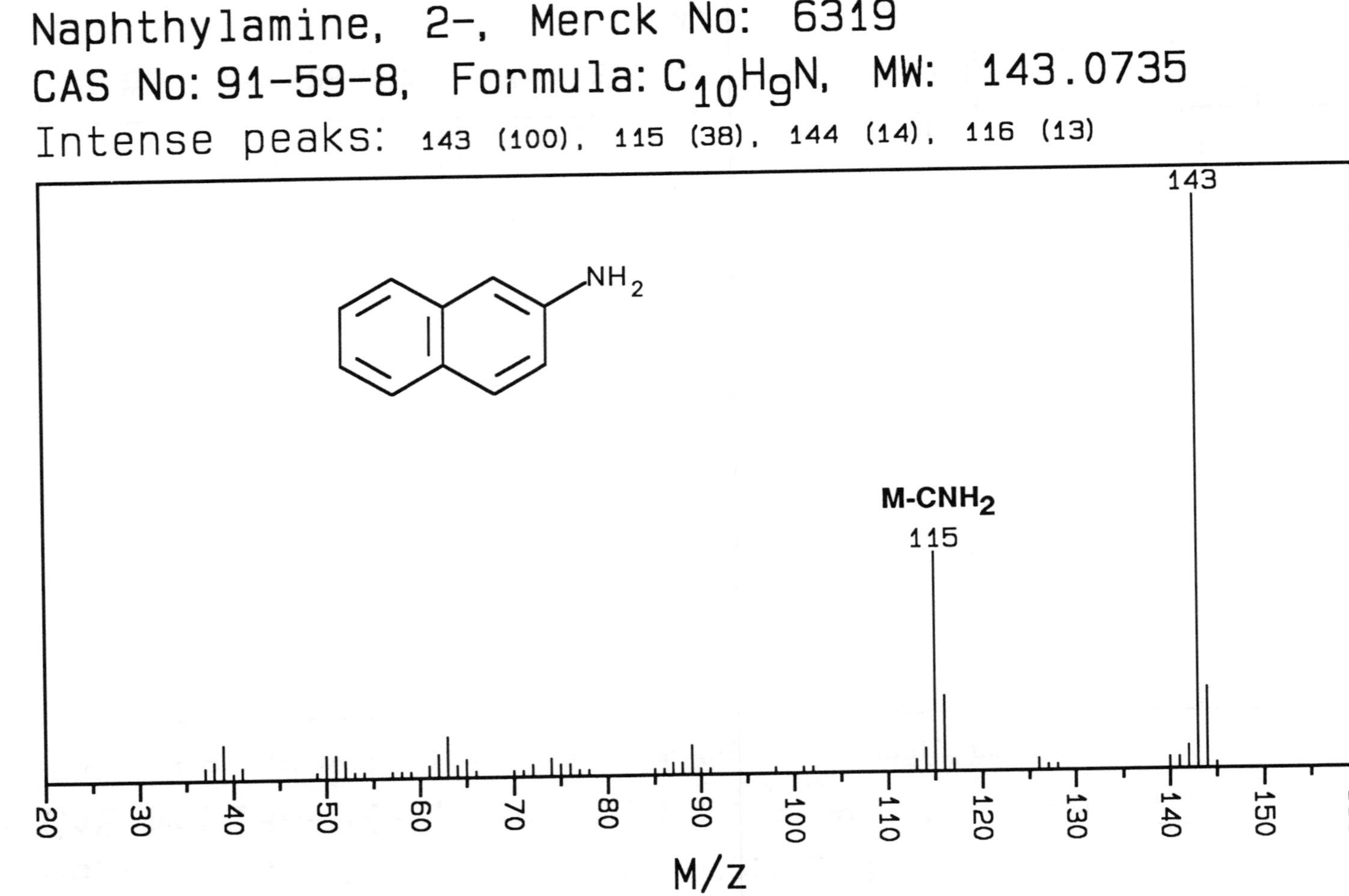
Naphthylamine, 2-, Merck No: 6319
CAS No: 91-59-8, Formula: $C_{10}H_9N$, MW: 143.0735
Intense peaks: 143 (100), 115 (38), 144 (14), 116 (13)
$NH_2$
143
M-$CNH_2$
115
20 30 40 50 60 70 80 90 100 110 120 130 140 150 160
M/Z

Diphenylamine, Merck No: 3317

CAS No: 122-39-4, Formula: $C_{12}H_{11}N$, MW: 169.0891

Intense peaks: 169 (100), 168 (47), 167 (28), 51 (14)

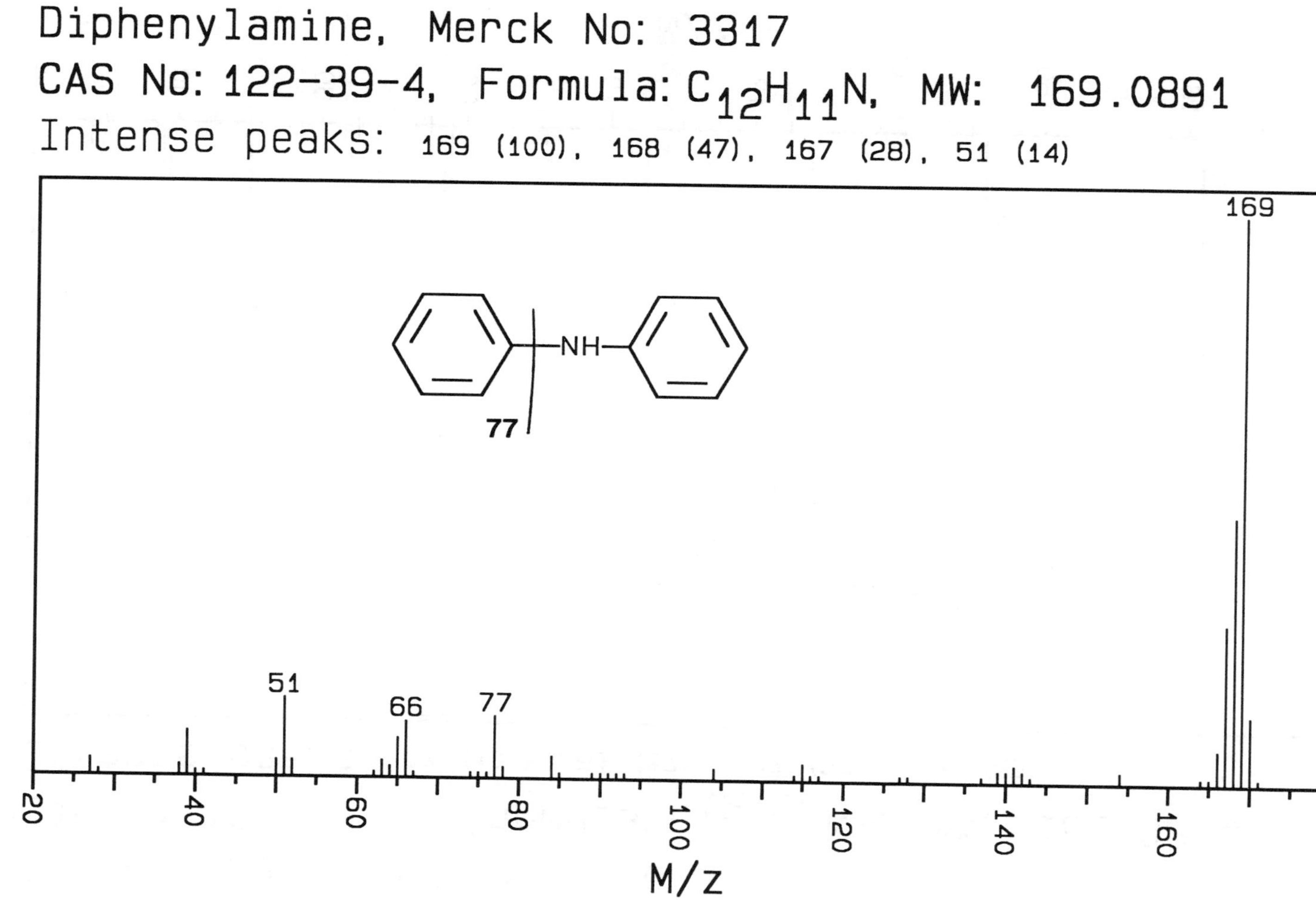

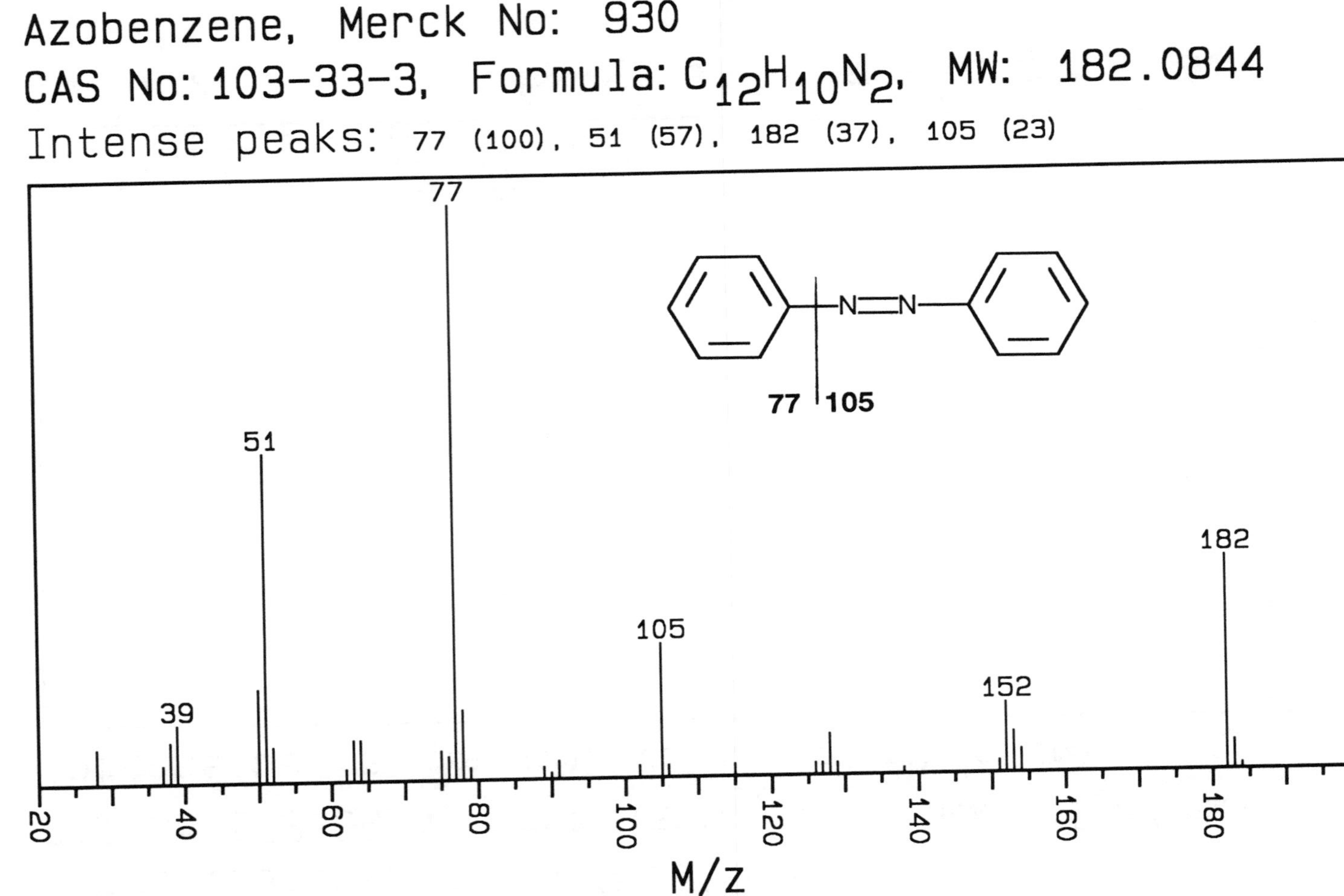

Azobenzene, Merck No: 930
CAS No: 103-33-3, Formula: $C_{12}H_{10}N_2$, MW: 182.0844
Intense peaks: 77 (100), 51 (57), 182 (37), 105 (23)
N=N
77
105
77
51
182
105
152
39
20
40
60
80
100
120
140
160
180
200
M/Z

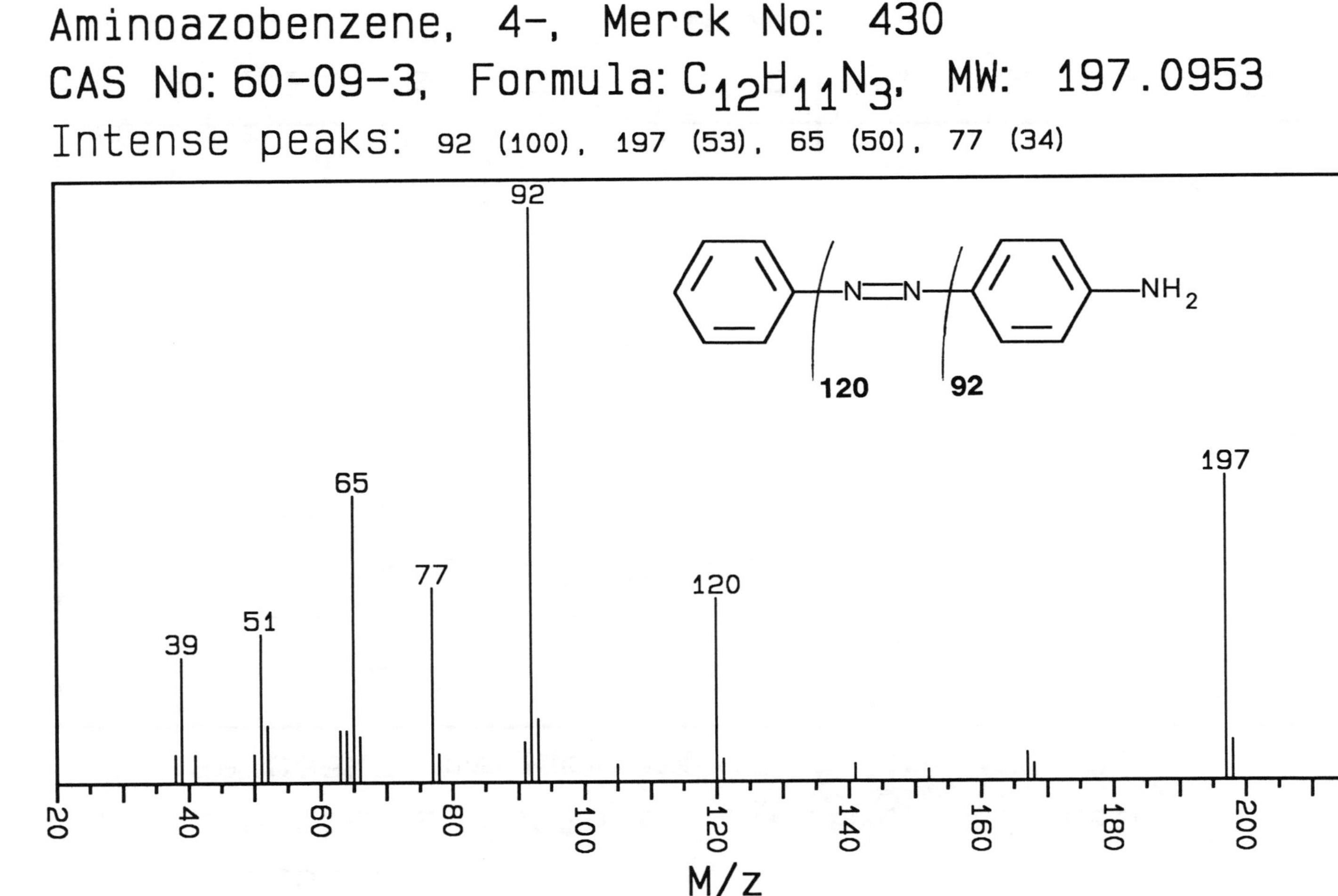
Aminoazobenzene, 4-, Merck No: 430
CAS No: 60-09-3, Formula: $C_{12}H_{11}N_3$, MW: 197.0953
Intense peaks: 92 (100), 197 (53), 65 (50), 77 (34)
92
65
77
51
39
120
197
N=N
$NH_2$
120
92
20
40
60
80
100
120
140
160
180
200
220
M/Z

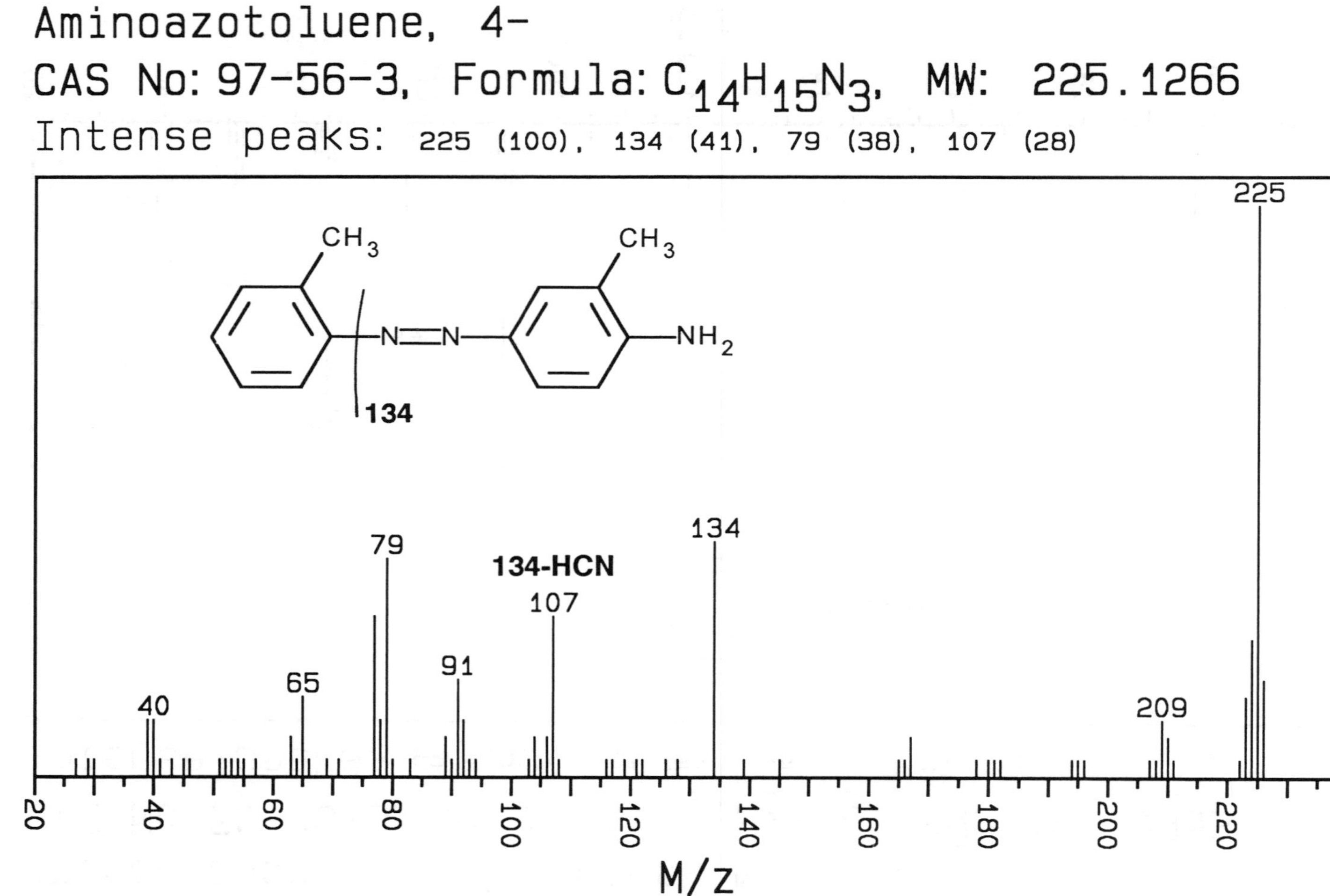
Aminoazotoluene, 4-
CAS No: 97-56-3, Formula: $C_{14}H_{15}N_3$, MW: 225.1266
Intense peaks: 225 (100), 134 (41), 79 (38), 107 (28)
CH3
CH3
NH2
134
225
134
79
134-HCN
107
91
65
40
209
20
40
60
80
100
120
140
160
180
200
220
240
M/Z

Aminoazobenzene, N,N-dimethyl-, Merck No: 3218
CAS No: 60-11-7, Formula: $C_{14}H_{15}N_3$, MW: 225.1266
Intense peaks: 120 (100), 225 (73), 77 (58), 106 (30)

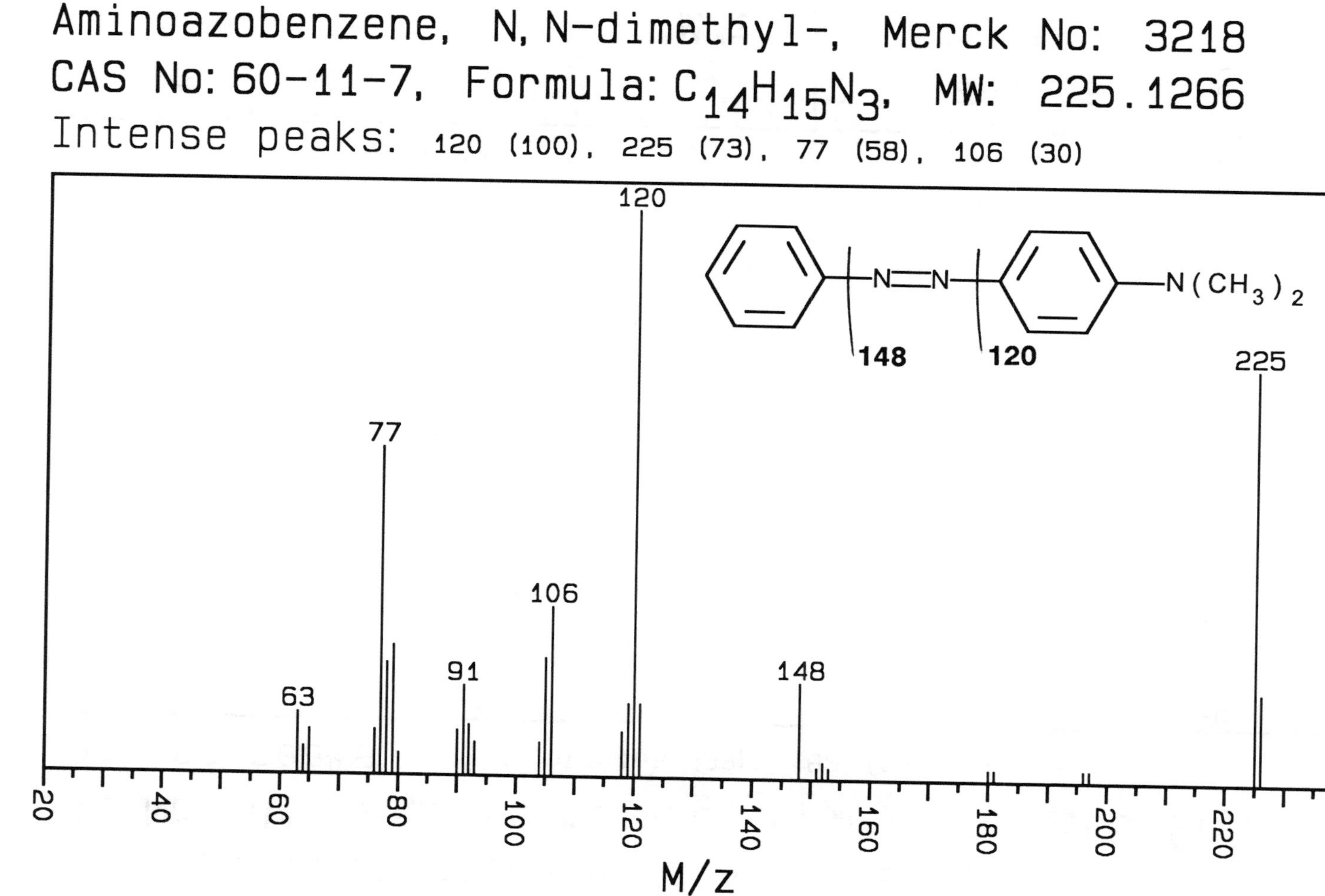

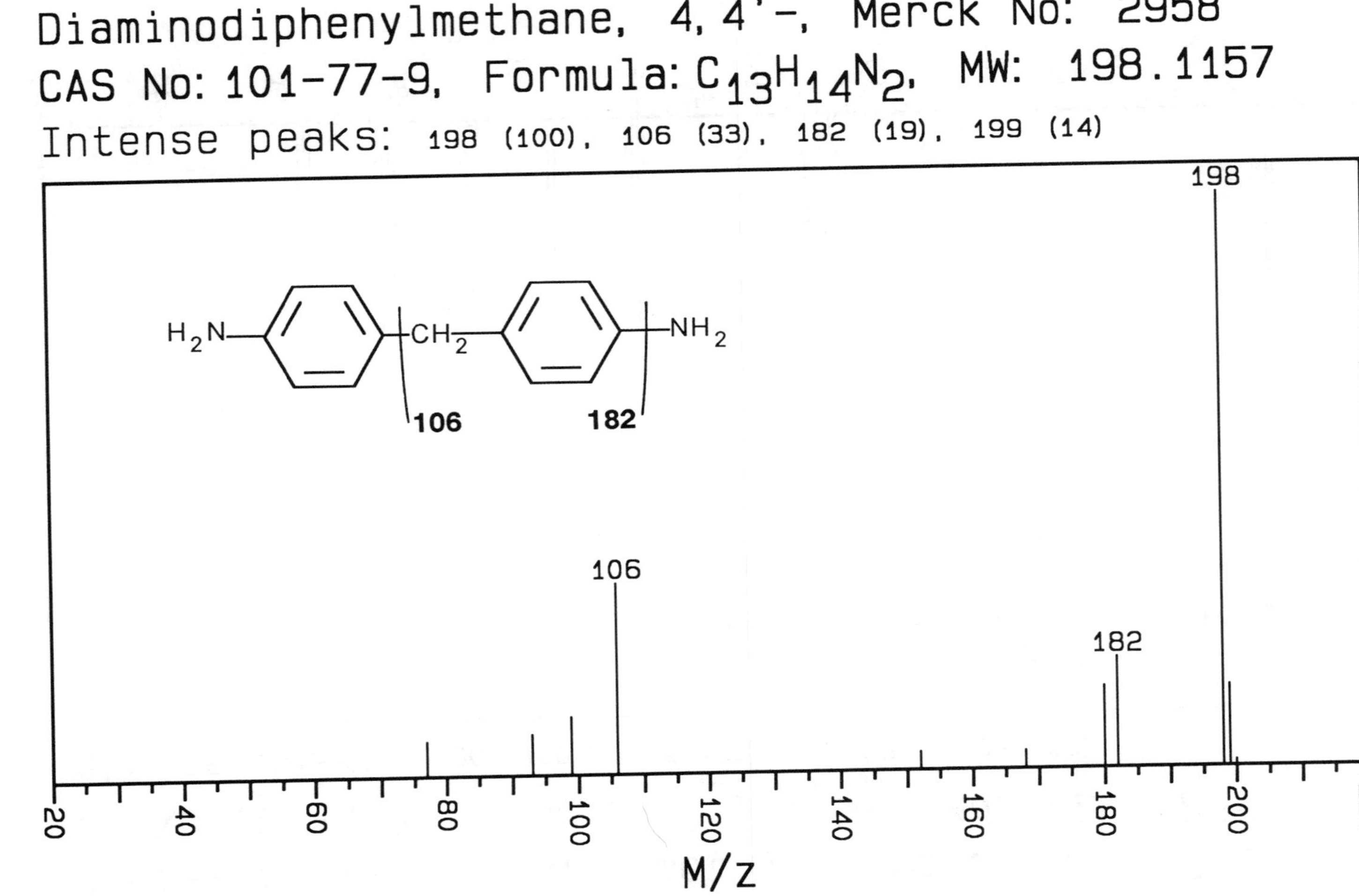
Diaminodiphenylmethane, 4,4'-, Merck No: 2958
CAS No: 101-77-9, Formula: $C_{13}H_{14}N_2$, MW: 198.1157
Intense peaks: 198 (100), 106 (33), 182 (19), 199 (14)
198
$H_2N$
$CH_2$
$NH_2$
106
182
106
182
20
40
60
80
100
120
140
160
180
200
220
M/Z

Biphenylamine, 4-, Merck No: 1248
CAS No: 92-67-1, Formula: $C_{12}H_{11}N$, MW: 169.0891
Intense peaks: 169 (100), 168 (35), 170 (25), 167 (22)

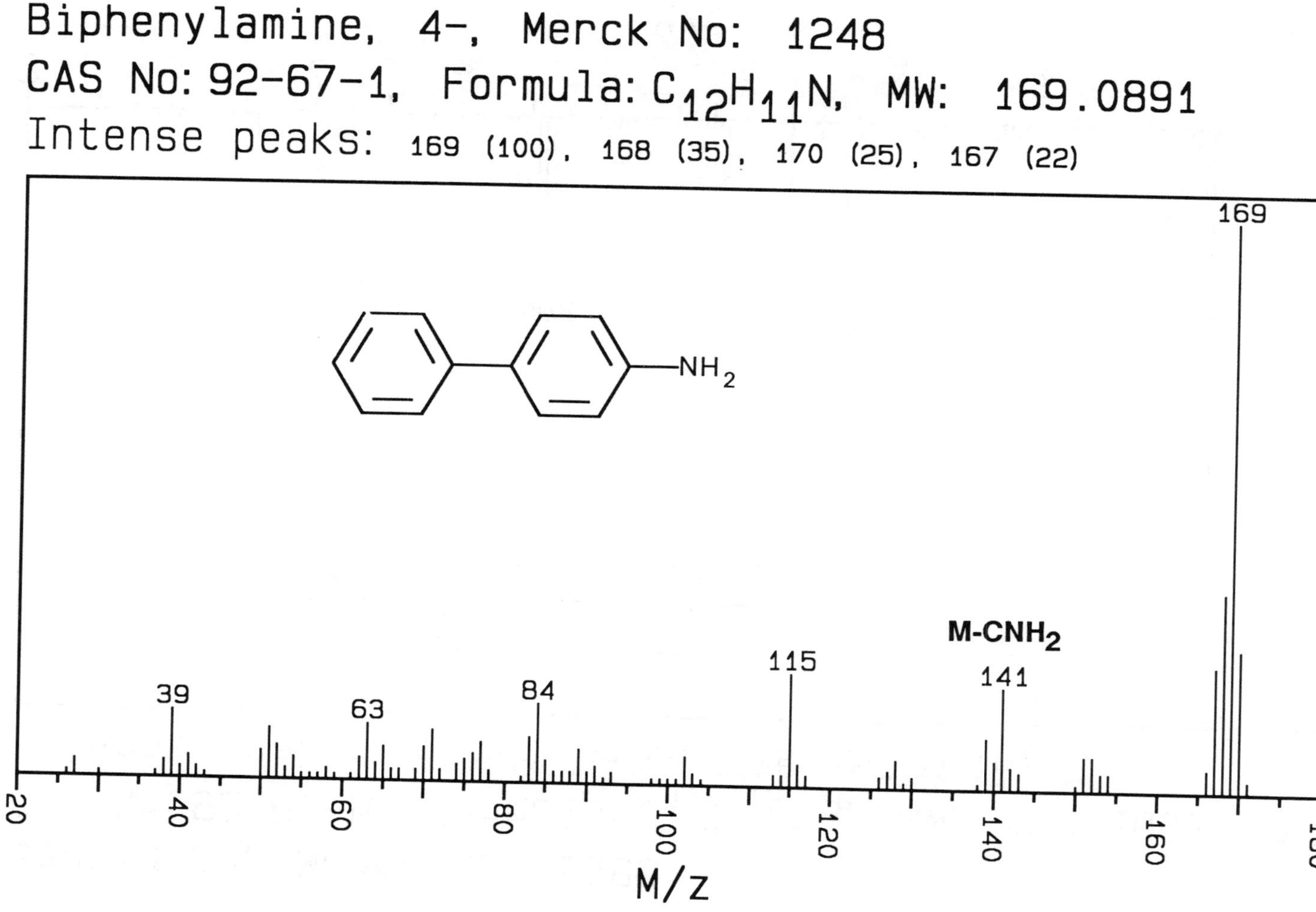

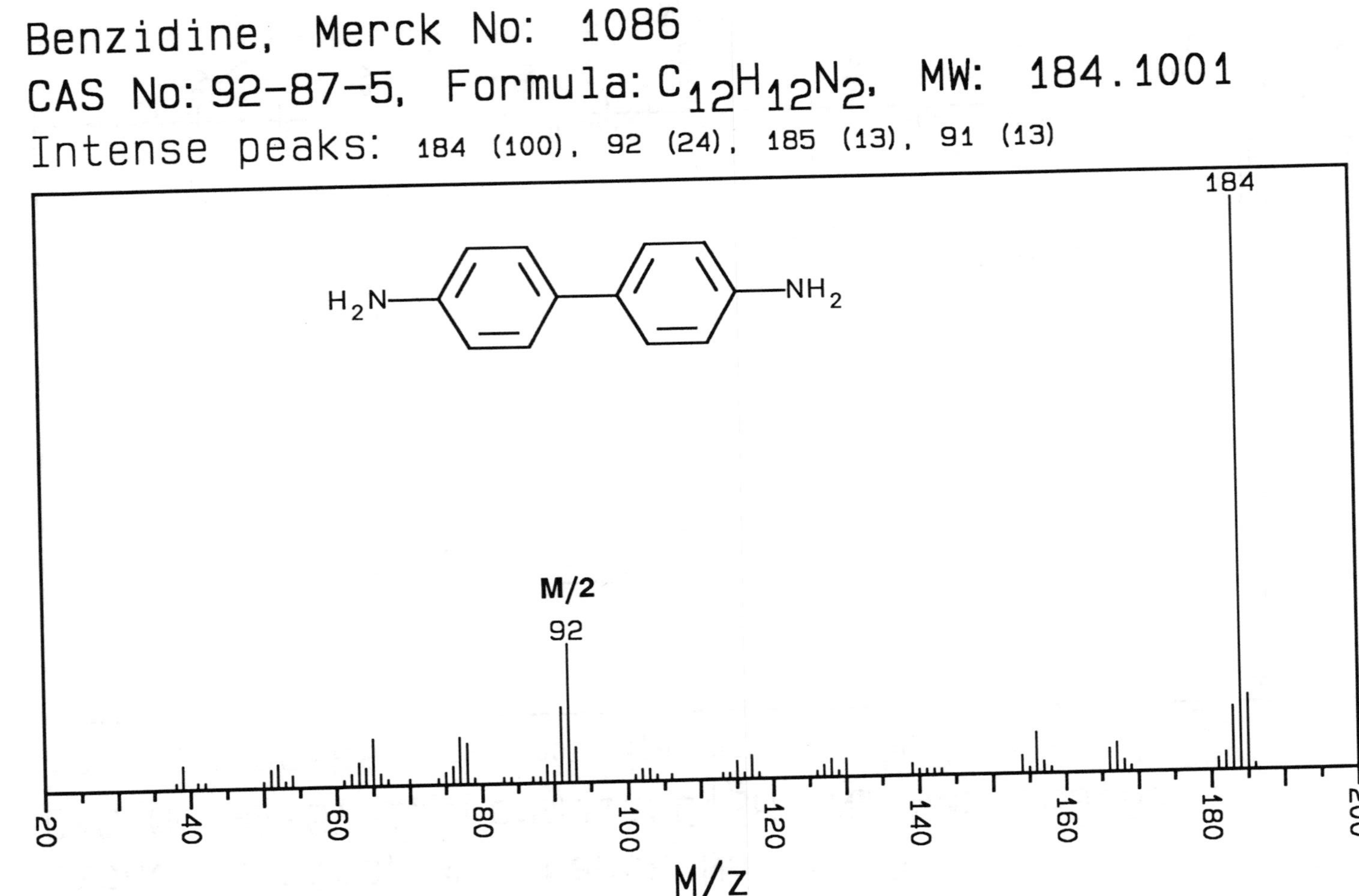
Benzidine, Merck No: 1086
CAS No: 92-87-5, Formula: $C_{12}H_{12}N_2$, MW: 184.1001
Intense peaks: 184 (100), 92 (24), 185 (13), 91 (13)
$H_2N$
$NH_2$
184
M/2
92
20
40
60
80
100
120
140
160
180
200
M/Z

Tolidine, 2-, Merck No: 9437

CAS No: 119-93-7, Formula: $C_{14}H_{16}N_2$, MW: 212.1313

Intense peaks: 212 (100), 106 (83), 213 (16), 211 (16)

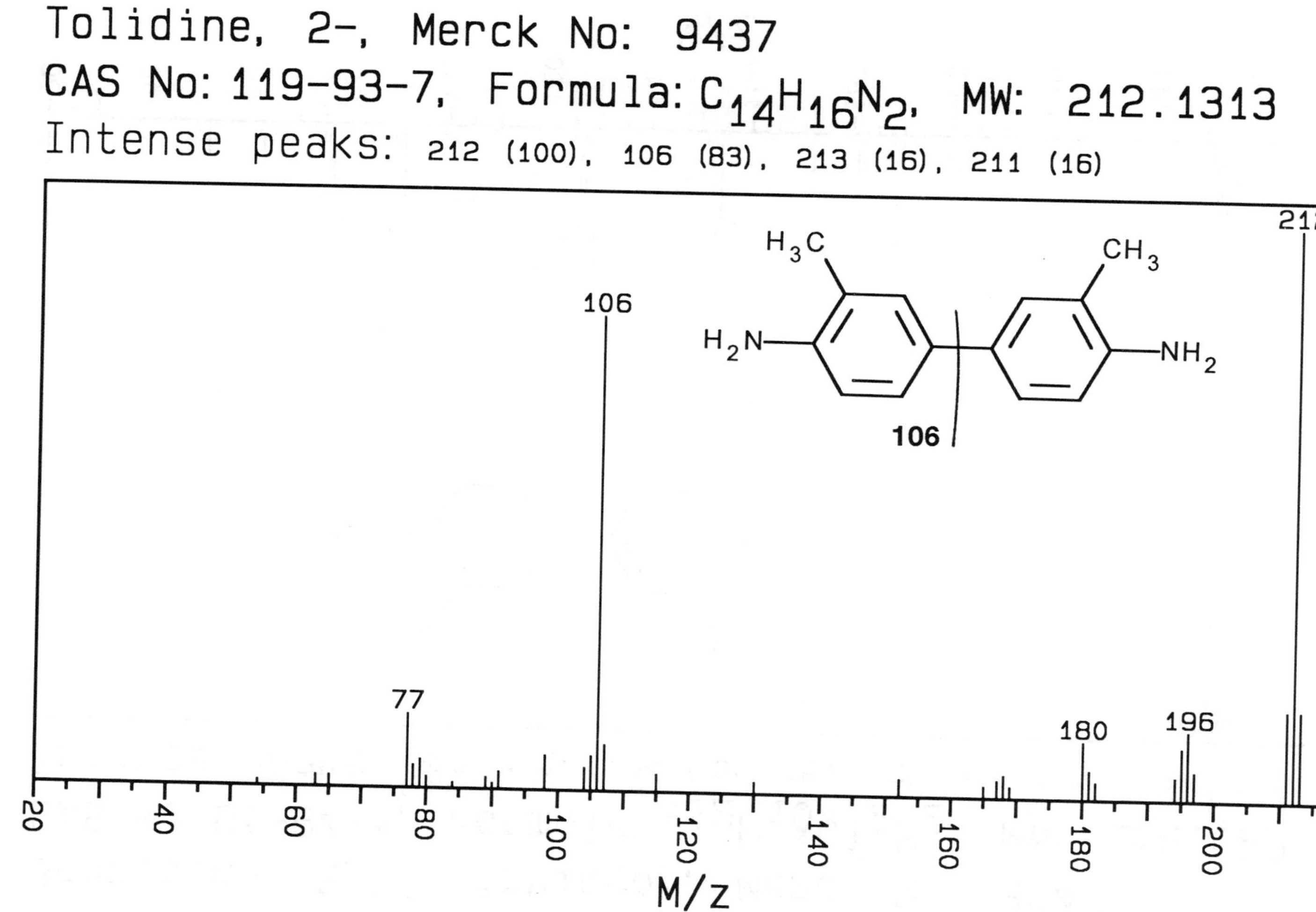

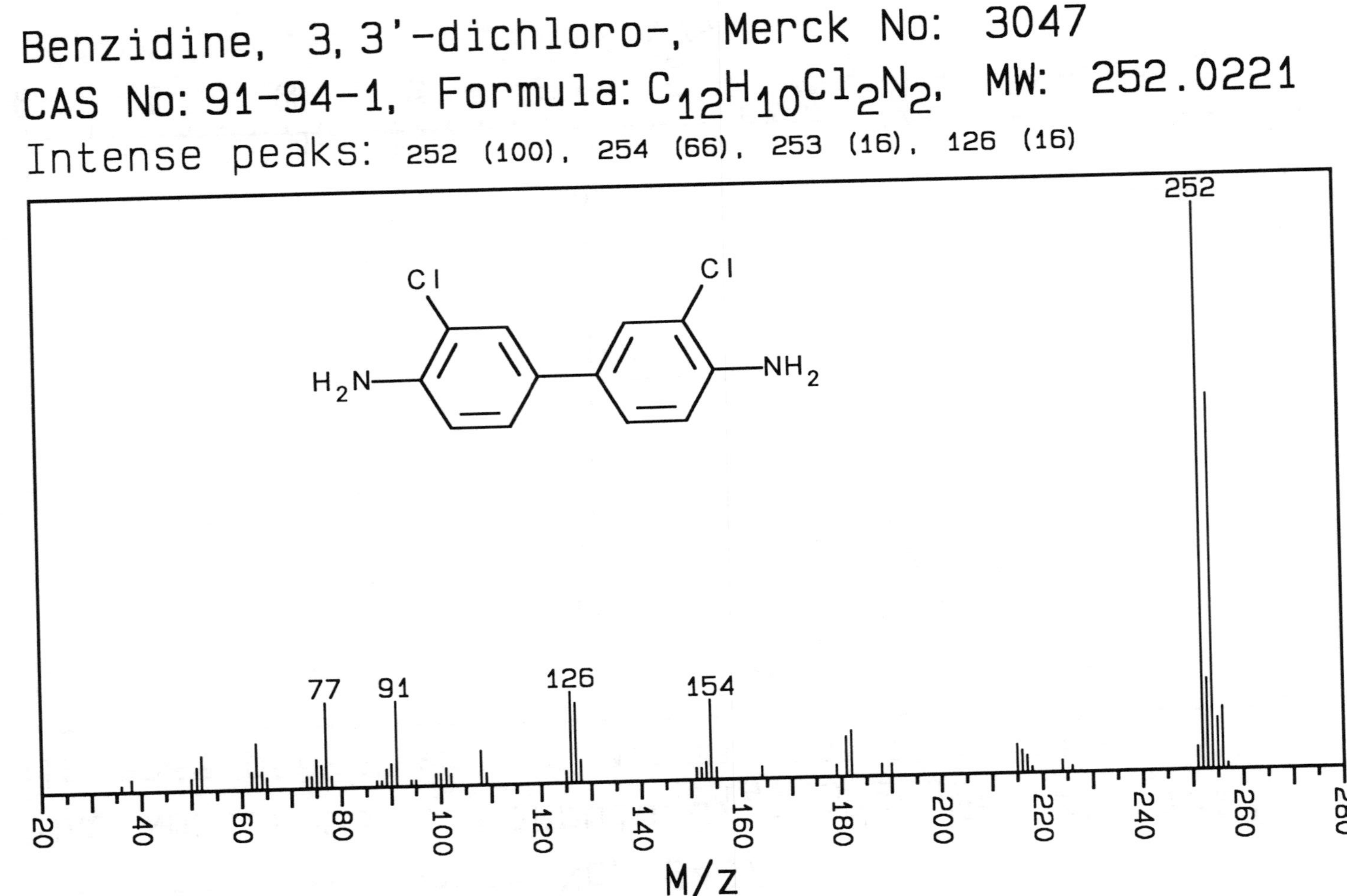
Benzidine, 3,3'-dichloro-, Merck No: 3047
CAS No: 91-94-1, Formula: $C_{12}H_{10}Cl_2N_2$, MW: 252.0221
Intense peaks: 252 (100), 254 (66), 253 (16), 126 (16)
252
Cl
Cl
$H_2N$
$NH_2$
77
91
126
154
20
40
60
80
100
120
140
160
180
200
220
240
260
280
M/Z

Picloram, Merck No: 7370

CAS No: 1918-02-1, Formula: $C_6H_3Cl_3N_2O_2$, MW: 239.9261

Intense peaks: 196 (100), 198 (95), 161 (40), 200 (31)

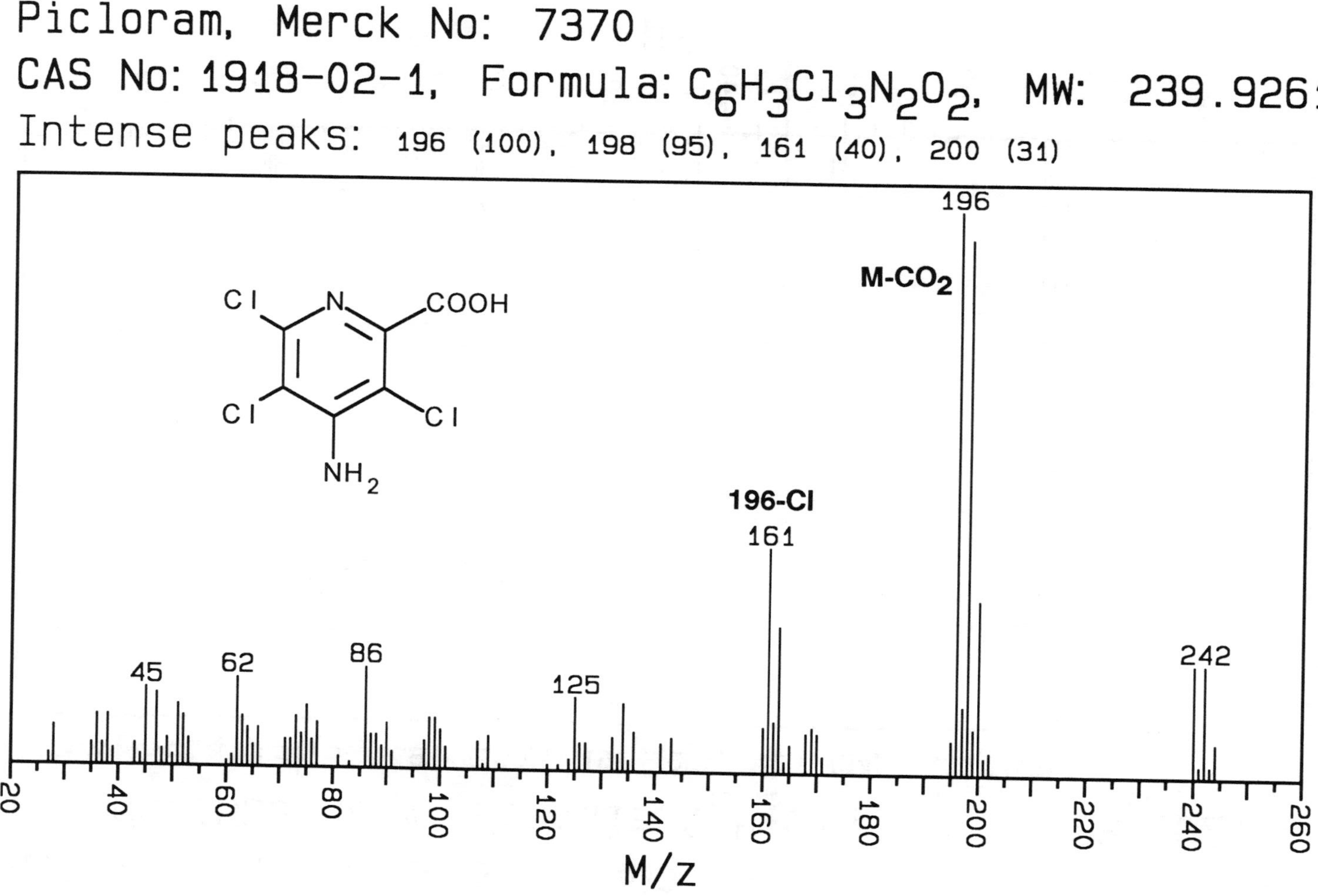

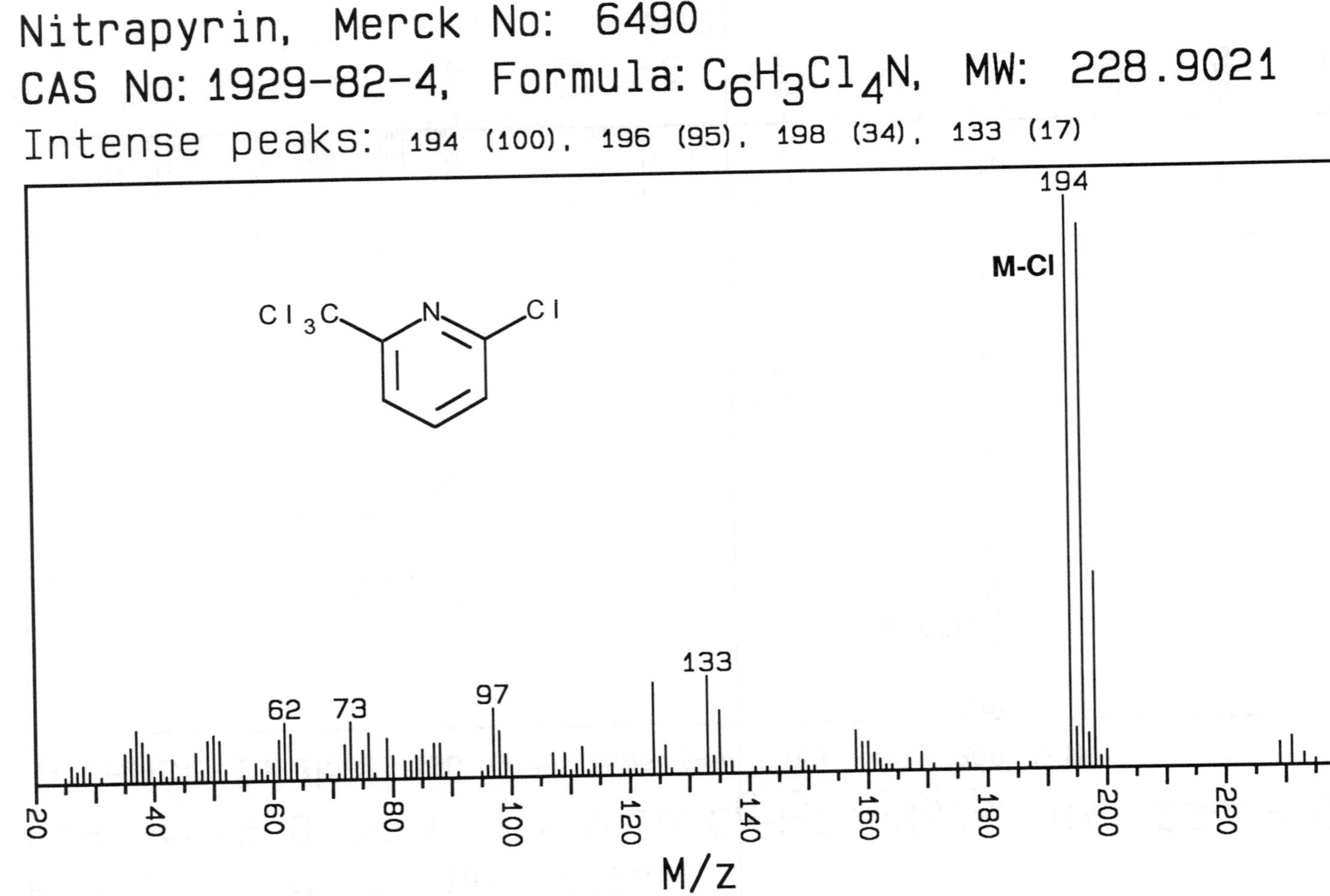

Nitrapyrin, Merck No: 6490
CAS No: 1929-82-4, Formula: $C_6H_3Cl_4N$, MW: 228.9021
Intense peaks: 194 (100), 196 (95), 198 (34), 133 (17)
194
M-Cl
$Cl_3C$
N
Cl
133
97
62
73
20
40
60
80
100
120
140
160
180
200
220
240
M/Z

Nornicotine, Merck No: 6631

CAS No: 494-97-3, Formula: $C_9H_{12}N_2$, MW: 148.1001

Intense peaks: 119 (100), 70 (80), 147 (34), 120 (33)

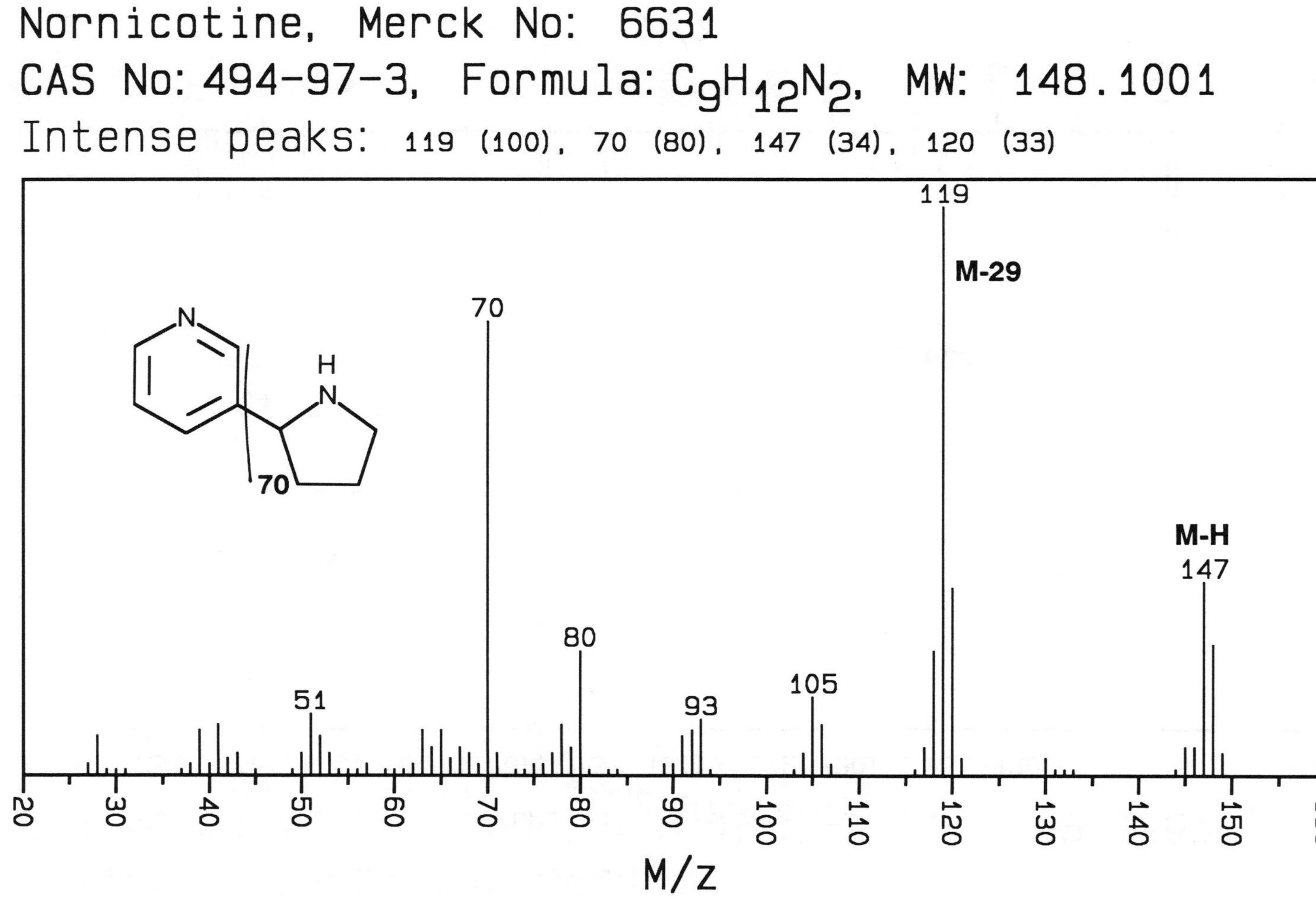

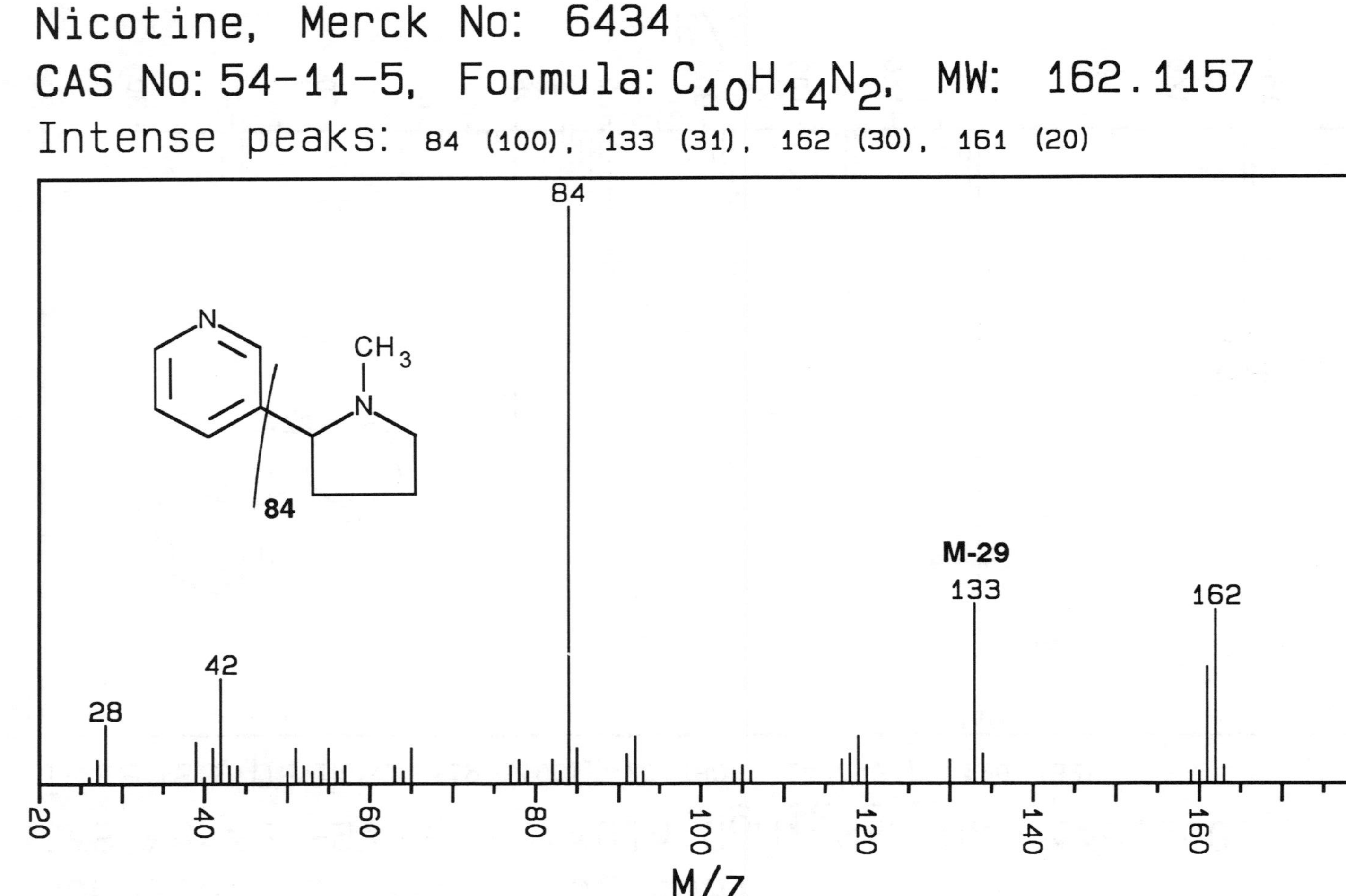
Nicotine, Merck No: 6434
CAS No: 54-11-5, Formula: $C_{10}H_{14}N_2$, MW: 162.1157
Intense peaks: 84 (100), 133 (31), 162 (30), 161 (20)
84
N
$CH_3$
N
84
M-29
133
162
42
28
20
40
60
80
100
120
140
160
180
M/Z

Anabasine, Merck No: 655

CAS No: 494-52-0, Formula: $C_{10}H_{14}N_2$, MW: 162.1157

Intense peaks: 84 (100), 105 (58), 106 (45), 133 (42)

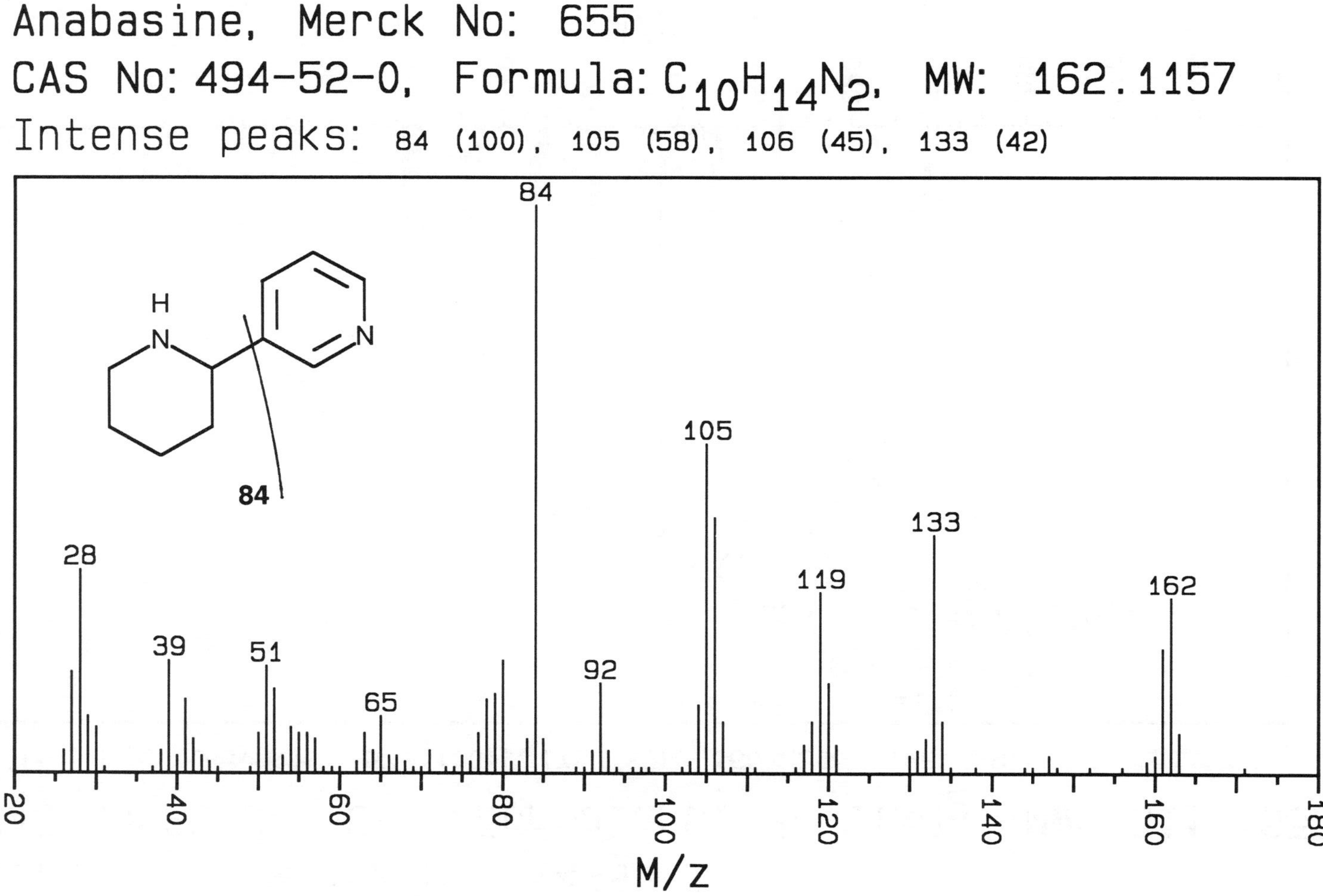

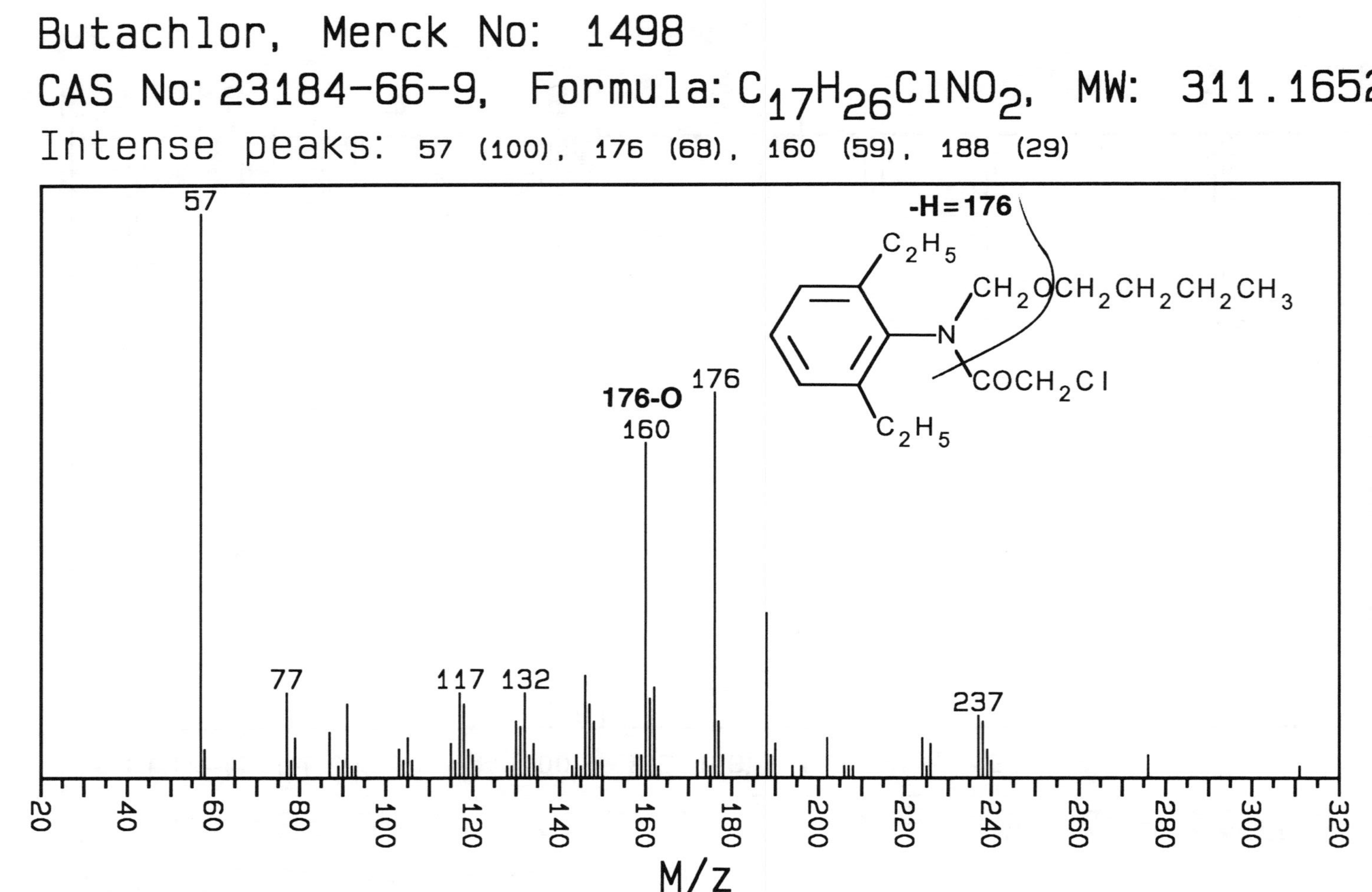
Butachlor, Merck No: 1498
CAS No: 23184-66-9, Formula: C17H26ClNO2, MW: 311.1652
Intense peaks: 57 (100), 176 (68), 160 (59), 188 (29)
-H=176
C2H5
CH2OCH2CH2CH2CH3
N
COCH2Cl
C2H5
57
77
117
132
176-O
160
176
237
20
40
60
80
100
120
140
160
180
200
220
240
260
280
300
320
M/Z

Chlordimeform, Merck No: 2083

CAS No: 6164-98-3, Formula: $C_{10}H_{13}ClN_2$, MW: 196.0767

Intense peaks: 196 (100), 44 (90), 181 (65), 117 (64)

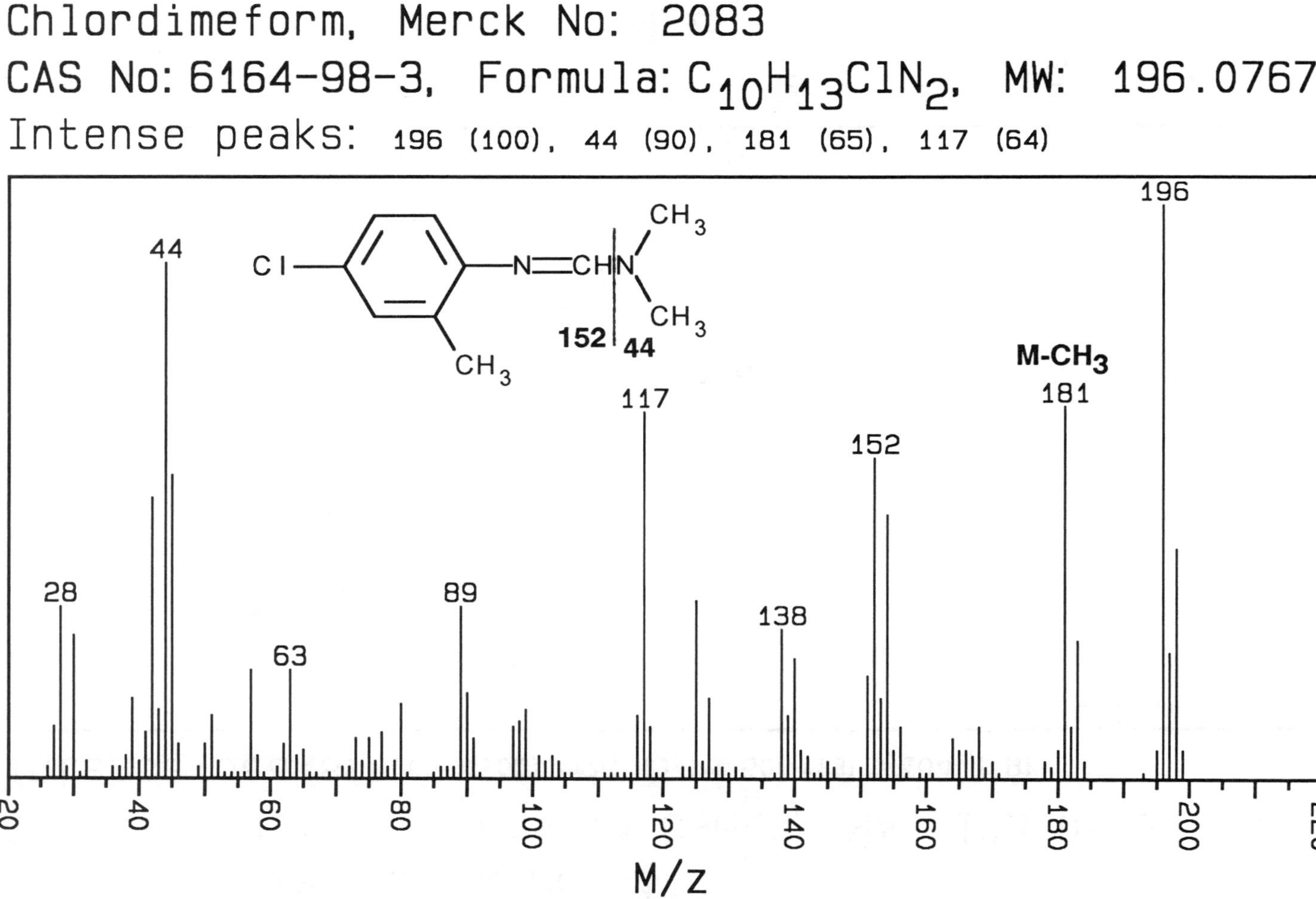

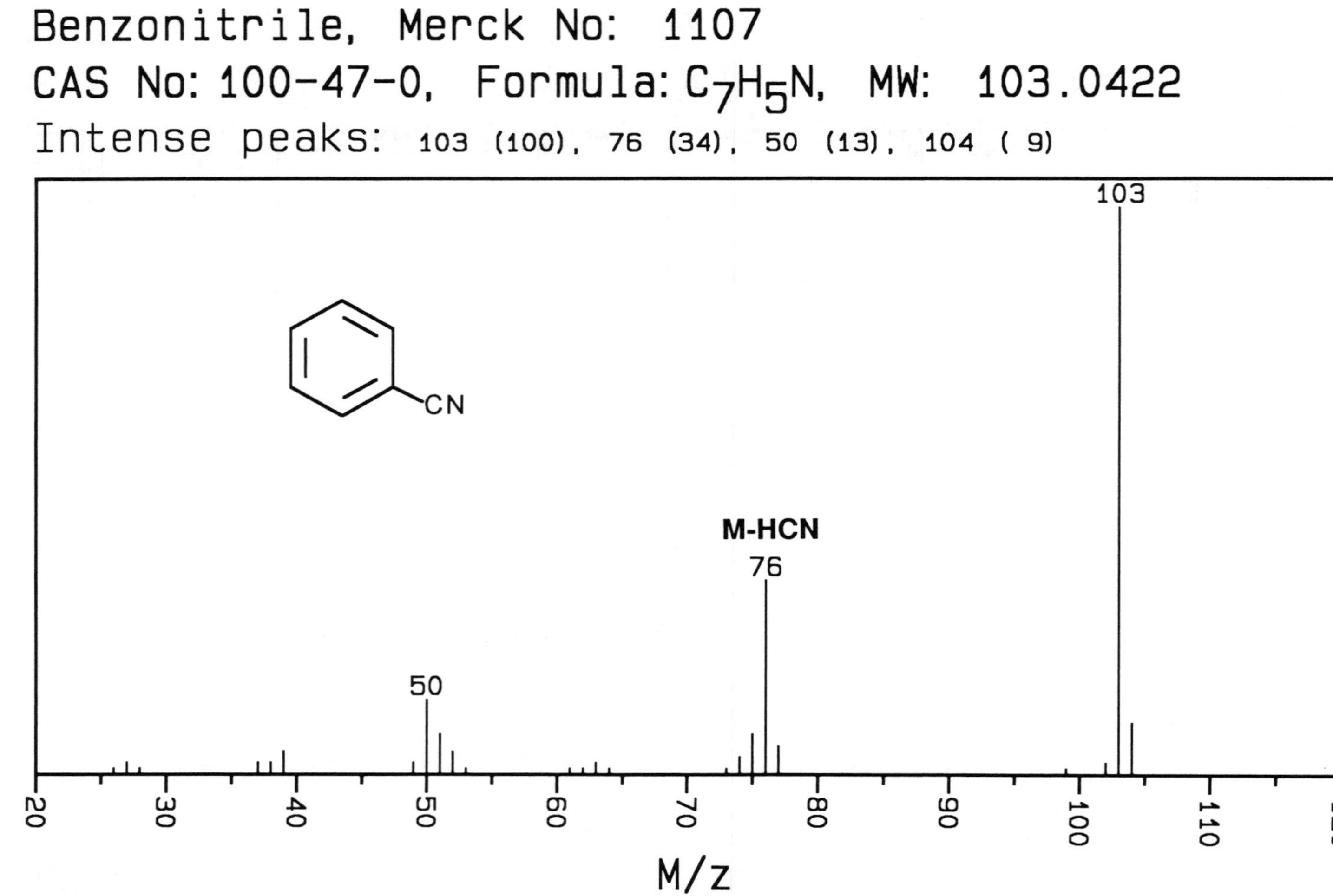
Benzonitrile, Merck No: 1107
CAS No: 100-47-0, Formula: $C_7H_5N$, MW: 103.0422
Intense peaks: 103 (100), 76 (34), 50 (13), 104 ( 9)
103
CN
M-HCN
76
50
20
30
40
50
60
70
80
90
100
110
120
M/Z

Dichlobenil, Merck No: 3029

CAS No: 1194-65-6, Formula: $C_7H_3Cl_2N$, MW: 170.9642

Intense peaks: 171 (100), 100 (71), 173 (69), 75 (56)

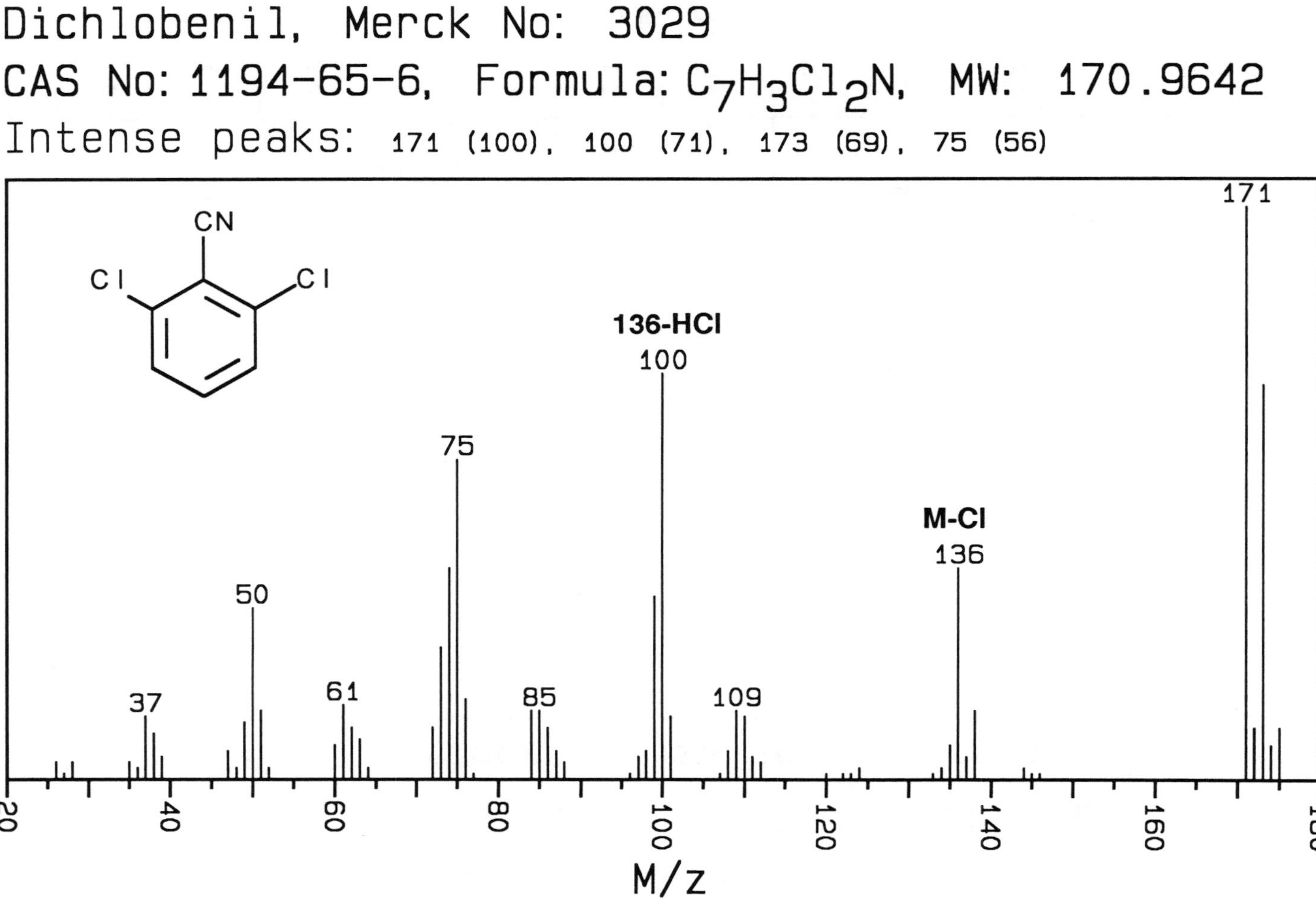

Chlorothalonil, Merck No: 2167

CAS No: 1897-45-6, Formula: $C_8Cl_4N_2$, MW: 263.8816

Intense peaks: 266 (100), 264 (85), 268 (52), 109 (30)

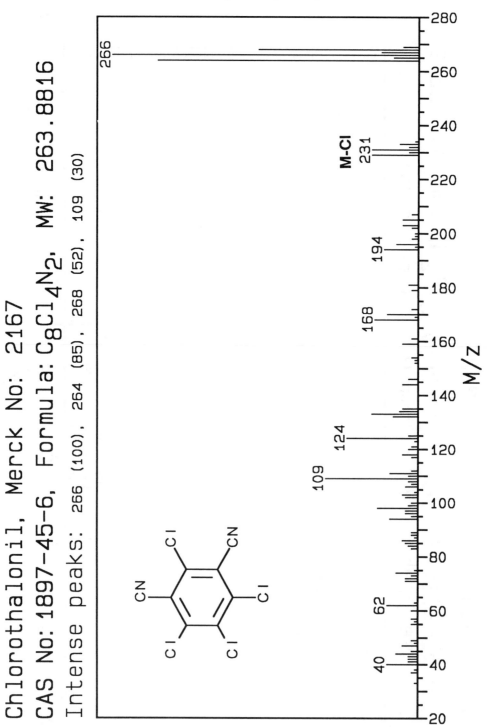

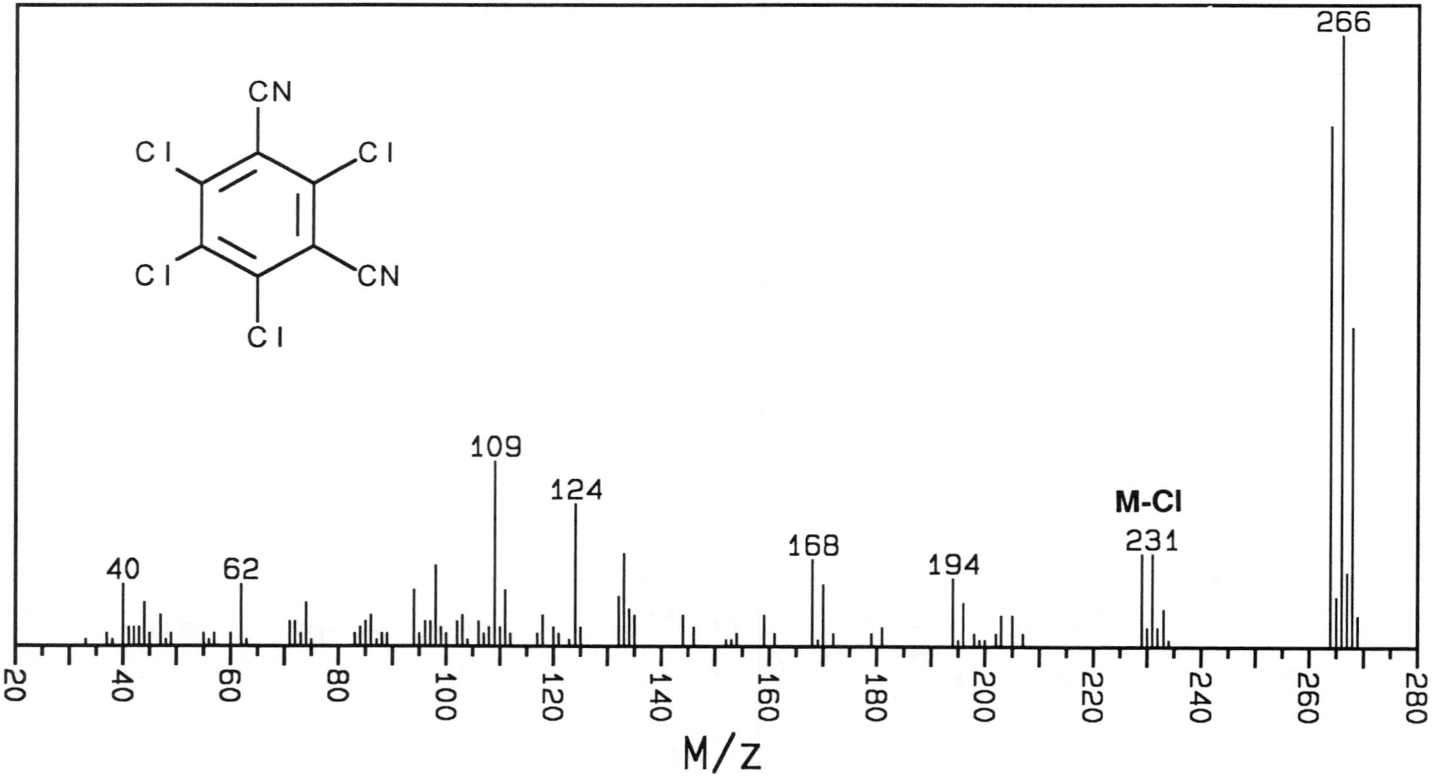

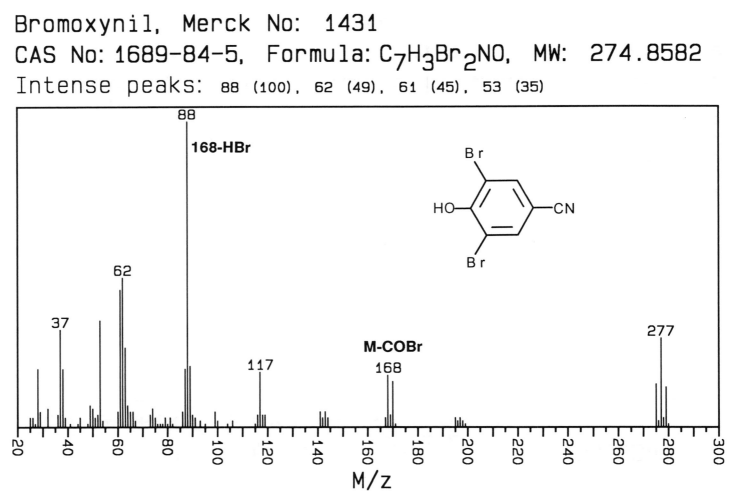

Bromoxynil, Merck No: 1431

CAS No: 1689-84-5, Formula: $C_7H_3Br_2NO$, MW: 274.8582

Intense peaks: 88 (100), 62 (49), 61 (45), 53 (35)

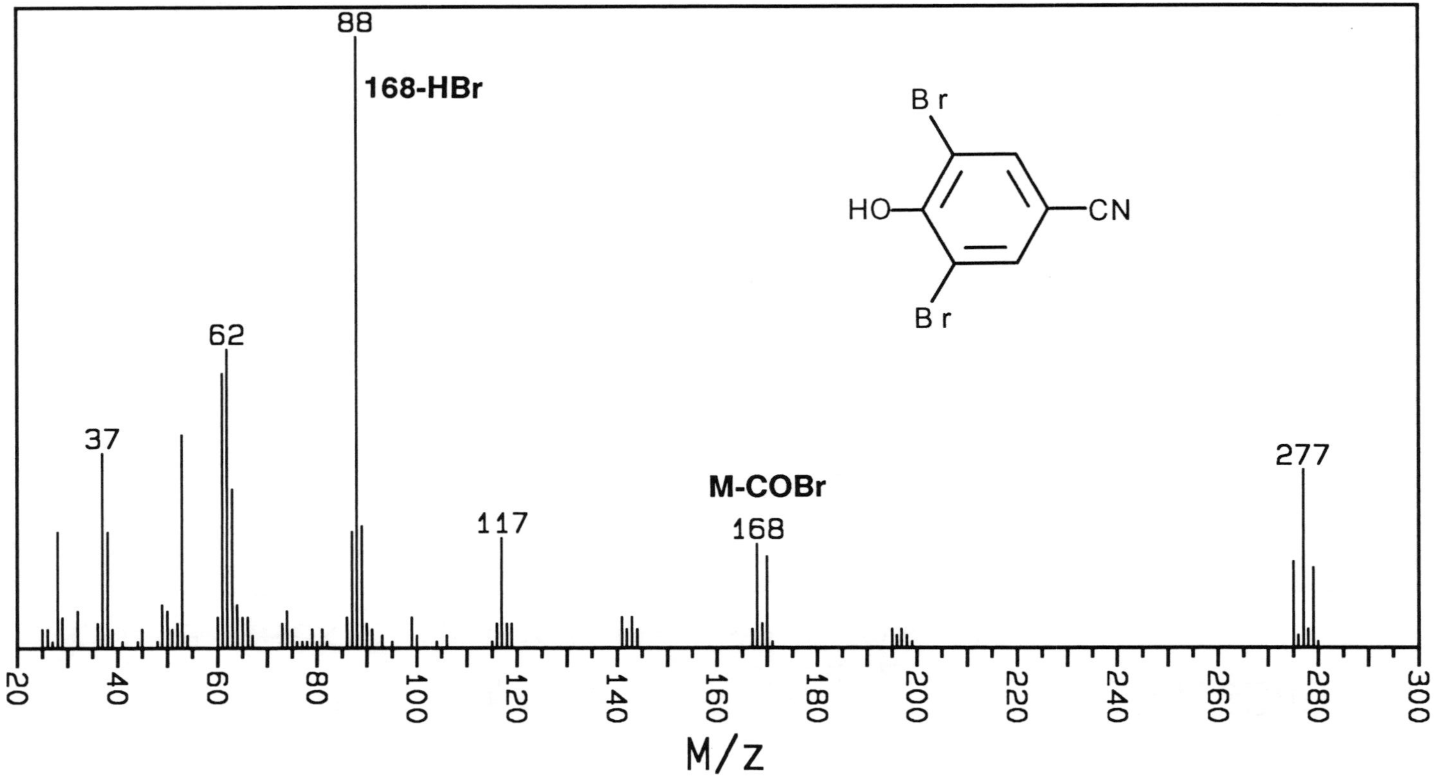

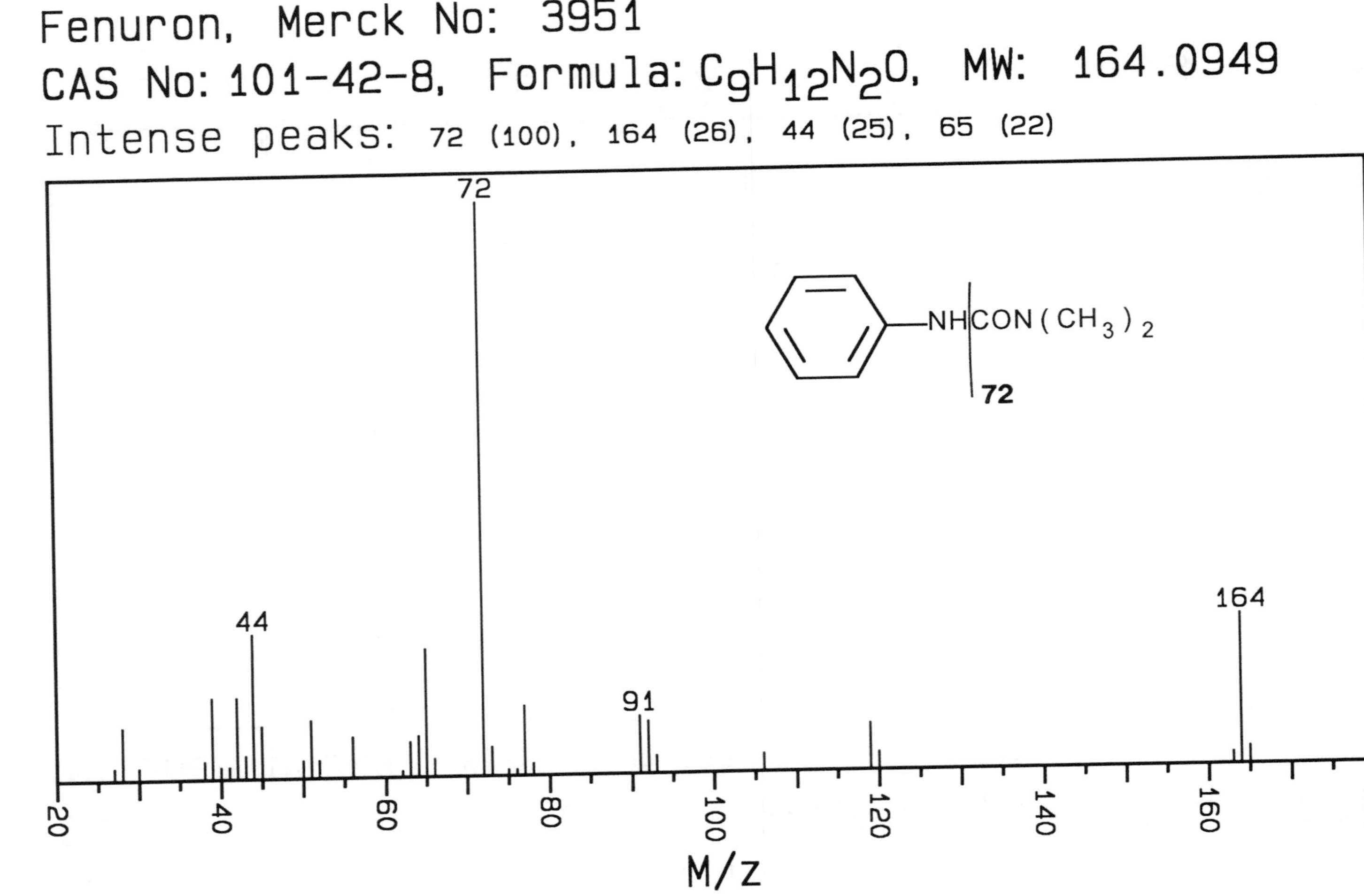
Fenuron, Merck No: 3951
CAS No: 101-42-8, Formula: $C_9H_{12}N_2O$, MW: 164.0949
Intense peaks: 72 (100), 164 (26), 44 (25), 65 (22)
72
NHCON(CH3)2
72
44
91
164
20
40
60
80
100
120
140
160
180
M/Z

Monuron, Merck No: 6169

CAS No: 150-68-5, Formula: $C_9H_{11}ClN_2O$, MW: 198.0561

Intense peaks: 72 (100), 40 (21), 198 (19), 28 (15)

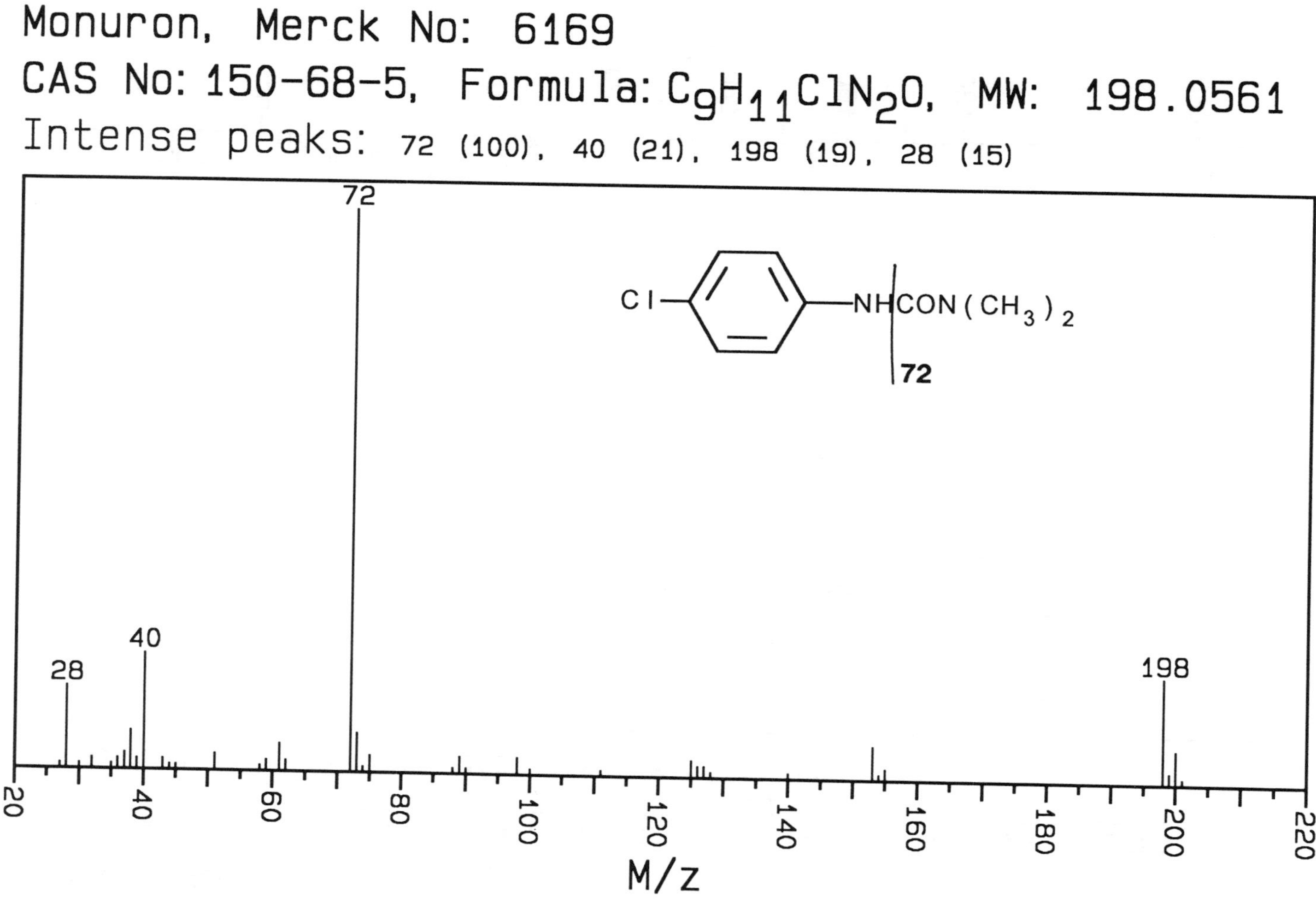

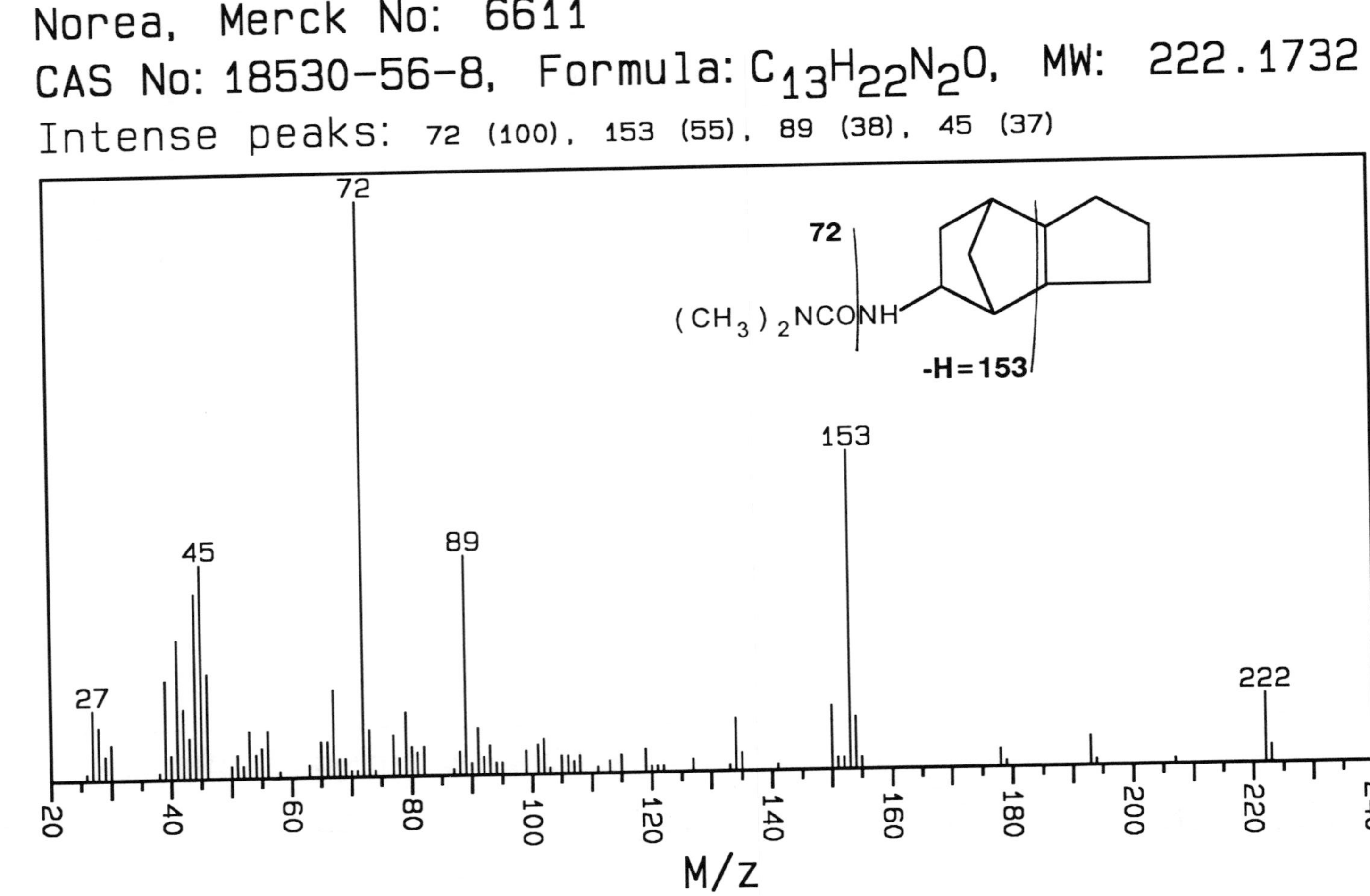
Norea, Merck No: 6611
CAS No: 18530-56-8, Formula: $C_{13}H_{22}N_2O$, MW: 222.1732
Intense peaks: 72 (100), 153 (55), 89 (38), 45 (37)
72
$(CH_3)_2NCONH$
-H=153
M/Z

Fluometuron, Merck No: 4080

CAS No: 2164-17-2, Formula: $C_{10}H_{11}F_3N_2O$, MW: 232.0823

Intense peaks: 72 (100), 232 (25), 44 (22), 28 (12)

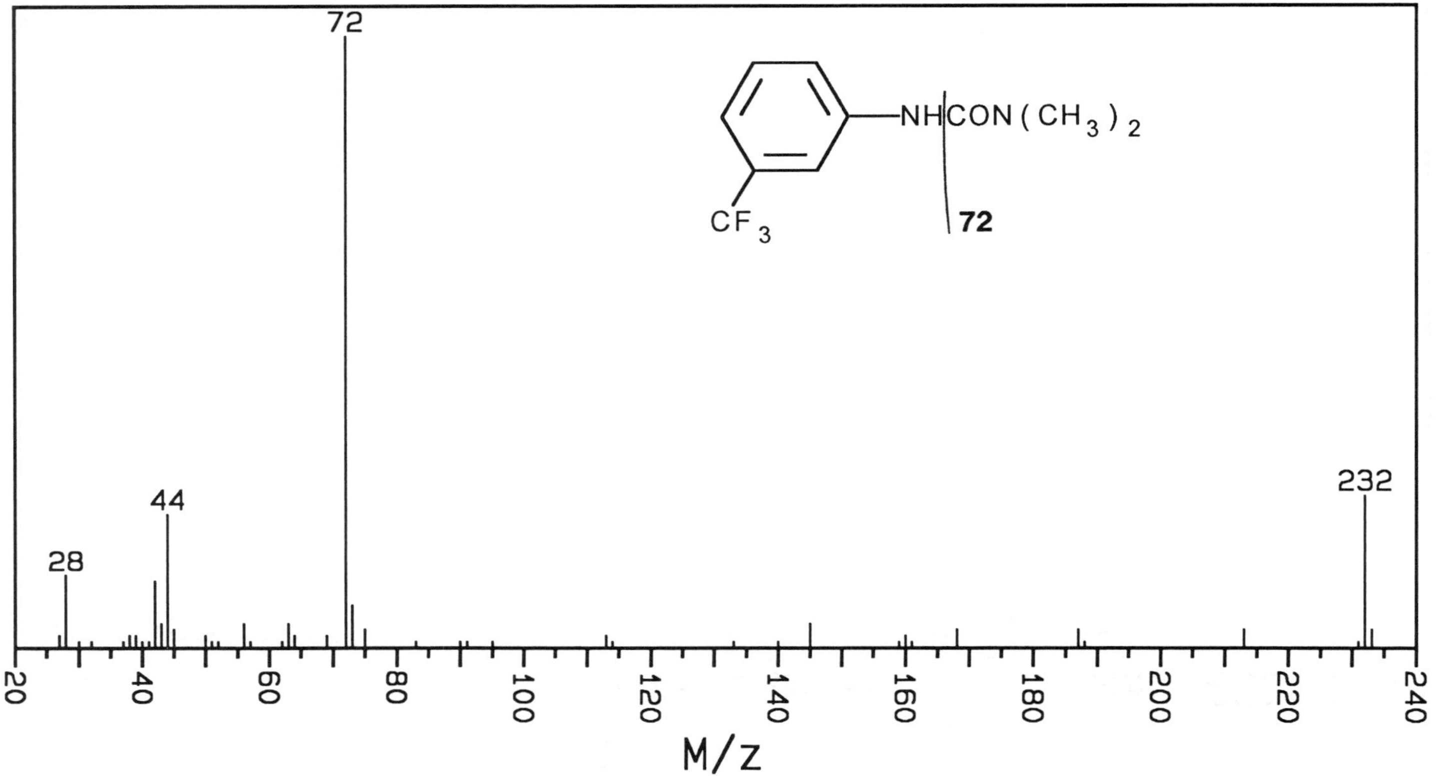

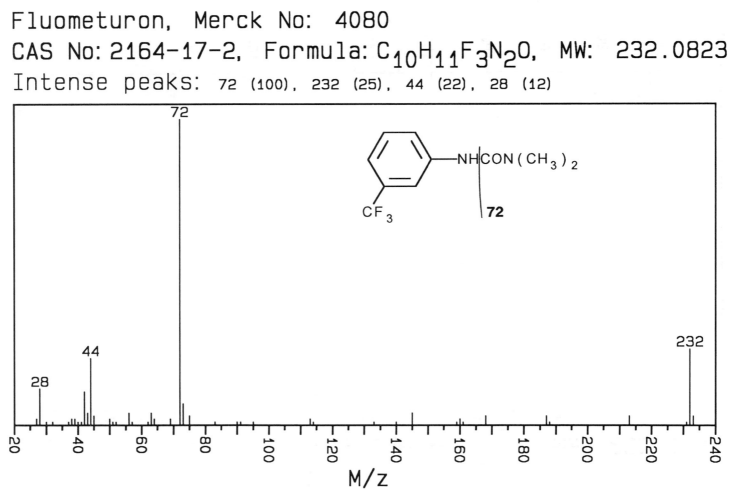

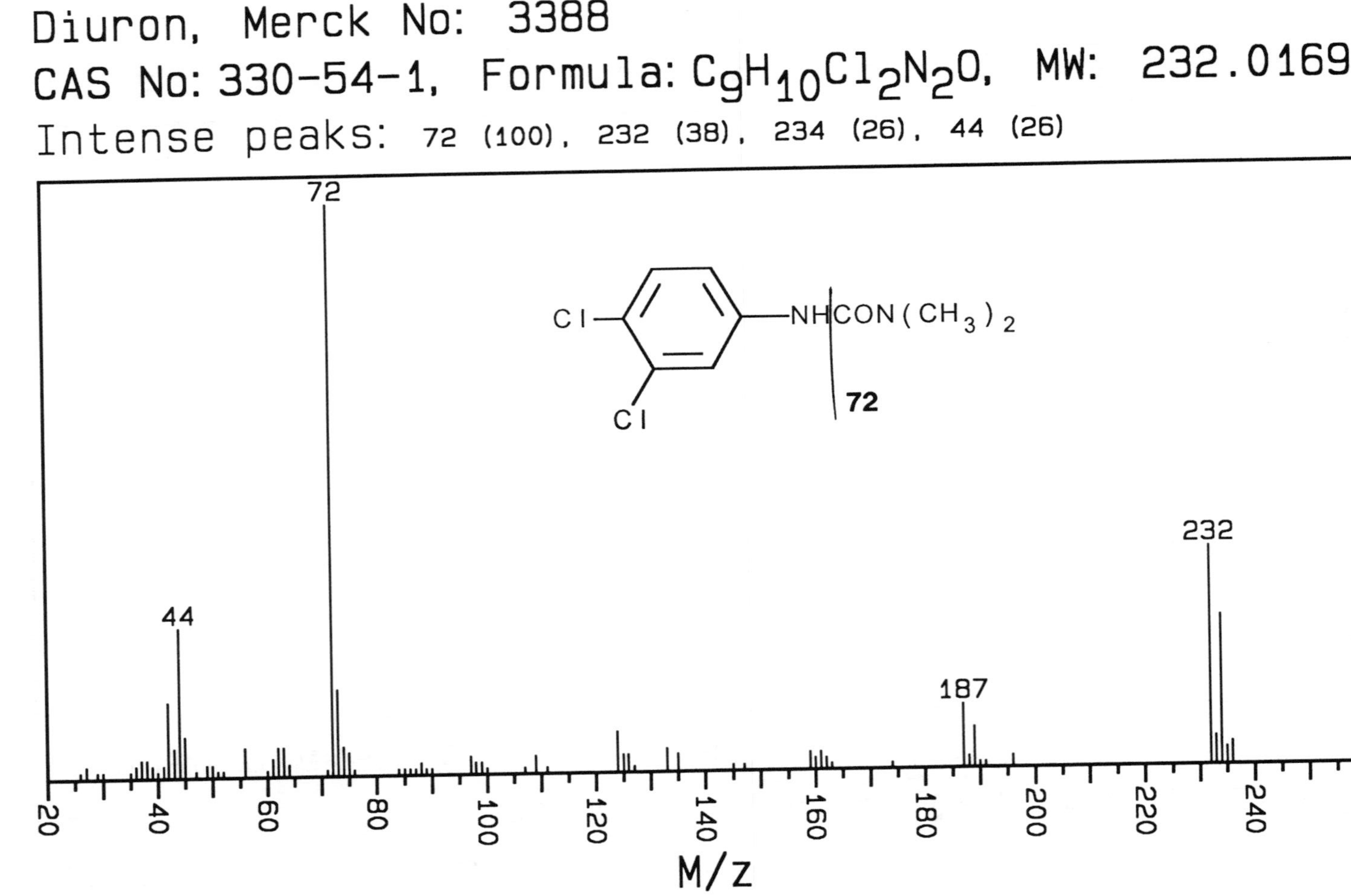
Diuron, Merck No: 3388
CAS No: 330-54-1, Formula: $C_9H_{10}Cl_2N_2O$, MW: 232.0169
Intense peaks: 72 (100), 232 (38), 234 (26), 44 (26)
72
Cl
Cl
NHCON(CH$_3$)$_2$
72
44
187
232
20
40
60
80
100
120
140
160
180
200
220
240
260
M/Z

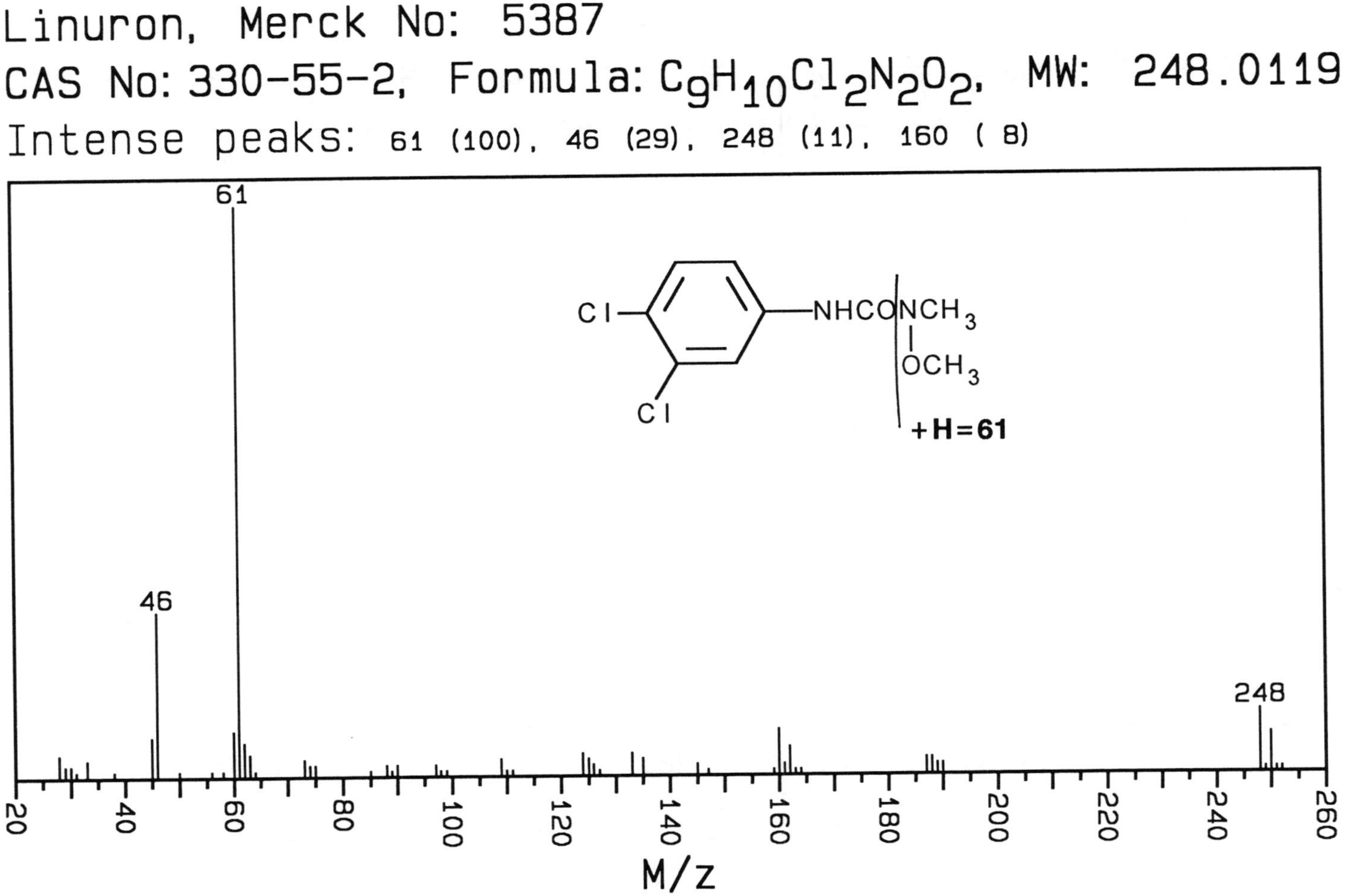
Linuron, Merck No: 5387
CAS No: 330-55-2, Formula: $C_9H_{10}Cl_2N_2O_2$, MW: 248.0119
Intense peaks: 61 (100), 46 (29), 248 (11), 160 ( 8)
61
Cl
NHCONCH3
OCH3
Cl
+H=61
46
248
20
40
60
80
100
120
140
160
180
200
220
240
260
M/Z

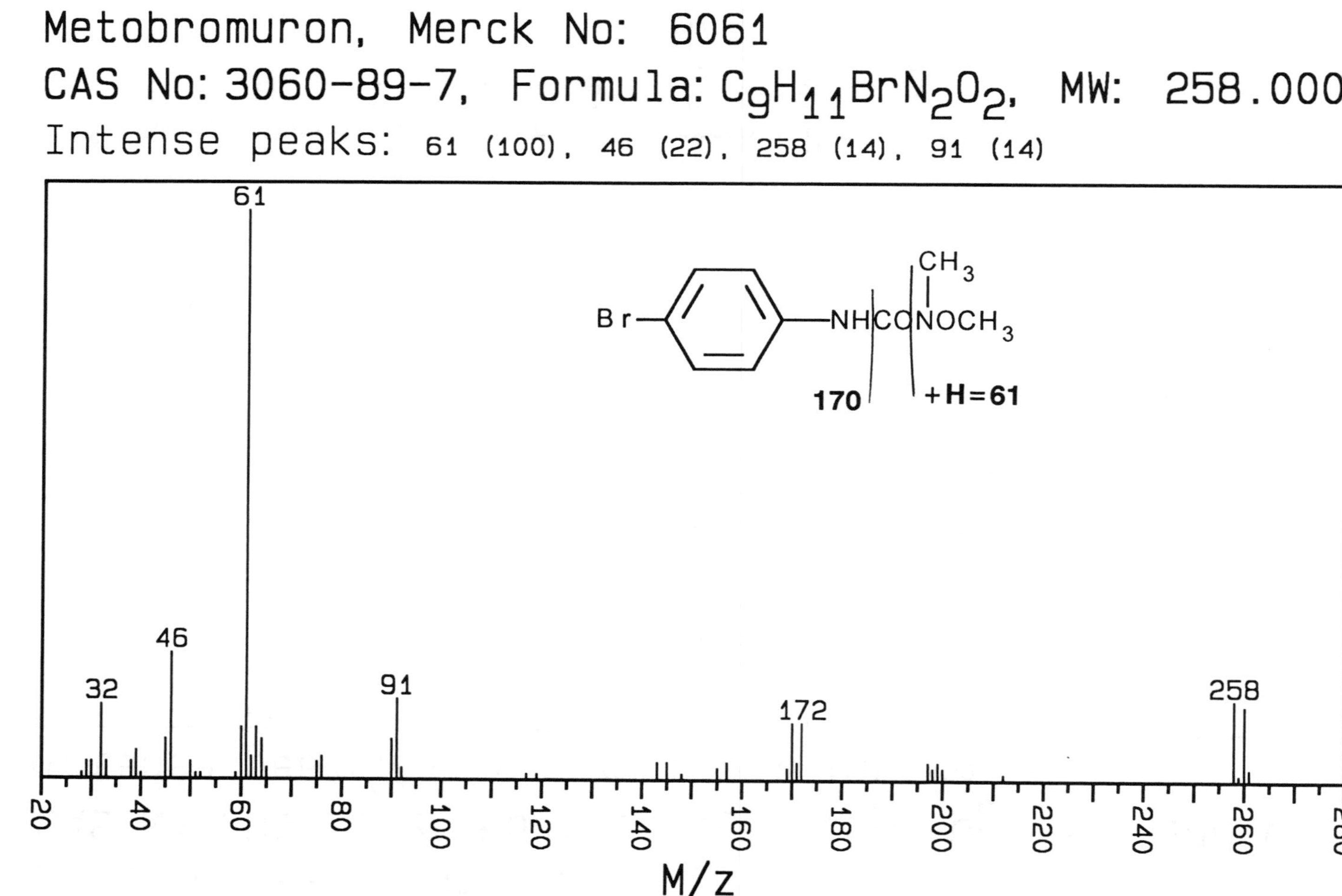
Metobromuron, Merck No: 6061
CAS No: 3060-89-7, Formula: $C_9H_{11}BrN_2O_2$, MW: 258.0004
Intense peaks: 61 (100), 46 (22), 258 (14), 91 (14)
61
Br
NHCONOCH3
CH3
170
+H=61
32
46
91
172
258
20
40
60
80
100
120
140
160
180
200
220
240
260
280
M/Z

Neburon, Merck No: 6354

CAS No: 555-37-3, Formula: $C_{12}H_{16}Cl_2N_2O$, MW: 274.0641

Intense peaks: 40 (100), 114 (75), 38 (48), 29 (37)

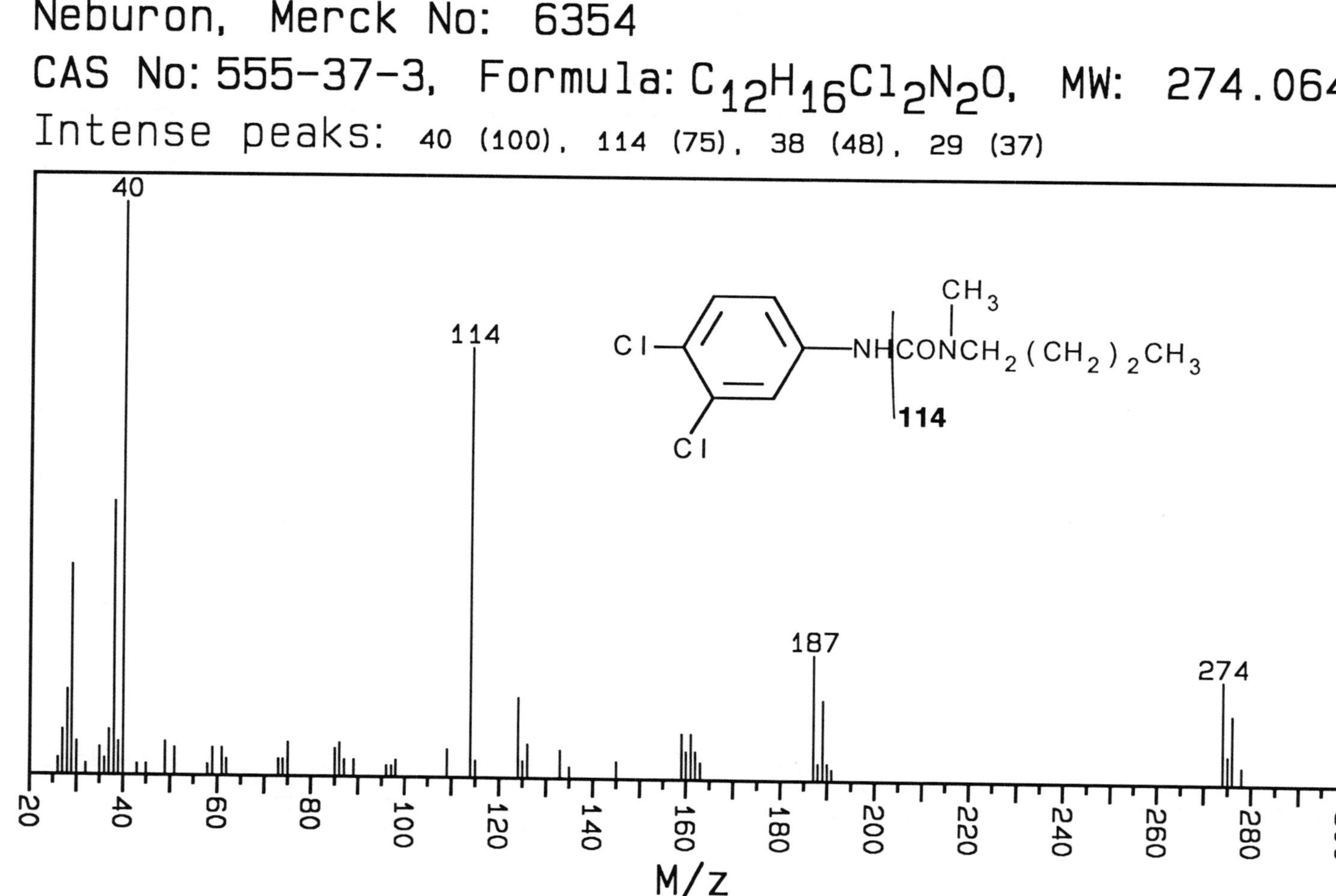

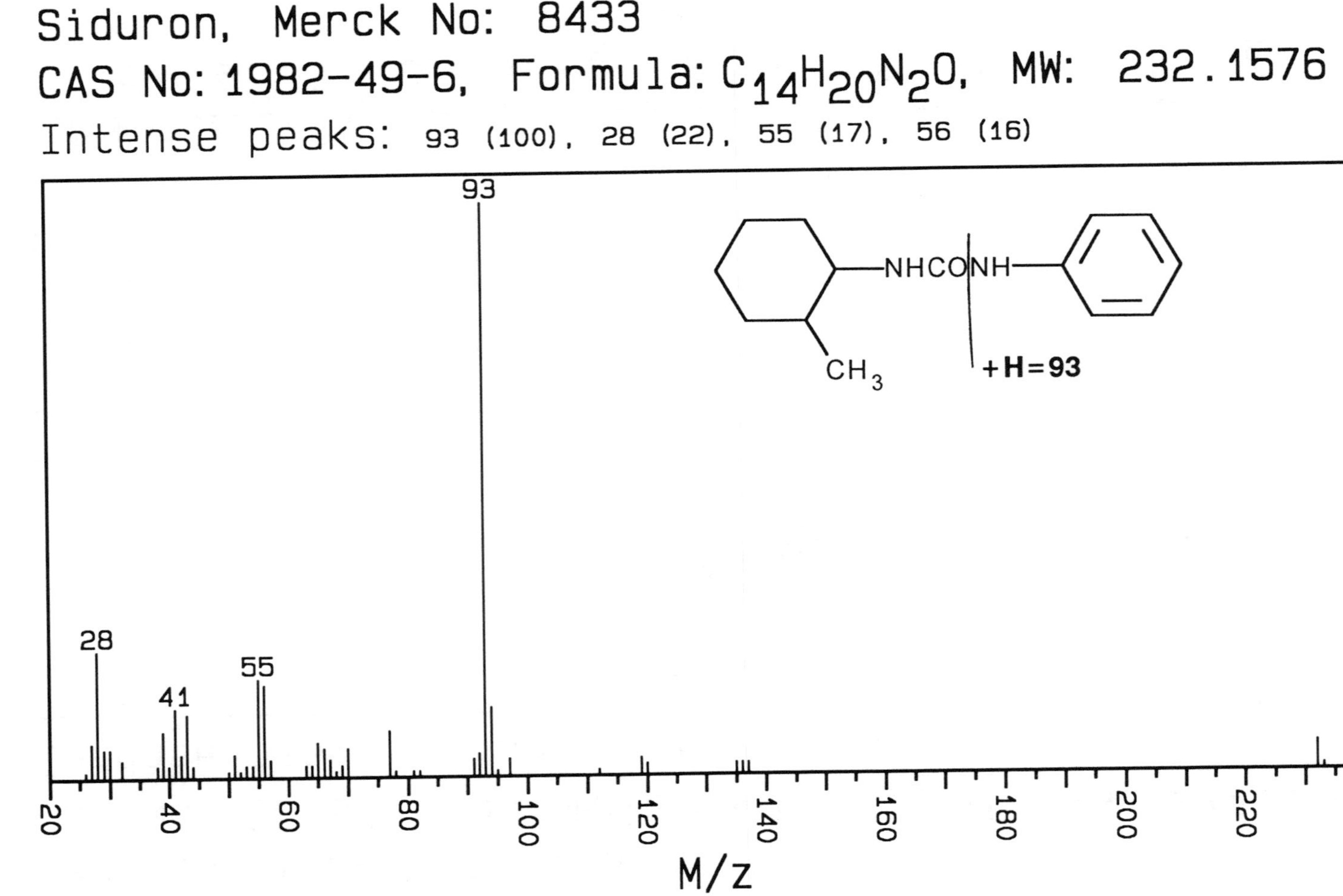
Siduron, Merck No: 8433
CAS No: 1982-49-6, Formula: $C_{14}H_{20}N_2O$, MW: 232.1576
Intense peaks: 93 (100), 28 (22), 55 (17), 56 (16)
93
NHCONH
$CH_3$
+H=93
28
41
55
20
40
60
80
100
120
140
160
180
200
220
240
M/Z

Diflubenzuron, Merck No: 3128

CAS No: 35367-38-5, Formula: $C_{14}H_9ClF_2N_2O_2$, MW: 310.0321

Intense peaks: 153 (100), 141 (84), 155 (46), 63 (38)

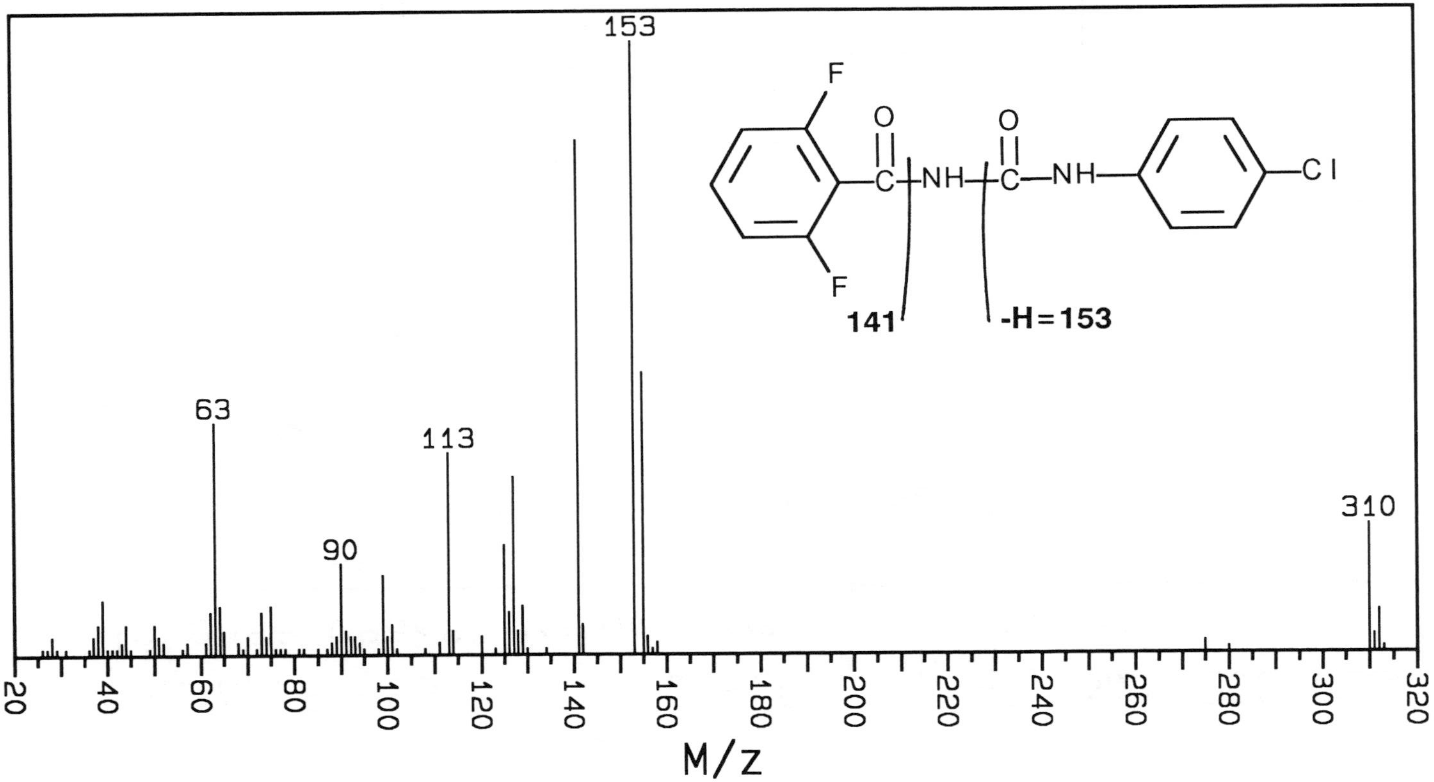

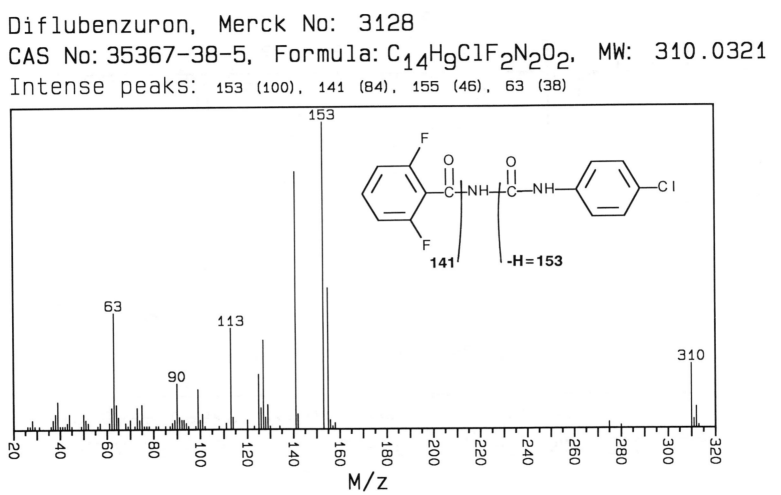

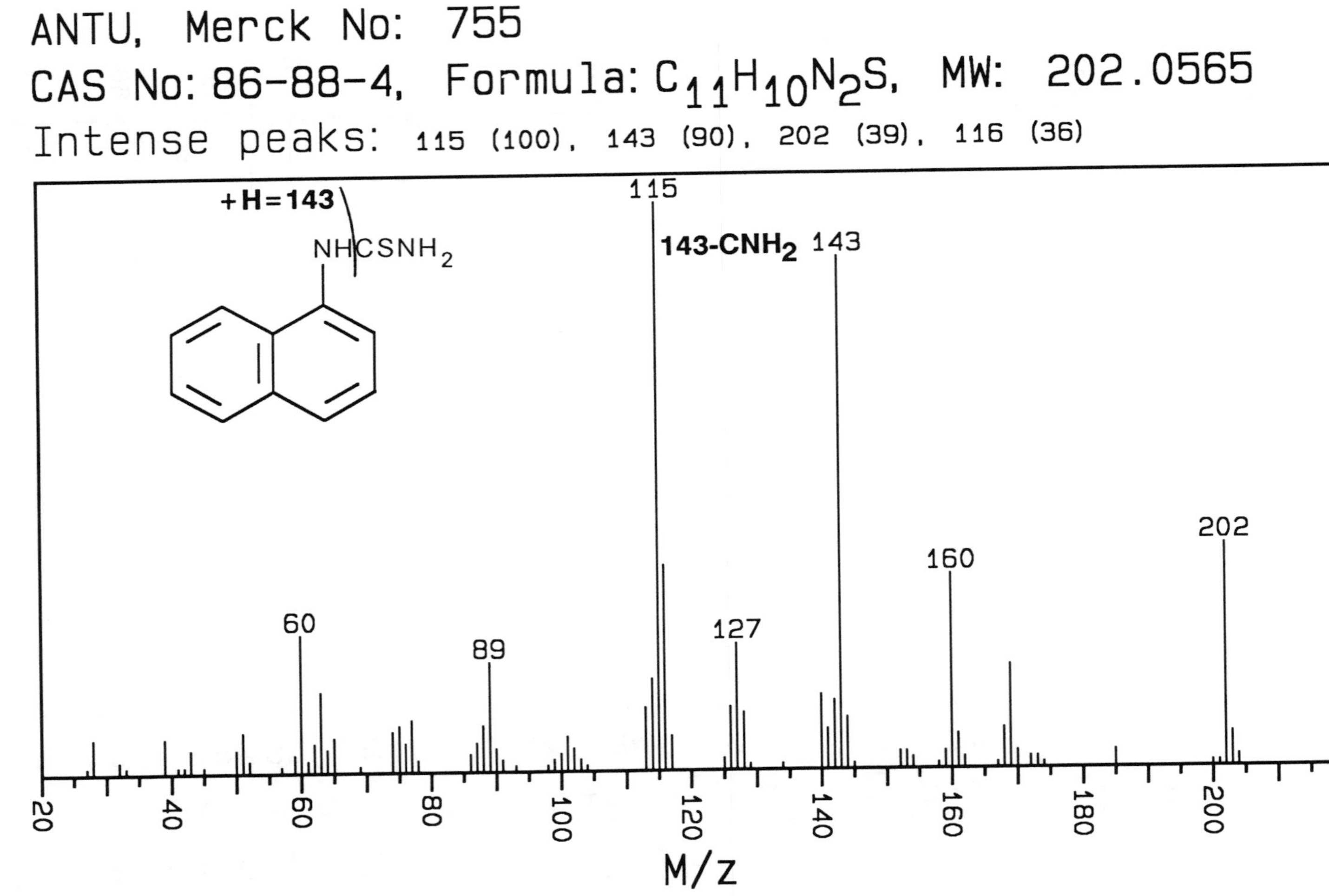
ANTU, Merck No: 755
CAS No: 86-88-4, Formula: $C_{11}H_{10}N_2S$, MW: 202.0565
Intense peaks: 115 (100), 143 (90), 202 (39), 116 (36)
+H=143
$NHCSNH_2$
115
143-$CNH_2$ 143
60
89
127
160
202
20
40
60
80
100
120
140
160
180
200
220
M/Z

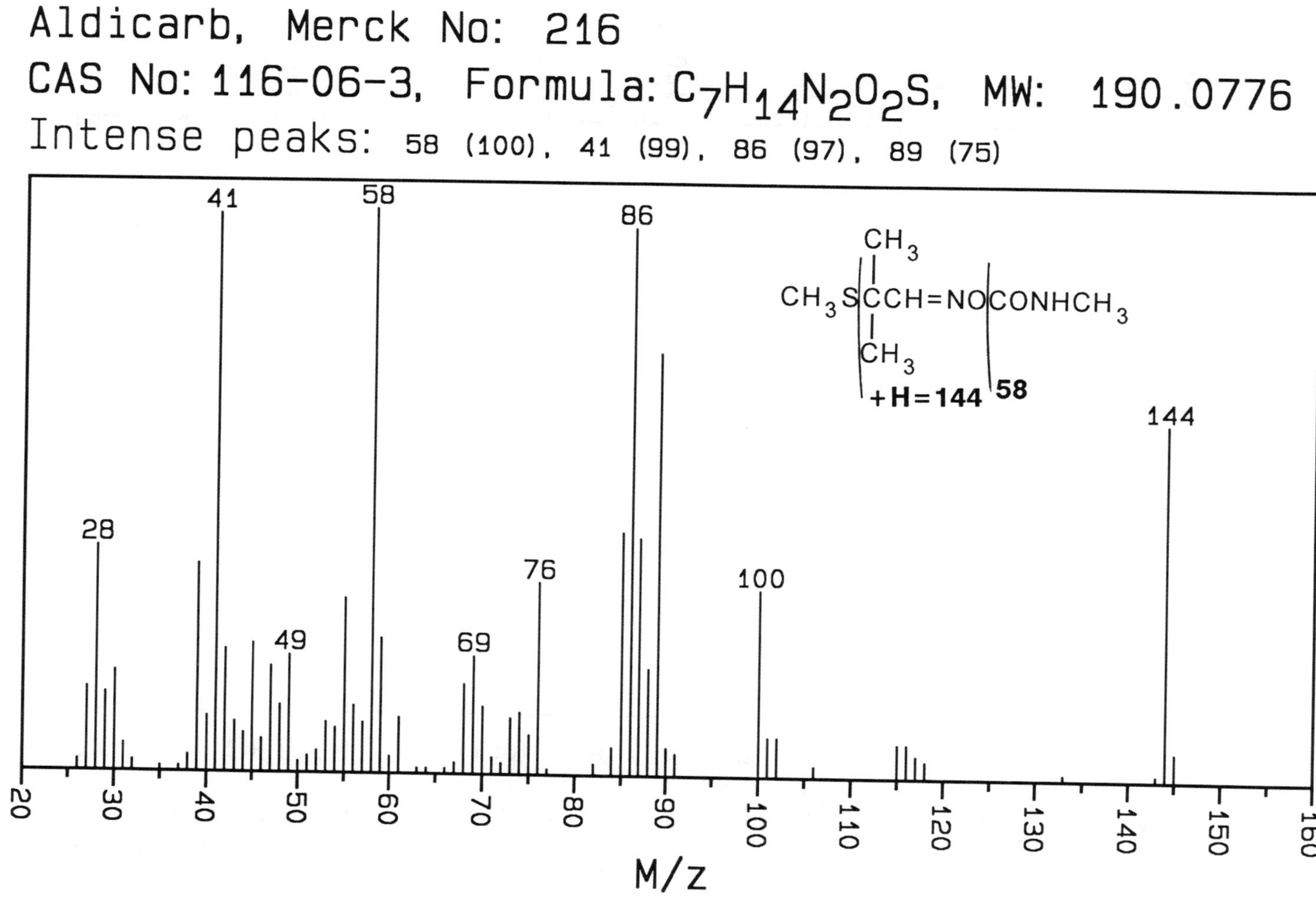

Aldicarb, Merck No: 216
CAS No: 116-06-3, Formula: $C_7H_{14}N_2O_2S$, MW: 190.0776
Intense peaks: 58 (100), 41 (99), 86 (97), 89 (75)
$CH_3$
$CH_3SCCH=NOCONHCH_3$
$CH_3$
+H=144
58
28
41
49
58
69
76
86
100
144
20
30
40
50
60
70
80
90
100
110
120
130
140
150
160
M/Z

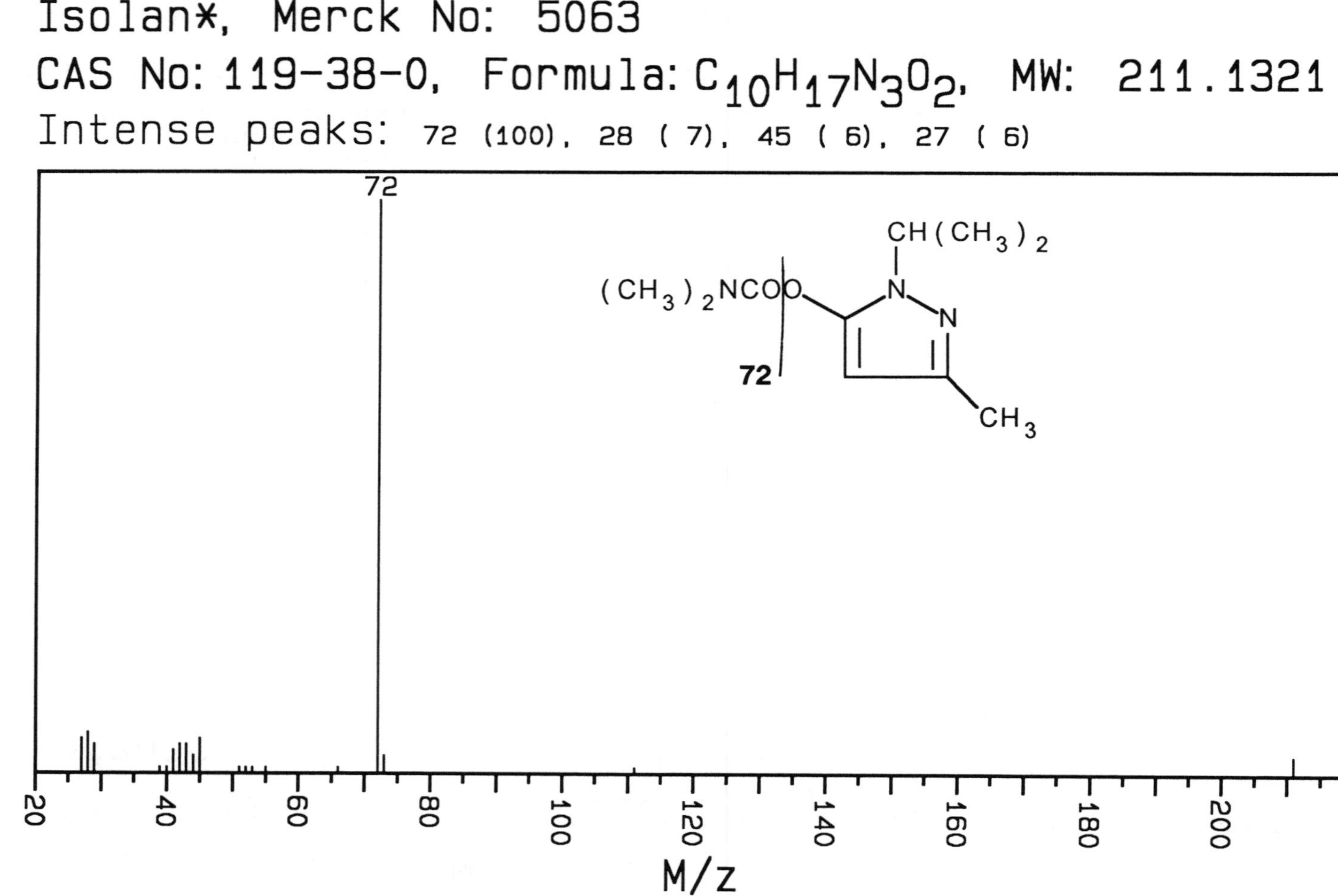
Isolan*, Merck No: 5063
CAS No: 119-38-0, Formula: $C_{10}H_{17}N_3O_2$, MW: 211.1321
Intense peaks: 72 (100), 28 ( 7), 45 ( 6), 27 ( 6)
72
CH(CH3)2
(CH3)2NCOO
N
N
72
CH3
20
40
60
80
100
120
140
160
180
200
220
M/Z

Pyrolan*, Merck No: 8013

CAS No: 87-47-8, Formula: $C_{13}H_{15}N_3O_2$, MW: 245.1164

Intense peaks: 72 (100), 77 (37), 39 (21), 51 (20)

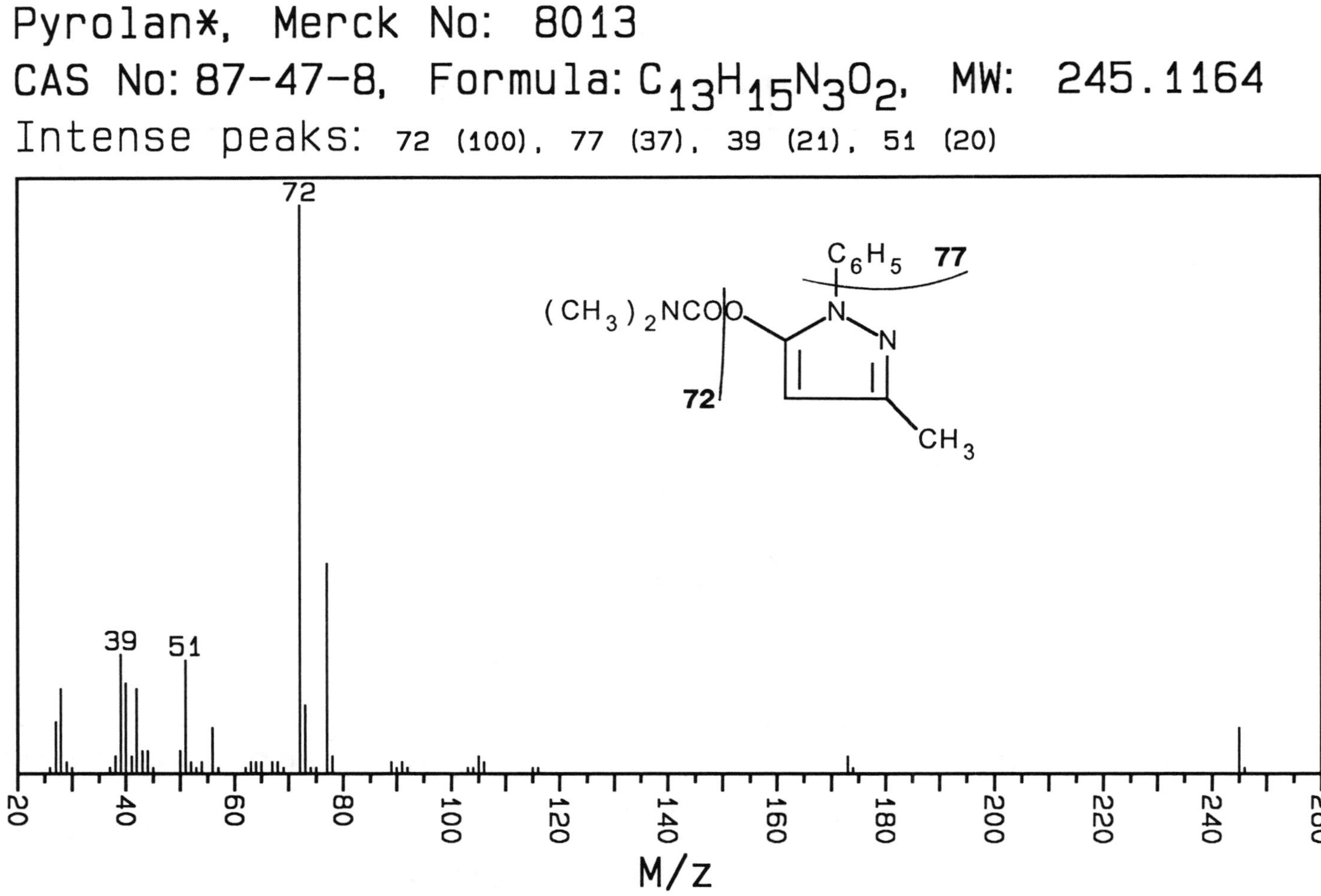

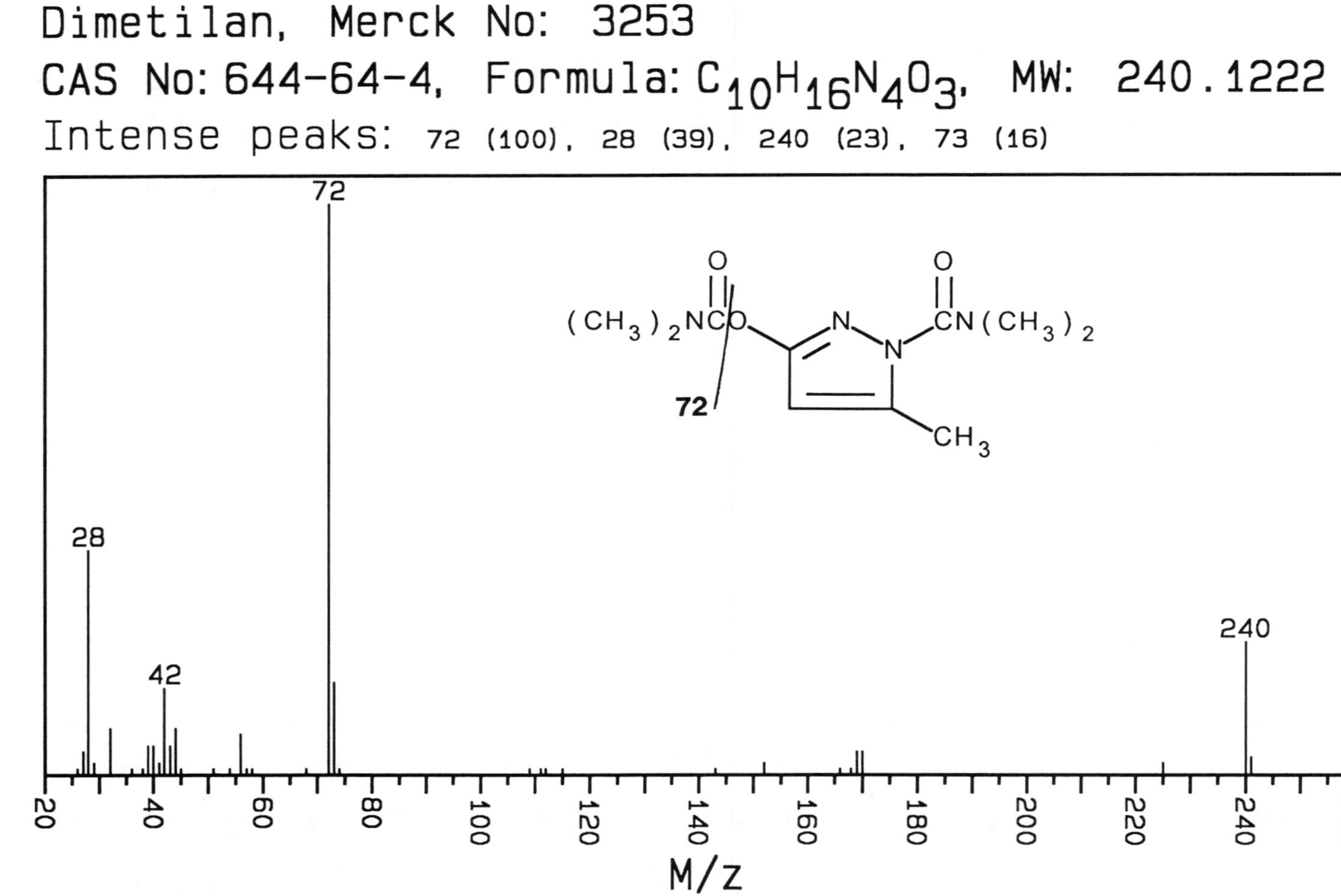
Dimetilan, Merck No: 3253
CAS No: 644-64-4, Formula: $C_{10}H_{16}N_4O_3$, MW: 240.1222
Intense peaks: 72 (100), 28 (39), 240 (23), 73 (16)
72
28
42
240
$(CH_3)_2NCO$
$CN(CH_3)_2$
$CH_3$
20
40
60
80
100
120
140
160
180
200
220
240
260
M/Z

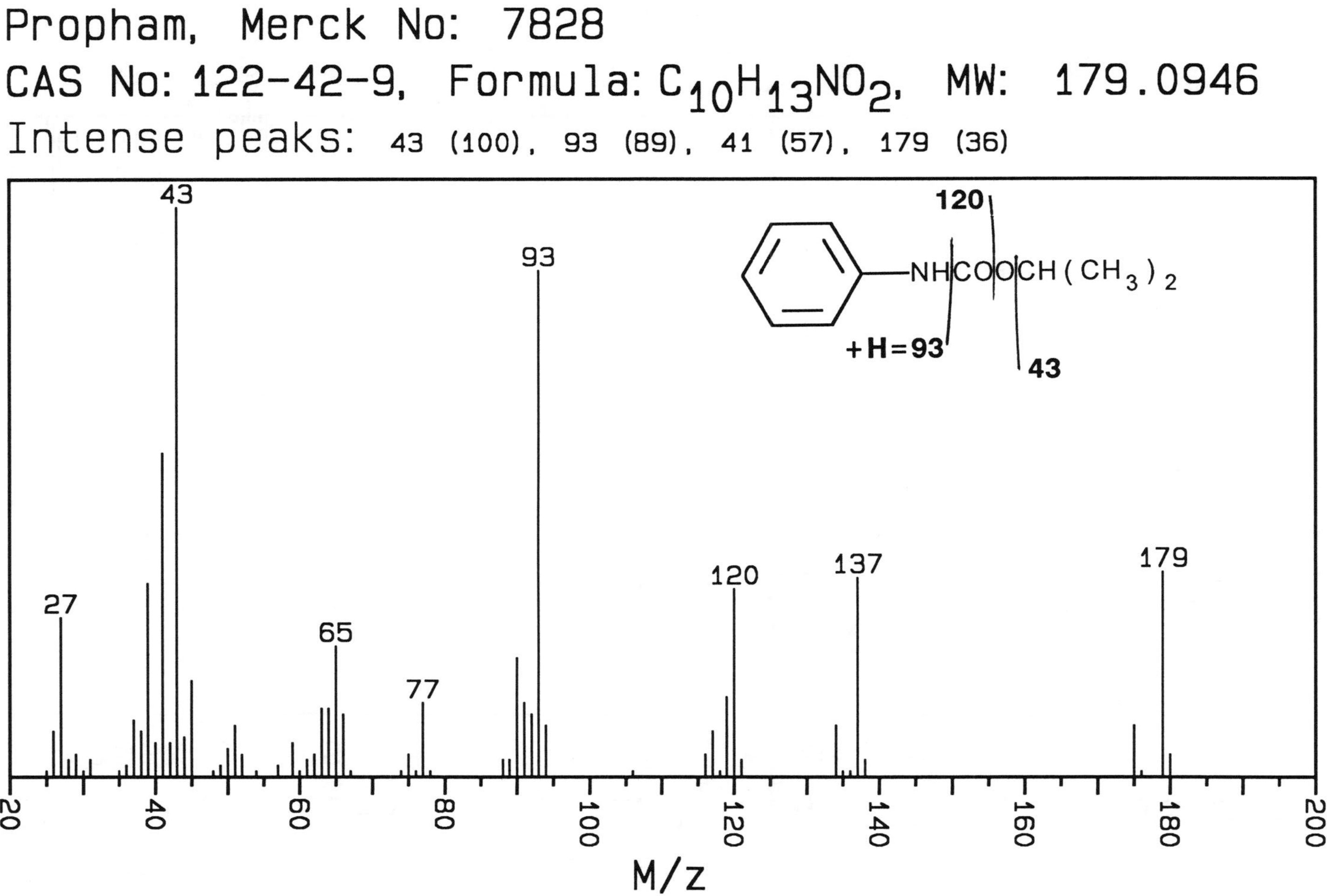
Propham, Merck No: 7828
CAS No: 122-42-9, Formula: $C_{10}H_{13}NO_2$, MW: 179.0946
Intense peaks: 43 (100), 93 (89), 41 (57), 179 (36)
120
NHCOOCH(CH3)2
+H=93
43
27
43
65
77
93
120
137
179
20
40
60
80
100
120
140
160
180
200
M/Z

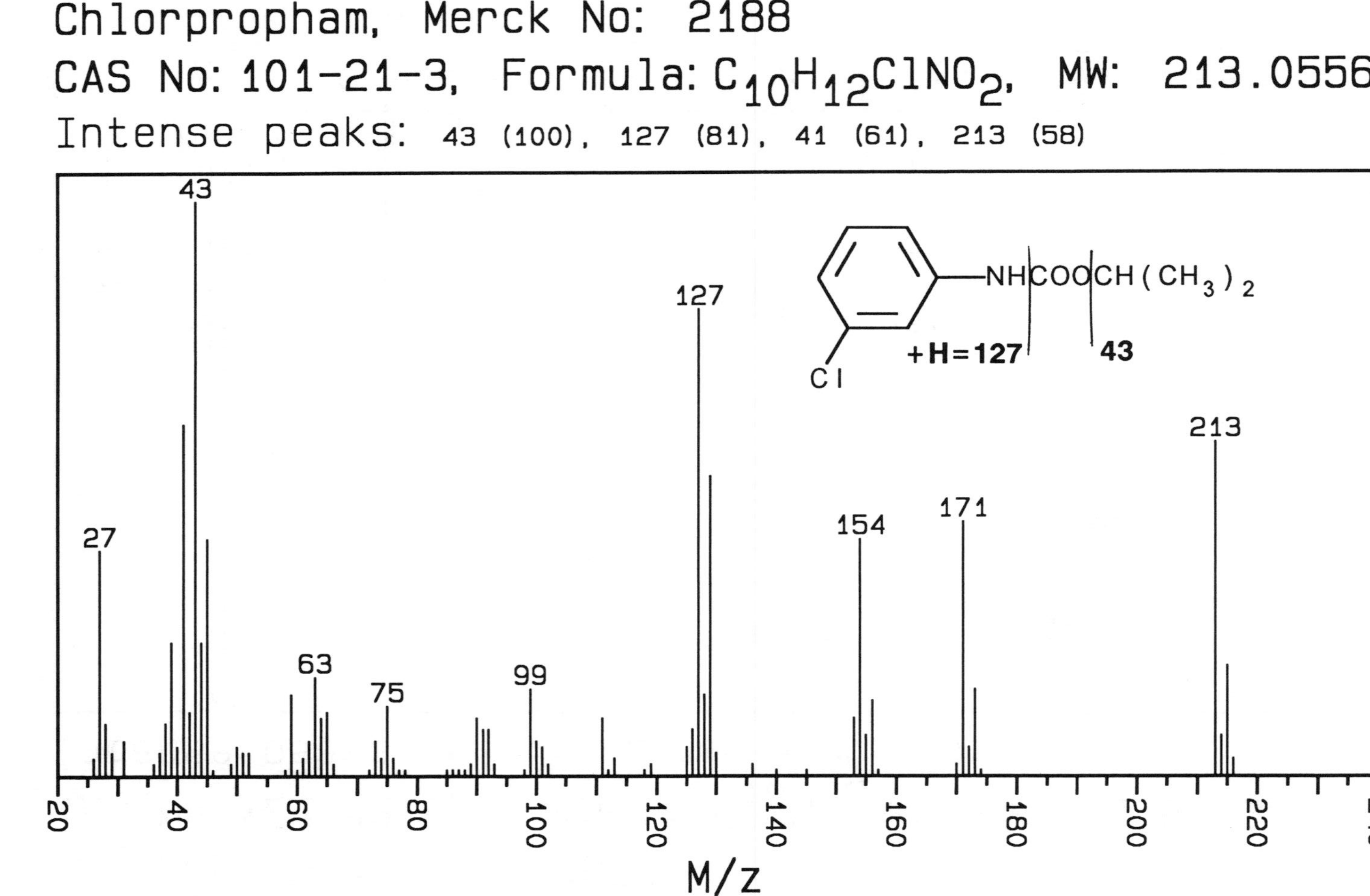

Chlorpropham, Merck No: 2188
CAS No: 101-21-3, Formula: $C_{10}H_{12}ClNO_2$, MW: 213.0556
Intense peaks: 43 (100), 127 (81), 41 (61), 213 (58)
NHCOOCH(CH$_3$)$_2$
Cl
+H=127
43
27
43
63
75
99
127
154
171
213
20
40
60
80
100
120
140
160
180
200
220
240
M/Z

Barban, Merck No: 969

CAS No: 101-27-9, Formula: $C_{11}H_9Cl_2NO_2$, MW: 257.0011

Intense peaks: 222 (100), 51 (74), 87 (67), 143 (40)

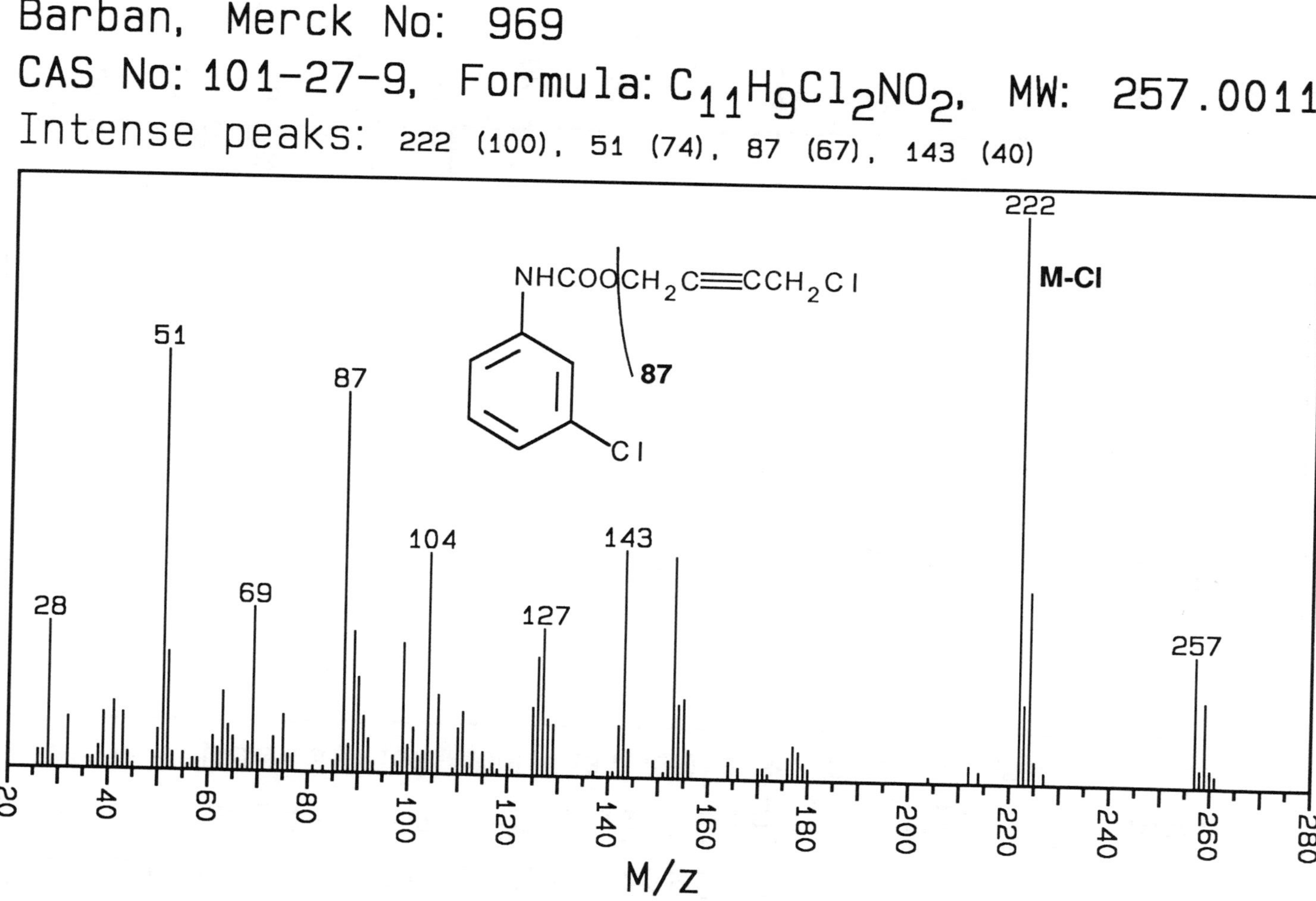

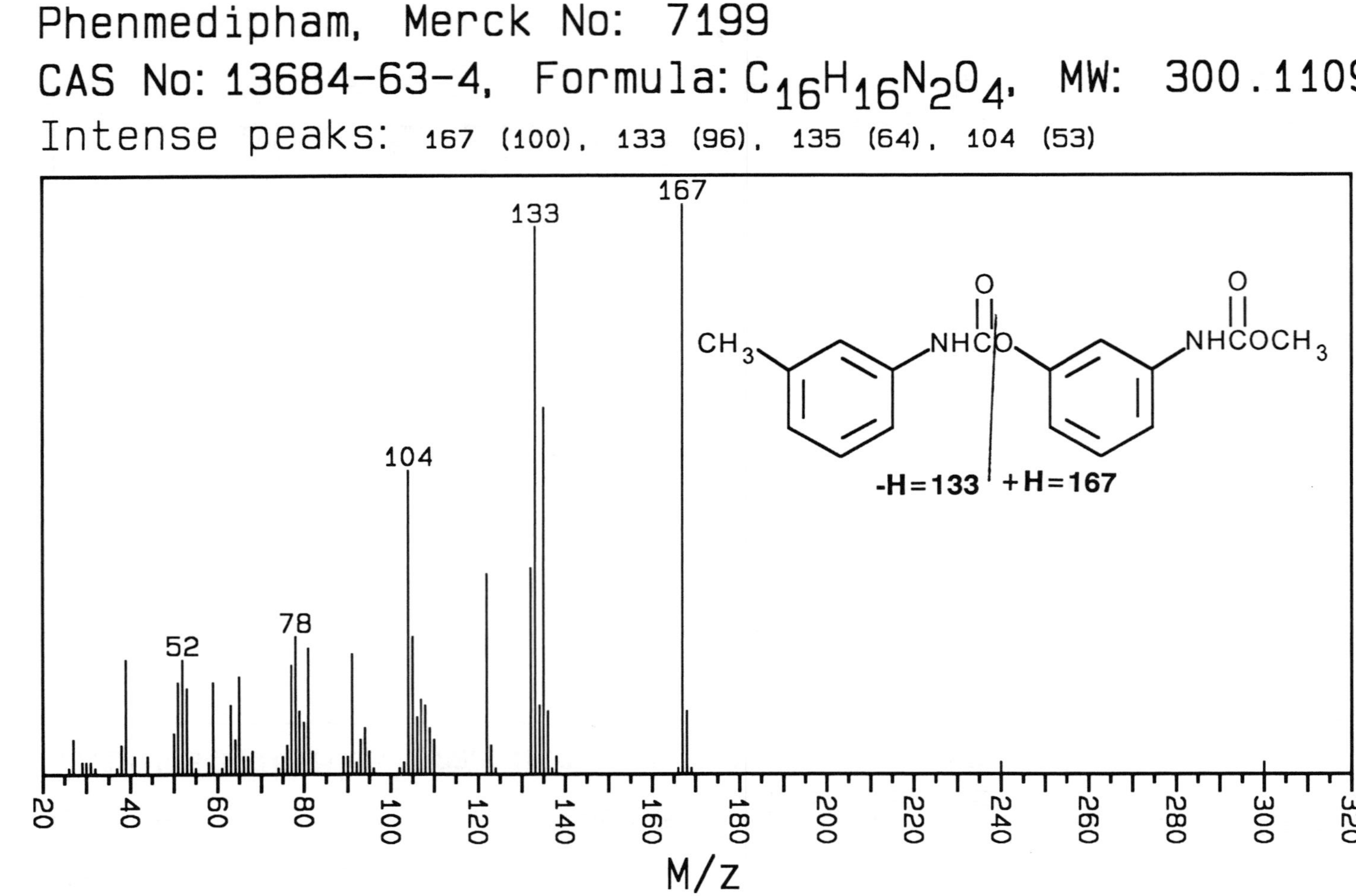

Phenmedipham, Merck No: 7199
CAS No: 13684-63-4, Formula: $C_{16}H_{16}N_2O_4$, MW: 300.1109
Intense peaks: 167 (100), 133 (96), 135 (64), 104 (53)
167
133
104
78
52
$CH_3$
NHCO
O
NHCOCH$_3$
O
-H=133
+H=167
M/Z
20
40
60
80
100
120
140
160
180
200
220
240
260
280
300
320

Carbendazim, Merck No: 1794

CAS No: 10605-21-7, Formula: $C_9H_9N_3O_2$, MW: 191.0695

Intense peaks: 191 (100), 105 (39), 159 (38), 132 (33)

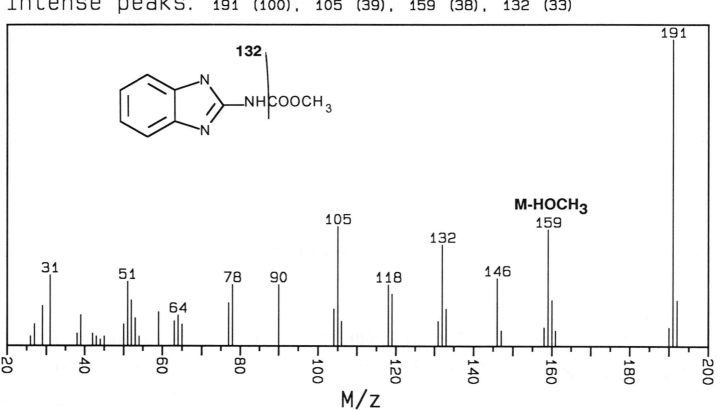

Benomyl, Merck No: 1053

CAS No: 17804-35-2, Formula: $C_{14}H_{18}N_4O_3$, MW: 290.1379

Intense peaks: 191 (100), 159 (99), 40 (46), 105 (44)

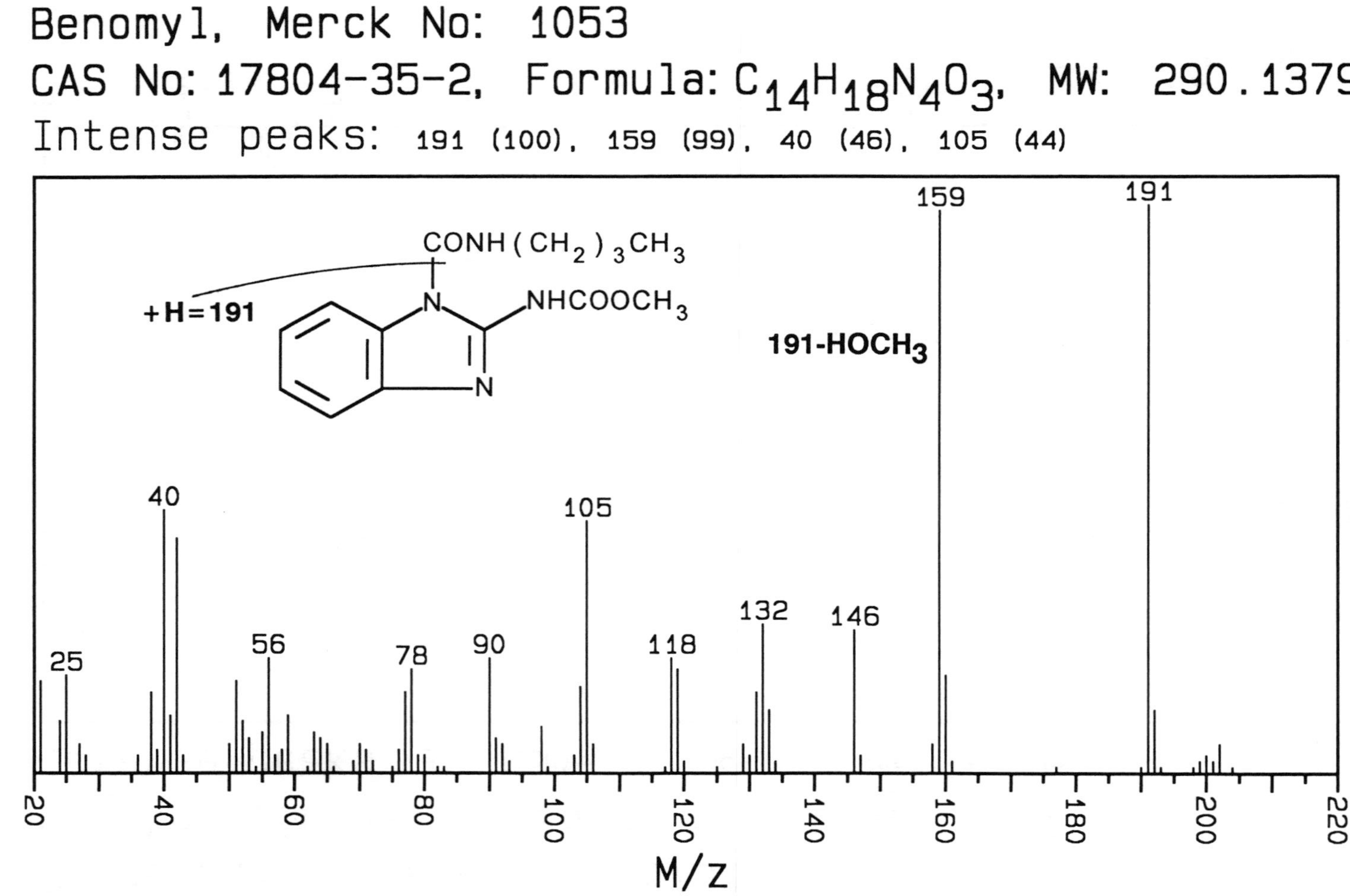

Thiophanate, Merck No: 9282

CAS No: 23564-06-9, Formula: $C_{14}H_{18}N_4O_4S_2$, MW: 370.0769

Intense peaks: 206 (100), 29 (78), 133 (69), 150 (56)

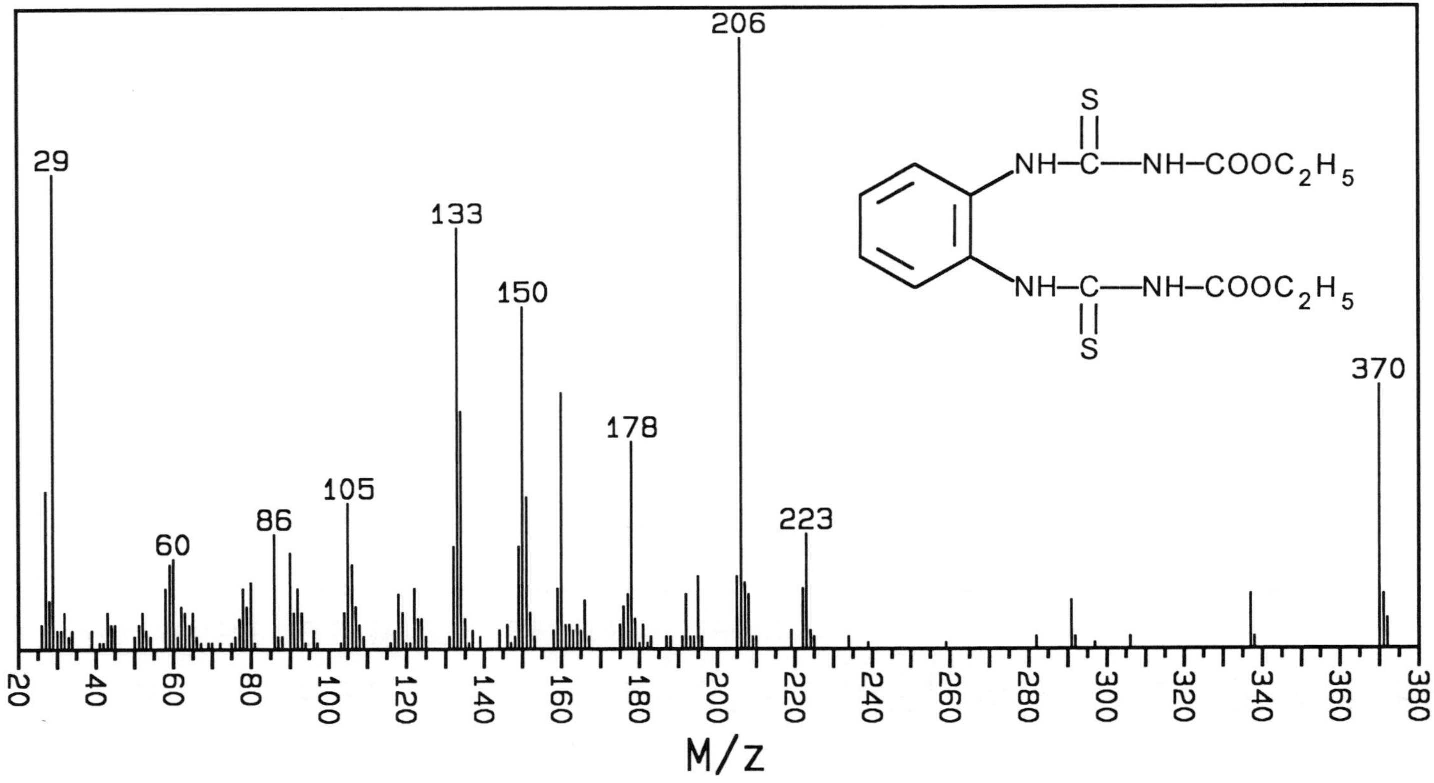

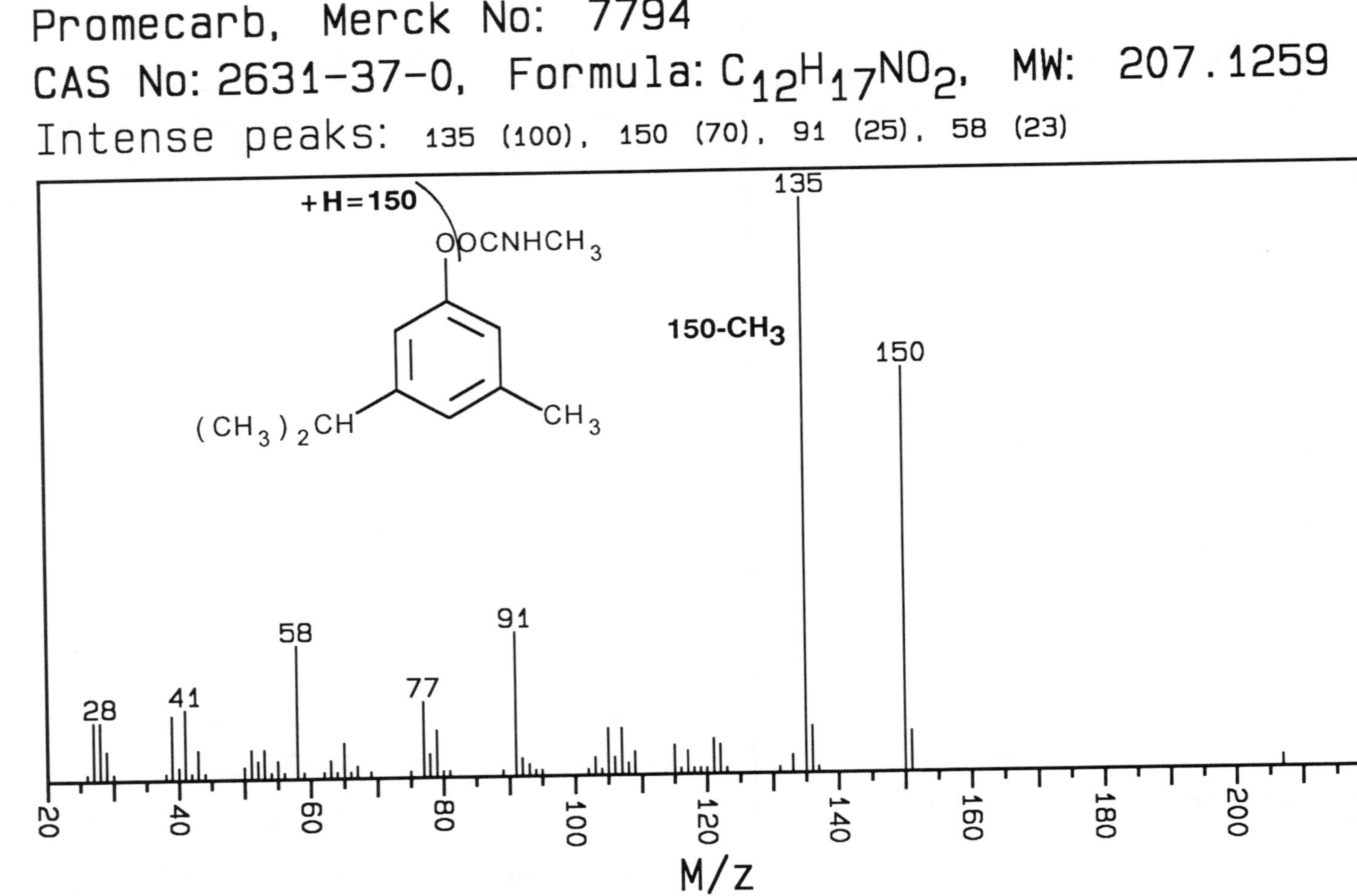
Promecarb, Merck No: 7794
CAS No: 2631-37-0, Formula: C12H17NO2, MW: 207.1259
Intense peaks: 135 (100), 150 (70), 91 (25), 58 (23)
+H=150
OOCNHCH3
(CH3)2CH
CH3
150-CH3
135
150
91
77
58
41
28
20
40
60
80
100
120
140
160
180
200
220
M/Z

Mexacarbate, Merck No: 6090

CAS No: 315-18-4, Formula: $C_{12}H_{18}N_2O_2$, MW: 222.1368

Intense peaks: 165 (100), 164 (67), 150 (66), 134 (30)

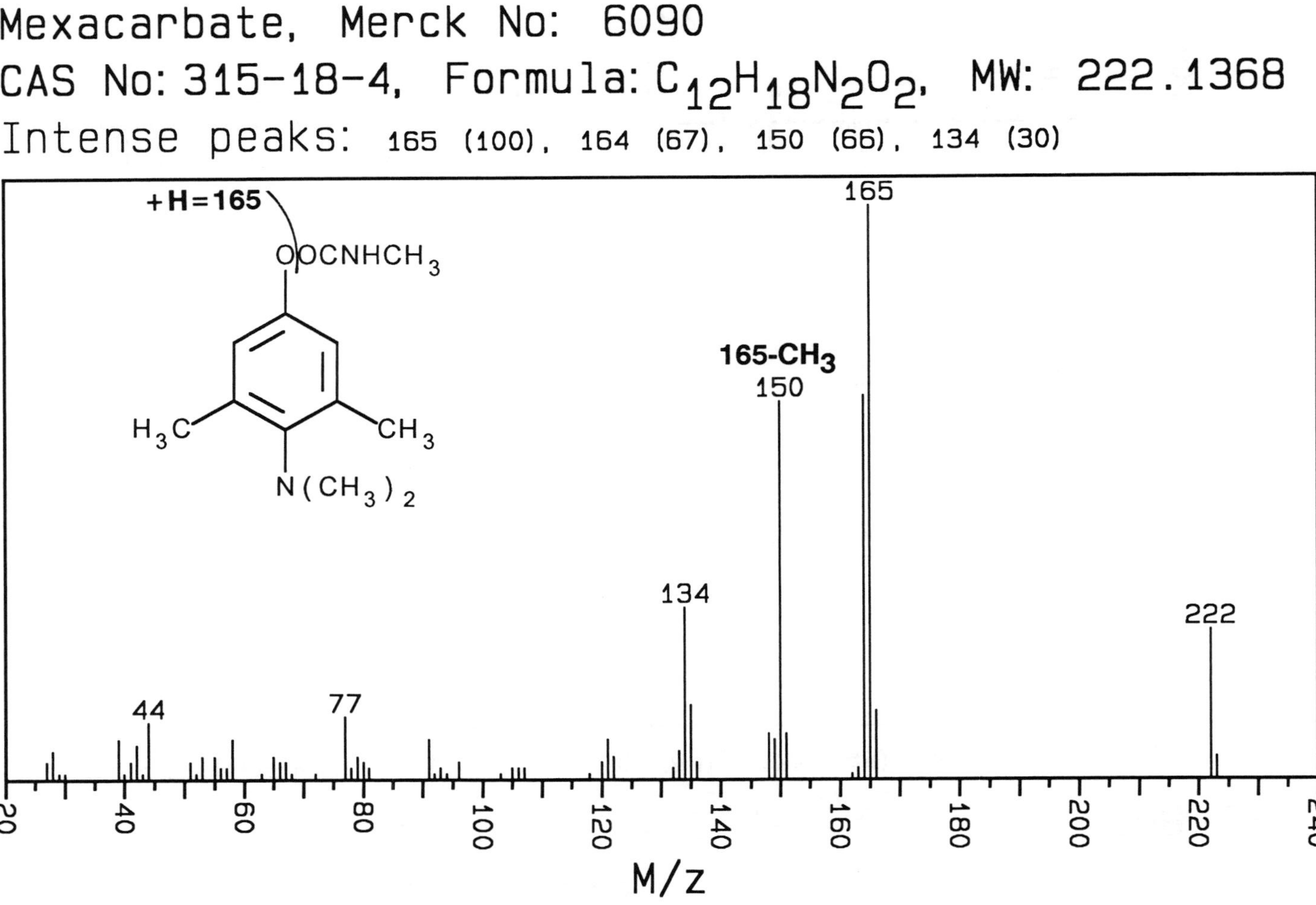

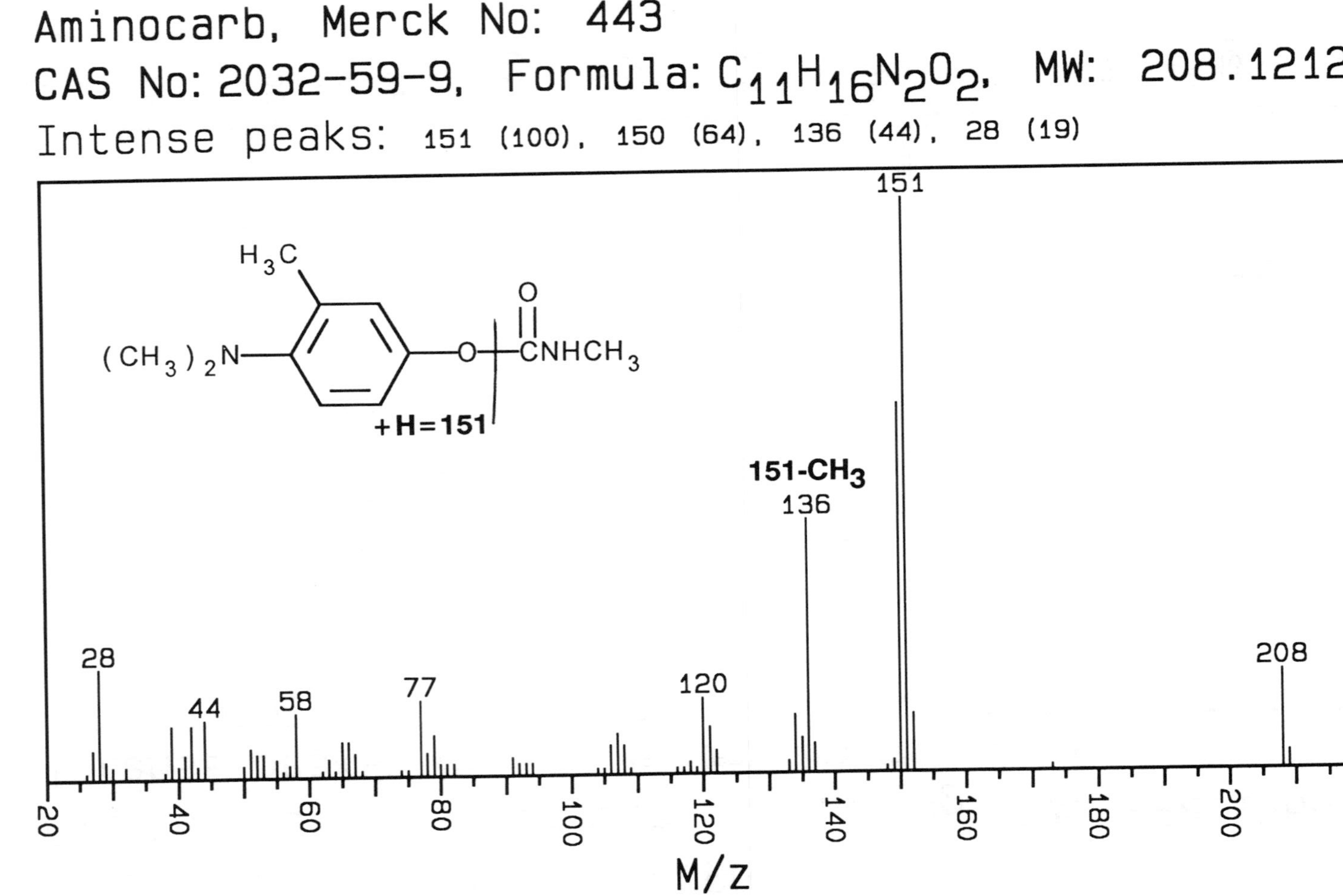
Aminocarb, Merck No: 443
CAS No: 2032-59-9, Formula: $C_{11}H_{16}N_2O_2$, MW: 208.1212
Intense peaks: 151 (100), 150 (64), 136 (44), 28 (19)
H3C
(CH3)2N
O
CNHCH3
+H=151
151
151-CH3
136
28
44
58
77
120
208
20
40
60
80
100
120
140
160
180
200
220
M/Z

Methiocarb, Merck No: 5893

CAS No: 2032-65-7, Formula: $C_{11}H_{15}NO_2S$, MW: 225.0824

Intense peaks: 168 (100), 153 (69), 109 (21), 91 (16)

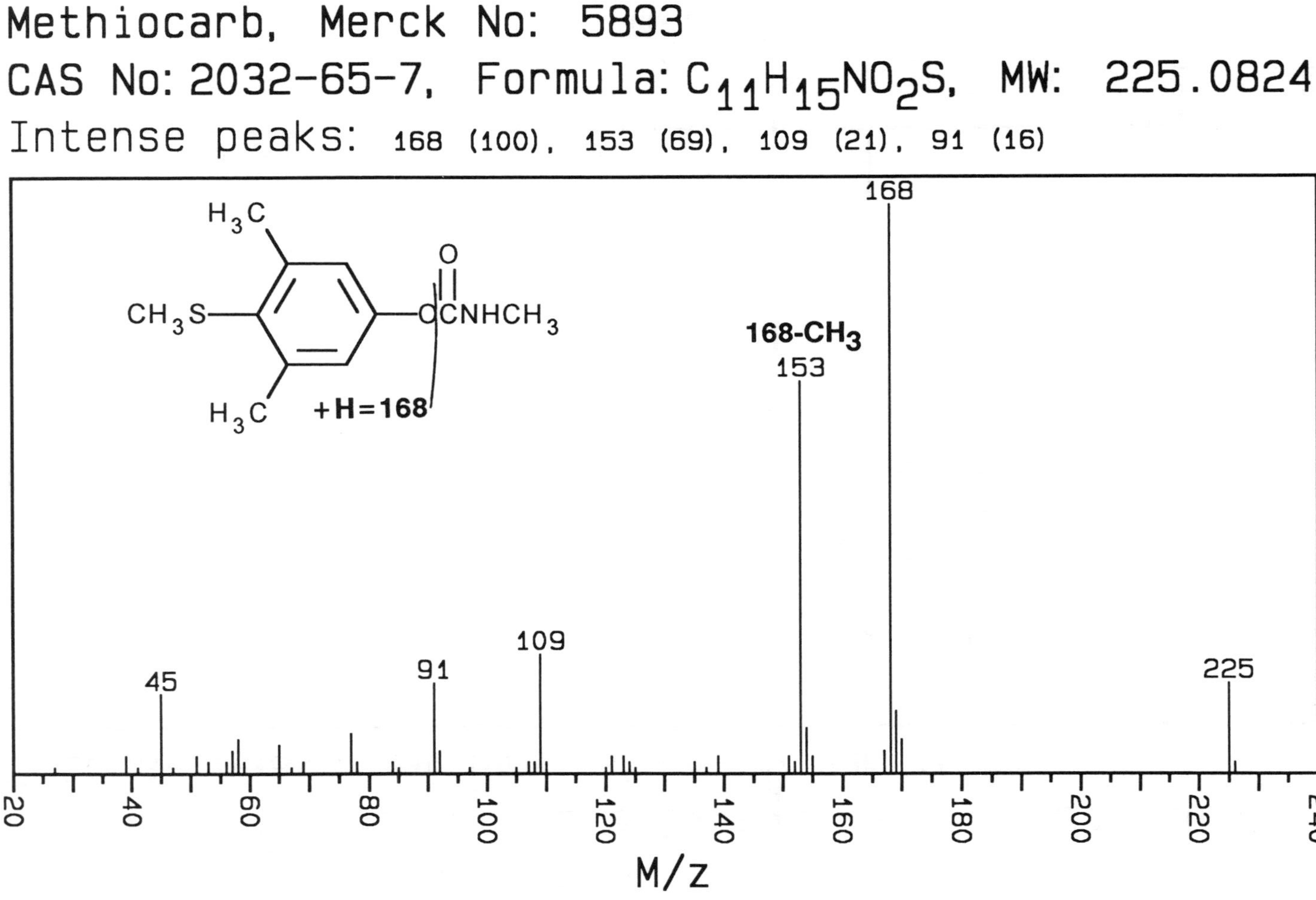

Propoxur, Merck No: 7849

CAS No: 114-26-1, Formula: $C_{11}H_{15}NO_3$, MW: 209.1052

Intense peaks: 110 (100), 152 (18), 27 (13), 111 (10)

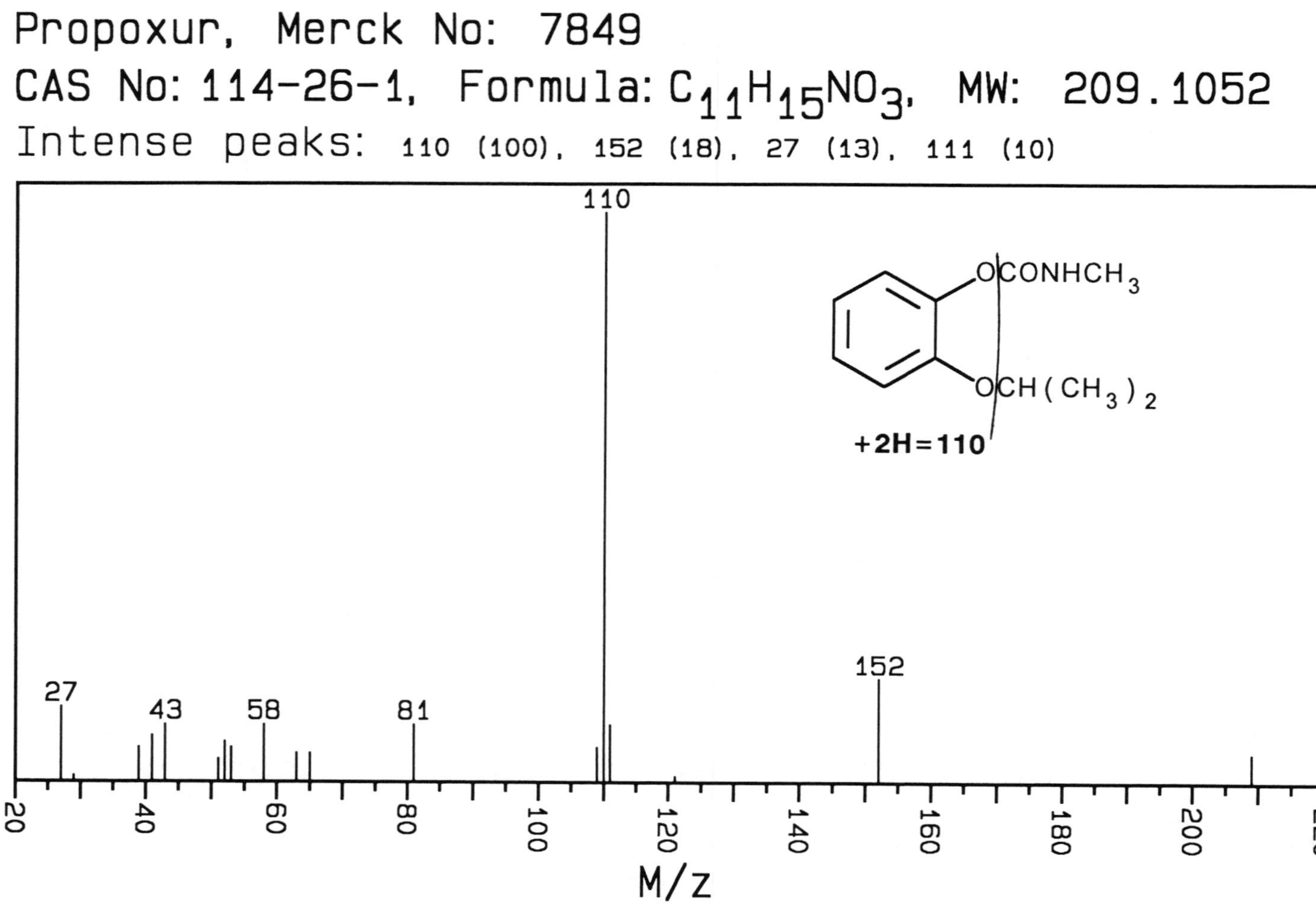

Carbaryl, Merck No: 1789

CAS No: 63-25-2, Formula: $C_{12}H_{11}NO_2$, MW: 201.0791

Intense peaks: 144 (100), 115 (44), 116 (37), 145 (14)

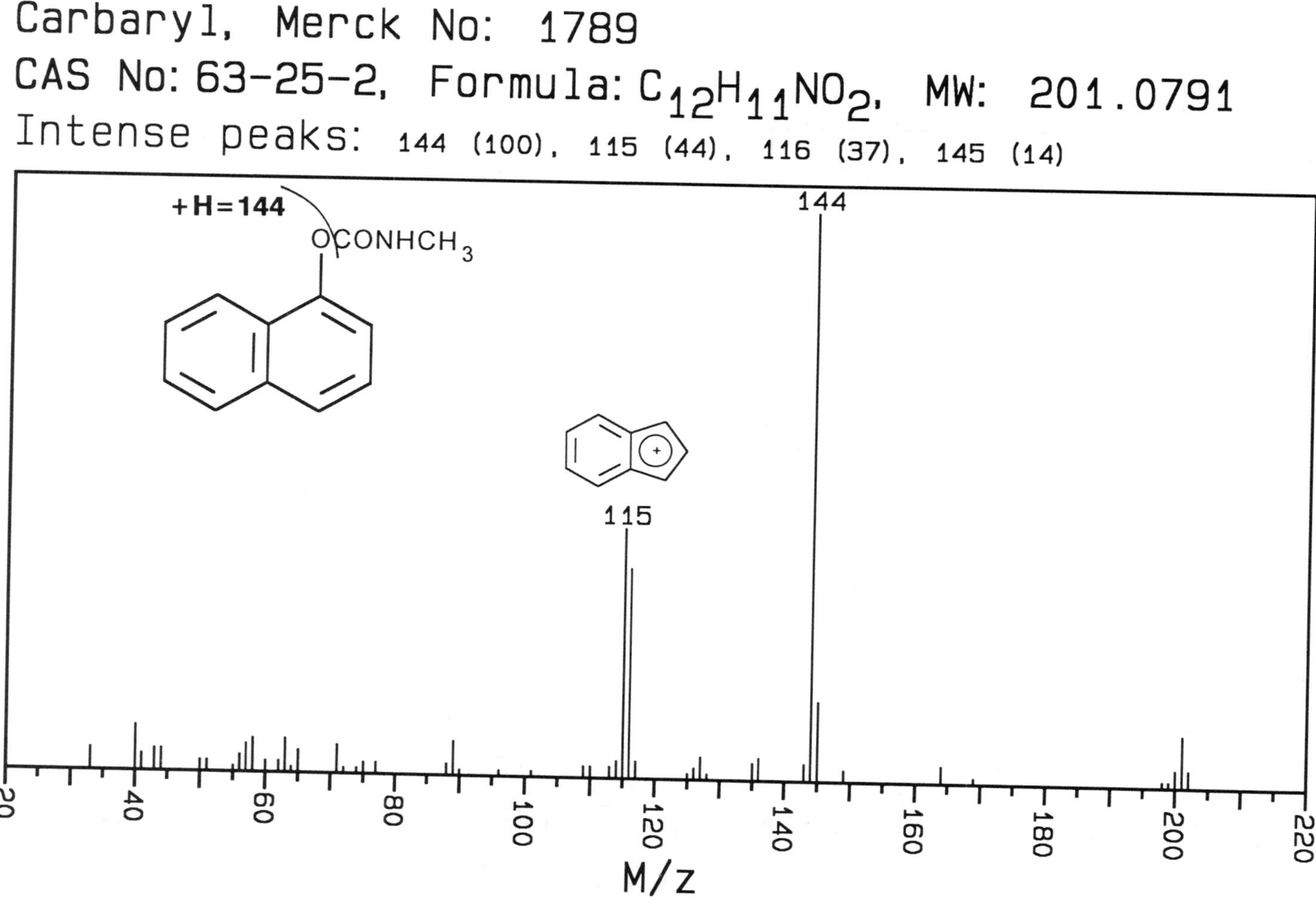

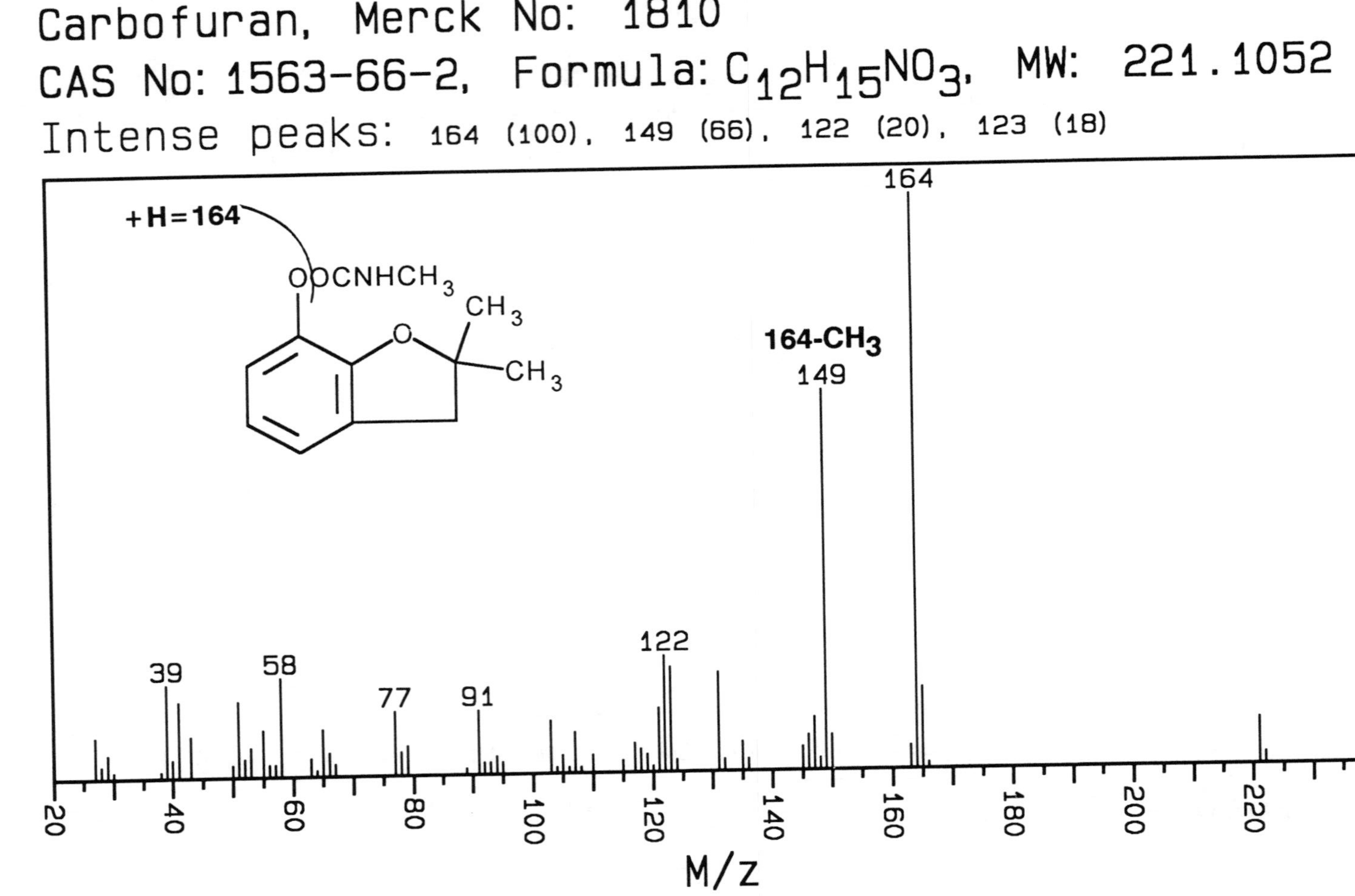
Carbofuran, Merck No: 1810
CAS No: 1563-66-2, Formula: $C_{12}H_{15}NO_3$, MW: 221.1052
Intense peaks: 164 (100), 149 (66), 122 (20), 123 (18)
+H=164
OOCNHCH3
CH3
O
CH3
164-CH3
149
164
39
58
77
91
122
20
40
60
80
100
120
140
160
180
200
220
240
M/Z

Bendiocarb, Merck No: 1044

CAS No: 22781-23-3, Formula: $C_{11}H_{13}NO_4$, MW: 223.0844

Intense peaks: 151 (100), 166 (69), 126 (61), 58 (60)

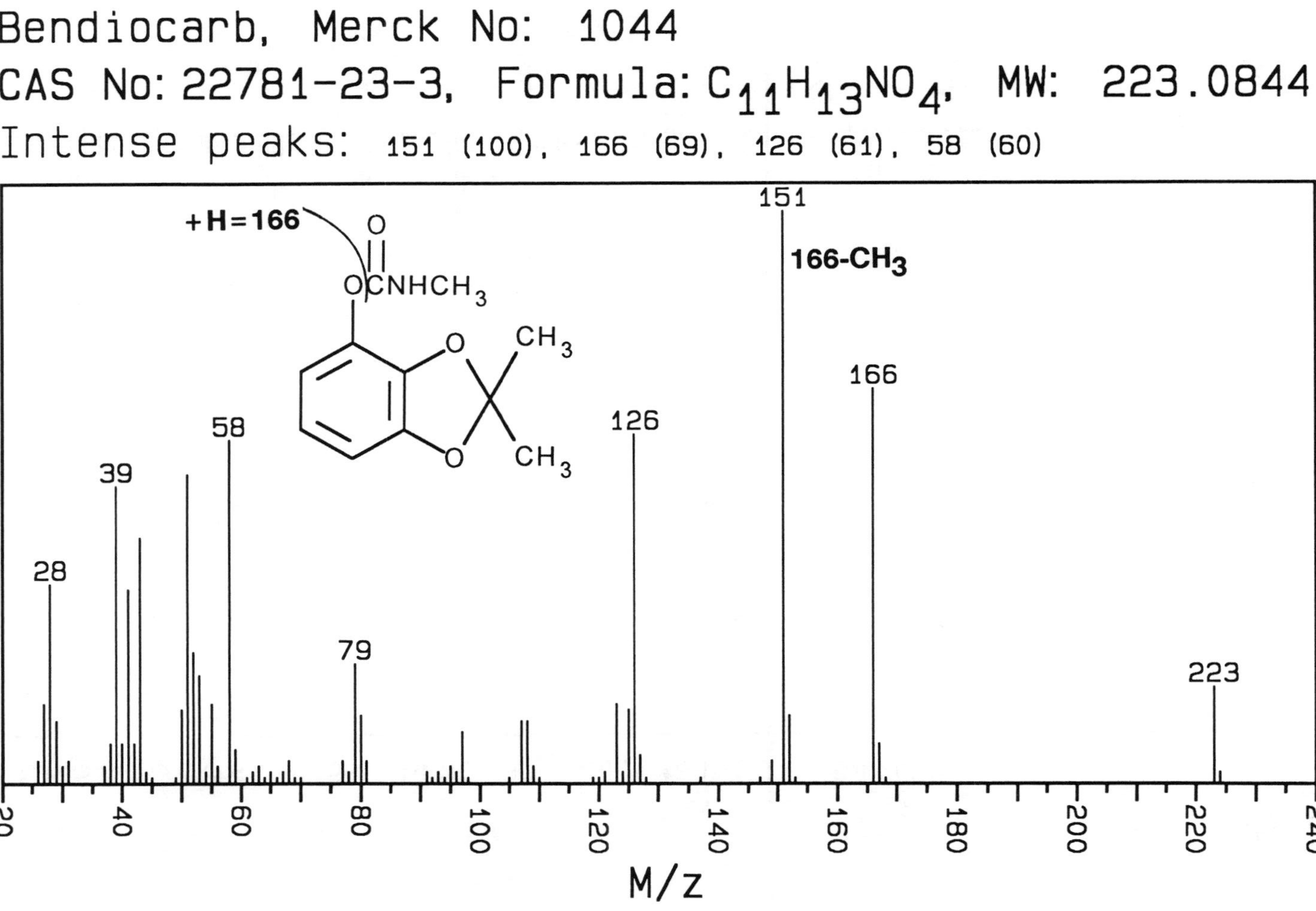

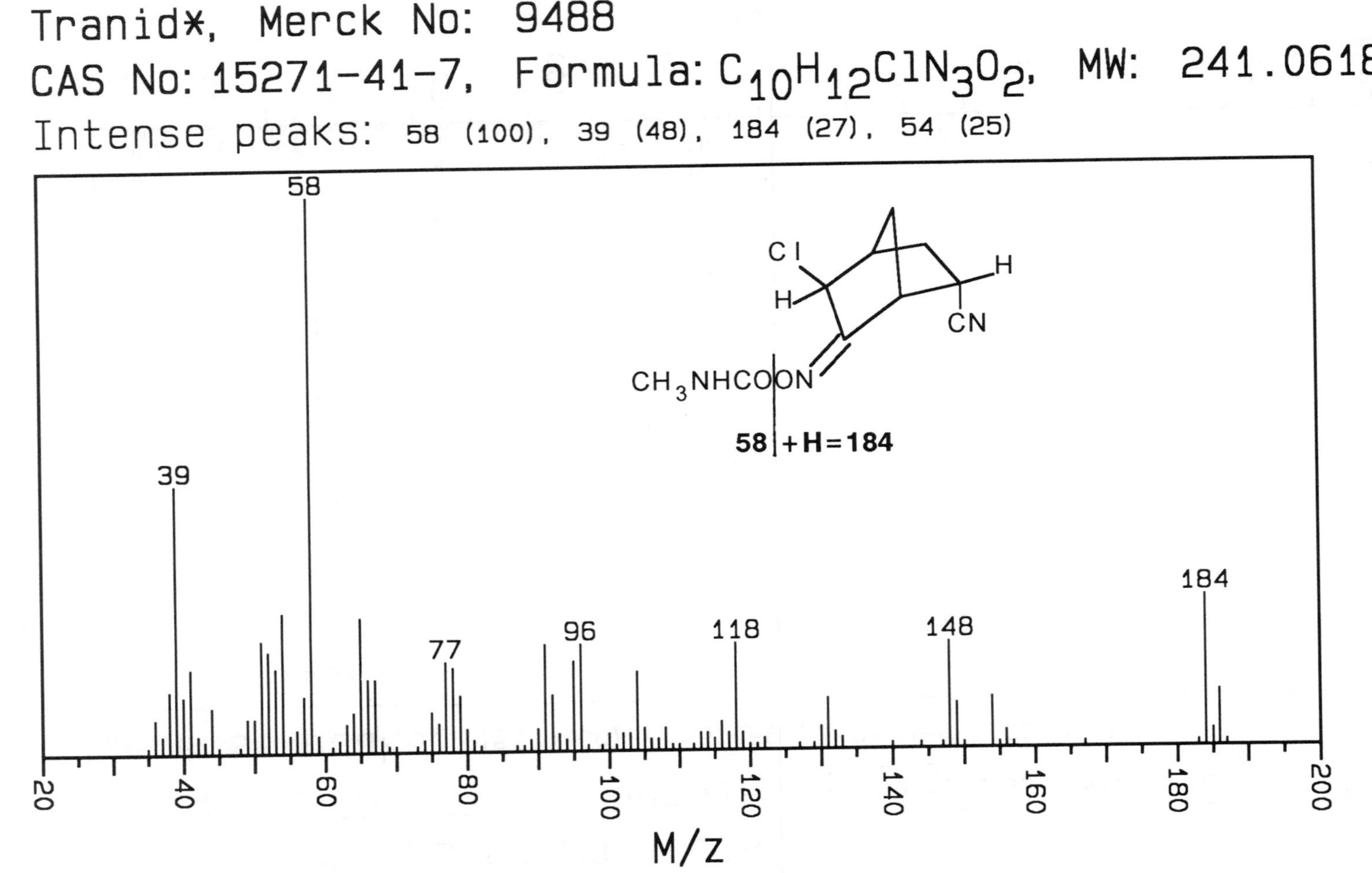
Tranid*, Merck No: 9488
CAS No: 15271-41-7, Formula: $C_{10}H_{12}ClN_3O_2$, MW: 241.0618
Intense peaks: 58 (100), 39 (48), 184 (27), 54 (25)
Cl
H
H
CN
$CH_3NHCOON$
58 | +H=184
58
39
77
96
118
148
184
20
40
60
80
100
120
140
160
180
200
M/Z

EPTC, Merck No: 3580

CAS No: 759-94-4, Formula: $C_9H_{19}NOS$, MW: 189.1187

Intense peaks: 43 (100), 128 (76), 86 (54), 29 (22)

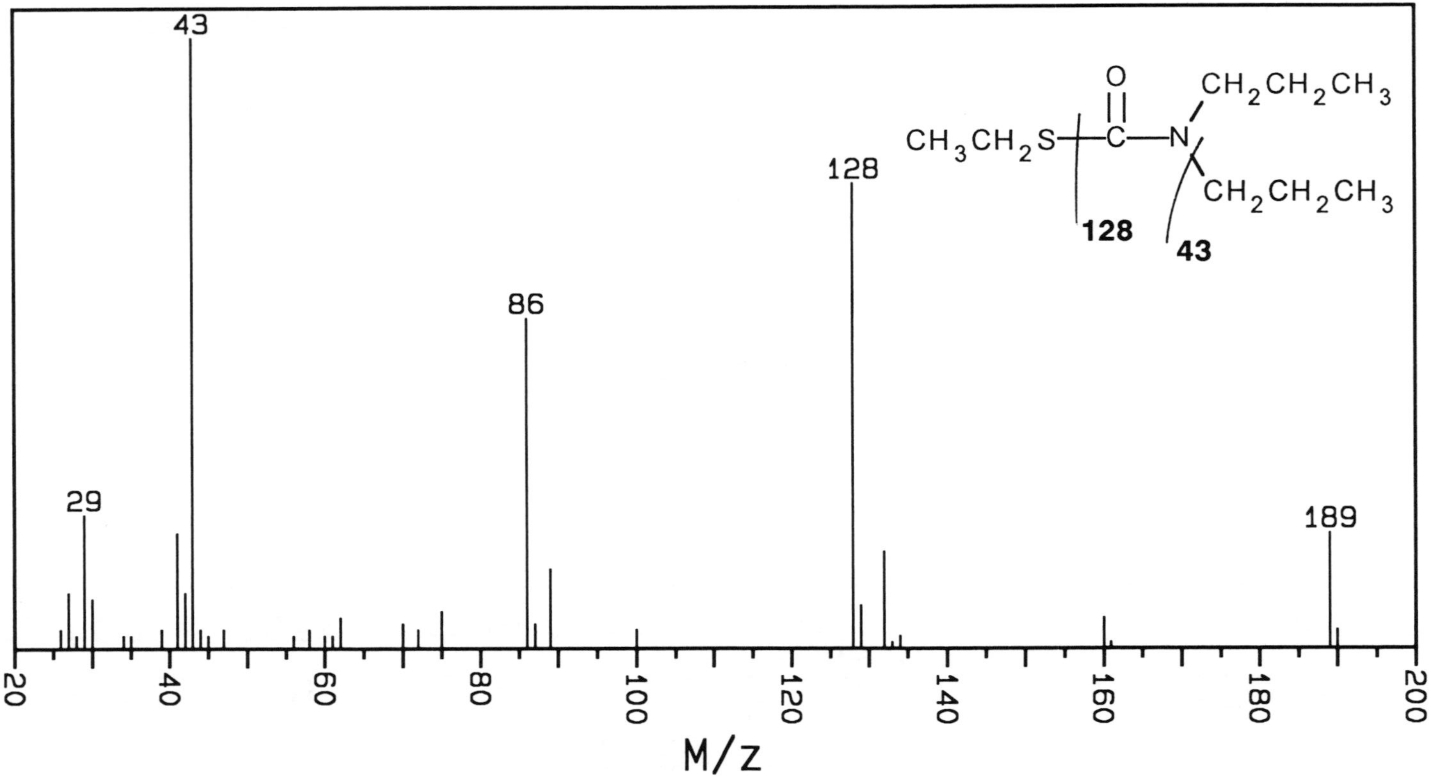

$CH_3CH_2S-C(=O)-N(CH_2CH_2CH_3)_2$

128

43

29

43

86

128

189

20 40 60 80 100 120 140 160 180 200

M/Z

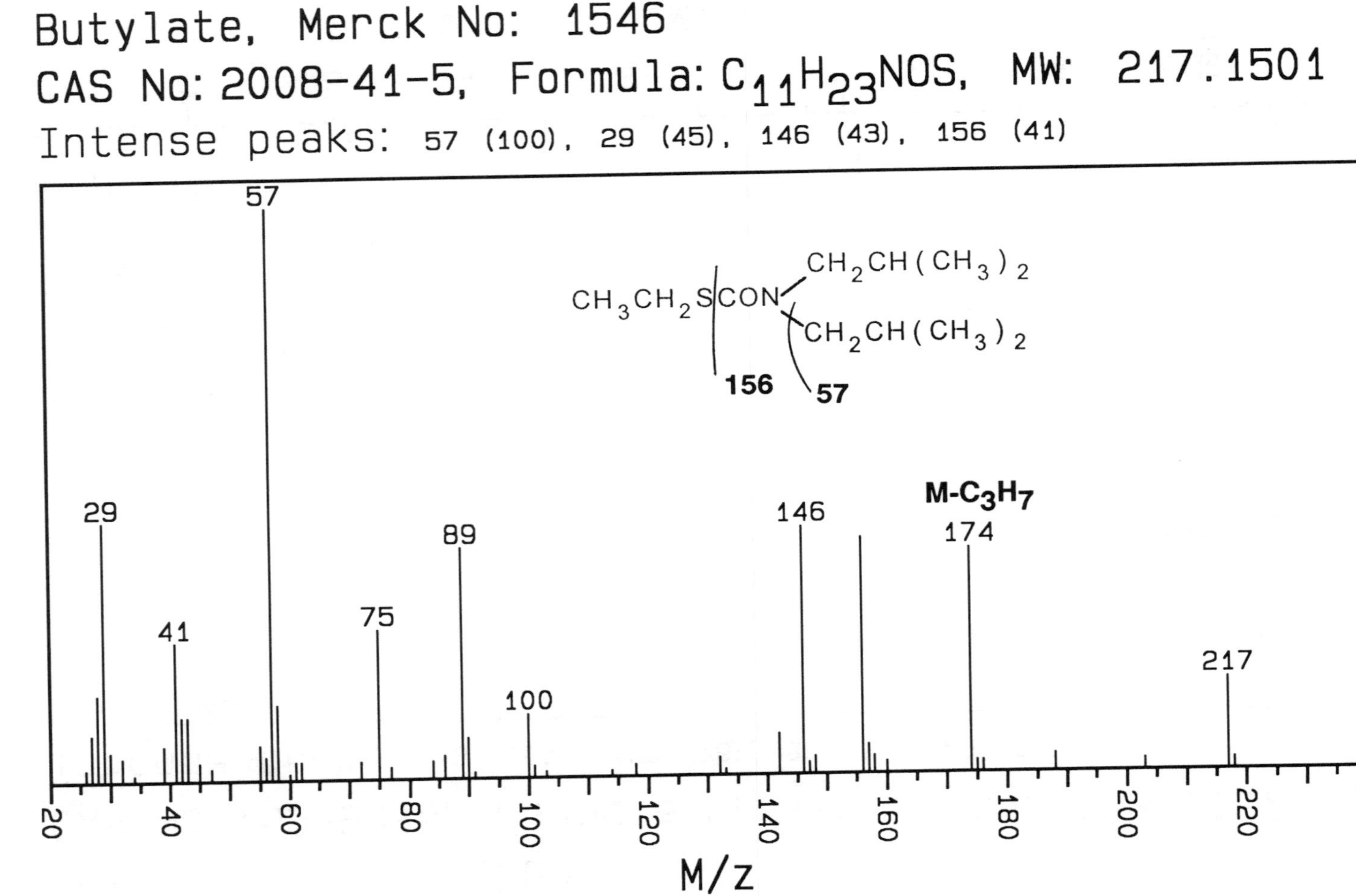

Butylate, Merck No: 1546
CAS No: 2008-41-5, Formula: C11H23NOS, MW: 217.1501
Intense peaks: 57 (100), 29 (45), 146 (43), 156 (41)
CH3CH2SCON CH2CH(CH3)2 CH2CH(CH3)2
156
57
29
41
57
75
89
100
146
M-C3H7
174
217
20
40
60
80
100
120
140
160
180
200
220
240
M/Z

Vernolate, Merck No: 9866

CAS No: 1929-77-7, Formula: $C_{10}H_{21}NOS$, MW: 203.1344

Intense peaks: 43 (100), 86 (56), 128 (48), 40 (47)

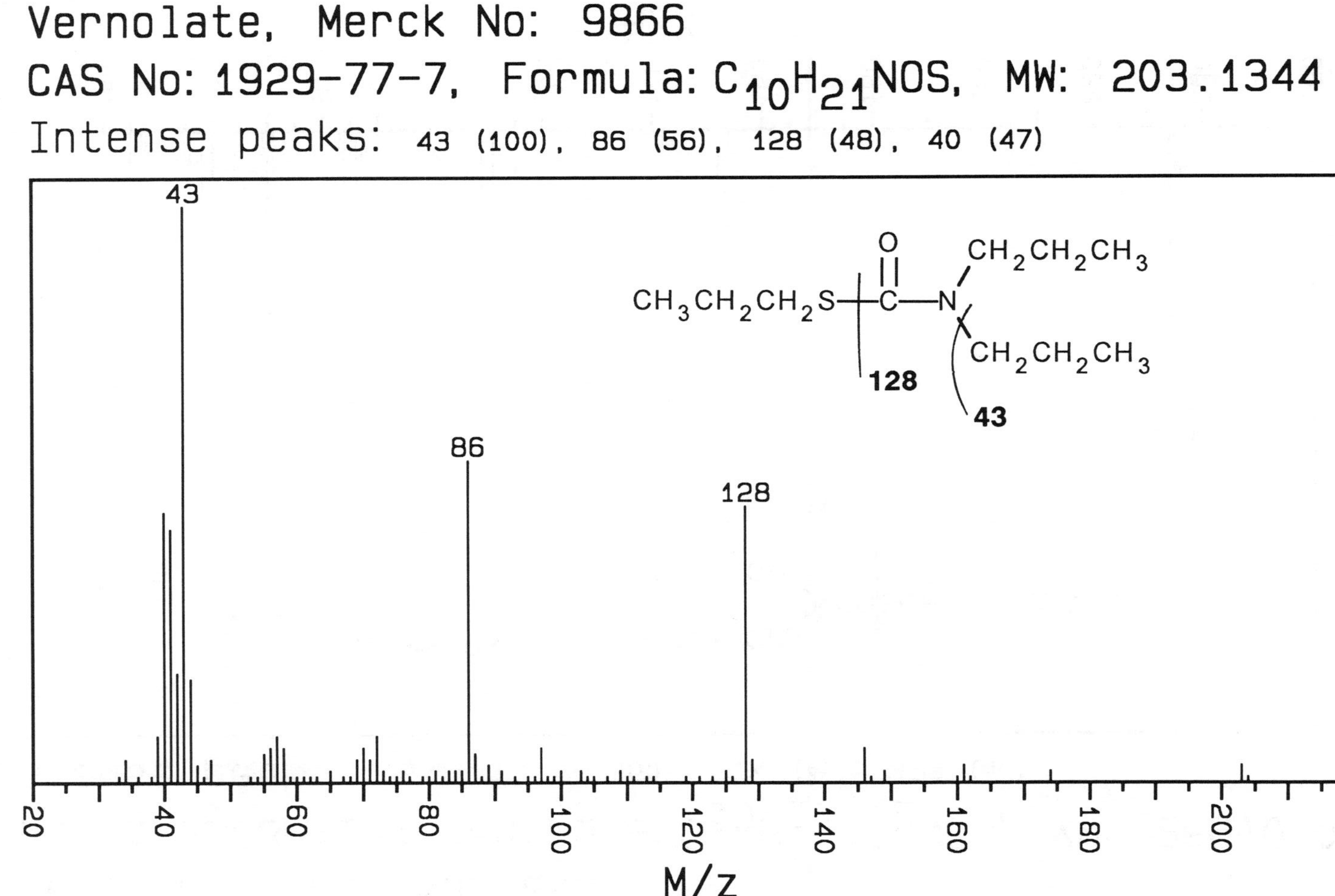

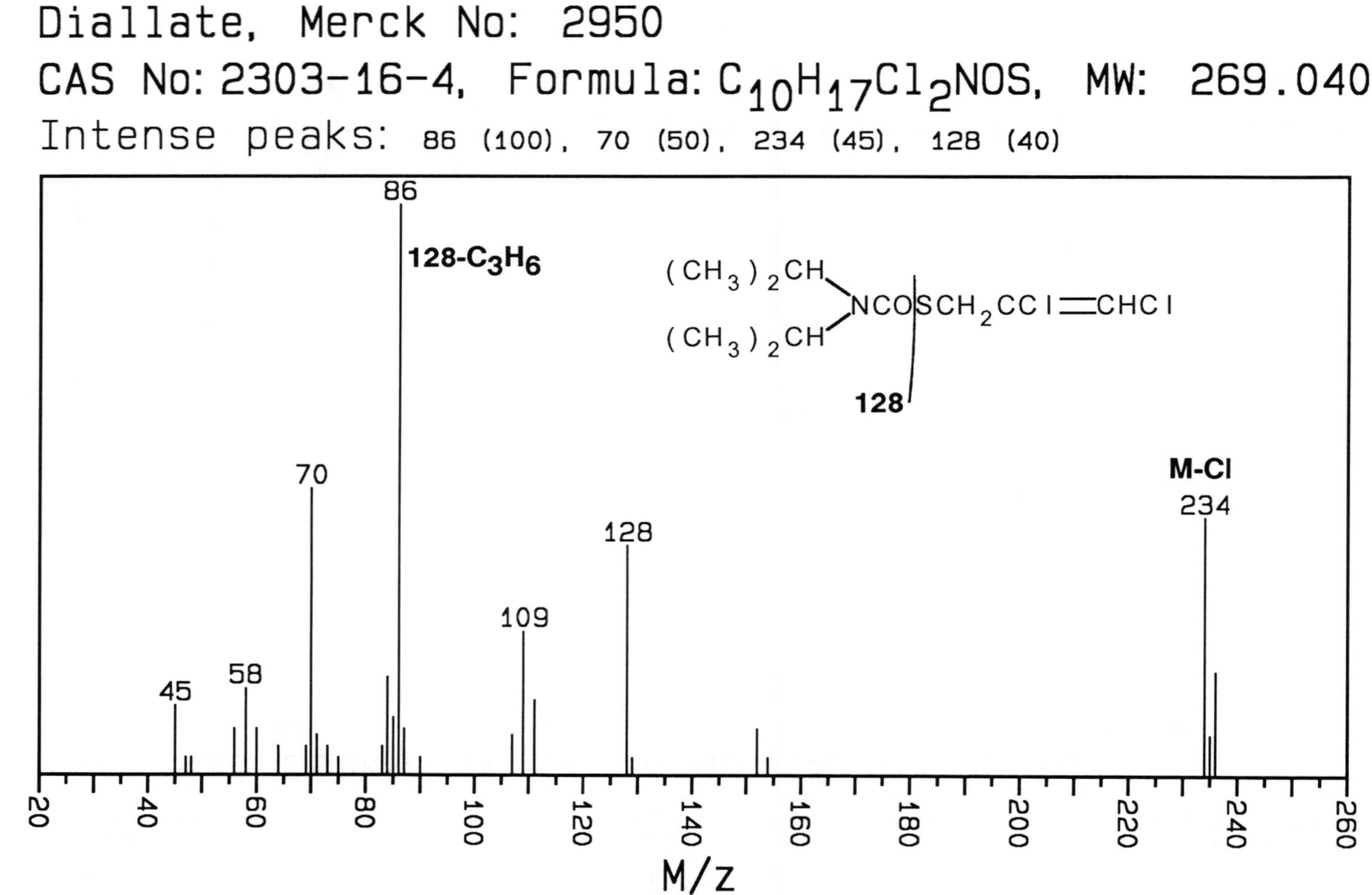

Diallate, Merck No: 2950
CAS No: 2303-16-4, Formula: $C_{10}H_{17}Cl_2NOS$, MW: 269.0408
Intense peaks: 86 (100), 70 (50), 234 (45), 128 (40)
86
128-$C_3H_6$
$(CH_3)_2CH$
$NCOSCH_2CCl{=}CHCl$
$(CH_3)_2CH$
128
70
M-Cl
234
128
109
58
45
20 40 60 80 100 120 140 160 180 200 220 240 260
M/Z

Triallate, Merck No: 9510

CAS No: 2303-17-5, Formula: $C_{10}H_{16}Cl_3NOS$, MW: 303.0018

Intense peaks: 86 (100), 268 (22), 128 (18), 270 (15)

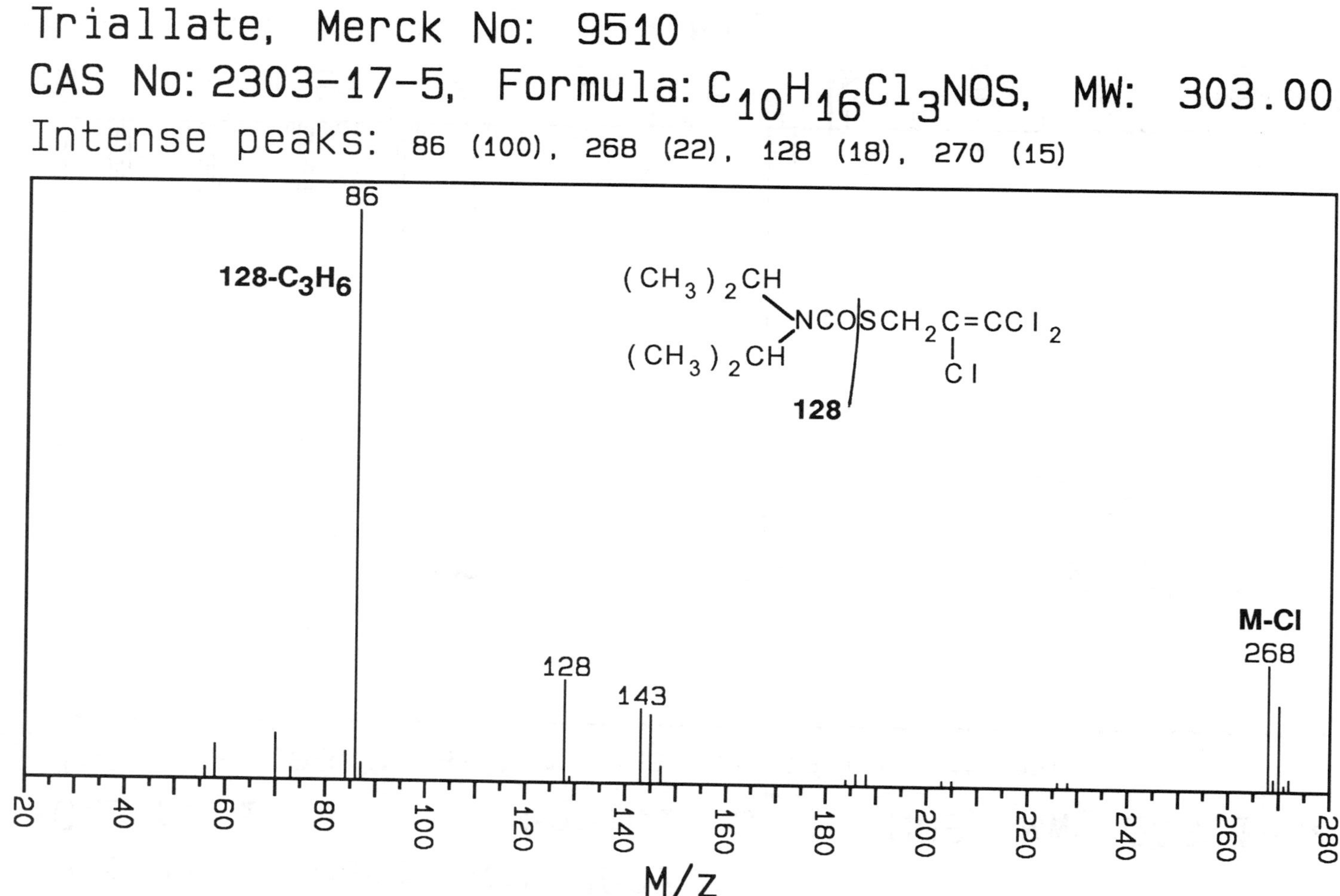

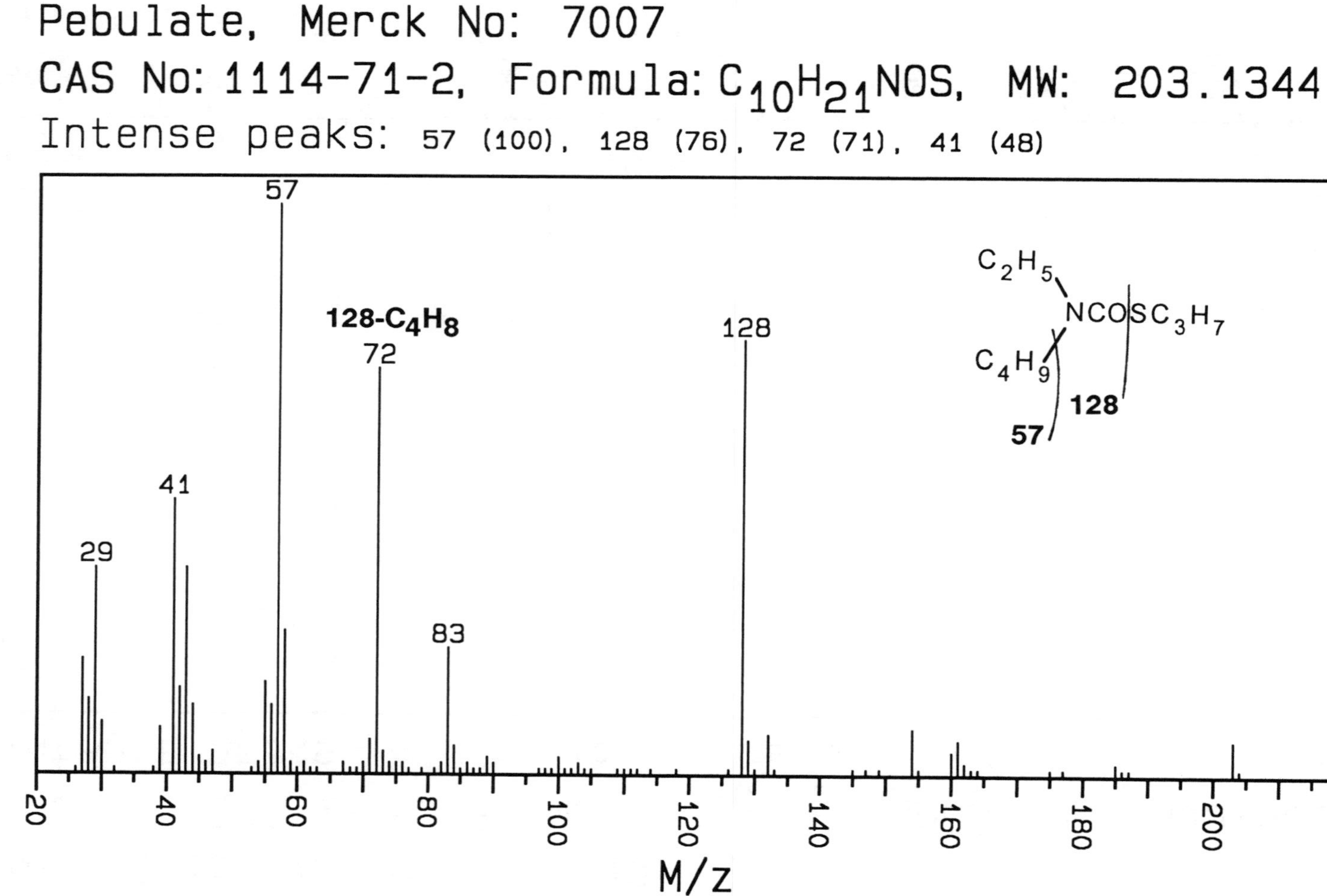
Pebulate, Merck No: 7007
CAS No: 1114-71-2, Formula: $C_{10}H_{21}NOS$, MW: 203.1344
Intense peaks: 57 (100), 128 (76), 72 (71), 41 (48)
$C_2H_5$
$NCOSC_3H_7$
$C_4H_9$
128
57
$128-C_4H_8$
72
57
41
29
83
128
M/Z
20
40
60
80
100
120
140
160
180
200
220

Sulfallate, Merck No: 8882

CAS No: 95-06-7, Formula: $C_8H_{14}ClNS_2$, MW: 223.0256

Intense peaks: 188 (100), 72 (33), 29 (33), 44 (30)

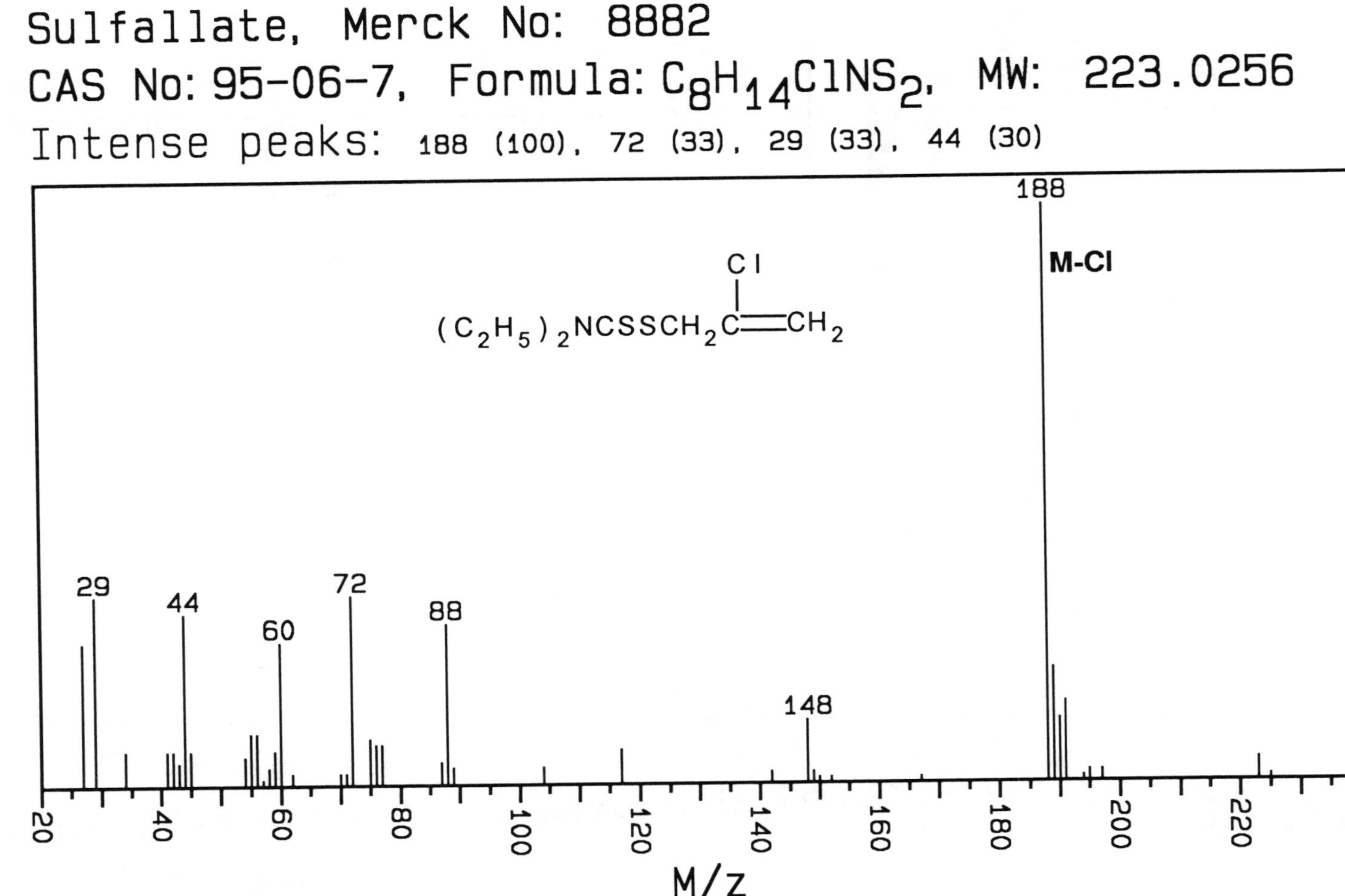

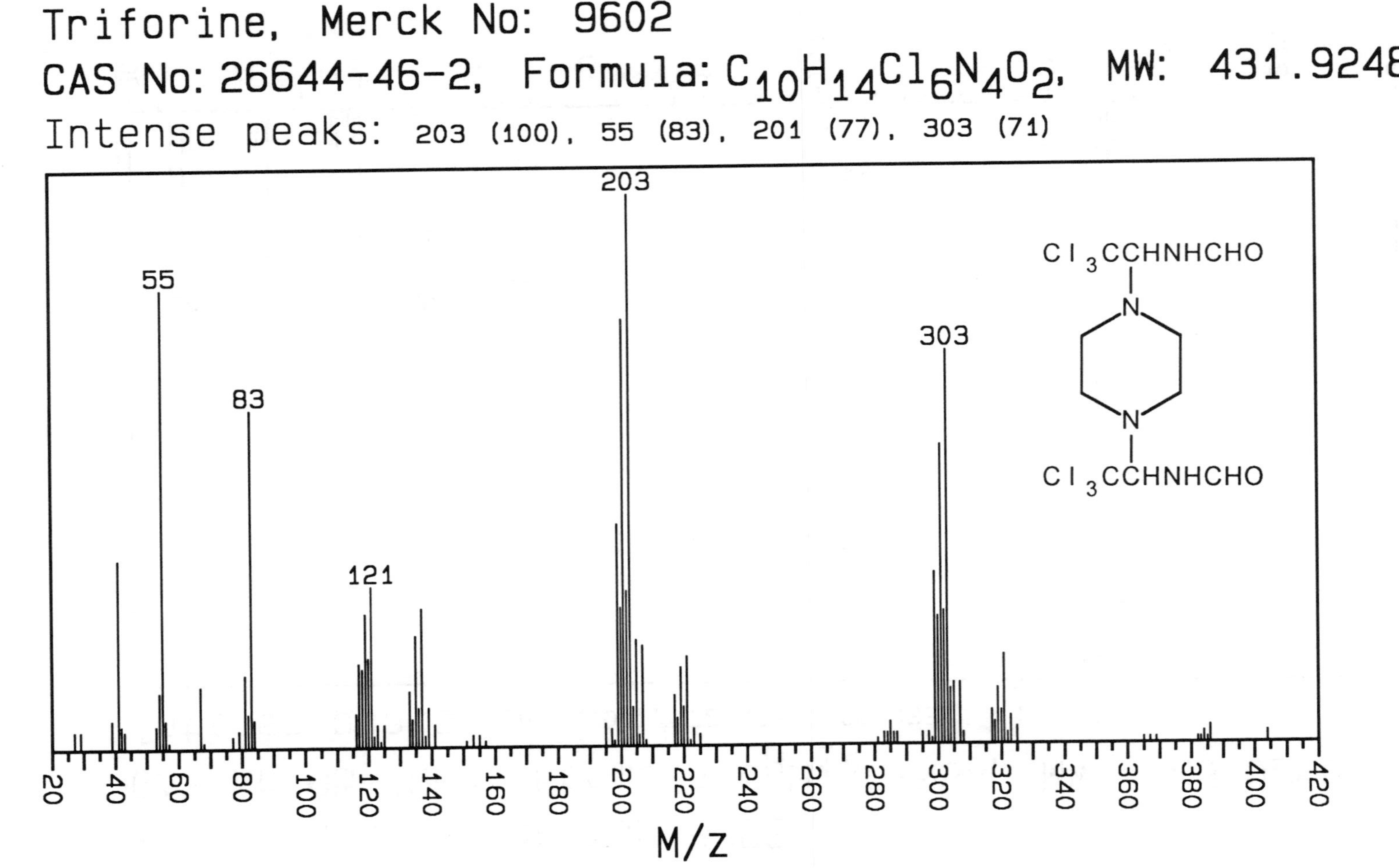
Triforine, Merck No: 9602
CAS No: 26644-46-2, Formula: $C_{10}H_{14}Cl_6N_4O_2$, MW: 431.9248
Intense peaks: 203 (100), 55 (83), 201 (77), 303 (71)
203
55
83
121
303
$Cl_3CCHNHCHO$
N
N
$Cl_3CCHNHCHO$
20
40
60
80
100
120
140
160
180
200
220
240
260
280
300
320
340
360
380
400
420
M/Z

Daminozide, Merck No: 2810

CAS No: 1596-84-5, Formula: $C_6H_{12}N_2O_3$, MW: 160.0848

Intense peaks: 59 (100), 60 (57), 43 (30), 45 (28)

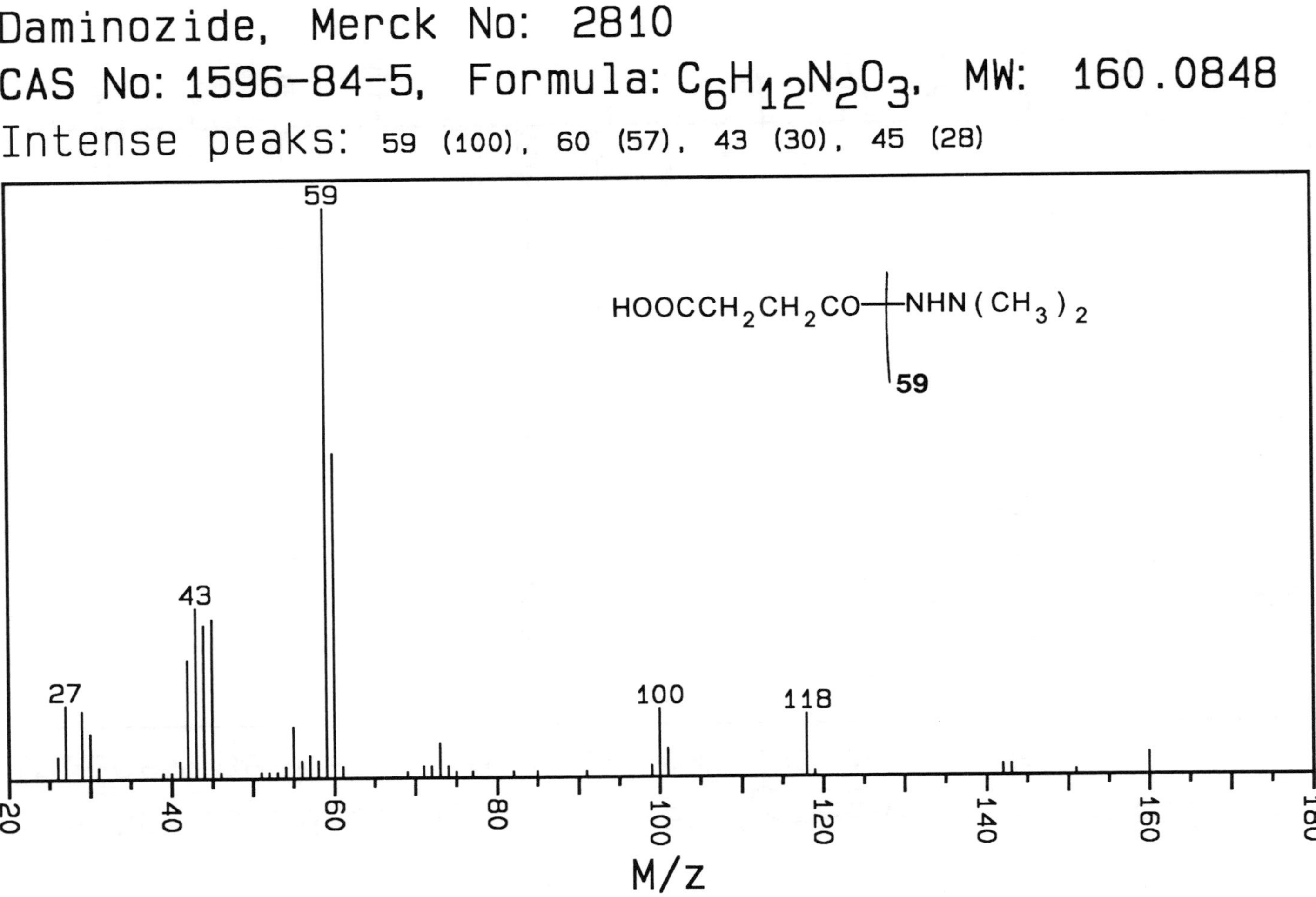

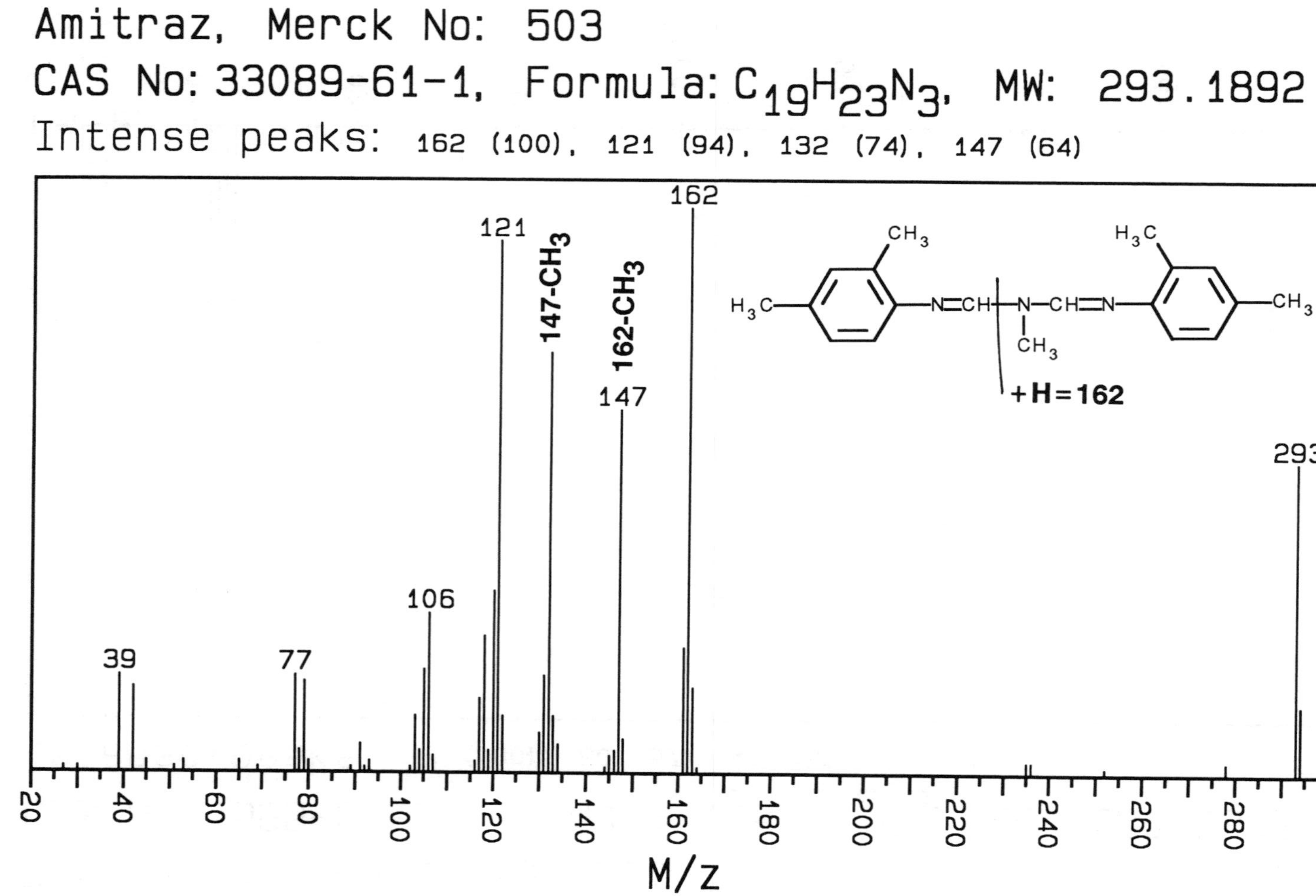
Amitraz, Merck No: 503
CAS No: 33089-61-1, Formula: $C_{19}H_{23}N_3$, MW: 293.1892
Intense peaks: 162 (100), 121 (94), 132 (74), 147 (64)
162
121
147-$CH_3$
162-$CH_3$
147
106
39
77
293
+H=162
M/Z

Oxamyl, Merck No: 6873

CAS No: 23135-22-0, Formula: $C_7H_{13}N_3O_3S$, MW: 219.0678

Intense peaks: 72 (100), 44 (79), 32 (51), 30 (45)

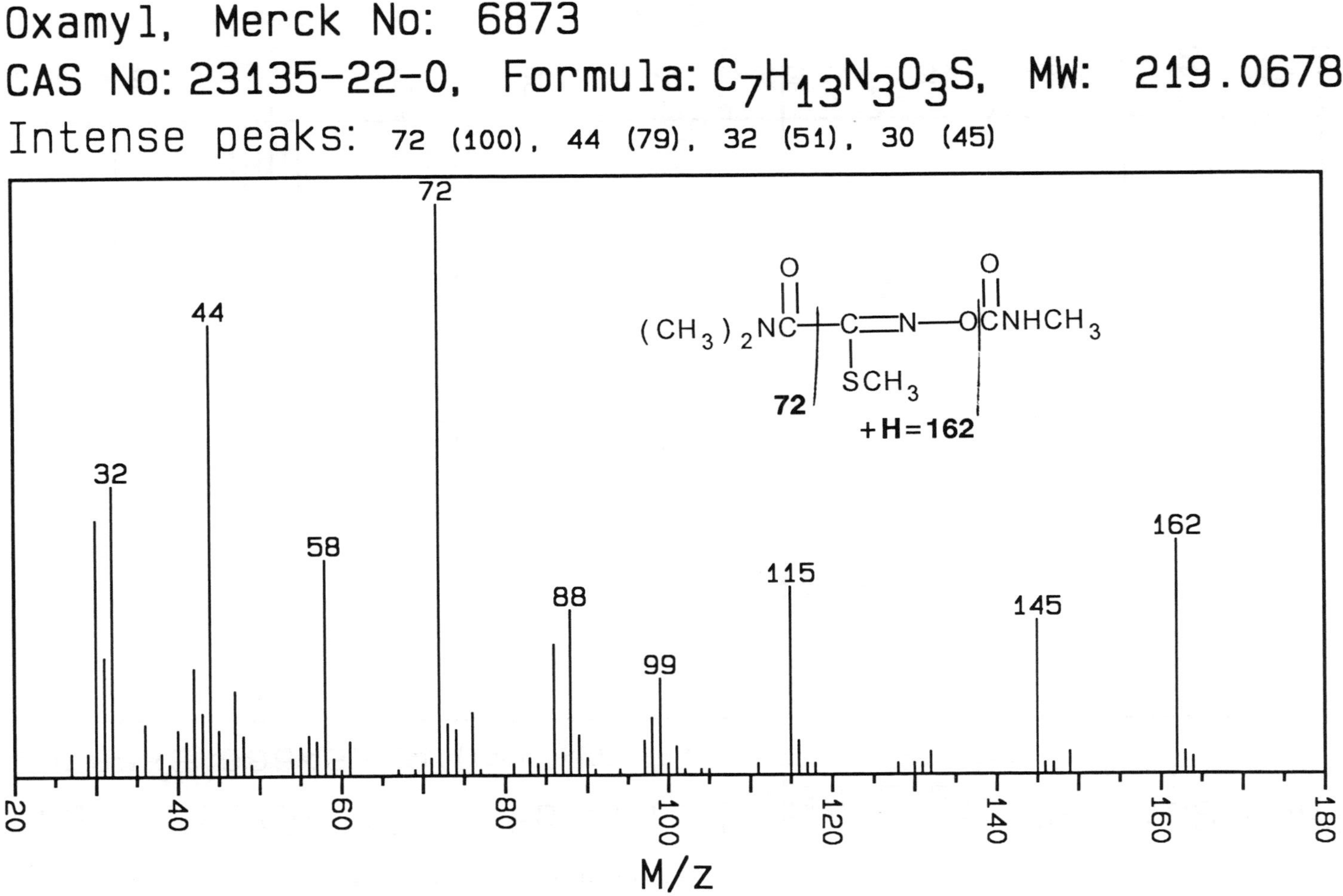

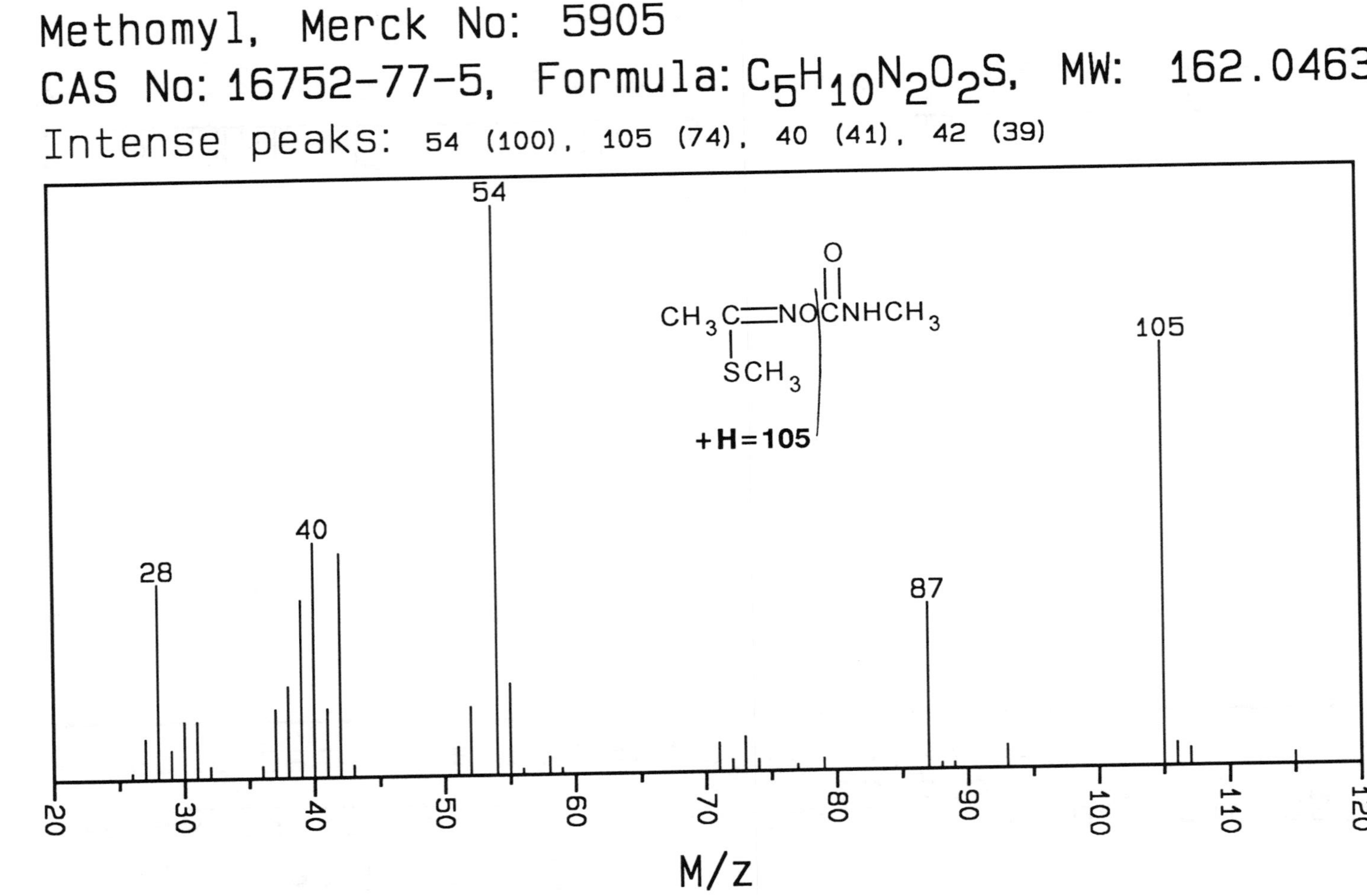
Methomyl, Merck No: 5905
CAS No: 16752-77-5, Formula: $C_5H_{10}N_2O_2S$, MW: 162.0463
Intense peaks: 54 (100), 105 (74), 40 (41), 42 (39)
54
O
CH3C=NOCNHCH3
SCH3
+H=105
105
40
28
87
20
30
40
50
60
70
80
90
100
110
120
M/Z

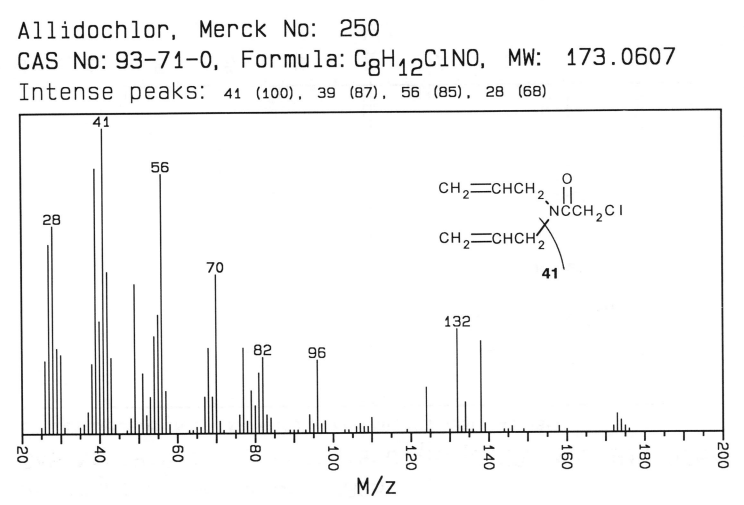

Allidochlor, Merck No: 250

CAS No: 93-71-0, Formula: $C_8H_{12}ClNO$, MW: 173.0607

Intense peaks: 41 (100), 39 (87), 56 (85), 28 (68)

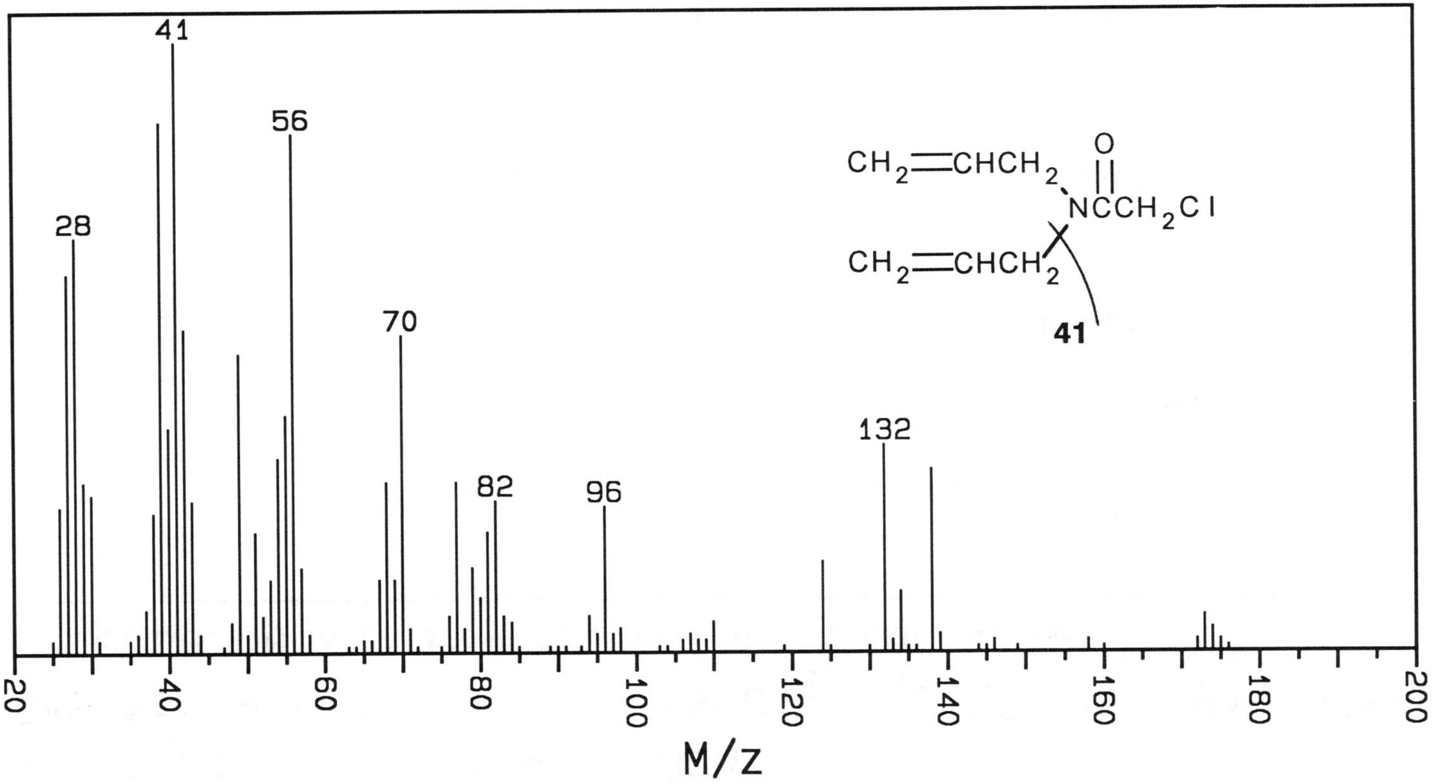

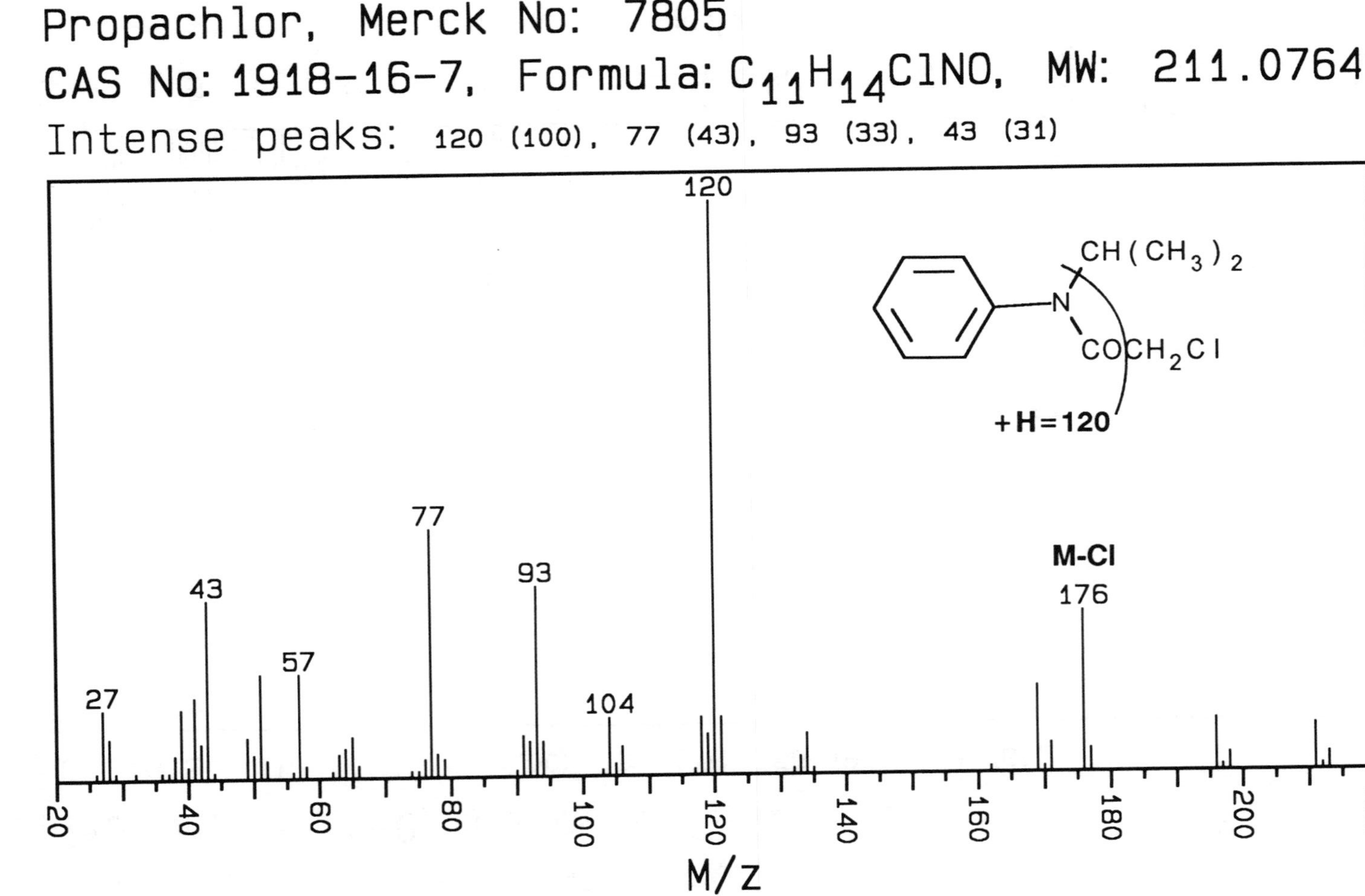
Propachlor, Merck No: 7805
CAS No: 1918-16-7, Formula: $C_{11}H_{14}ClNO$, MW: 211.0764
Intense peaks: 120 (100), 77 (43), 93 (33), 43 (31)
120
$CH(CH_3)_2$
N
$COCH_2Cl$
+H=120
77
93
43
57
27
104
M-Cl
176
20
40
60
80
100
120
140
160
180
200
220
M/Z

Metolachlor, Merck No: 6067

CAS No: 51218-45-2, Formula: $C_{15}H_{22}ClNO_2$, MW: 283.1339

Intense peaks: 162 (100), 45 (76), 41 (33), 238 (26)

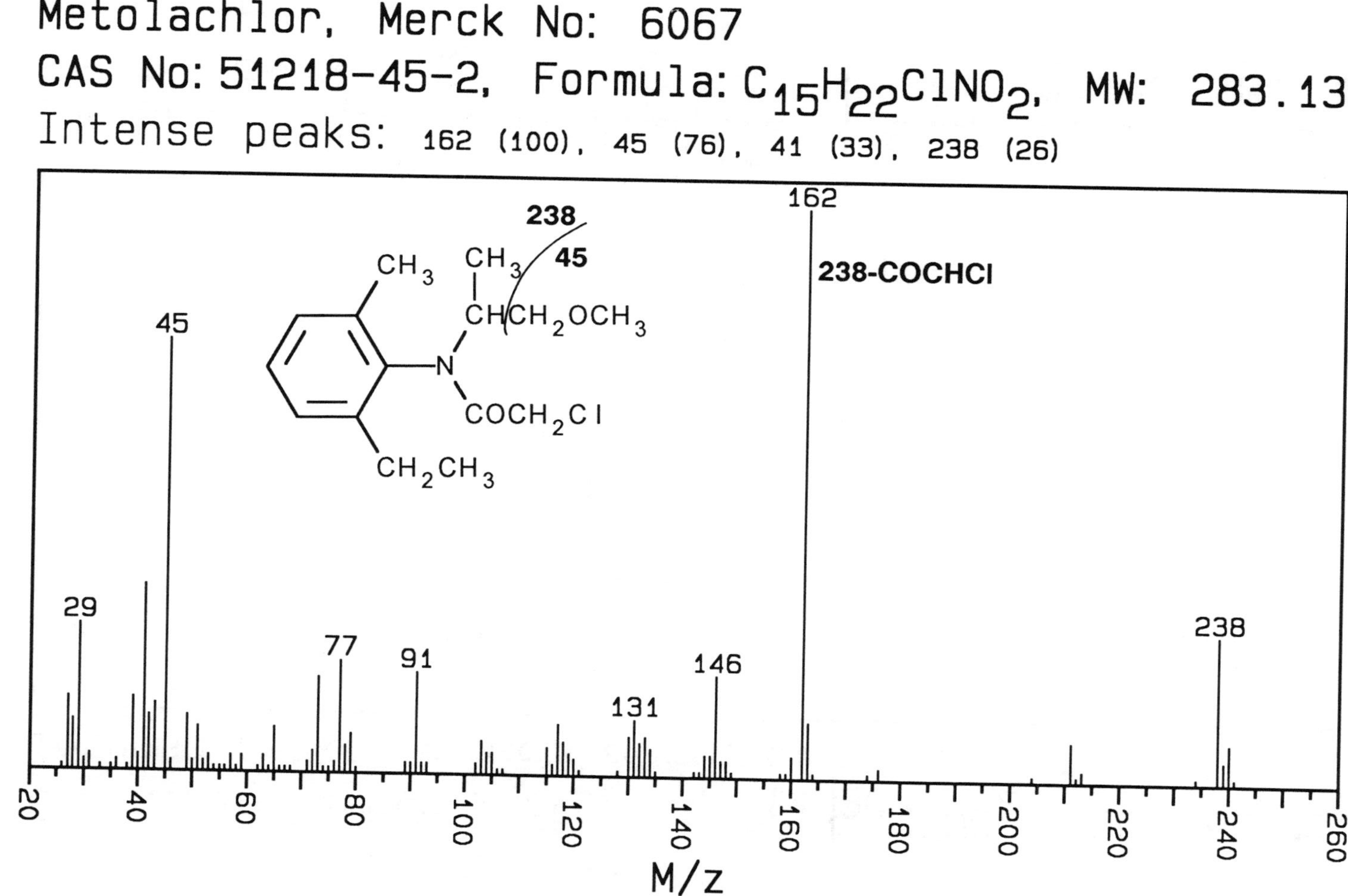

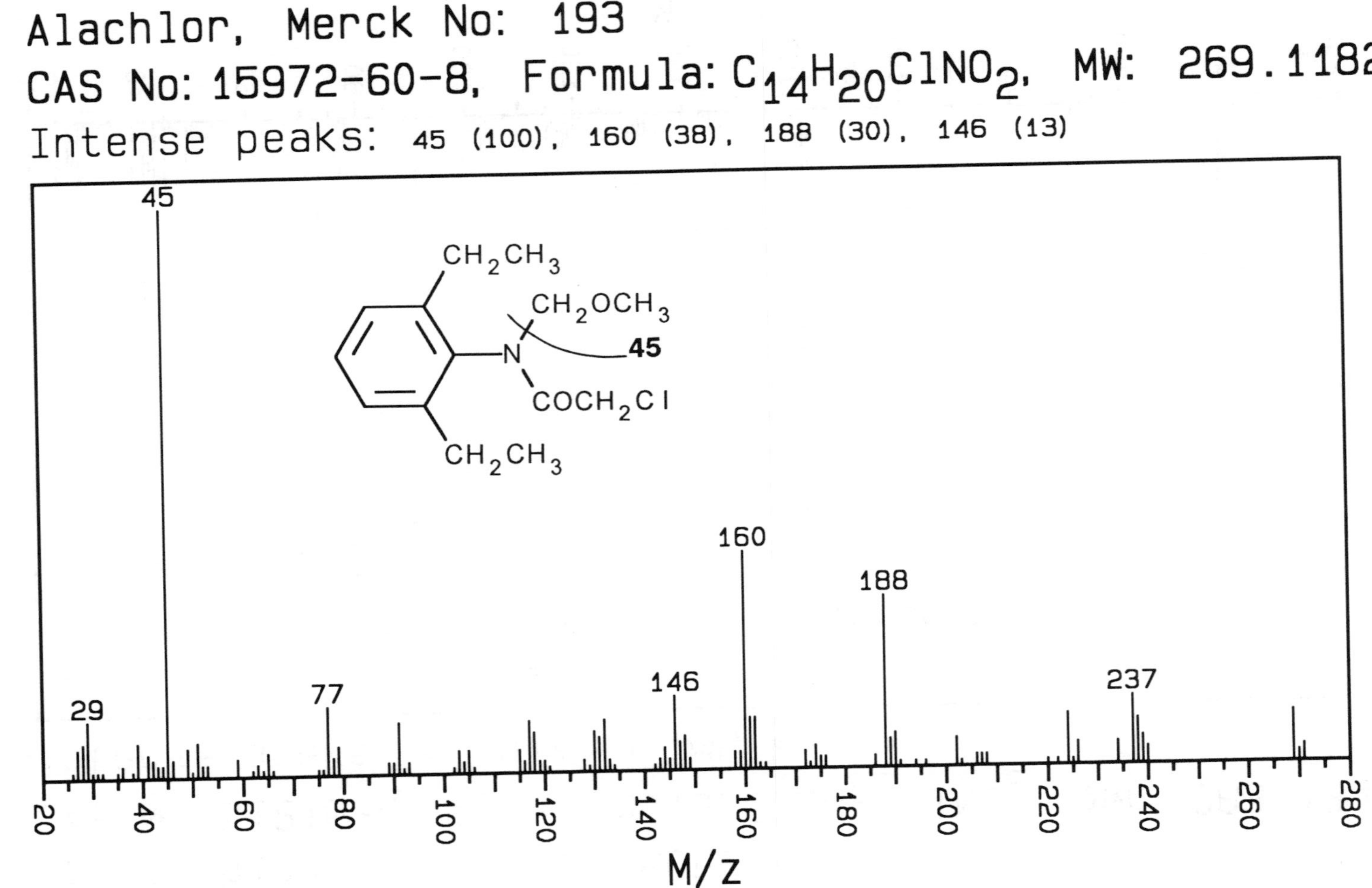
Alachlor, Merck No: 193
CAS No: 15972-60-8, Formula: $C_{14}H_{20}ClNO_2$, MW: 269.1182
Intense peaks: 45 (100), 160 (38), 188 (30), 146 (13)
45
$CH_2CH_3$
$CH_2OCH_3$
N
45
$COCH_2Cl$
$CH_2CH_3$
29
77
146
160
188
237
20
40
60
80
100
120
140
160
180
200
220
240
260
280
M/Z

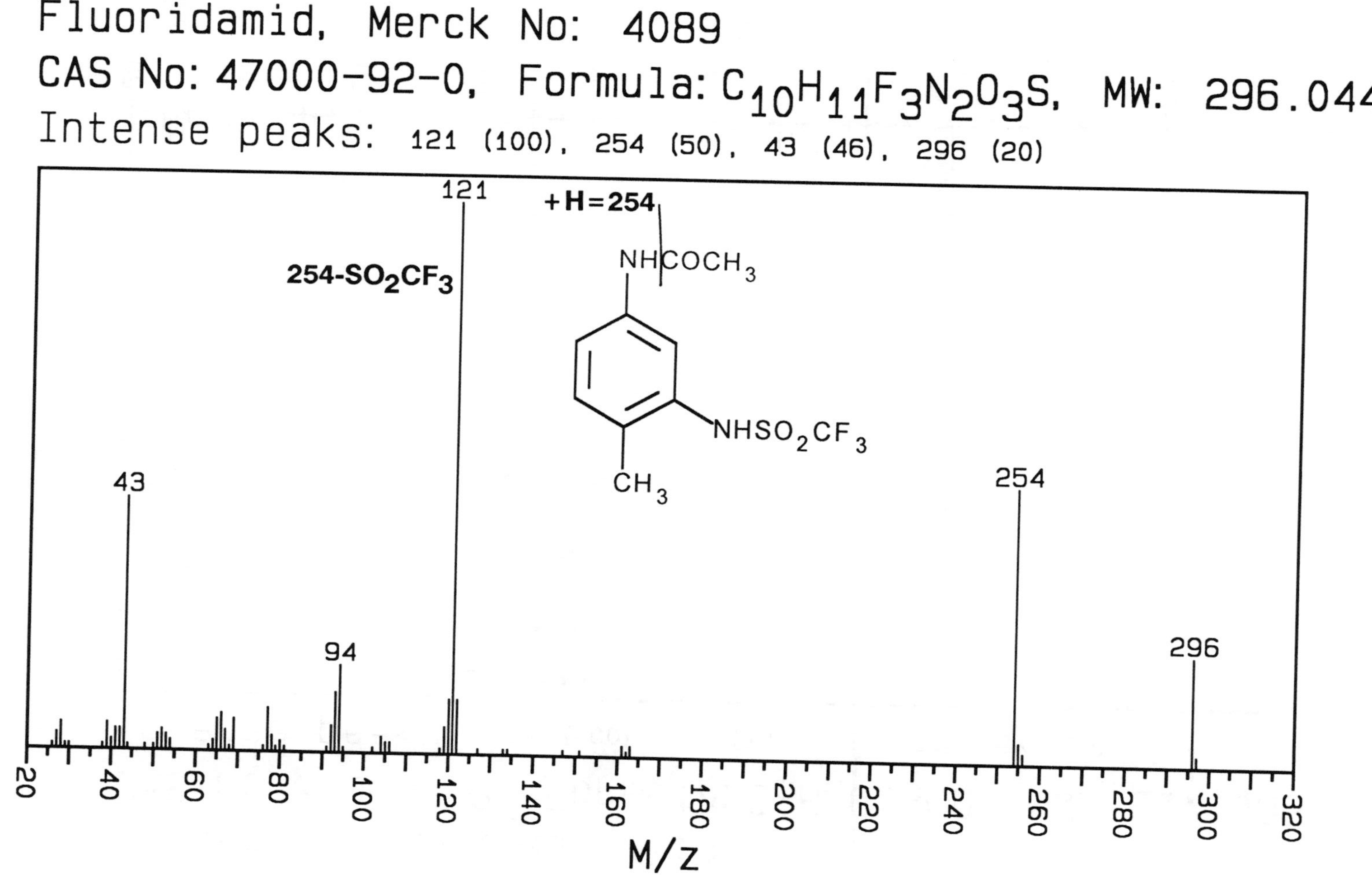
Fluoridamid, Merck No: 4089
CAS No: 47000-92-0, Formula: $C_{10}H_{11}F_3N_2O_3S$, MW: 296.0442
Intense peaks: 121 (100), 254 (50), 43 (46), 296 (20)
121
+H=254
254-SO2CF3
NHCOCH3
NHSO2CF3
CH3
43
94
254
296
20
40
60
80
100
120
140
160
180
200
220
240
260
280
300
320
M/Z

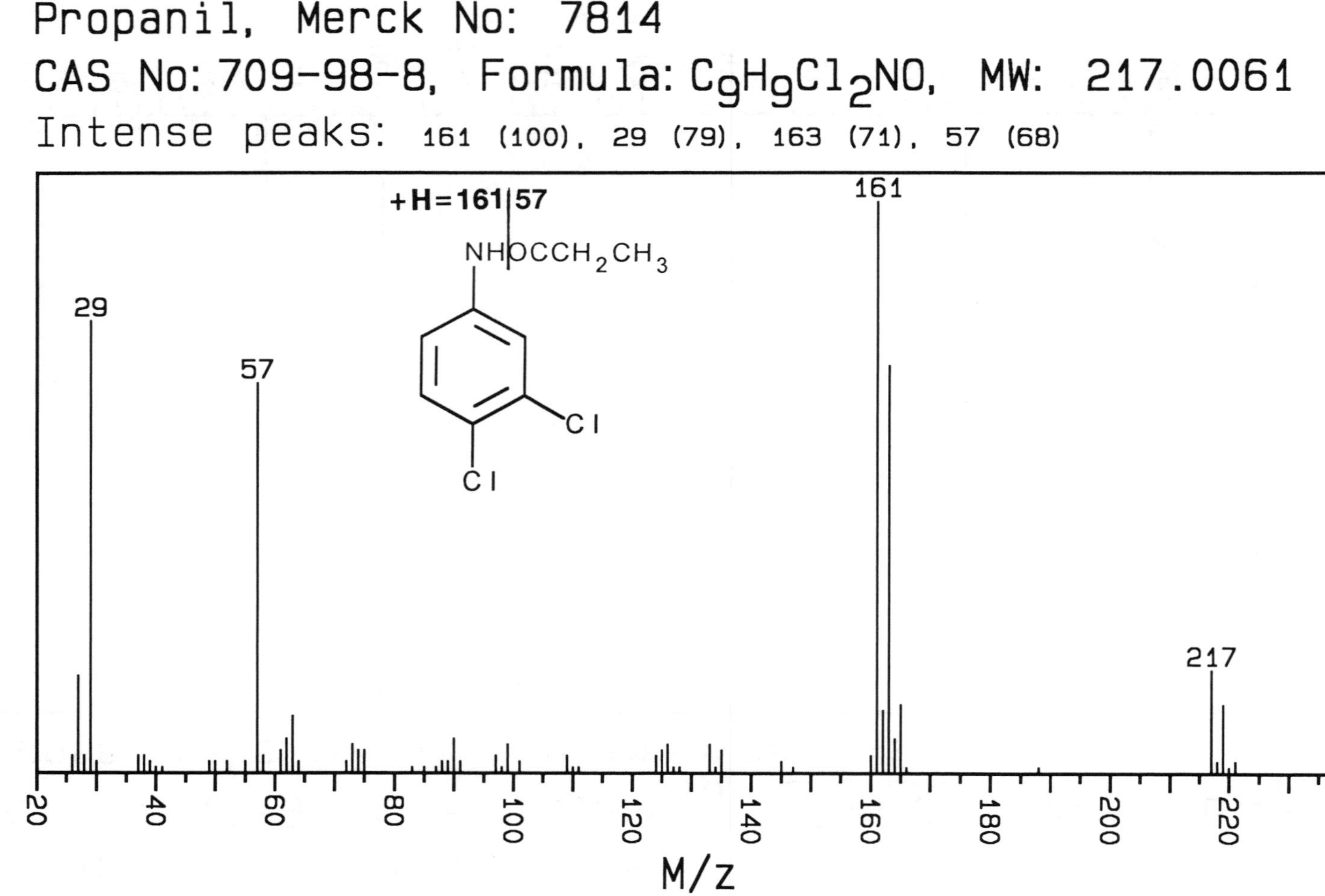
Propanil, Merck No: 7814
CAS No: 709-98-8, Formula: $C_9H_9Cl_2NO$, MW: 217.0061
Intense peaks: 161 (100), 29 (79), 163 (71), 57 (68)
+H=161|57
$NHOCCH_2CH_3$
Cl
Cl
29
57
161
217
20
40
60
80
100
120
140
160
180
200
220
240
M/Z

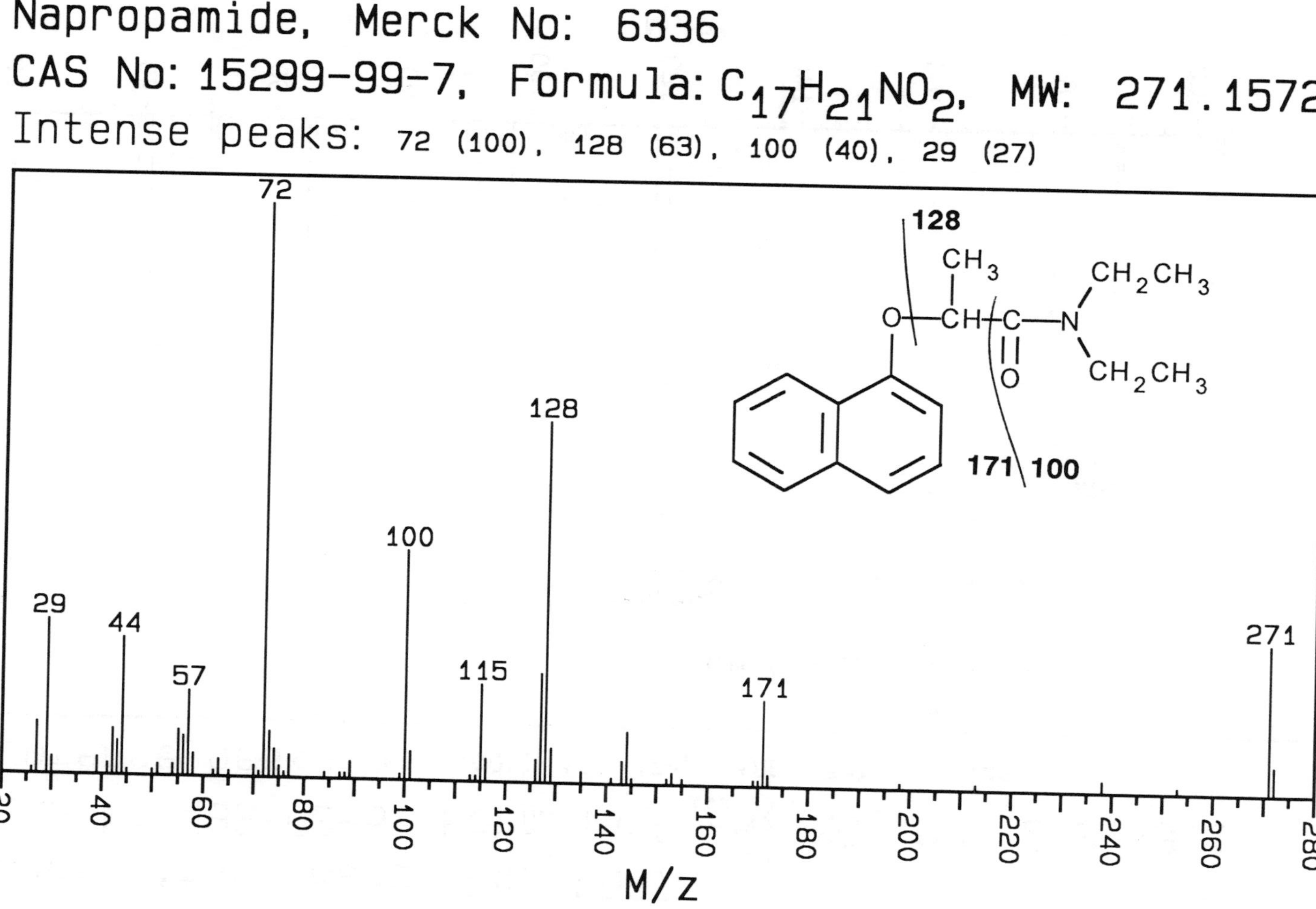
Napropamide, Merck No: 6336
CAS No: 15299-99-7, Formula: $C_{17}H_{21}NO_2$, MW: 271.1572
Intense peaks: 72 (100), 128 (63), 100 (40), 29 (27)
72
128
100
29
44
57
115
171
271
128
$CH_3$
$CH_2CH_3$
O
CH
C
N
O
$CH_2CH_3$
171
100
20
40
60
80
100
120
140
160
180
200
220
240
260
280
M/Z

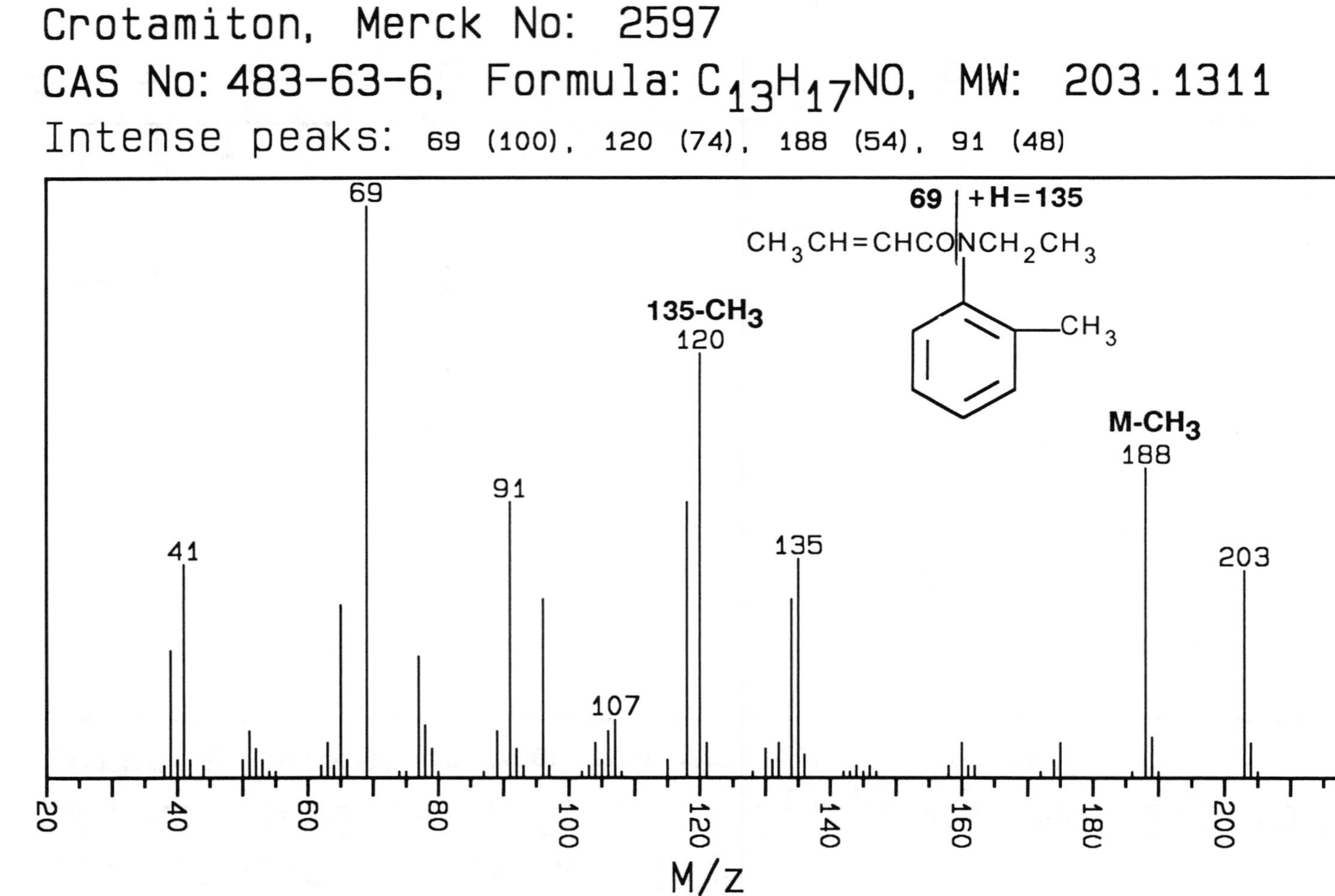
Crotamiton, Merck No: 2597
CAS No: 483-63-6, Formula: C13H17NO, MW: 203.1311
Intense peaks: 69 (100), 120 (74), 188 (54), 91 (48)
69 | +H=135
CH3CH=CHCONCH2CH3
CH3
135-CH3
120
M-CH3
188
41
69
91
107
135
203
20
40
60
80
100
120
140
160
180
200
220
M/Z

Dicryl, Merck No: 3078

CAS No: 2164-09-2, Formula: $C_{10}H_9Cl_2NO$, MW: 229.0061

Intense peaks: 69 (100), 41 (97), 39 (49), 63 (12)

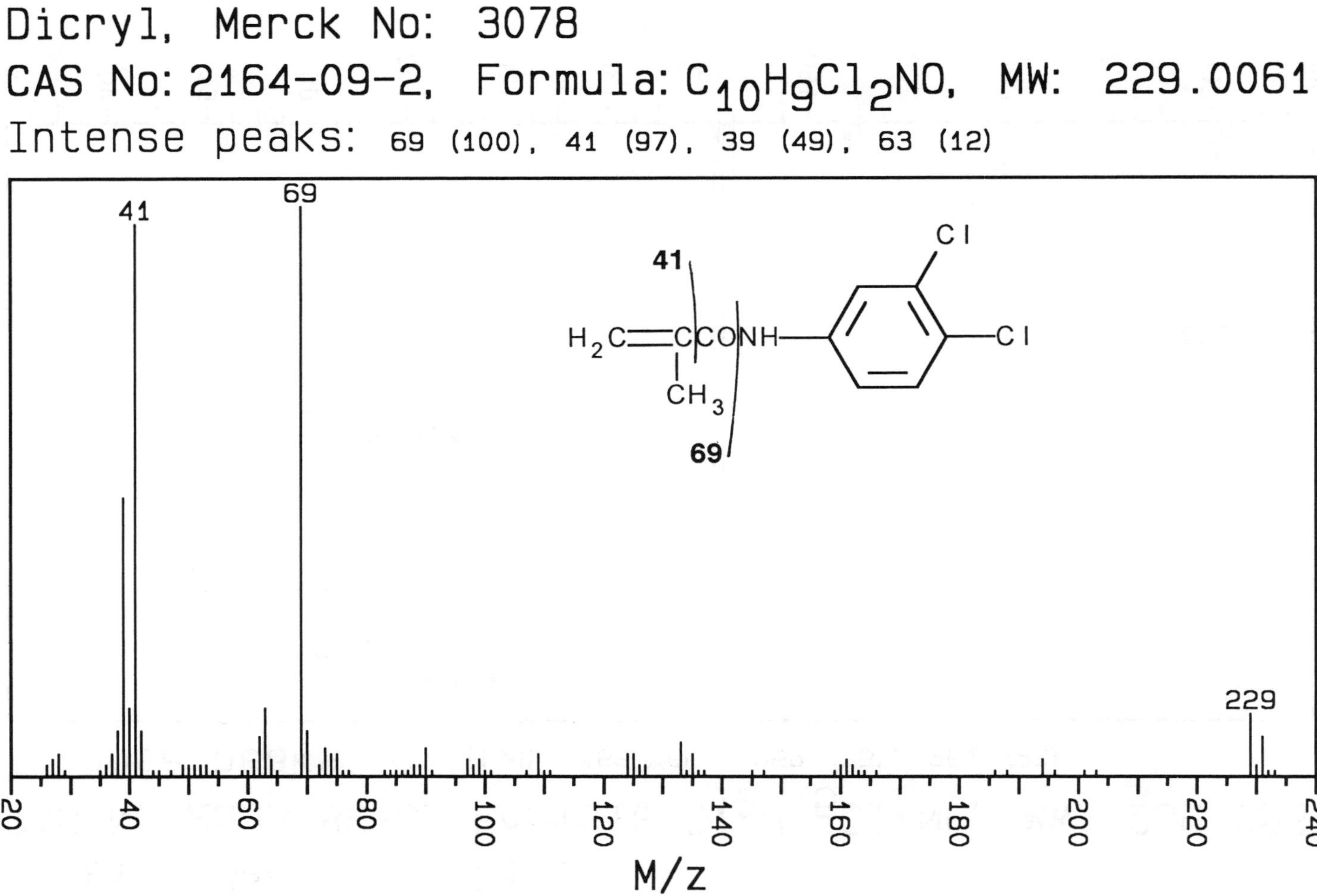

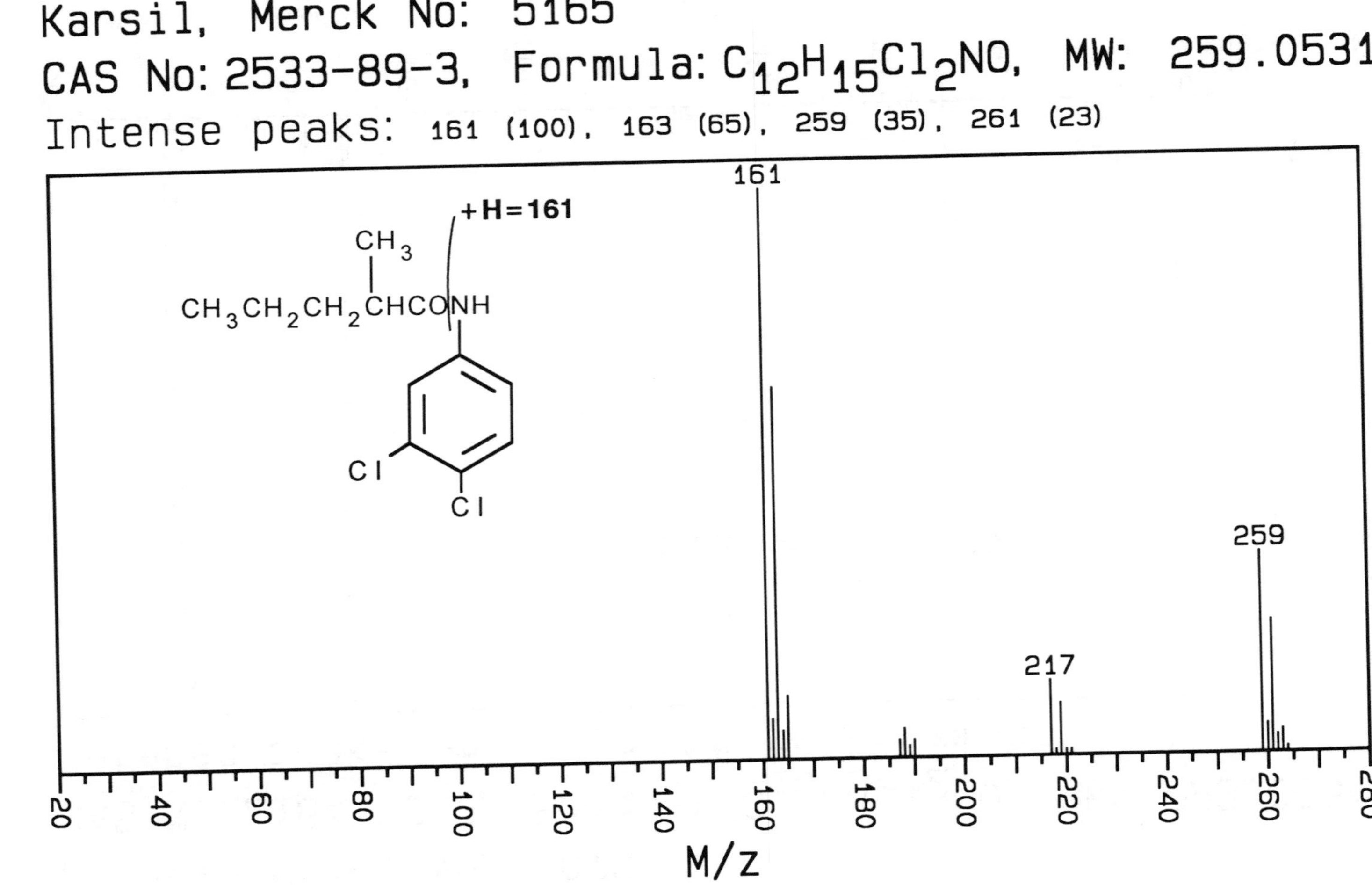
Karsil, Merck No: 5165
CAS No: 2533-89-3, Formula: $C_{12}H_{15}Cl_2NO$, MW: 259.0531
Intense peaks: 161 (100), 163 (65), 259 (35), 261 (23)
161
+H=161
CH3
CH3CH2CH2CHCONH
Cl
Cl
259
217
20
40
60
80
100
120
140
160
180
200
220
240
260
280
M/Z

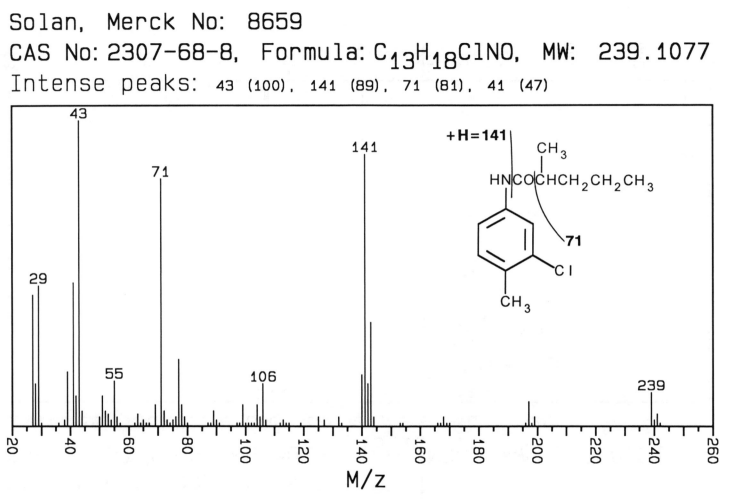

Solan, Merck No: 8659

CAS No: 2307-68-8, Formula: $C_{13}H_{18}ClNO$, MW: 239.1077

Intense peaks: 43 (100), 141 (89), 71 (81), 41 (47)

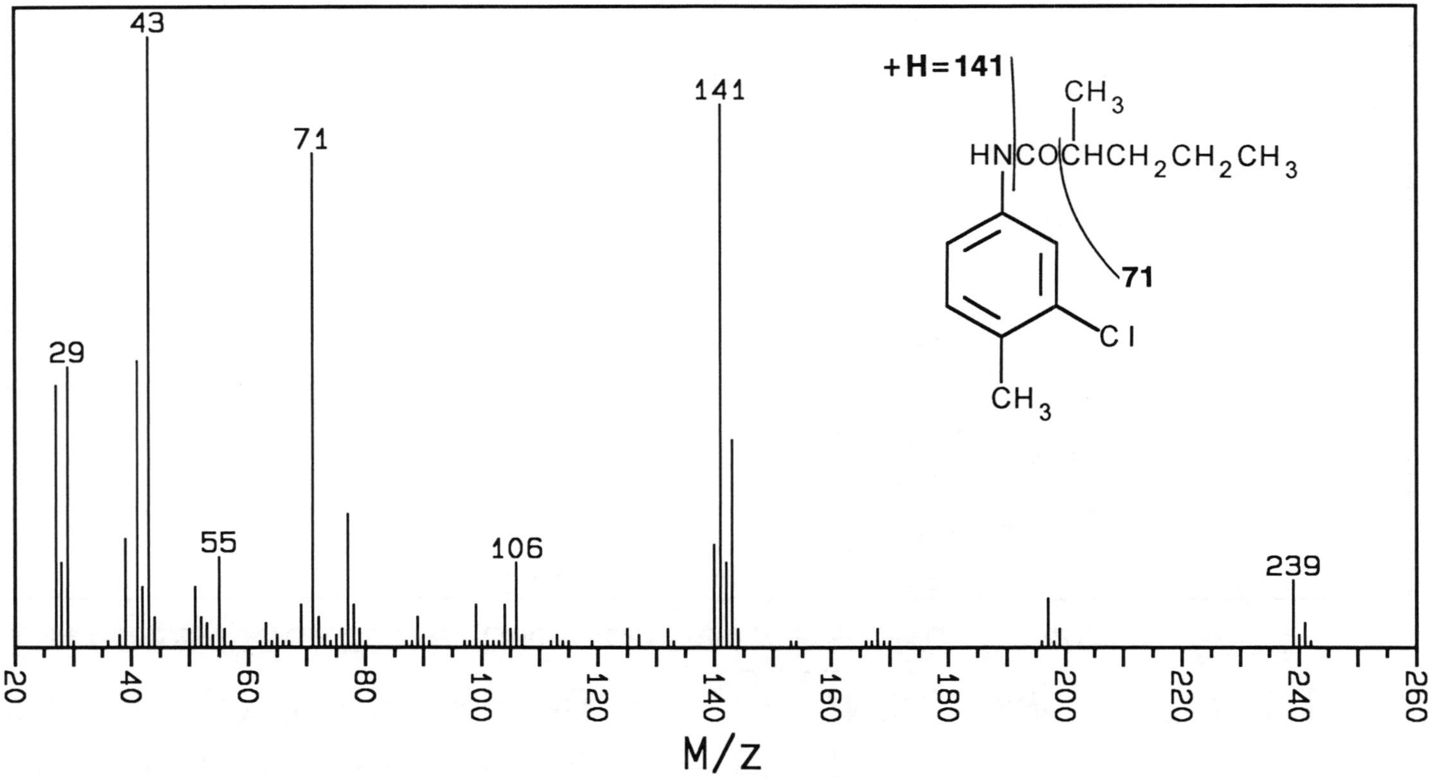

29
43
55
71
106
141
239
+H=141
71
$HNCOCH(CH_3)CH_2CH_2CH_3$
Cl
$CH_3$
20 40 60 80 100 120 140 160 180 200 220 240 260
M/Z

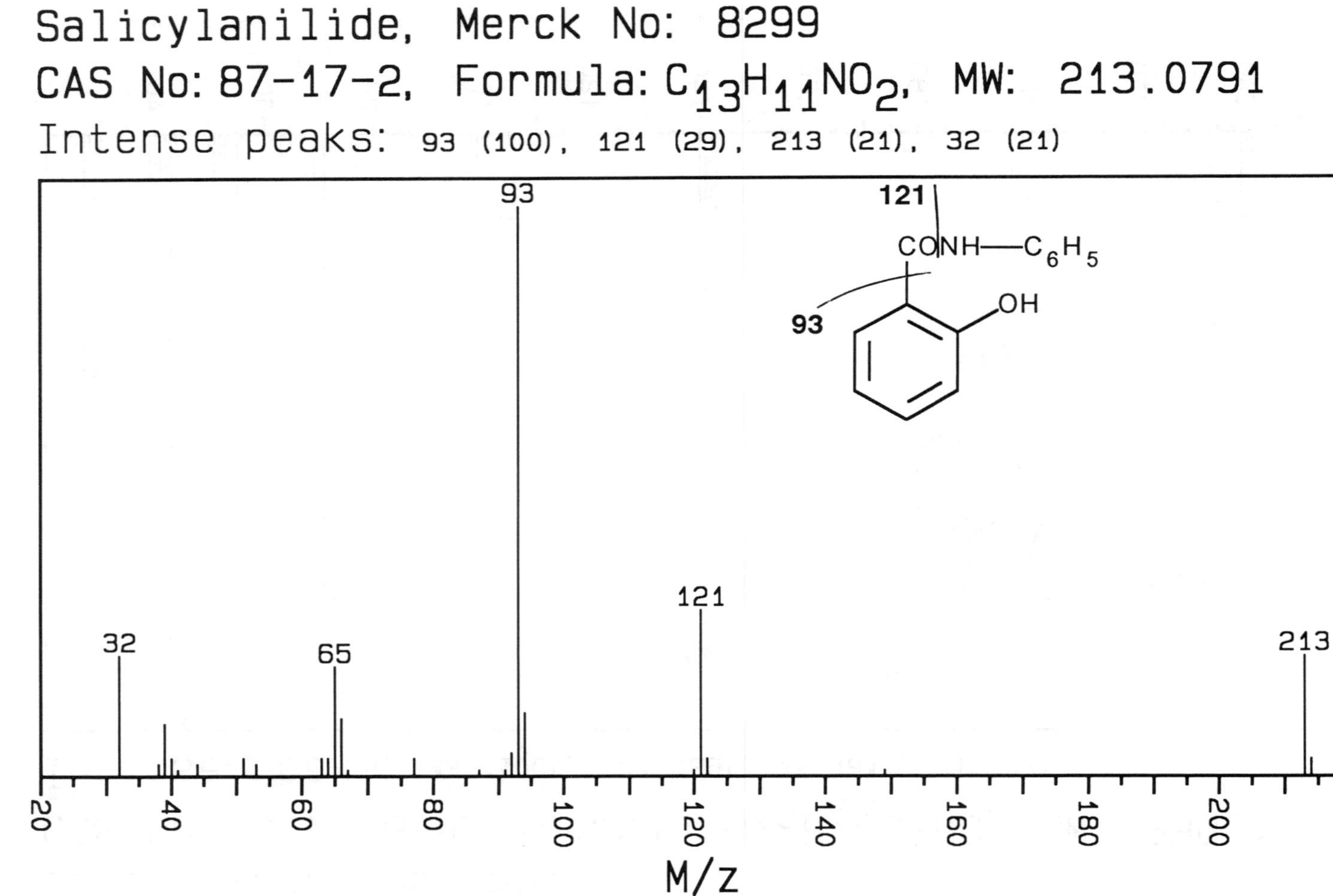
Salicylanilide, Merck No: 8299
CAS No: 87-17-2, Formula: C13H11NO2, MW: 213.0791
Intense peaks: 93 (100), 121 (29), 213 (21), 32 (21)
93
121
CONH—C6H5
93
OH
32
65
121
213
20
40
60
80
100
120
140
160
180
200
220
M/Z

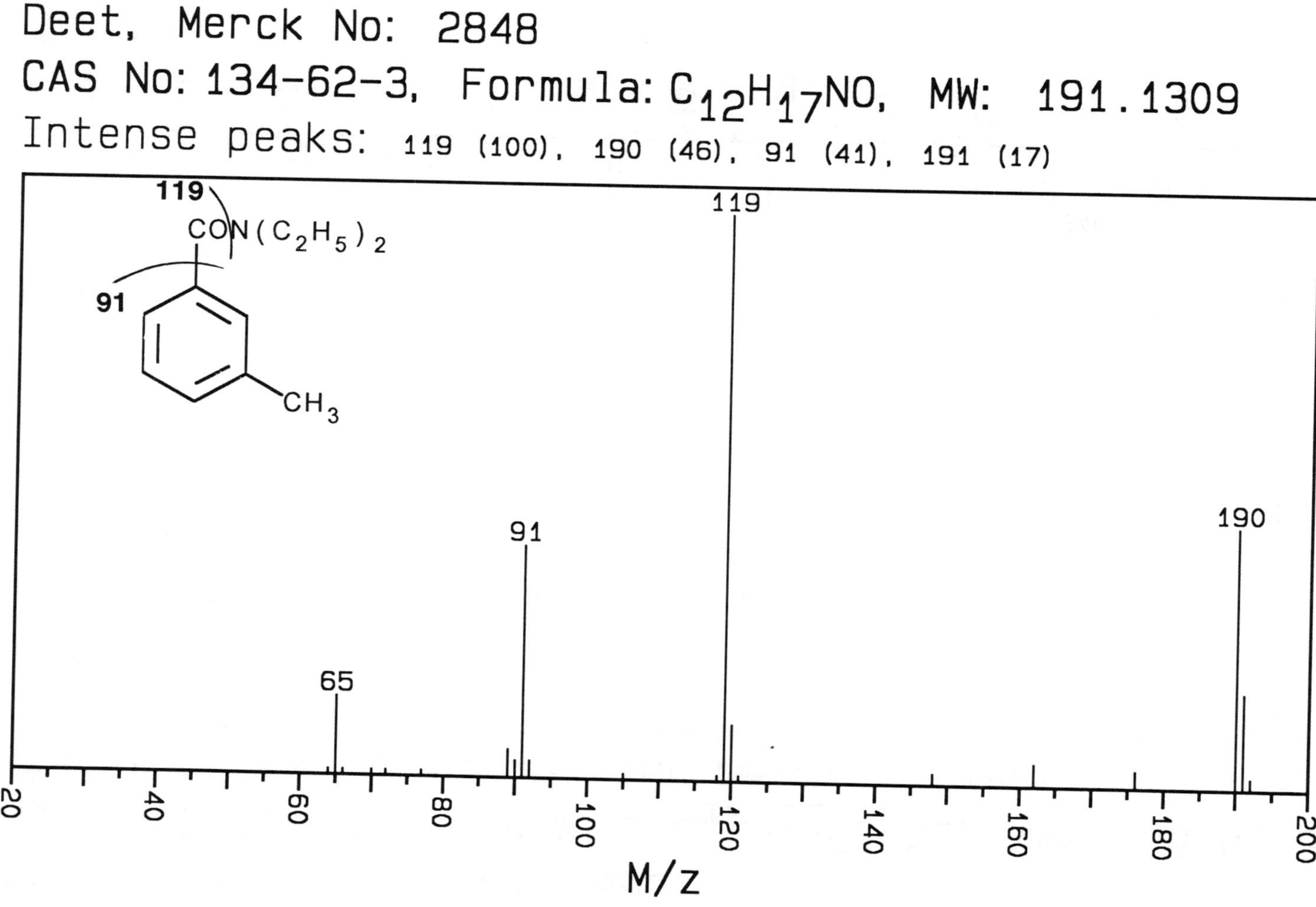

Deet, Merck No: 2848
CAS No: 134-62-3, Formula: $C_{12}H_{17}NO$, MW: 191.1309
Intense peaks: 119 (100), 190 (46), 91 (41), 191 (17)
119
$CON(C_2H_5)_2$
91
$CH_3$
119
91
65
190
20
40
60
80
100
120
140
160
180
200
M/Z

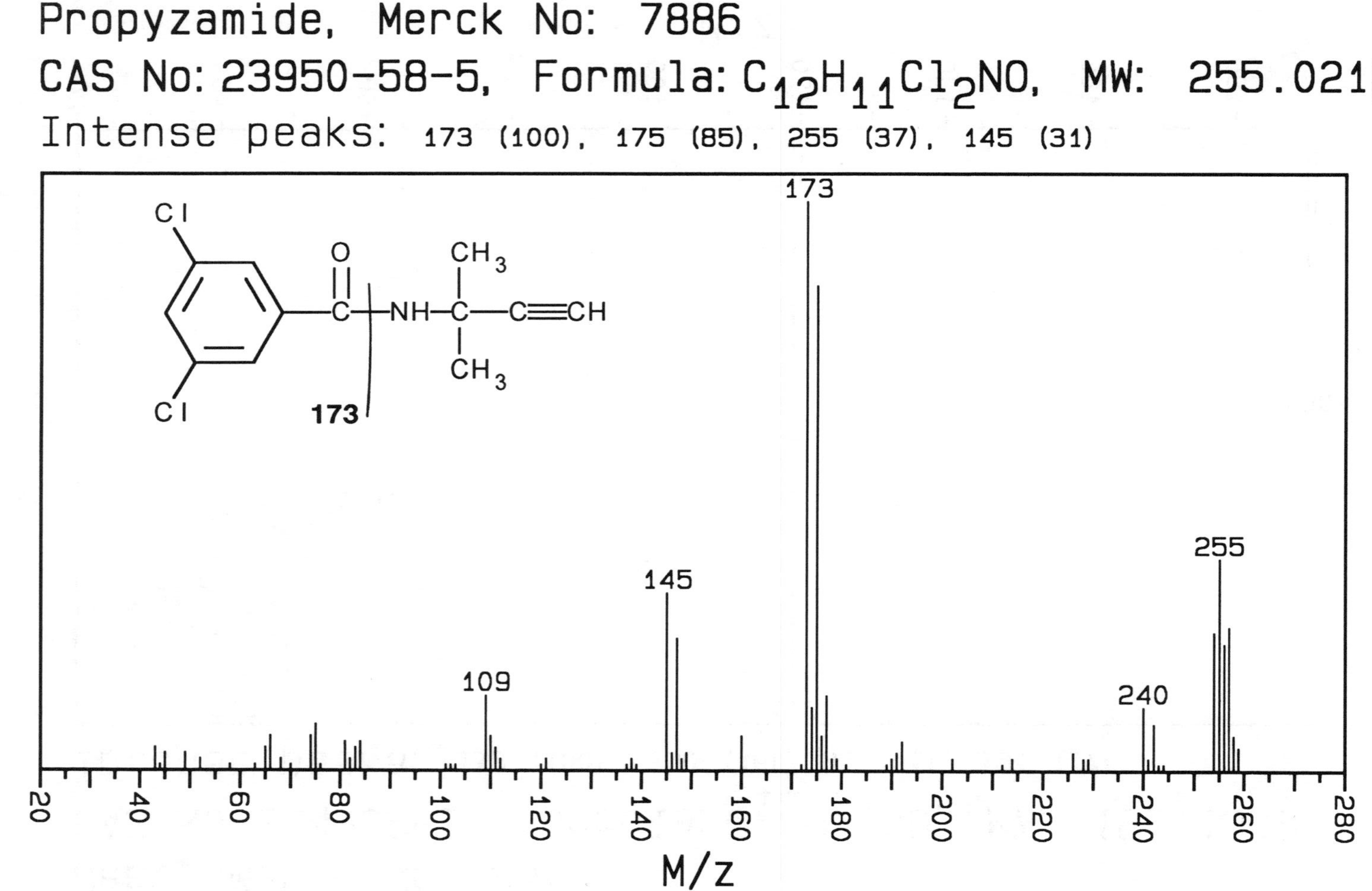
Propyzamide, Merck No: 7886
CAS No: 23950-58-5, Formula: $C_{12}H_{11}Cl_2NO$, MW: 255.0218
Intense peaks: 173 (100), 175 (85), 255 (37), 145 (31)
Cl
Cl
O
C
NH
C
CH3
CH3
C≡CH
173
173
145
109
255
240
20
40
60
80
100
120
140
160
180
200
220
240
260
280
M/Z

Carboxin, Merck No: 1832

CAS No: 5234-68-4, Formula: $C_{12}H_{13}NO_2S$, MW: 235.0667

Intense peaks: 43 (100), 143 (96), 87 (80), 235 (33)

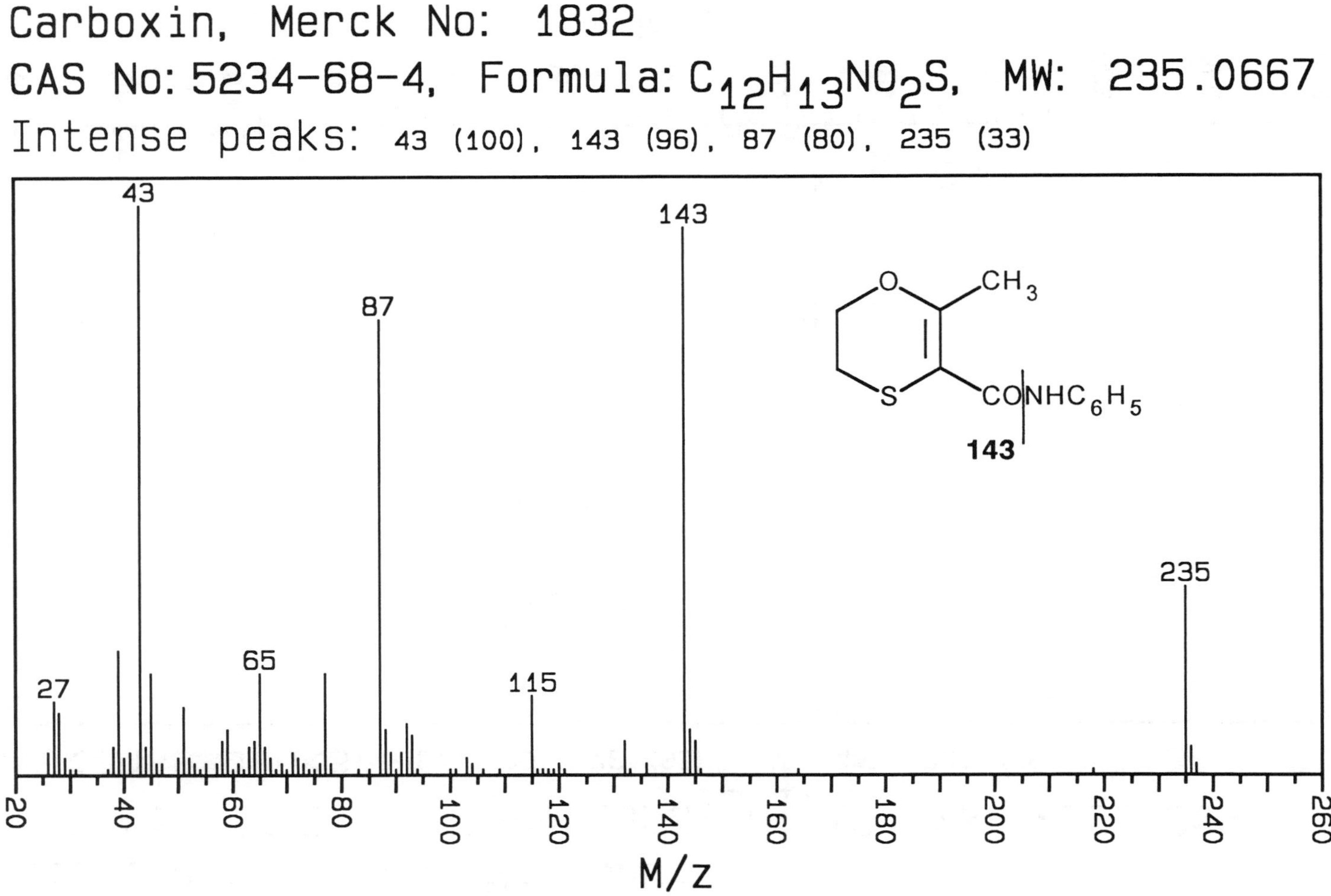

Diphenamid, Merck No: 3305

CAS No: 957-51-7, Formula: $C_{16}H_{17}NO$, MW: 239.1309

Intense peaks: 167 (100), 72 (95), 165 (44), 239 (28)

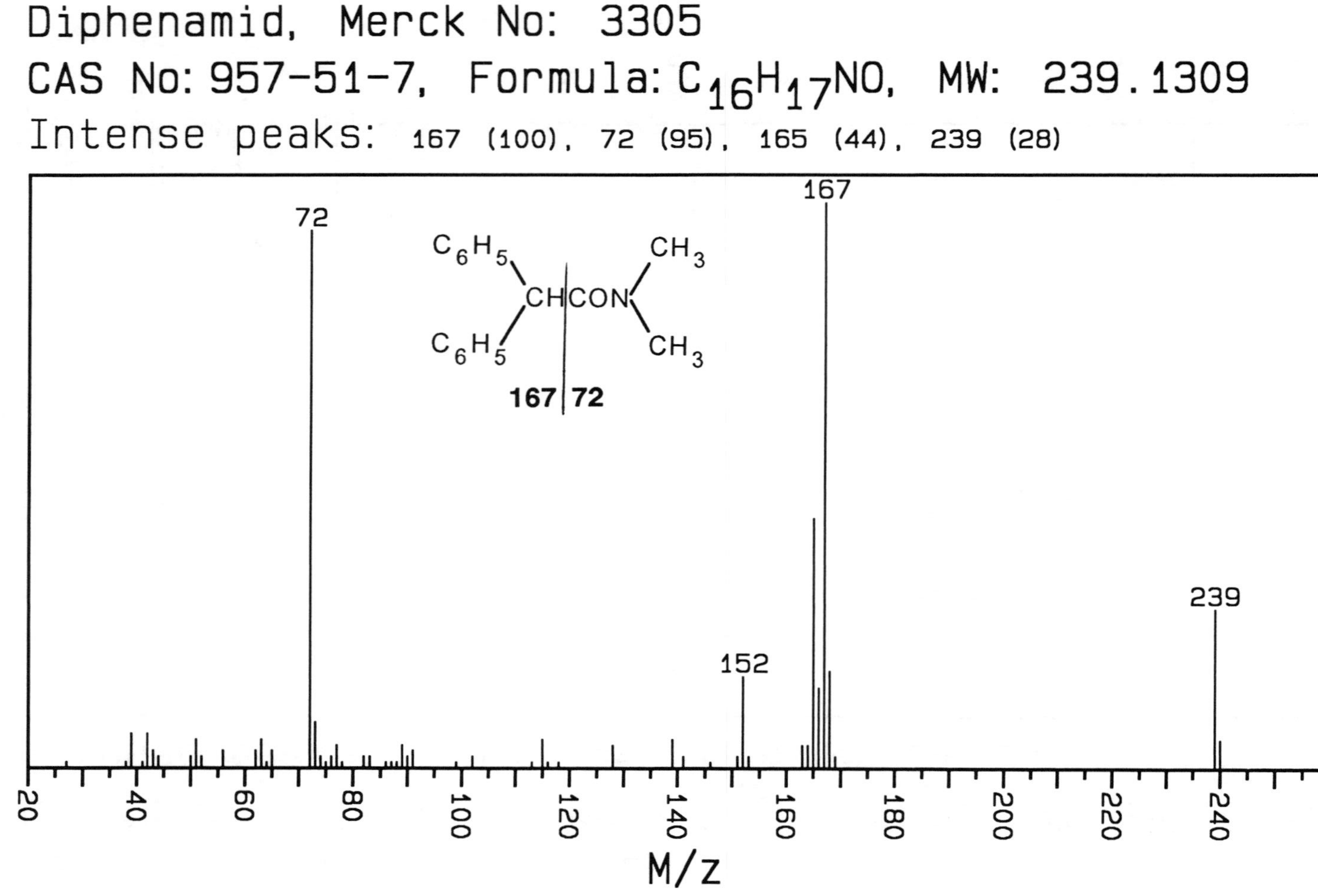

Maleanilic acid, Merck No: 5584

CAS No: 555-59-9, Formula: $C_{10}H_9NO_3$, MW: 191.0582

Intense peaks: 93 (100), 191 (33), 99 (14), 146 (11)

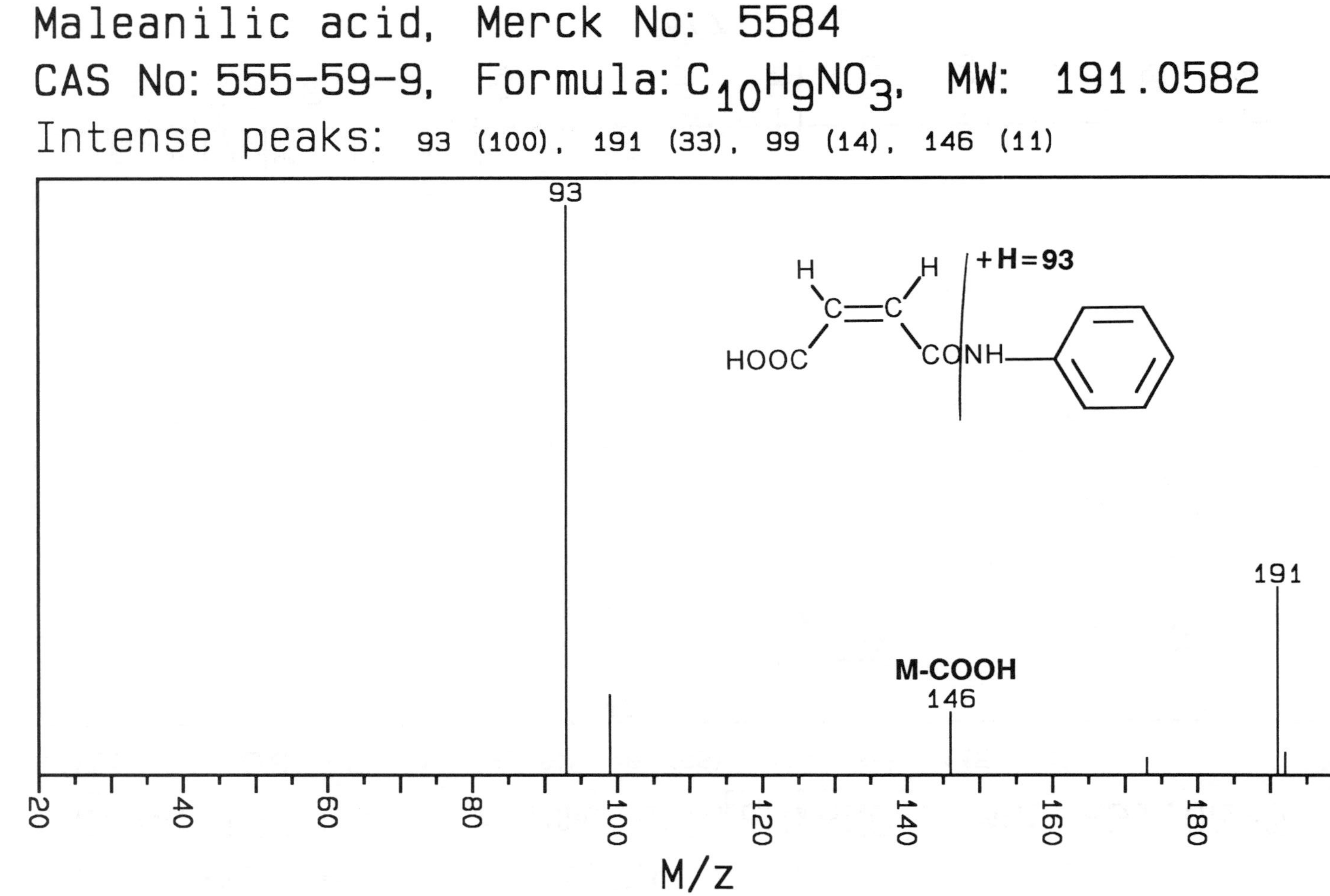

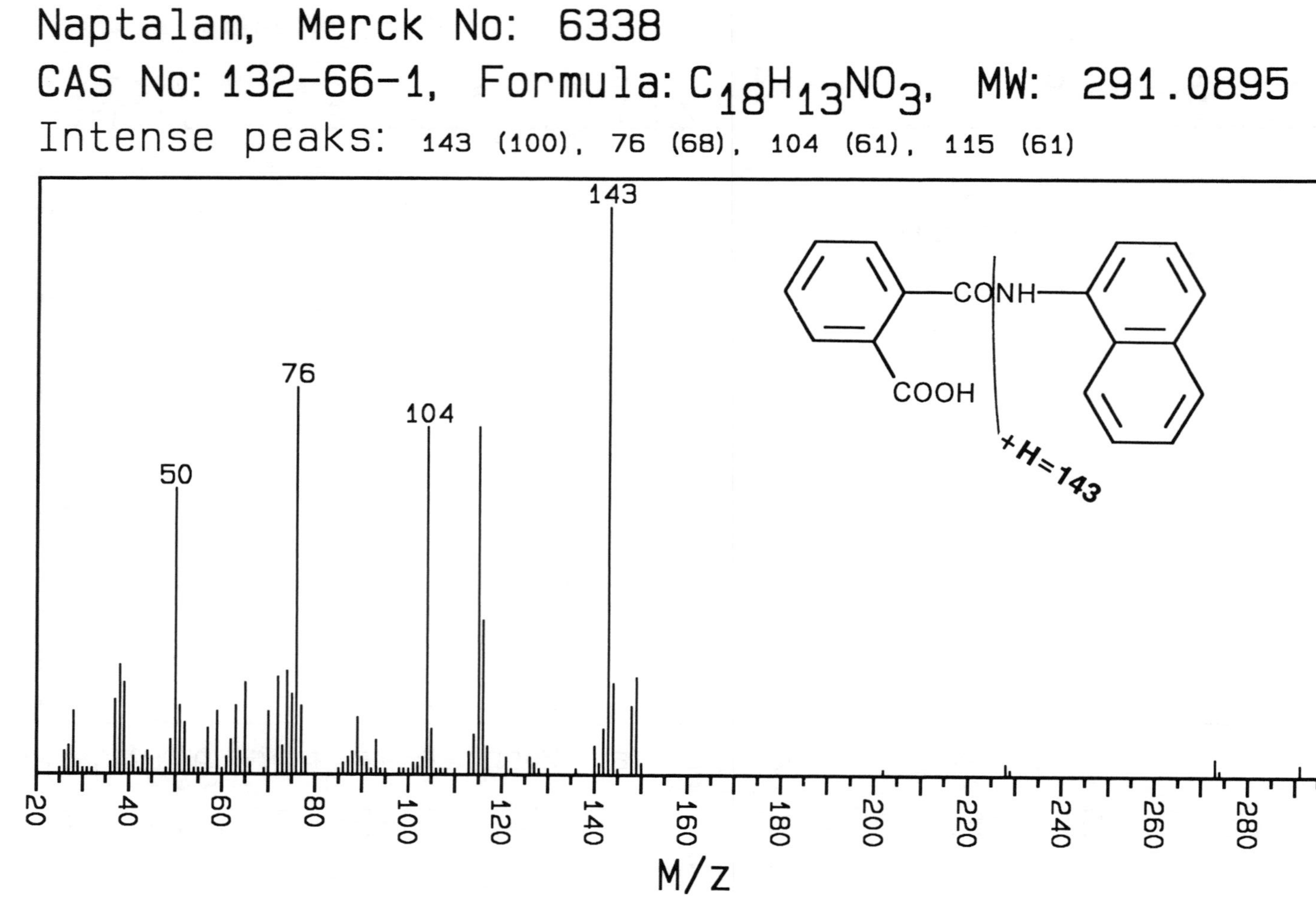

Naptalam, Merck No: 6338
CAS No: 132-66-1, Formula: $C_{18}H_{13}NO_3$, MW: 291.0895
Intense peaks: 143 (100), 76 (68), 104 (61), 115 (61)
143
76
104
50
CONH
COOH
+H=143
20
40
60
80
100
120
140
160
180
200
220
240
260
280
300
M/Z

Duraset*, Merck No: 6027

CAS No: 85-72-3, Formula: $C_{15}H_{13}NO_3$, MW: 255.0895

Intense peaks: 107 (100), 106 (77), 76 (65), 104 (61)

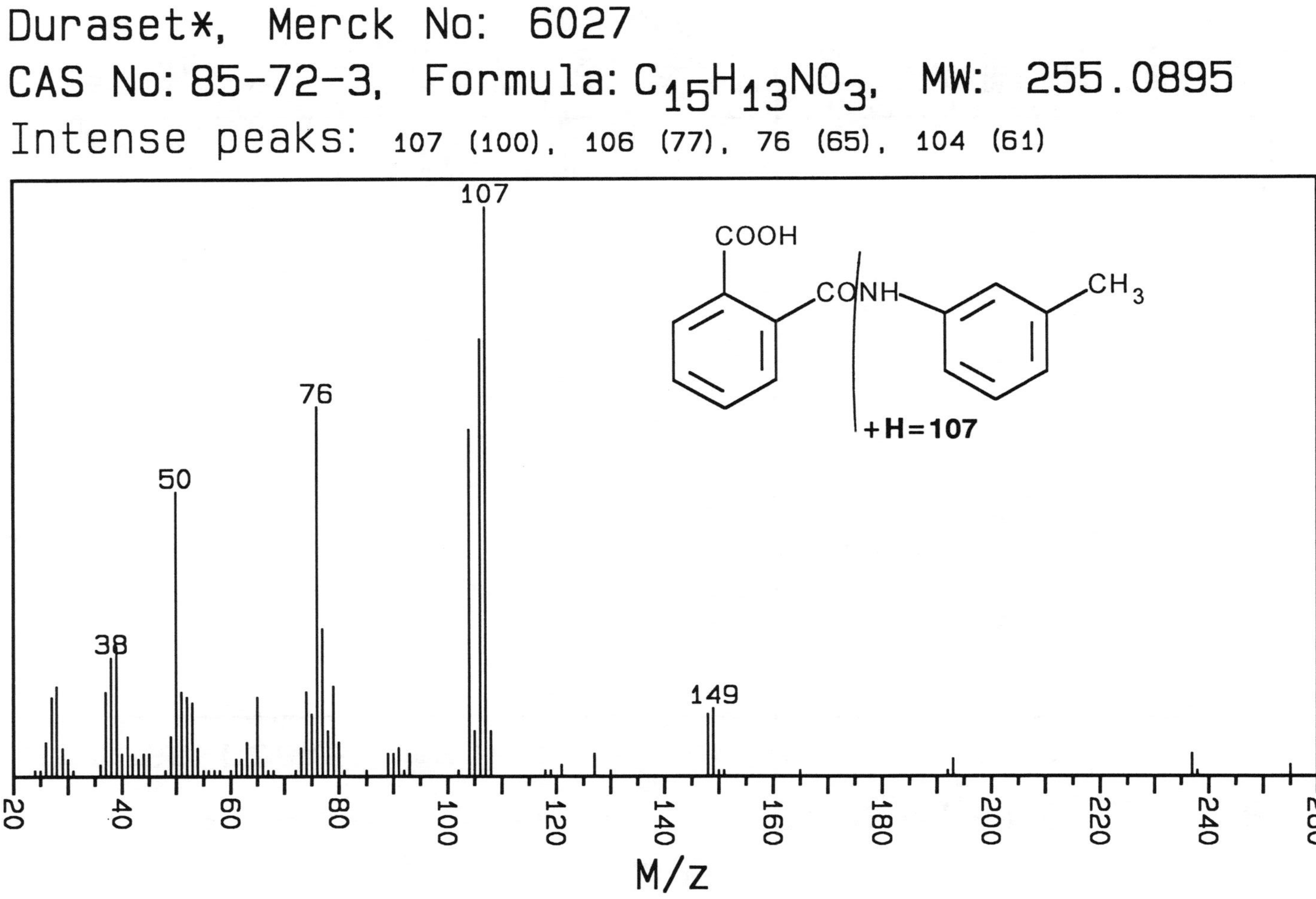

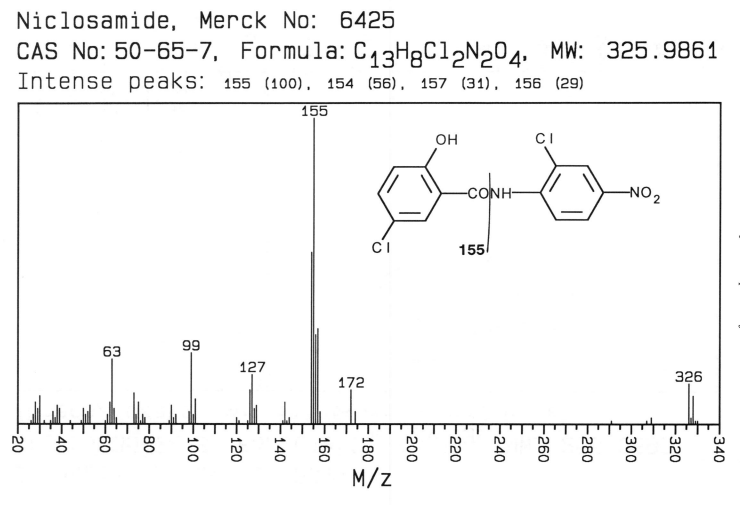

Niclosamide, Merck No: 6425

CAS No: 50-65-7, Formula: $C_{13}H_8Cl_2N_2O_4$, MW: 325.9861

Intense peaks: 155 (100), 154 (56), 157 (31), 156 (29)

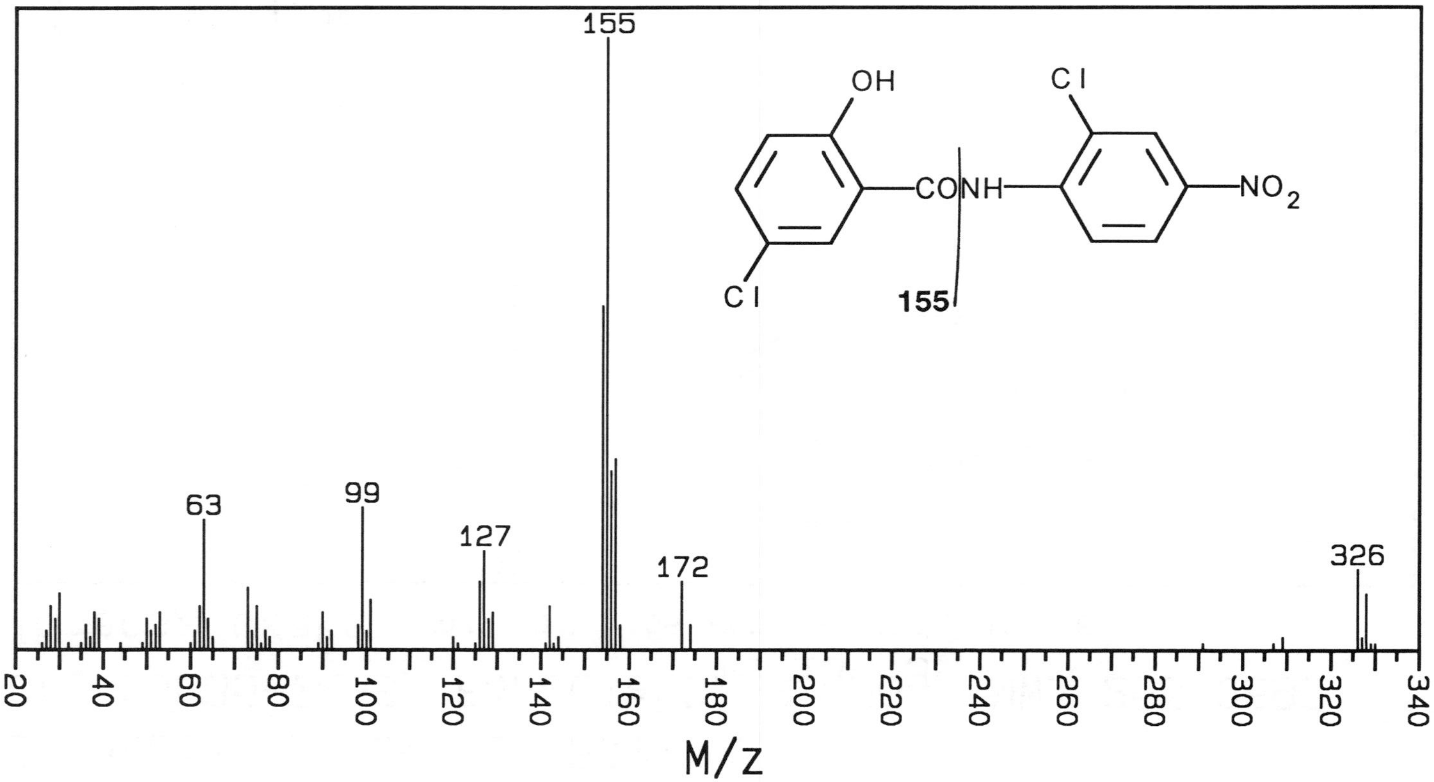

155

OH

Cl

CONH

$NO_2$

Cl

155

63

99

127

172

326

20 40 60 80 100 120 140 160 180 200 220 240 260 280 300 320 340

M/Z

Terbacil, Merck No: 9085

CAS No: 5902-51-2, Formula: $C_9H_{13}ClN_2O_2$, MW: 216.0665

Intense peaks: 161 (100), 160 (72), 41 (63), 56 (55)

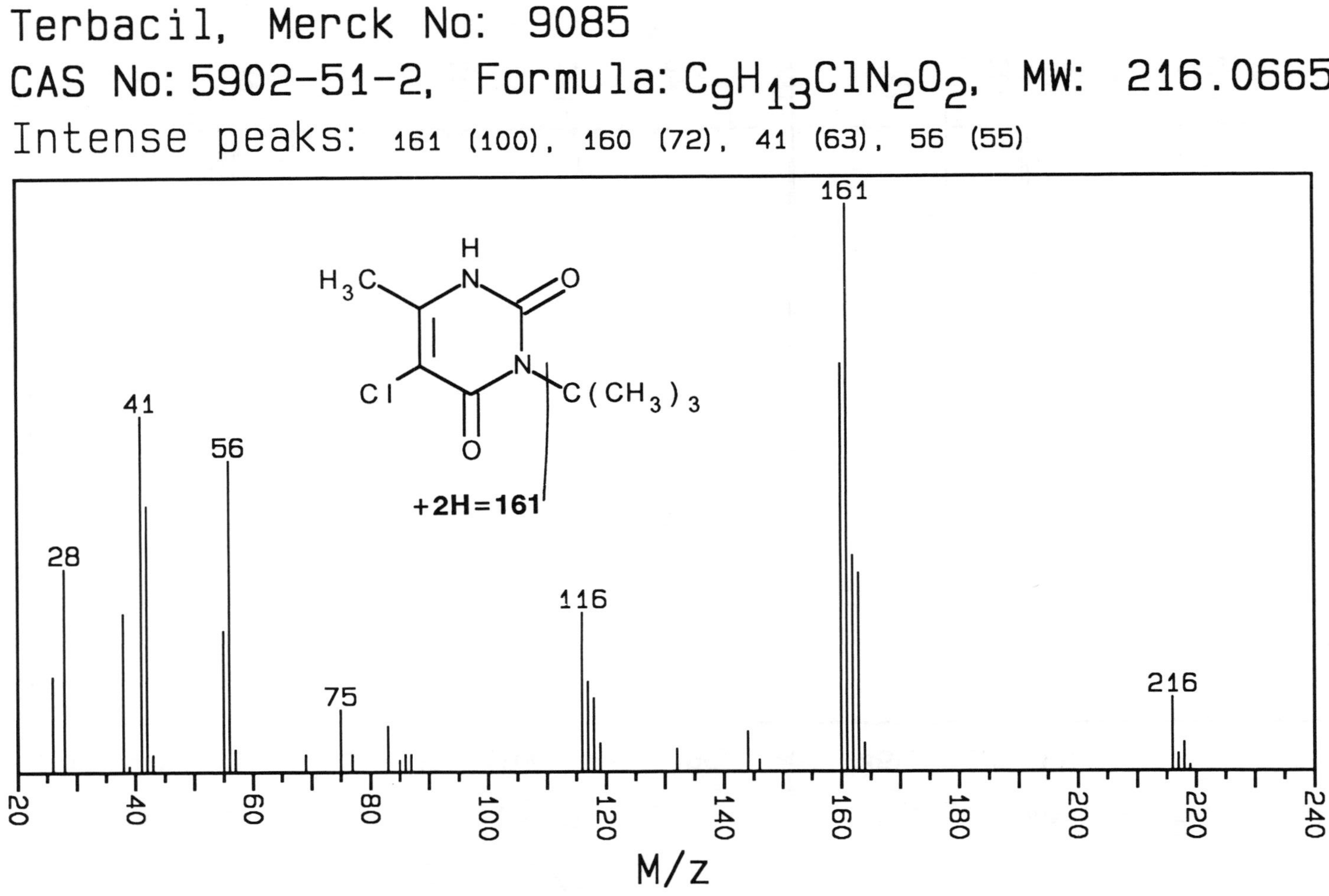

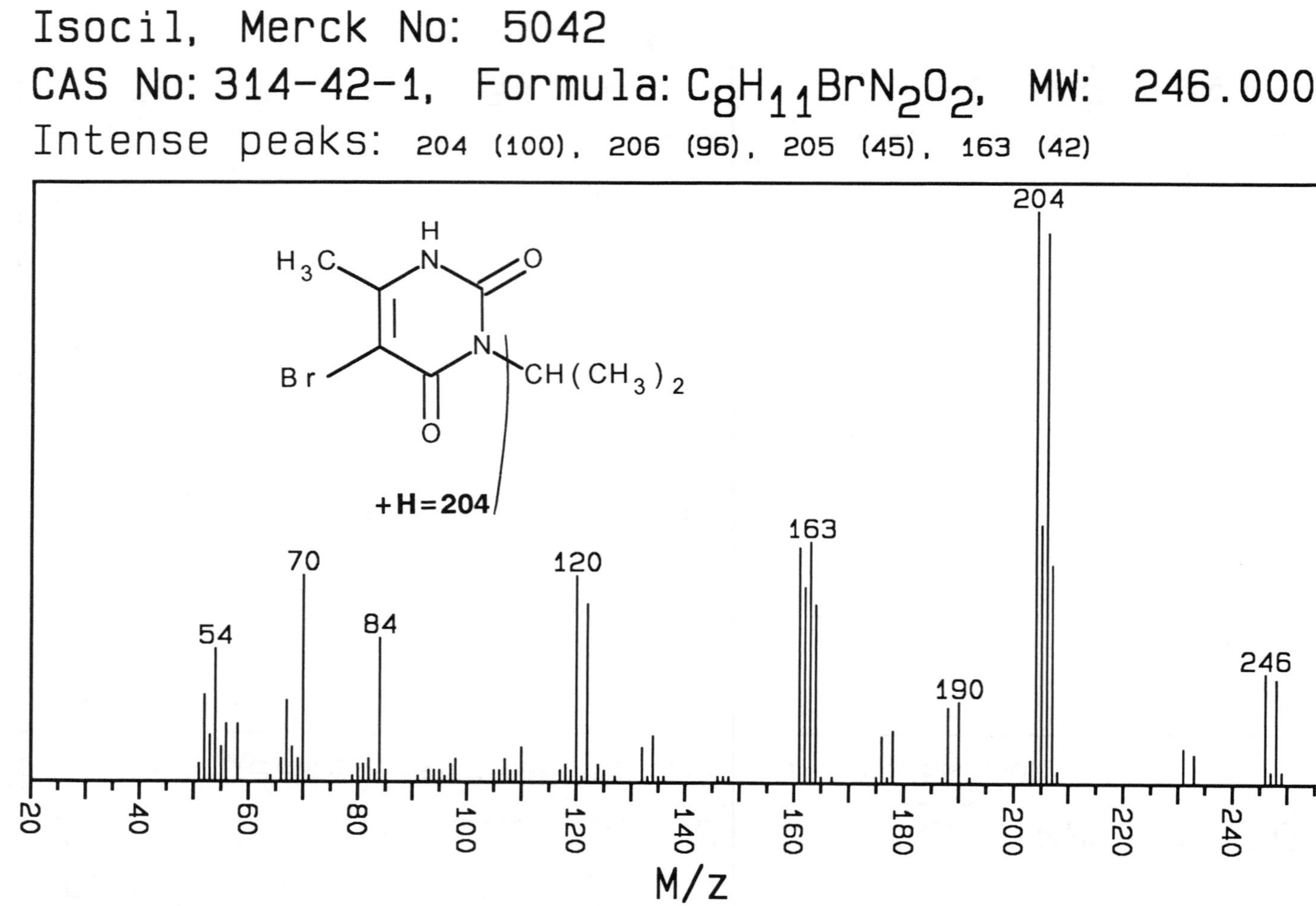
Isocil, Merck No: 5042
CAS No: 314-42-1, Formula: $C_8H_{11}BrN_2O_2$, MW: 246.0004
Intense peaks: 204 (100), 206 (96), 205 (45), 163 (42)
H
$H_3C$
N
O
N
CH(CH$_3$)$_2$
Br
O
+H=204
204
163
70
120
54
84
190
246
20
40
60
80
100
120
140
160
180
200
220
240
260
M/Z

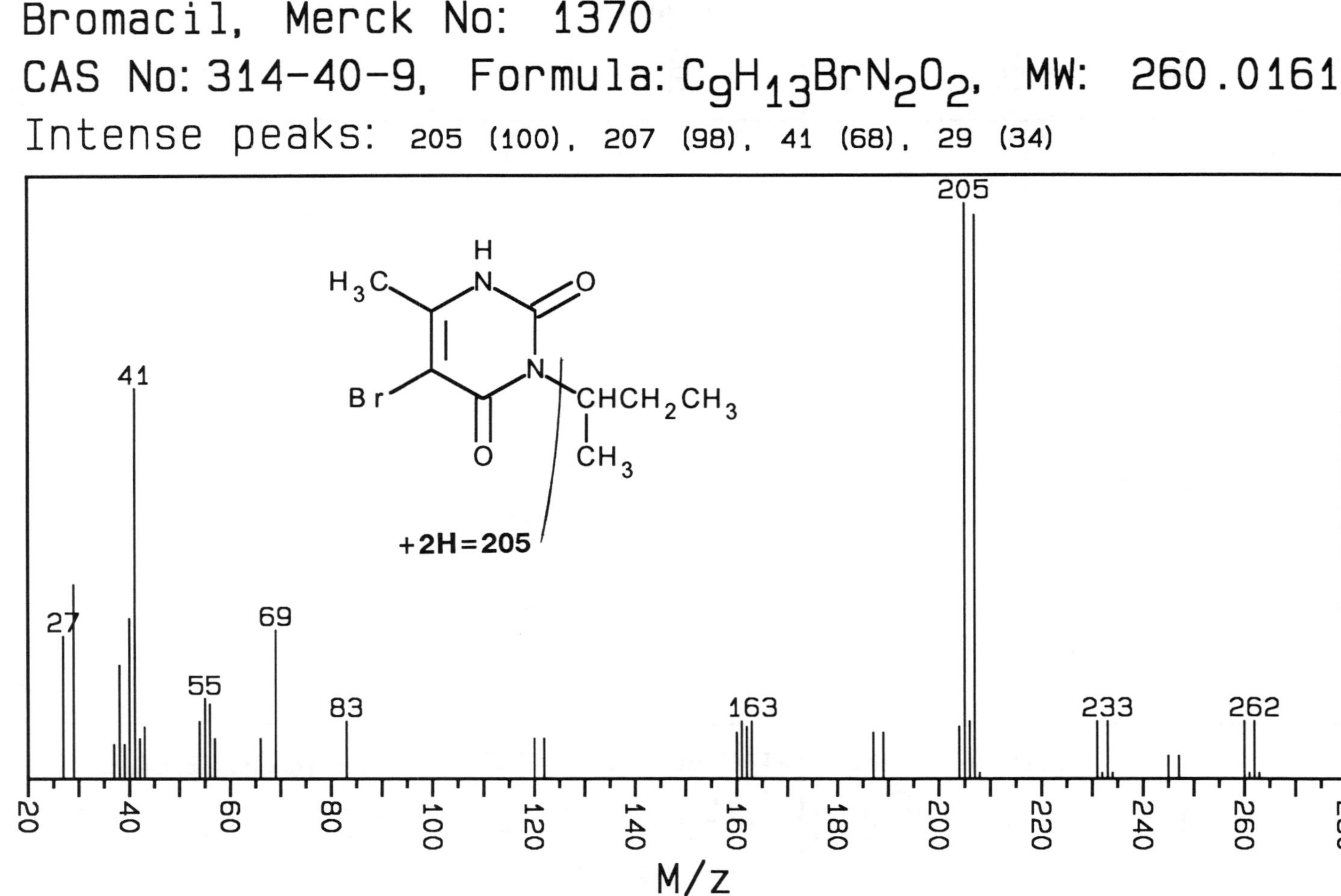

Bromacil, Merck No: 1370
CAS No: 314-40-9, Formula: C9H13BrN2O2, MW: 260.0161
Intense peaks: 205 (100), 207 (98), 41 (68), 29 (34)
205
41
27
55
69
83
163
233
262
H
H3C
N
O
N
Br
CHCH2CH3
CH3
O
+2H=205
20
40
60
80
100
120
140
160
180
200
220
240
260
280
M/Z

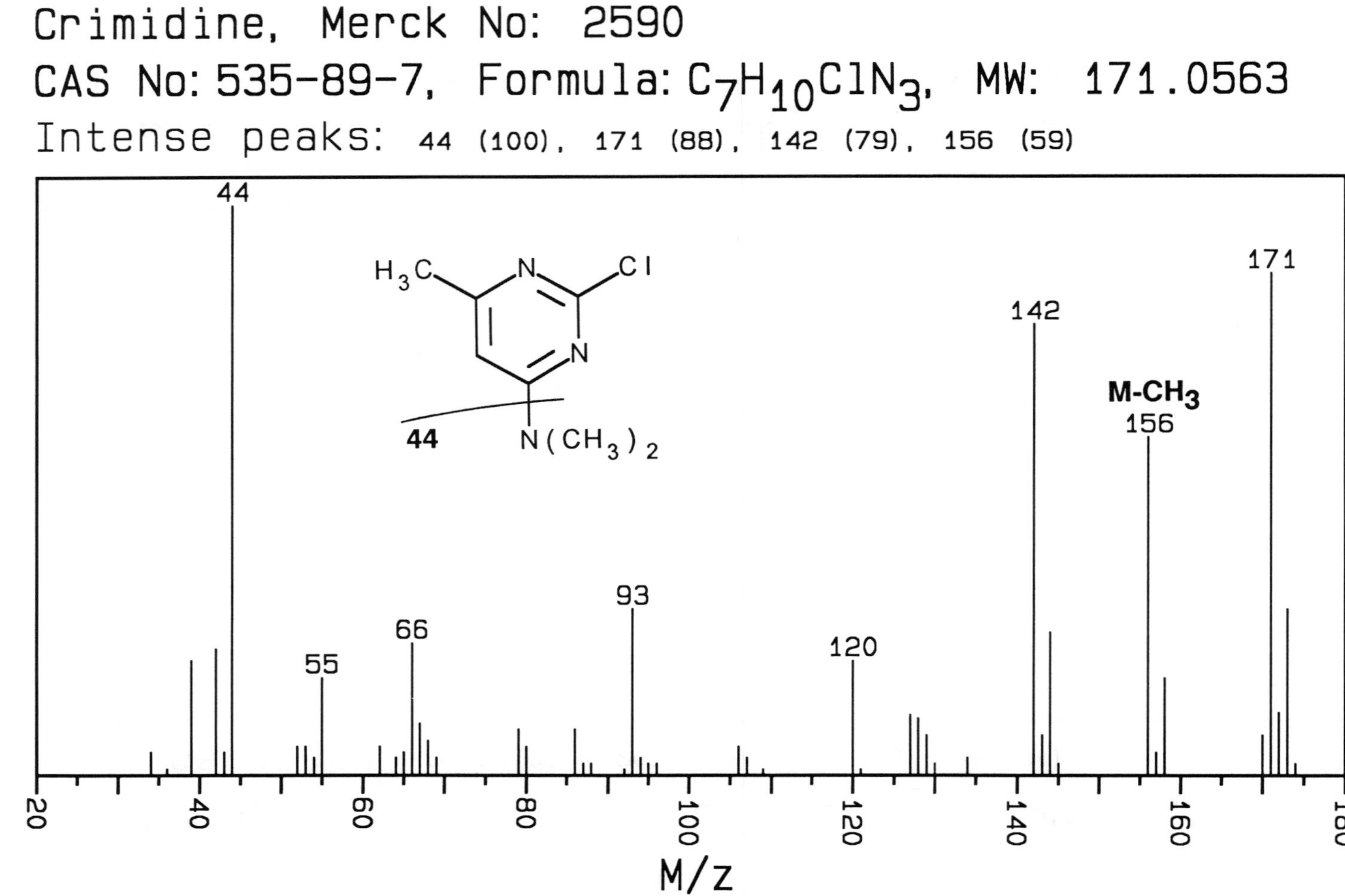
Crimidine, Merck No: 2590
CAS No: 535-89-7, Formula: $C_7H_{10}ClN_3$, MW: 171.0563
Intense peaks: 44 (100), 171 (88), 142 (79), 156 (59)
$H_3C$
N
Cl
N
44
$N(CH_3)_2$
44
55
66
93
120
142
M-$CH_3$
156
171
20
40
60
80
100
120
140
160
180
M/Z

Dimethirimol, Merck No: 3206

CAS No: 5221-53-4, Formula: $C_{11}H_{19}N_3O$, MW: 209.1528

Intense peaks: 166 (100), 32 (18), 209 (14), 96 (13)

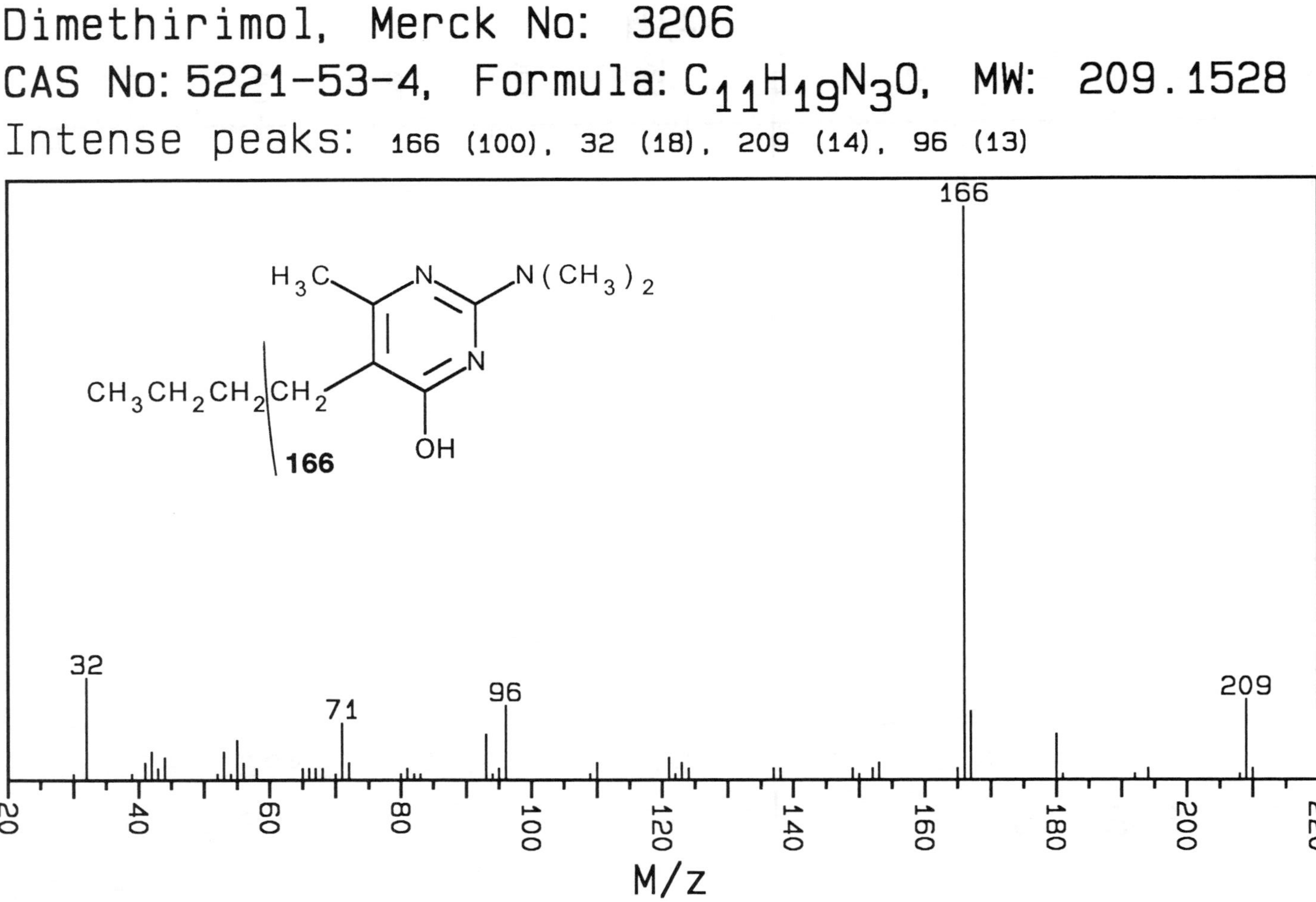

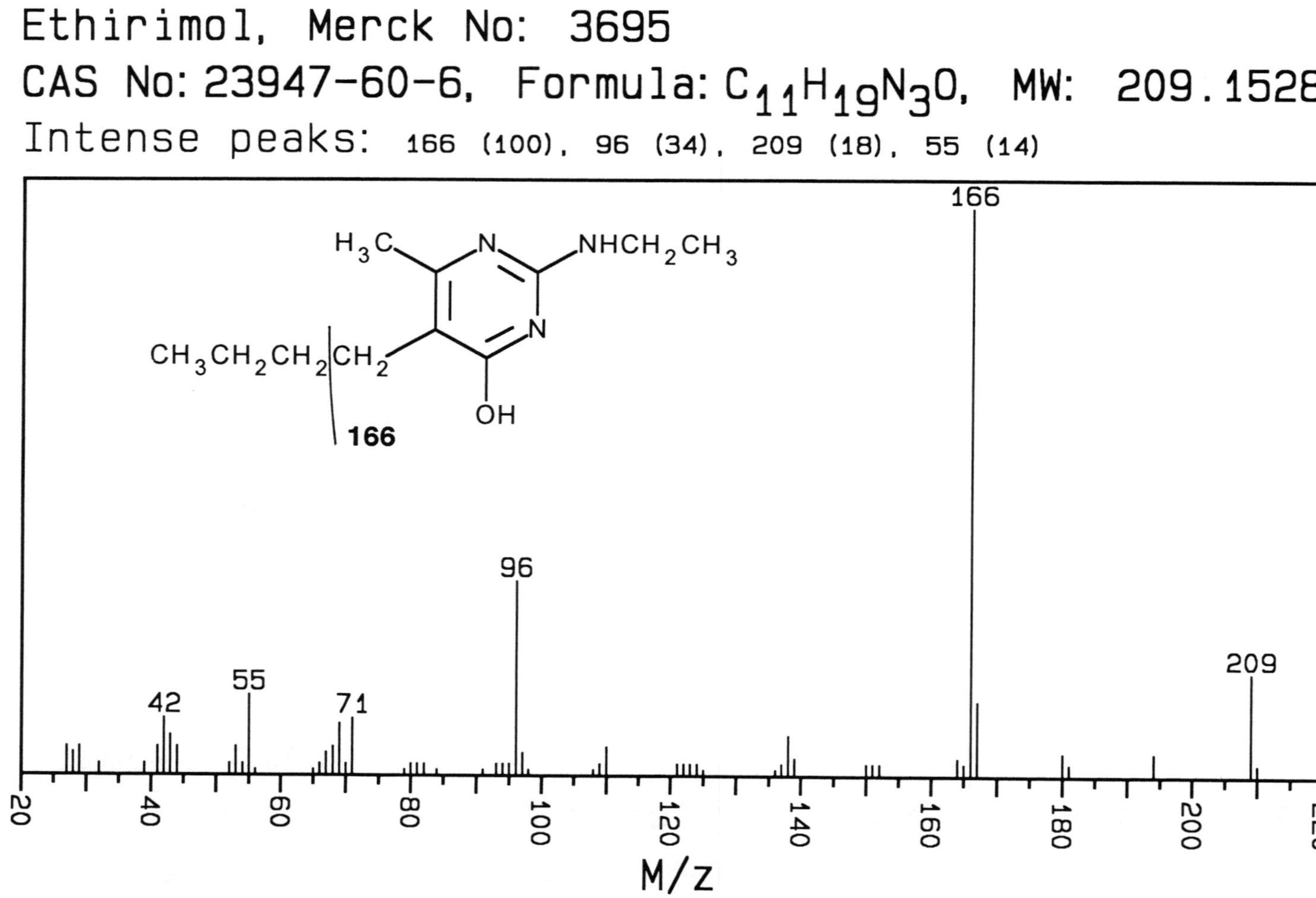
Ethirimol, Merck No: 3695
CAS No: 23947-60-6, Formula: C11H19N3O, MW: 209.1528
Intense peaks: 166 (100), 96 (34), 209 (18), 55 (14)
H3C
N
NHCH2CH3
N
CH3CH2CH2CH2
OH
166
166
96
209
42
55
71
20
40
60
80
100
120
140
160
180
200
220
M/Z

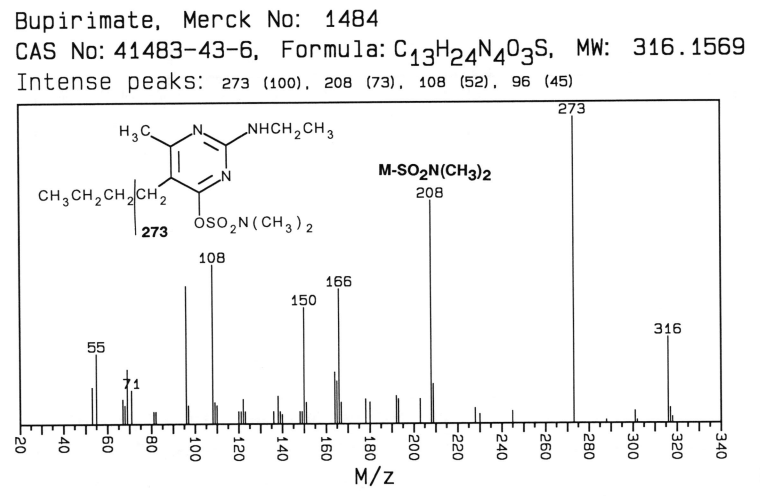

Bupirimate, Merck No: 1484

CAS No: 41483-43-6, Formula: $C_{13}H_{24}N_4O_3S$, MW: 316.1569

Intense peaks: 273 (100), 208 (73), 108 (52), 96 (45)

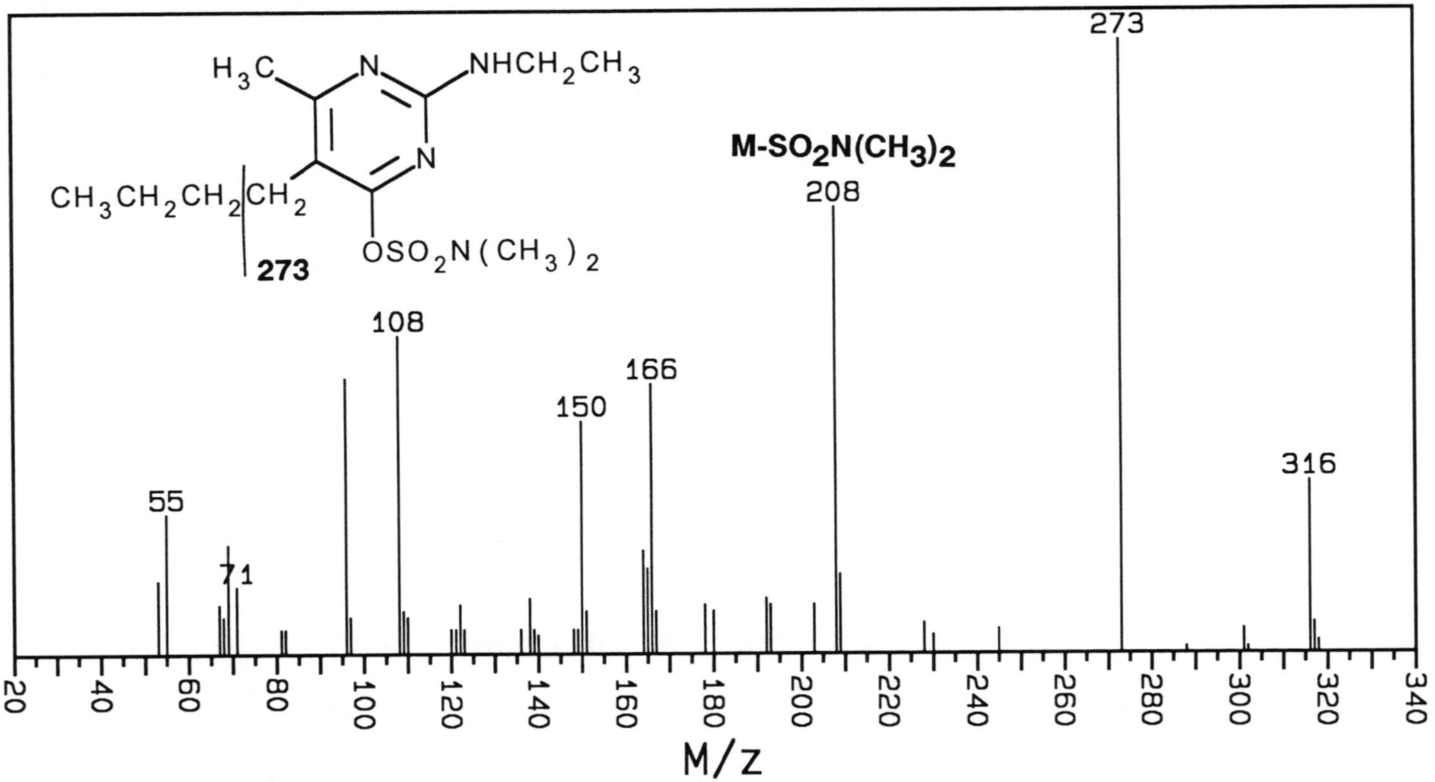

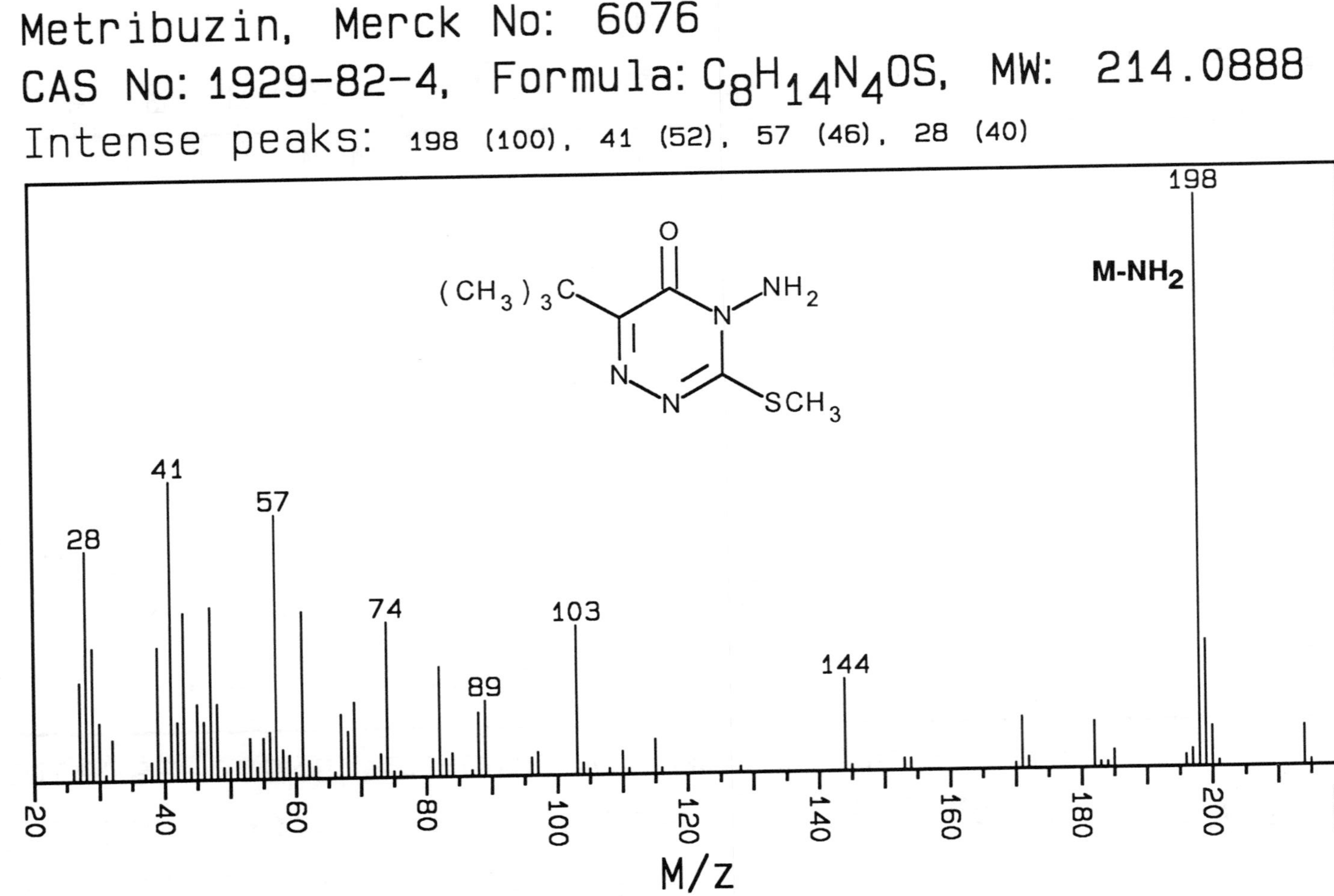
Metribuzin, Merck No: 6076
CAS No: 1929-82-4, Formula: $C_8H_{14}N_4OS$, MW: 214.0888
Intense peaks: 198 (100), 41 (52), 57 (46), 28 (40)
198
M-NH2
O
$(CH_3)_3C$
$NH_2$
N
N
N
$SCH_3$
41
57
28
74
103
89
144
20
40
60
80
100
120
140
160
180
200
220
M/Z

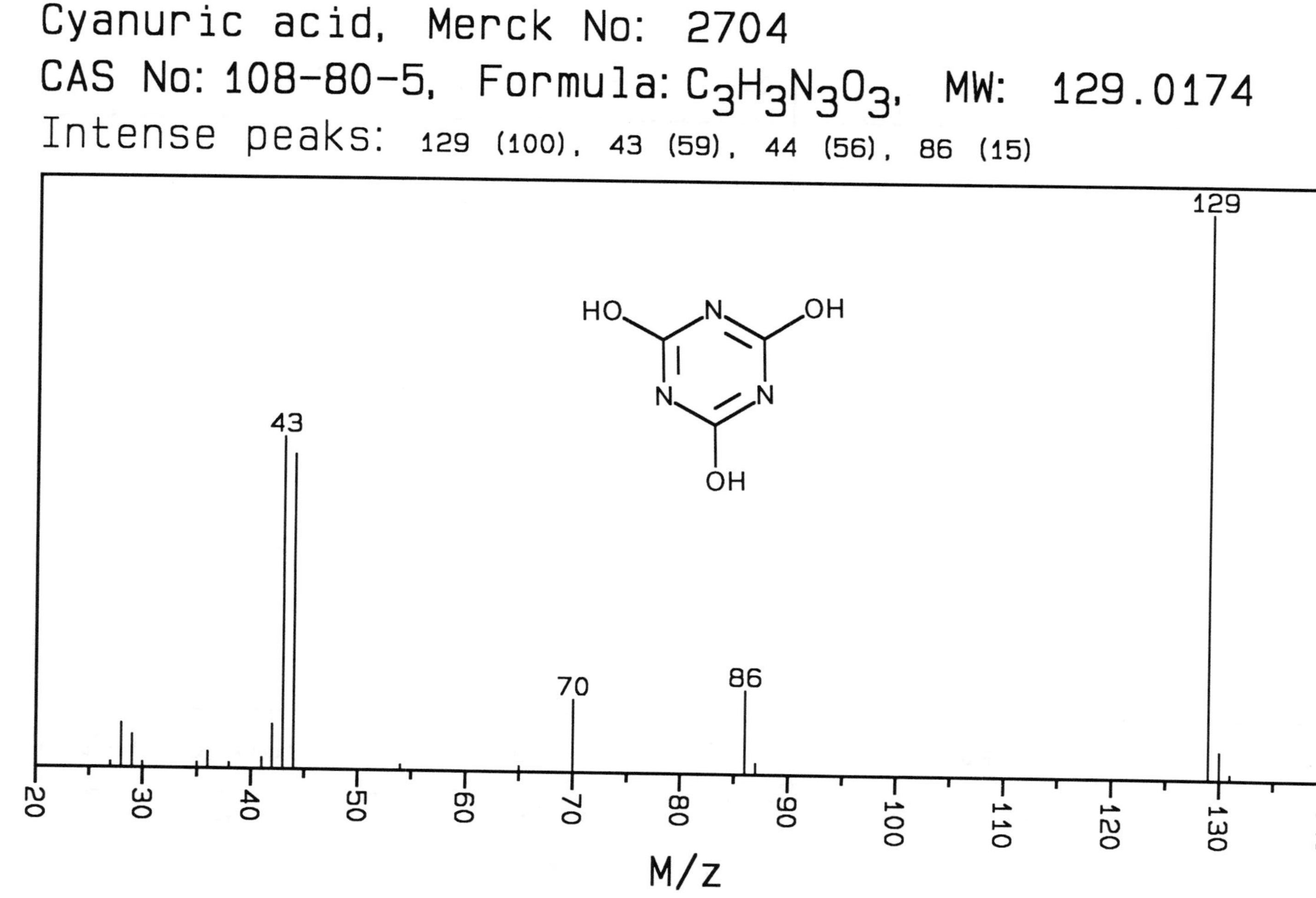
Cyanuric acid, Merck No: 2704
CAS No: 108-80-5, Formula: $C_3H_3N_3O_3$, MW: 129.0174
Intense peaks: 129 (100), 43 (59), 44 (56), 86 (15)
129
43
70
86
HO
N
OH
N
N
OH
20
30
40
50
60
70
80
90
100
110
120
130
140
M/Z

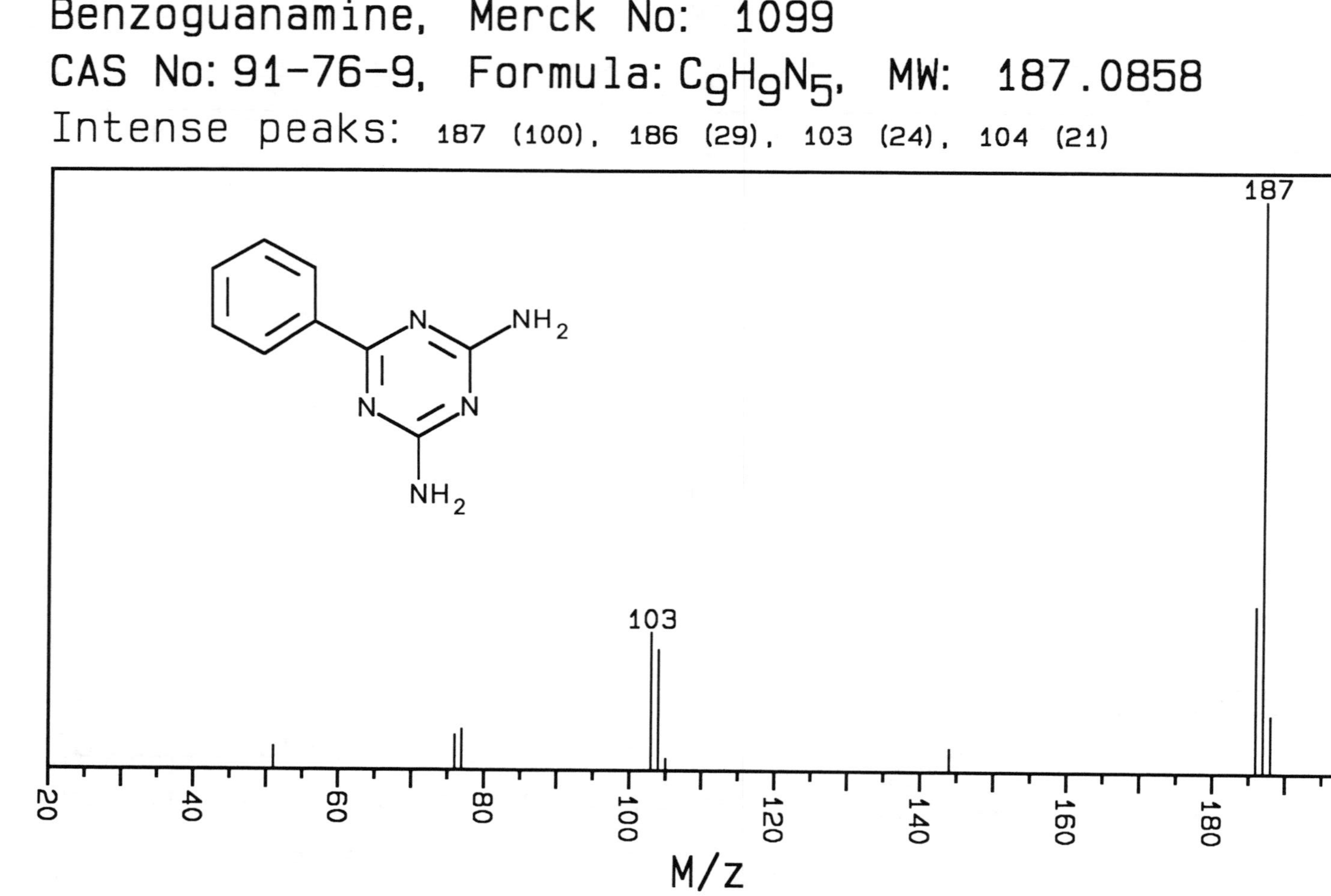
Benzoguanamine, Merck No: 1099
CAS No: 91-76-9, Formula: $C_9H_9N_5$, MW: 187.0858
Intense peaks: 187 (100), 186 (29), 103 (24), 104 (21)
187
103
N
N
N
NH2
NH2
20
40
60
80
100
120
140
160
180
200
M/Z

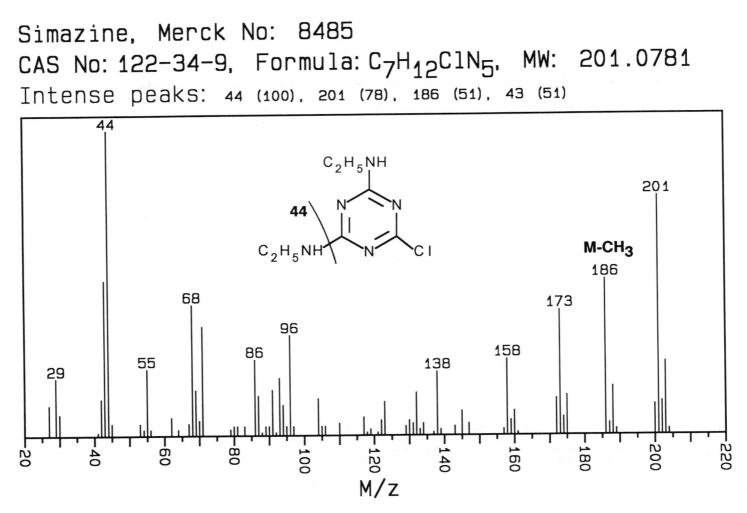

Simazine, Merck No: 8485

CAS No: 122-34-9, Formula: $C_7H_{12}ClN_5$, MW: 201.0781

Intense peaks: 44 (100), 201 (78), 186 (51), 43 (51)

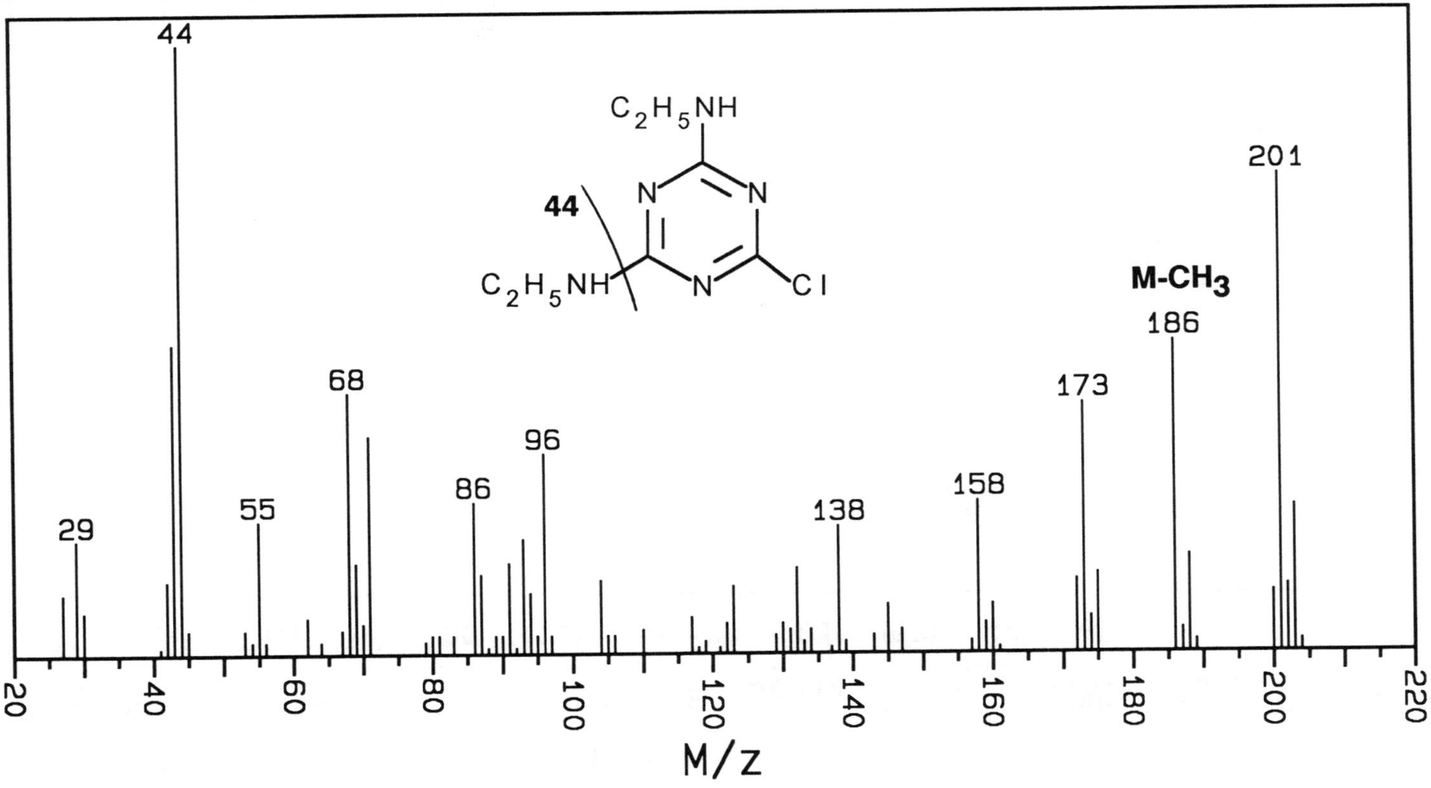

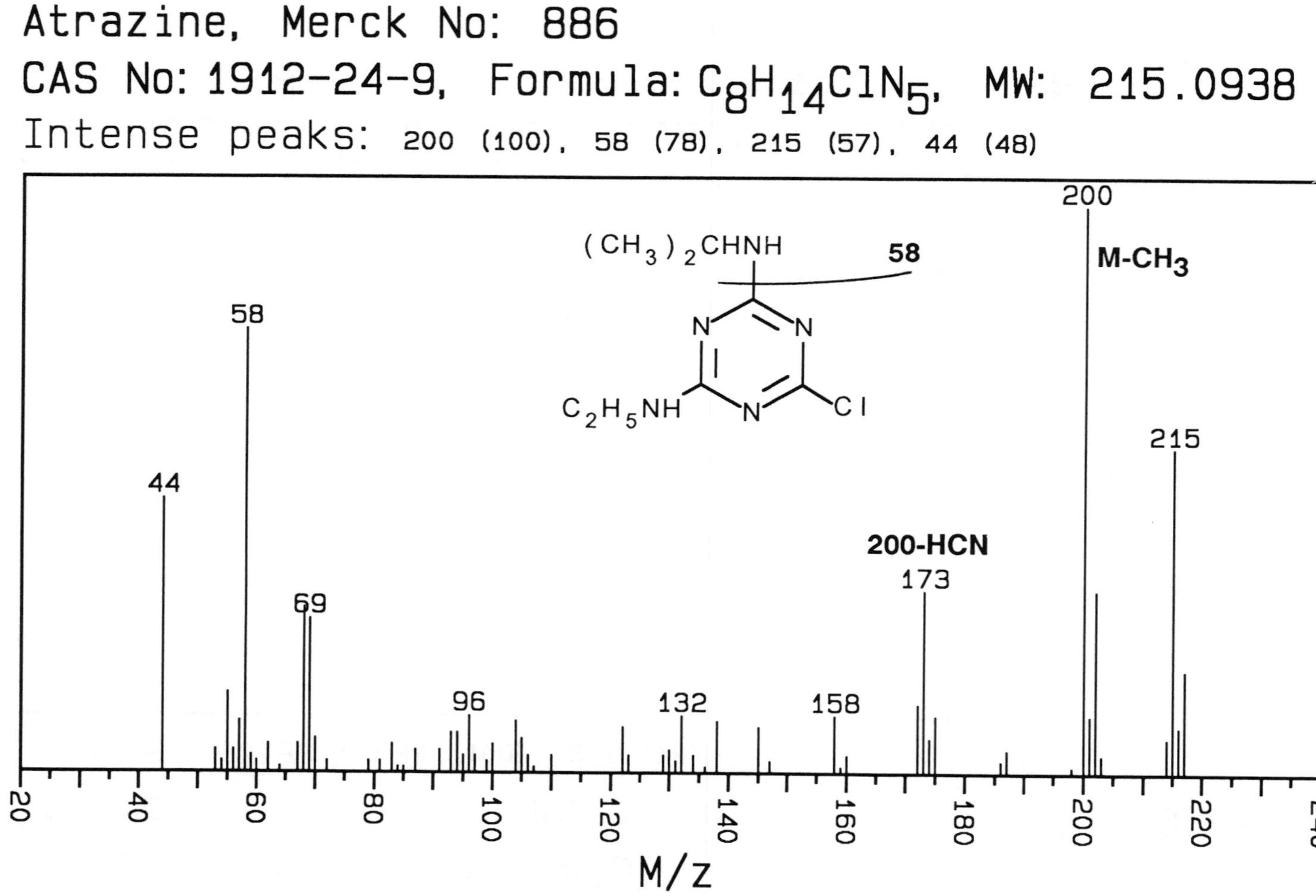

Atrazine, Merck No: 886
CAS No: 1912-24-9, Formula: $C_8H_{14}ClN_5$, MW: 215.0938
Intense peaks: 200 (100), 58 (78), 215 (57), 44 (48)
(CH3)2CHNH
58
N
N
N
C2H5NH
Cl
M-CH3
200-HCN
44
58
69
96
132
158
173
200
215
20
40
60
80
100
120
140
160
180
200
220
240
M/Z

Simetryne*, Merck No: 8488

CAS No: 1014-70-6, Formula: $C_8H_{15}N_5S$, MW: 213.1048

Intense peaks: 213 (100), 68 (74), 71 (59), 170 (58)

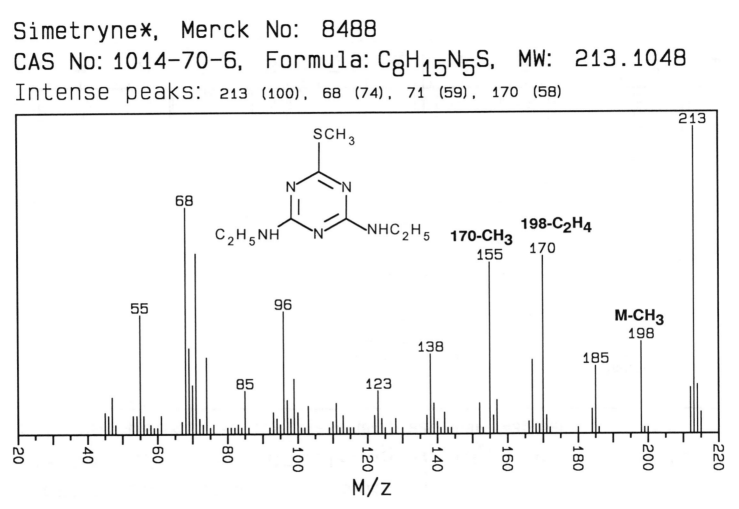

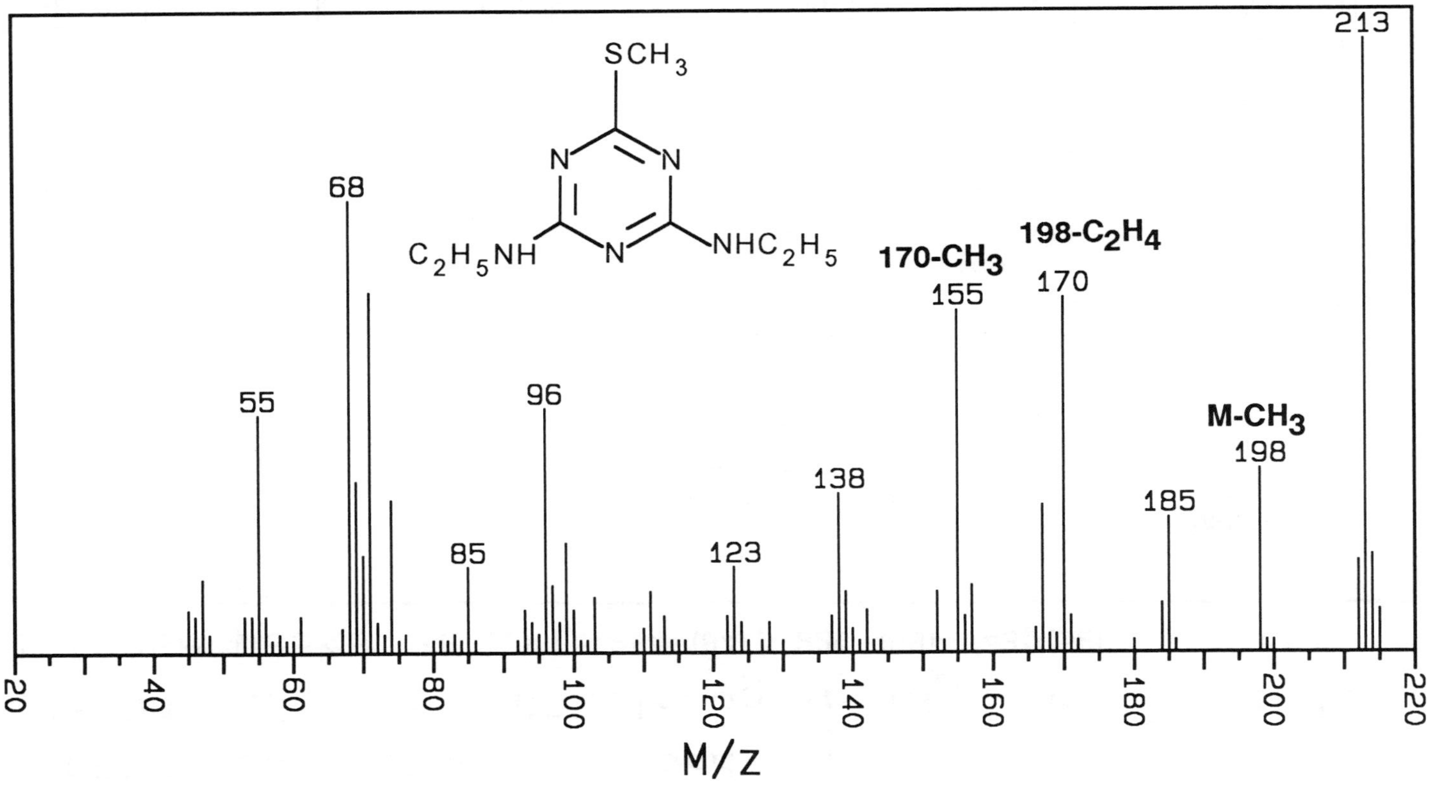

$SCH_3$

N N

$C_2H_5NH$ N $NHC_2H_5$

213

68

55

96

85

123

138

**170-$CH_3$**
155

**198-$C_2H_4$**
170

185

**M-$CH_3$**
198

20 40 60 80 100 120 140 160 180 200 220

M/Z

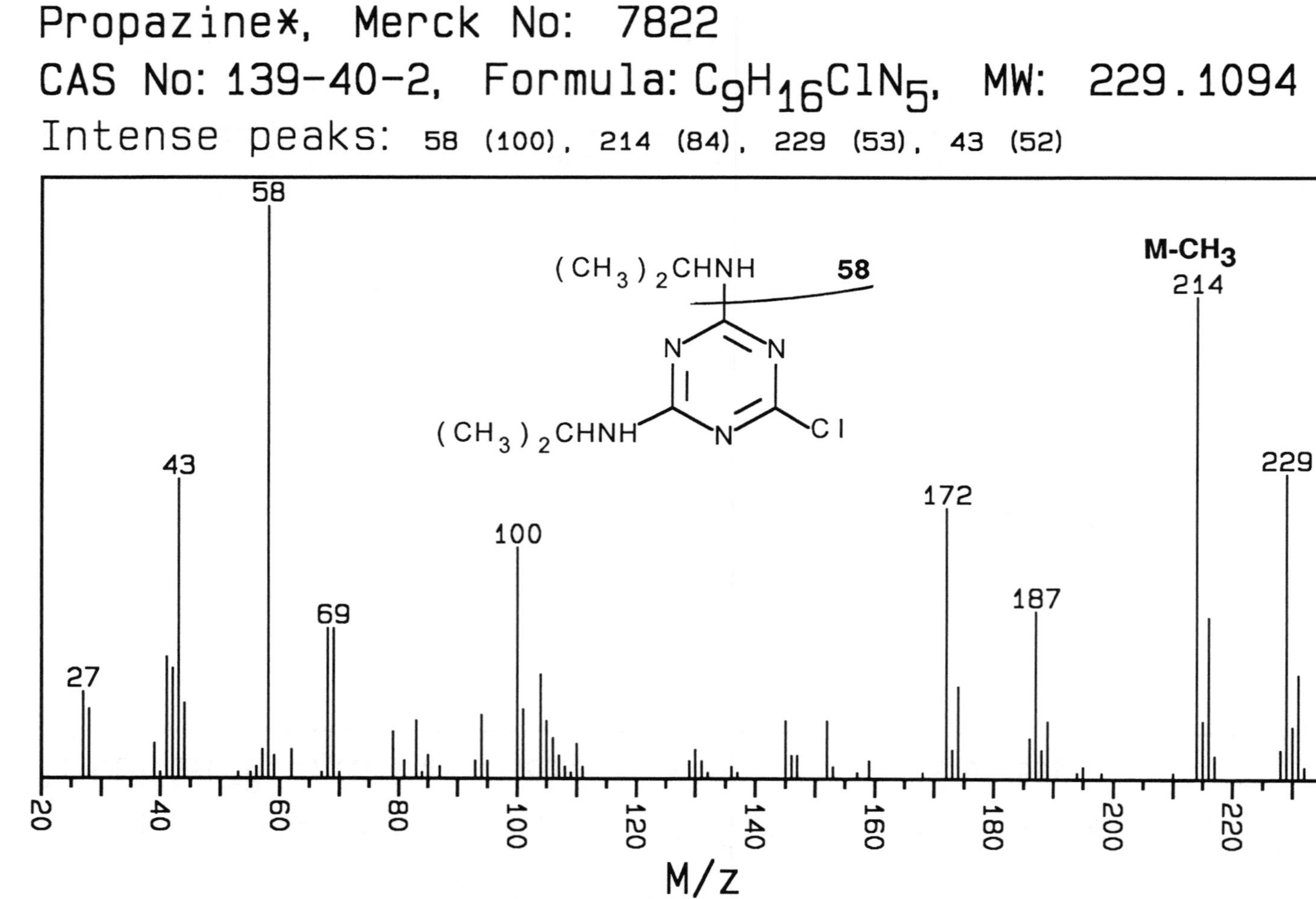
Propazine*, Merck No: 7822
CAS No: 139-40-2, Formula: C9H16ClN5, MW: 229.1094
Intense peaks: 58 (100), 214 (84), 229 (53), 43 (52)
58
(CH3)2CHNH
58
M-CH3
214
N
N
(CH3)2CHNH
N
Cl
43
229
172
100
187
69
27
20
40
60
80
100
120
140
160
180
200
220
240
M/Z

Trietazine, Merck No: 9580

CAS No: 1912-26-1, Formula: $C_9H_{16}ClN_5$, MW: 229.1094

Intense peaks: 200 (100), 186 (61), 229 (60), 214 (59)

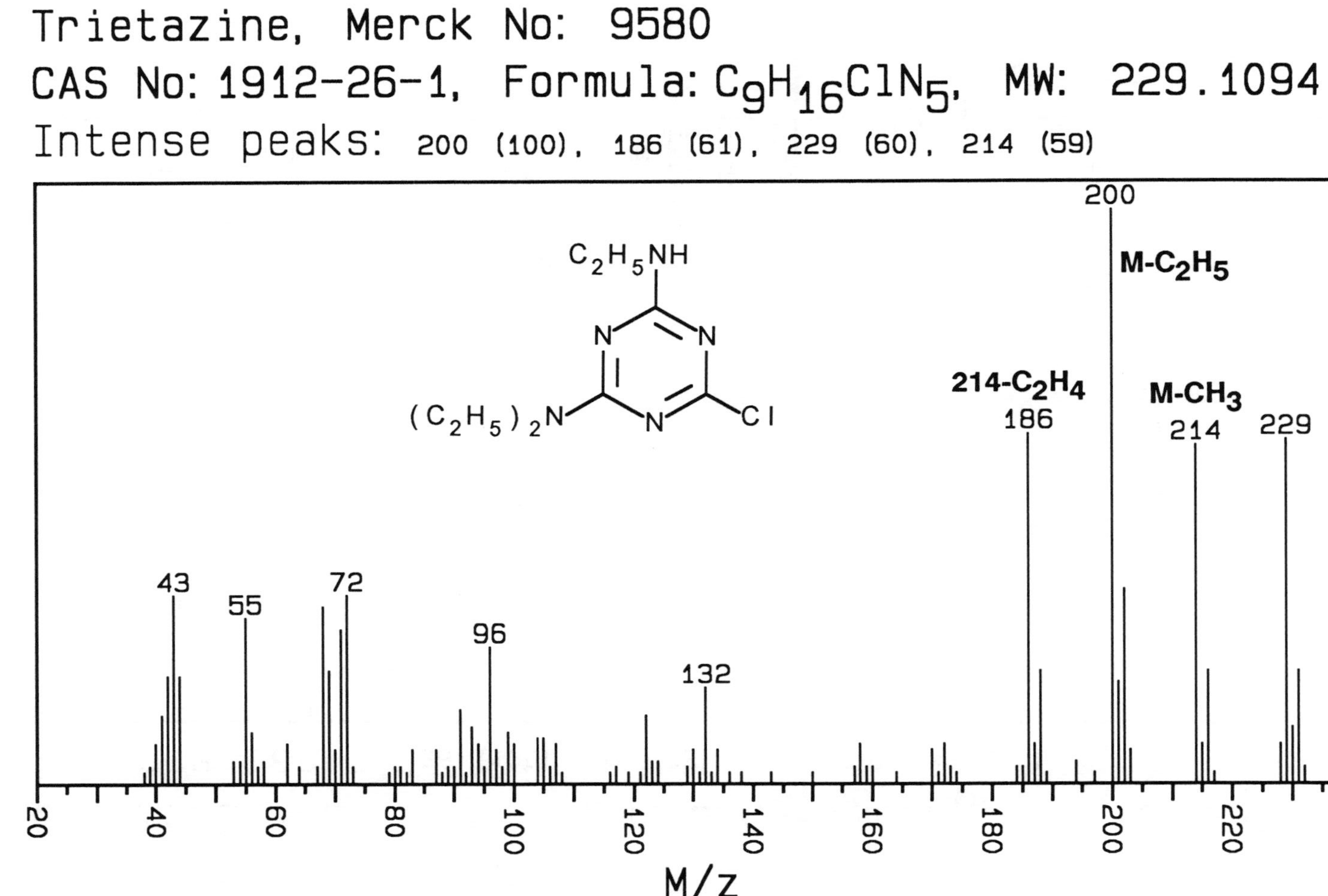

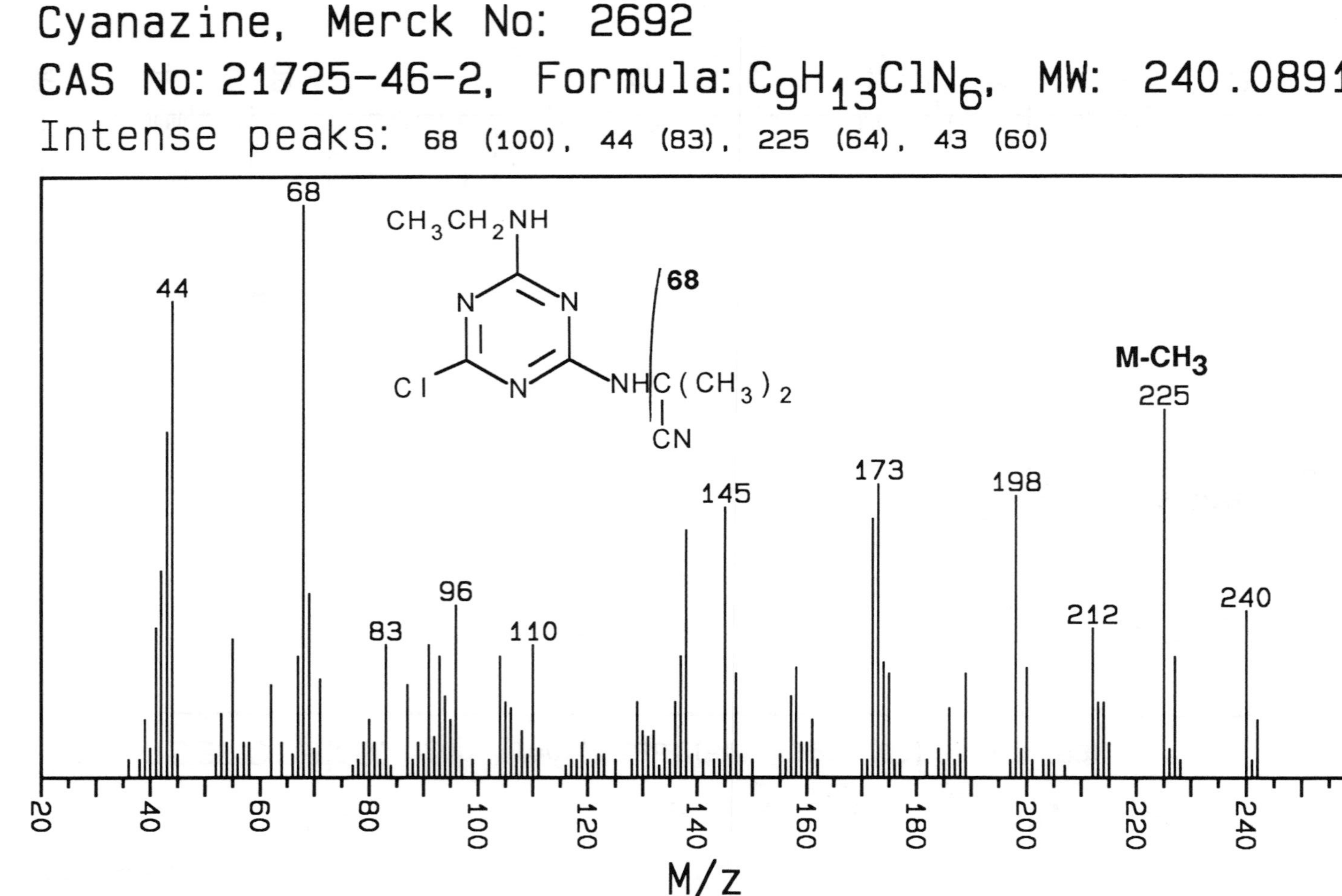
Cyanazine, Merck No: 2692
CAS No: 21725-46-2, Formula: $C_9H_{13}ClN_6$, MW: 240.0891
Intense peaks: 68 (100), 44 (83), 225 (64), 43 (60)
$CH_3CH_2NH$
N
N
N
Cl
NHC(CH$_3$)$_2$
CN
68
M-CH$_3$
44
68
83
96
110
145
173
198
212
225
240
20
40
60
80
100
120
140
160
180
200
220
240
260
M/Z

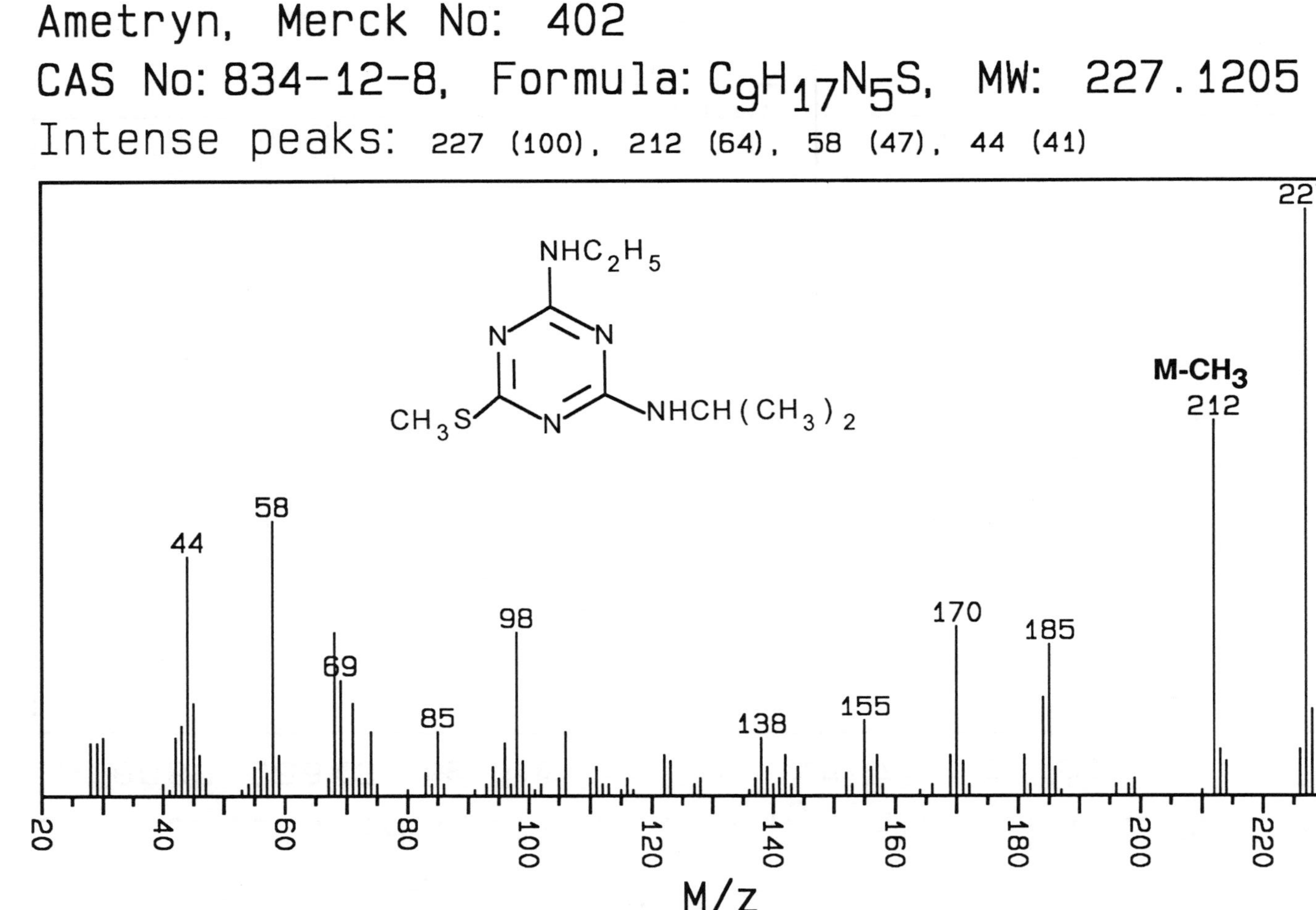
Ametryn, Merck No: 402
CAS No: 834-12-8, Formula: $C_9H_{17}N_5S$, MW: 227.1205
Intense peaks: 227 (100), 212 (64), 58 (47), 44 (41)
$NHC_2H_5$
N
N
N
$CH_3S$
$NHCH(CH_3)_2$
227
M-$CH_3$
212
44
58
69
85
98
138
155
170
185
20
40
60
80
100
120
140
160
180
200
220
240
M/Z

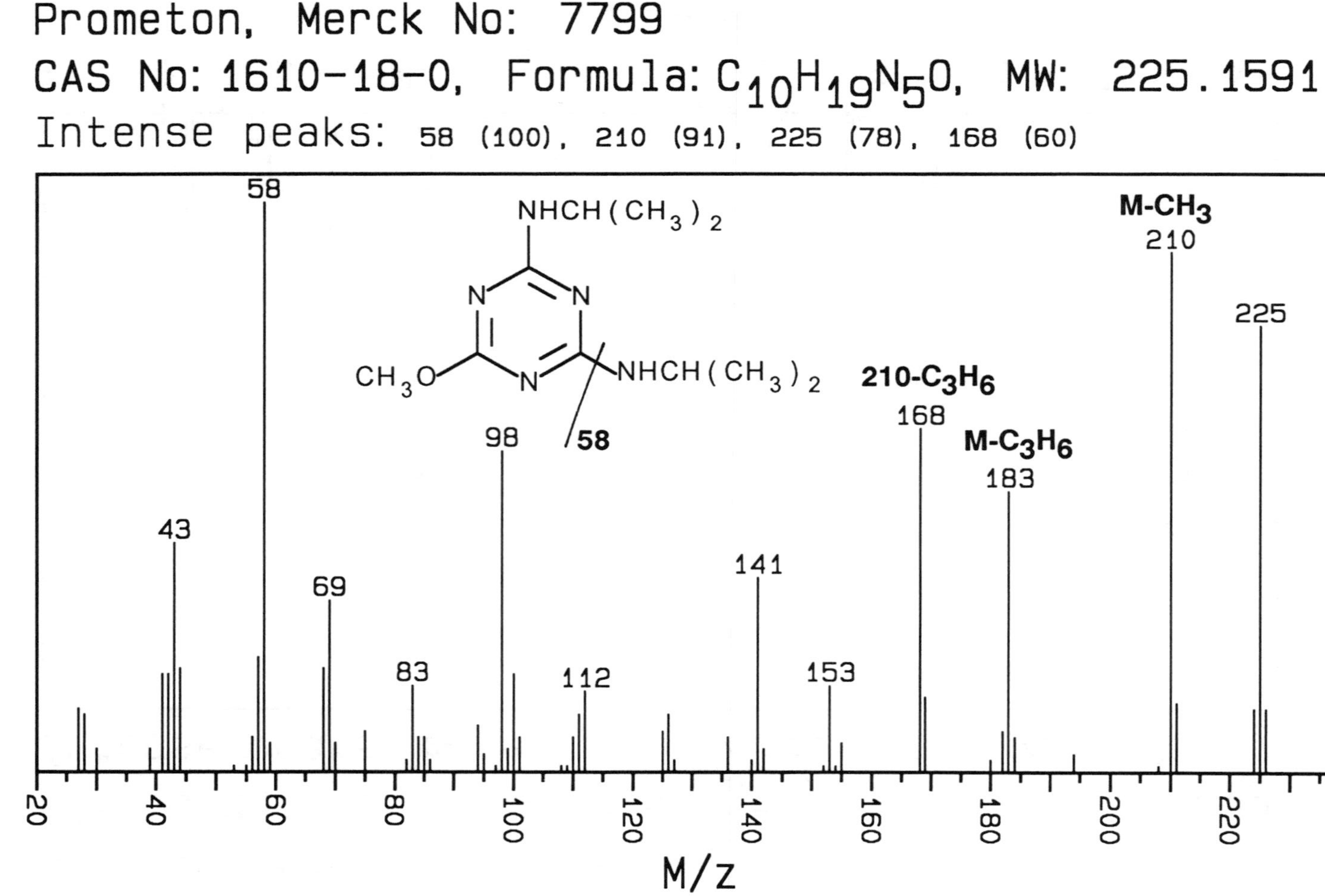

Prometon, Merck No: 7799
CAS No: 1610-18-0, Formula: $C_{10}H_{19}N_5O$, MW: 225.1591
Intense peaks: 58 (100), 210 (91), 225 (78), 168 (60)
NHCH(CH3)2
N
N
N
CH3O
NHCH(CH3)2
58
M-CH3
210-C3H6
M-C3H6
43
58
69
83
98
112
141
153
168
183
210
225
20
40
60
80
100
120
140
160
180
200
220
240
M/Z

Prometryn, Merck No: 7800

CAS No: 7287-19-6, Formula: $C_{10}H_{19}N_5S$, MW: 241.1361

Intense peaks: 241 (100), 58 (84), 184 (72), 226 (62)

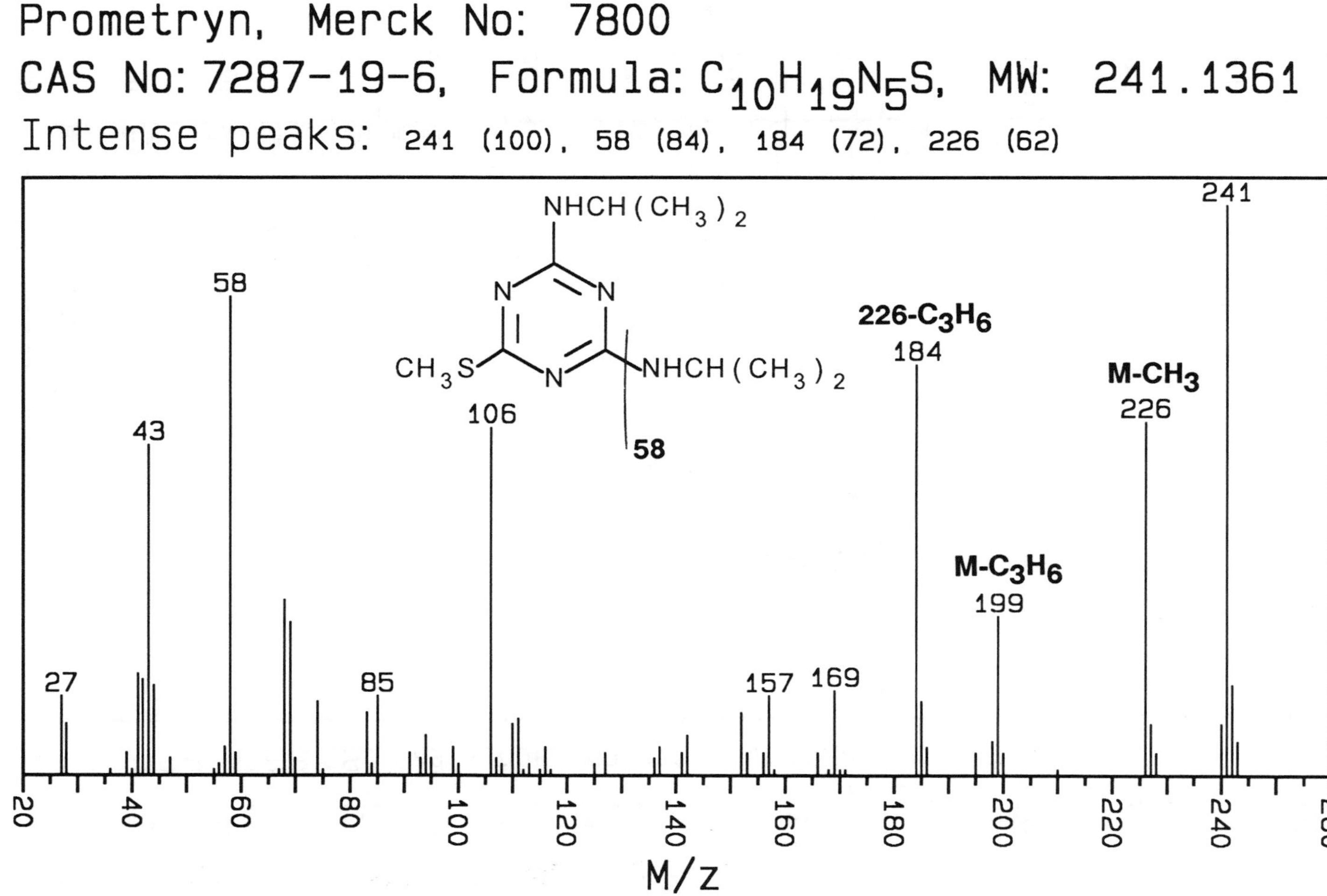

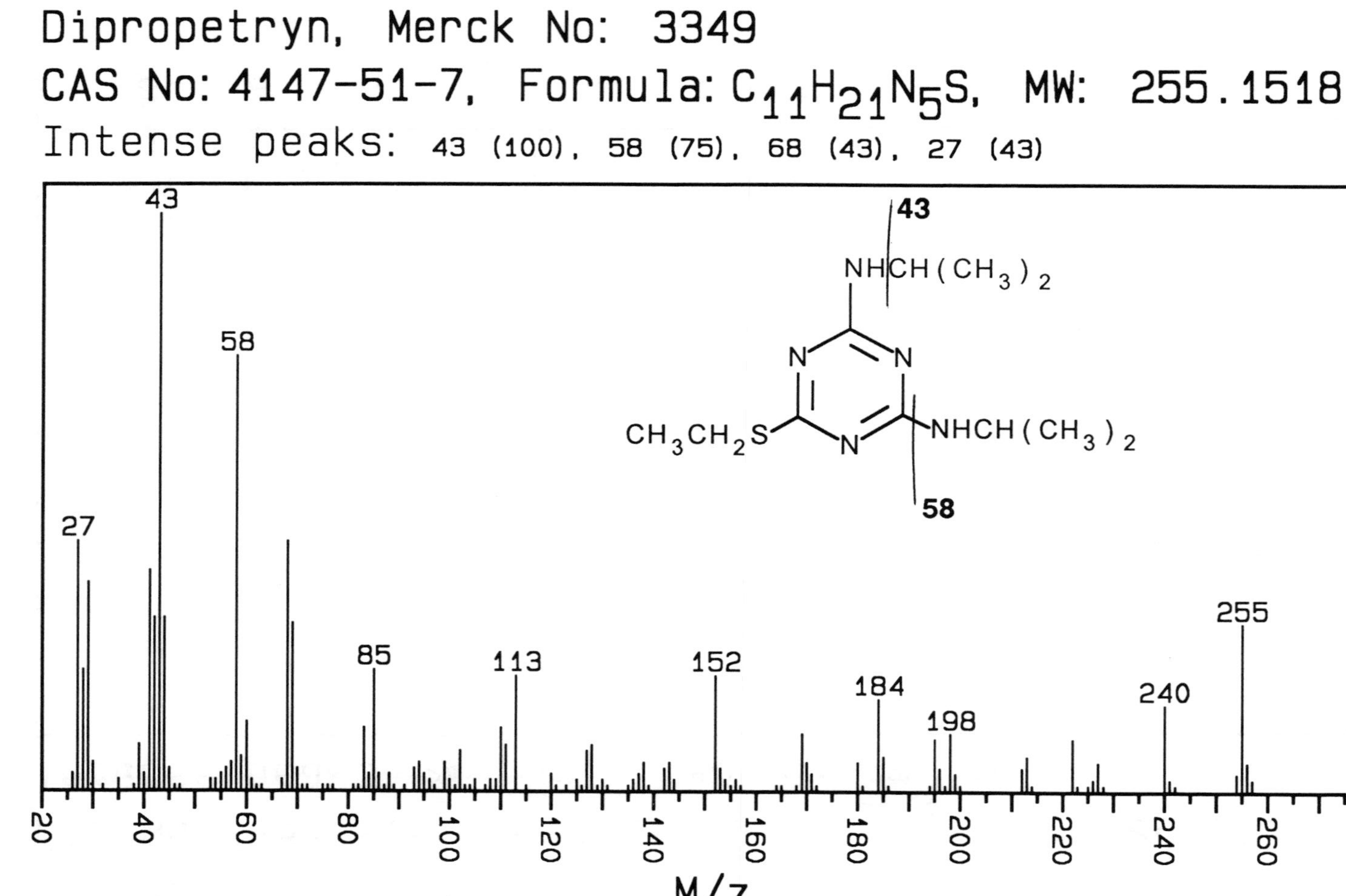
Dipropetryn, Merck No: 3349
CAS No: 4147-51-7, Formula: $C_{11}H_{21}N_5S$, MW: 255.1518
Intense peaks: 43 (100), 58 (75), 68 (43), 27 (43)
43
NHCH(CH$_3$)$_2$
N
N
$CH_3CH_2S$
N
NHCH(CH$_3$)$_2$
58
27
43
58
85
113
152
184
198
240
255
20
40
60
80
100
120
140
160
180
200
220
240
260
280
M/Z

Anilazine, Merck No: 685

CAS No: 101-05-3, Formula: $C_9H_5Cl_3N_4$, MW: 273.9581

Intense peaks: 239 (100), 241 (66), 178 (35), 143 (33)

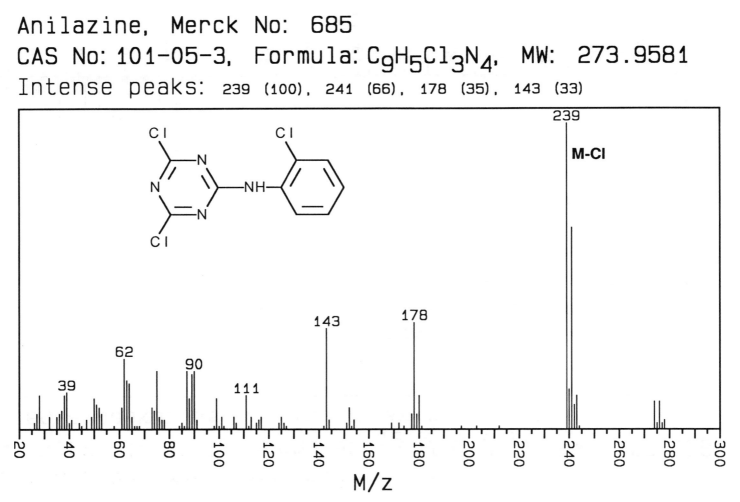

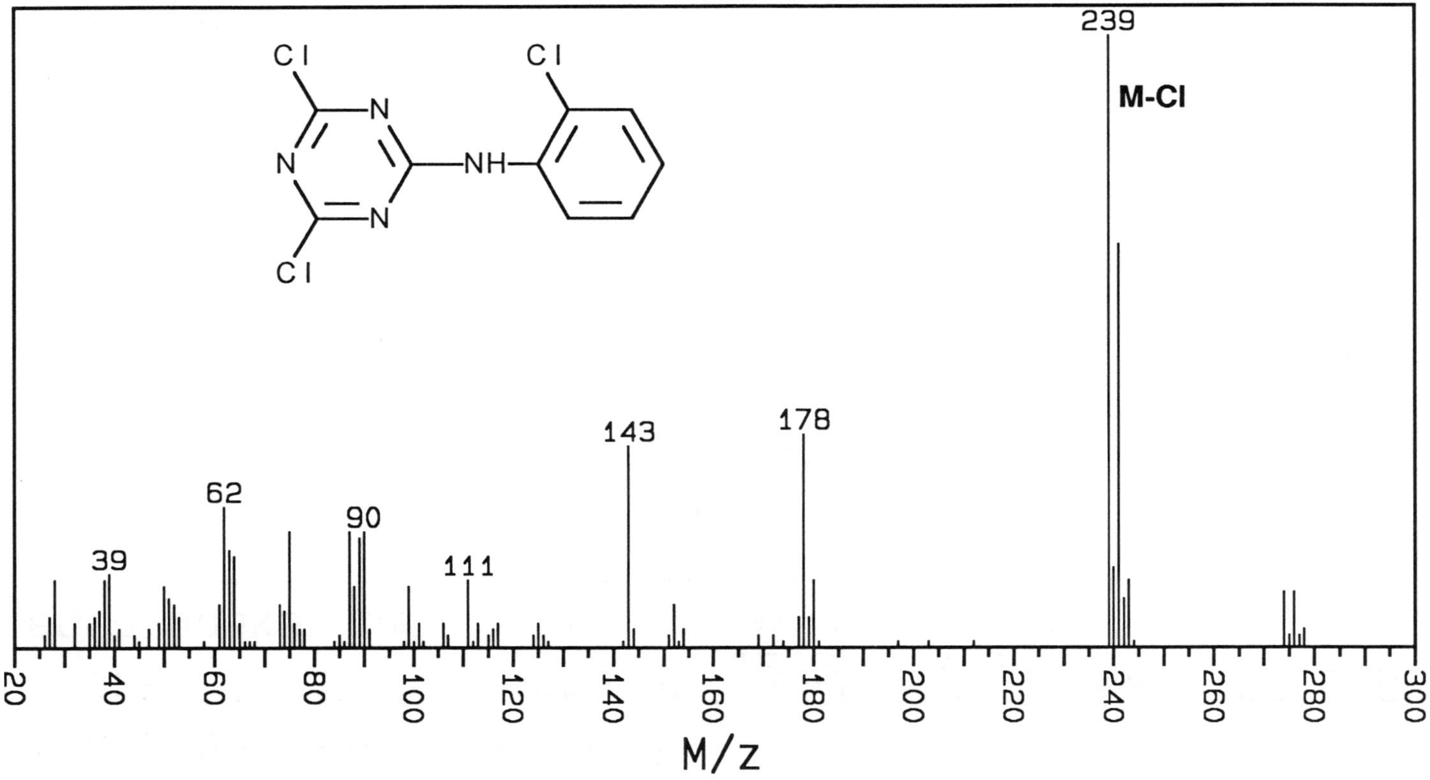

Cl
N
N
N
NH
Cl
Cl
Cl
239
M-Cl
143
178
62
90
39
111
20
40
60
80
100
120
140
160
180
200
220
240
260
280
300
M/Z

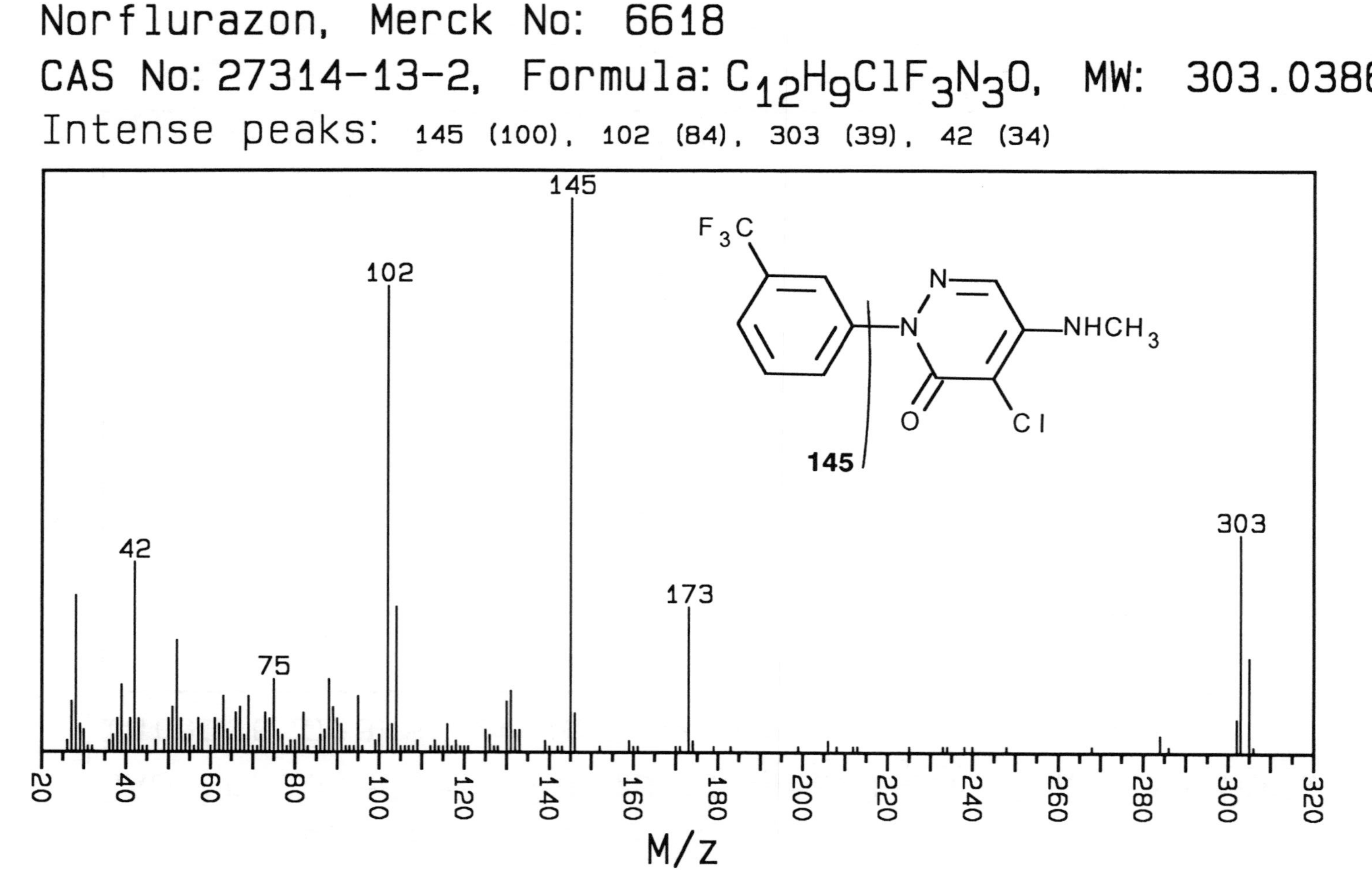
Norflurazon, Merck No: 6618
CAS No: 27314-13-2, Formula: $C_{12}H_9ClF_3N_3O$, MW: 303.0386
Intense peaks: 145 (100), 102 (84), 303 (39), 42 (34)
145
$F_3C$
102
N
N
NHCH$_3$
O
Cl
145
303
42
173
75
20
40
60
80
100
120
140
160
180
200
220
240
260
280
300
320
M/Z

Oxadiazon, Merck No: 6860

CAS No: 19666-30-9, Formula: $C_{15}H_{18}Cl_2N_2O_3$, MW: 344.0694

Intense peaks: 41 (100), 43 (71), 175 (60), 57 (59)

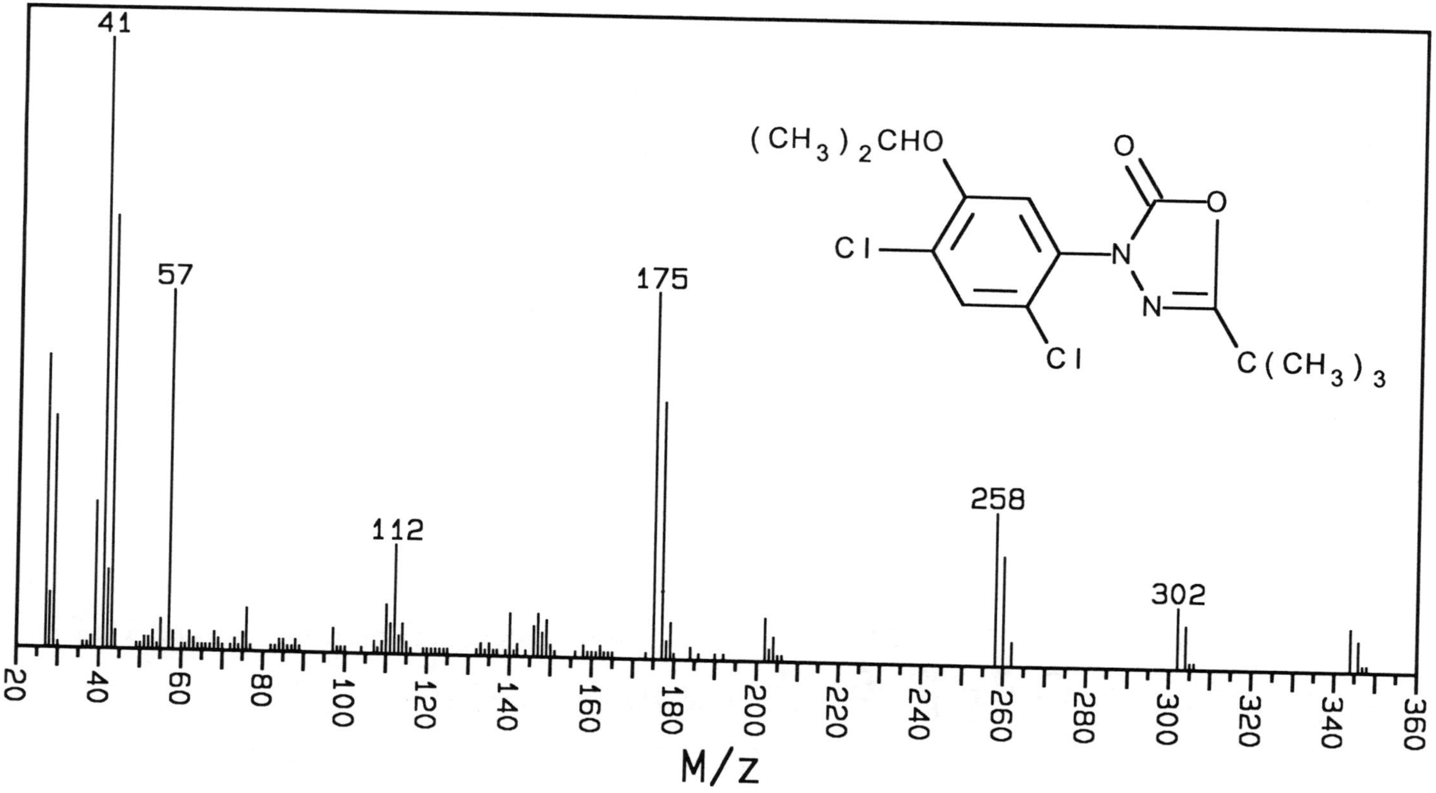

Methazole, Merck No: 5876

CAS No: 20354-26-1, Formula: $C_9H_6Cl_2N_2O_3$, MW: 259.9755

Intense peaks: 159 (100), 124 (97), 260 (71), 161 (71)

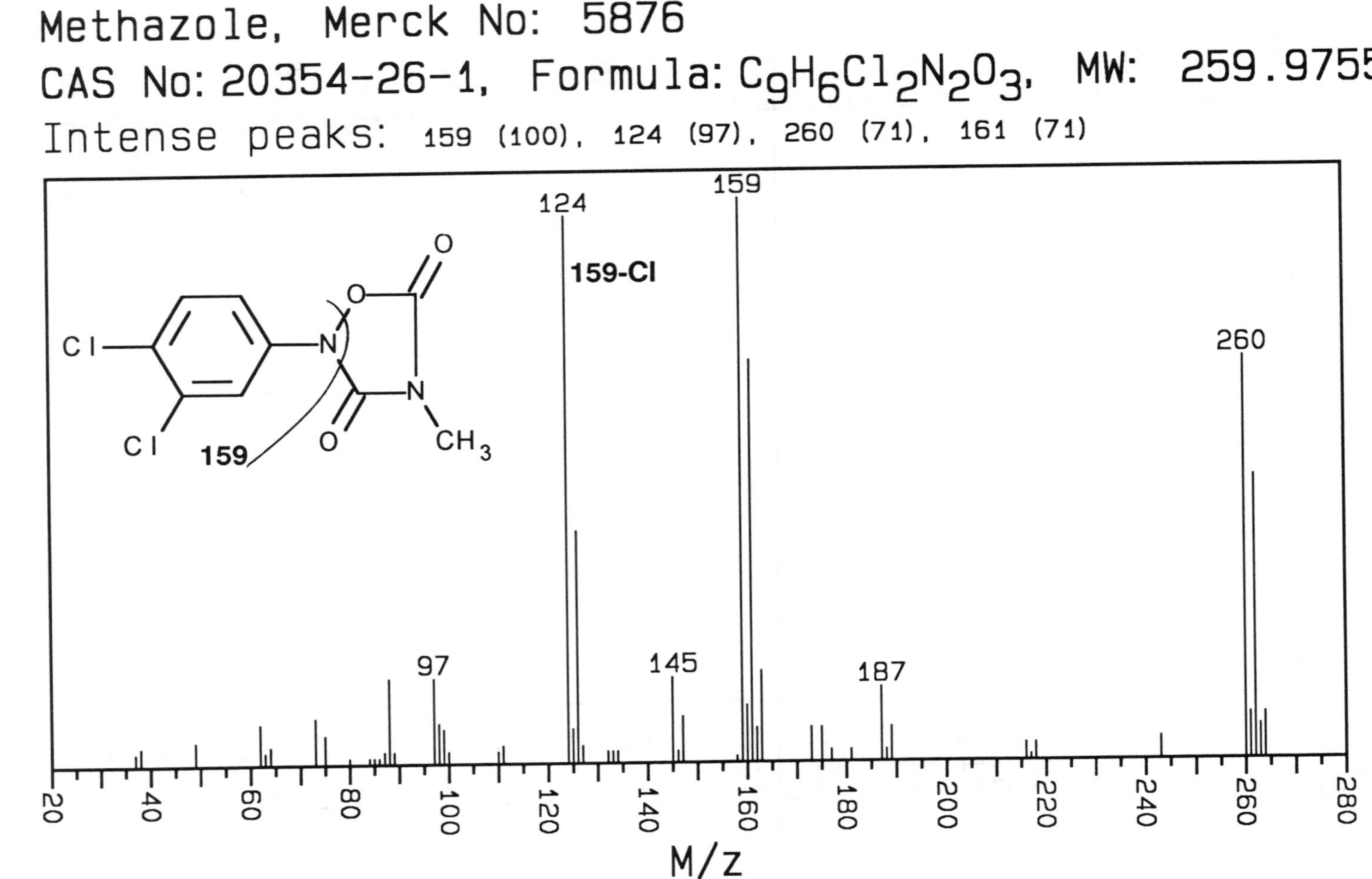

Iprodione, Merck No: 4964

CAS No: 36734-19-7, Formula: $C_{13}H_{13}Cl_2N_3O_3$, MW: 329.0334

Intense peaks: 56 (100), 43 (93), 58 (61), 314 (41)

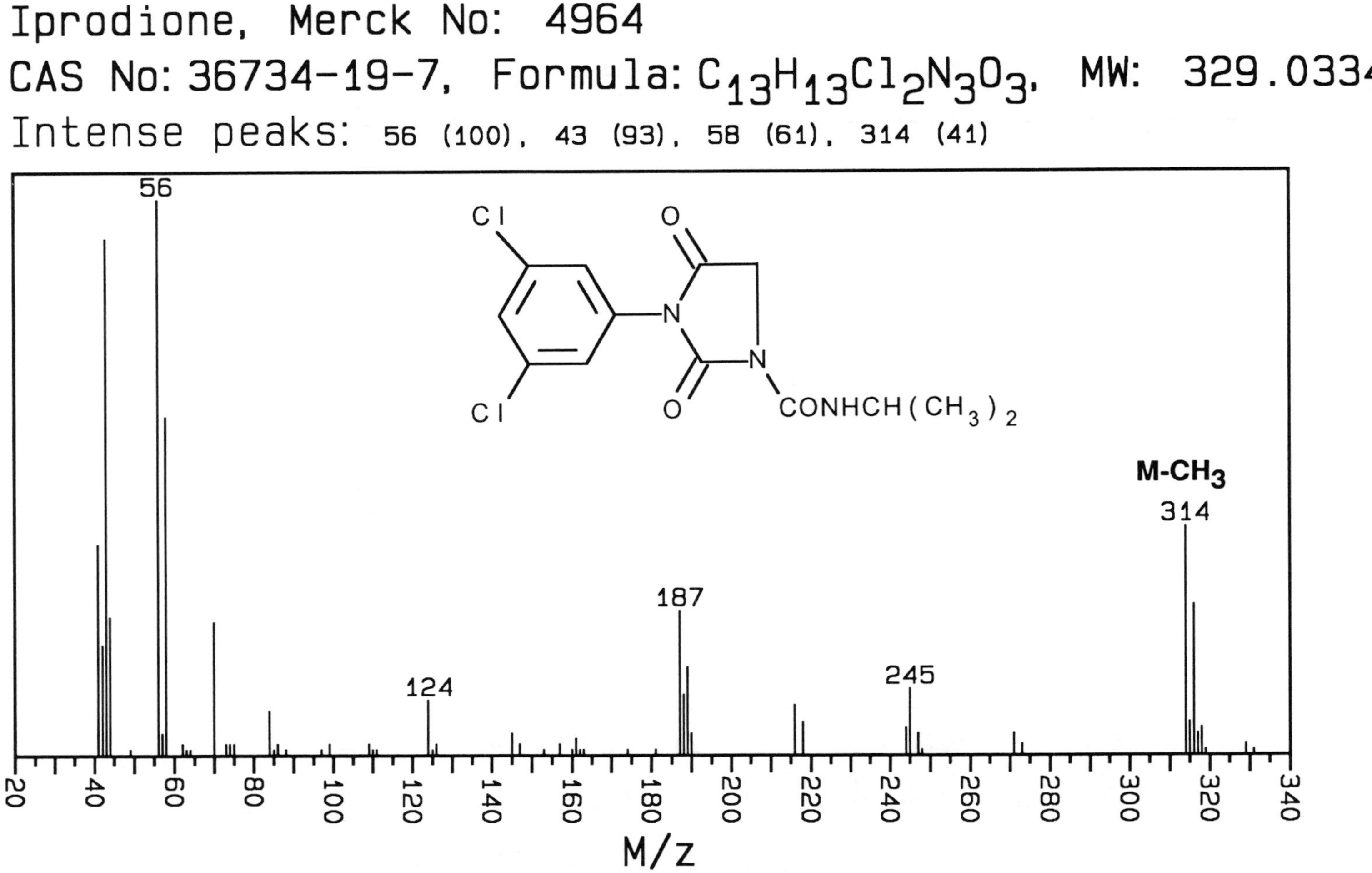

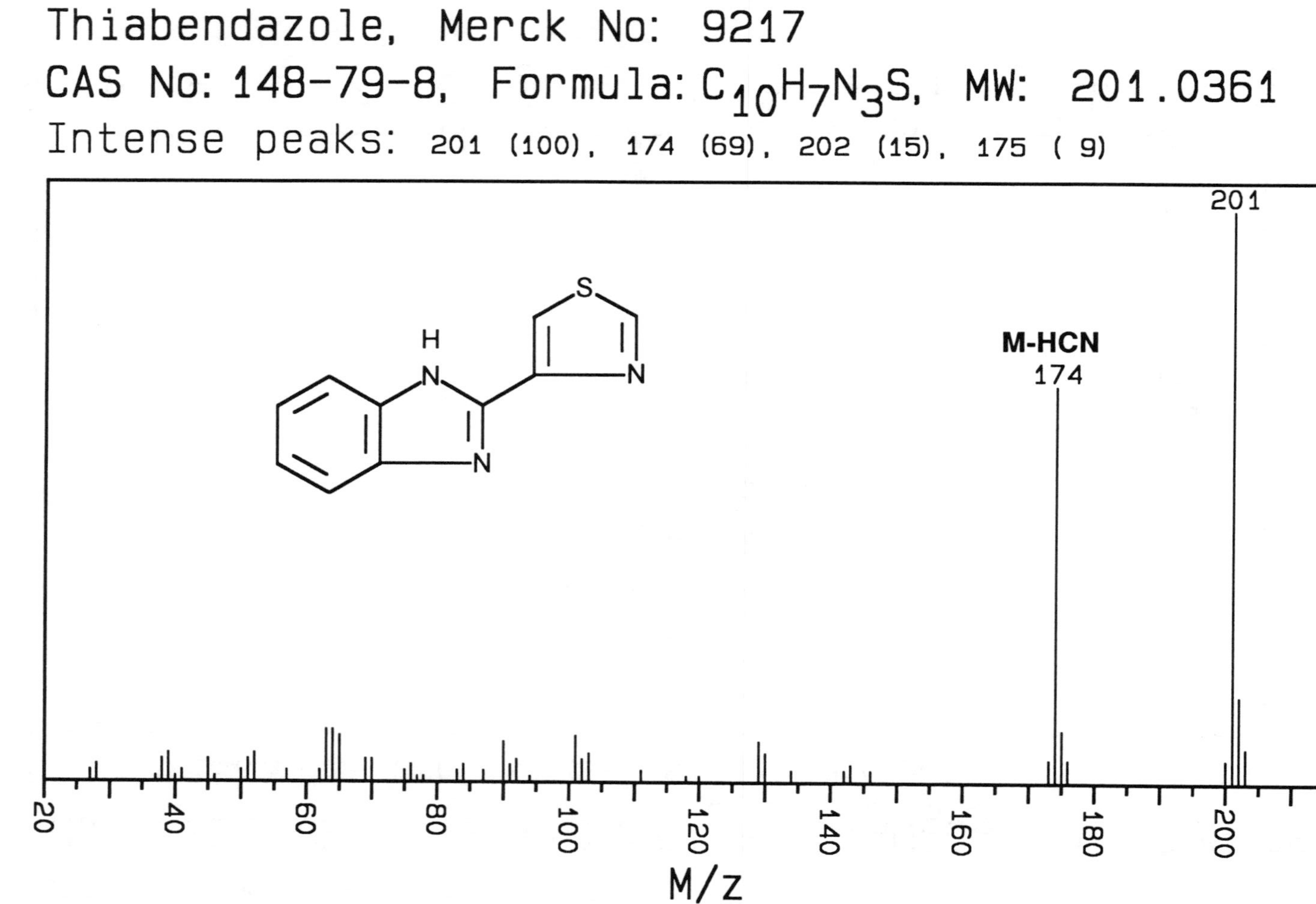
Thiabendazole, Merck No: 9217
CAS No: 148-79-8, Formula: $C_{10}H_7N_3S$, MW: 201.0361
Intense peaks: 201 (100), 174 (69), 202 (15), 175 ( 9)
201
M-HCN
174
H
N
N
S
N
20
40
60
80
100
120
140
160
180
200
220
M/Z

Piperine, Merck No: 7442

CAS No: 94-62-2, Formula: $C_{17}H_{19}NO_3$, MW: 285.1365

Intense peaks: 201 (100), 115 (92), 285 (65), 173 (42)

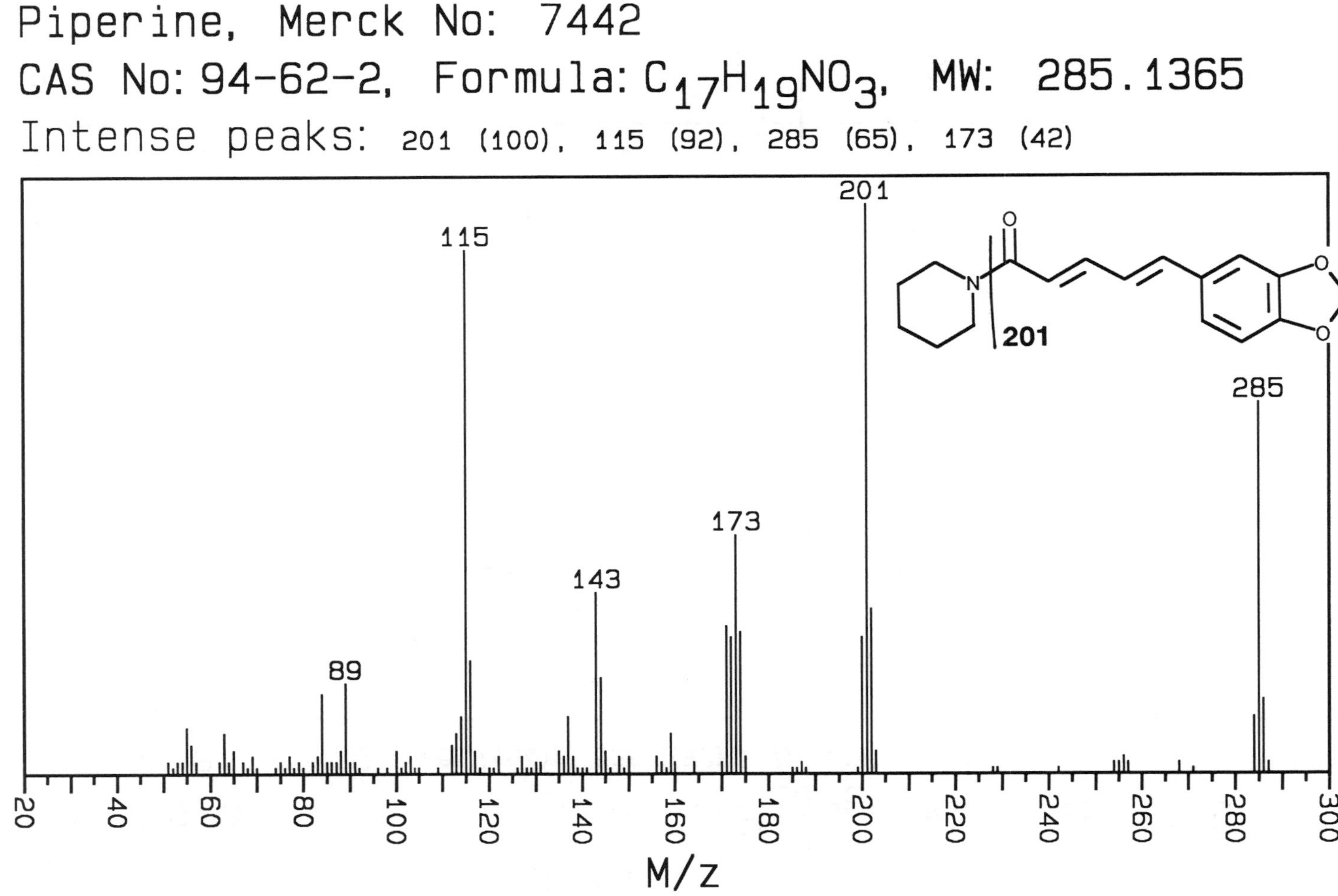

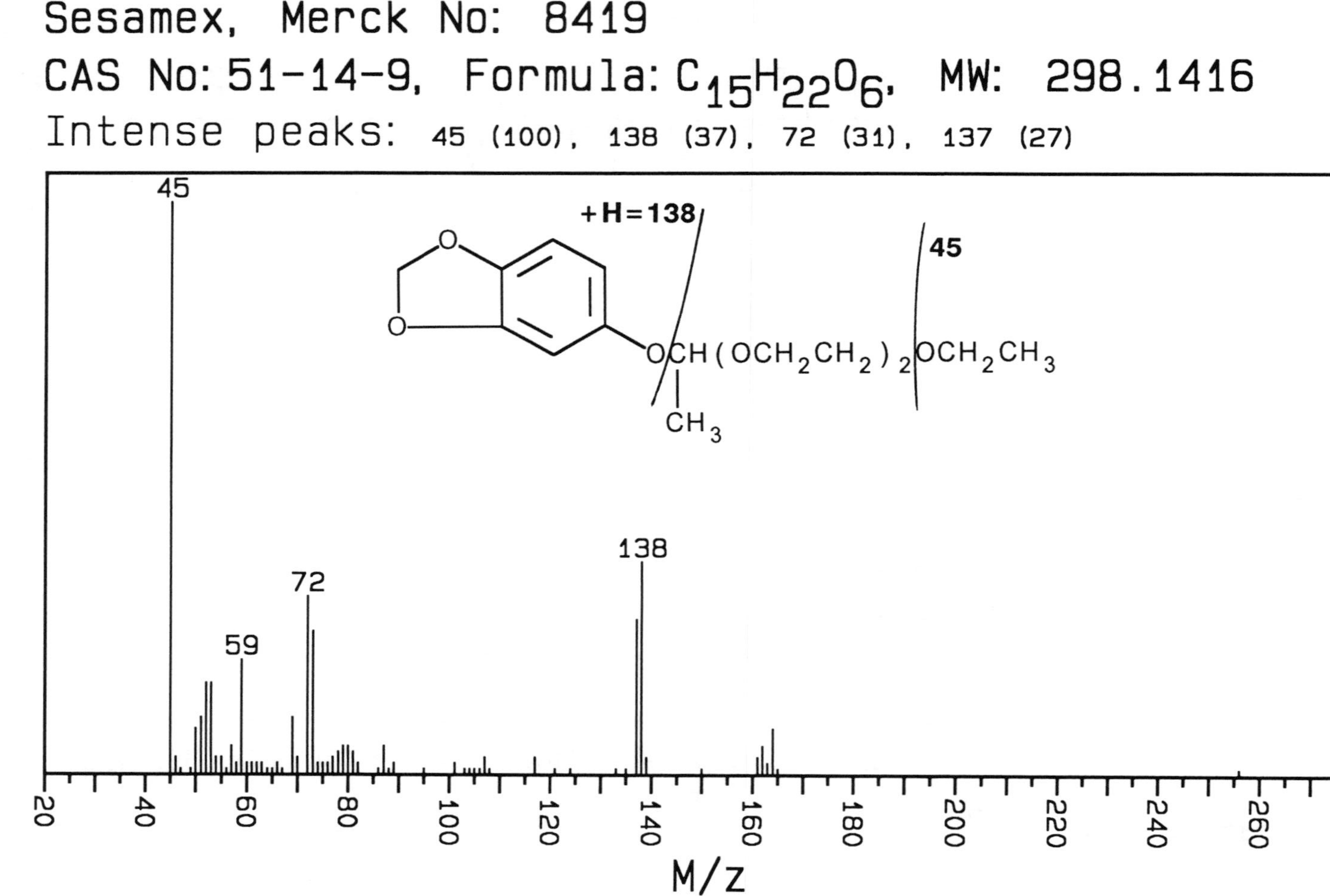
Sesamex, Merck No: 8419
CAS No: 51-14-9, Formula: $C_{15}H_{22}O_6$, MW: 298.1416
Intense peaks: 45 (100), 138 (37), 72 (31), 137 (27)
+H=138
45
OCH(OCH2CH2)2OCH2CH3
CH3
45
59
72
138
M/Z
20
40
60
80
100
120
140
160
180
200
220
240
260
280

Piperonyl butoxide, Merck No: 7446

CAS No: 51-03-6, Formula: $C_{19}H_{30}O_5$, MW: 338.2093

Intense peaks: 176 (100), 177 (41), 194 (26), 57 (19)

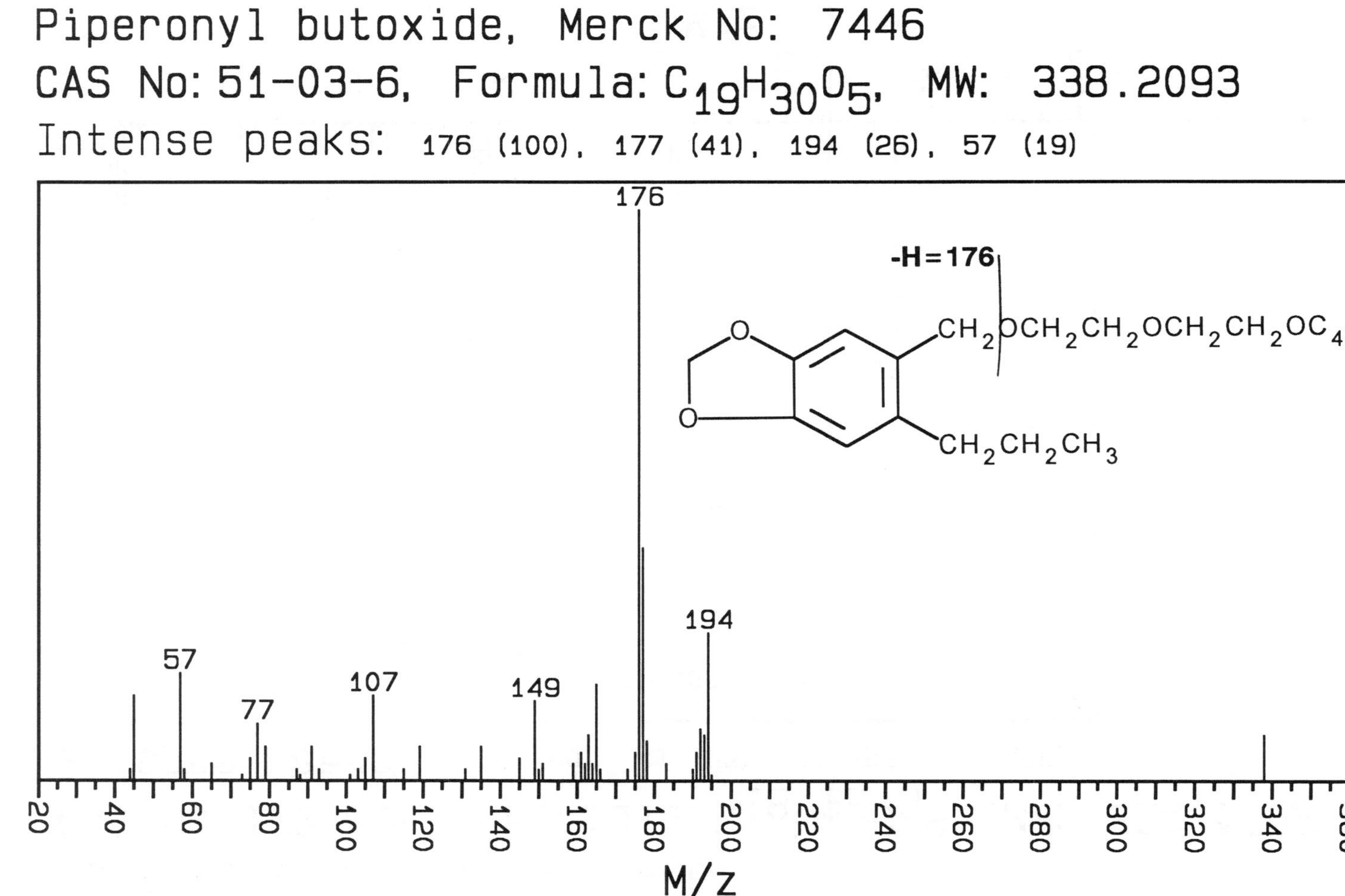

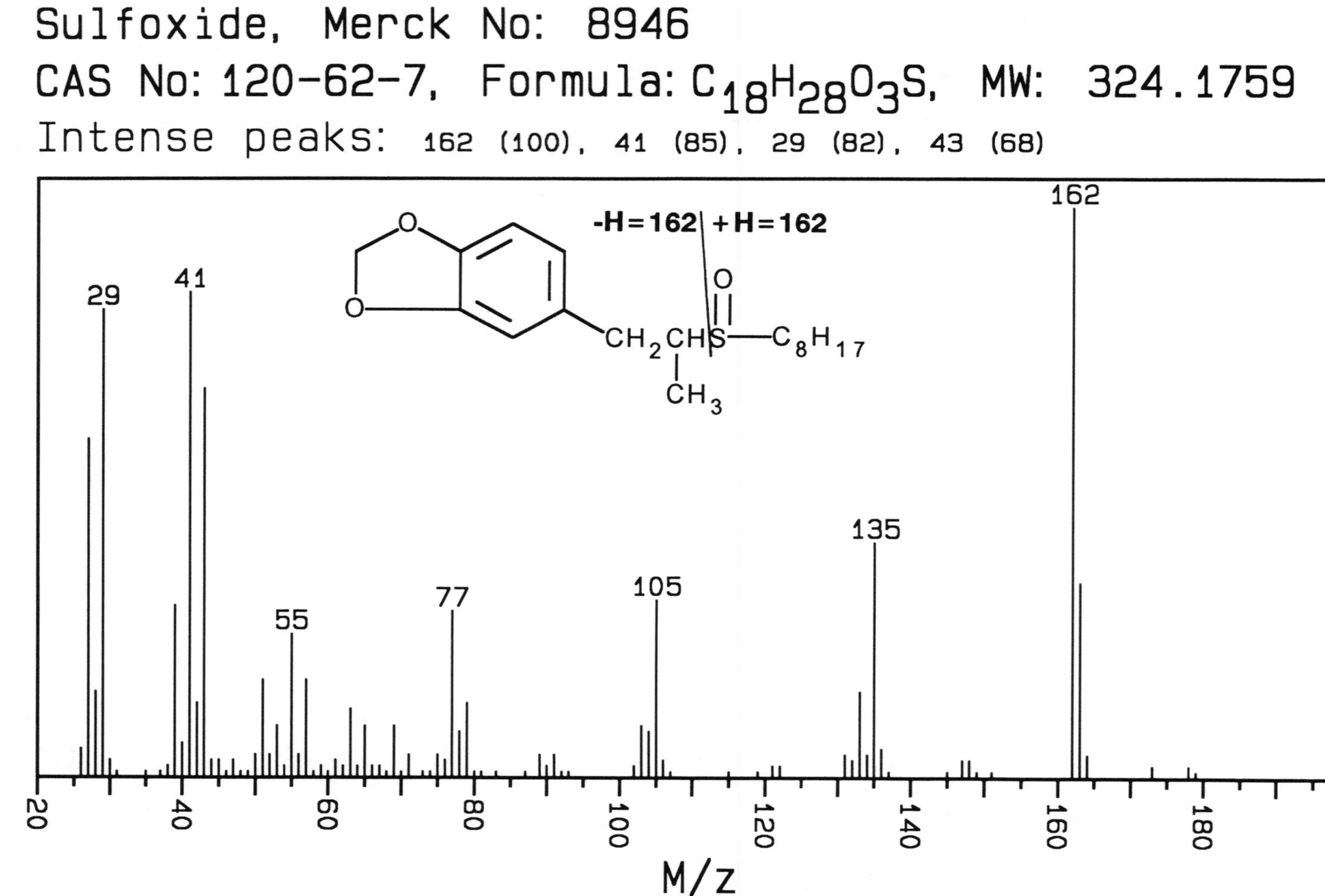
Sulfoxide, Merck No: 8946
CAS No: 120-62-7, Formula: $C_{18}H_{28}O_3S$, MW: 324.1759
Intense peaks: 162 (100), 41 (85), 29 (82), 43 (68)
-H=162 +H=162
O
O
O
$CH_2CHS$—$C_8H_{17}$
$CH_3$
162
29
41
55
77
105
135
20
40
60
80
100
120
140
160
180
200
M/Z

Warfarin, Merck No: 9950

CAS No: 81-81-2, Formula: $C_{19}H_{16}O_4$, MW: 308.1041

Intense peaks: 265 (100), 43 (46), 121 (38), 187 (27)

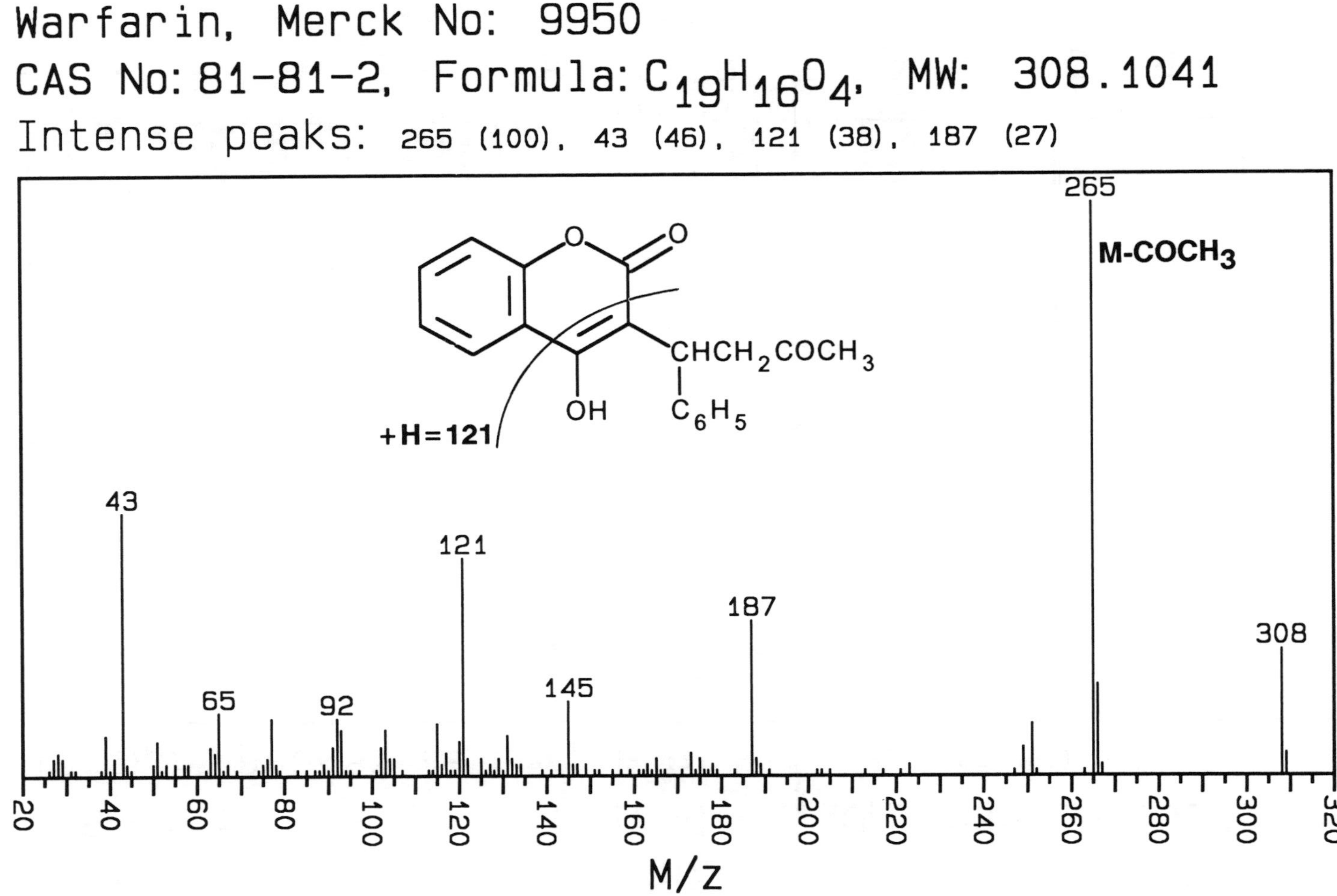

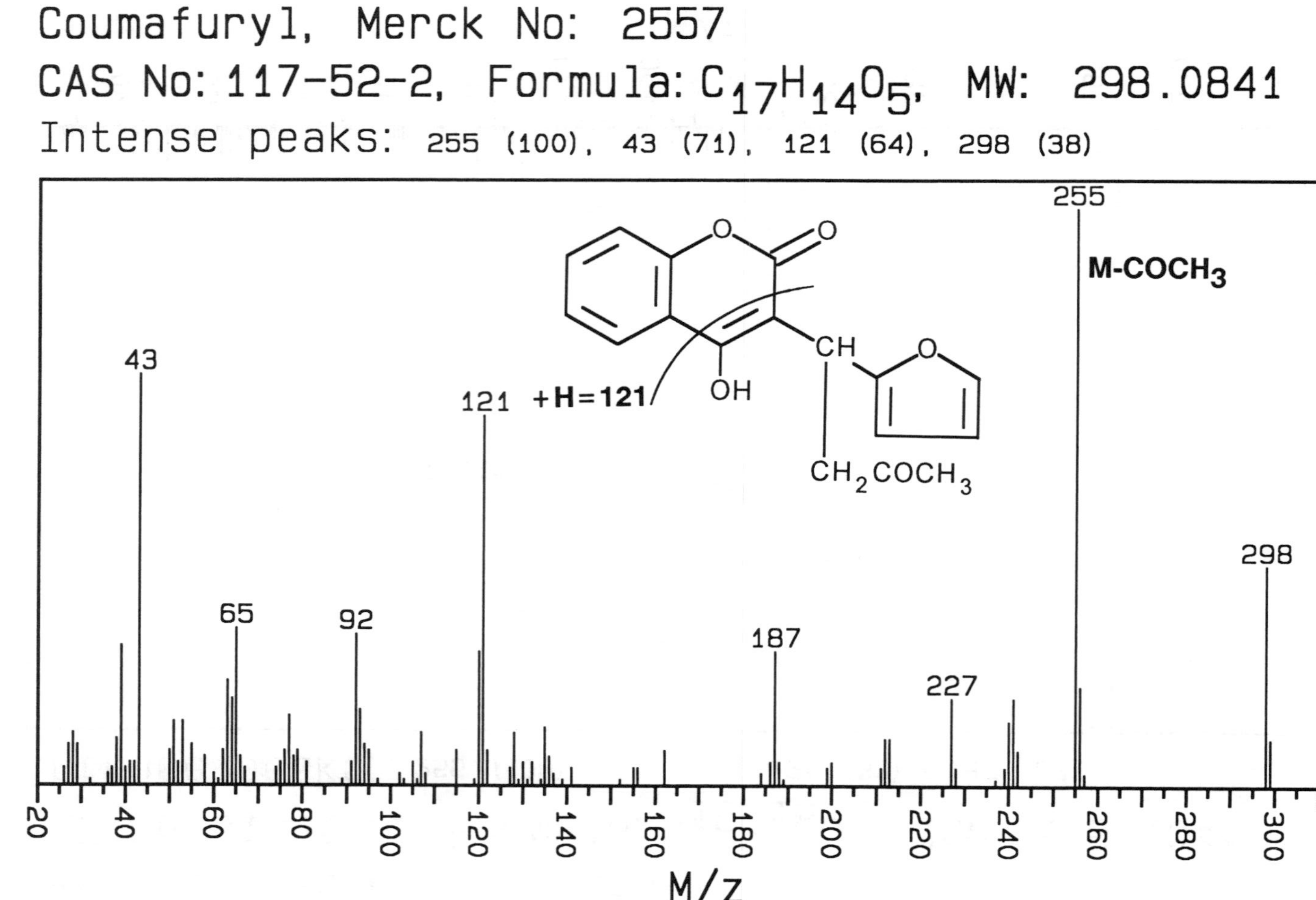
Coumafuryl, Merck No: 2557
CAS No: 117-52-2, Formula: $C_{17}H_{14}O_5$, MW: 298.0841
Intense peaks: 255 (100), 43 (71), 121 (64), 298 (38)
255
M-COCH$_3$
O
O
CH
O
OH
CH$_2$COCH$_3$
43
121 +H=121
65
92
187
227
298
20
40
60
80
100
120
140
160
180
200
220
240
260
280
300
320
M/Z

Coumachlor, Merck No: 2556

CAS No: 81-82-3, Formula: $C_{19}H_{15}ClO_4$, MW: 342.0659

Intense peaks: 299 (100), 43 (46), 121 (41), 301 (34)

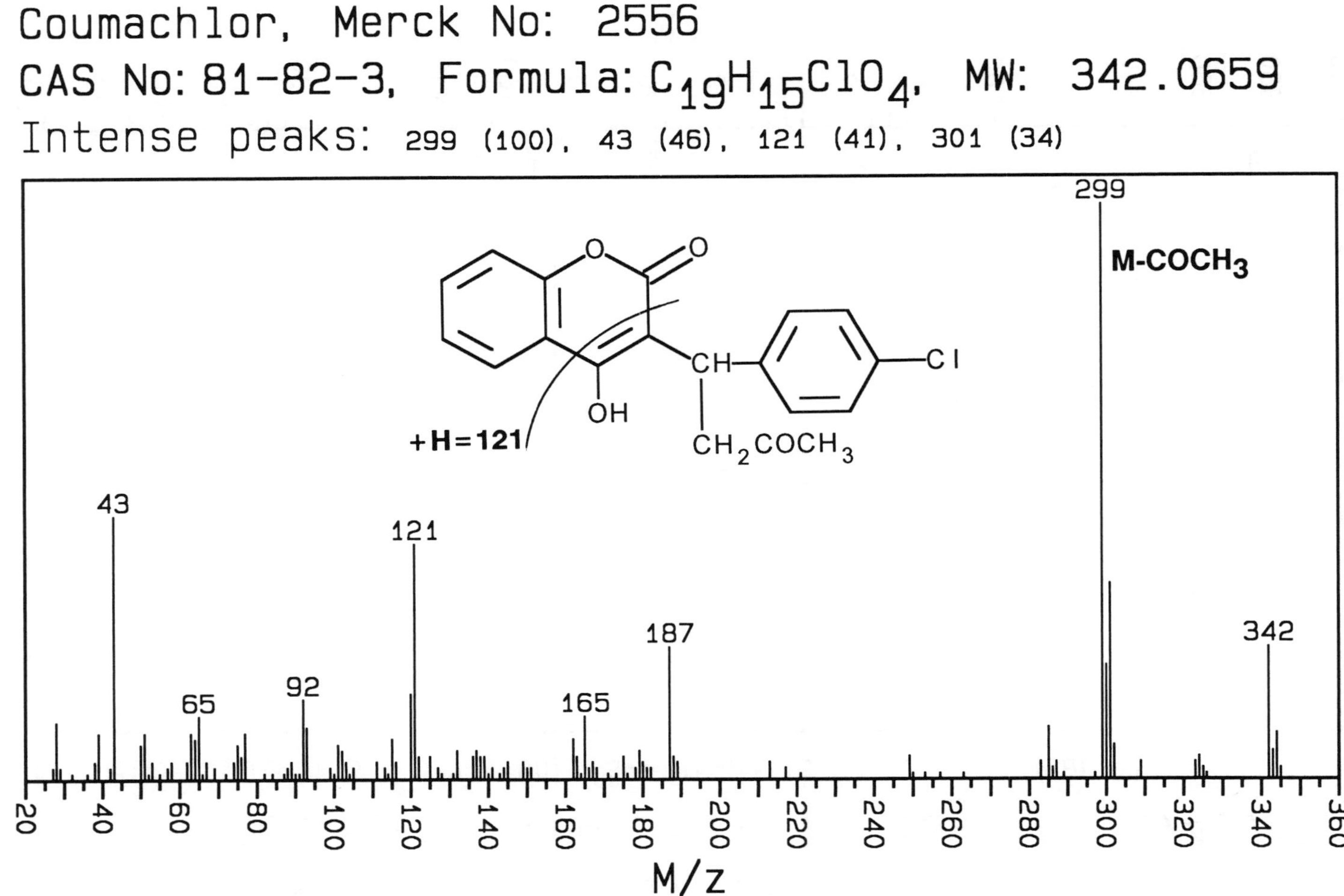

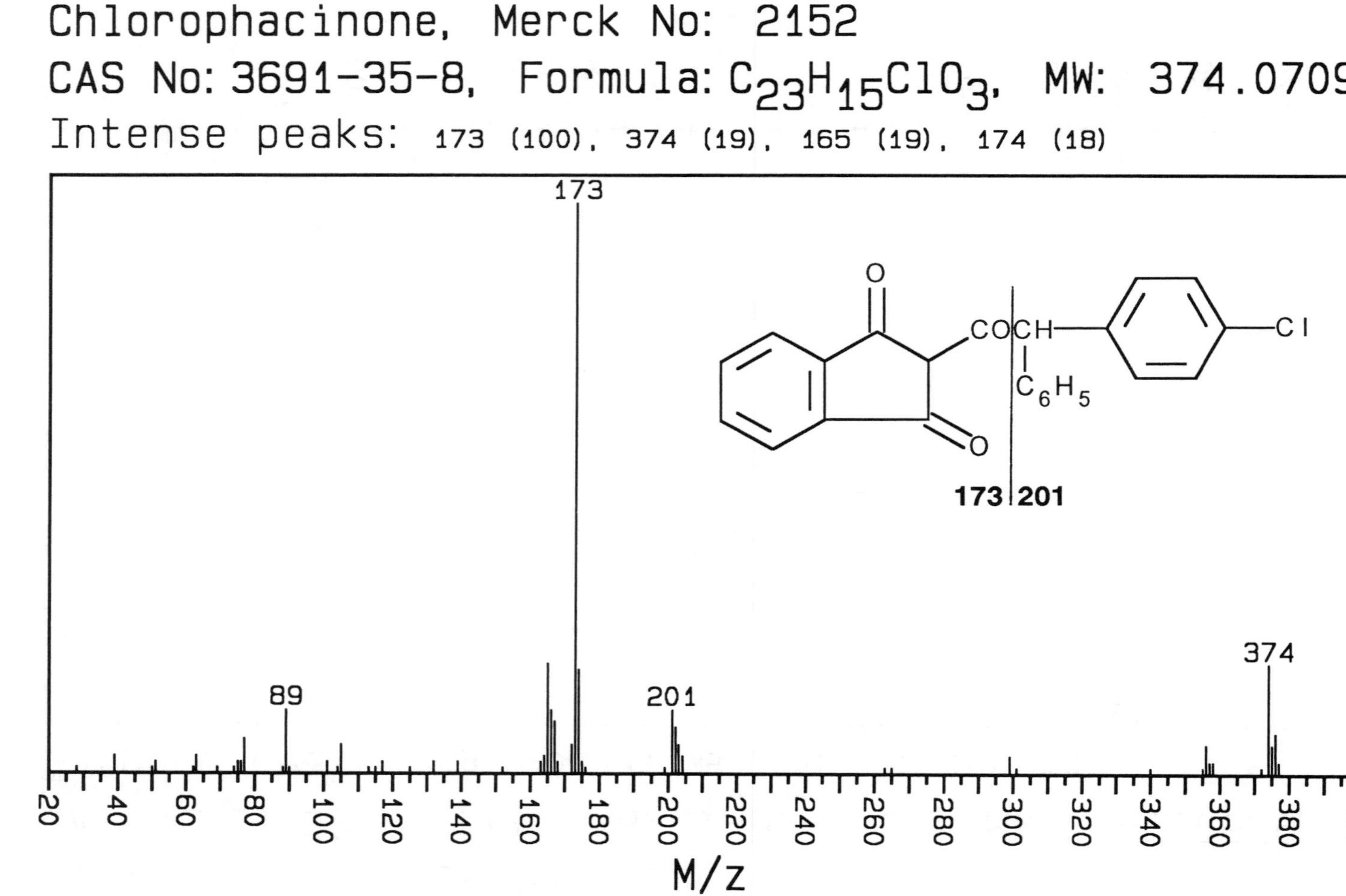
Chlorophacinone, Merck No: 2152
CAS No: 3691-35-8, Formula: C23H15ClO3, MW: 374.0709
Intense peaks: 173 (100), 374 (19), 165 (19), 174 (18)
173
O
COCH
Cl
C6H5
O
173 201
89
201
374
20
40
60
80
100
120
140
160
180
200
220
240
260
280
300
320
340
360
380
400
M/Z

Pindone, Merck No: 7413

CAS No: 83-26-1, Formula: $C_{14}H_{14}O_3$, MW: 230.0943

Intense peaks: 173 (100), 174 (81), 146 (41), 89 (32)

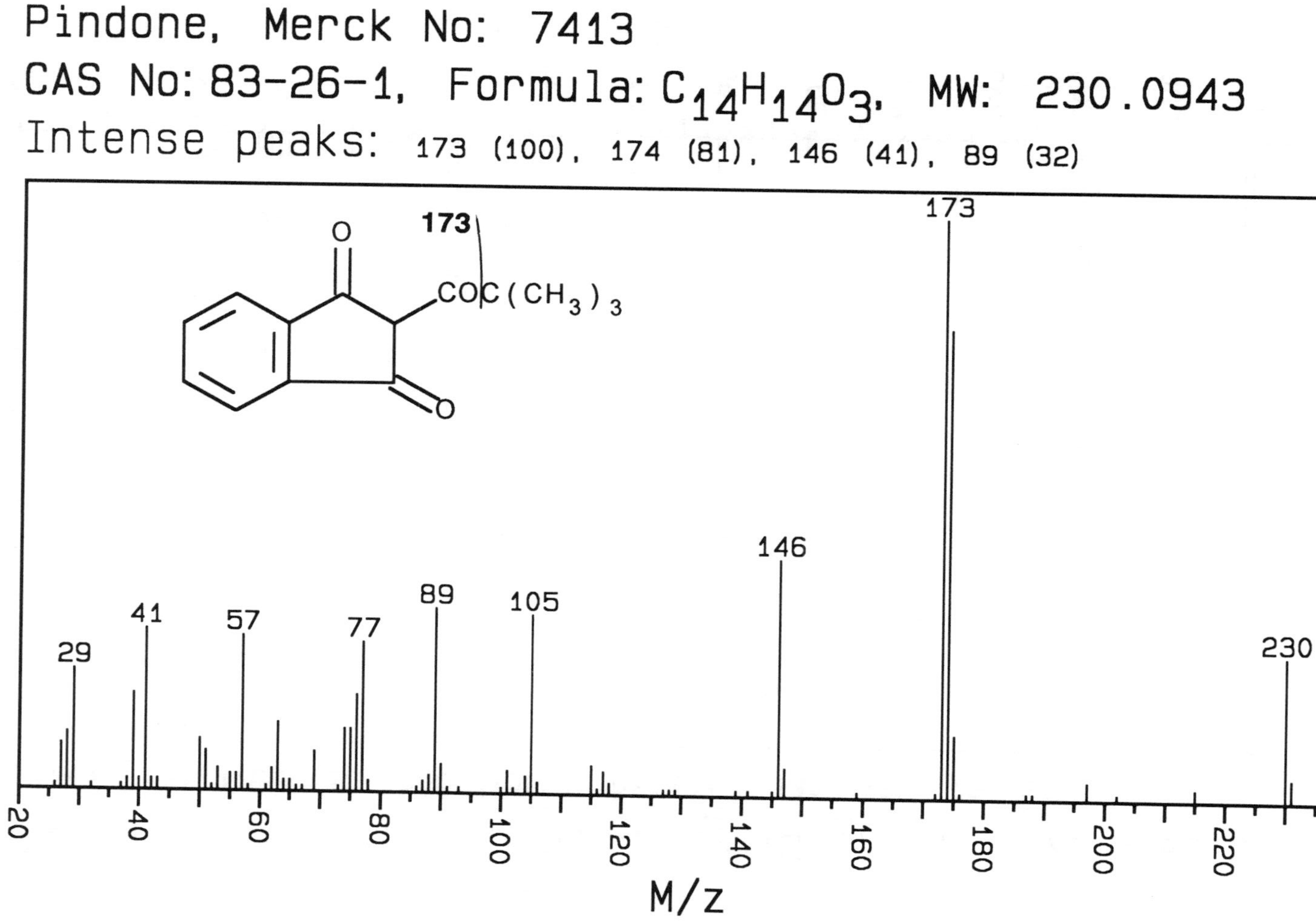

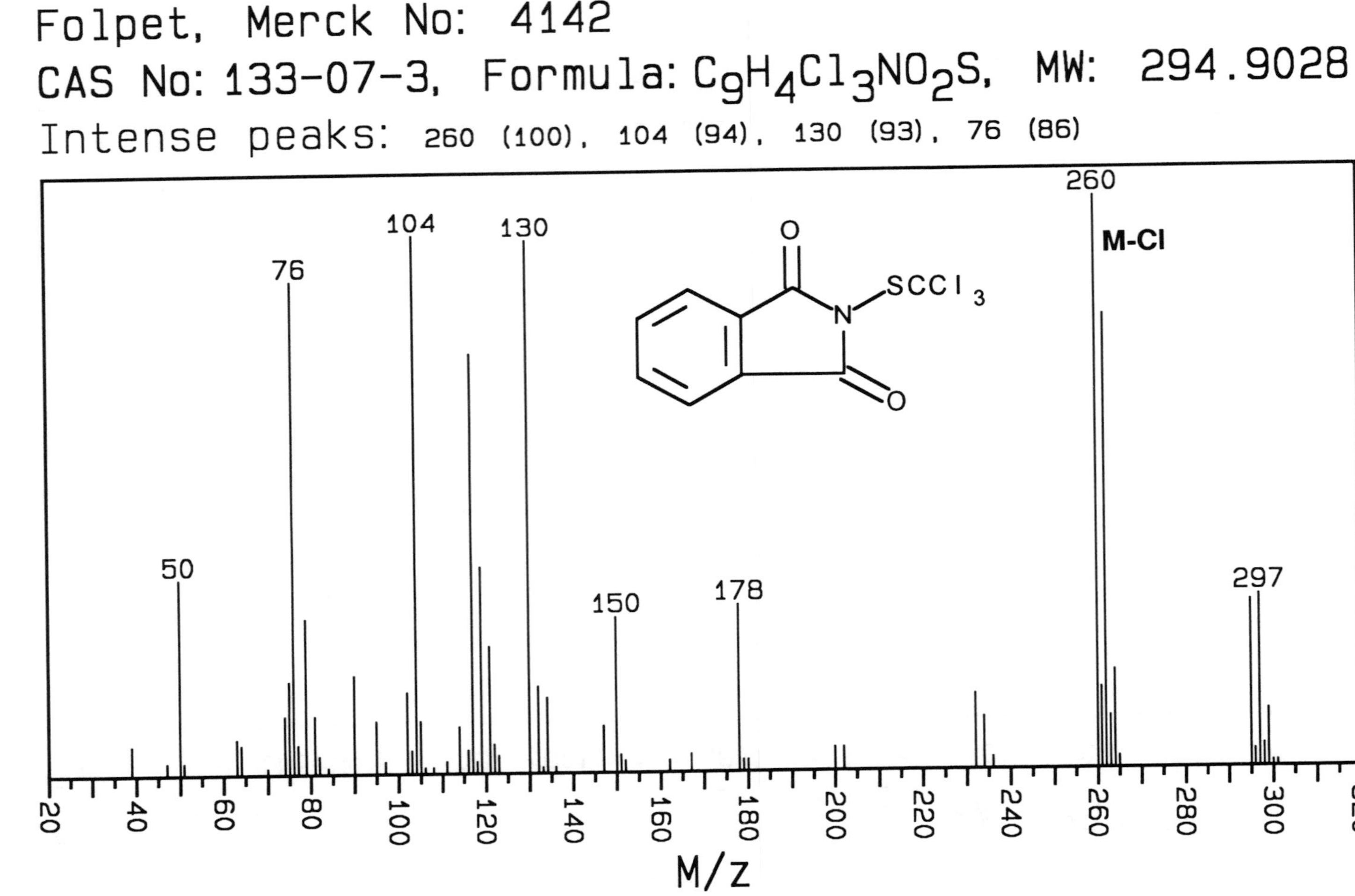
Folpet, Merck No: 4142
CAS No: 133-07-3, Formula: C9H4Cl3NO2S, MW: 294.9028
Intense peaks: 260 (100), 104 (94), 130 (93), 76 (86)
260
M-Cl
104
130
76
50
150
178
297
O
O
N
SCCl3
20
40
60
80
100
120
140
160
180
200
220
240
260
280
300
320
M/Z

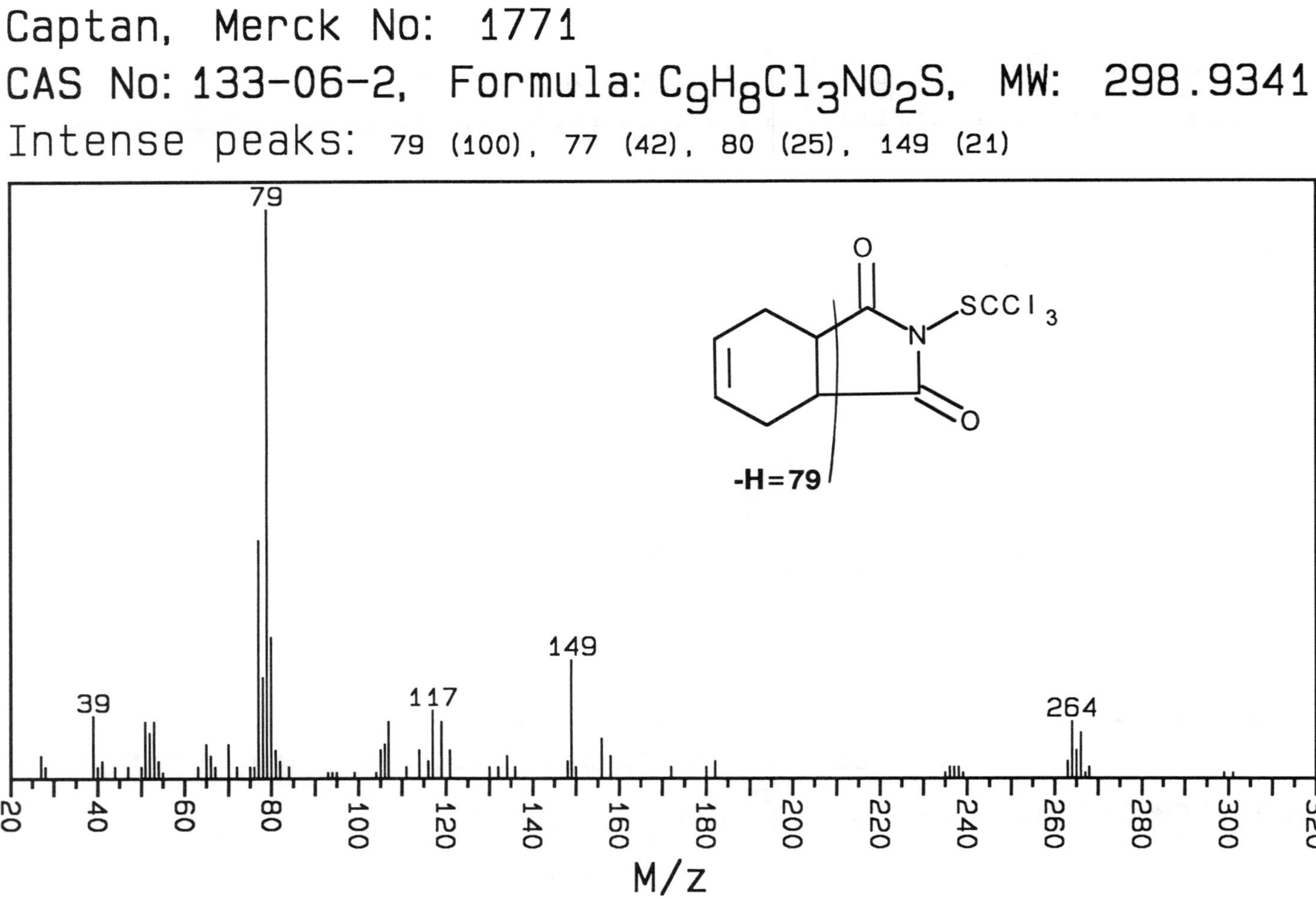

Captan, Merck No: 1771
CAS No: 133-06-2, Formula: C9H8Cl3NO2S, MW: 298.9341
Intense peaks: 79 (100), 77 (42), 80 (25), 149 (21)
79
39
117
149
264
O
SCCl3
N
O
-H=79
20
40
60
80
100
120
140
160
180
200
220
240
260
280
300
320
M/Z

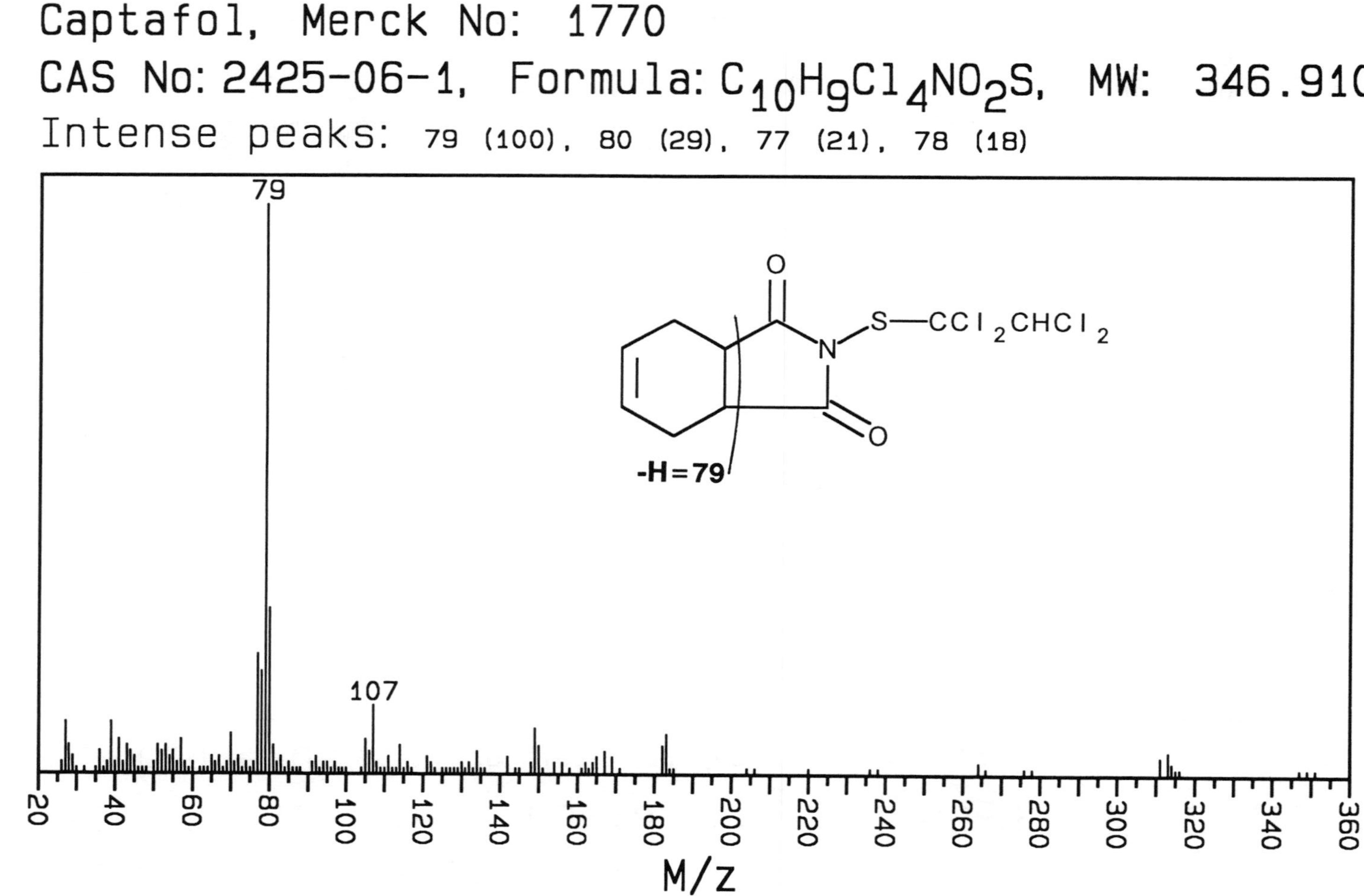
Captafol, Merck No: 1770
CAS No: 2425-06-1, Formula: $C_{10}H_9Cl_4NO_2S$, MW: 346.9108
Intense peaks: 79 (100), 80 (29), 77 (21), 78 (18)
79
107
O
O
N
S—CCl2CHCl2
-H=79
20
40
60
80
100
120
140
160
180
200
220
240
260
280
300
320
340
360
M/Z

Permethrin, Merck No: 7132

CAS No: 52645-53-1, Formula: $C_{21}H_{20}Cl_2O_3$, MW: 390.0789

Intense peaks: 183 (100), 163 (16), 165 (12), 184 ( 9)

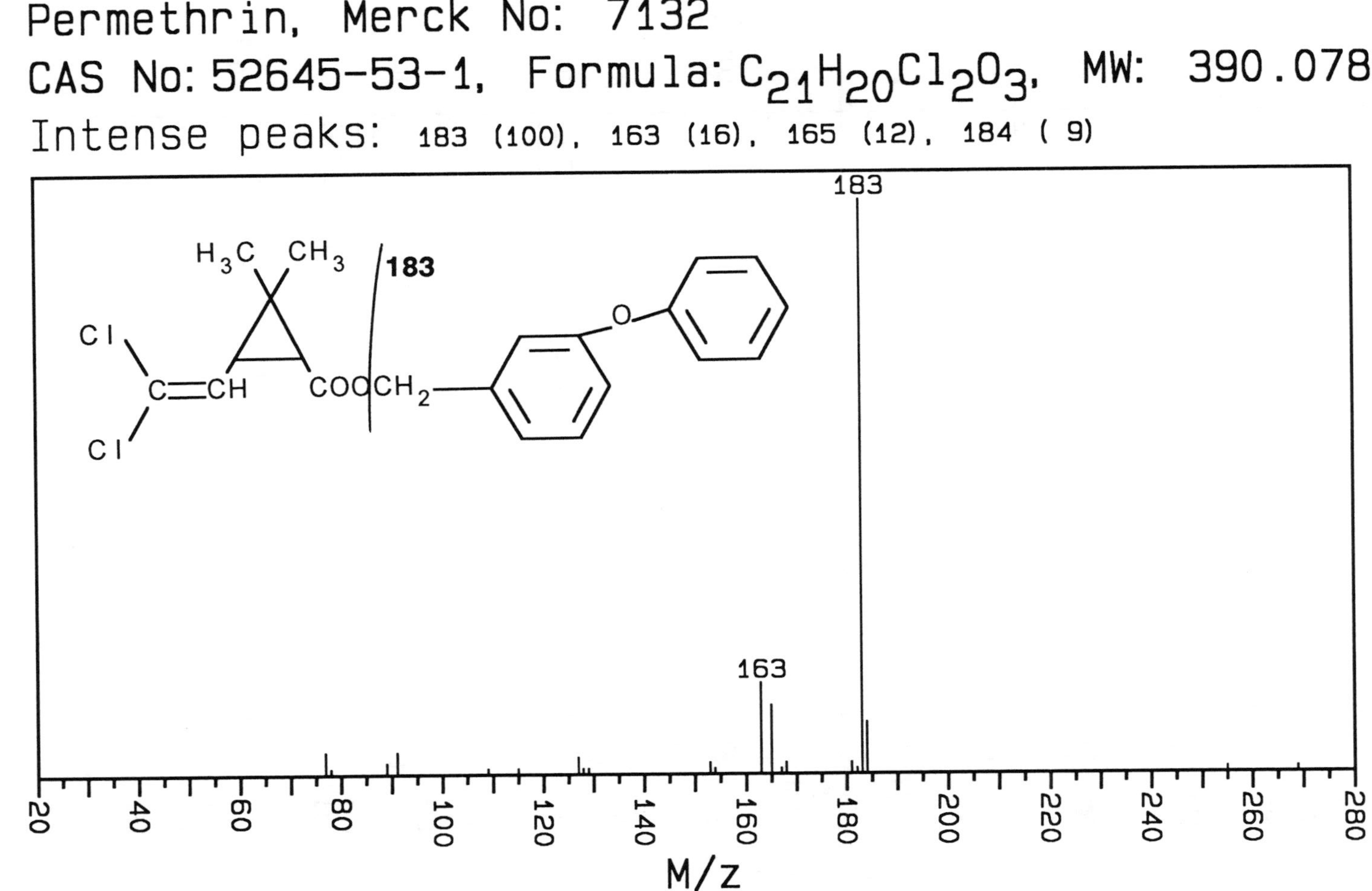

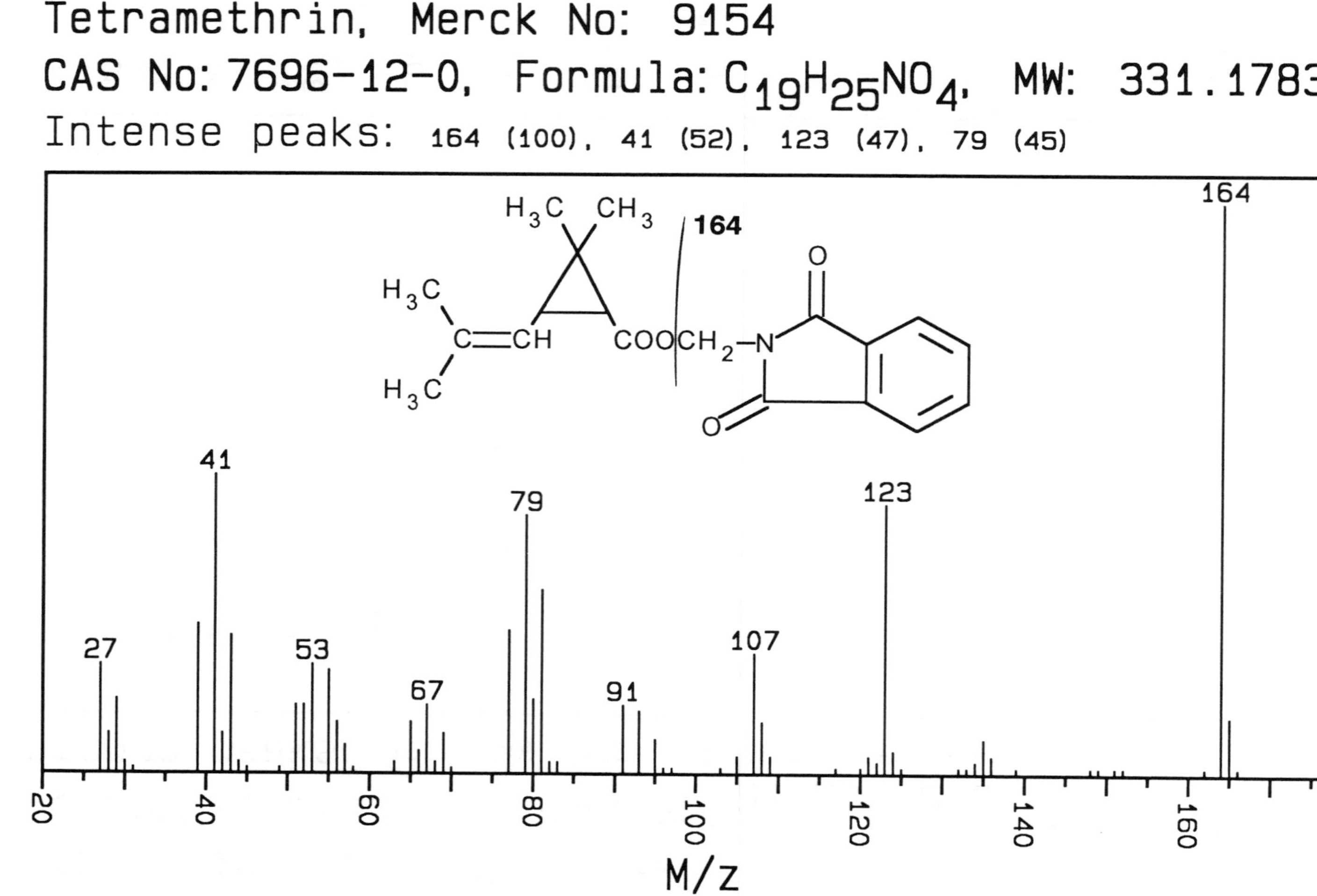
Tetramethrin, Merck No: 9154
CAS No: 7696-12-0, Formula: $C_{19}H_{25}NO_4$, MW: 331.1783
Intense peaks: 164 (100), 41 (52), 123 (47), 79 (45)
H3C CH3
164
H3C
C=CH
H3C
COOCH2–N
O
O
164
41
79
123
27
53
107
67
91
20
40
60
80
100
120
140
160
180
M/Z

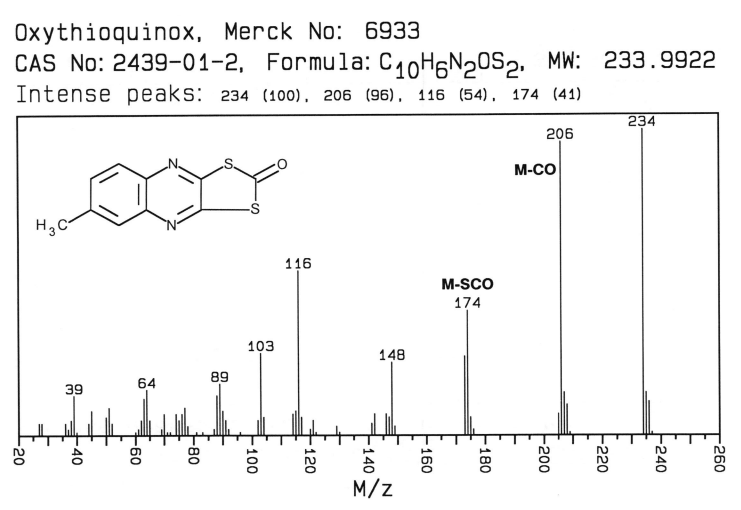

Oxythioquinox, Merck No: 6933

CAS No: 2439-01-2, Formula: $C_{10}H_6N_2OS_2$, MW: 233.9922

Intense peaks: 234 (100), 206 (96), 116 (54), 174 (41)

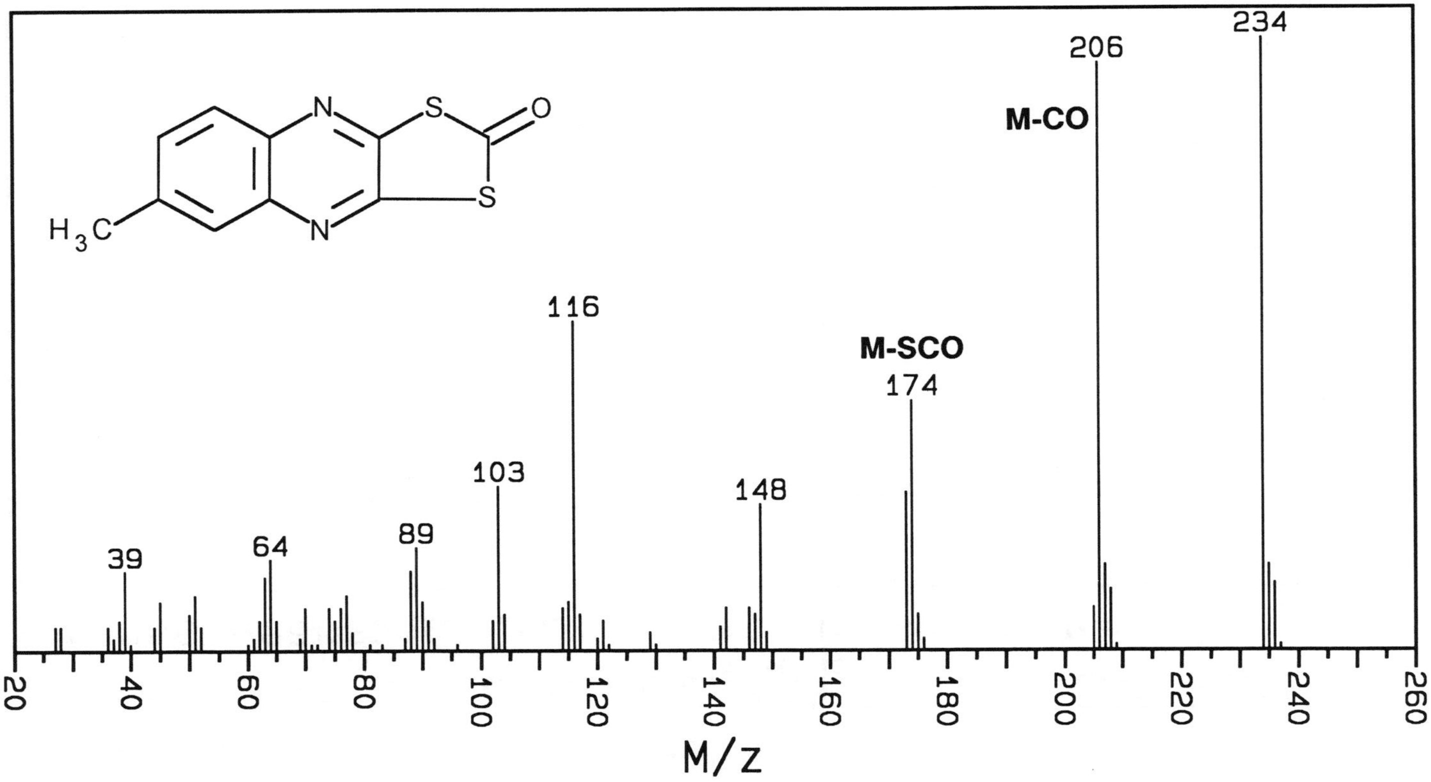

N S O S N $H_3C$

234
206
M-CO
116
M-SCO
174
103
148
39
64
89

20 40 60 80 100 120 140 160 180 200 220 240 260

M/Z

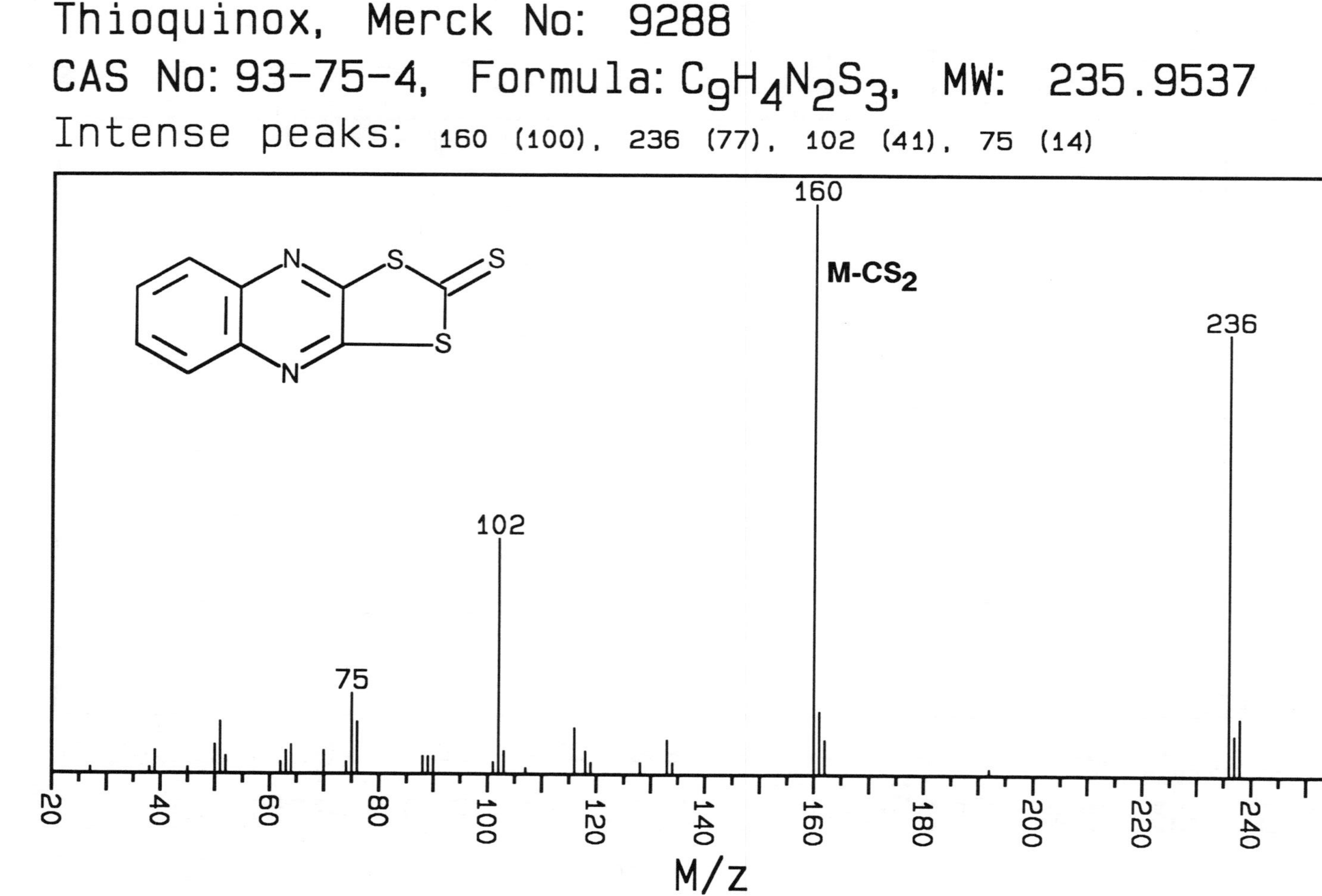
Thioquinox, Merck No: 9288
CAS No: 93-75-4, Formula: $C_9H_4N_2S_3$, MW: 235.9537
Intense peaks: 160 (100), 236 (77), 102 (41), 75 (14)
160
M-$CS_2$
236
102
75
N
N
S
S
S
20
40
60
80
100
120
140
160
180
200
220
240
260
M/Z

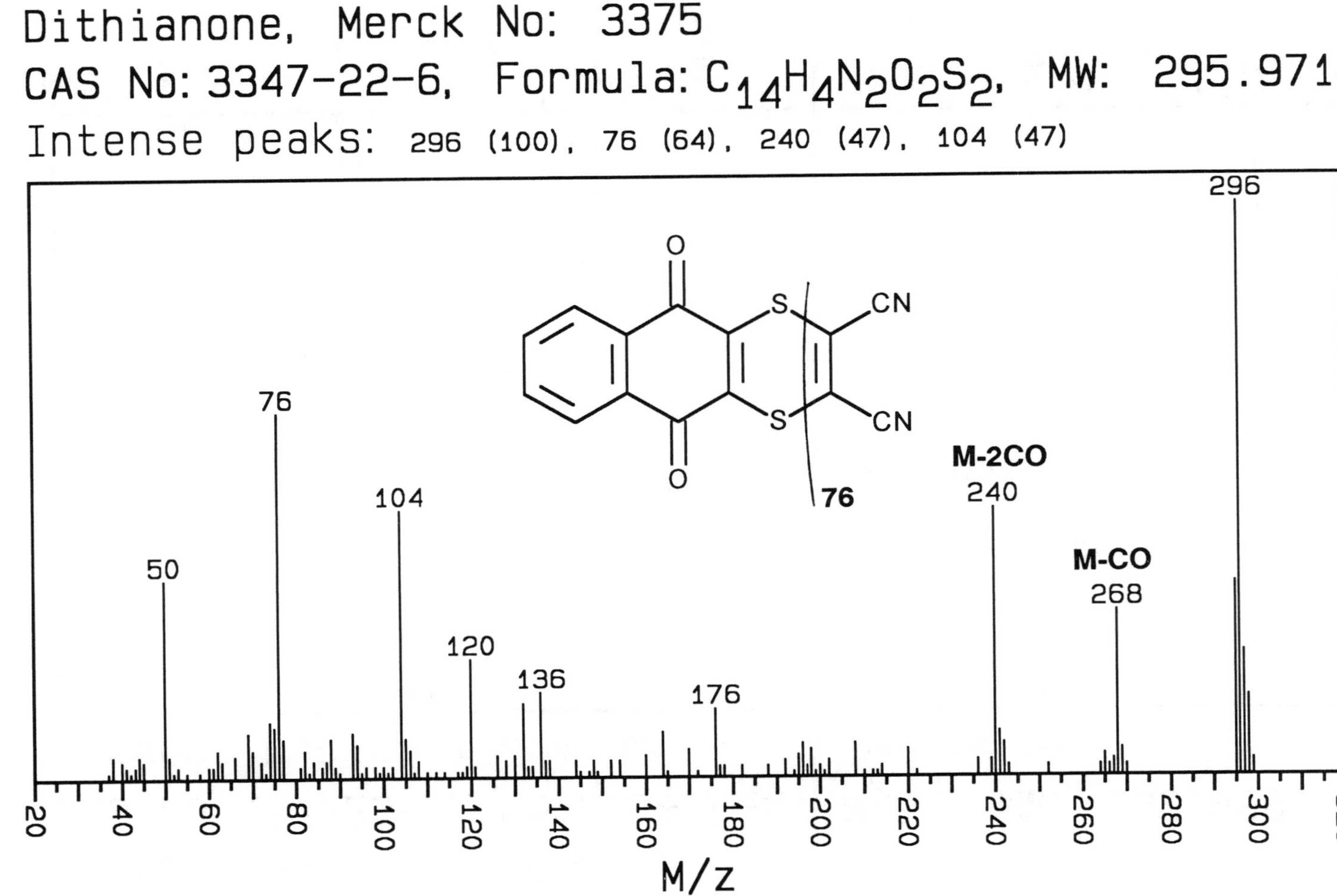
Dithianone, Merck No: 3375
CAS No: 3347-22-6, Formula: $C_{14}H_4N_2O_2S_2$, MW: 295.9714
Intense peaks: 296 (100), 76 (64), 240 (47), 104 (47)
O
S
CN
S
CN
O
76
296
76
104
50
120
136
176
M-2CO
240
M-CO
268
20
40
60
80
100
120
140
160
180
200
220
240
260
280
300
320
M/Z

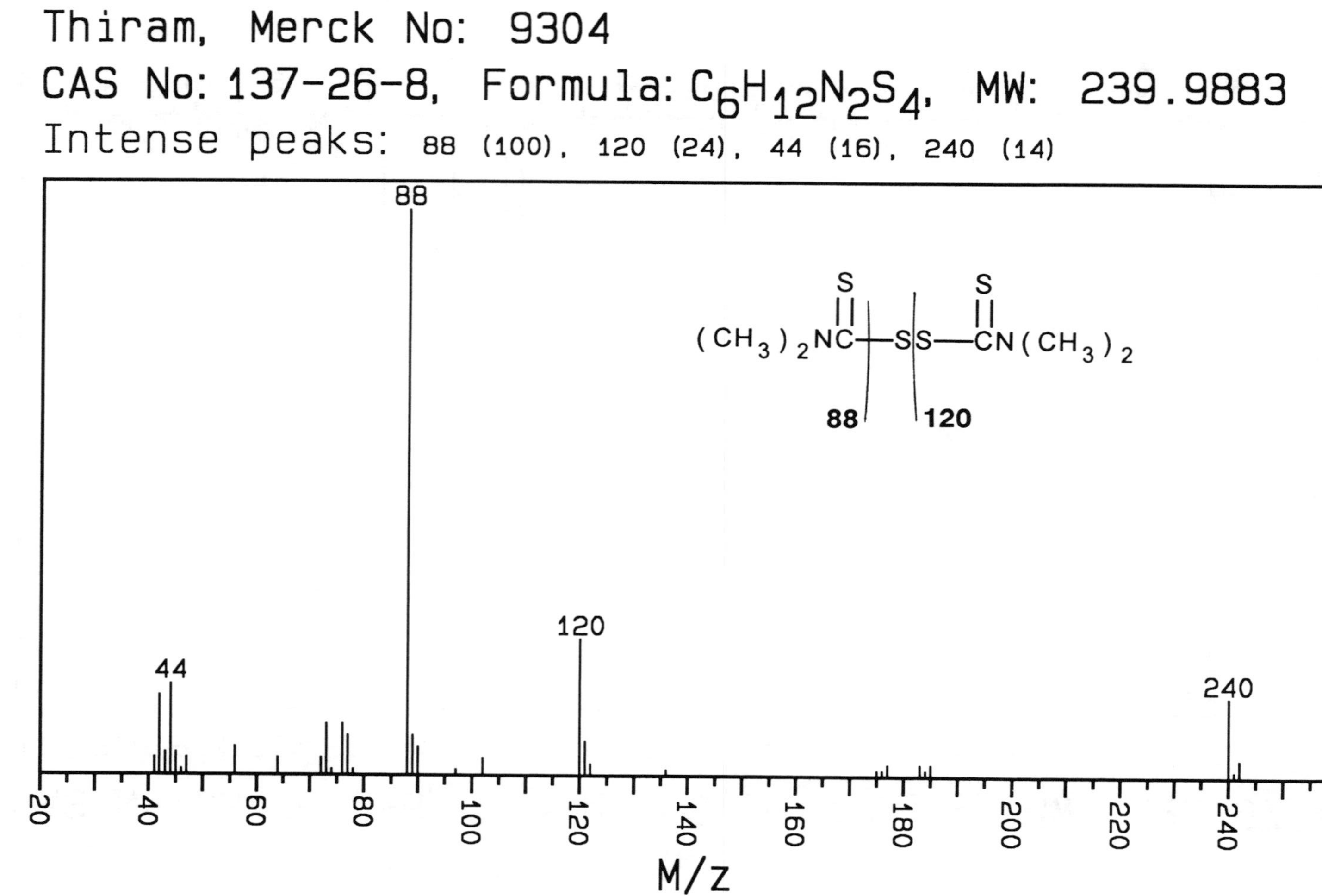
Thiram, Merck No: 9304
CAS No: 137-26-8, Formula: $C_6H_{12}N_2S_4$, MW: 239.9883
Intense peaks: 88 (100), 120 (24), 44 (16), 240 (14)
$(CH_3)_2NC(=S)-SS-C(=S)N(CH_3)_2$
88
120
88
44
120
240
20
40
60
80
100
120
140
160
180
200
220
240
260
M/Z

Dixanthogen, Merck No: 3390

CAS No: 502-55-6, Formula: $C_6H_{10}O_2S_4$, MW: 241.9564

Intense peaks: 29 (100), 89 (43), 27 (41), 94 (30)

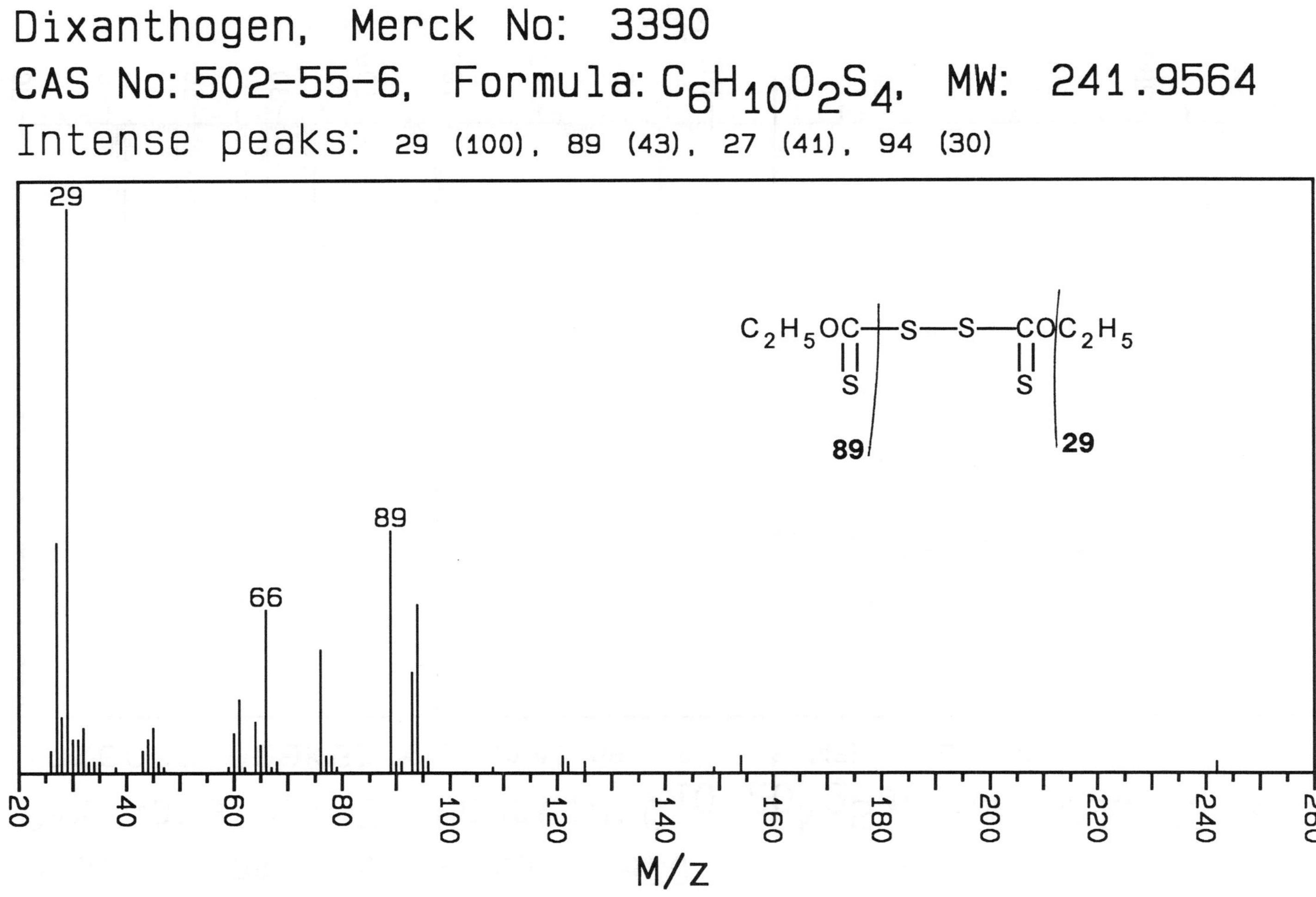

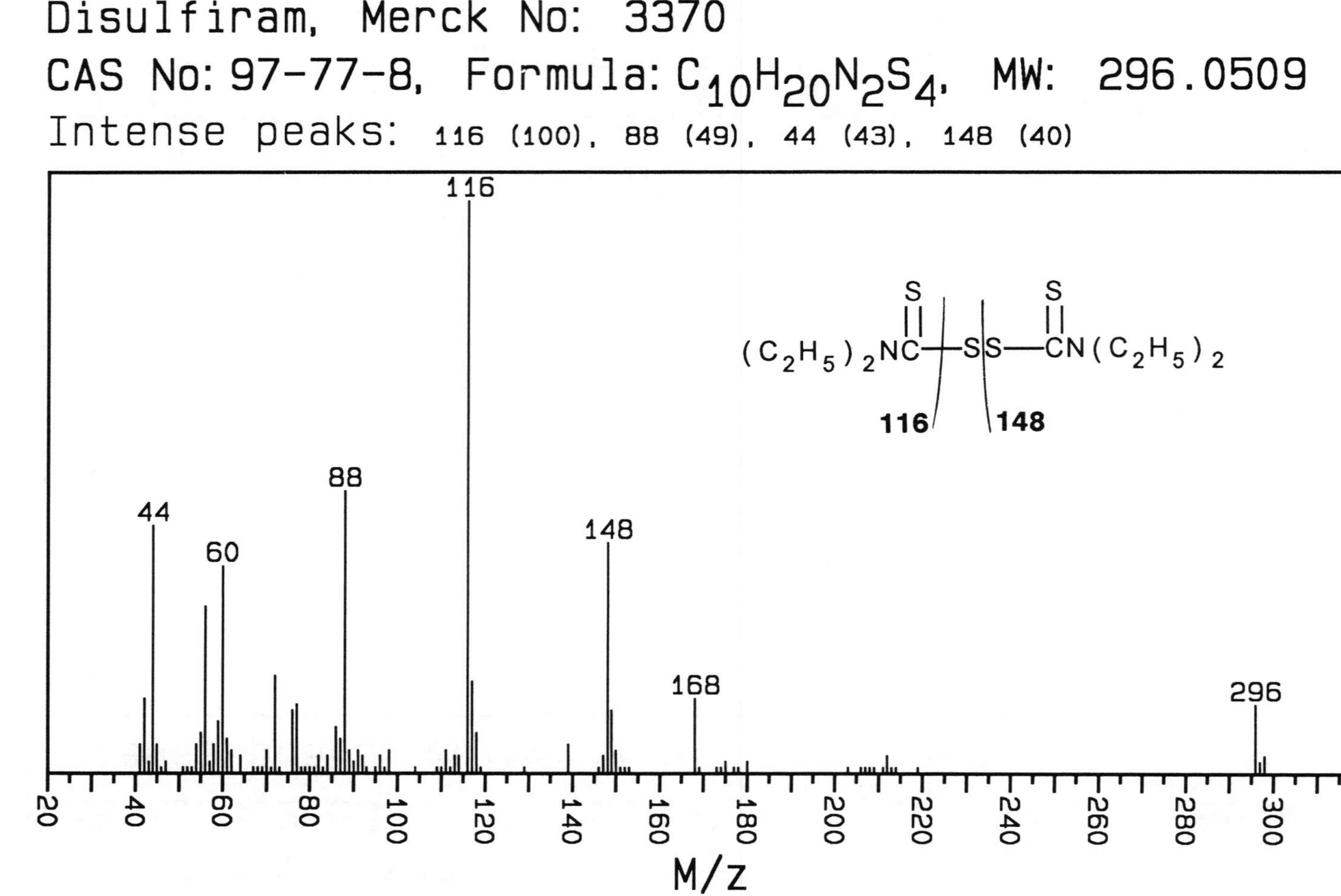
Disulfiram, Merck No: 3370
CAS No: 97-77-8, Formula: $C_{10}H_{20}N_2S_4$, MW: 296.0509
Intense peaks: 116 (100), 88 (49), 44 (43), 148 (40)
$(C_2H_5)_2NC(=S)$—SS—$C(=S)N(C_2H_5)_2$
116
148
116
88
44
60
148
168
296
20
40
60
80
100
120
140
160
180
200
220
240
260
280
300
320
M/Z

Chlorbenside, Merck No: 2074

CAS No: 103-17-3, Formula: $C_{13}H_{10}Cl_2S$, MW: 267.9881

Intense peaks: 125 (100), 127 (34), 89 (19), 63 (15)

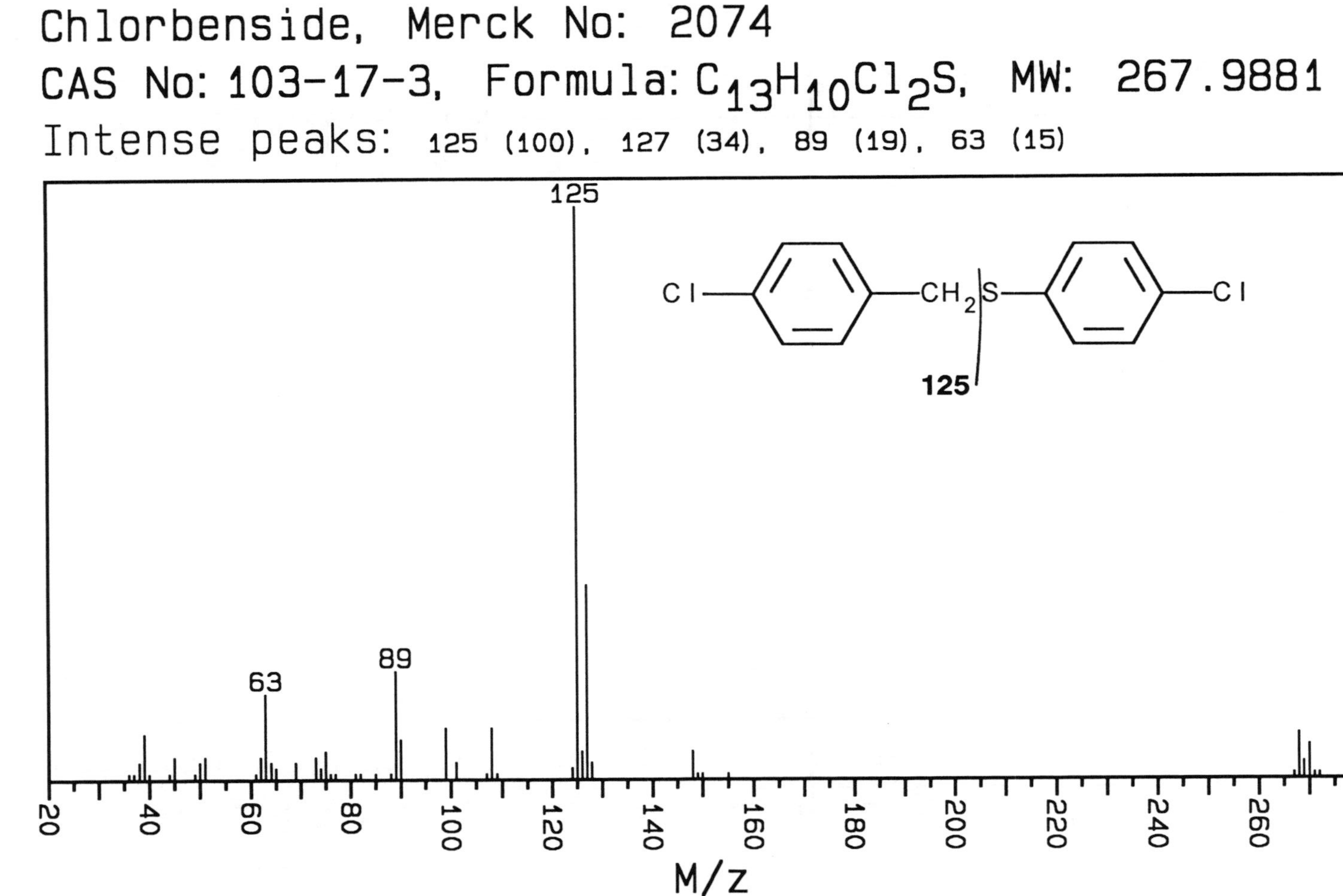

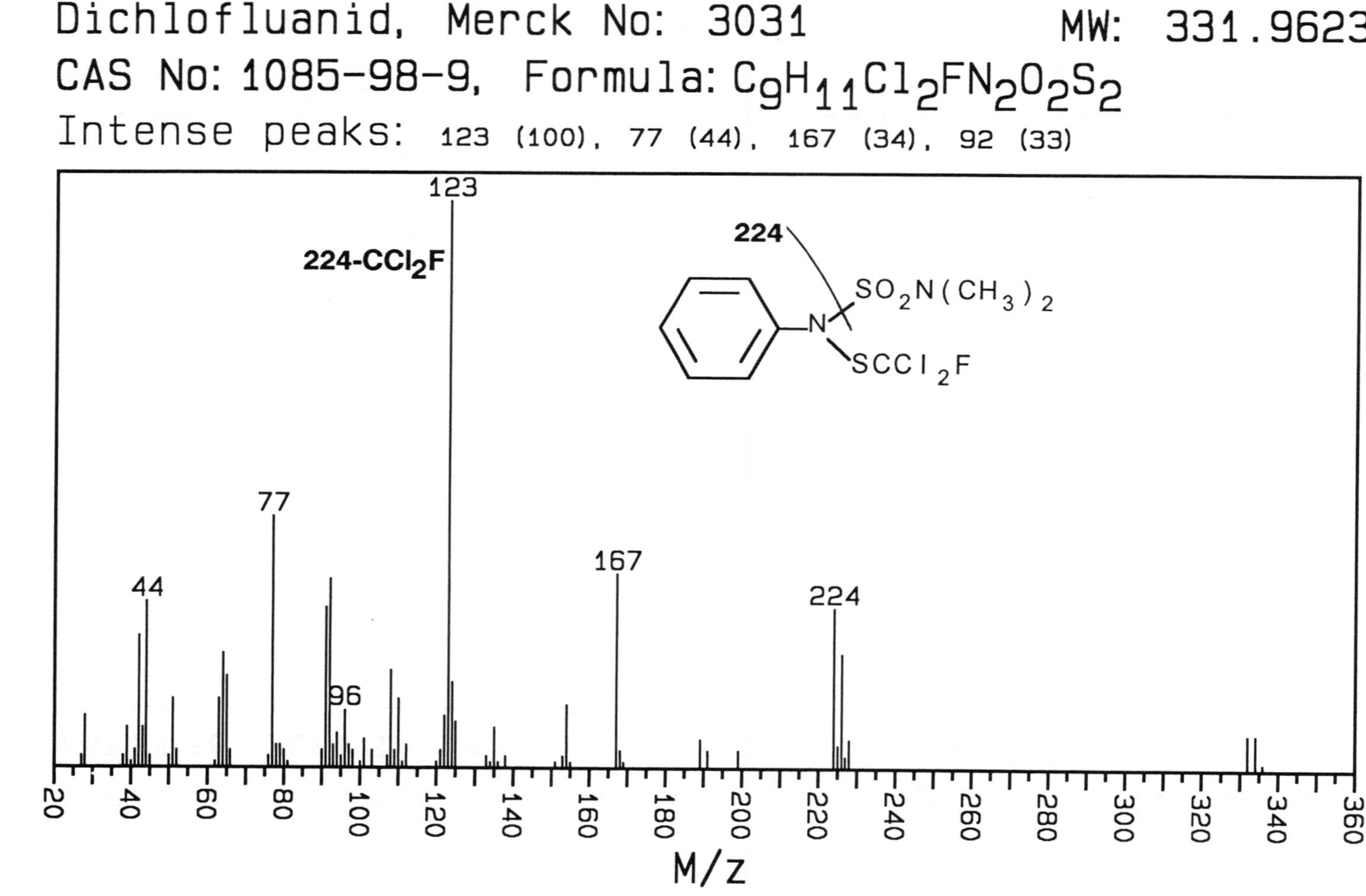
Dichlofluanid, Merck No: 3031 MW: 331.9623
CAS No: 1085-98-9, Formula: $C_9H_{11}Cl_2FN_2O_2S_2$
Intense peaks: 123 (100), 77 (44), 167 (34), 92 (33)
123
224-$CCl_2F$
224
$SO_2N(CH_3)_2$
N
$SCCl_2F$
77
44
96
167
224
20
40
60
80
100
120
140
160
180
200
220
240
260
280
300
320
340
360
M/Z

Propargite, Merck No: 7818

CAS No: 2312-35-8, Formula: $C_{19}H_{26}O_4S$, MW: 350.1552

Intense peaks: 135 (100), 81 (63), 173 (36), 39 (31)

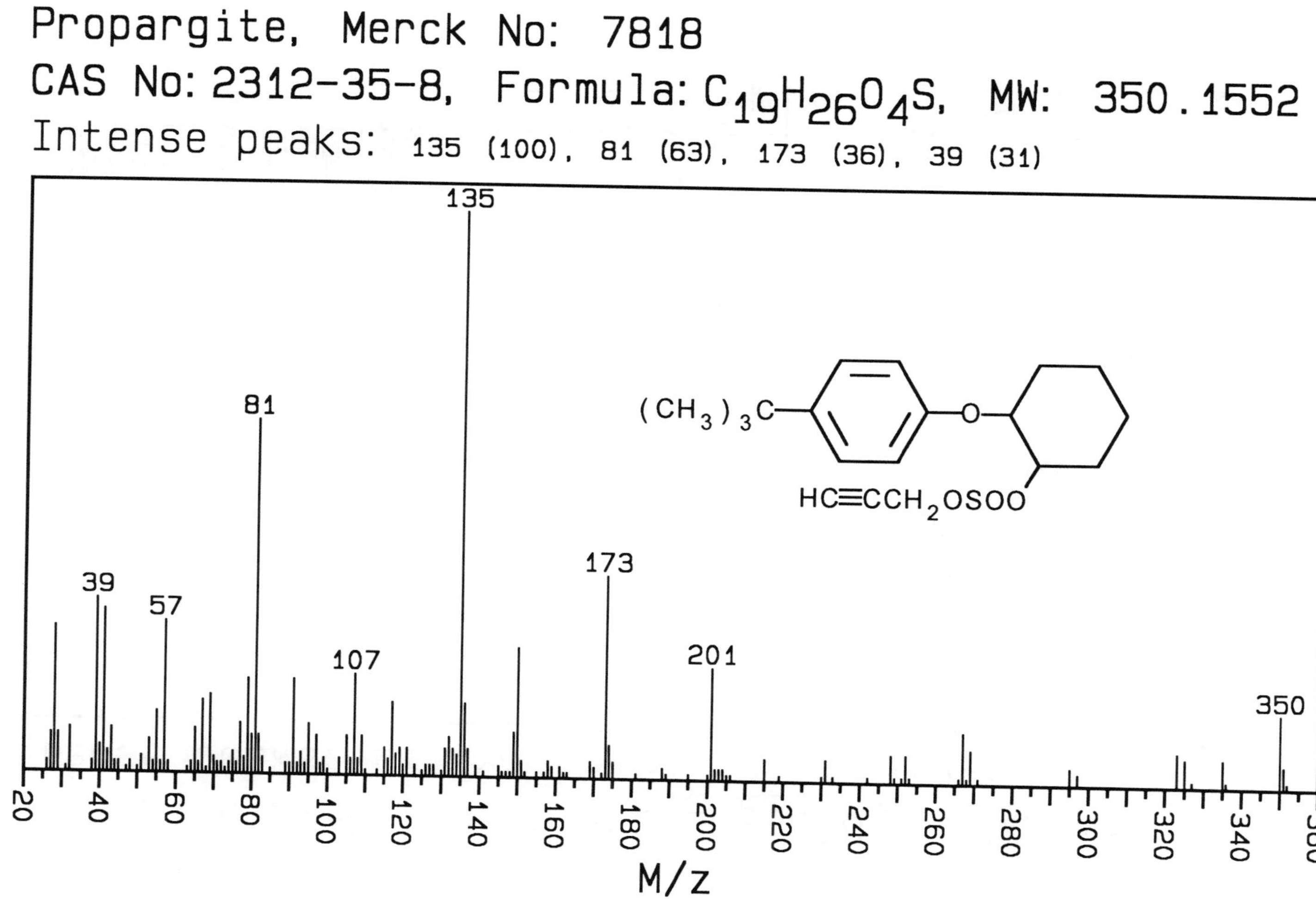

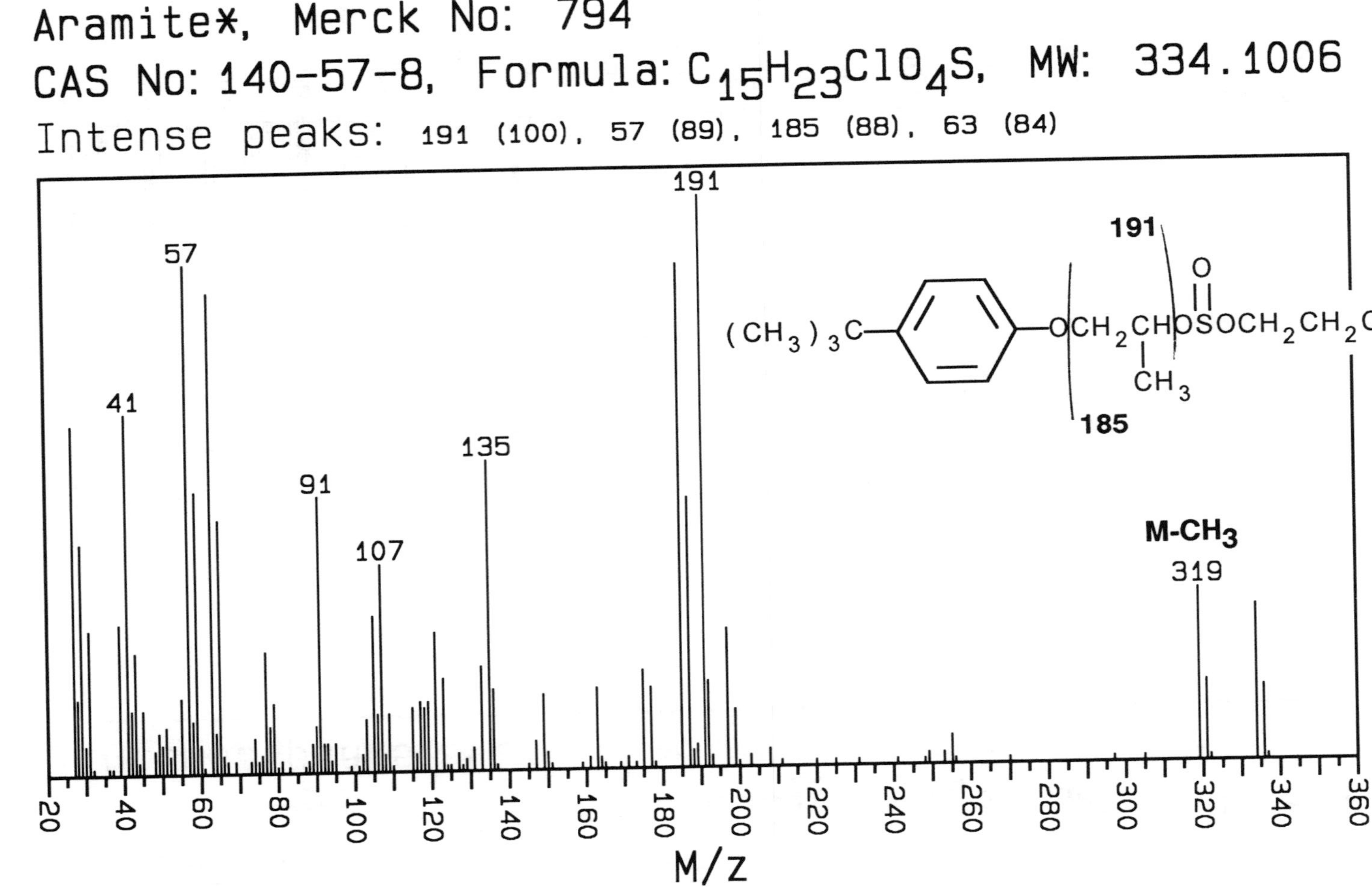
Aramite*, Merck No: 794
CAS No: 140-57-8, Formula: C15H23ClO4S, MW: 334.1006
Intense peaks: 191 (100), 57 (89), 185 (88), 63 (84)
(CH3)3C
OCH2CHOSOCH2CH2Cl
O
CH3
191
185
M-CH3
319
191
135
107
91
57
41
M/Z

Diphenyl sulfone, Merck No: 3336

CAS No: 127-63-9, Formula: $C_{12}H_{10}O_2S$, MW: 218.0401

Intense peaks: 125 (100), 77 (80), 51 (75), 218 (62)

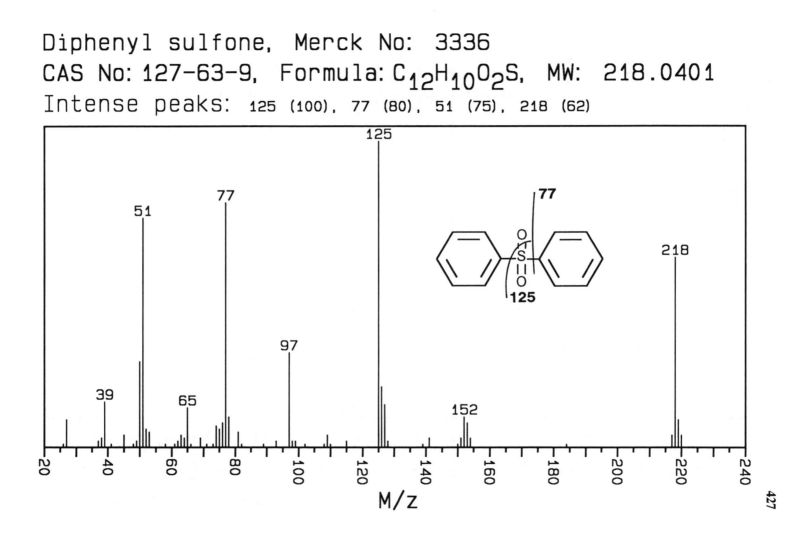

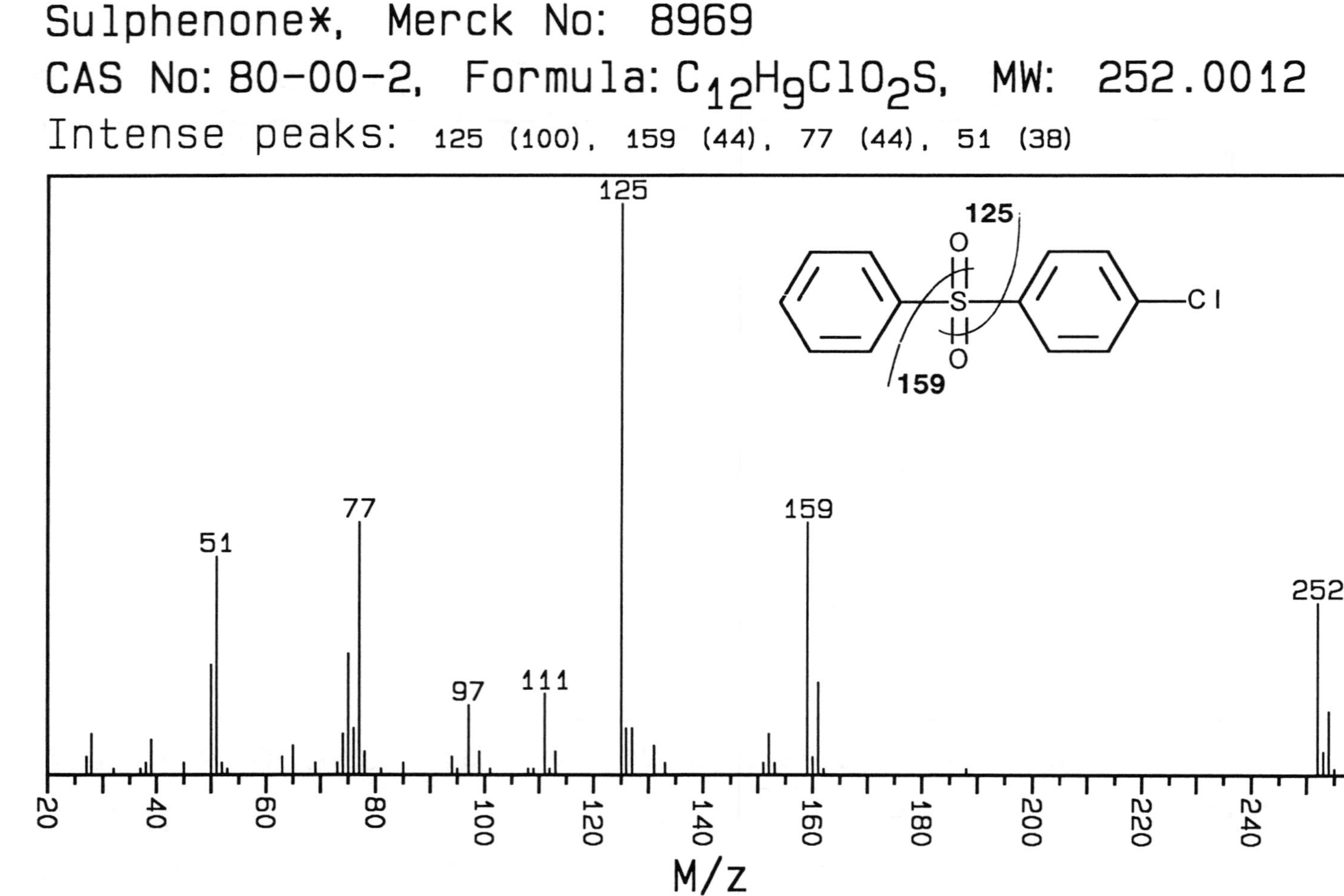
Sulphenone*, Merck No: 8969
CAS No: 80-00-2, Formula: $C_{12}H_9ClO_2S$, MW: 252.0012
Intense peaks: 125 (100), 159 (44), 77 (44), 51 (38)
125
125
O
S
O
Cl
159
51
77
97
111
159
252
20
40
60
80
100
120
140
160
180
200
220
240
260
M/Z

Perfluidone, Merck No: 7113

CAS No: 37924-13-3, Formula: $C_{14}H_{12}F_3NO_4S_2$, MW: 379.0161

Intense peaks: 125 (100), 77 (37), 51 (32), 97 (21)

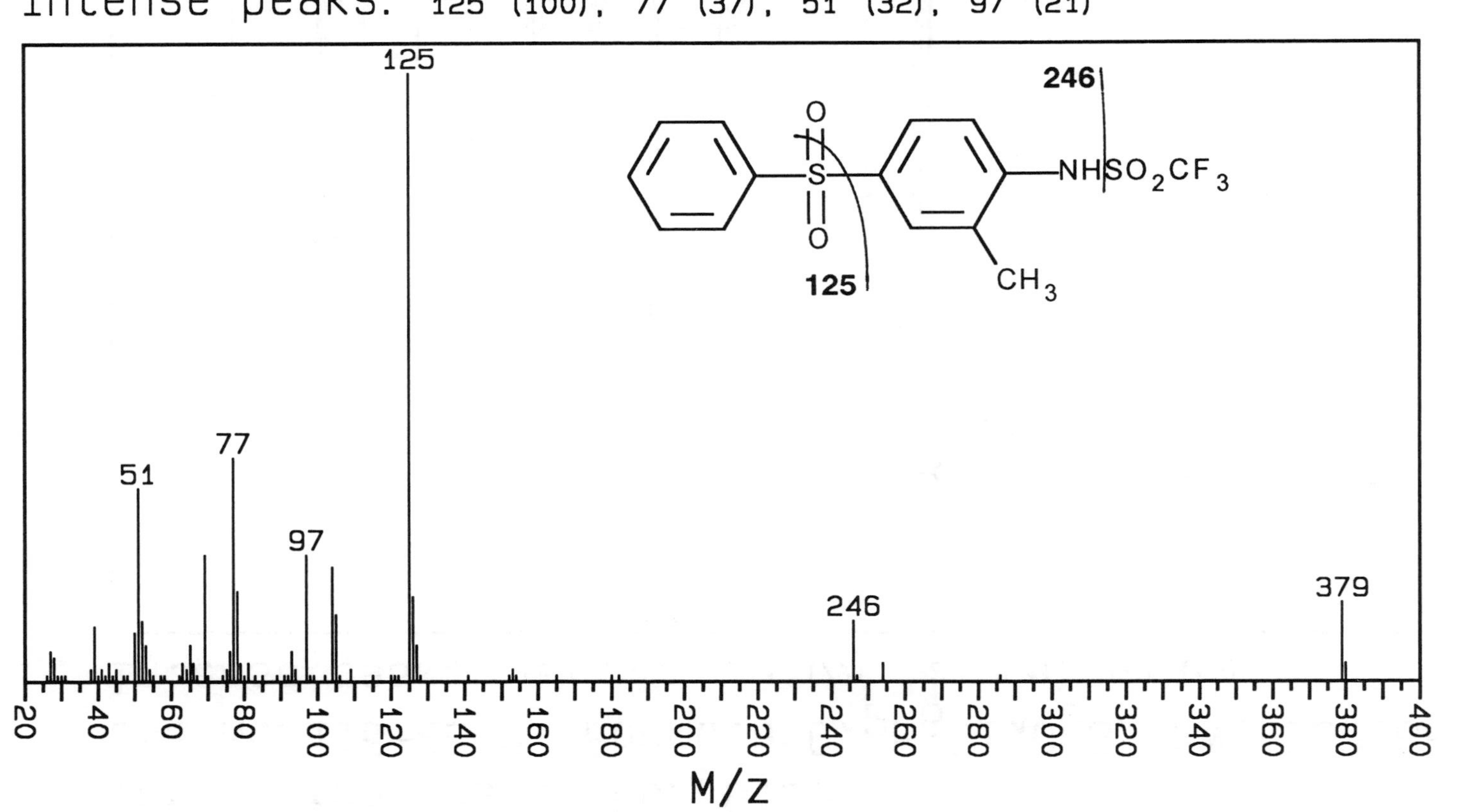

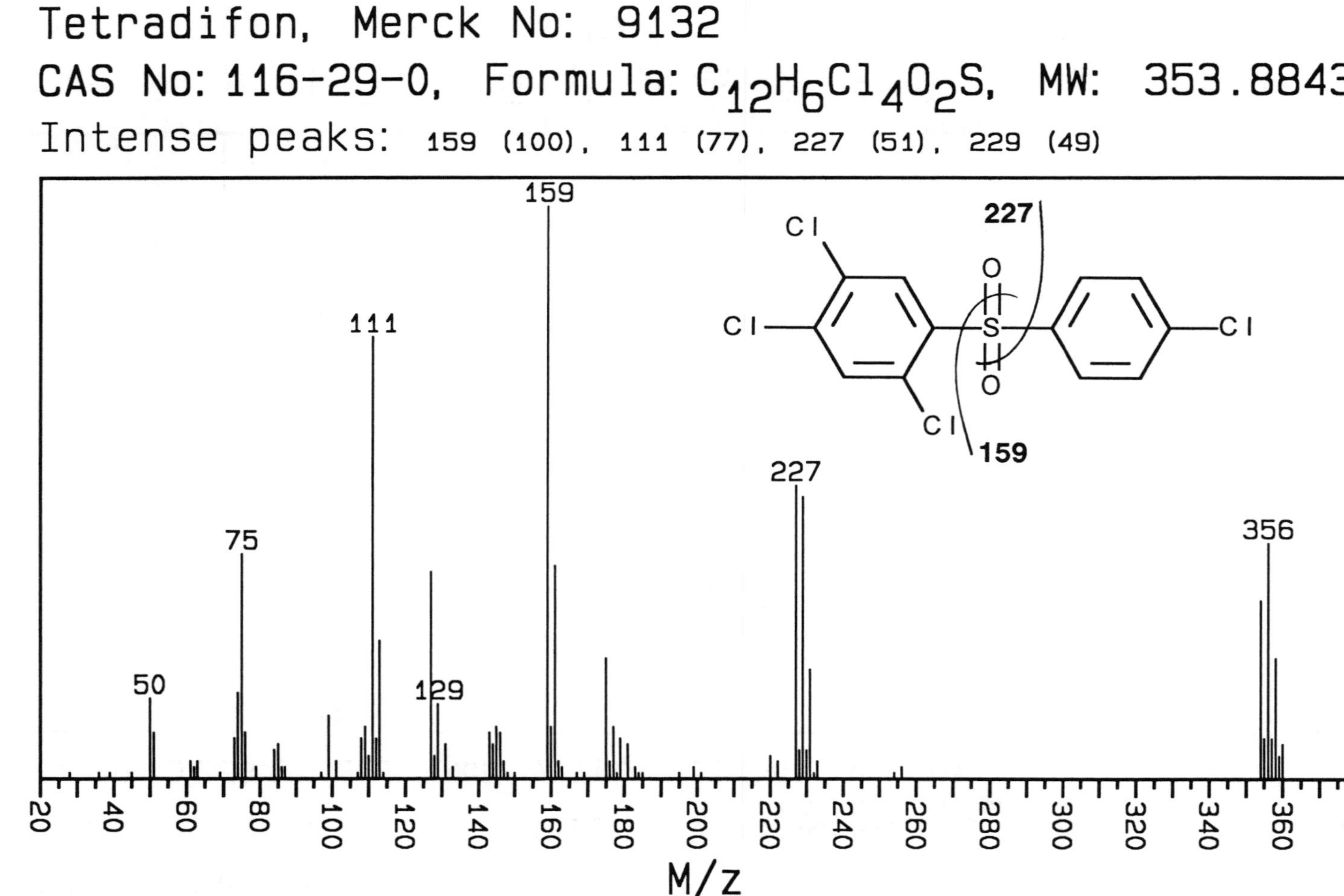

Tetradifon, Merck No: 9132
CAS No: 116-29-0, Formula: $C_{12}H_6Cl_4O_2S$, MW: 353.8843
Intense peaks: 159 (100), 111 (77), 227 (51), 229 (49)
159
111
75
50
129
227
356
Cl
O
S
227
159
20
40
60
80
100
120
140
160
180
200
220
240
260
280
300
320
340
360
380
M/Z

Ovex, Merck No: 6856

CAS No: 80-33-1, Formula: $C_{12}H_8Cl_2O_3S$, MW: 301.9571

Intense peaks: 111 (100), 175 (75), 75 (53), 99 (51)

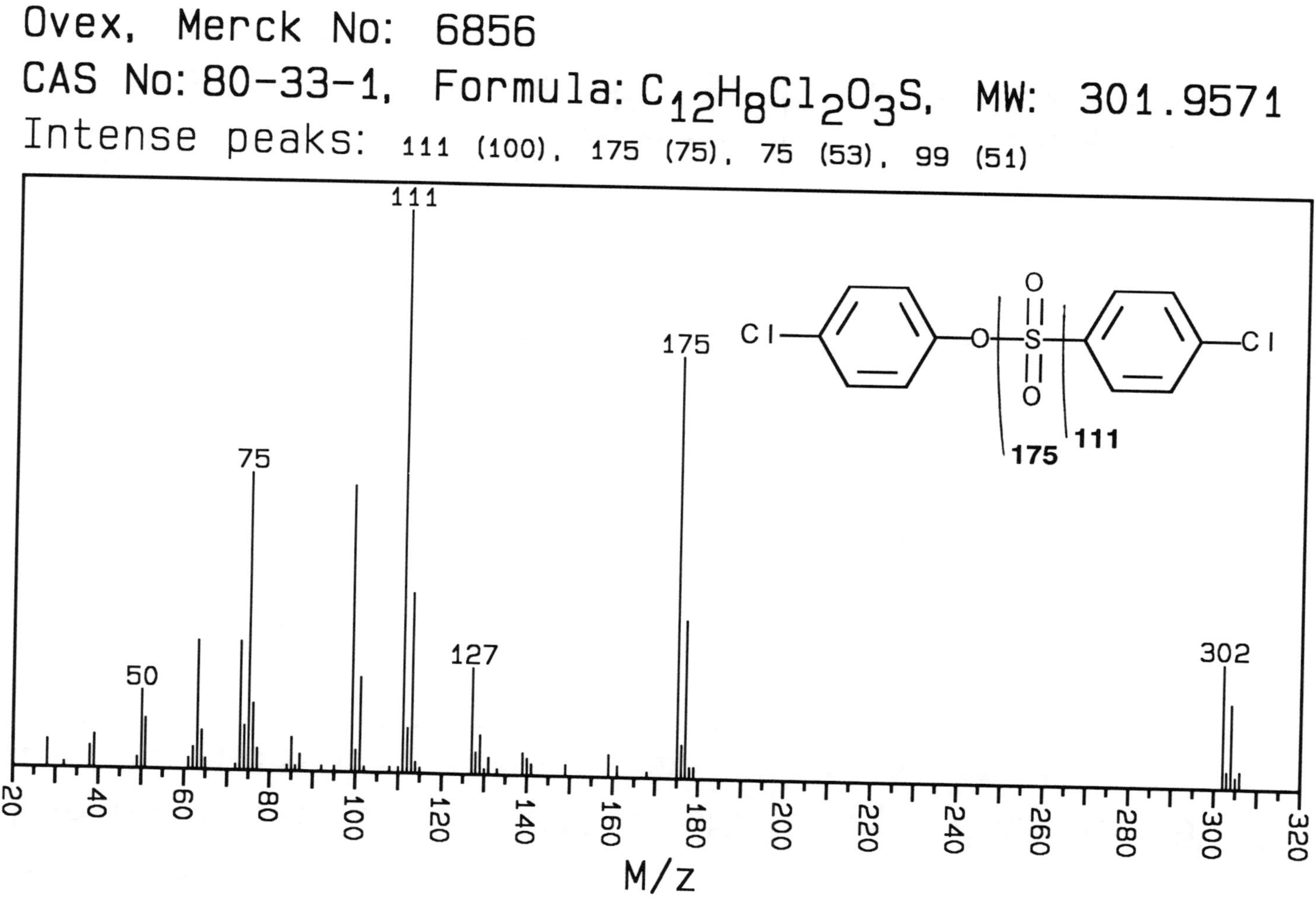

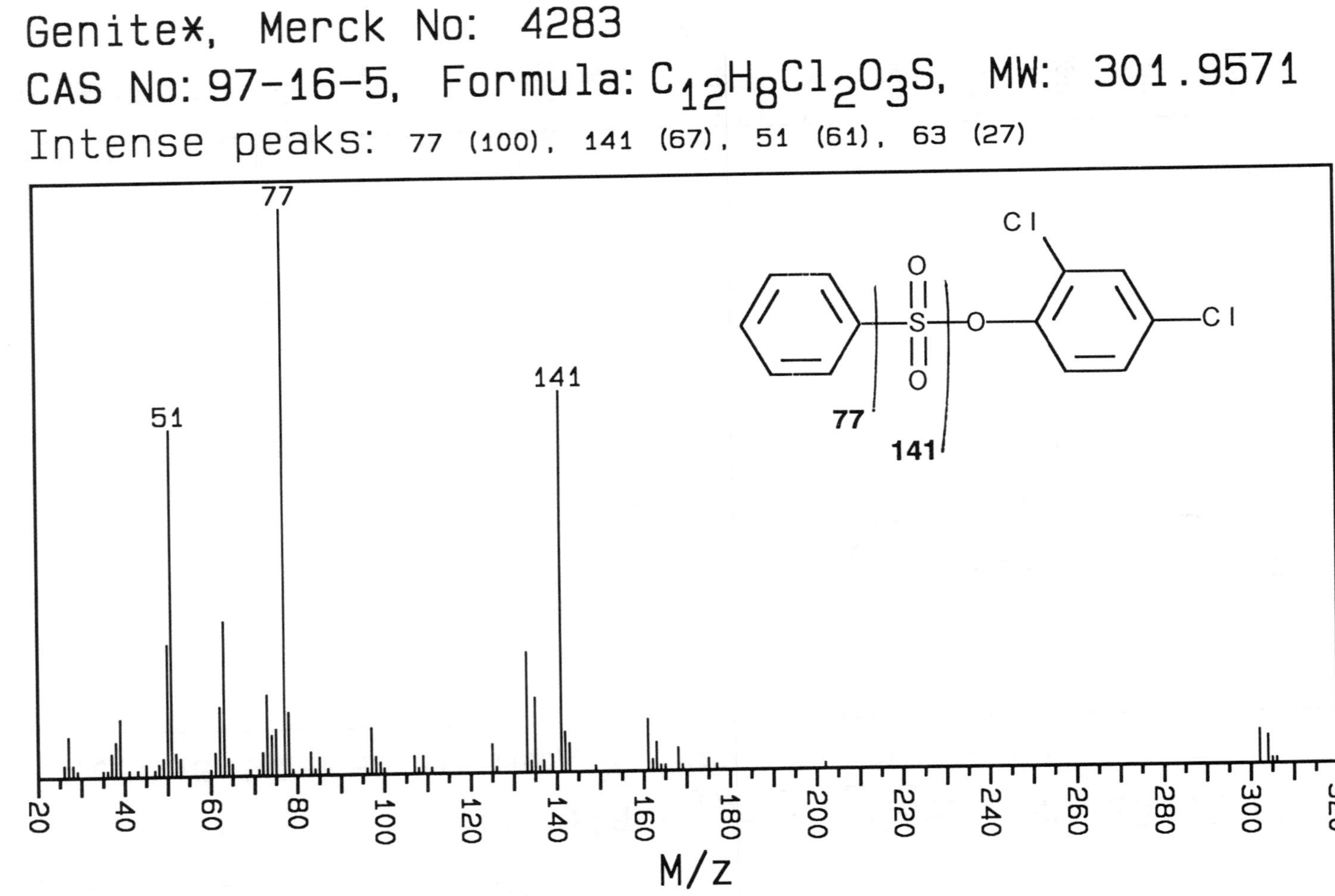
Genite*, Merck No: 4283
CAS No: 97-16-5, Formula: $C_{12}H_8Cl_2O_3S$, MW: 301.9571
Intense peaks: 77 (100), 141 (67), 51 (61), 63 (27)
77
141
51
Cl
Cl
O
S
O
O
77
141
20
40
60
80
100
120
140
160
180
200
220
240
260
280
300
320
M/Z

Bentazon, Merck No: 1060

CAS No: 25057-89-0, Formula: $C_{10}H_{12}N_2O_3S$, MW: 240.0569

Intense peaks: 119 (100), 198 (88), 92 (34), 161 (33)

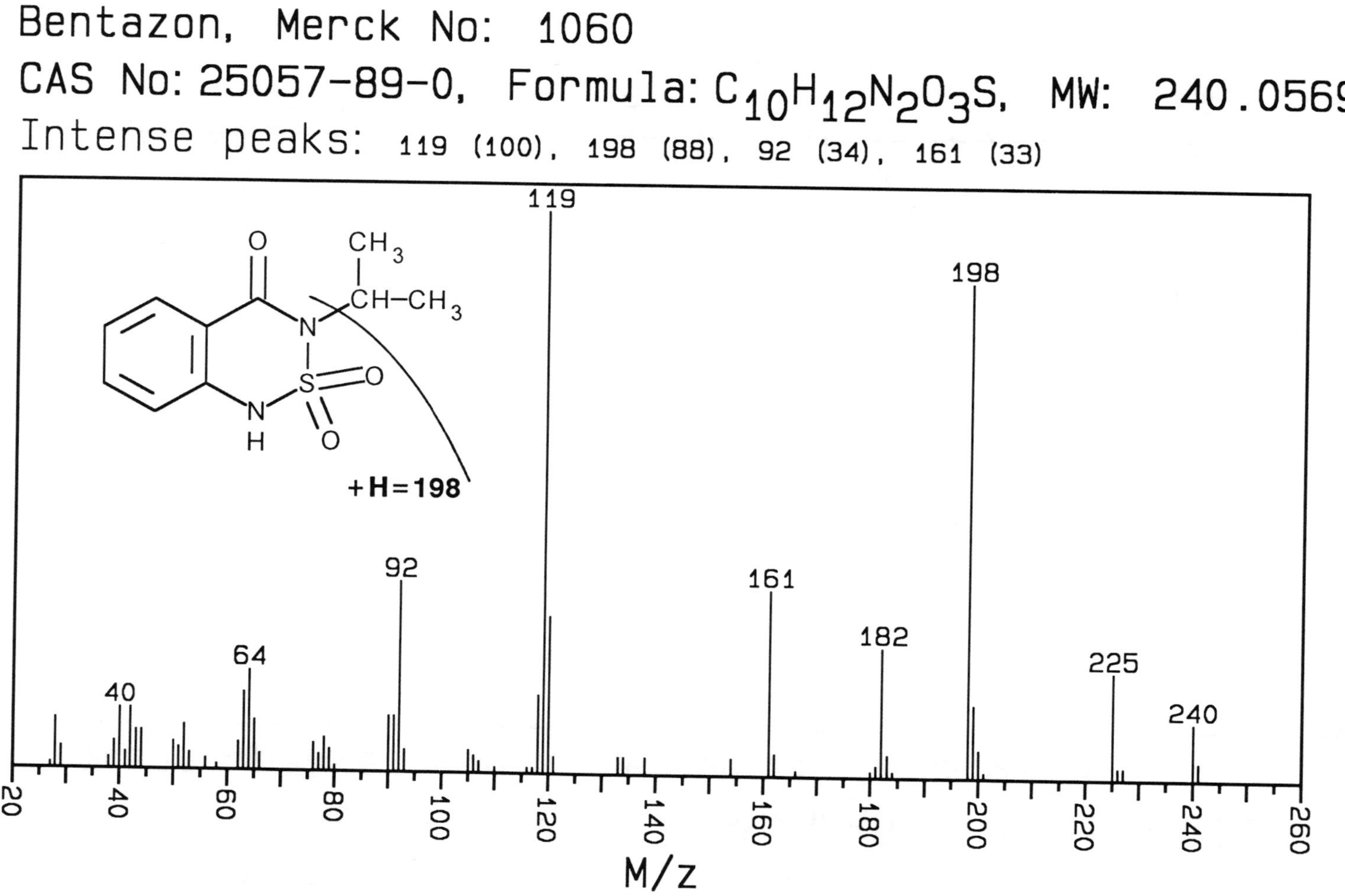

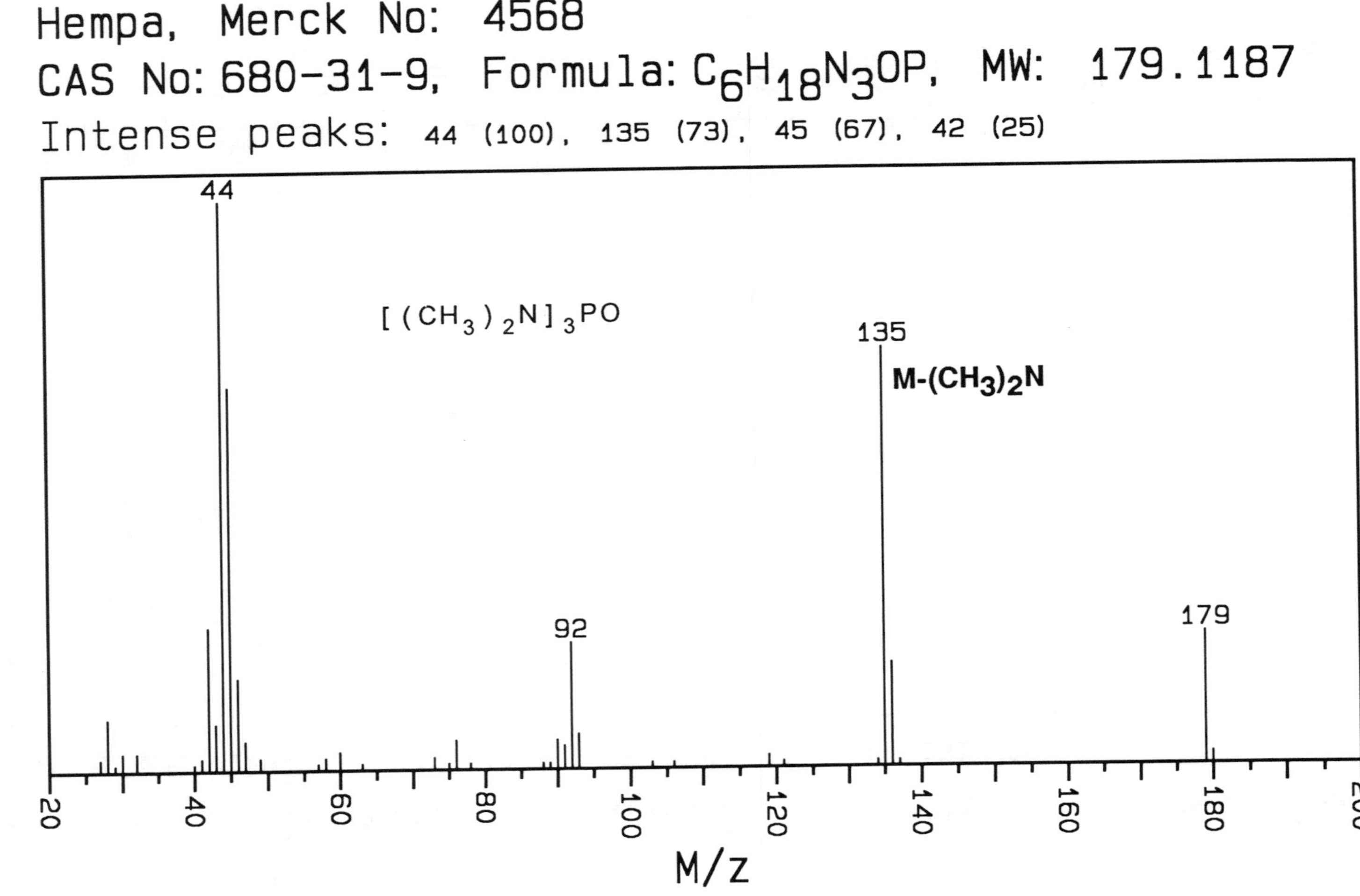
Hempa, Merck No: 4568
CAS No: 680-31-9, Formula: $C_6H_{18}N_3OP$, MW: 179.1187
Intense peaks: 44 (100), 135 (73), 45 (67), 42 (25)
$[(CH_3)_2N]_3PO$
44
135
M-$(CH_3)_2N$
92
179
20
40
60
80
100
120
140
160
180
200
M/Z

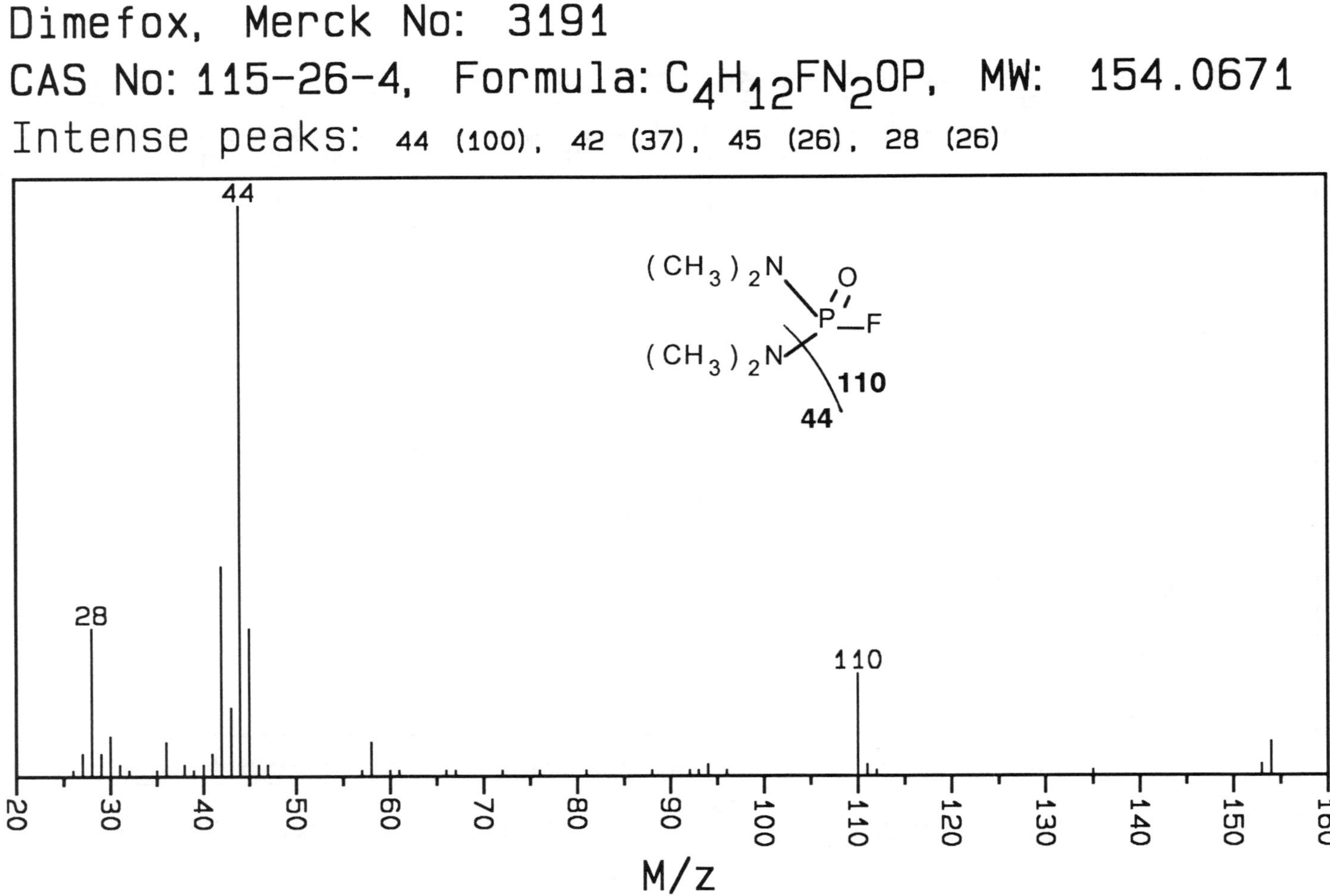

Dimefox, Merck No: 3191
CAS No: 115-26-4, Formula: $C_4H_{12}FN_2OP$, MW: 154.0671
Intense peaks: 44 (100), 42 (37), 45 (26), 28 (26)
44
28
110
$(CH_3)_2N$
$(CH_3)_2N$
O
P
F
110
44
20
30
40
50
60
70
80
90
100
110
120
130
140
150
160
M/Z

Crufomate, Merck No: 2607

CAS No: 299-86-5, Formula: $C_{12}H_{19}ClNO_3P$, MW: 291.0791

Intense peaks: 256 (100), 108 (80), 276 (71), 182 (65)

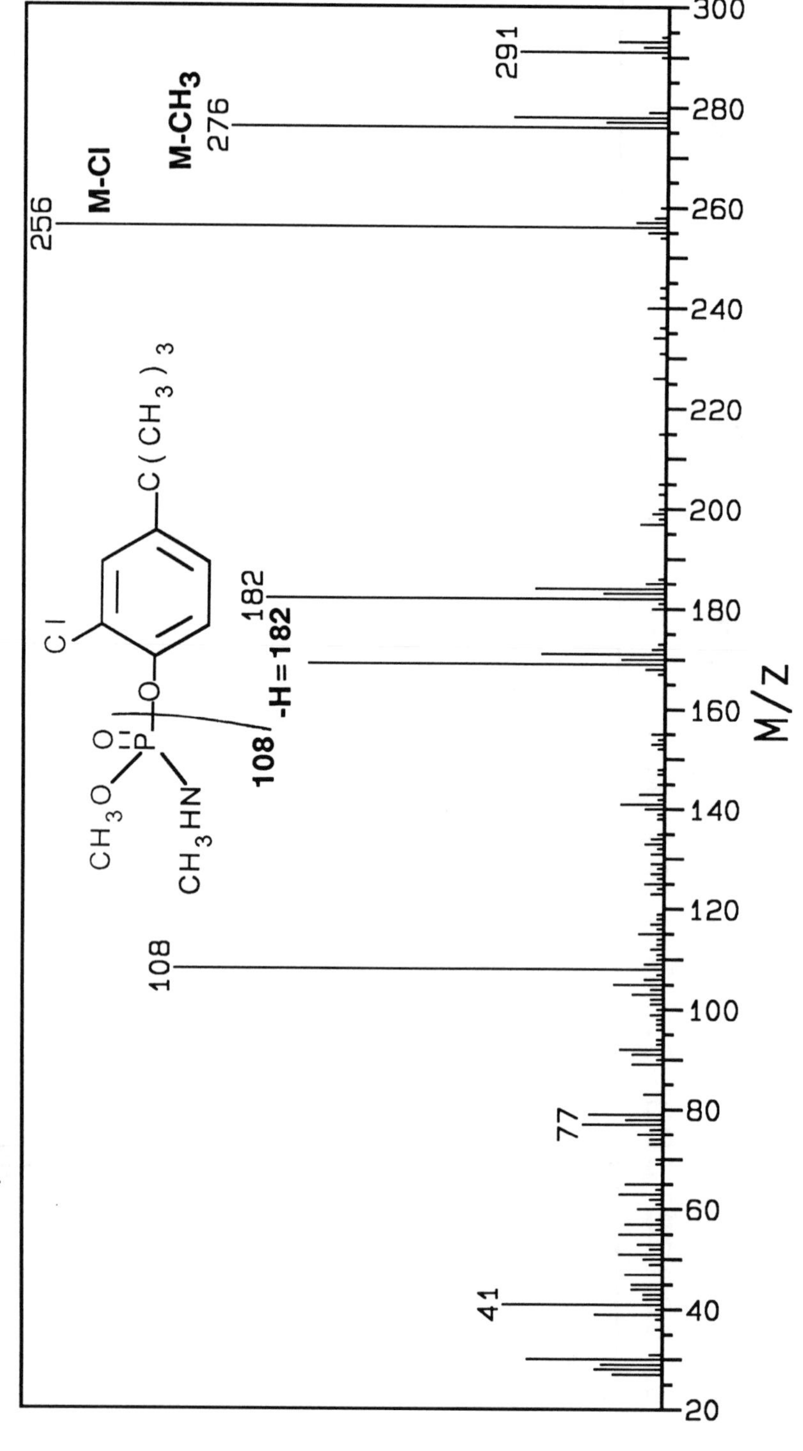

Ethephon, Merck No: 3686

CAS No: 16672-87-0, Formula: $C_2H_6ClO_3P$, MW: 143.9743

Intense peaks: 82 (100), 81 (55), 109 (41), 27 (31)

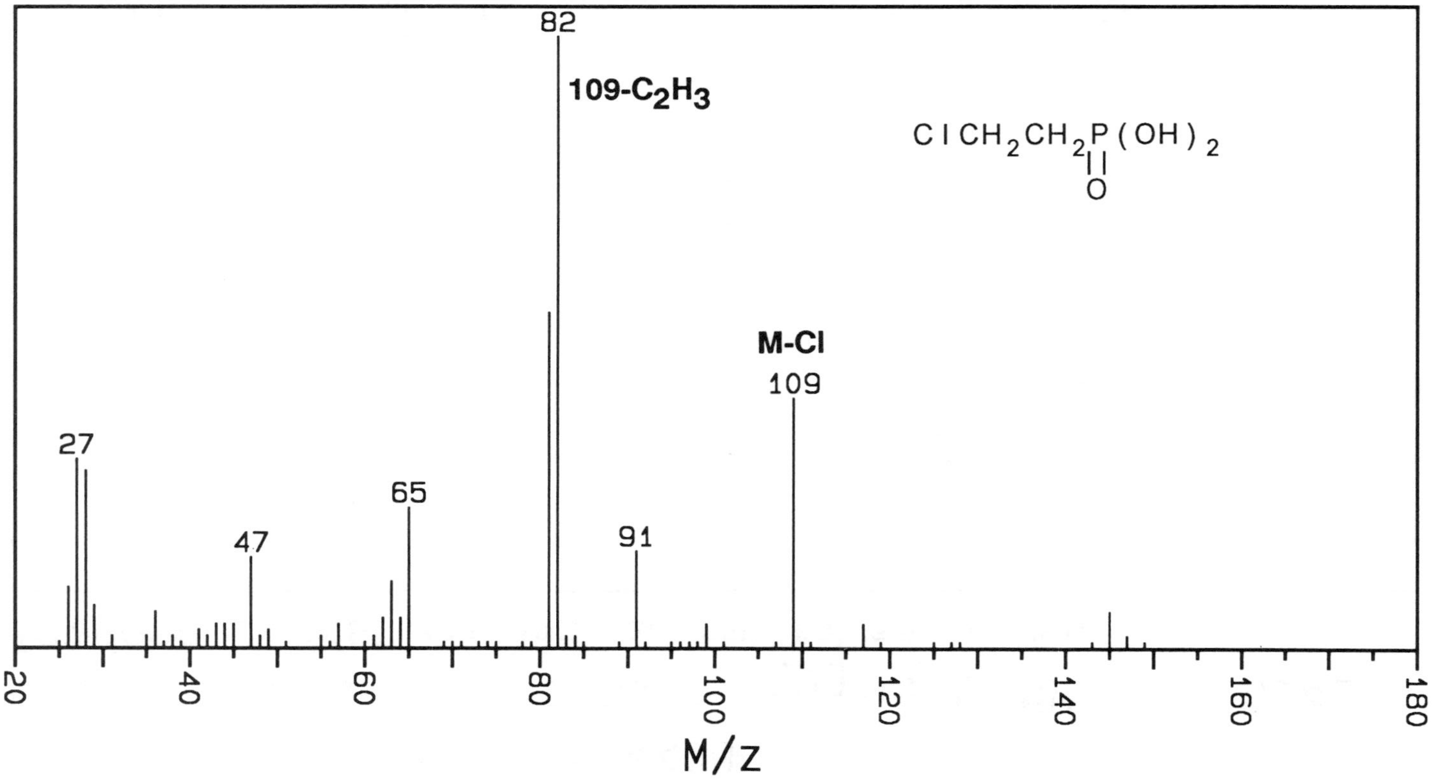

82

109-$C_2H_3$

$ClCH_2CH_2P(O)(OH)_2$

M-Cl

109

27

65

47

91

20 40 60 80 100 120 140 160 180

M/Z

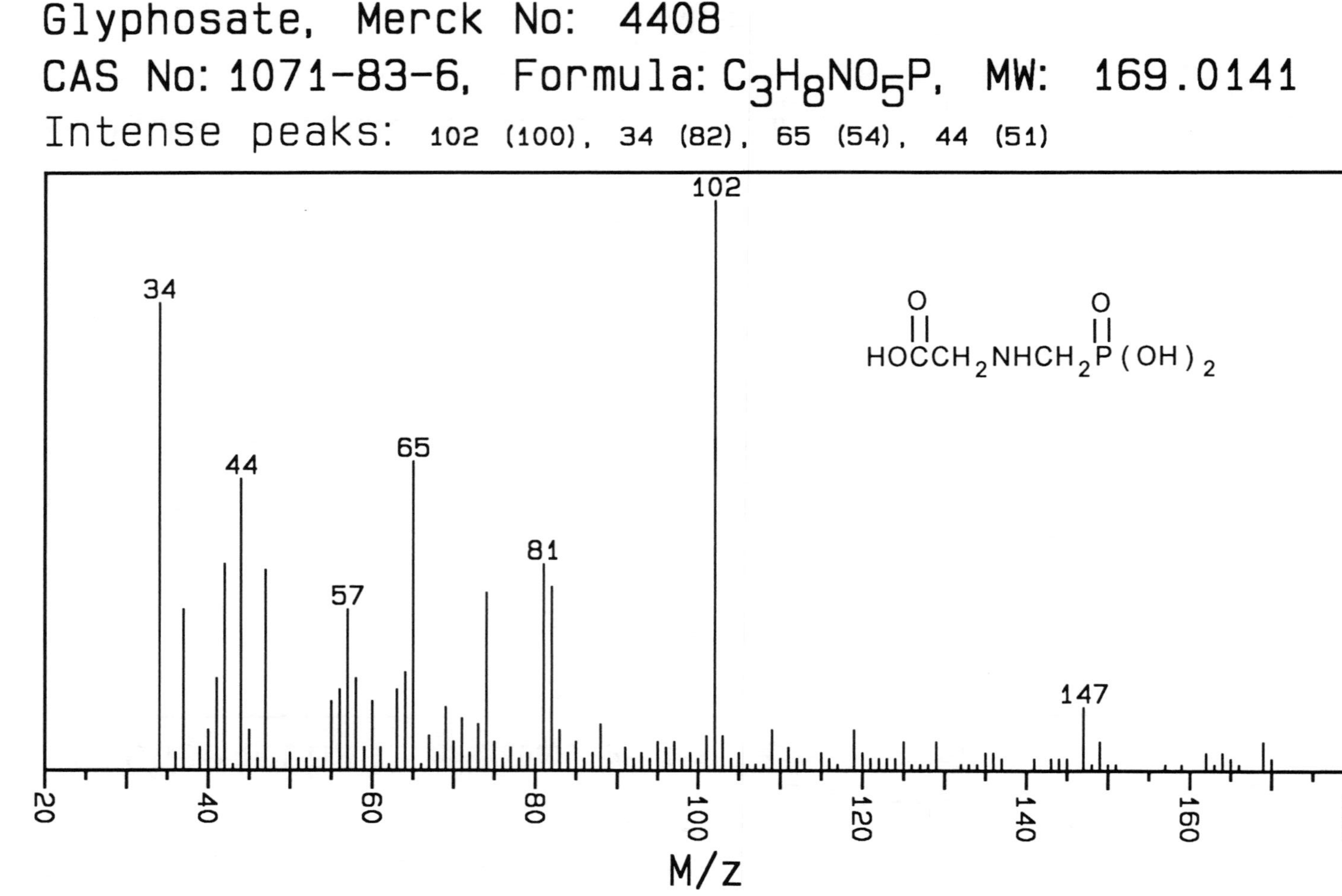
Glyphosate, Merck No: 4408
CAS No: 1071-83-6, Formula: C3H8NO5P, MW: 169.0141
Intense peaks: 102 (100), 34 (82), 65 (54), 44 (51)
102
34
O O
HOCCH2NHCH2P(OH)2
44
65
81
57
147
20
40
60
80
100
120
140
160
180
M/Z

Trichlorfon, Merck No: 9536

CAS No: 52-68-6, Formula: $C_4H_8Cl_3O_4P$, MW: 255.9226

Intense peaks: 79 (100), 109 (96), 110 (72), 139 (55)

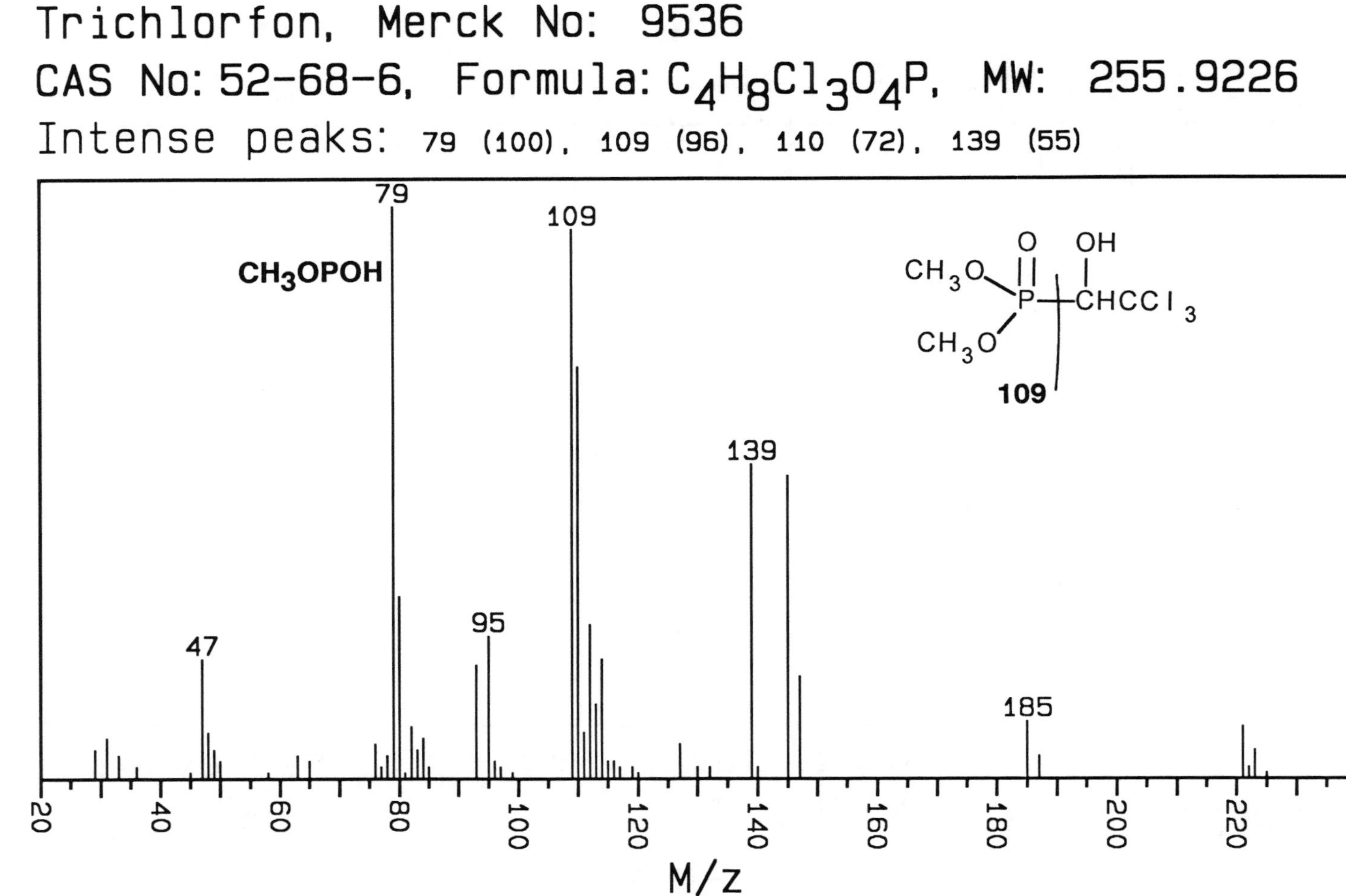

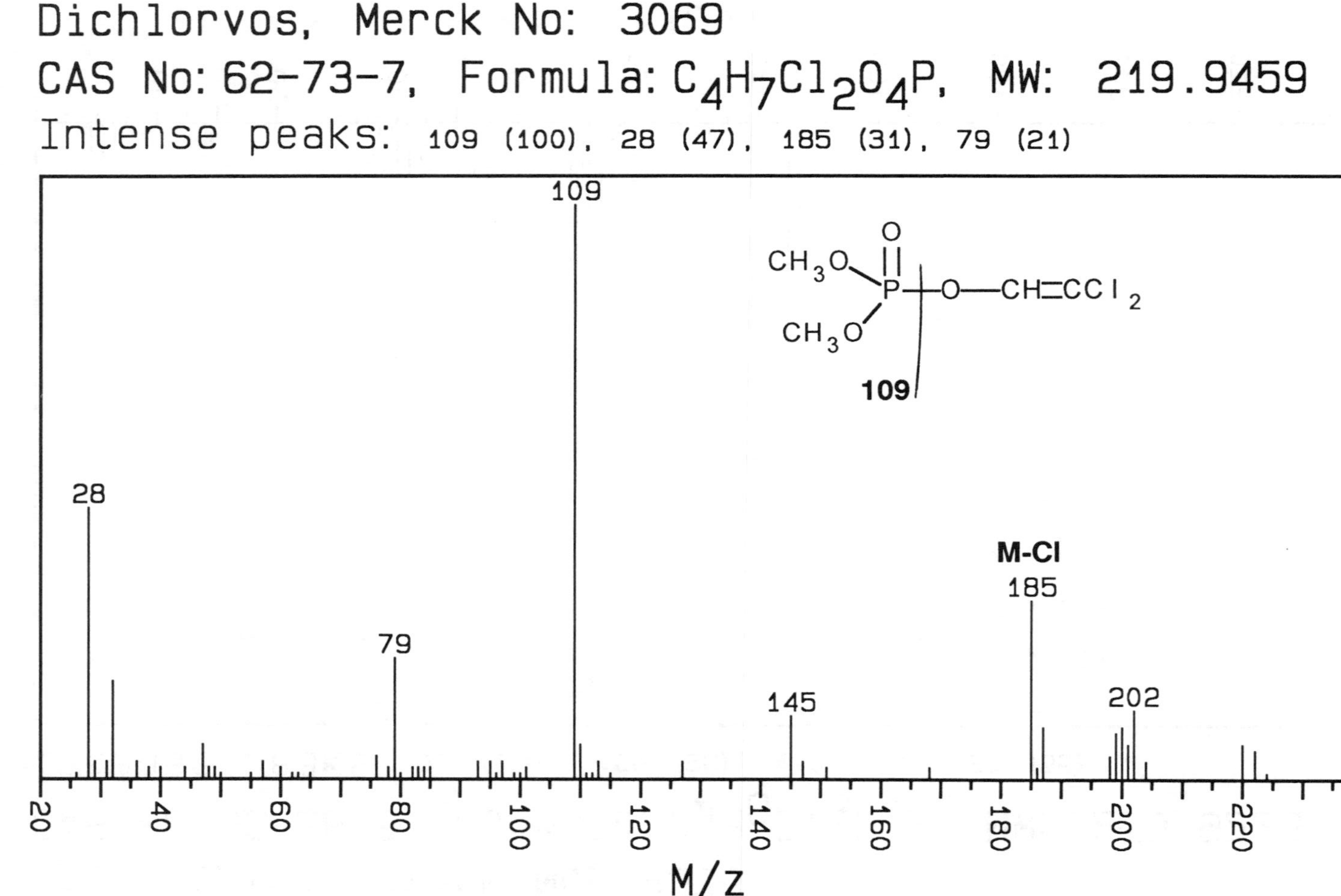
Dichlorvos, Merck No: 3069
CAS No: 62-73-7, Formula: C4H7Cl2O4P, MW: 219.9459
Intense peaks: 109 (100), 28 (47), 185 (31), 79 (21)
CH3O
CH3O
O
P
O—CH═CCl2
109
109
28
79
145
M-Cl
185
202
20
40
60
80
100
120
140
160
180
200
220
240
M/Z

Naled, Merck No: 6272

CAS No: 300-76-5, Formula: $C_4H_7Br_2Cl_2O_4P$, MW: 377.7826

Intense peaks: 109 (100), 145 (53), 79 (39), 47 (33)

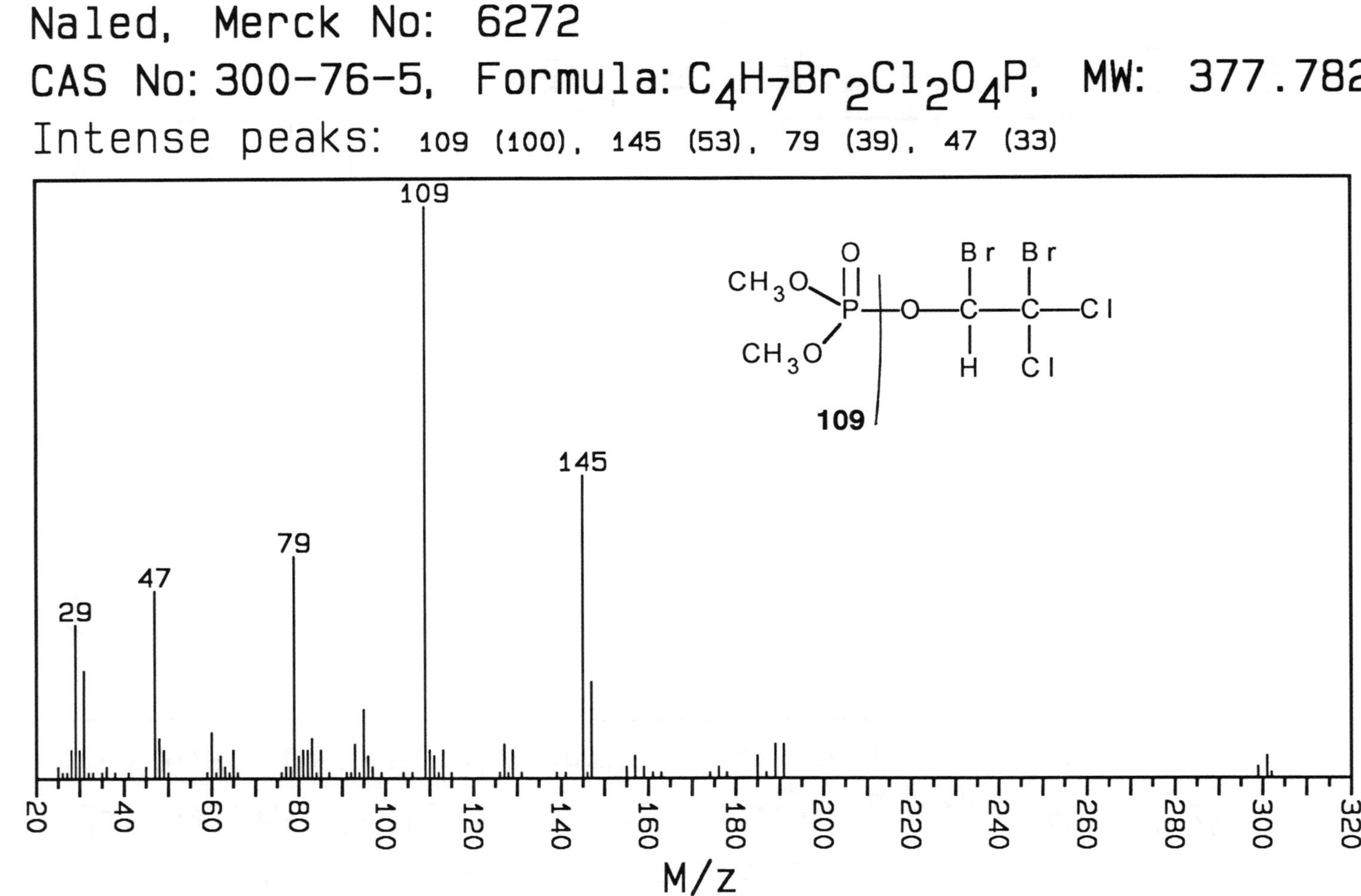

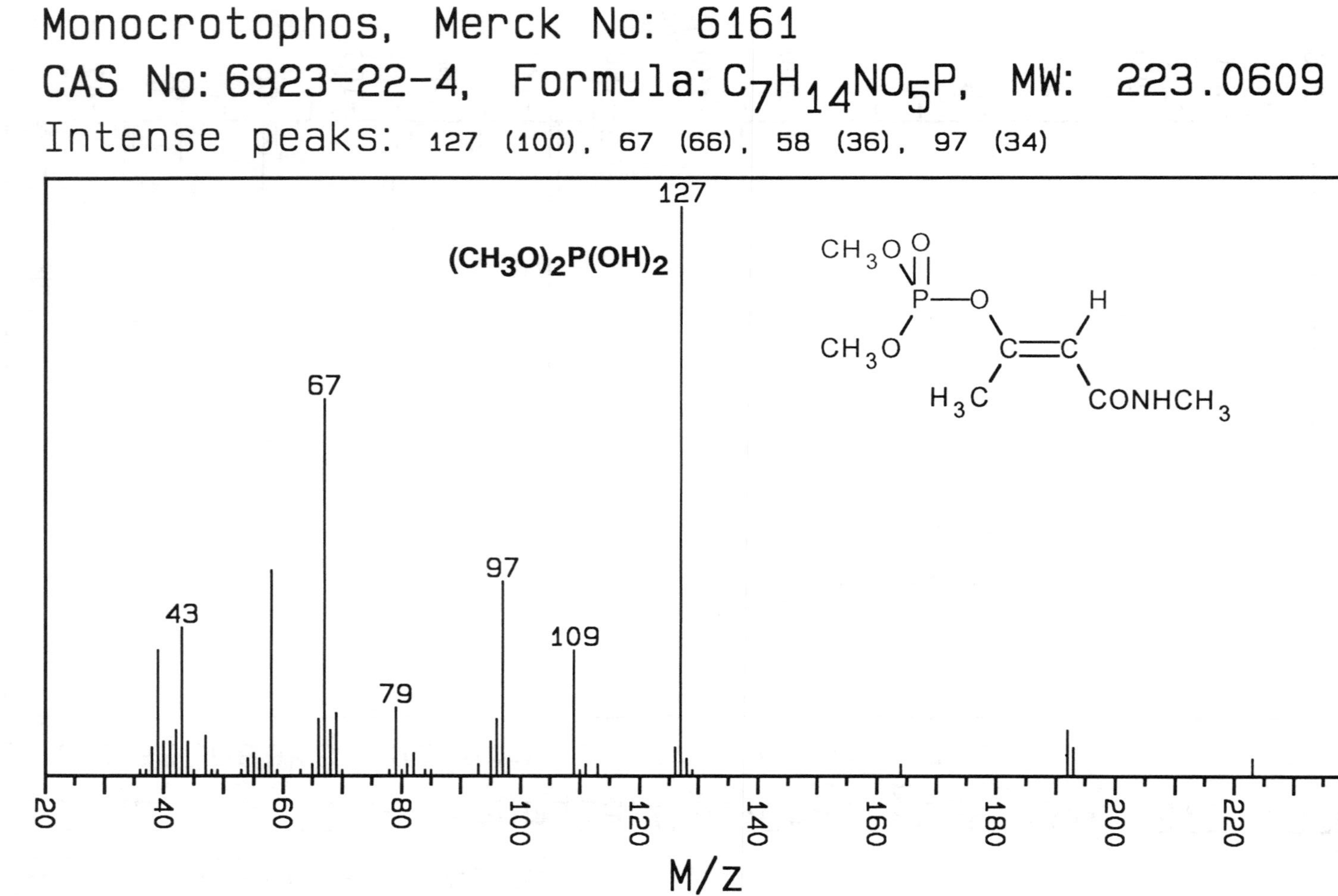
Monocrotophos, Merck No: 6161
CAS No: 6923-22-4, Formula: C7H14NO5P, MW: 223.0609
Intense peaks: 127 (100), 67 (66), 58 (36), 97 (34)
127
(CH3O)2P(OH)2
67
43
97
109
79
CH3O
CH3O
P
O
H
C=C
H3C
CONHCH3
20
40
60
80
100
120
140
160
180
200
220
240
M/Z

Mevinphos, Merck No: 6089

CAS No: 7786-34-7, Formula: $C_7H_{13}O_6P$, MW: 224.0451

Intense peaks: 127 (100), 192 (41), 67 (19), 70 (15)

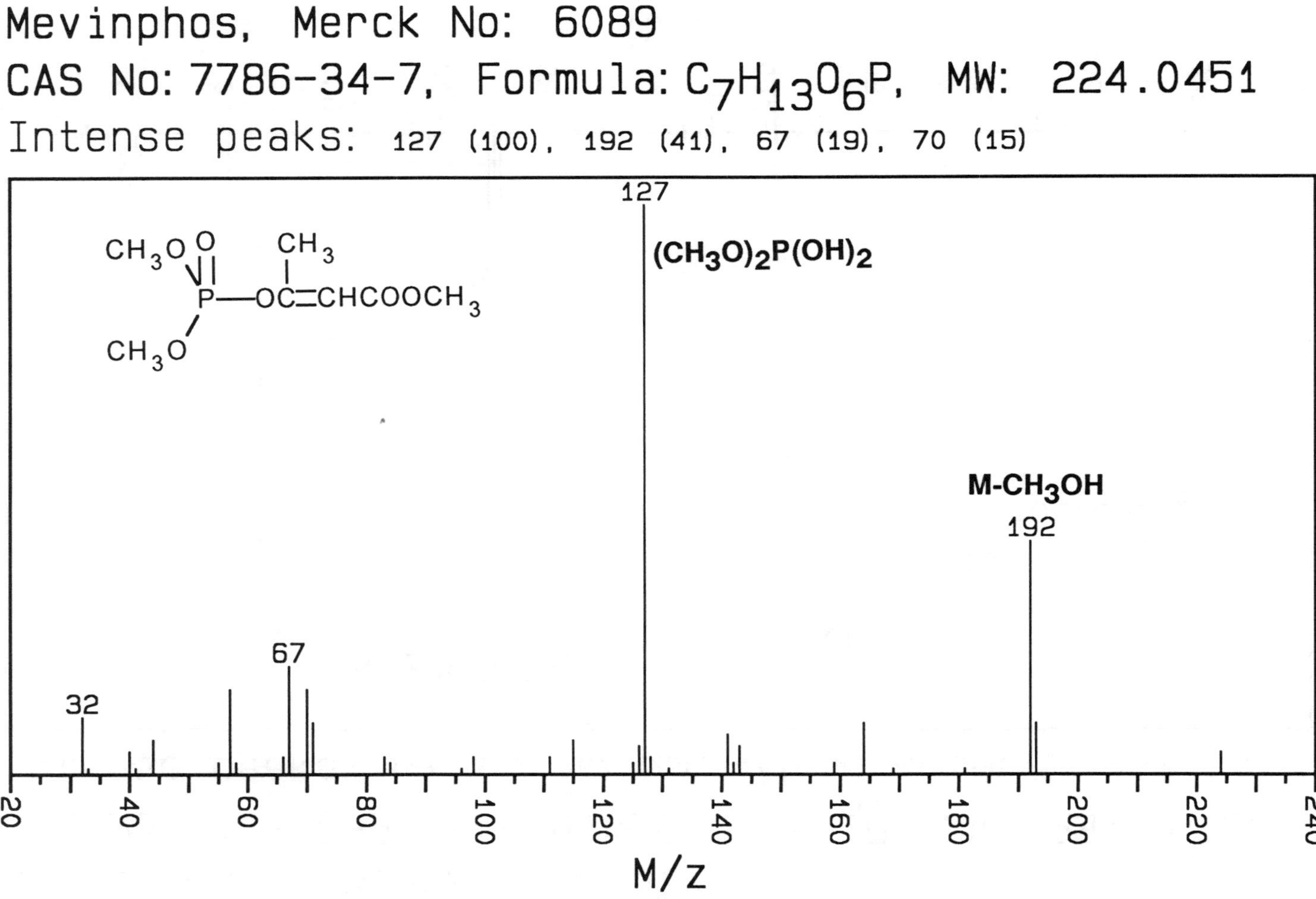

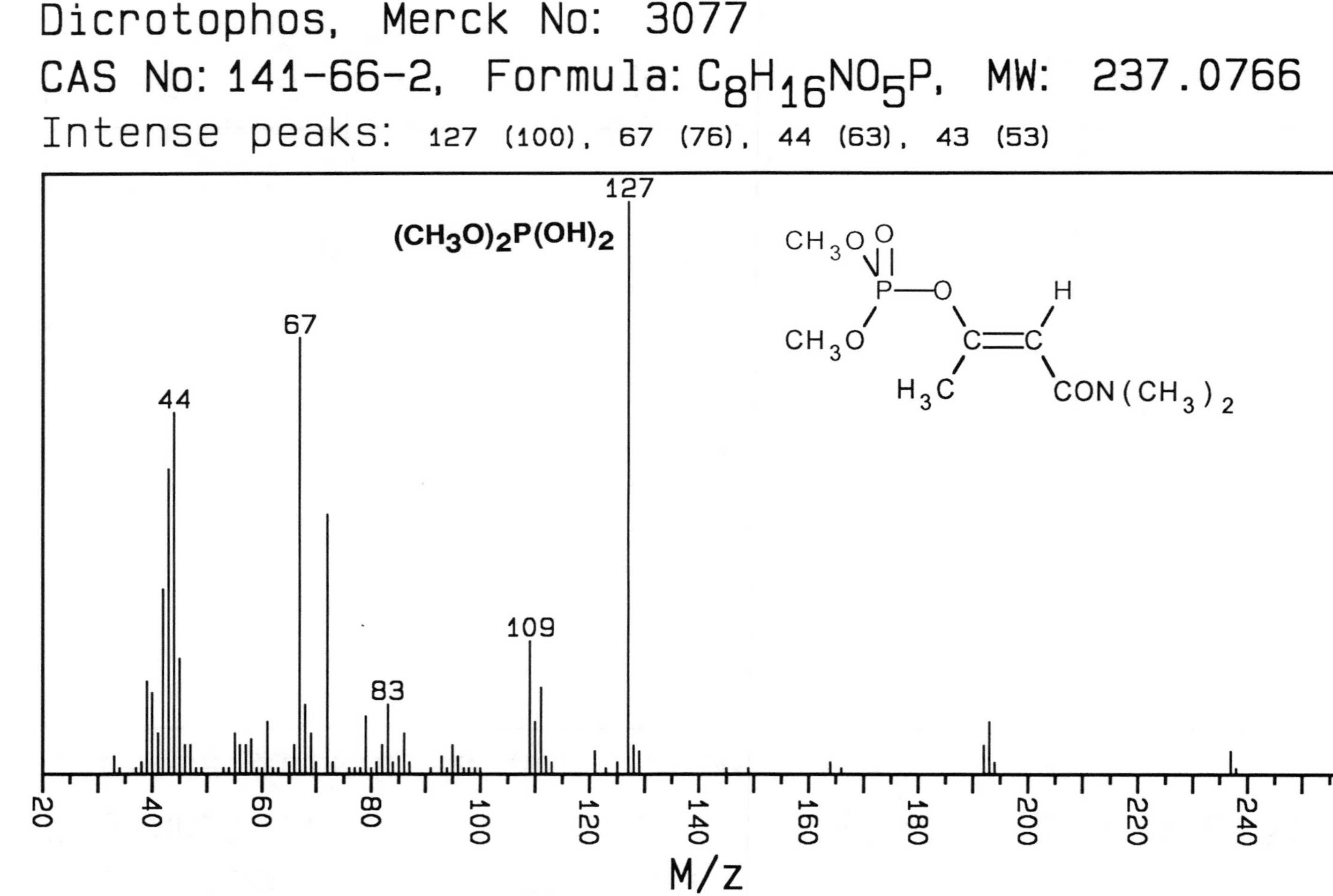
Dicrotophos, Merck No: 3077
CAS No: 141-66-2, Formula: $C_8H_{16}NO_5P$, MW: 237.0766
Intense peaks: 127 (100), 67 (76), 44 (63), 43 (53)
127
$(CH_3O)_2P(OH)_2$
67
44
109
83
$CH_3O$
$CH_3O$
P
O
H
C=C
$H_3C$
$CON(CH_3)_2$
20 40 60 80 100 120 140 160 180 200 220 240 260
M/Z

Phosphamidon, Merck No: 7312

CAS No: 13171-21-6, Formula: $C_{10}H_{19}ClNO_5P$, MW: 299.0689

Intense peaks: 127 (100), 72 (61), 264 (46), 138 (35)

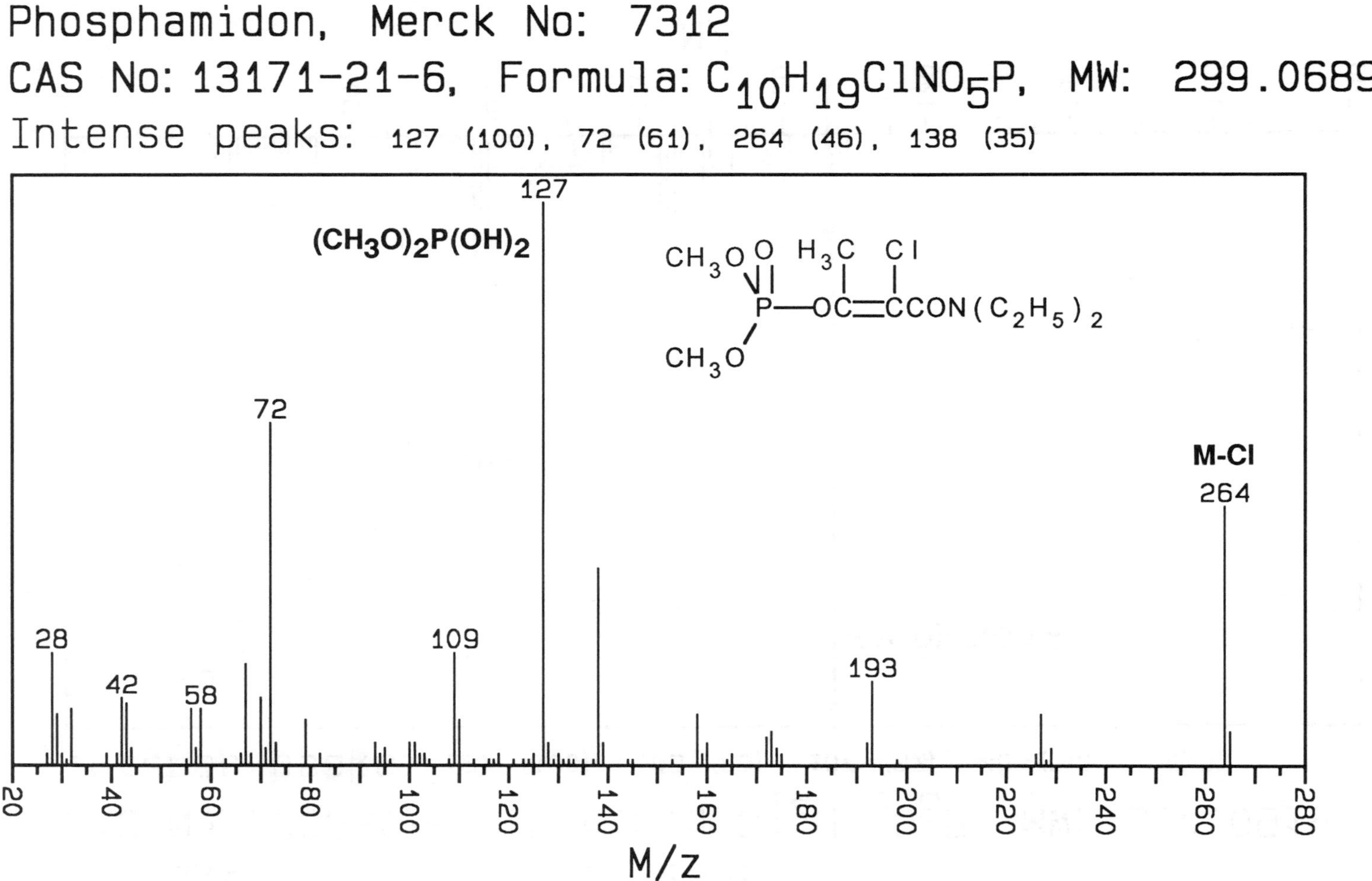

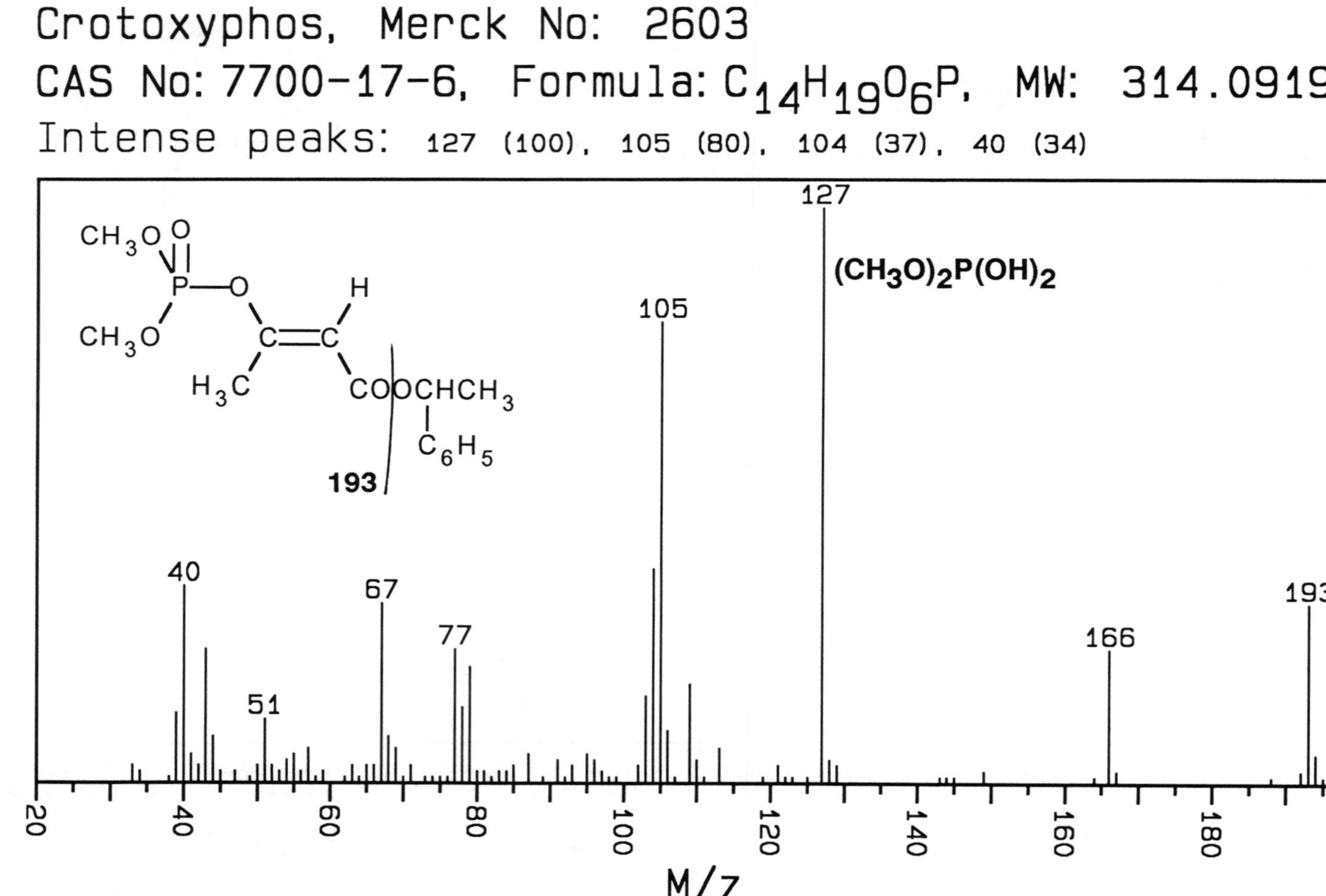

Crotoxyphos, Merck No: 2603
CAS No: 7700-17-6, Formula: $C_{14}H_{19}O_6P$, MW: 314.0919
Intense peaks: 127 (100), 105 (80), 104 (37), 40 (34)
$(CH_3O)_2P(OH)_2$
193
127
105
40
67
77
51
166
193
20
40
60
80
100
120
140
160
180
200
M/Z

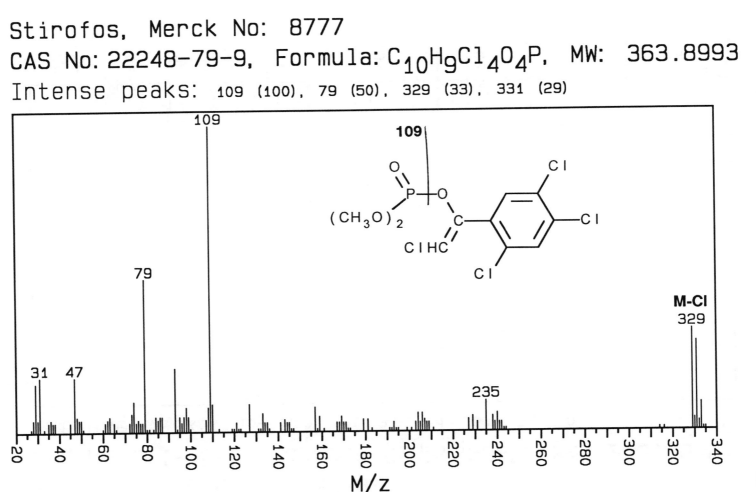

Stirofos, Merck No: 8777

CAS No: 22248-79-9, Formula: $C_{10}H_9Cl_4O_4P$, MW: 363.8993

Intense peaks: 109 (100), 79 (50), 329 (33), 331 (29)

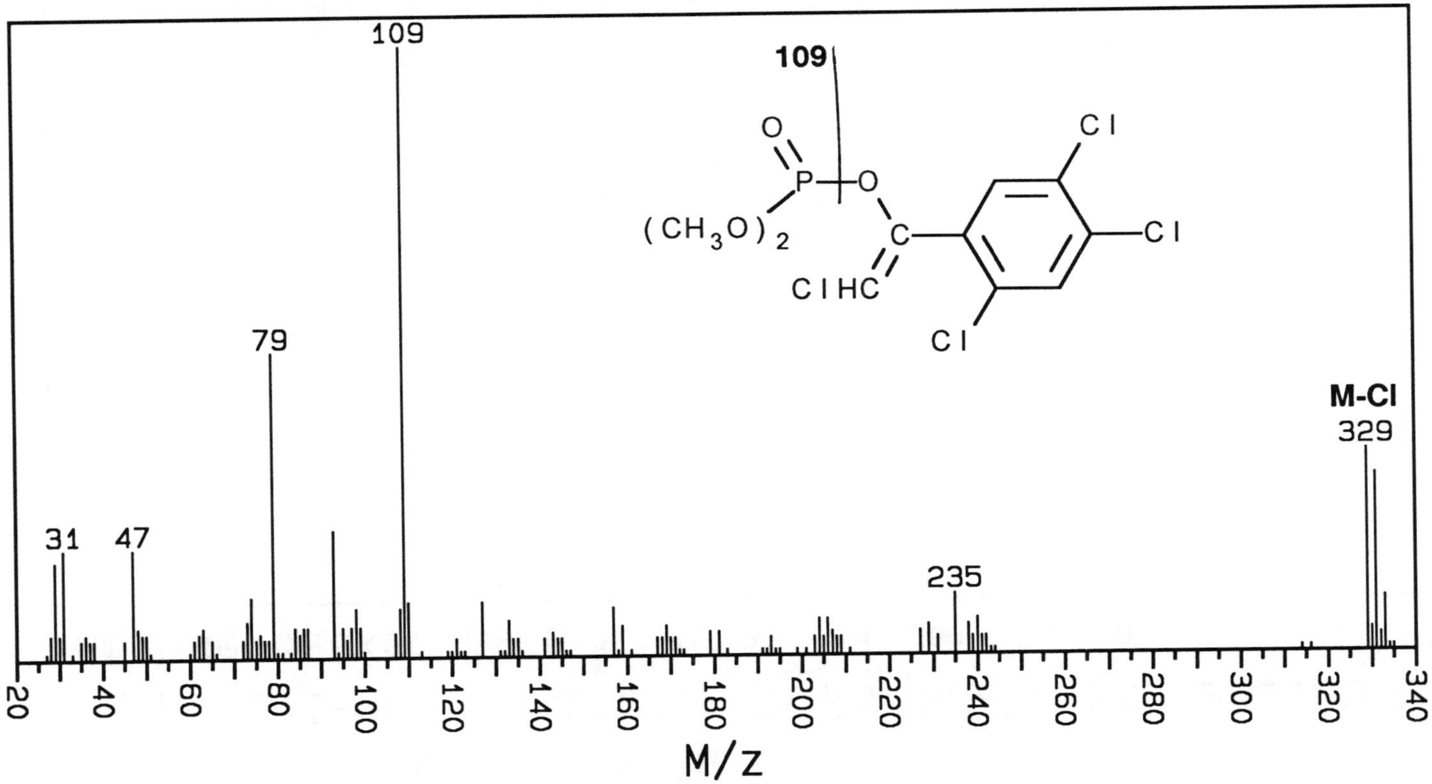

109
79
31
47
235
M-Cl
329
109
$(CH_3O)_2$
O
P
O
C
ClHC
Cl
Cl
Cl
20
40
60
80
100
120
140
160
180
200
220
240
260
280
300
320
340
M/Z

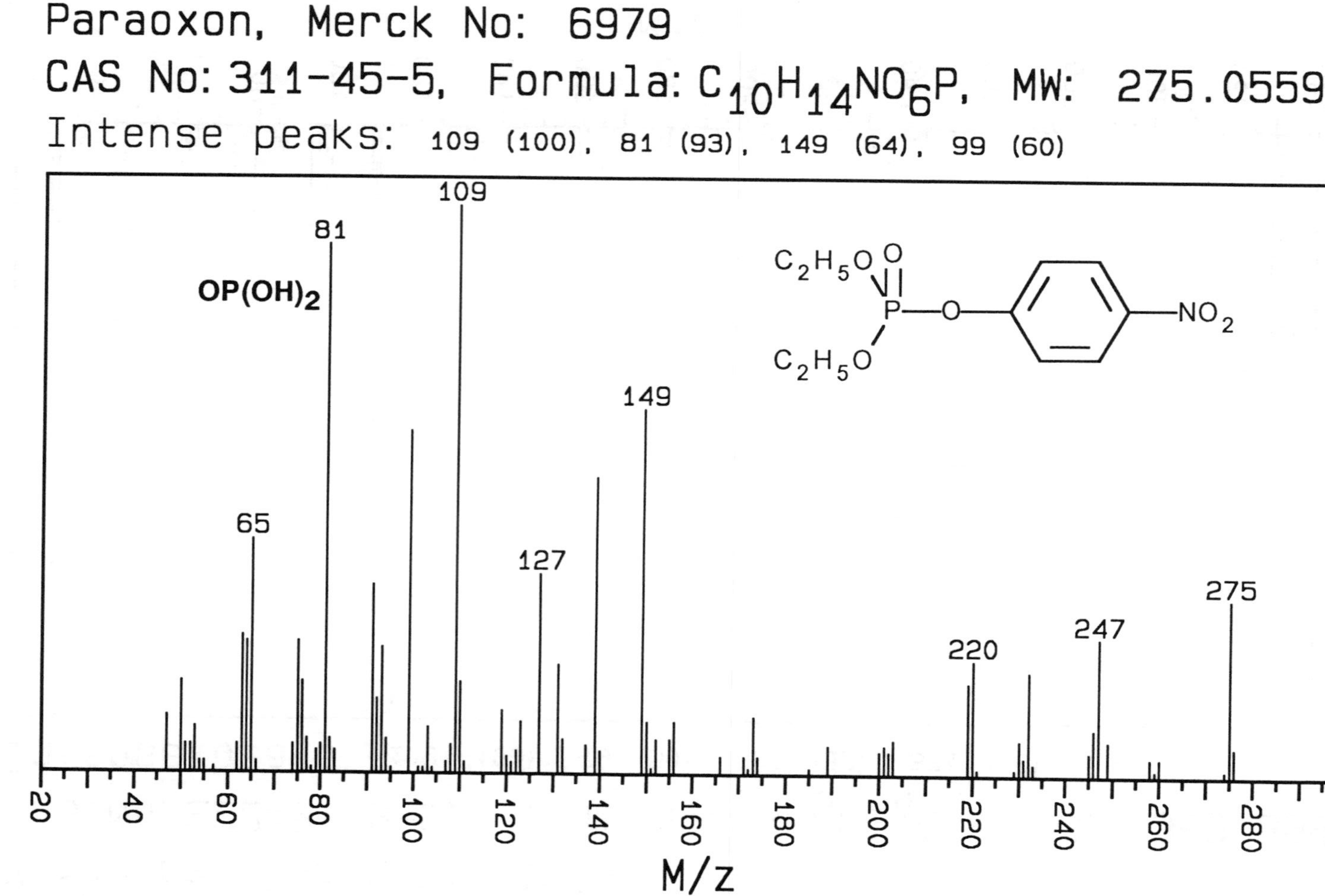

Paraoxon, Merck No: 6979
CAS No: 311-45-5, Formula: $C_{10}H_{14}NO_6P$, MW: 275.0559
Intense peaks: 109 (100), 81 (93), 149 (64), 99 (60)
$C_2H_5O$
O
P
O
$C_2H_5O$
$NO_2$
$OP(OH)_2$
65
81
109
127
149
220
247
275
20
40
60
80
100
120
140
160
180
200
220
240
260
280
300
M/Z

Chlorfenvinphos, Merck No: 2087

CAS No: 470-90-6, Formula: $C_{12}H_{14}Cl_3O_4P$, MW: 357.9695

Intense peaks: 267 (100), 323 (76), 269 (63), 325 (50)

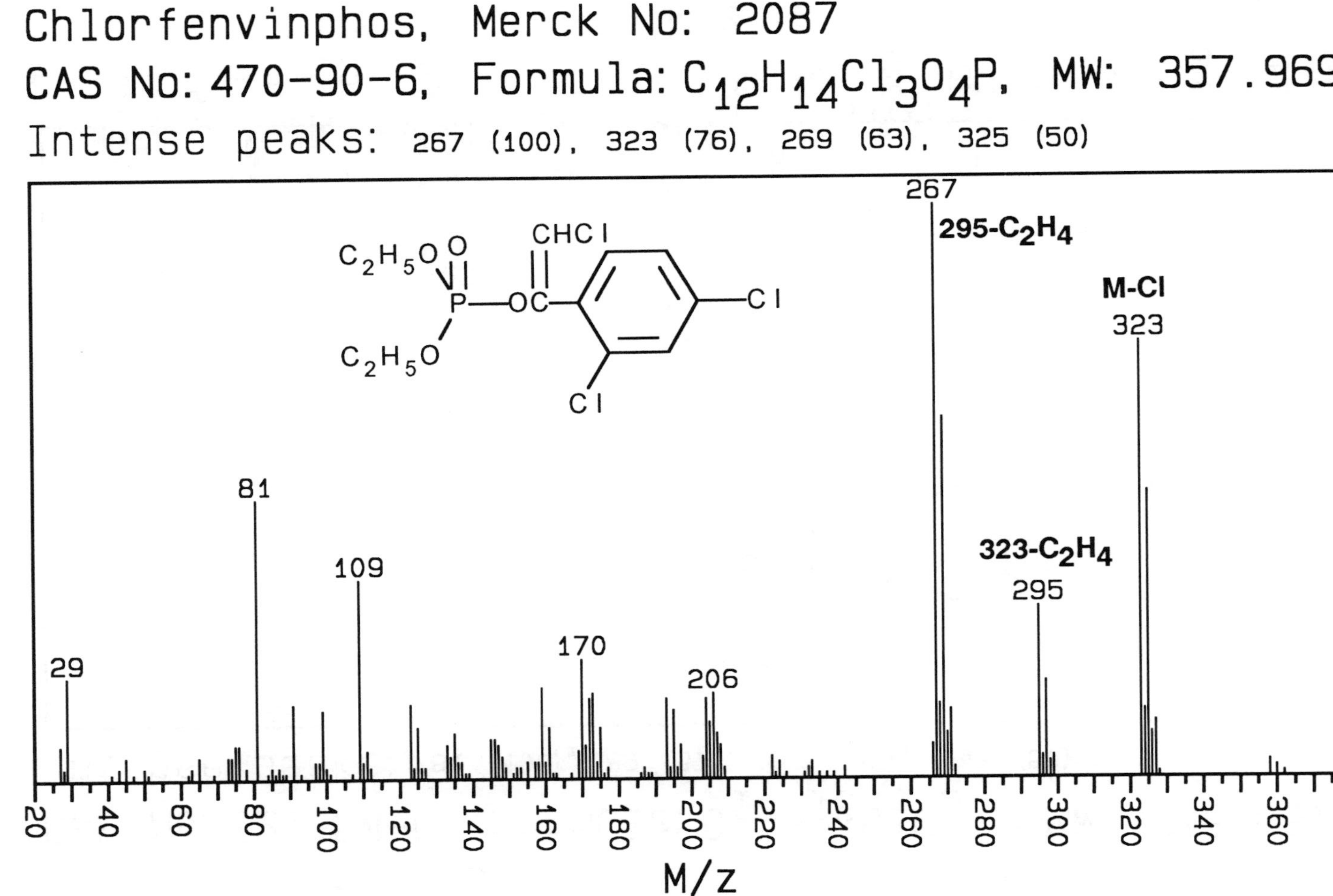

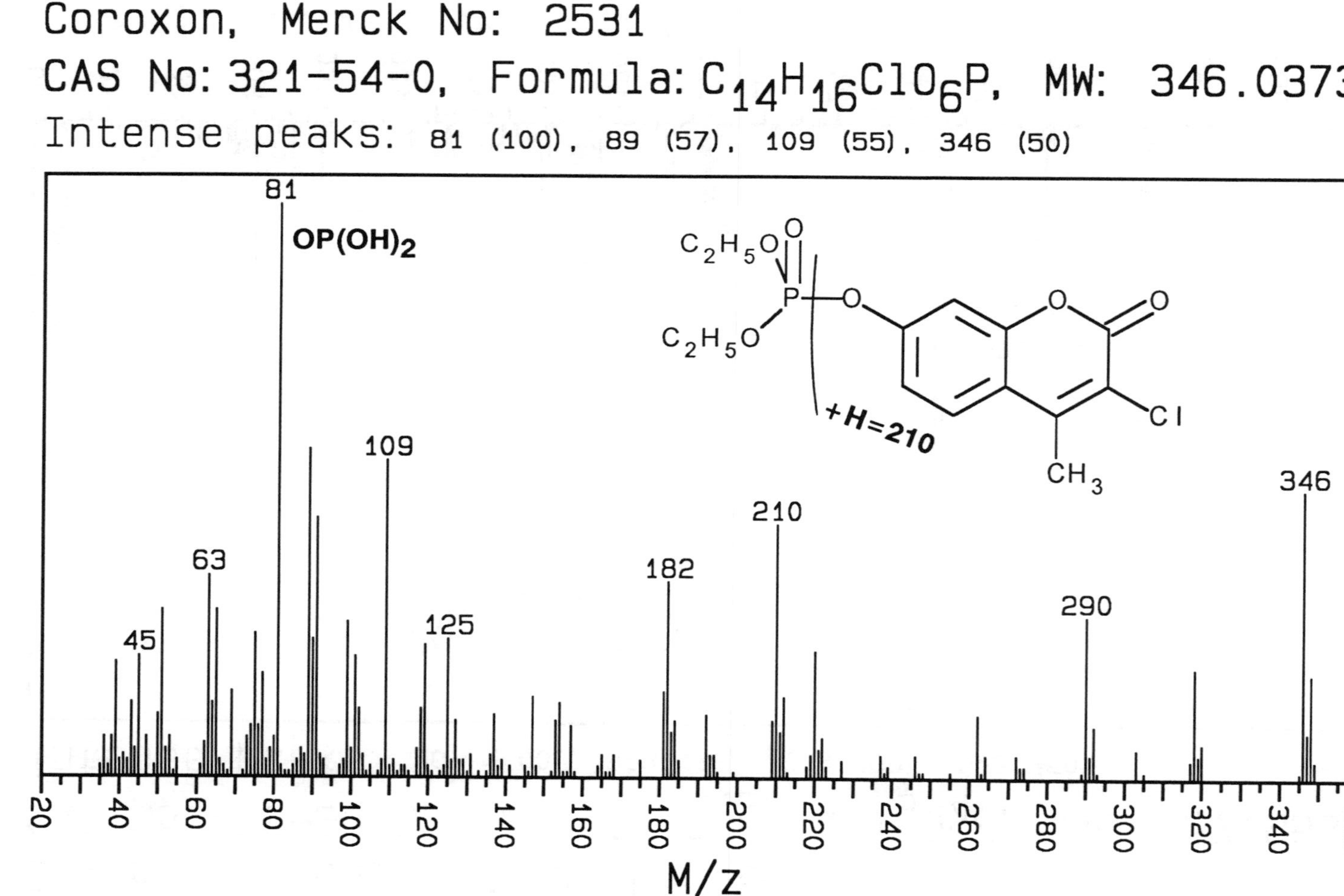
Coroxon, Merck No: 2531
CAS No: 321-54-0, Formula: C14H16ClO6P, MW: 346.0373
Intense peaks: 81 (100), 89 (57), 109 (55), 346 (50)
81
OP(OH)2
+H=210
45
63
109
125
182
210
290
346
M/Z

Triethyl phosphate, Merck No: 9589

CAS No: 78-40-0, Formula: $C_6H_{15}O_4P$, MW: 182.0708

Intense peaks: 99 (100), 81 (71), 155 (56), 82 (45)

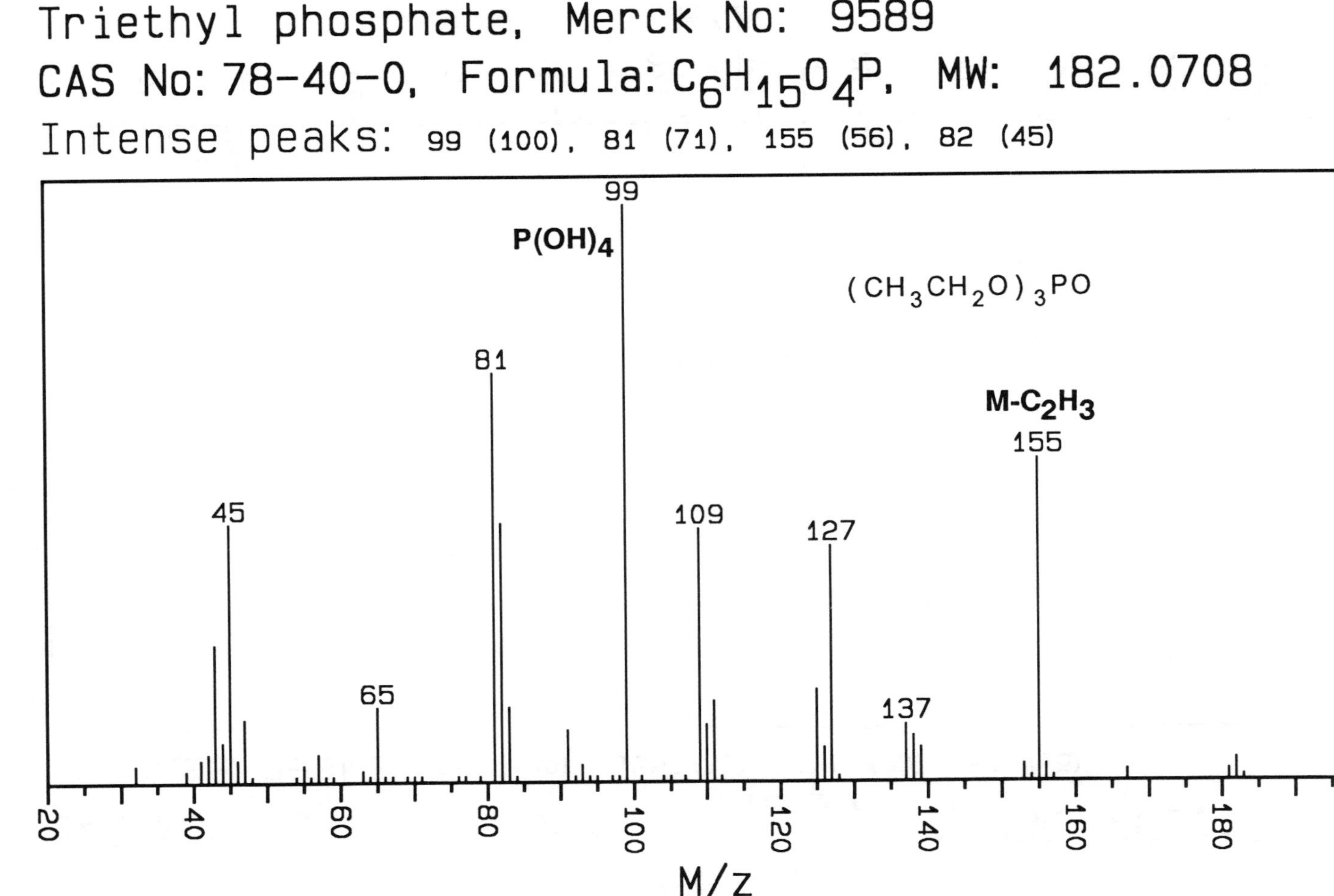

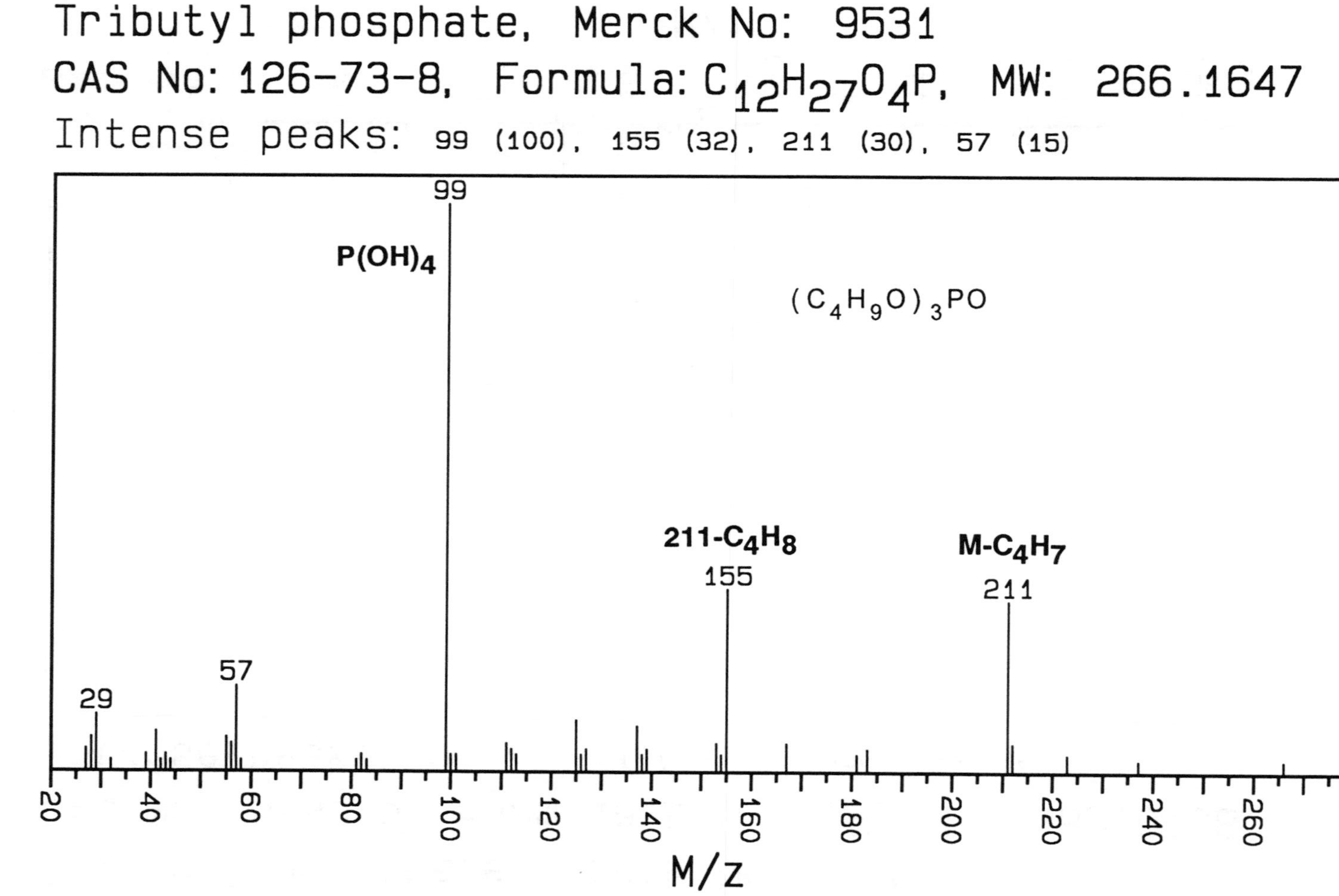
Tributyl phosphate, Merck No: 9531
CAS No: 126-73-8, Formula: $C_{12}H_{27}O_4P$, MW: 266.1647
Intense peaks: 99 (100), 155 (32), 211 (30), 57 (15)
99
$P(OH)_4$
$(C_4H_9O)_3PO$
$211-C_4H_8$
155
$M-C_4H_7$
211
57
29
20
40
60
80
100
120
140
160
180
200
220
240
260
280
M/Z

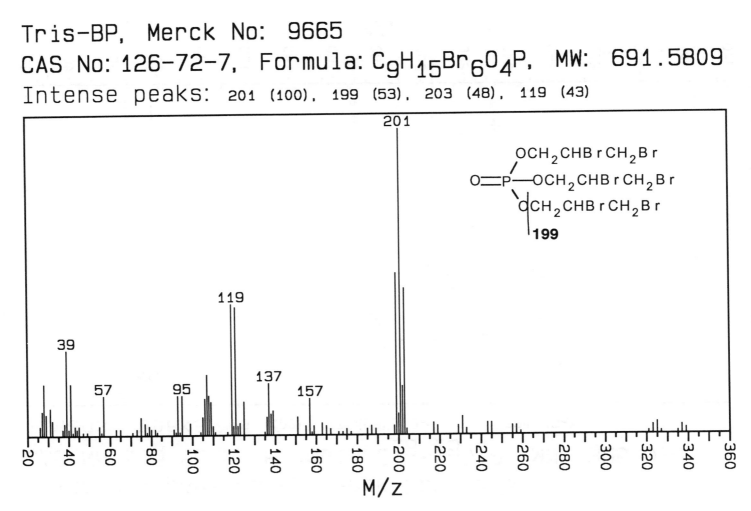

Tris-BP, Merck No: 9665

CAS No: 126-72-7, Formula: $C_9H_{15}Br_6O_4P$, MW: 691.5809

Intense peaks: 201 (100), 199 (53), 203 (48), 119 (43)

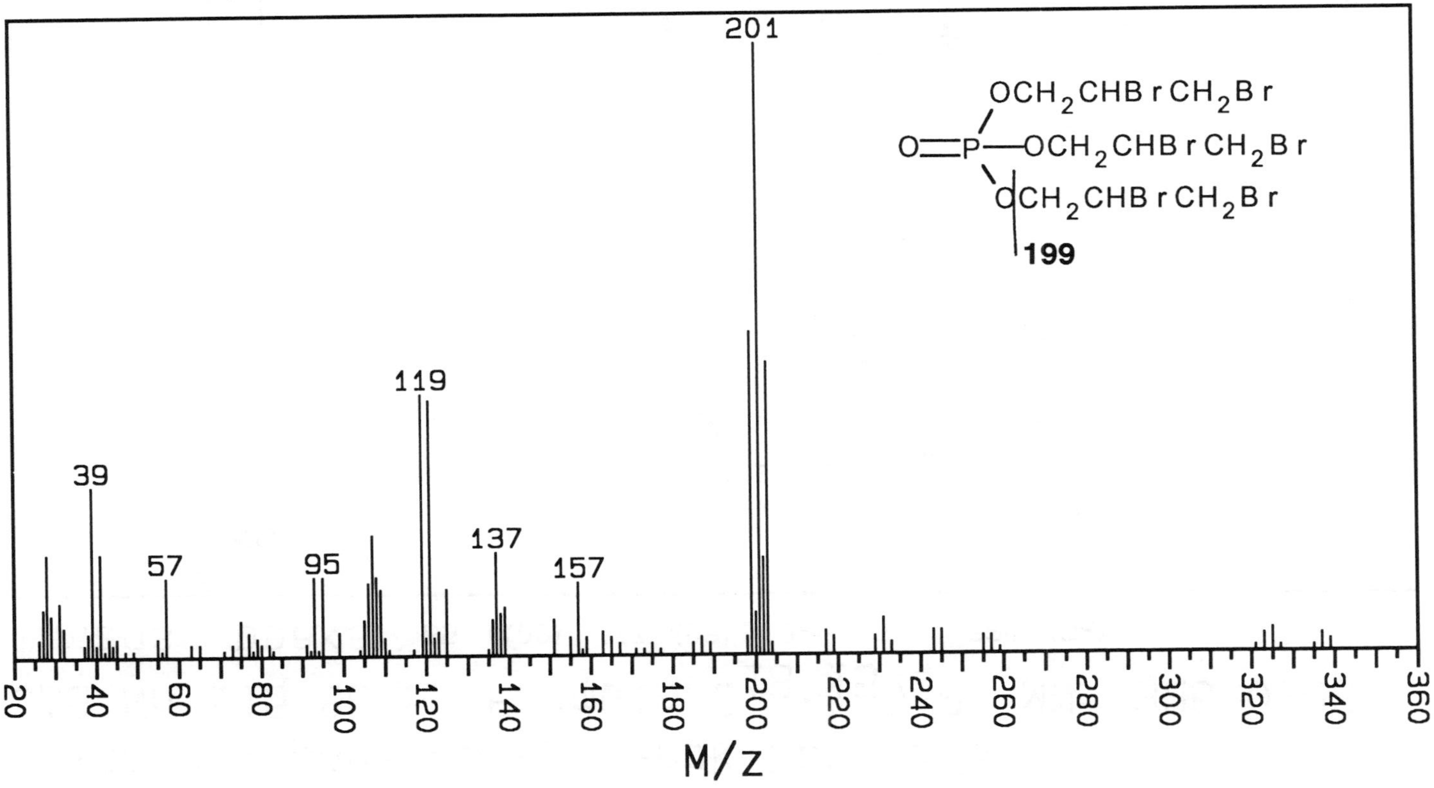

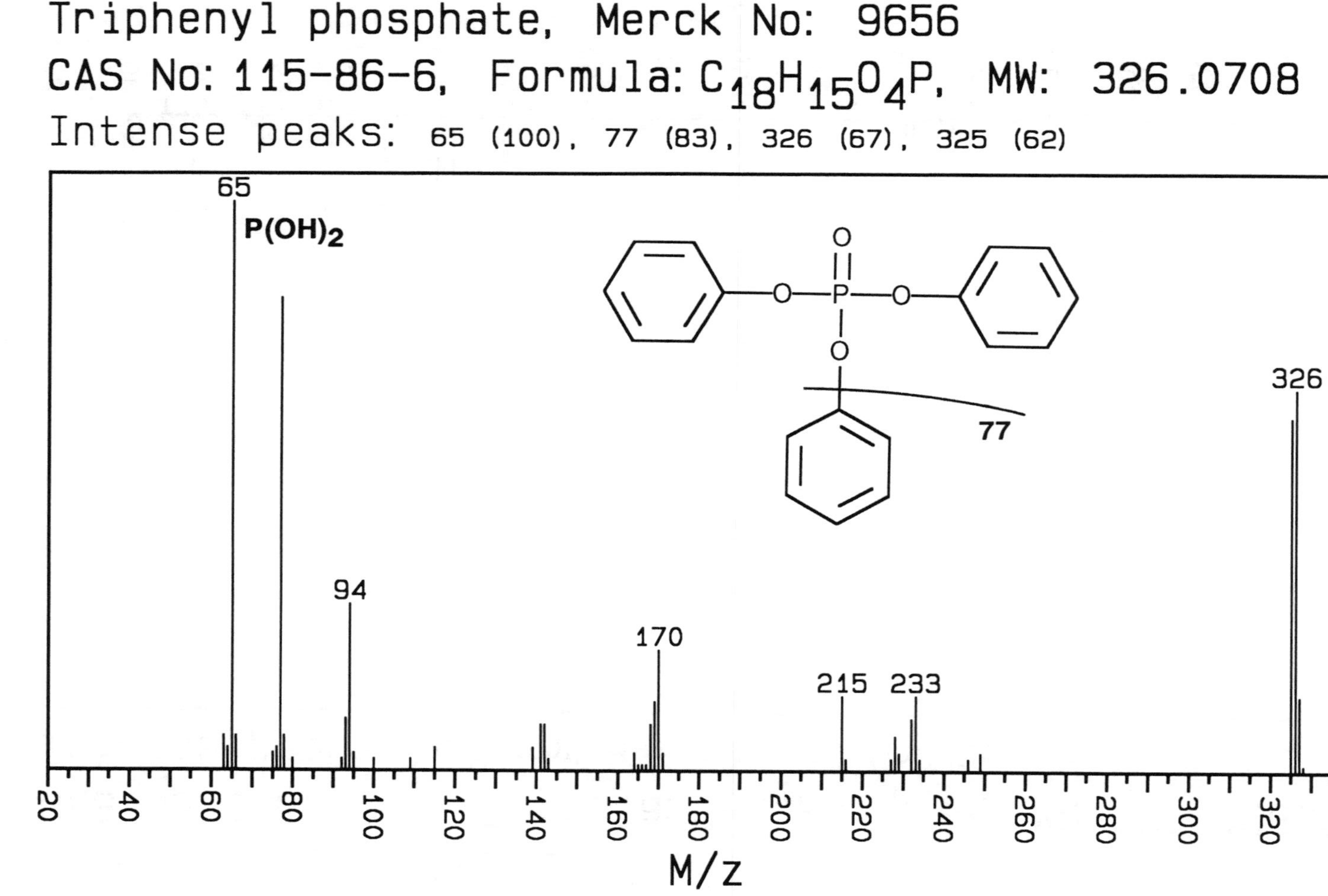
Triphenyl phosphate, Merck No: 9656
CAS No: 115-86-6, Formula: $C_{18}H_{15}O_4P$, MW: 326.0708
Intense peaks: 65 (100), 77 (83), 326 (67), 325 (62)
65
P(OH)2
77
94
170
215
233
326
20 40 60 80 100 120 140 160 180 200 220 240 260 280 300 320 340
M/Z

Tritolyl phosphate, Merck No: 9675

CAS No: 1330-78-5, Formula: $C_{21}H_{21}O_4P$, MW: 368.1177

Intense peaks: 368 (100), 367 (60), 91 (35), 165 (30)

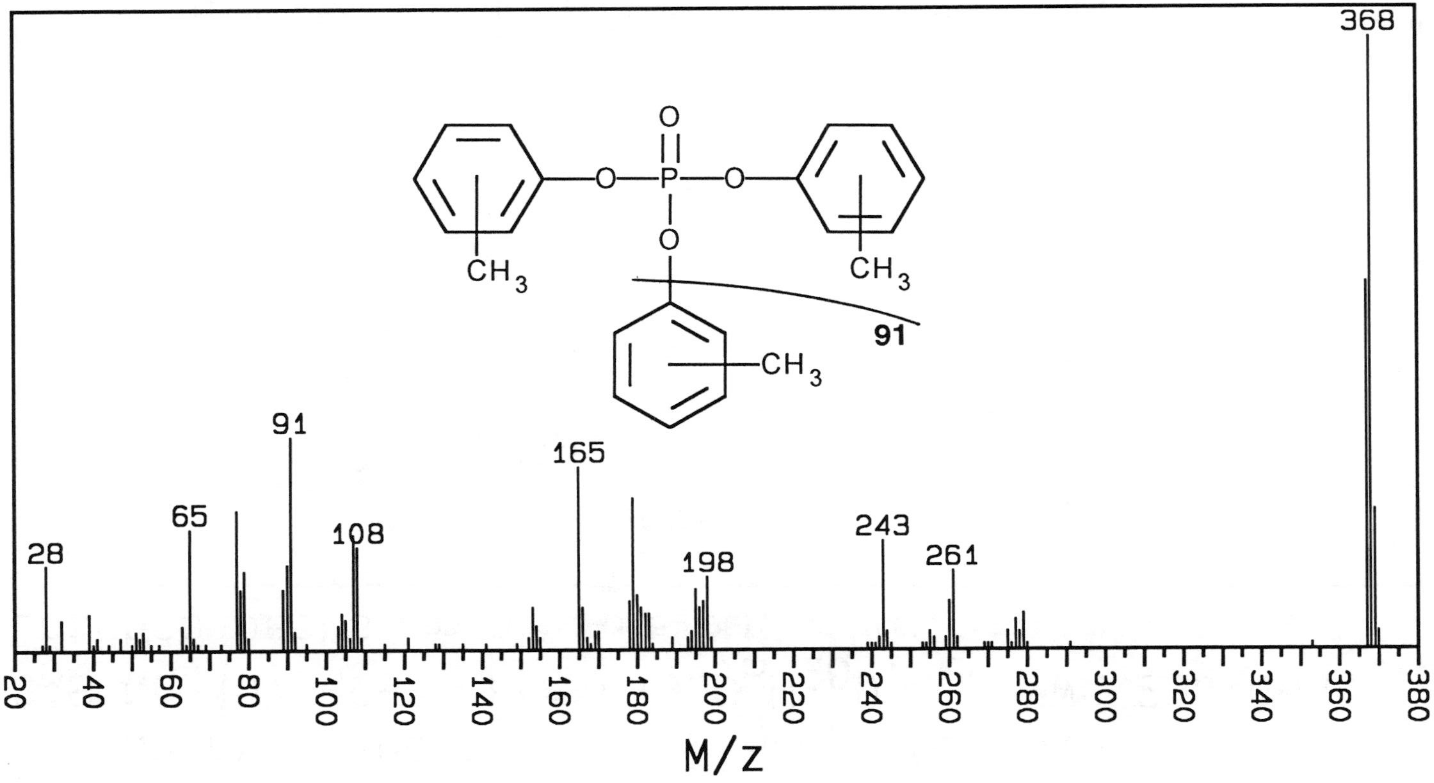

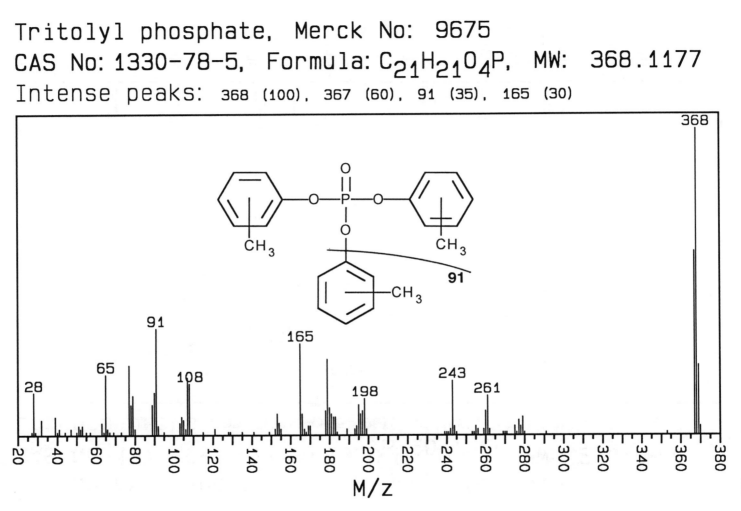

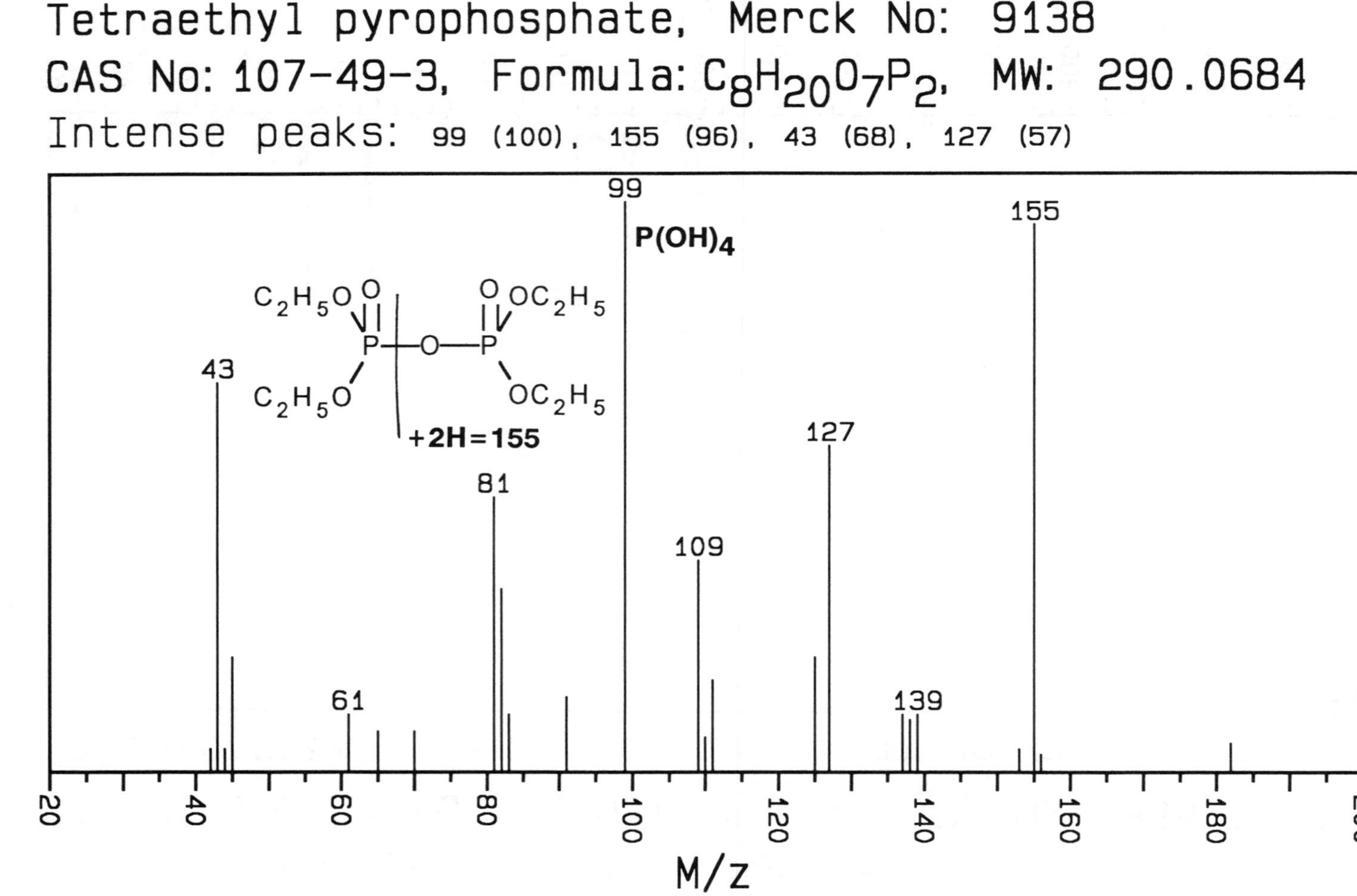

Tetraethyl pyrophosphate, Merck No: 9138
CAS No: 107-49-3, Formula: $C_8H_{20}O_7P_2$, MW: 290.0684
Intense peaks: 99 (100), 155 (96), 43 (68), 127 (57)
99
$P(OH)_4$
155
43
+2H=155
127
81
109
61
139
M/Z

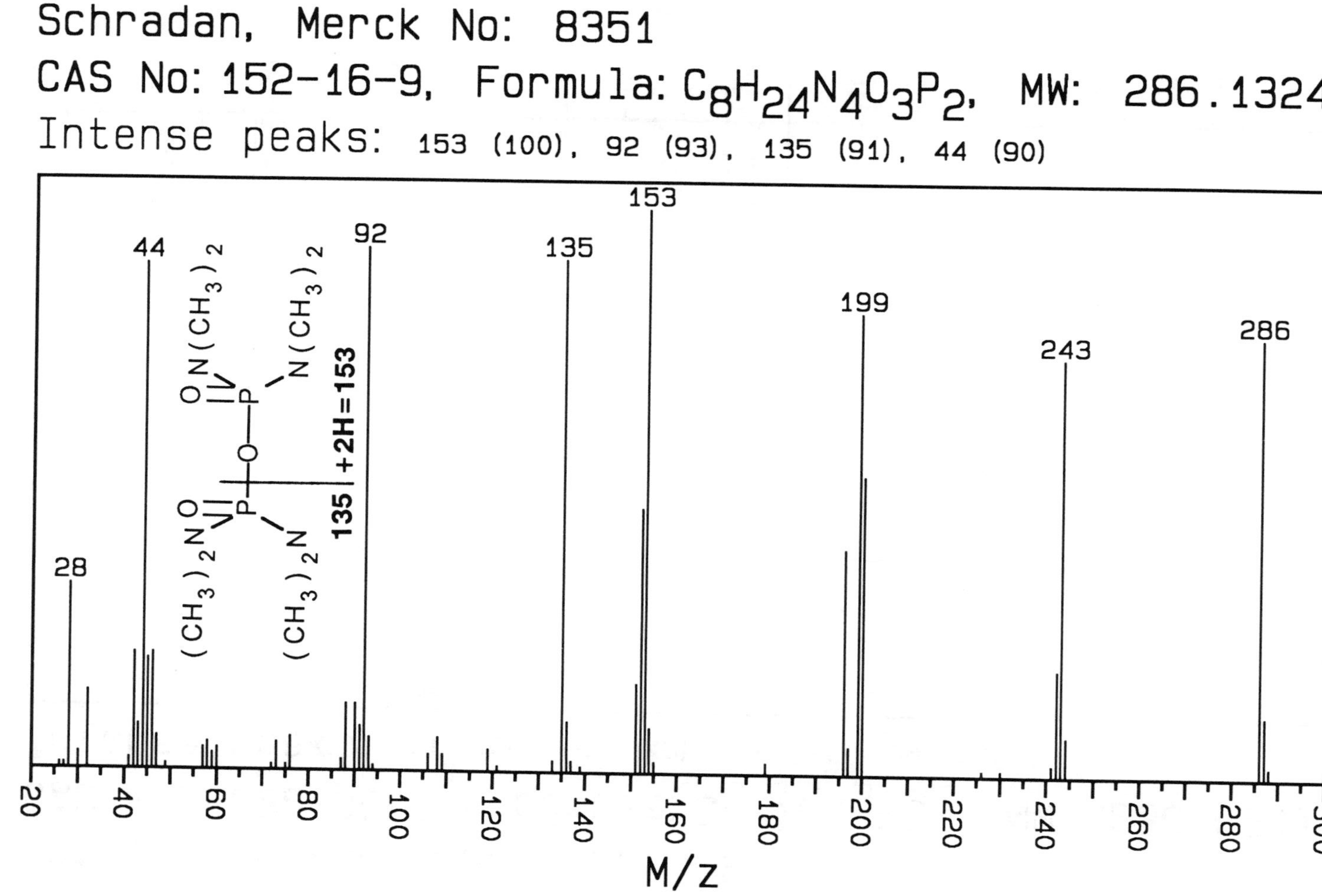
Schradan, Merck No: 8351
CAS No: 152-16-9, Formula: $C_8H_{24}N_4O_3P_2$, MW: 286.1324
Intense peaks: 153 (100), 92 (93), 135 (91), 44 (90)
28
44
92
135
153
199
243
286
135 | +2H=153
20
40
60
80
100
120
140
160
180
200
220
240
260
280
300
M/Z

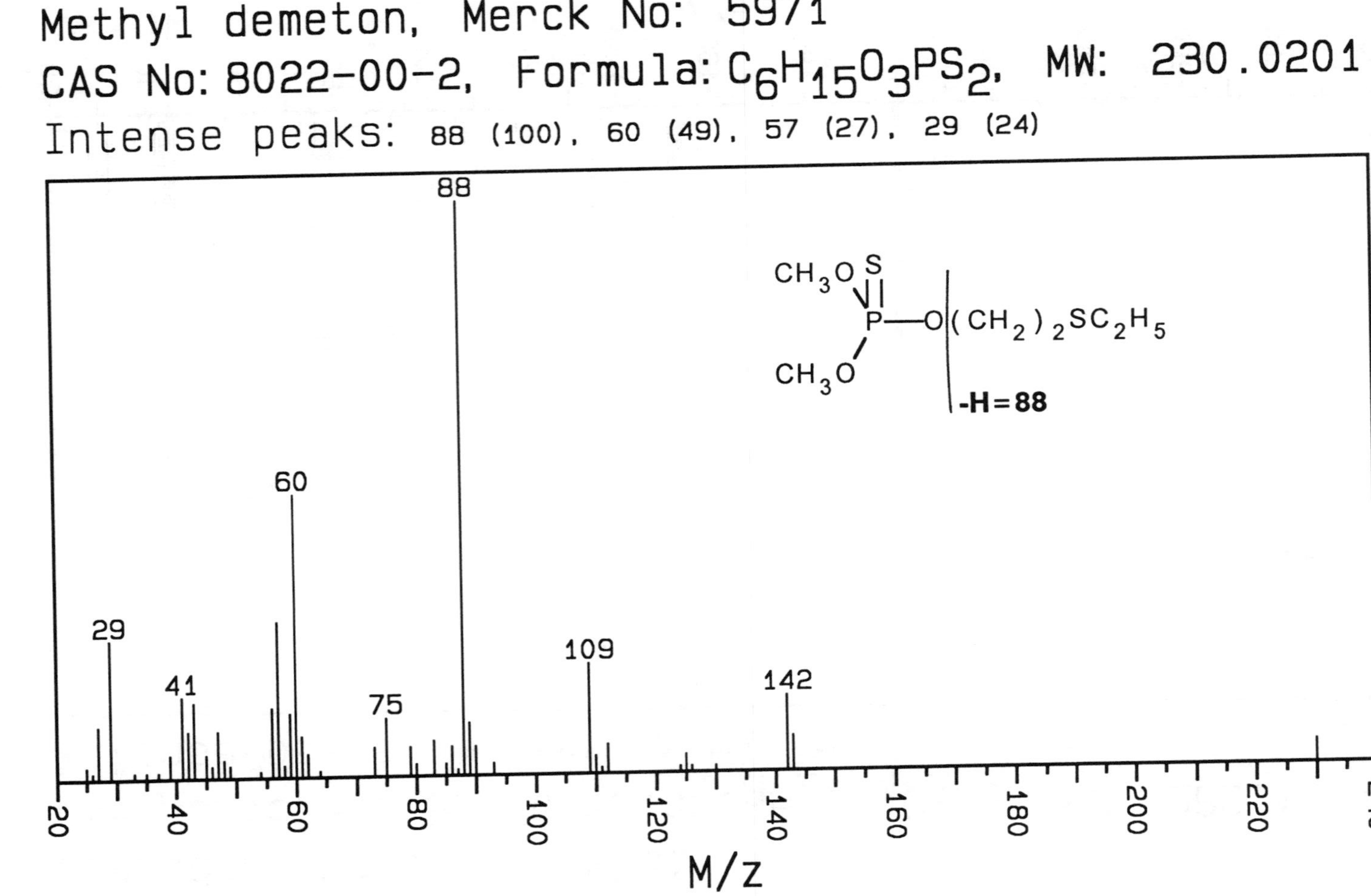
Methyl demeton, Merck No: 5971
CAS No: 8022-00-2, Formula: C6H15O3PS2, MW: 230.0201
Intense peaks: 88 (100), 60 (49), 57 (27), 29 (24)
CH3O
CH3O
S
P
O(CH2)2SC2H5
-H=88
29
41
60
75
88
109
142
20
40
60
80
100
120
140
160
180
200
220
240
M/Z

Methyl parathion, Merck No: 6022

CAS No: 298-00-0, Formula: $C_8H_{10}NO_5PS$, MW: 263.0017

Intense peaks: 263 (100), 125 (99), 109 (99), 79 (39)

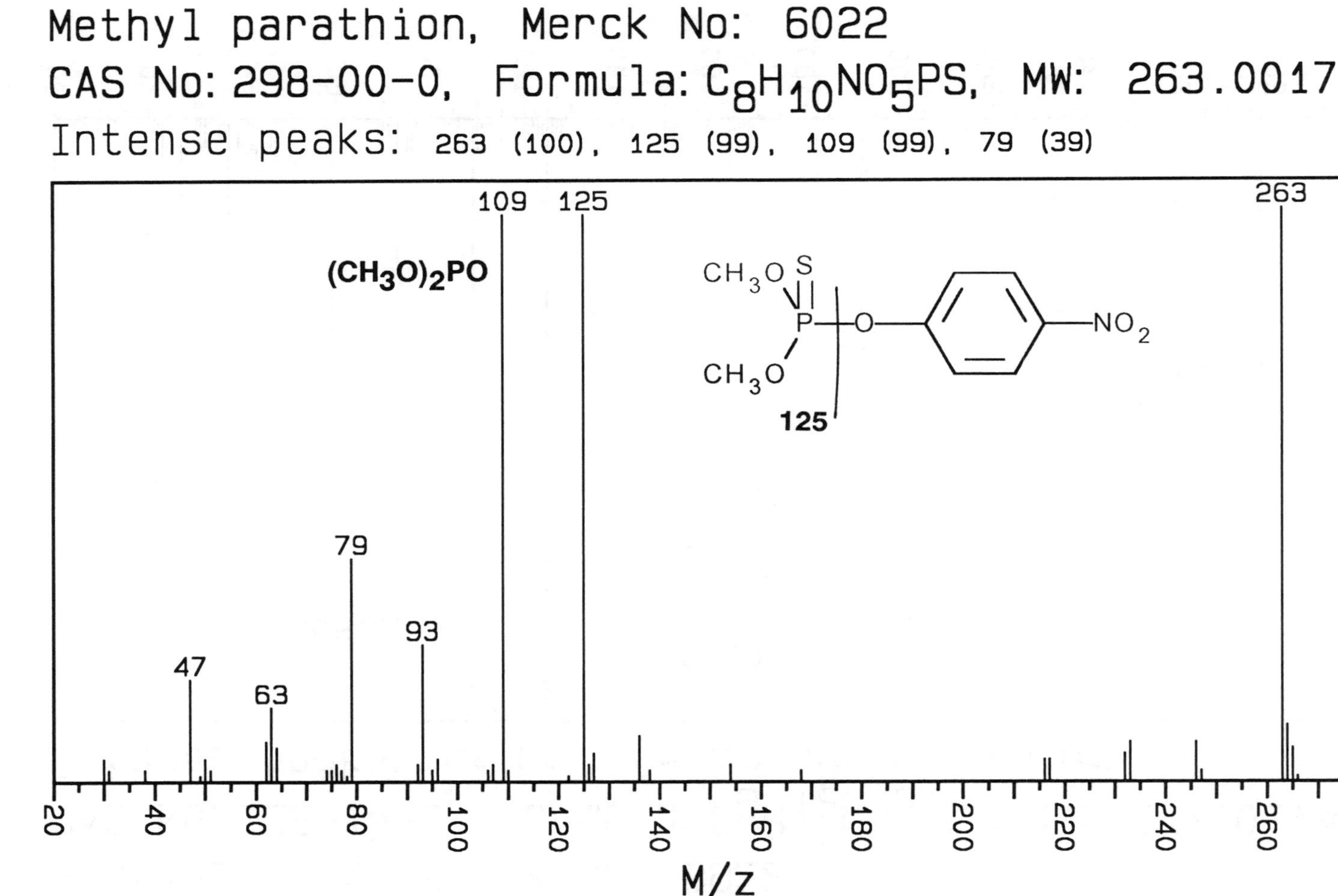

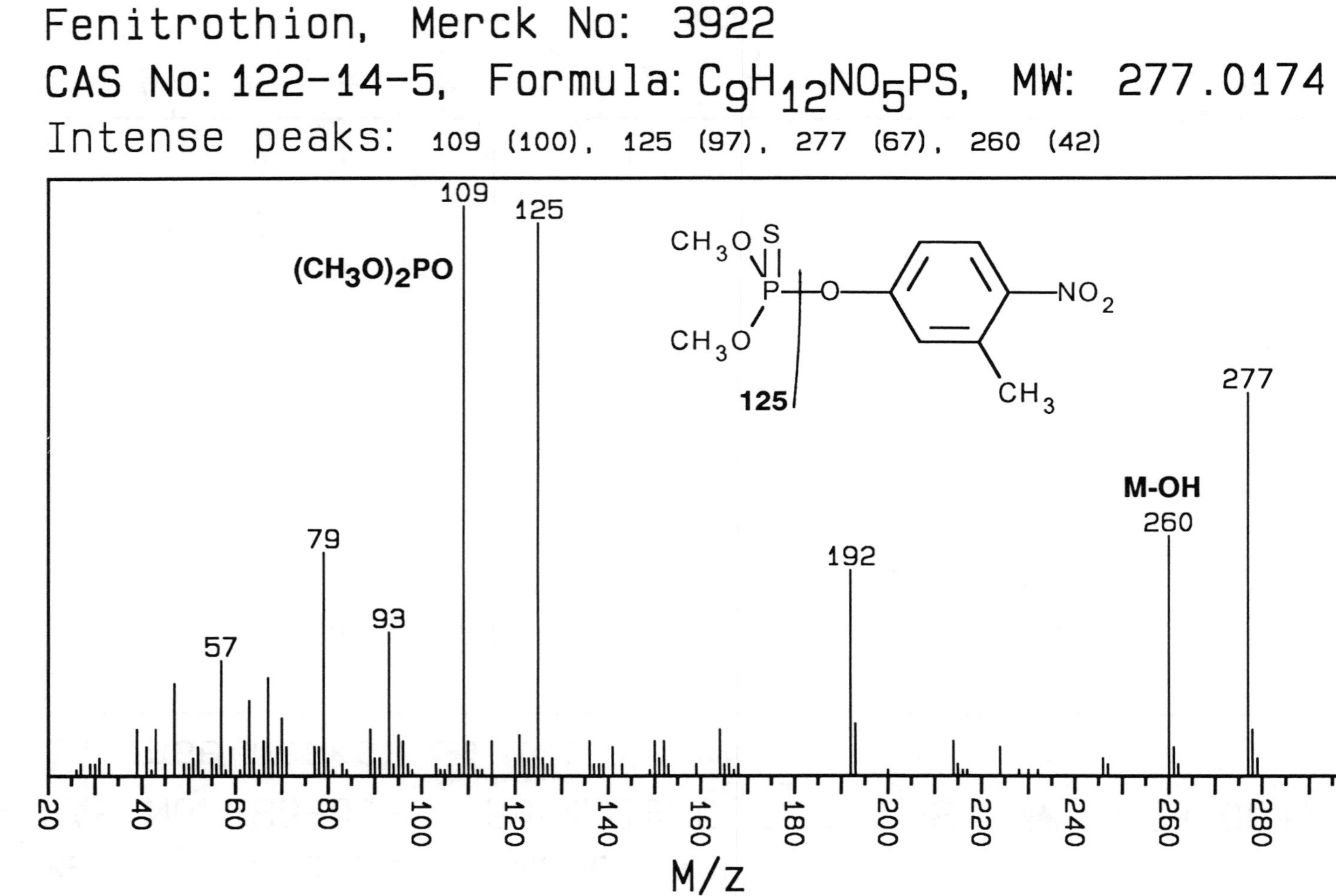
Fenitrothion, Merck No: 3922
CAS No: 122-14-5, Formula: C9H12NO5PS, MW: 277.0174
Intense peaks: 109 (100), 125 (97), 277 (67), 260 (42)
109
125
(CH3O)2PO
CH3O
S
P
O
NO2
CH3O
125
CH3
277
M-OH
260
79
192
93
57
20
40
60
80
100
120
140
160
180
200
220
240
260
280
300
M/Z

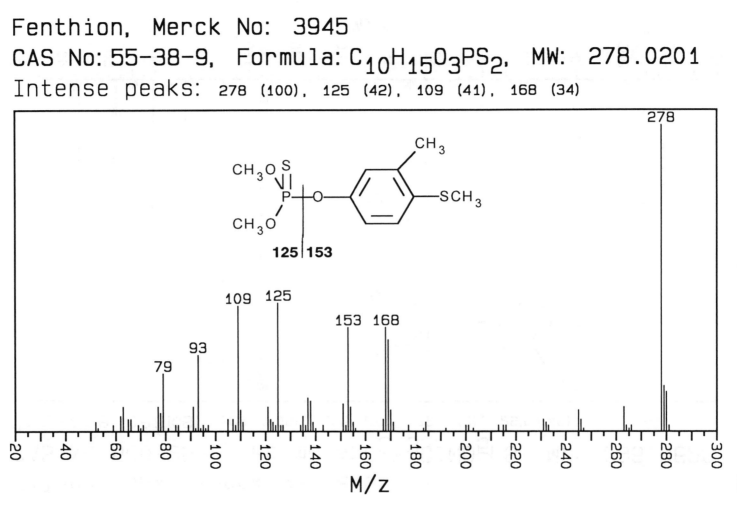

Fenthion, Merck No: 3945

CAS No: 55-38-9, Formula: $C_{10}H_{15}O_3PS_2$, MW: 278.0201

Intense peaks: 278 (100), 125 (42), 109 (41), 168 (34)

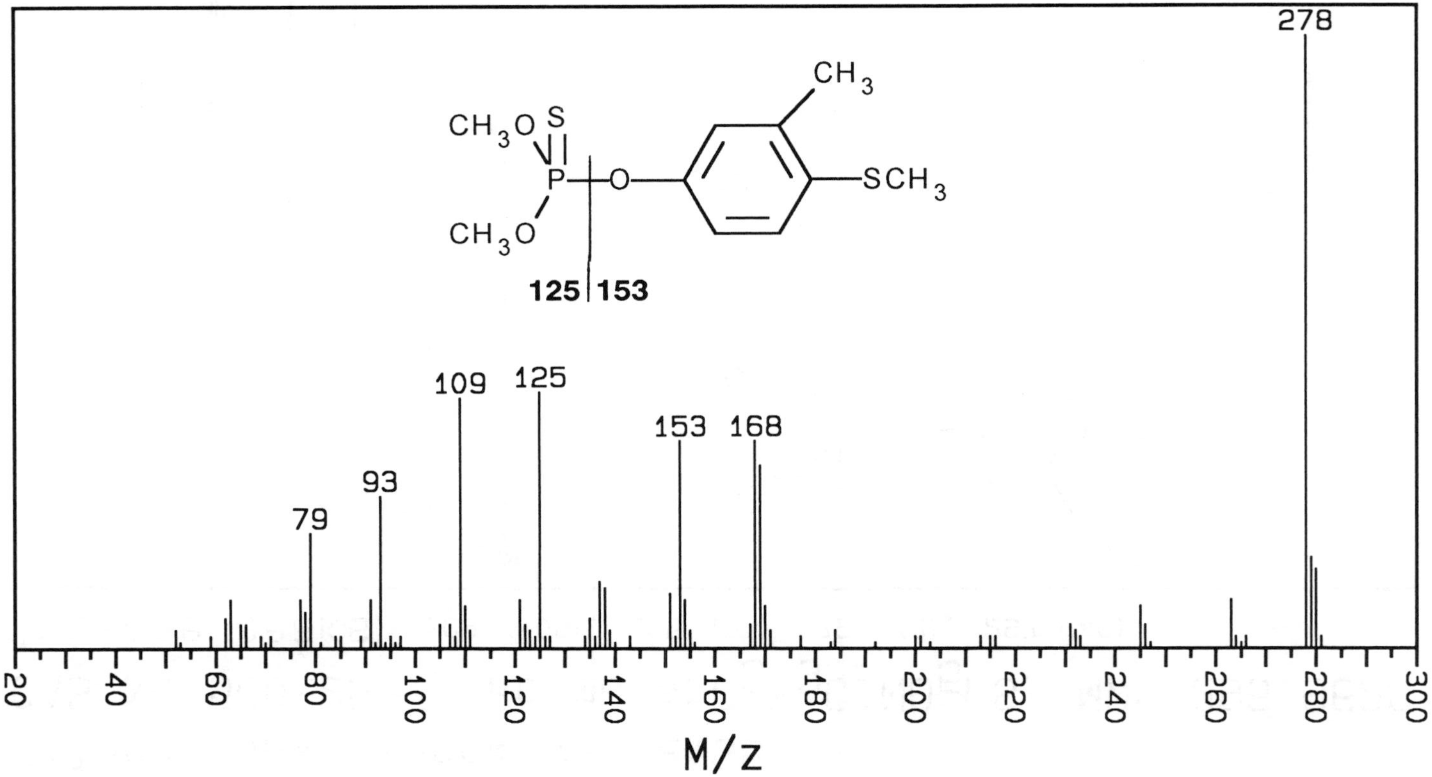

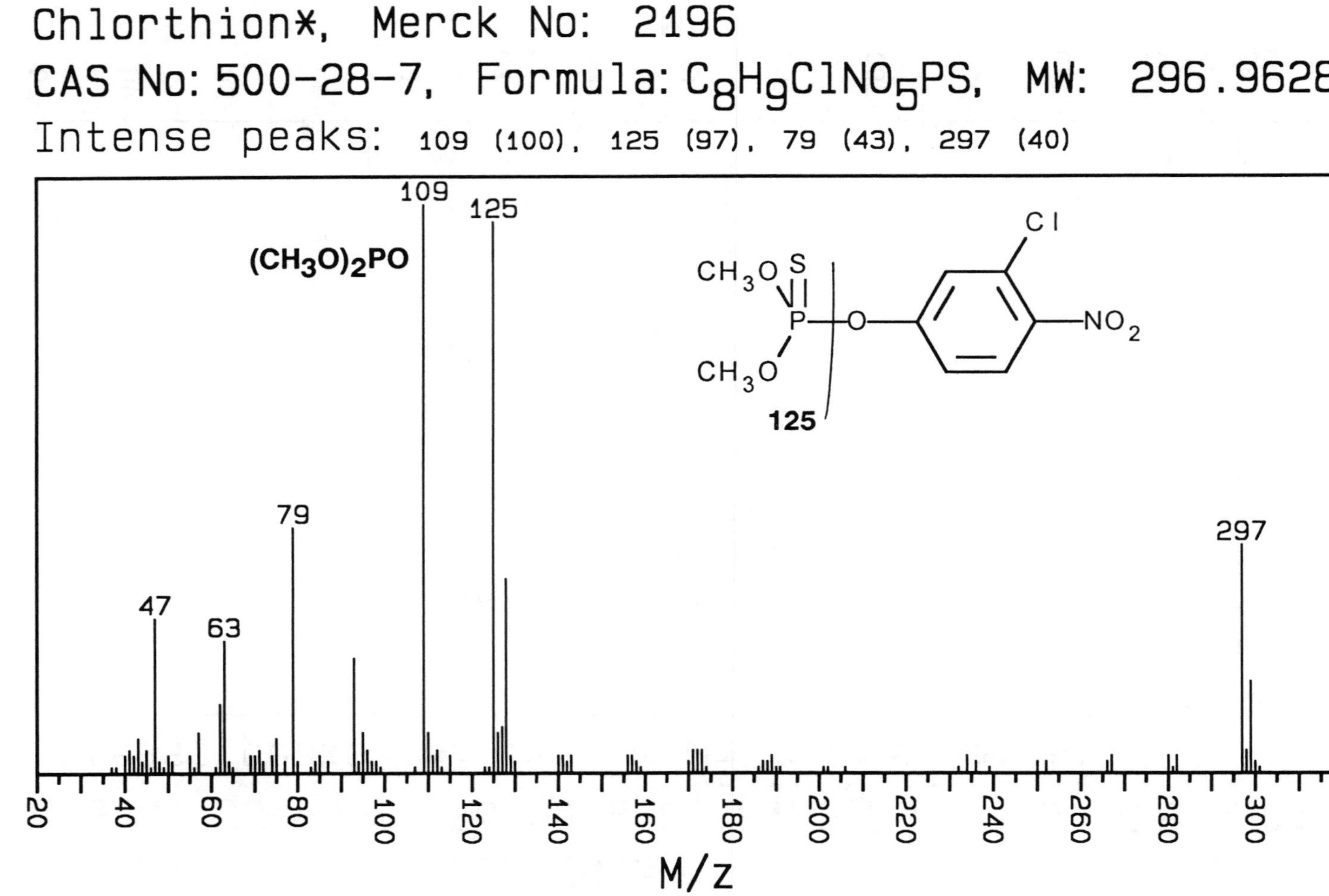
Chlorthion*, Merck No: 2196
CAS No: 500-28-7, Formula: $C_8H_9ClNO_5PS$, MW: 296.9628
Intense peaks: 109 (100), 125 (97), 79 (43), 297 (40)
$(CH_3O)_2PO$
109
125
79
47
63
297
Cl
$NO_2$
$CH_3O$
S
P
O
$CH_3O$
125
M/Z

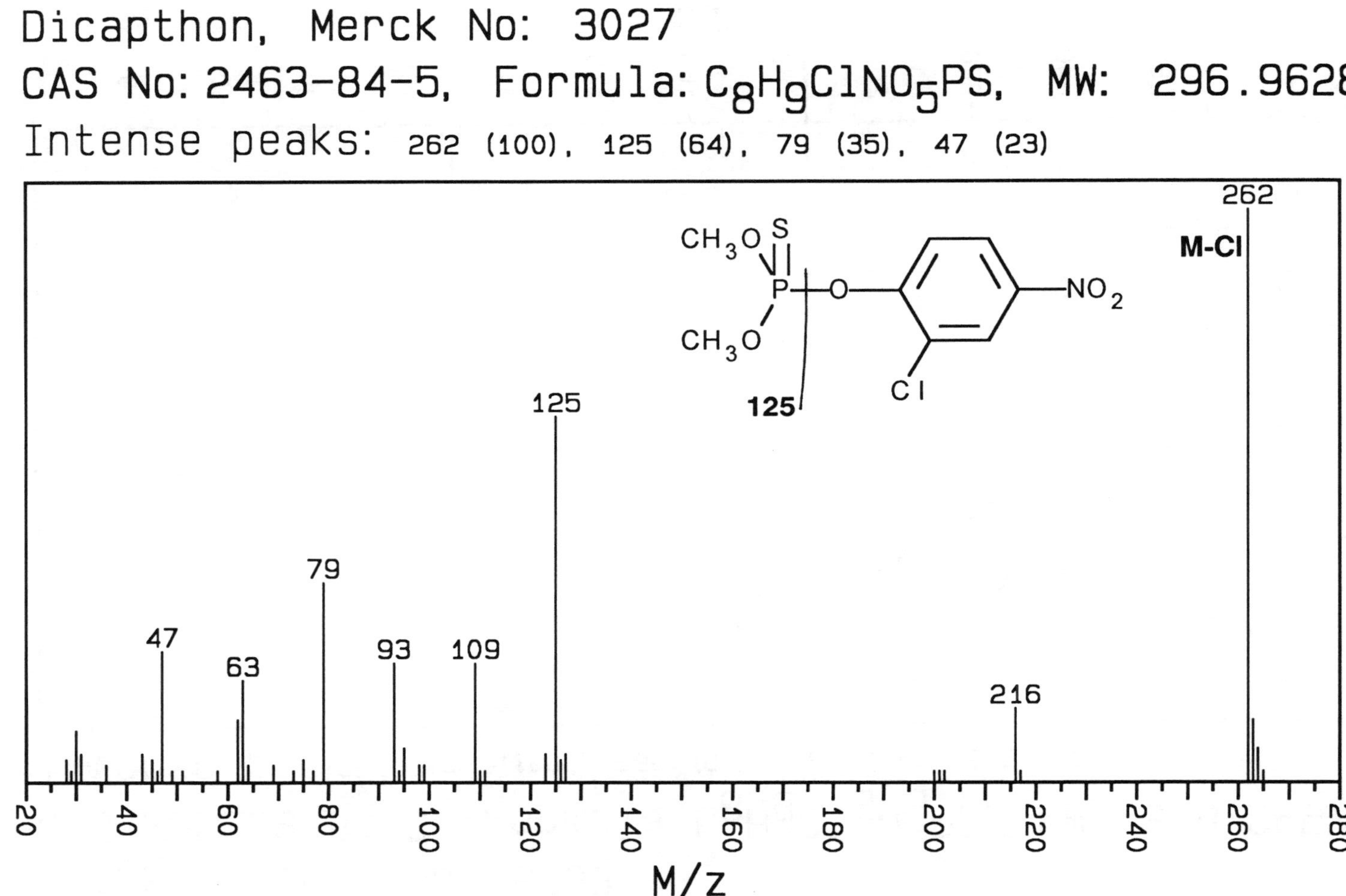
Dicapthon, Merck No: 3027
CAS No: 2463-84-5, Formula: $C_8H_9ClNO_5PS$, MW: 296.9628
Intense peaks: 262 (100), 125 (64), 79 (35), 47 (23)
262
M-Cl
$CH_3O$
S
P
O
$NO_2$
$CH_3O$
Cl
125
125
79
47
63
93
109
216
20
40
60
80
100
120
140
160
180
200
220
240
260
280
M/Z

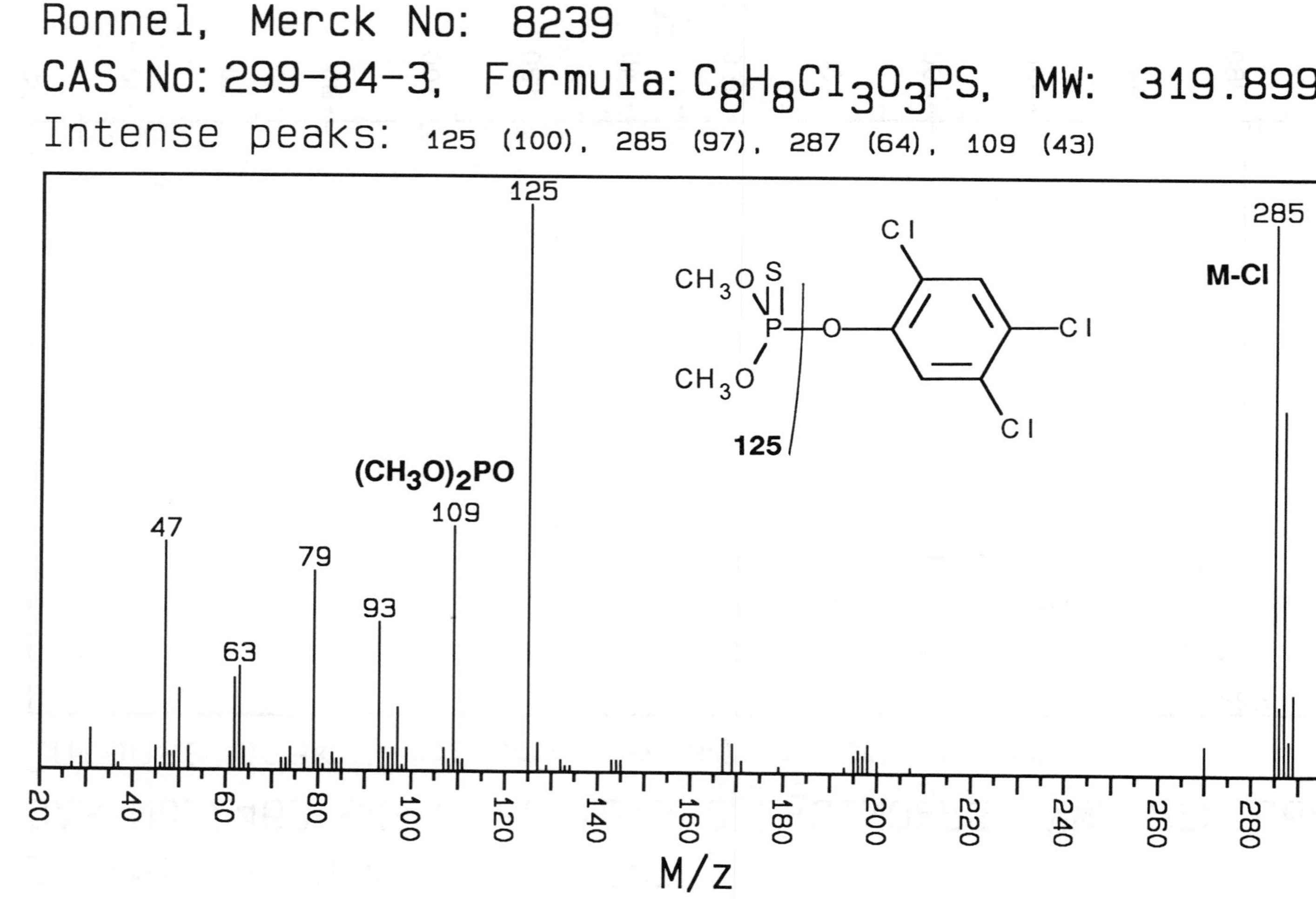
Ronnel, Merck No: 8239
CAS No: 299-84-3, Formula: $C_8H_8Cl_3O_3PS$, MW: 319.8997
Intense peaks: 125 (100), 285 (97), 287 (64), 109 (43)
125
285
M-Cl
$CH_3O$
S
P
O
$CH_3O$
Cl
Cl
Cl
125
$(CH_3O)_2PO$
109
47
79
93
63
20
40
60
80
100
120
140
160
180
200
220
240
260
280
300
M/Z

Bromophos, Merck No: 1419

CAS No: 2104-96-3, Formula: $C_8H_8BrCl_2O_3PS$, MW: 363.8492

Intense peaks: 331 (100), 125 (79), 329 (75), 47 (42)

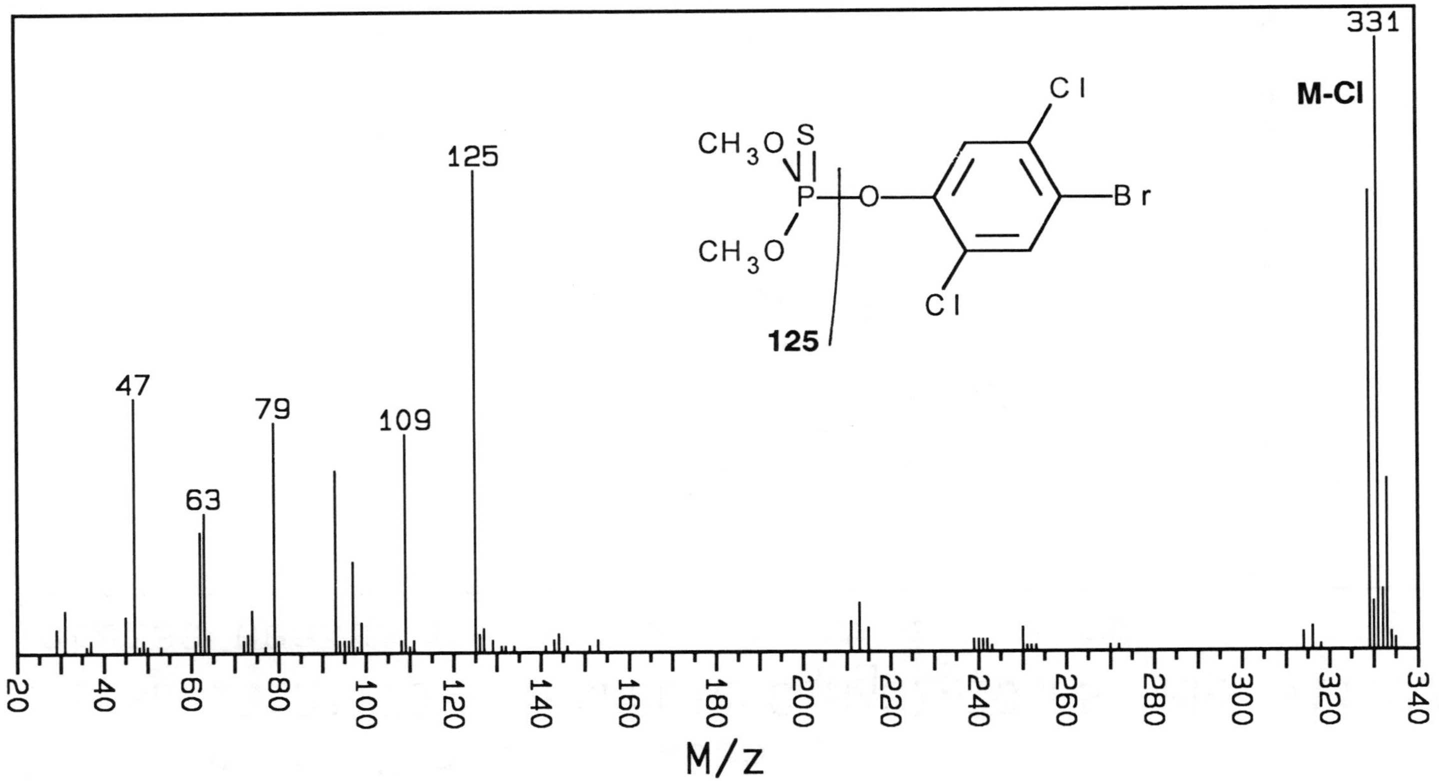

Iodofenphos, Merck No: 4925

CAS No: 18181-70-9, Formula: $C_8H_8Cl_2IO_3PS$, MW: 411.8352

Intense peaks: 377 (100), 125 (44), 379 (37), 42 (22)

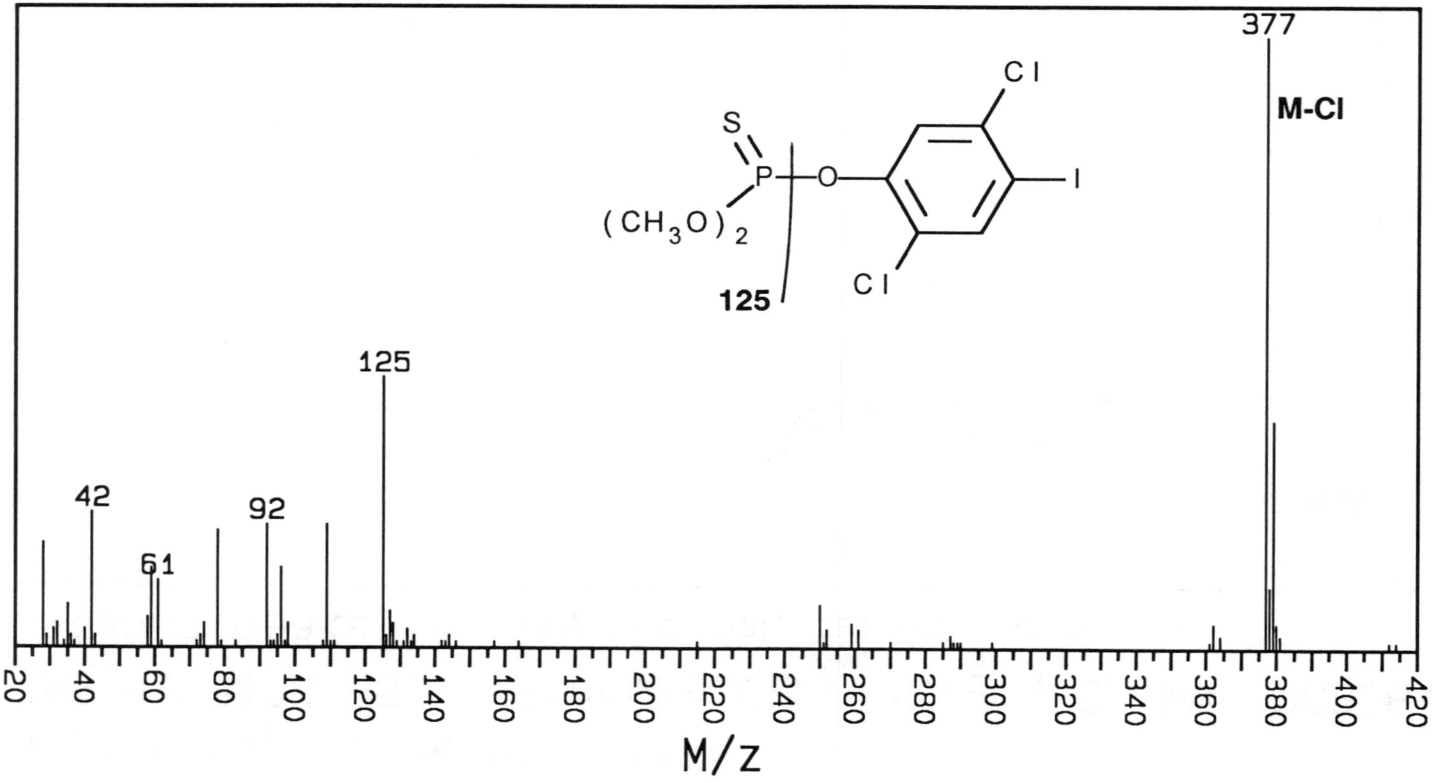

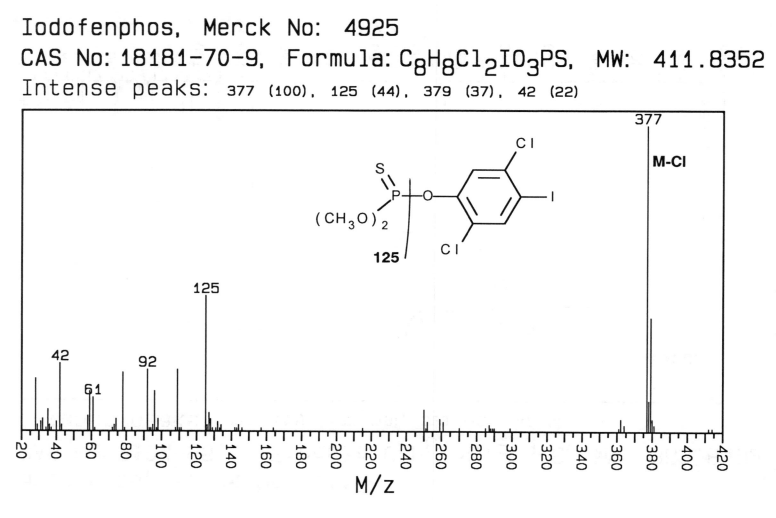

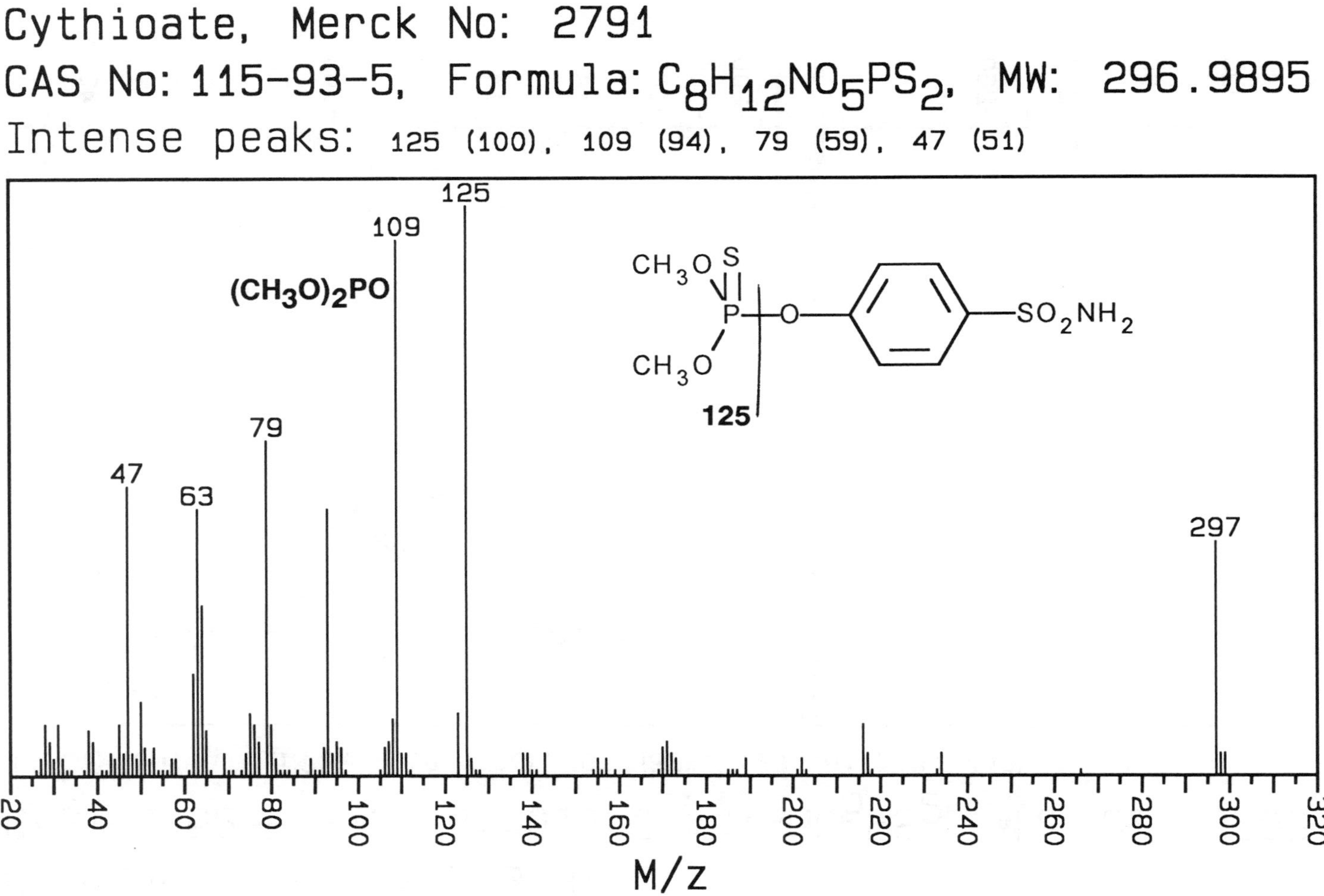

Cythioate, Merck No: 2791
CAS No: 115-93-5, Formula: $C_8H_{12}NO_5PS_2$, MW: 296.9895
Intense peaks: 125 (100), 109 (94), 79 (59), 47 (51)
125
109
$(CH_3O)_2PO$
79
47
63
297
$CH_3O$
S
P
O
$SO_2NH_2$
$CH_3O$
125
20
40
60
80
100
120
140
160
180
200
220
240
260
280
300
320
M/Z

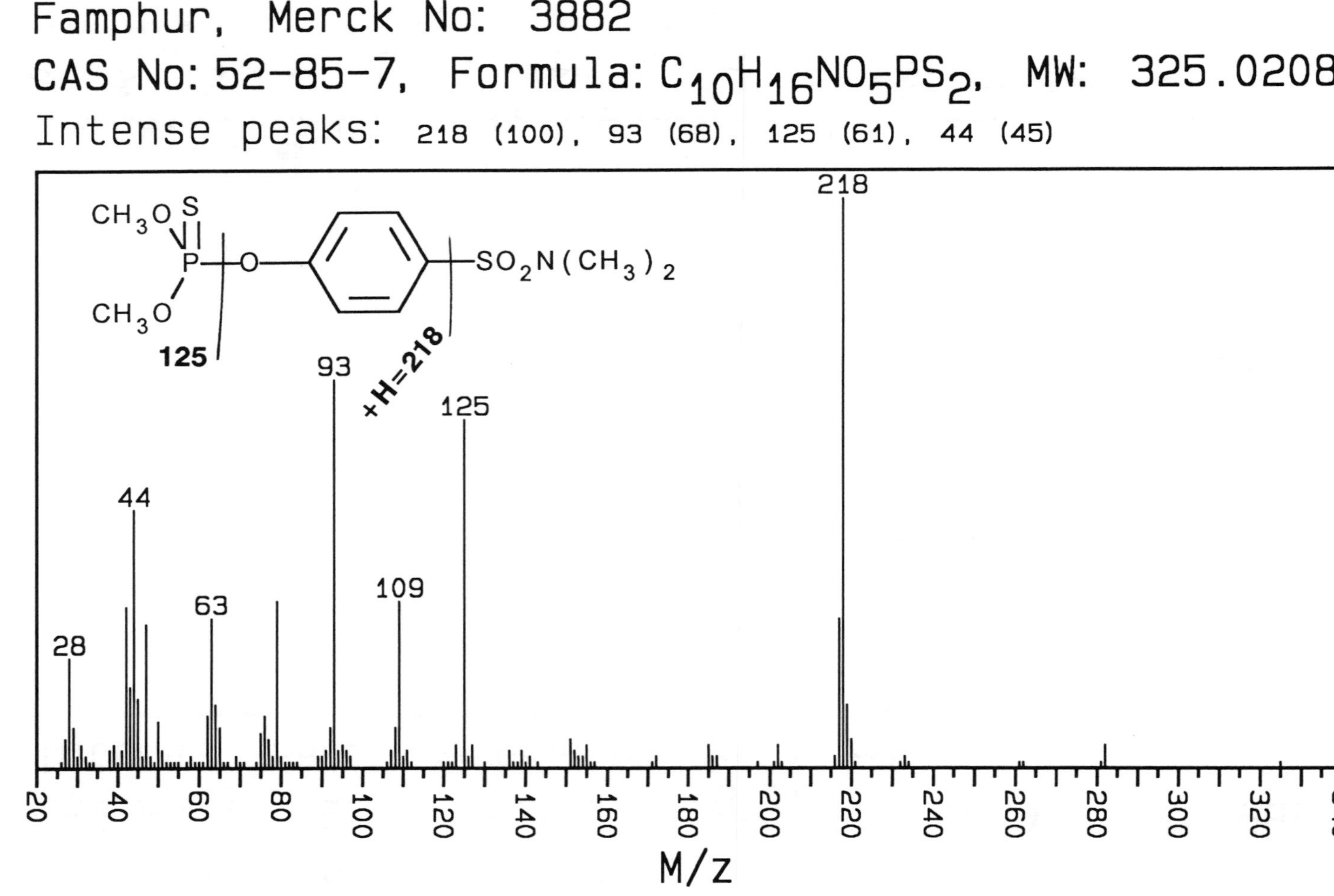
Famphur, Merck No: 3882
CAS No: 52-85-7, Formula: $C_{10}H_{16}NO_5PS_2$, MW: 325.0208
Intense peaks: 218 (100), 93 (68), 125 (61), 44 (45)
$CH_3O$
S
P
O
$CH_3O$
125
$SO_2N(CH_3)_2$
+H=218
218
93
125
44
109
63
28
20
40
60
80
100
120
140
160
180
200
220
240
260
280
300
320
340
M/Z

Temephos, Merck No: 9075

CAS No: 3383-96-8, Formula: $C_{16}H_{20}O_6P_2S_3$, MW: 465.9897

Intense peaks: 466 (100), 125 (51), 93 (38), 47 (35)

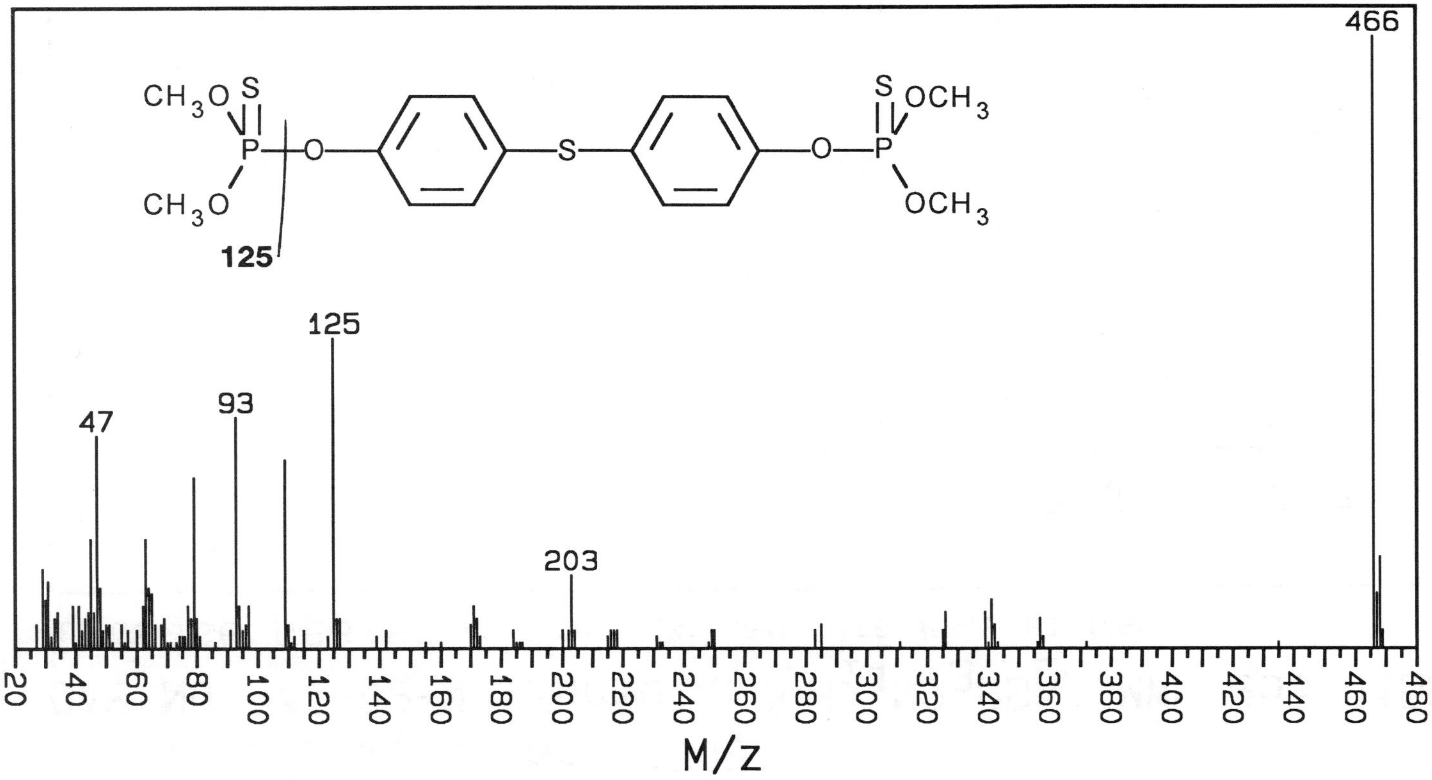

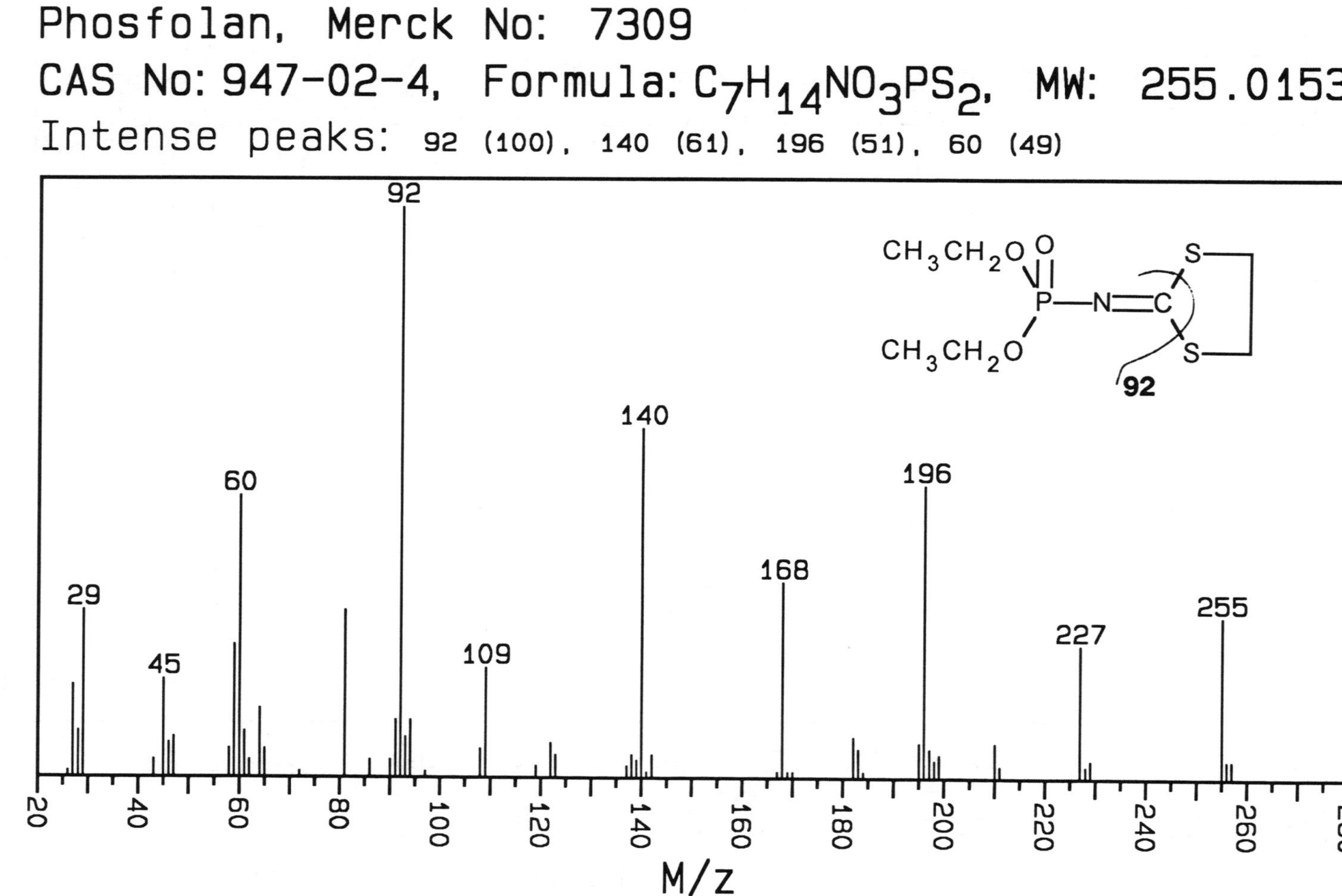
Phosfolan, Merck No: 7309
CAS No: 947-02-4, Formula: C7H14NO3PS2, MW: 255.0153
Intense peaks: 92 (100), 140 (61), 196 (51), 60 (49)
CH3CH2O
CH3CH2O
O
P
N
C
S
S
92
29
45
60
92
109
140
168
196
227
255
20
40
60
80
100
120
140
160
180
200
220
240
260
280
M/Z

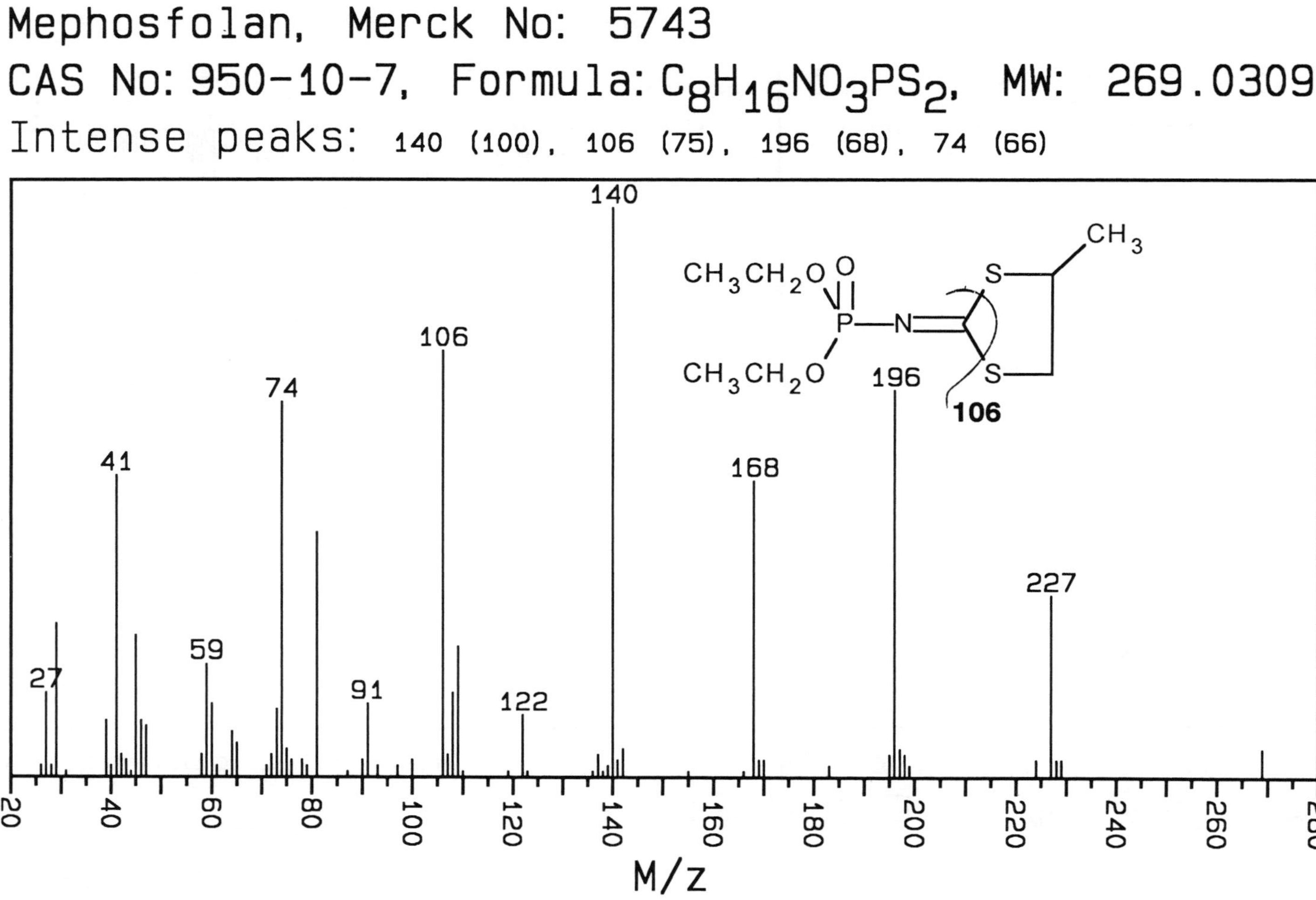
Mephosfolan, Merck No: 5743
CAS No: 950-10-7, Formula: $C_8H_{16}NO_3PS_2$, MW: 269.0309
Intense peaks: 140 (100), 106 (75), 196 (68), 74 (66)
140
106
74
41
59
27
91
122
168
196
227
$CH_3CH_2O$
$CH_3CH_2O$
O
P
N
S
S
$CH_3$
106
20
40
60
80
100
120
140
160
180
200
220
240
260
280
M/Z

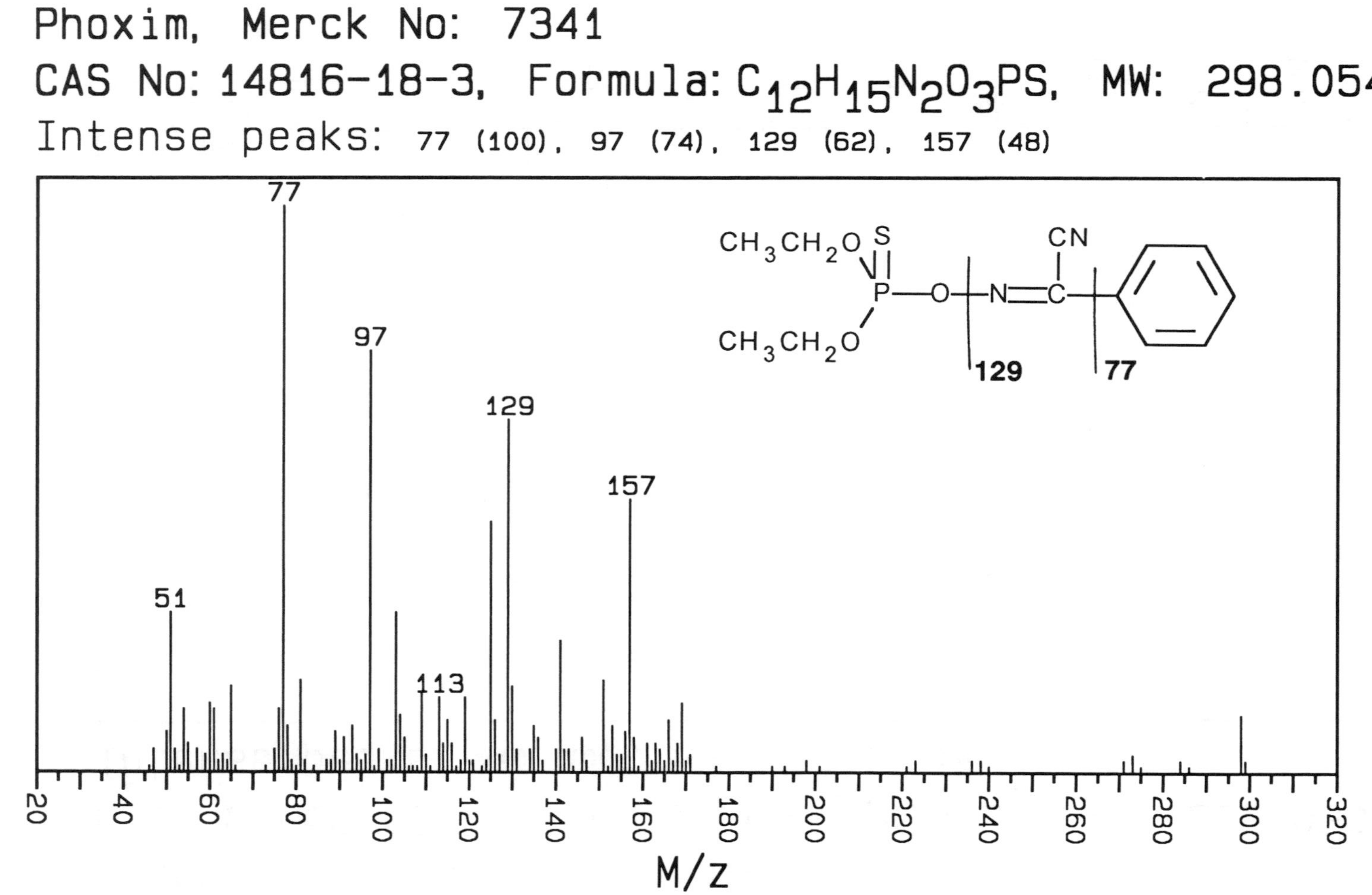
Phoxim, Merck No: 7341
CAS No: 14816-18-3, Formula: C12H15N2O3PS, MW: 298.0541
Intense peaks: 77 (100), 97 (74), 129 (62), 157 (48)
CH3CH2O
CH3CH2O
S
P
O
N
C
CN
129
77
51
77
97
113
129
157
20
40
60
80
100
120
140
160
180
200
220
240
260
280
300
320
M/Z

Thionazin, Merck No: 9275

CAS No: 297-97-2, Formula: $C_8H_{13}N_2O_3PS$, MW: 248.0384

Intense peaks: 107 (100), 96 (98), 106 (73), 97 (64)

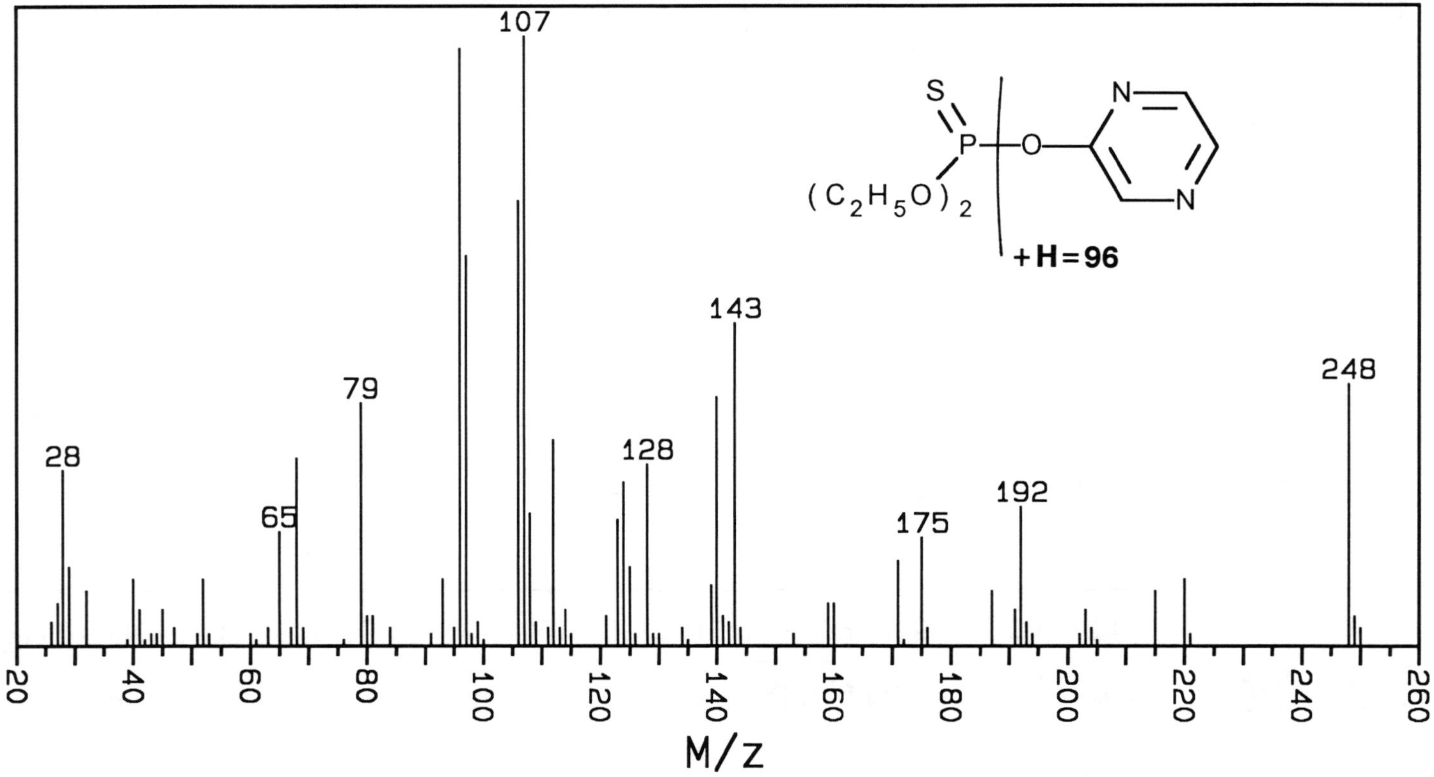

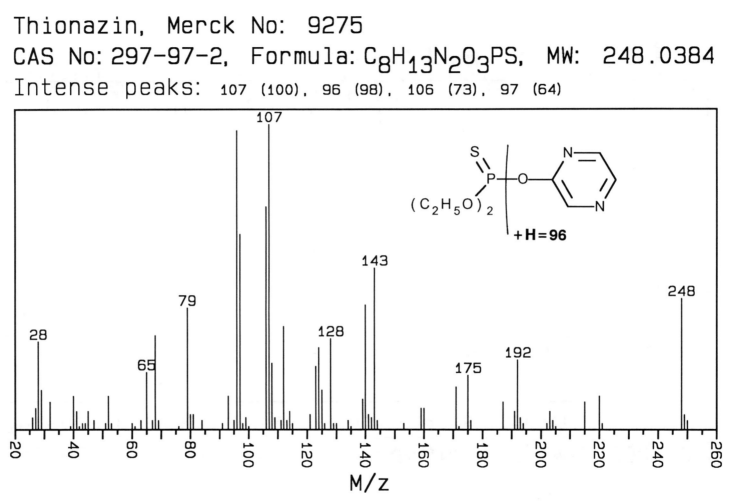

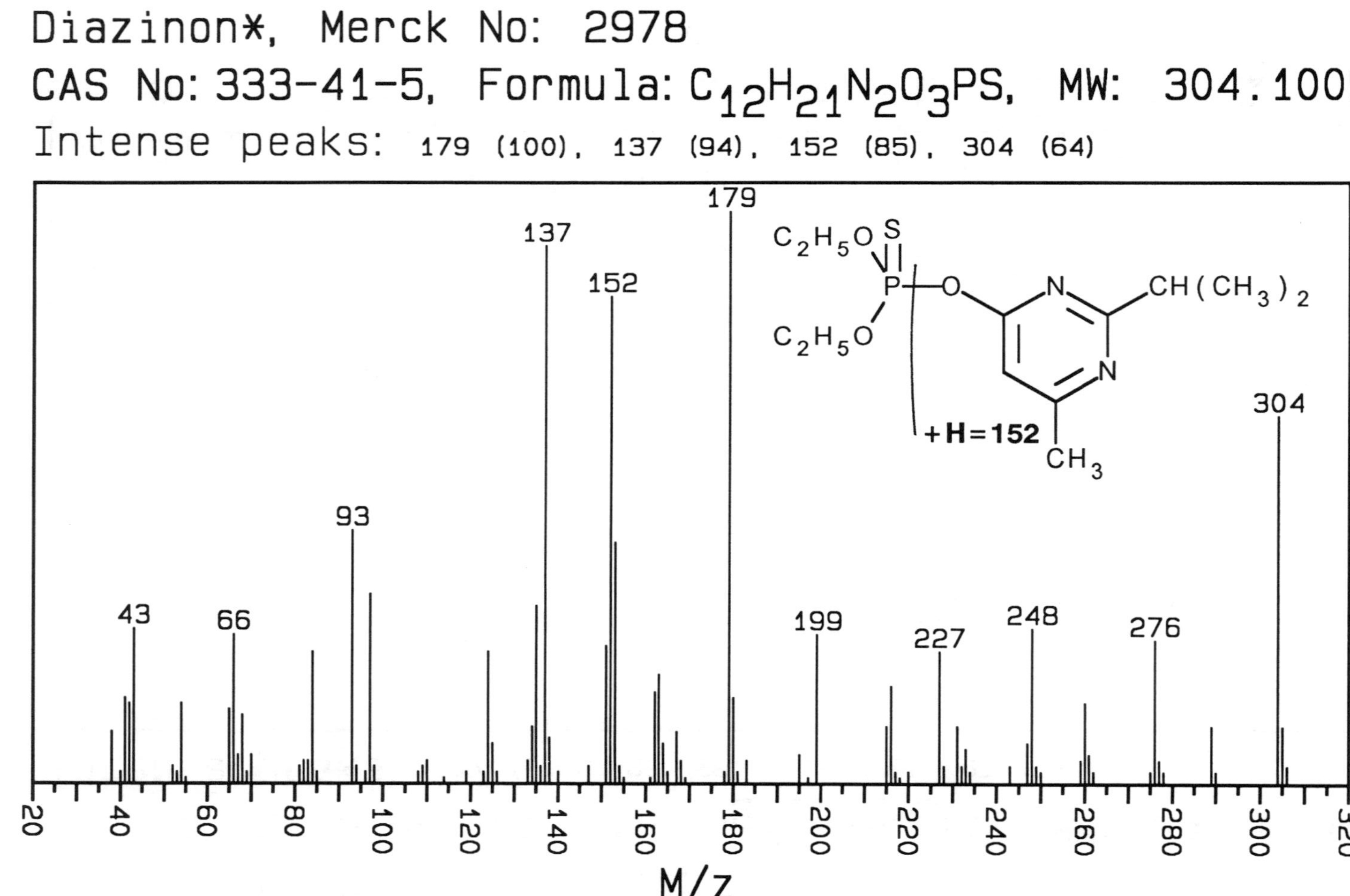

Diazinon*, Merck No: 2978
CAS No: 333-41-5, Formula: $C_{12}H_{21}N_2O_3PS$, MW: 304.1009
Intense peaks: 179 (100), 137 (94), 152 (85), 304 (64)
$C_2H_5O$
S
P
O
$C_2H_5O$
N
N
$CH(CH_3)_2$
$CH_3$
+H=152
43
66
93
137
152
179
199
227
248
276
304
M/Z

Pirimiphos-ethyl, Merck No: 7469

CAS No: 23505-41-1, Formula: $C_{13}H_{24}N_3O_3PS$, MW: 333.1276

Intense peaks: 333 (100), 318 (82), 57 (60), 304 (57)

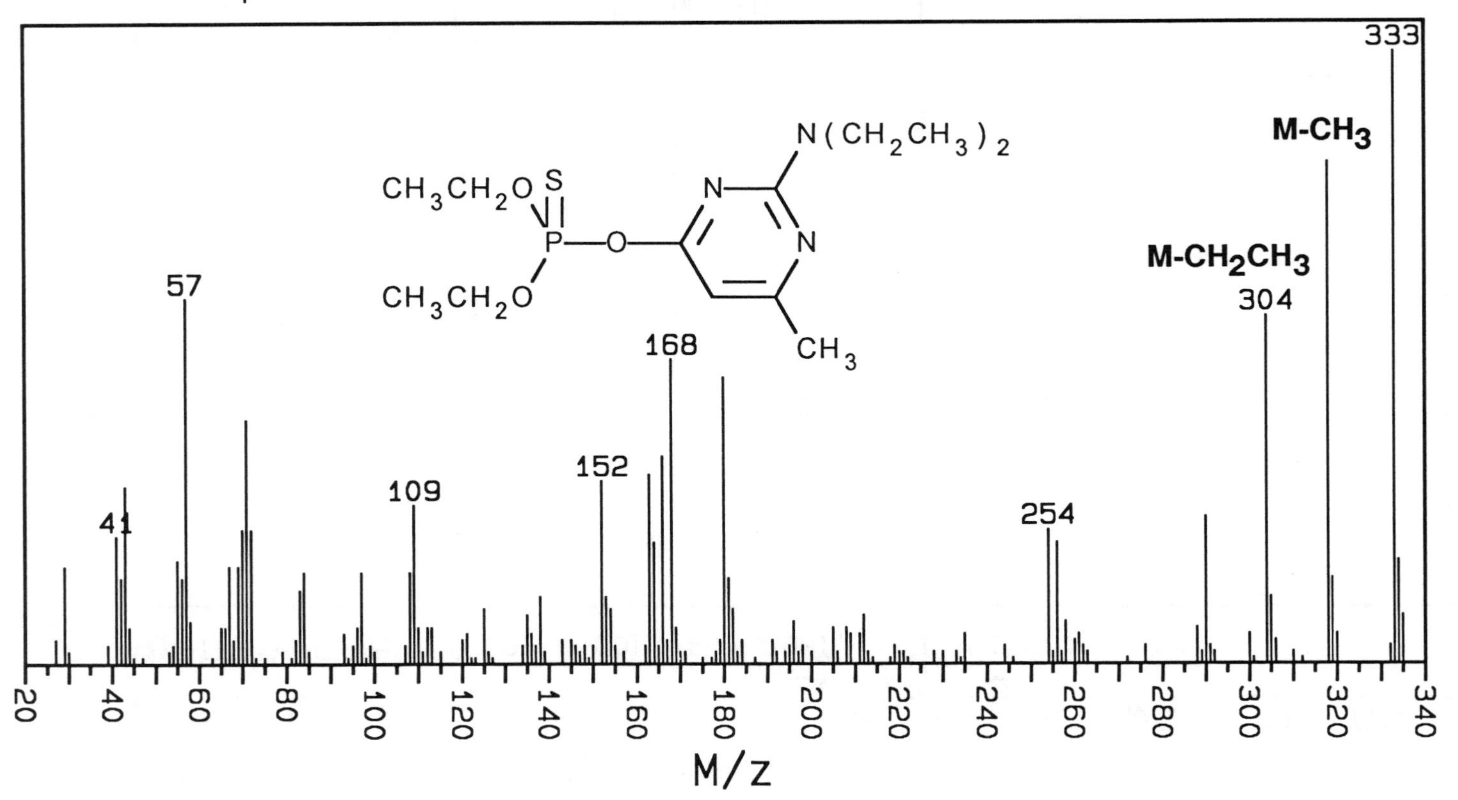

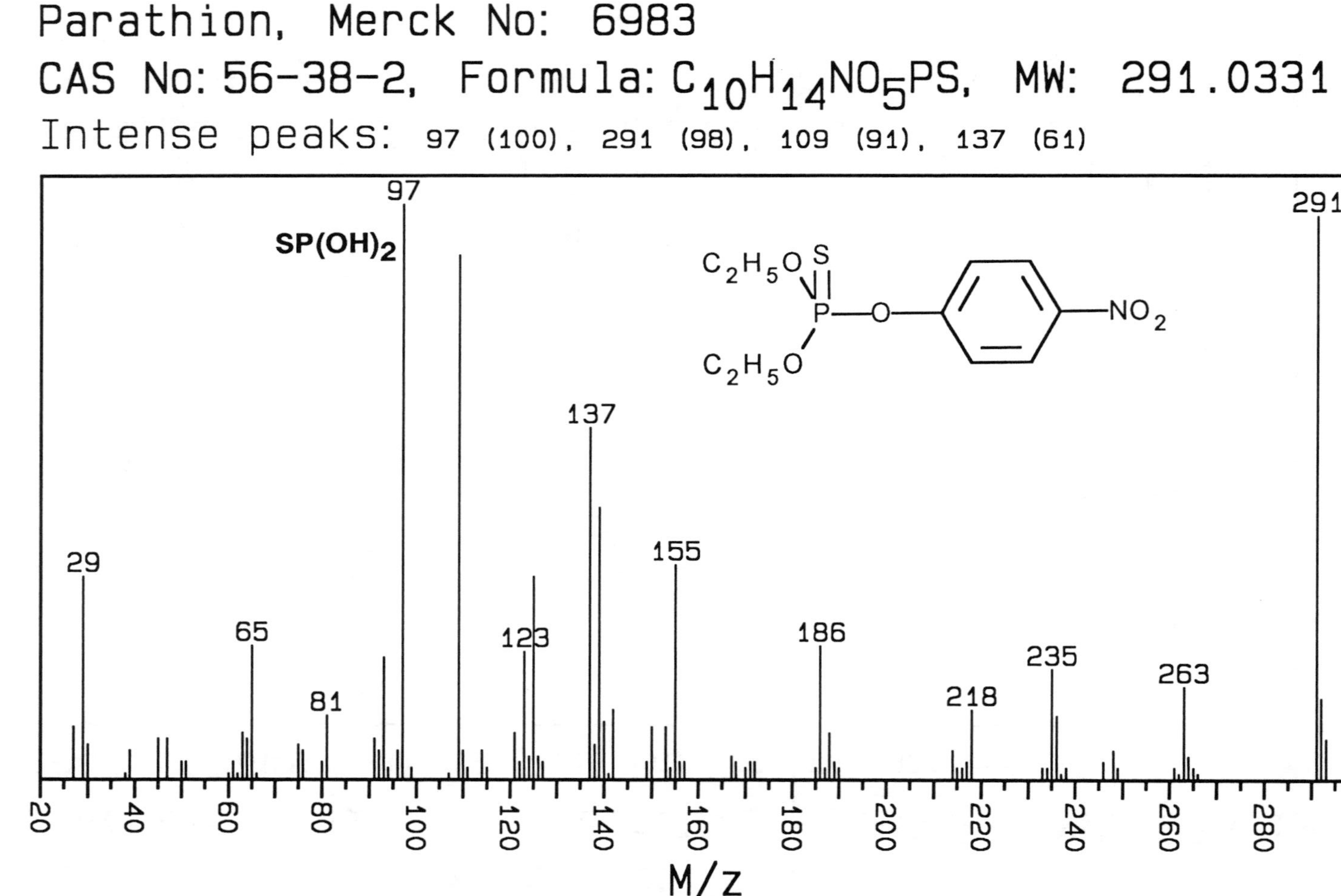

Parathion, Merck No: 6983
CAS No: 56-38-2, Formula: $C_{10}H_{14}NO_5PS$, MW: 291.0331
Intense peaks: 97 (100), 291 (98), 109 (91), 137 (61)
97
SP(OH)2
291
$C_2H_5O$
S
P
O
$NO_2$
$C_2H_5O$
137
29
155
65
123
186
81
235
263
218
20
40
60
80
100
120
140
160
180
200
220
240
260
280
300
M/Z

Fensulfothion, Merck No: 3943

CAS No: 115-90-2, Formula: $C_{11}H_{17}O_4PS_2$, MW: 308.0306

Intense peaks: 29 (100), 97 (65), 125 (49), 27 (44)

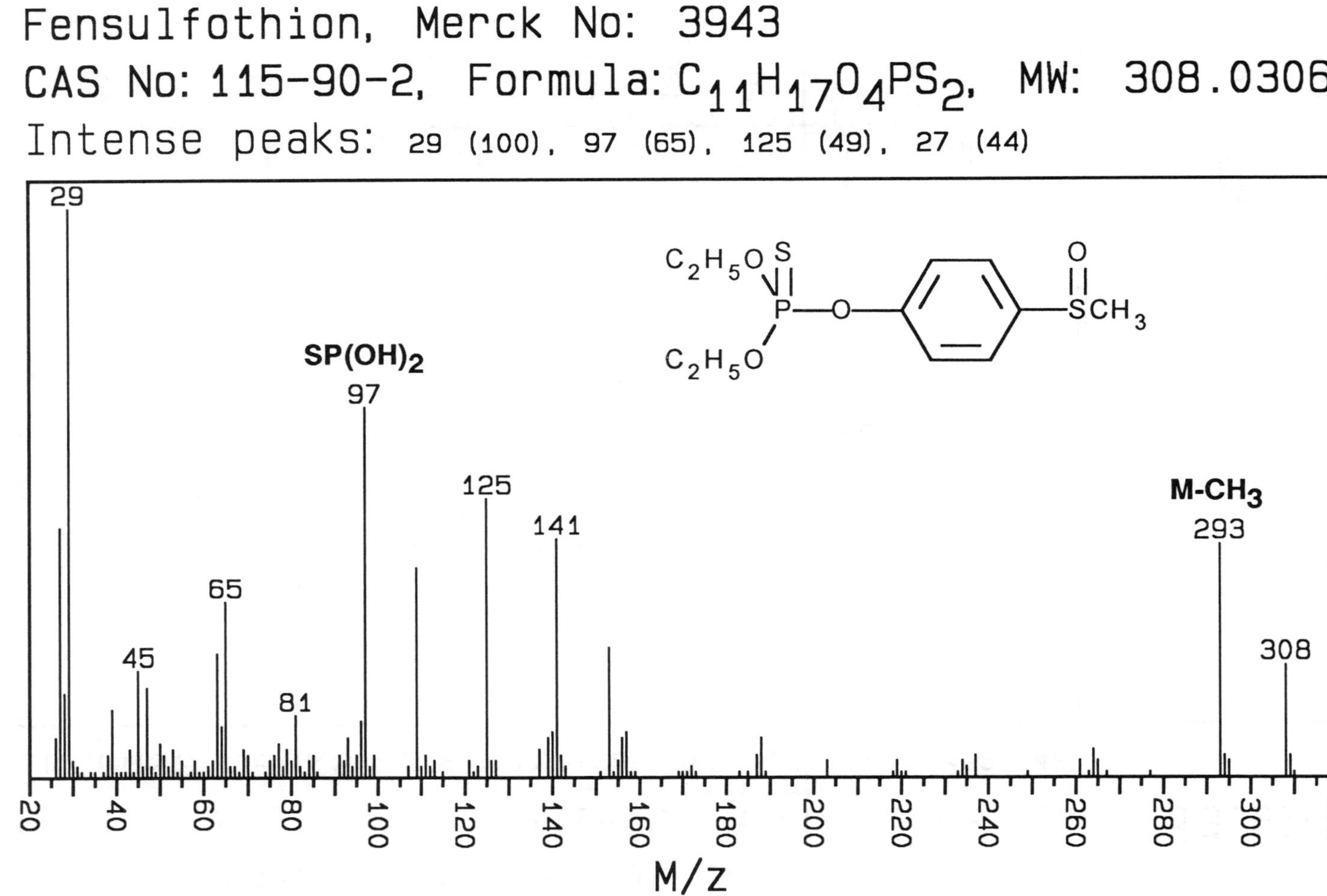

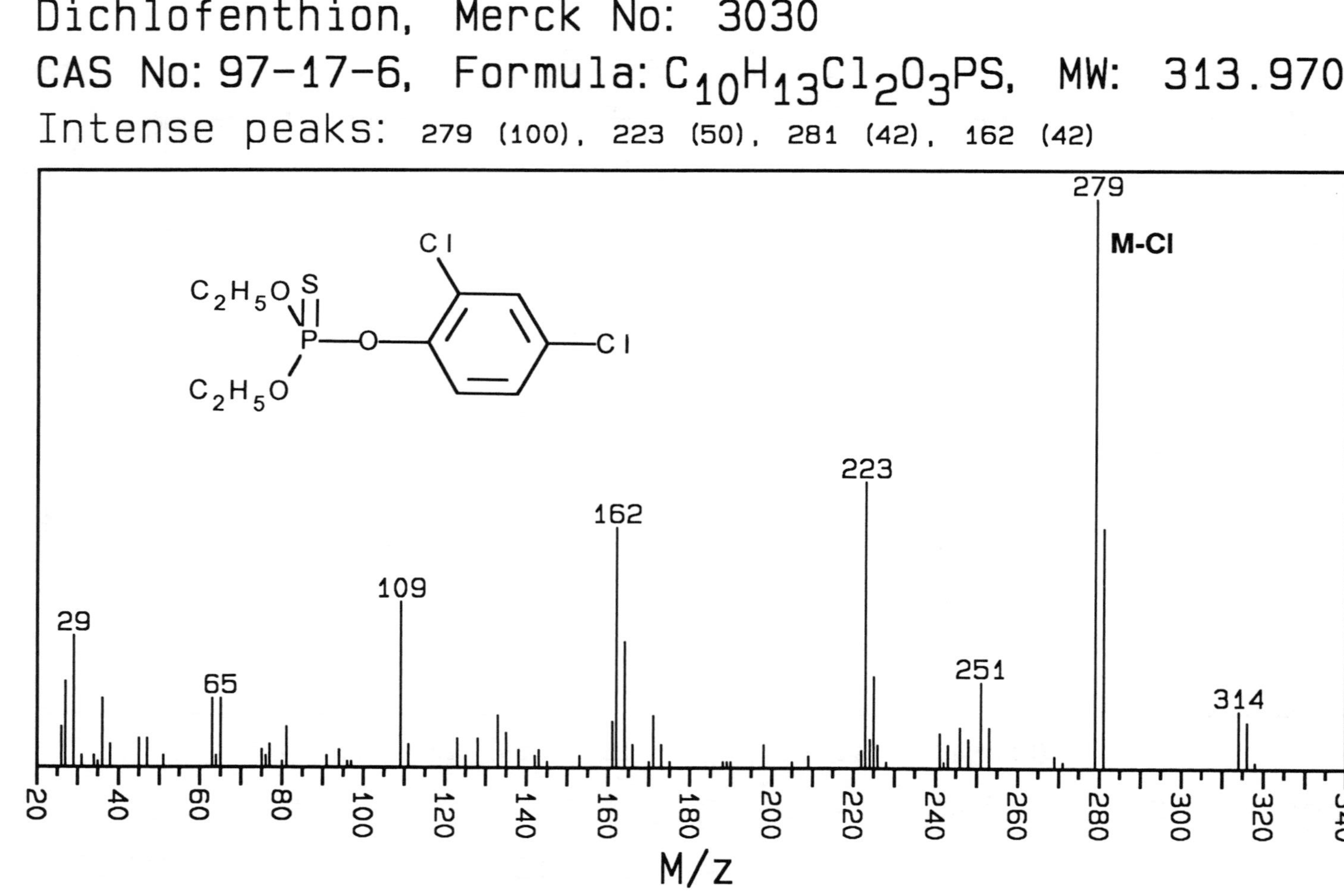

Dichlofenthion, Merck No: 3030
CAS No: 97-17-6, Formula: $C_{10}H_{13}Cl_2O_3PS$, MW: 313.9701
Intense peaks: 279 (100), 223 (50), 281 (42), 162 (42)
279
M-Cl
Cl
$C_2H_5O$
S
P
O
Cl
$C_2H_5O$
223
162
109
29
251
65
314
20
40
60
80
100
120
140
160
180
200
220
240
260
280
300
320
340
M/Z

Chlorpyrifos, Merck No: 2190

CAS No: 2921-88-2, Formula: $C_9H_{11}Cl_3NO_3PS$, MW: 348.9263

Intense peaks: 97 (100), 197 (97), 199 (94), 314 (64)

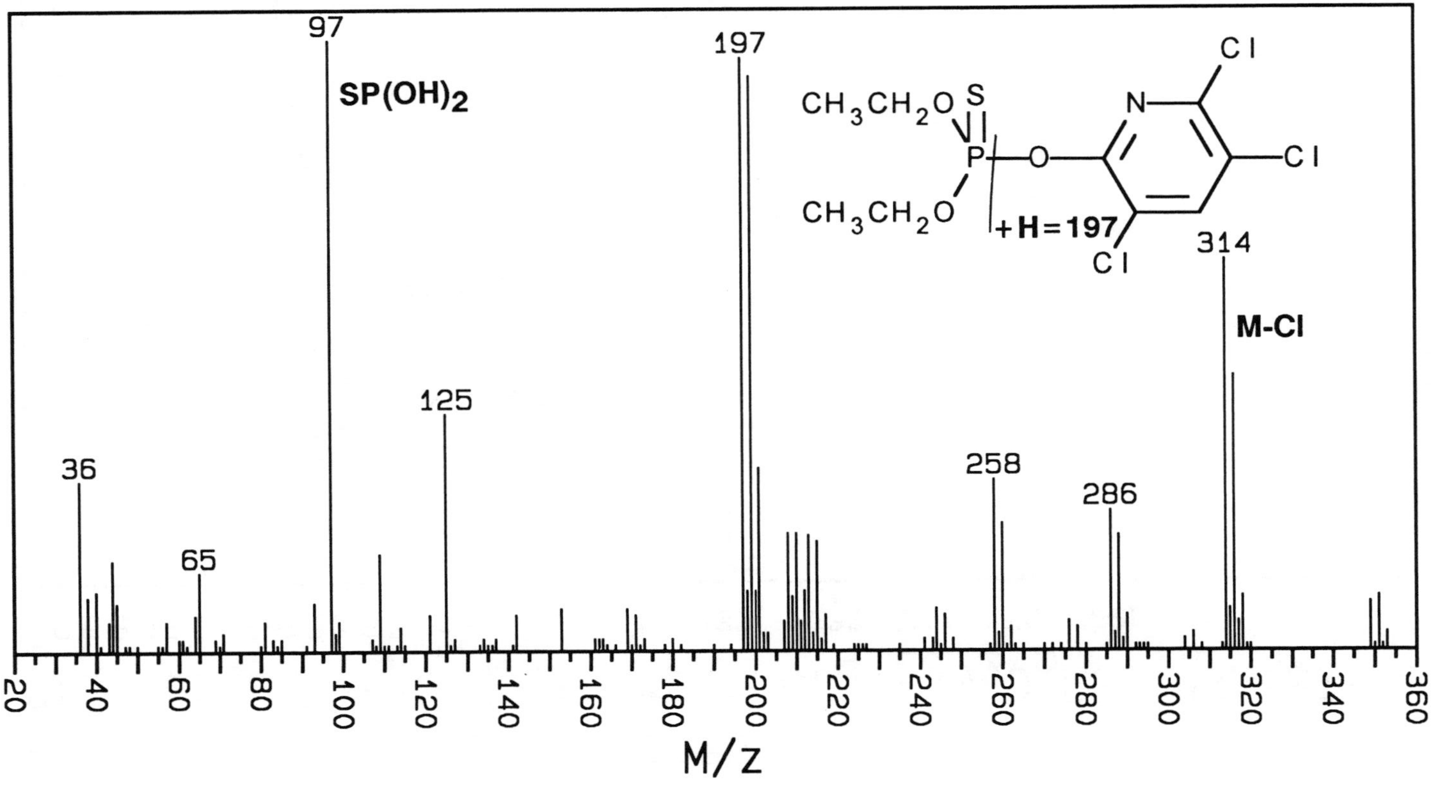

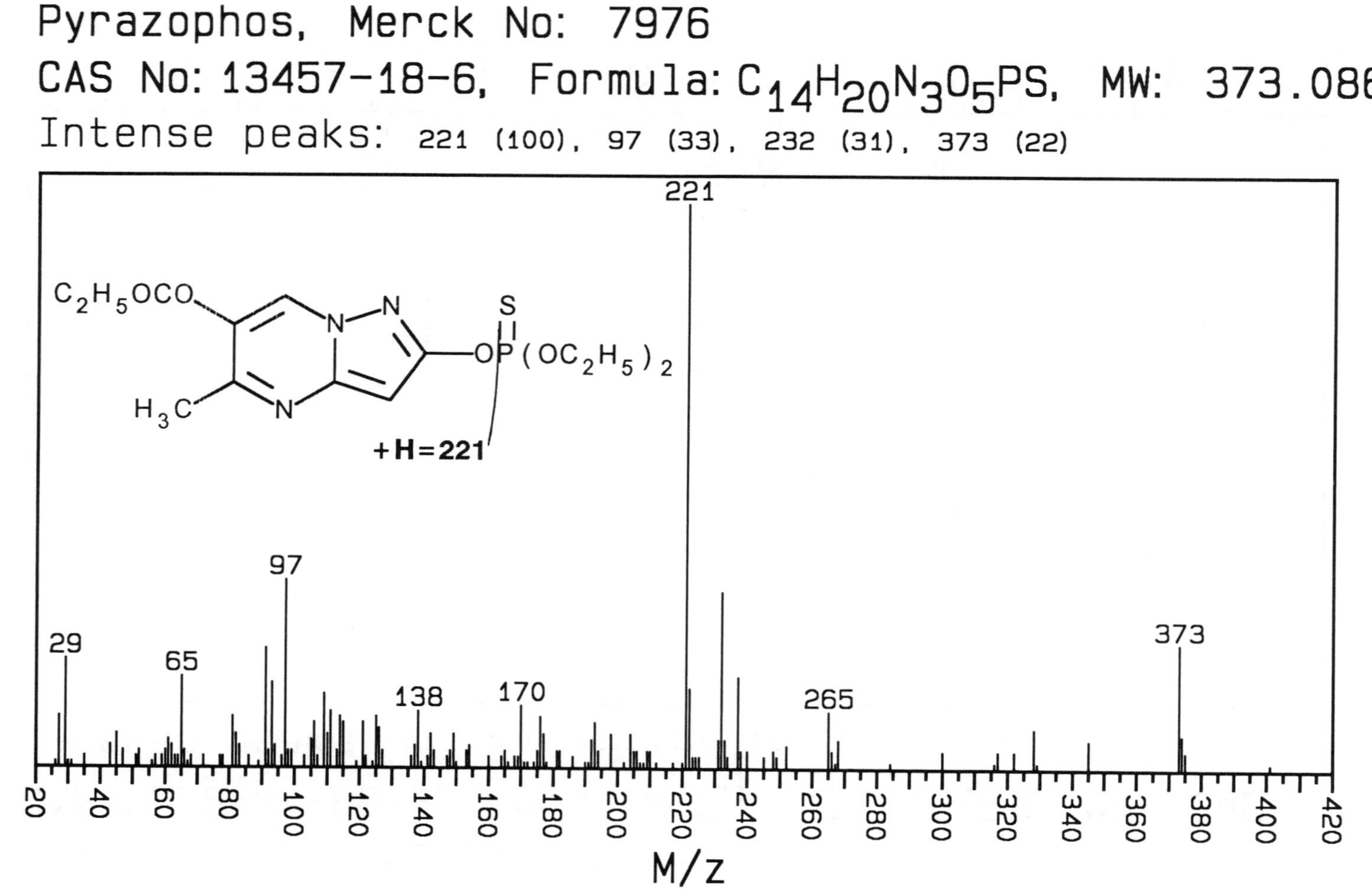

Pyrazophos, Merck No: 7976
CAS No: 13457-18-6, Formula: $C_{14}H_{20}N_3O_5PS$, MW: 373.0861
Intense peaks: 221 (100), 97 (33), 232 (31), 373 (22)
221
$C_2H_5OCO$
N
N
S
$OP(OC_2H_5)_2$
$H_3C$
N
+H=221
97
29
65
138
170
265
373
20
40
60
80
100
120
140
160
180
200
220
240
260
280
300
320
340
360
380
400
420
M/Z

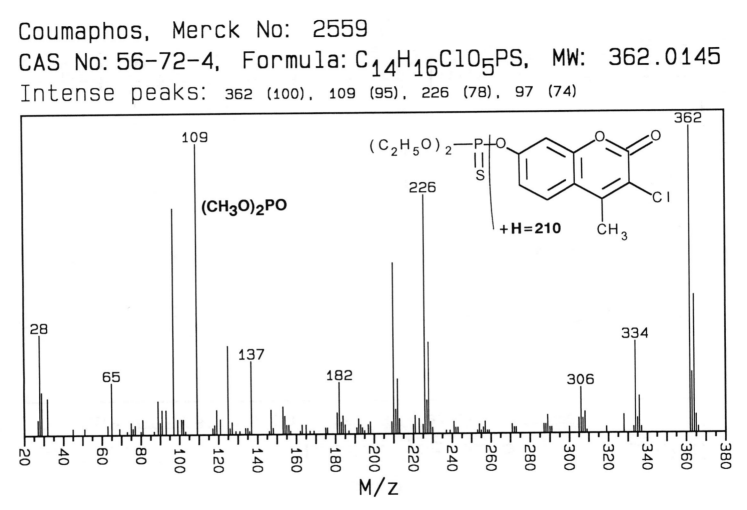

Coumaphos, Merck No: 2559

CAS No: 56-72-4, Formula: $C_{14}H_{16}ClO_5PS$, MW: 362.0145

Intense peaks: 362 (100), 109 (95), 226 (78), 97 (74)

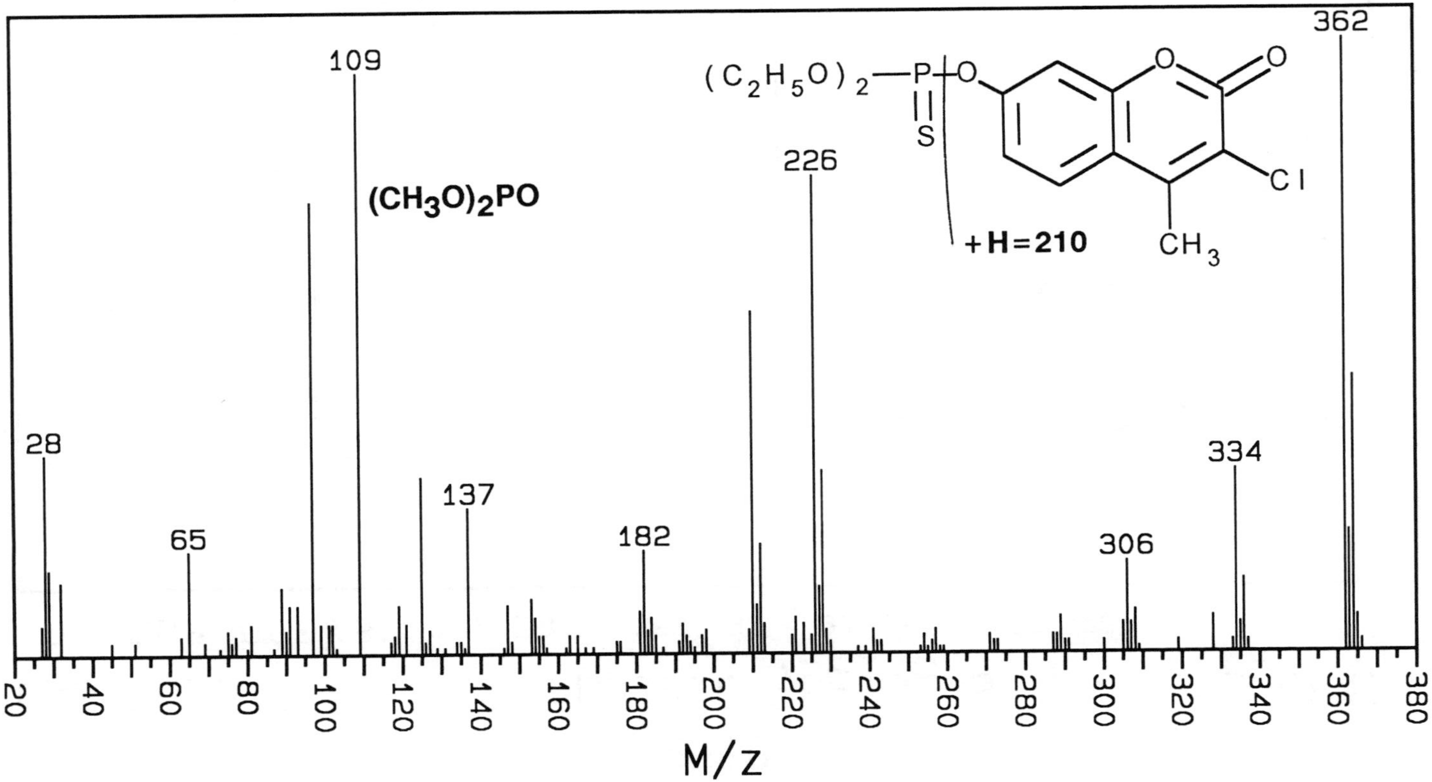

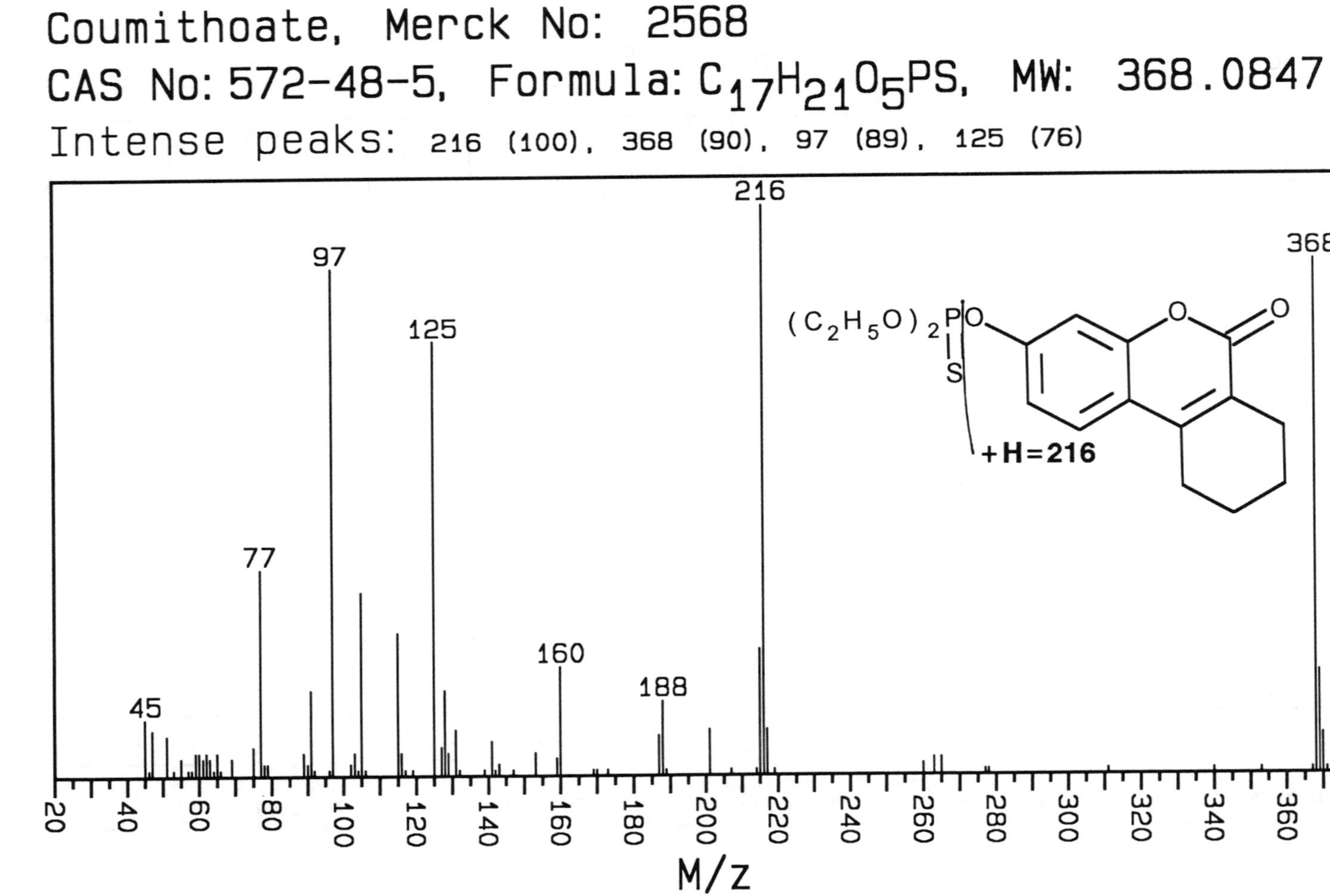
Coumithoate, Merck No: 2568
CAS No: 572-48-5, Formula: $C_{17}H_{21}O_5PS$, MW: 368.0847
Intense peaks: 216 (100), 368 (90), 97 (89), 125 (76)
216
368
97
125
77
160
188
45
$(C_2H_5O)_2P$
S
O
O
O
+H=216
20
40
60
80
100
120
140
160
180
200
220
240
260
280
300
320
340
360
380
M/Z

Leptophos, Merck No: 5327 MW: 409.8701

CAS No: 21609-90-5, Formula: $C_{13}H_{10}BrCl_2O_2PS$

Intense peaks: 171 (100), 377 (62), 28 (60), 375 (45)

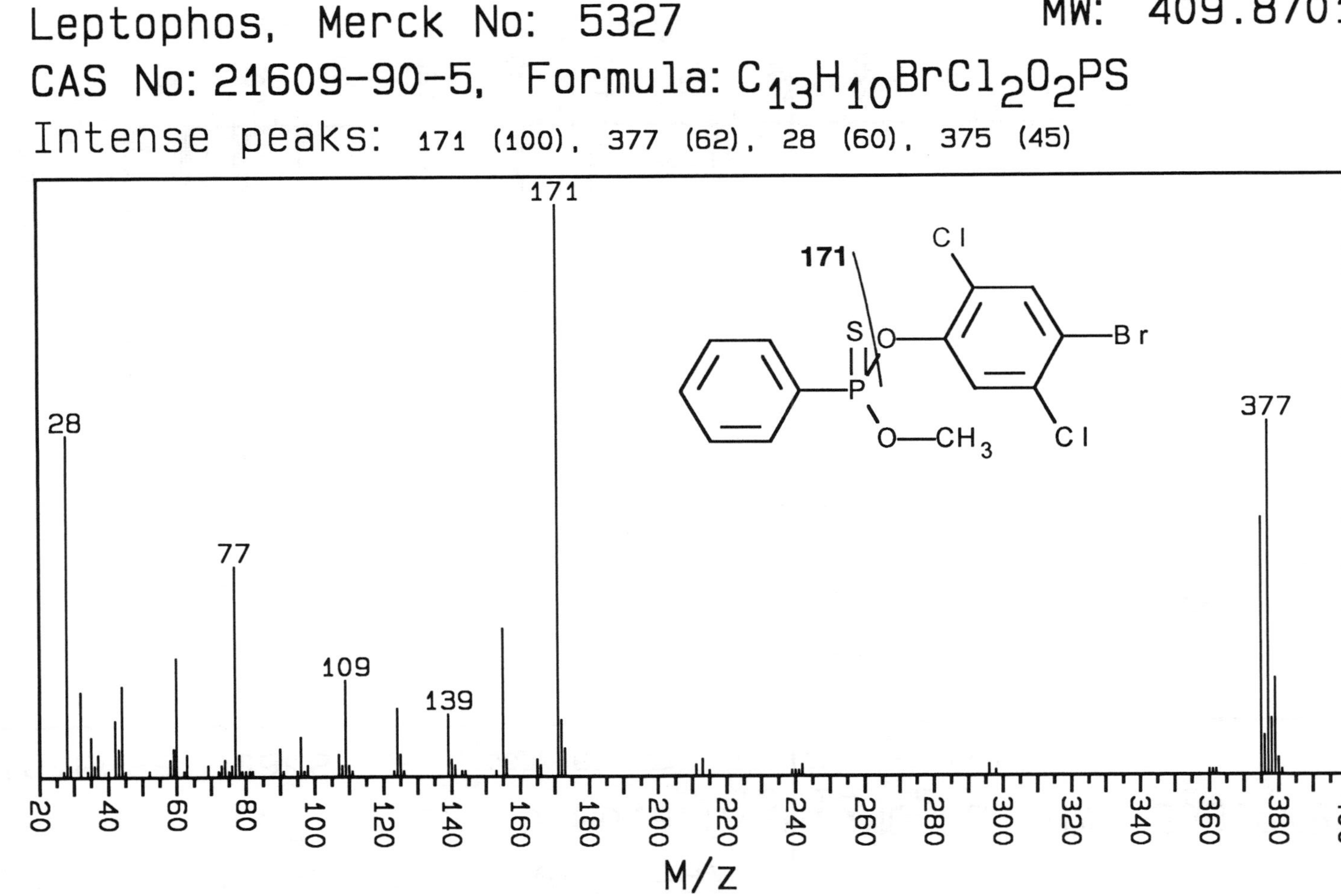

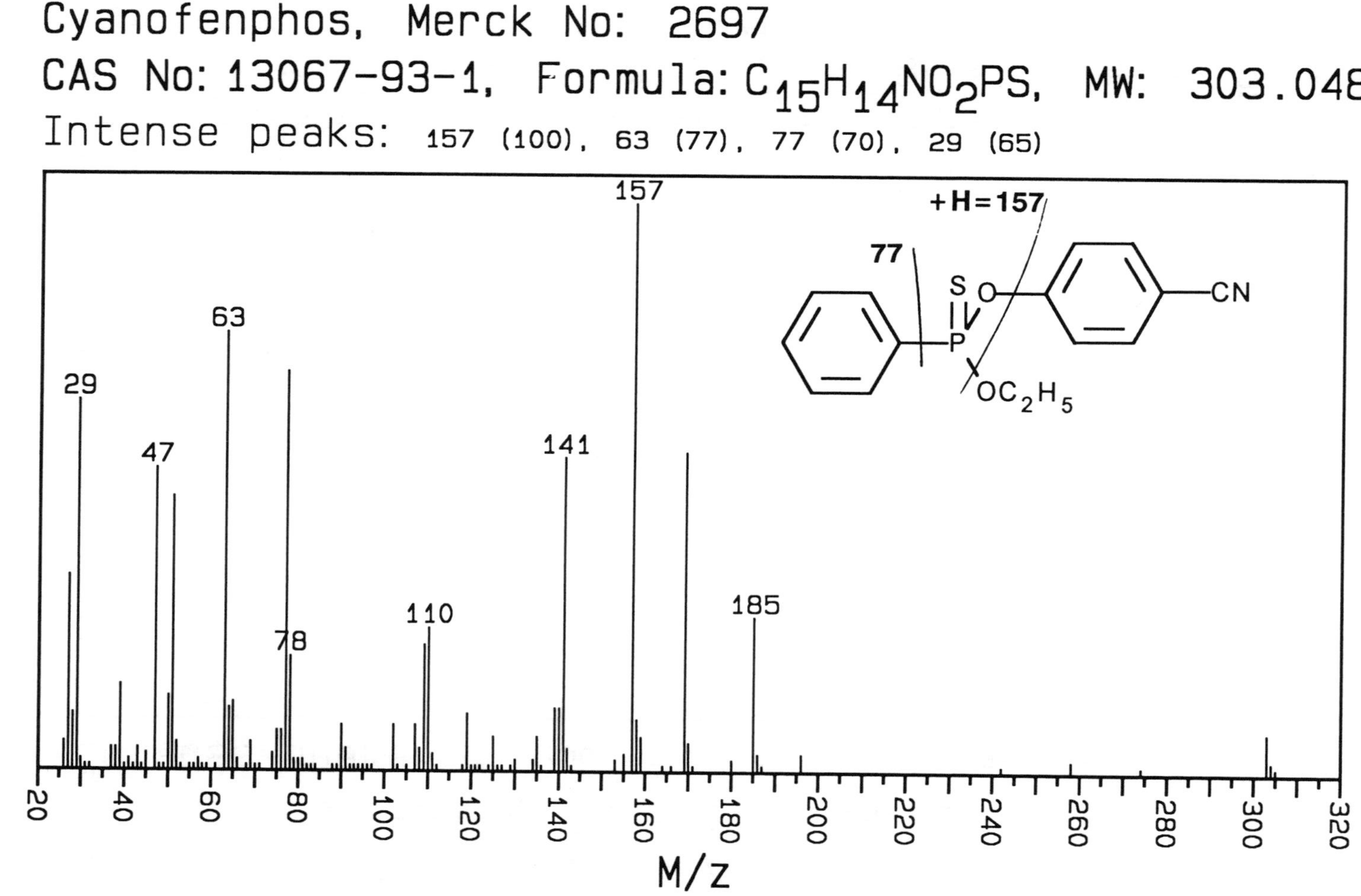

Cyanofenphos, Merck No: 2697
CAS No: 13067-93-1, Formula: C15H14NO2PS, MW: 303.0483
Intense peaks: 157 (100), 63 (77), 77 (70), 29 (65)
+H=157
77
S
P
O
OC2H5
CN
29
47
63
78
110
141
157
185
20
40
60
80
100
120
140
160
180
200
220
240
260
280
300
320
M/Z

EPN, Merck No: 3576

CAS No: 2104-64-5, Formula: $C_{14}H_{14}NO_4PS$, MW: 323.0381

Intense peaks: 323 (100), 185 (98), 157 (76), 77 (67)

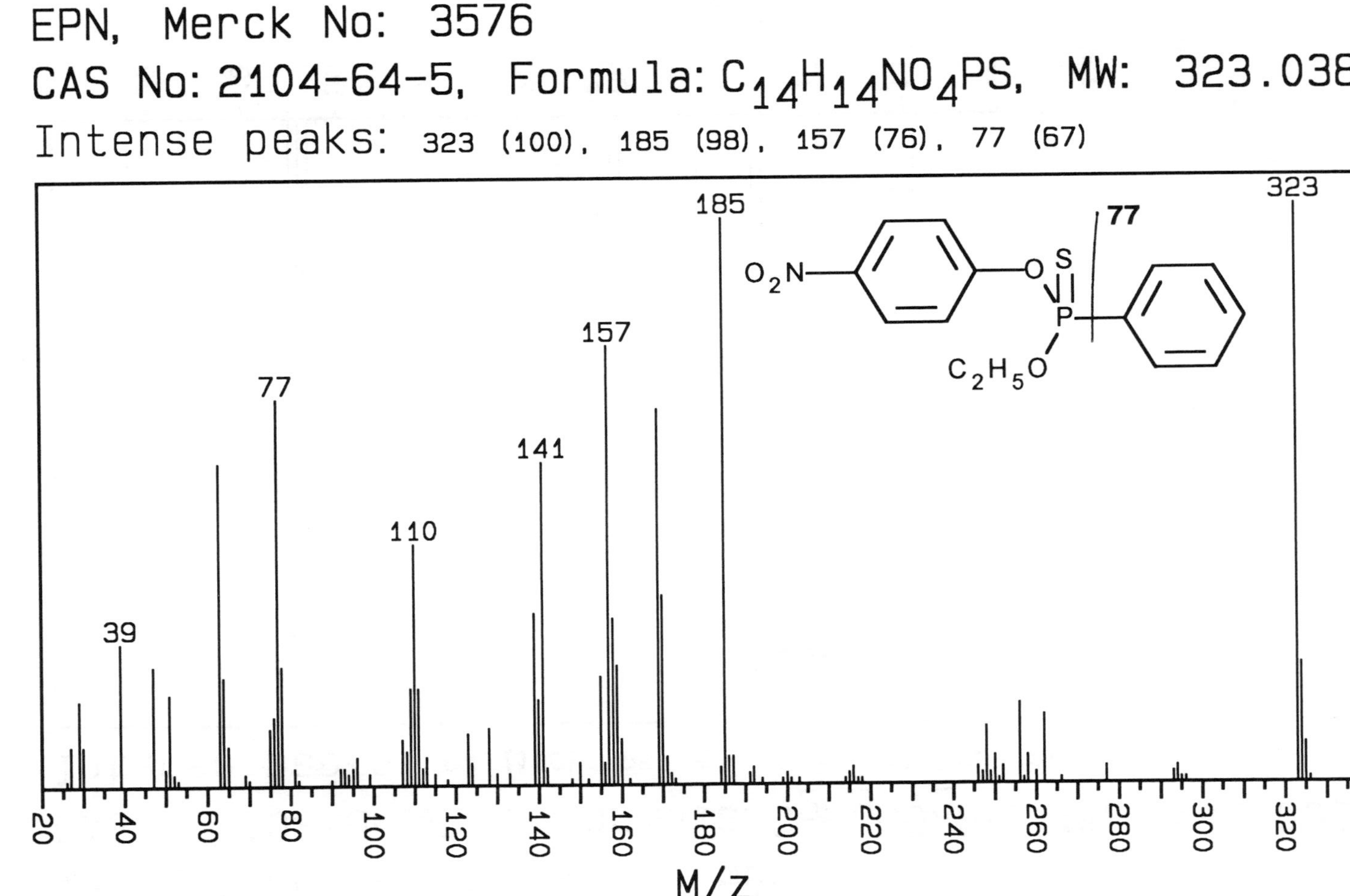

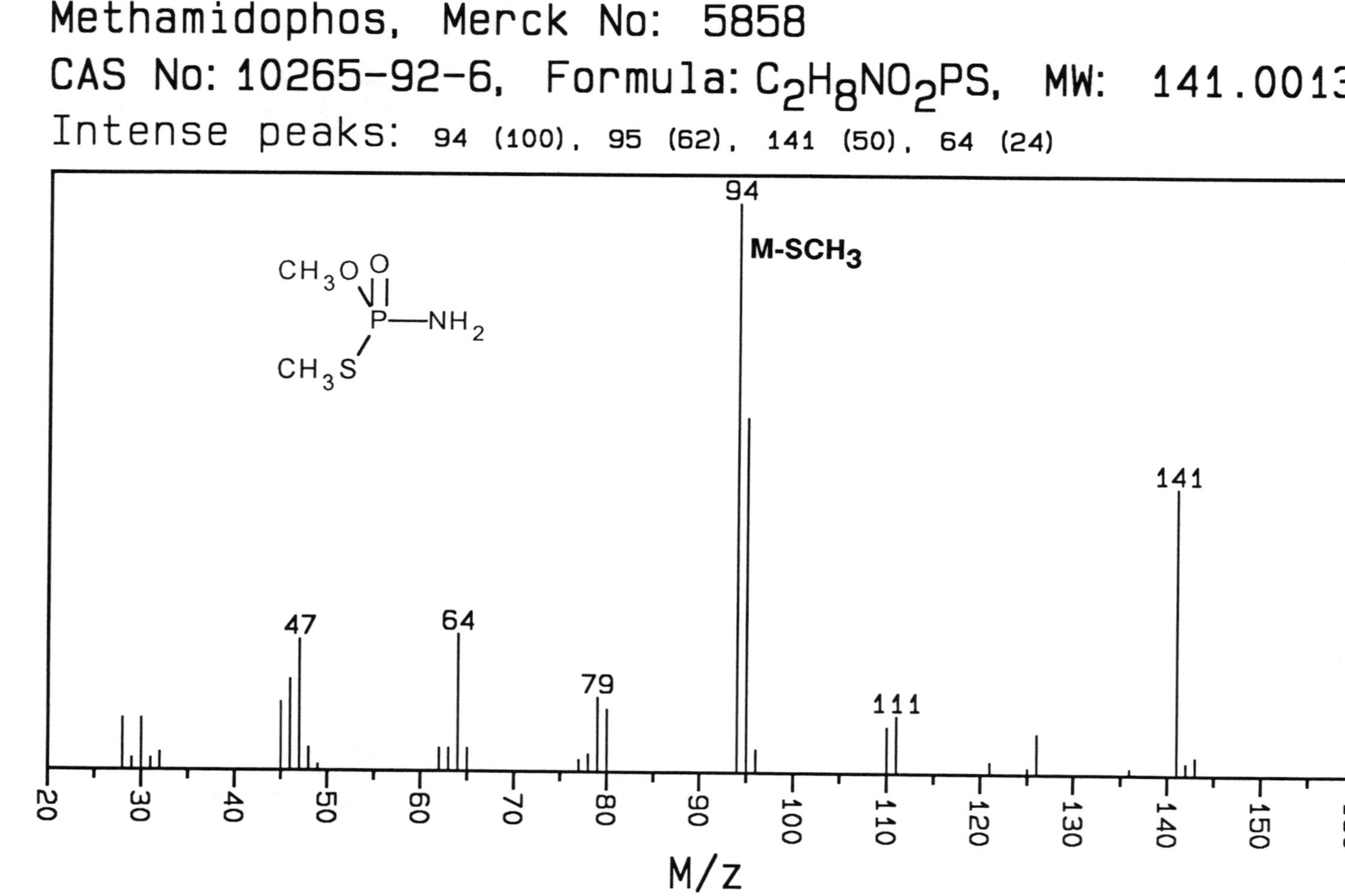
Methamidophos, Merck No: 5858
CAS No: 10265-92-6, Formula: $C_2H_8NO_2PS$, MW: 141.0013
Intense peaks: 94 (100), 95 (62), 141 (50), 64 (24)
94
M-SCH3
$CH_3O$
O
P
$NH_2$
$CH_3S$
141
47
64
79
111
20
30
40
50
60
70
80
90
100
110
120
130
140
150
160
M/Z

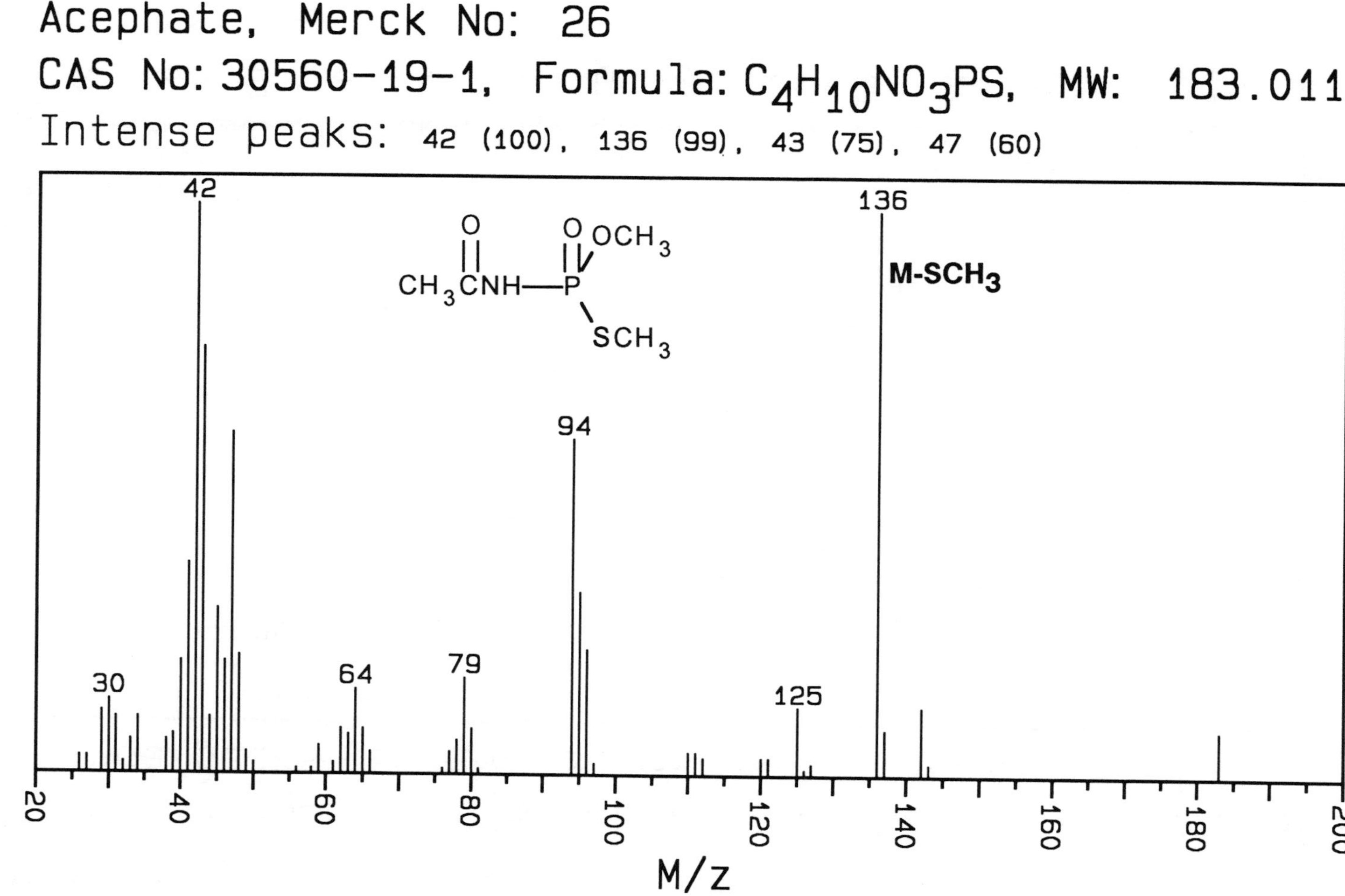
Acephate, Merck No: 26
CAS No: 30560-19-1, Formula: C4H10NO3PS, MW: 183.0119
Intense peaks: 42 (100), 136 (99), 43 (75), 47 (60)
42
136
M-SCH3
CH3CNH—P(=O)(OCH3)SCH3
94
30
64
79
125
20
40
60
80
100
120
140
160
180
200
M/Z

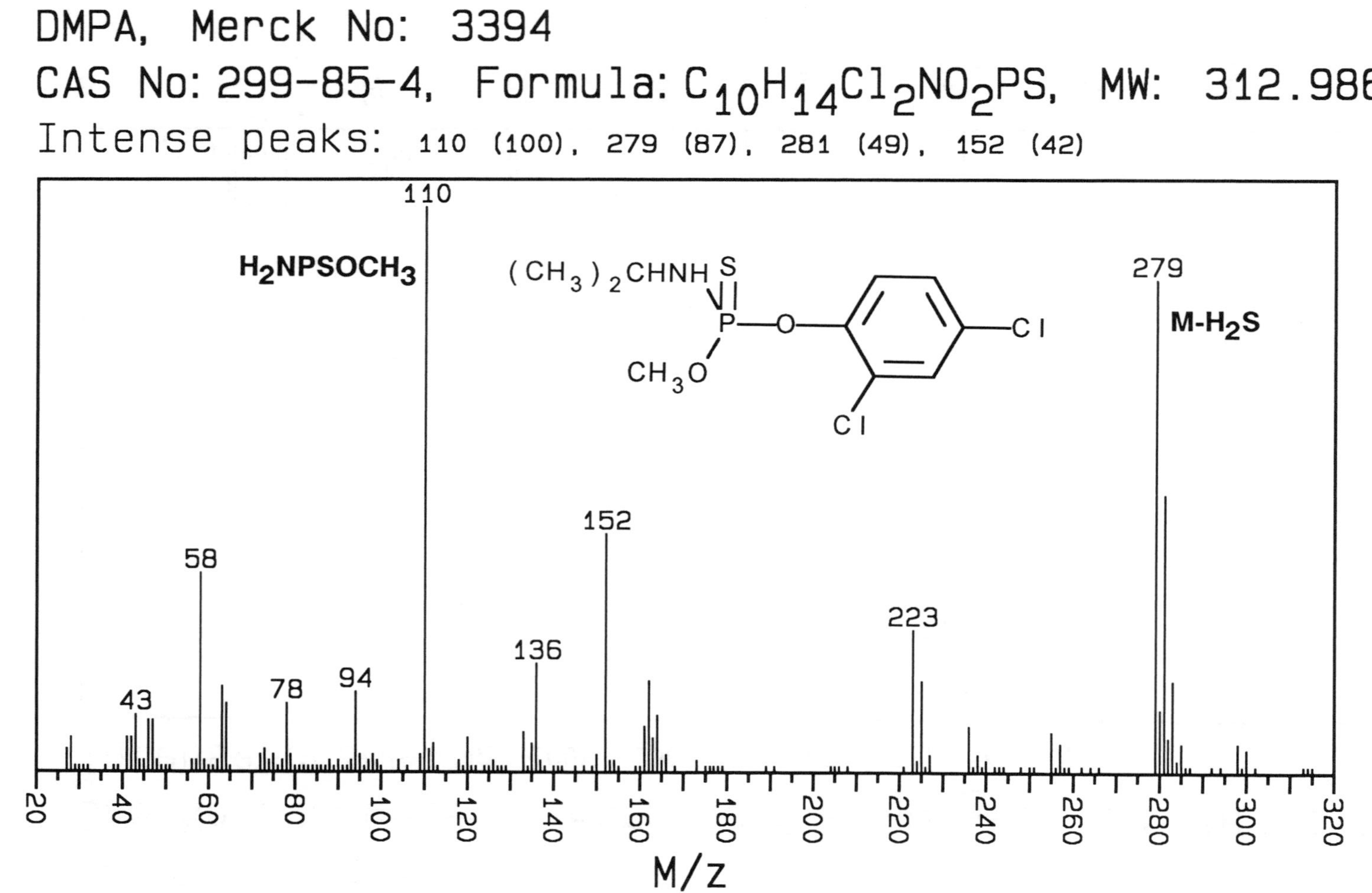
DMPA, Merck No: 3394
CAS No: 299-85-4, Formula: $C_{10}H_{14}Cl_2NO_2PS$, MW: 312.9861
Intense peaks: 110 (100), 279 (87), 281 (49), 152 (42)
110
$H_2NPSOCH_3$
$(CH_3)_2CHNH$
S
P
O
$CH_3O$
Cl
Cl
279
M-$H_2S$
58
152
223
136
43
78
94
20
40
60
80
100
120
140
160
180
200
220
240
260
280
300
320
M/Z

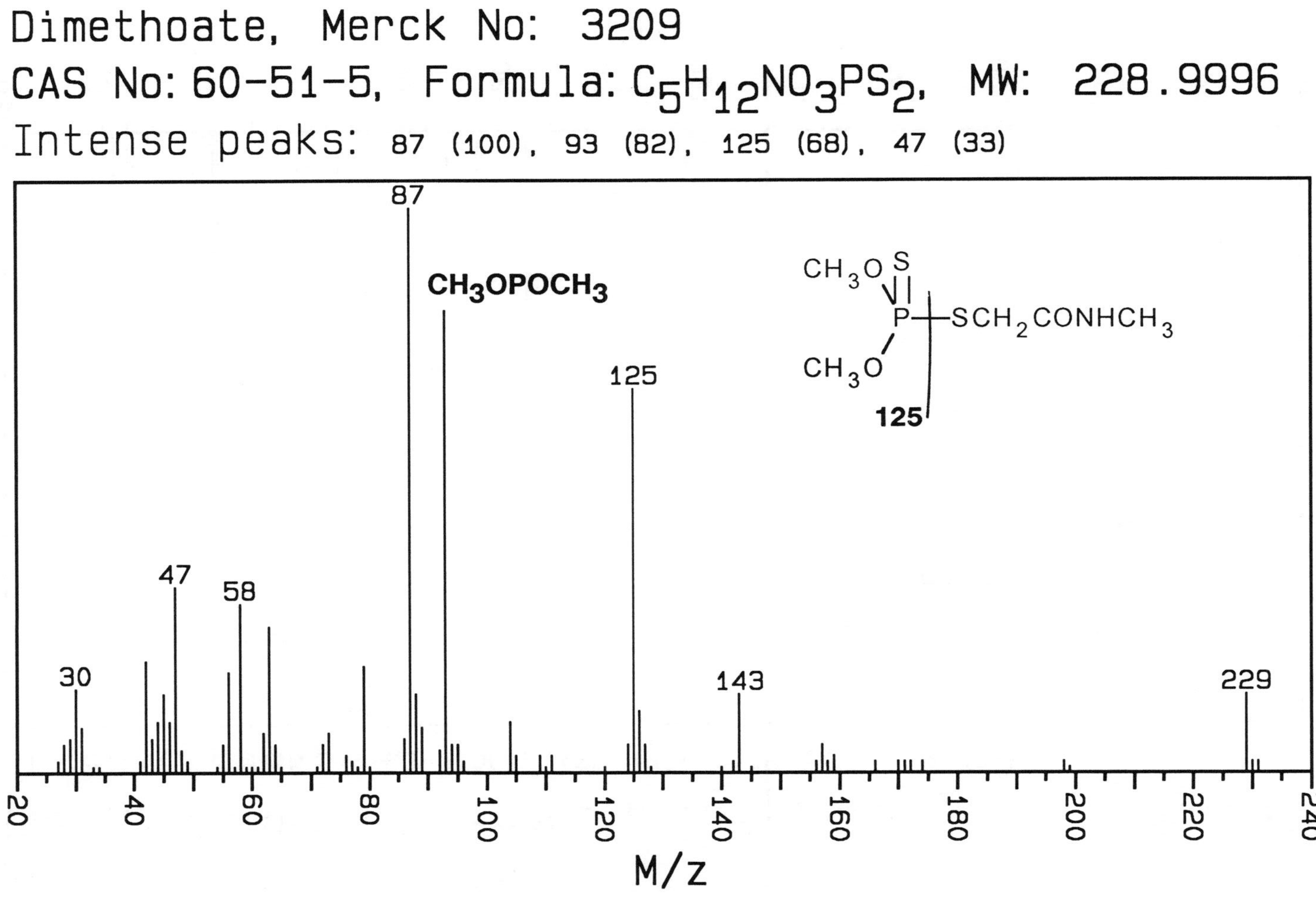
Dimethoate, Merck No: 3209
CAS No: 60-51-5, Formula: $C_5H_{12}NO_3PS_2$, MW: 228.9996
Intense peaks: 87 (100), 93 (82), 125 (68), 47 (33)
87
$CH_3OPOCH_3$
125
30
47
58
143
229
$CH_3O$
S
P
$SCH_2CONHCH_3$
$CH_3O$
125
20
40
60
80
100
120
140
160
180
200
220
240
M/Z

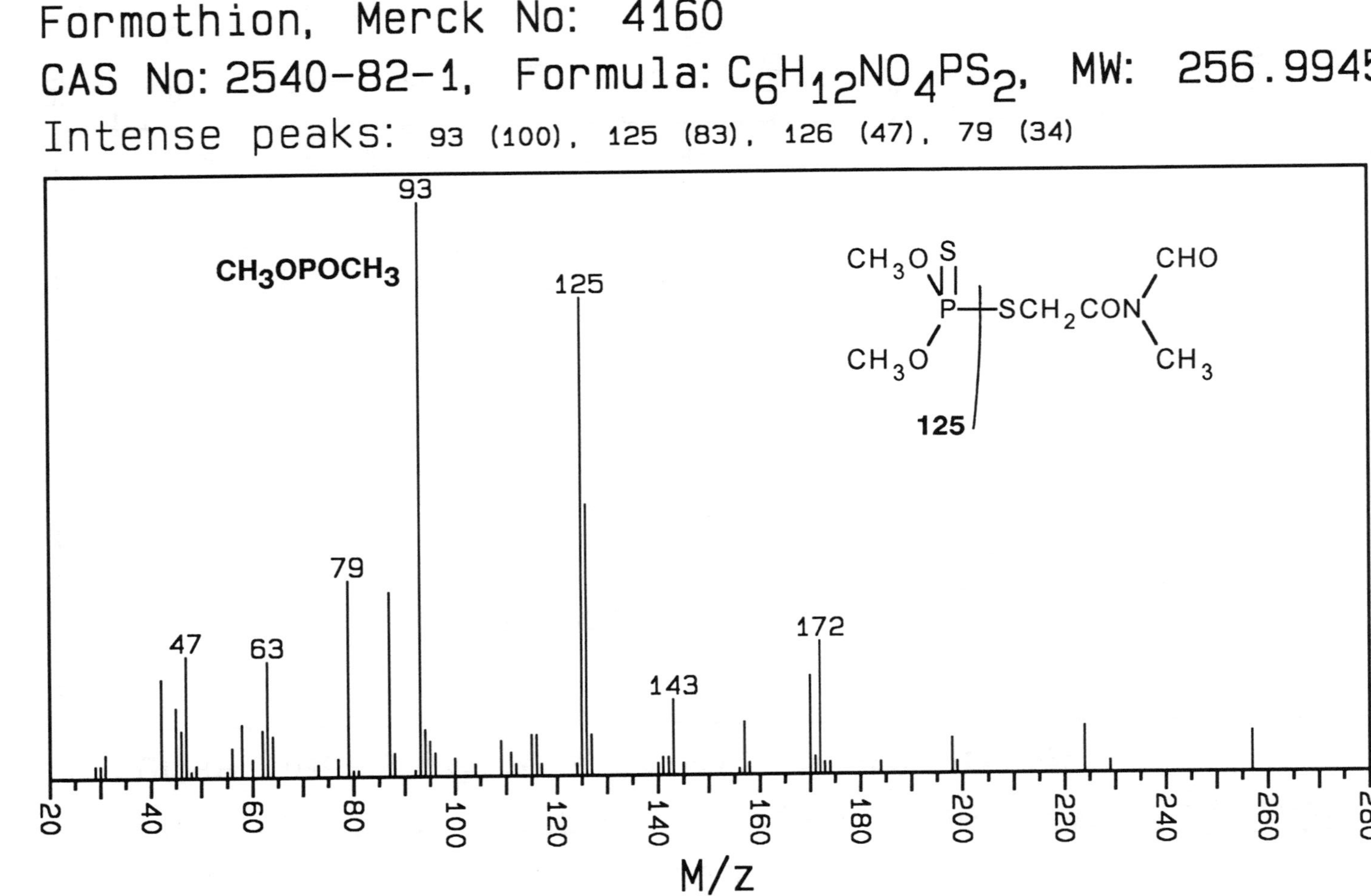
Formothion, Merck No: 4160
CAS No: 2540-82-1, Formula: $C_6H_{12}NO_4PS_2$, MW: 256.9945
Intense peaks: 93 (100), 125 (83), 126 (47), 79 (34)
93
$CH_3OPOCH_3$
125
$CH_3O$
S
P
$SCH_2CON$
CHO
$CH_3O$
$CH_3$
125
79
47
63
172
143
20
40
60
80
100
120
140
160
180
200
220
240
260
280
M/Z

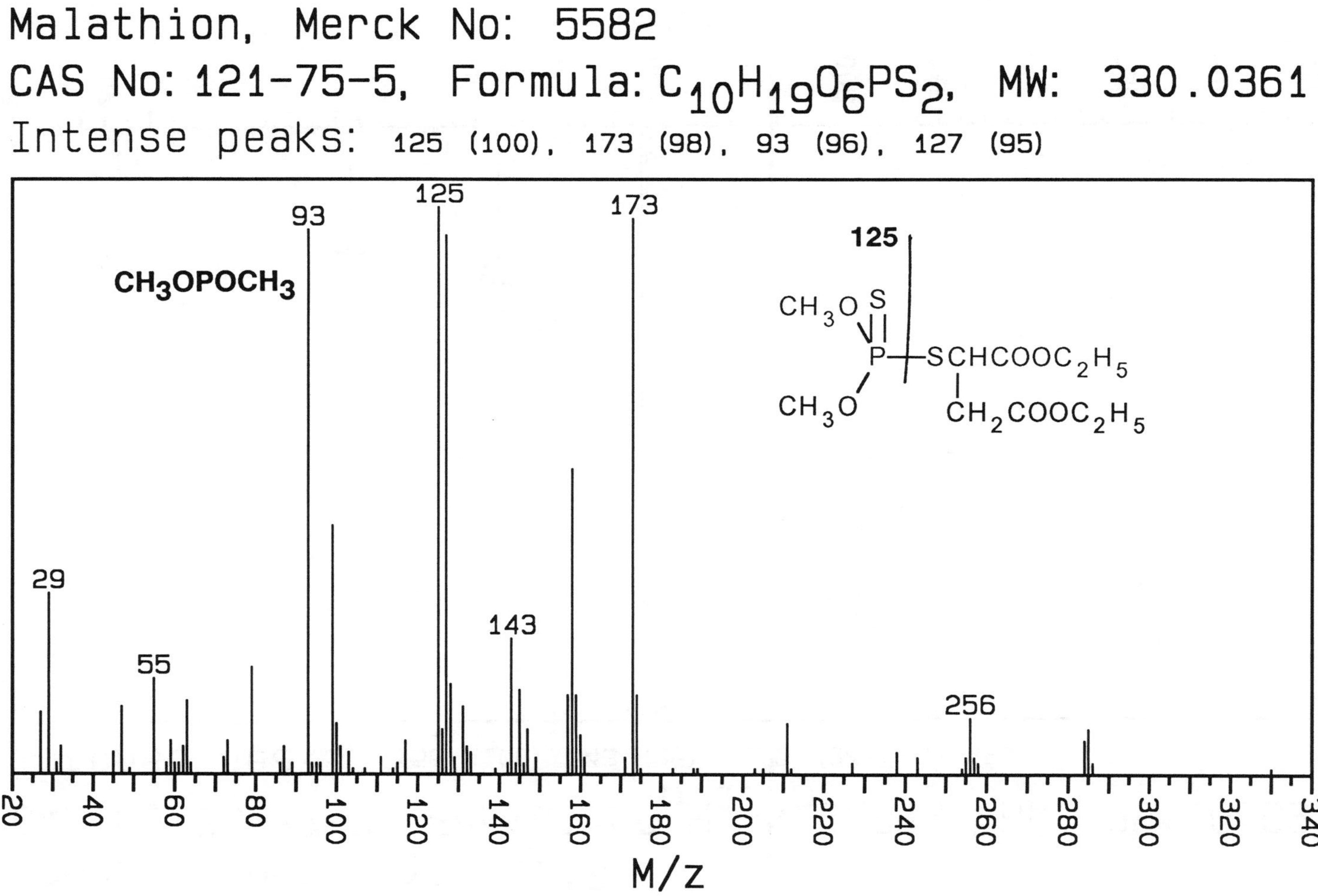
Malathion, Merck No: 5582
CAS No: 121-75-5, Formula: $C_{10}H_{19}O_6PS_2$, MW: 330.0361
Intense peaks: 125 (100), 173 (98), 93 (96), 127 (95)
29
55
93
125
143
173
256
$CH_3OPOCH_3$
125
$CH_3O$
S
P
$SCHCOOC_2H_5$
$CH_3O$
$CH_2COOC_2H_5$
20
40
60
80
100
120
140
160
180
200
220
240
260
280
300
320
340
M/Z

Menazon, Merck No: 5718

CAS No: 78-57-9, Formula: $C_6H_{12}N_5O_2PS_2$, MW: 281.0169

Intense peaks: 156 (100), 43 (25), 125 (24), 93 (20)

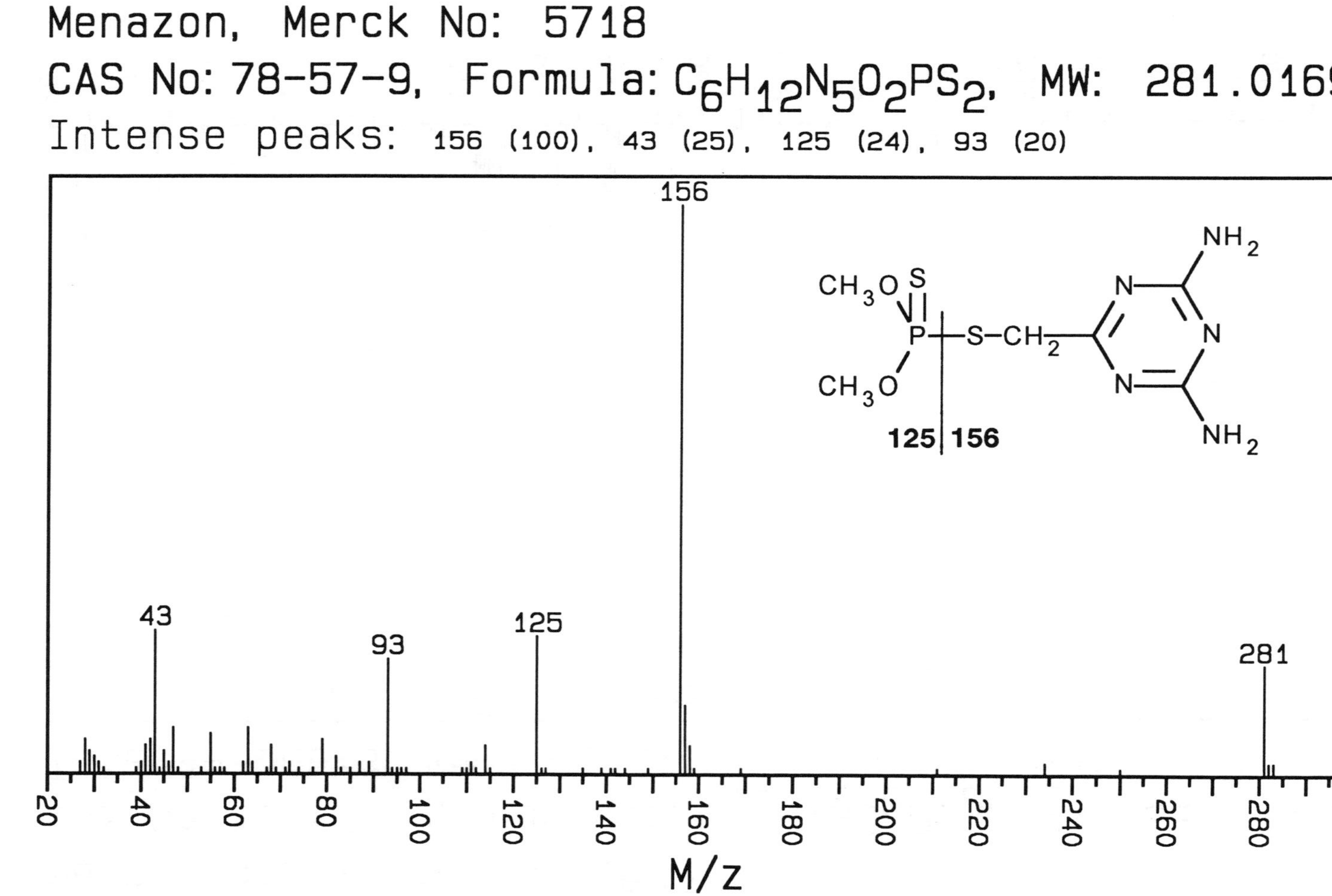

Azinphos-methyl, Merck No: 926

CAS No: 86-50-0, Formula: $C_{10}H_{12}N_3O_3PS_2$, MW: 317.0058

Intense peaks: 132 (100), 77 (91), 160 (45), 104 (33)

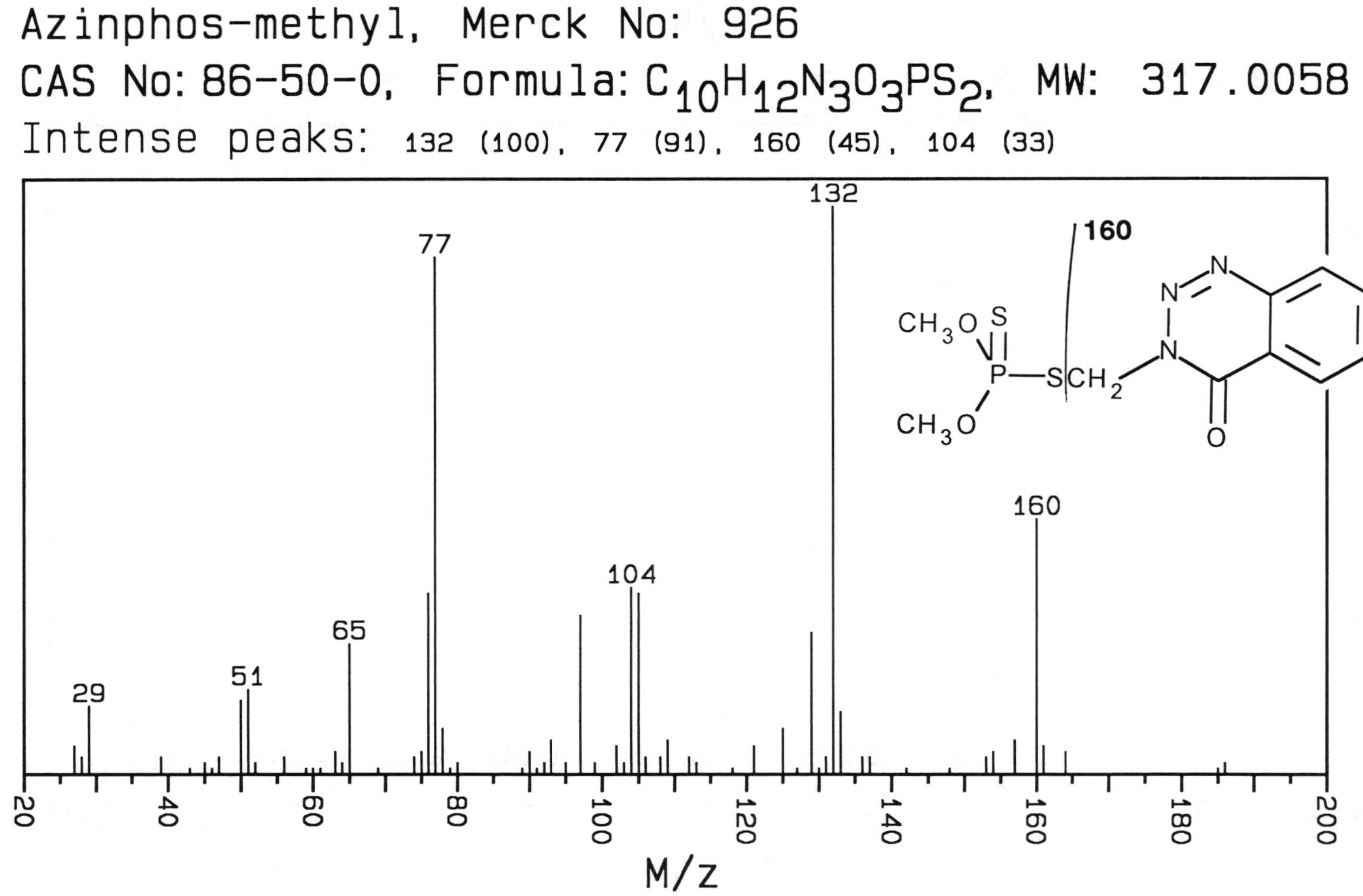

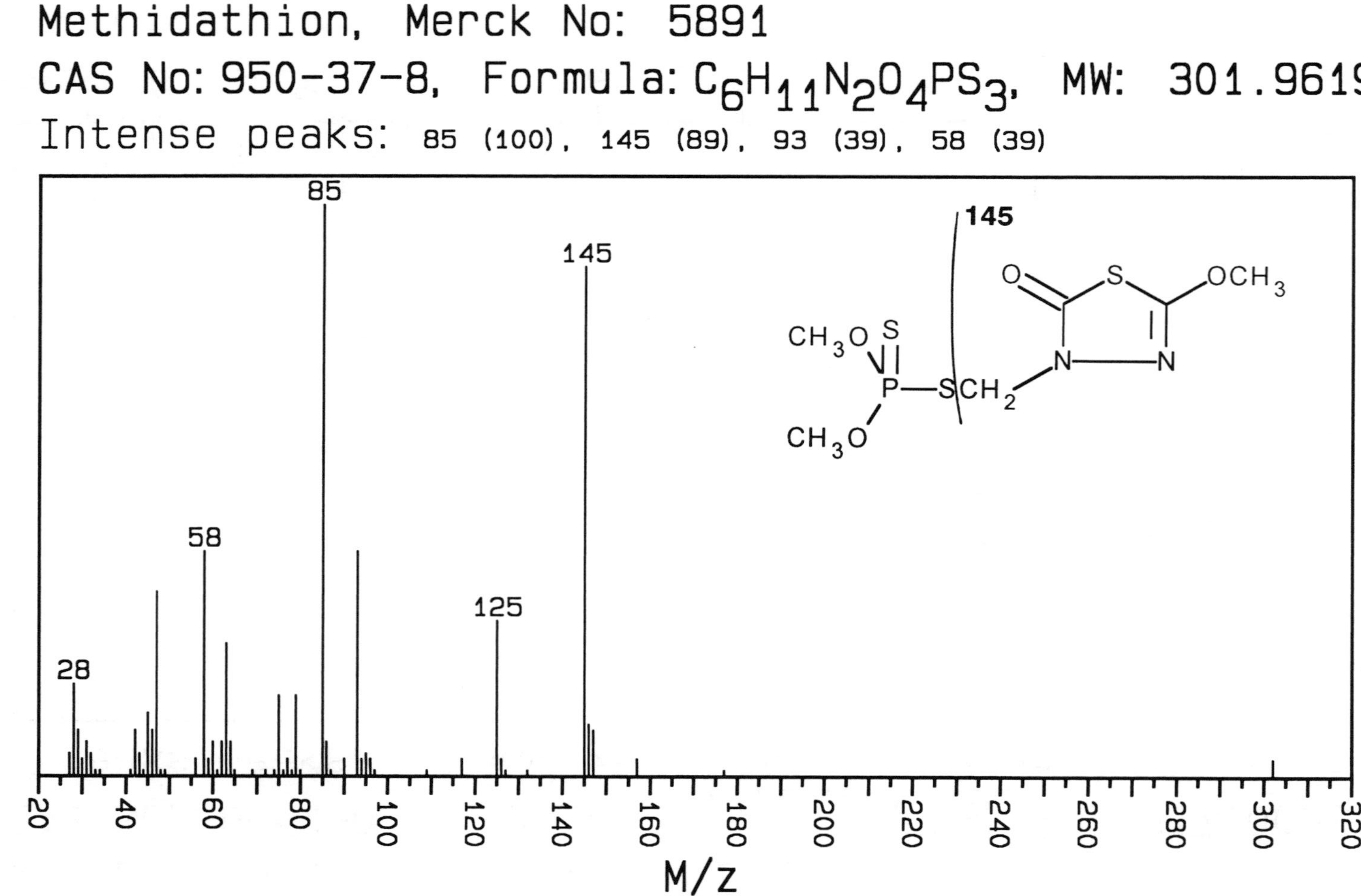

Methidathion, Merck No: 5891
CAS No: 950-37-8, Formula: $C_6H_{11}N_2O_4PS_3$, MW: 301.9619
Intense peaks: 85 (100), 145 (89), 93 (39), 58 (39)
85
145
145
$CH_3O$
S
P
S
$CH_2$
$CH_3O$
O
S
$OCH_3$
N
N
58
125
28
20
40
60
80
100
120
140
160
180
200
220
240
260
280
300
320
M/Z

Phosmet, Merck No: 7311

CAS No: 732-11-6, Formula: $C_{11}H_{12}NO_4PS_2$, MW: 316.9945

Intense peaks: 160 (100), 77 (34), 161 (33), 28 (31)

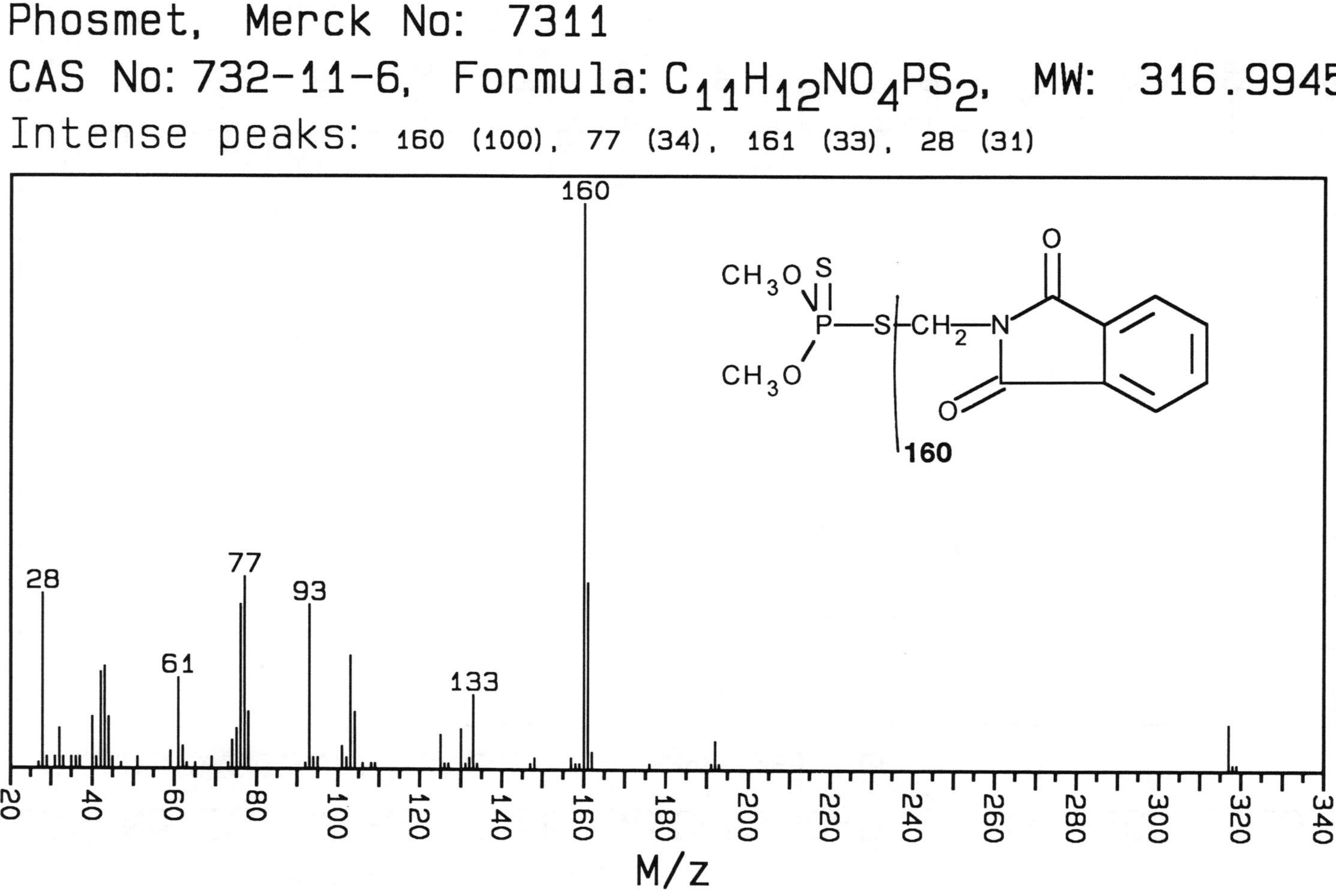

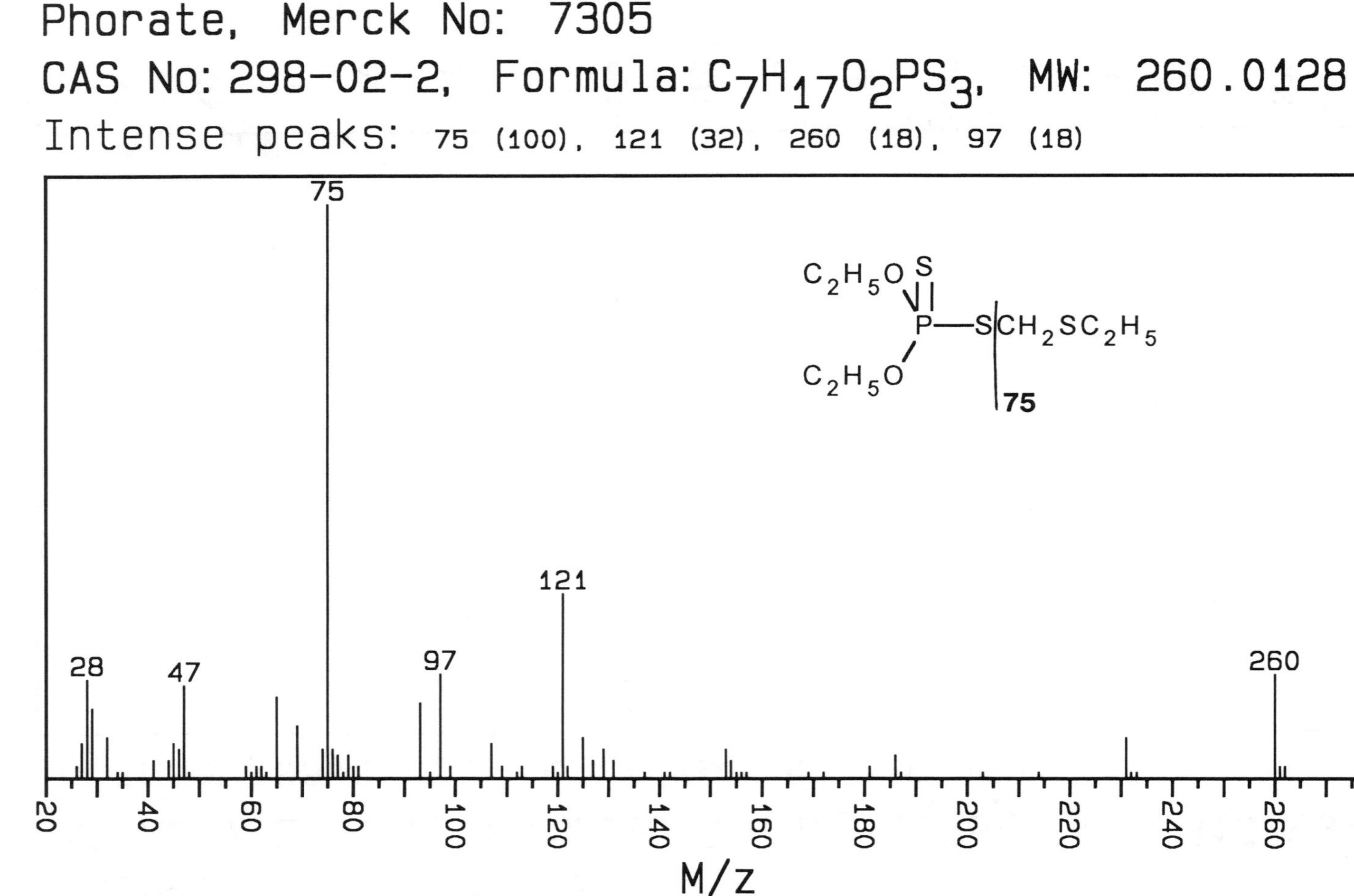
Phorate, Merck No: 7305
CAS No: 298-02-2, Formula: $C_7H_{17}O_2PS_3$, MW: 260.0128
Intense peaks: 75 (100), 121 (32), 260 (18), 97 (18)
$C_2H_5O$
S
P
S
$CH_2SC_2H_5$
$C_2H_5O$
75
75
28
47
97
121
260
20
40
60
80
100
120
140
160
180
200
220
240
260
280
M/Z

Disulfoton, Merck No: 3371

CAS No: 298-04-4, Formula: $C_8H_{19}O_2PS_3$, MW: 274.0285

Intense peaks: 88 (100), 89 (37), 60 (24), 61 (23)

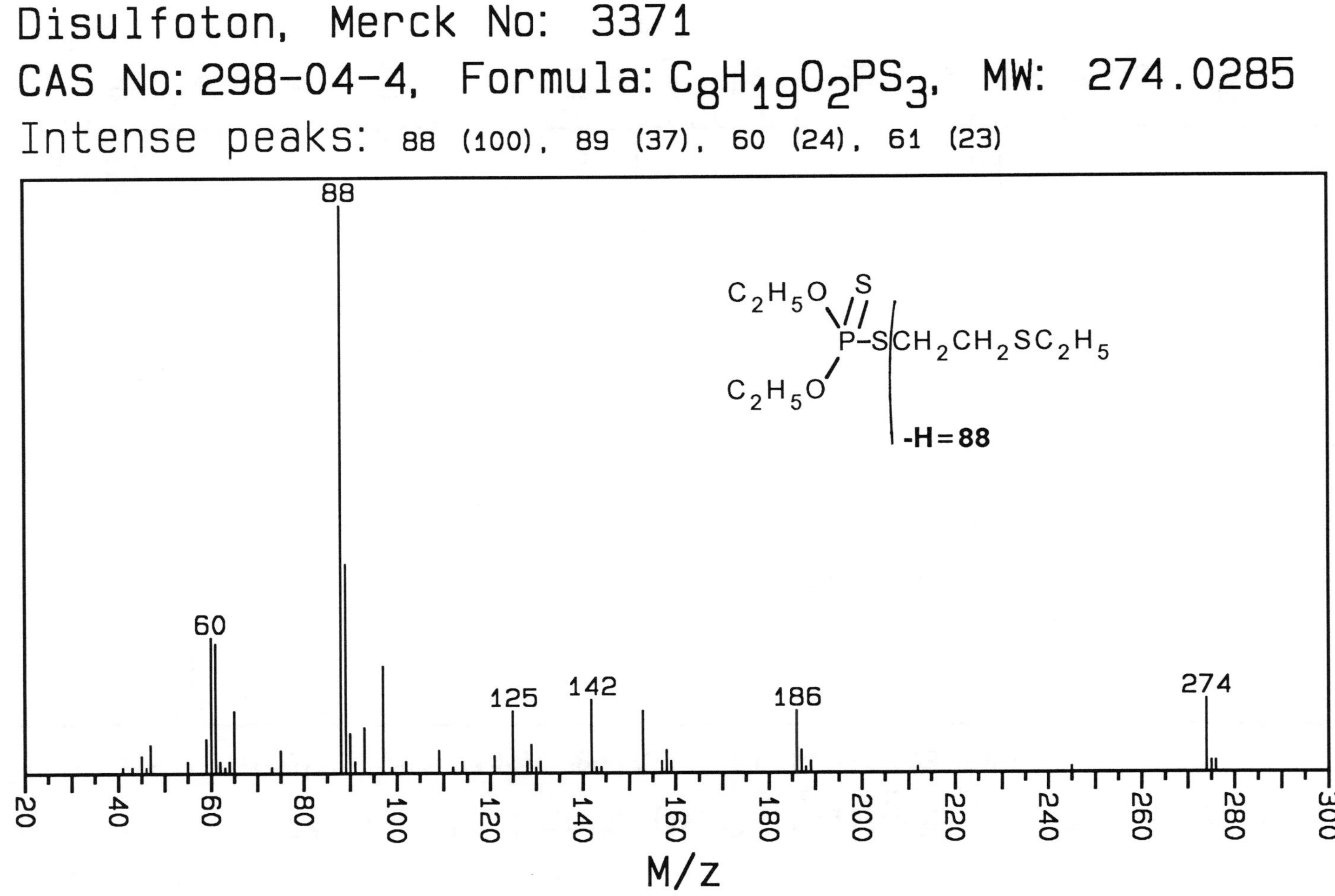

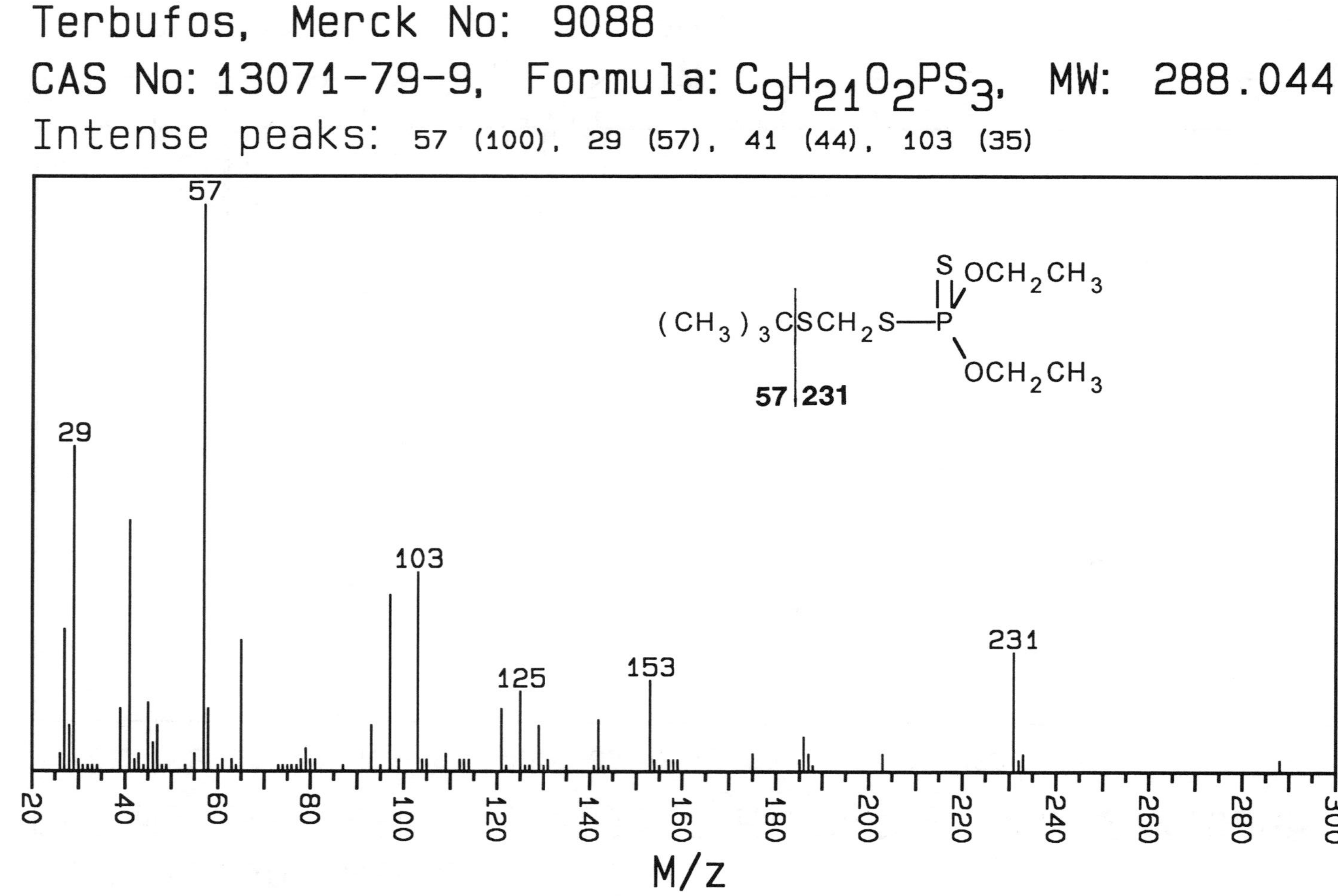
Terbufos, Merck No: 9088
CAS No: 13071-79-9, Formula: $C_9H_{21}O_2PS_3$, MW: 288.0441
Intense peaks: 57 (100), 29 (57), 41 (44), 103 (35)
$(CH_3)_3CSCH_2S-P(=S)(OCH_2CH_3)(OCH_2CH_3)$
57 | 231
57
29
103
125
153
231
20
40
60
80
100
120
140
160
180
200
220
240
260
280
300
M/Z

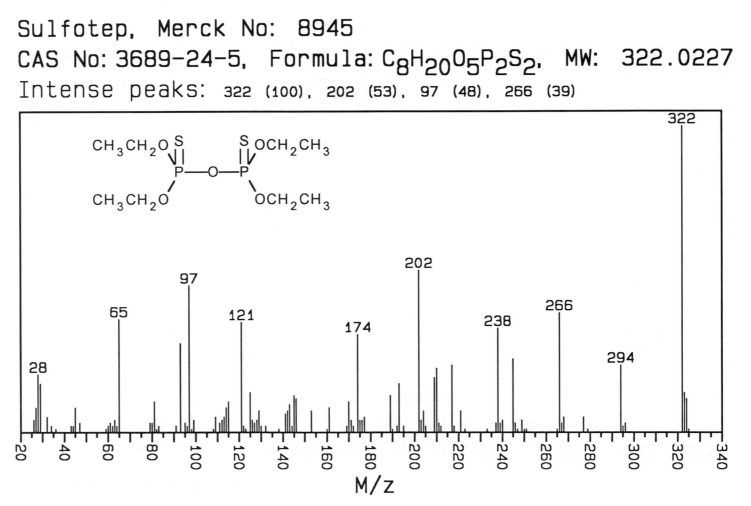

Sulfotep, Merck No: 8945

CAS No: 3689-24-5, Formula: $C_8H_{20}O_5P_2S_2$, MW: 322.0227

Intense peaks: 322 (100), 202 (53), 97 (48), 266 (39)

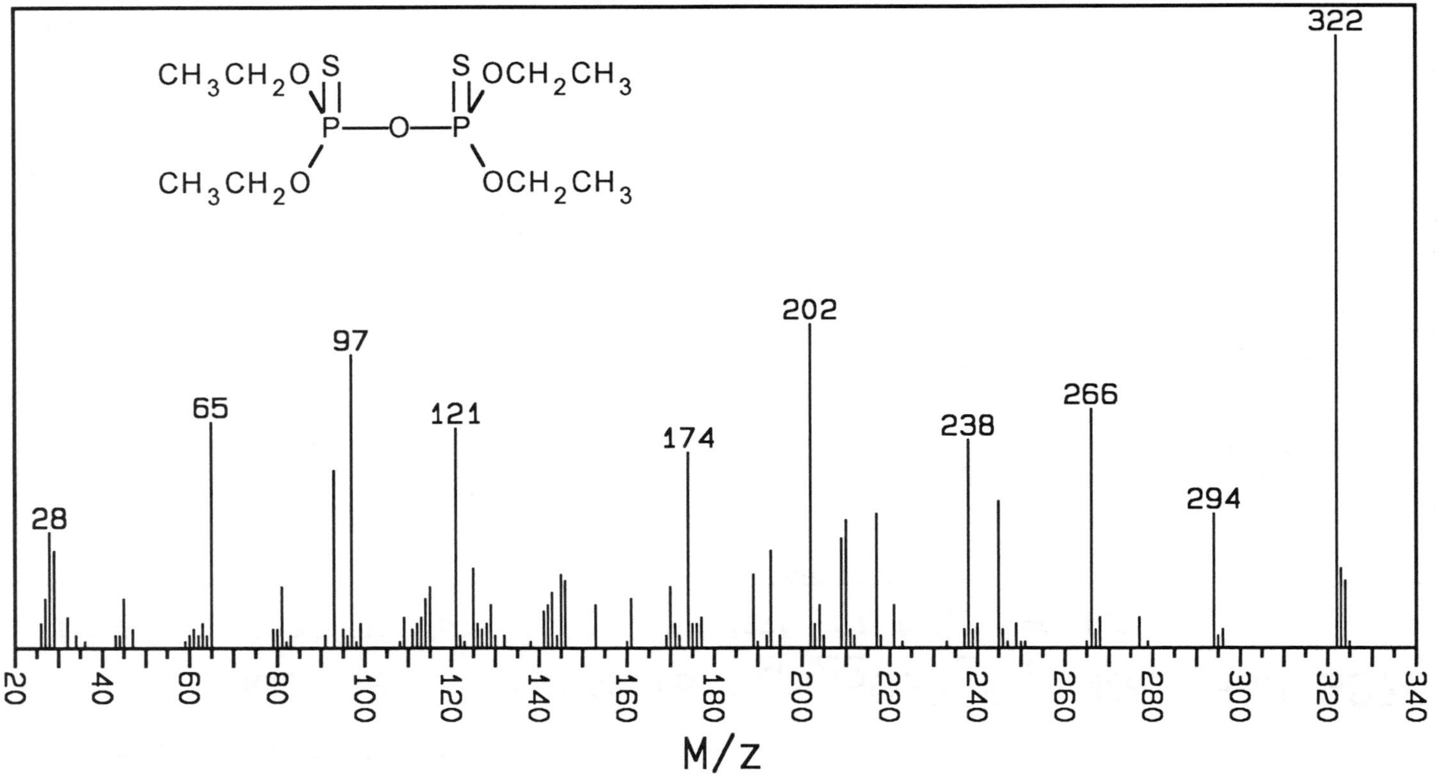

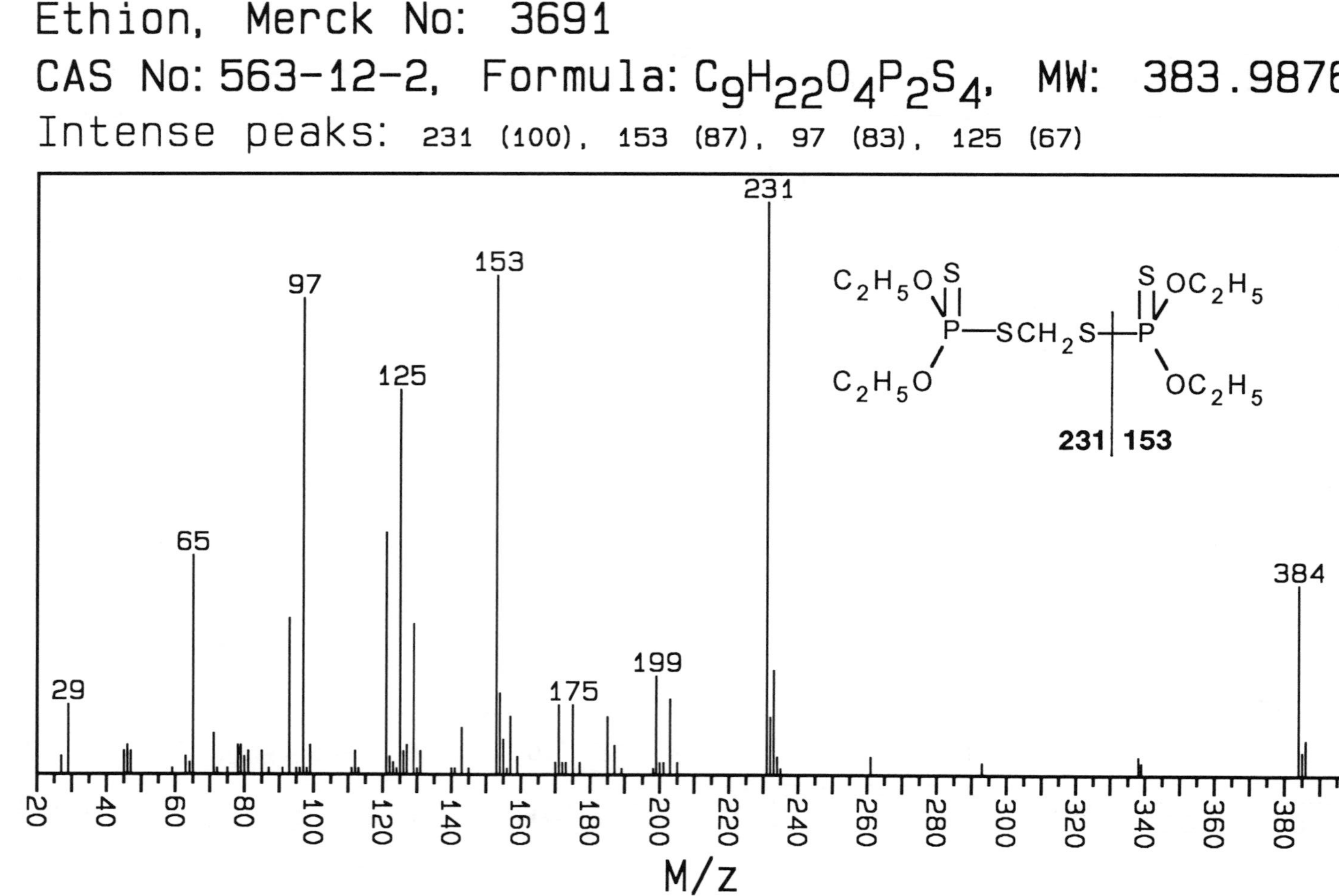
Ethion, Merck No: 3691
CAS No: 563-12-2, Formula: C9H22O4P2S4, MW: 383.9876
Intense peaks: 231 (100), 153 (87), 97 (83), 125 (67)
231
153
97
125
65
29
175
199
384
C2H5O
S
P—SCH2S—P
S
OC2H5
C2H5O
OC2H5
231
153
20
40
60
80
100
120
140
160
180
200
220
240
260
280
300
320
340
360
380
400
M/Z

Carbophenothion, Merck No: 1827

CAS No: 786-19-6, Formula: $C_{11}H_{16}ClO_2PS_3$, MW: 341.9739

Intense peaks: 157 (100), 45 (57), 97 (55), 121 (48)

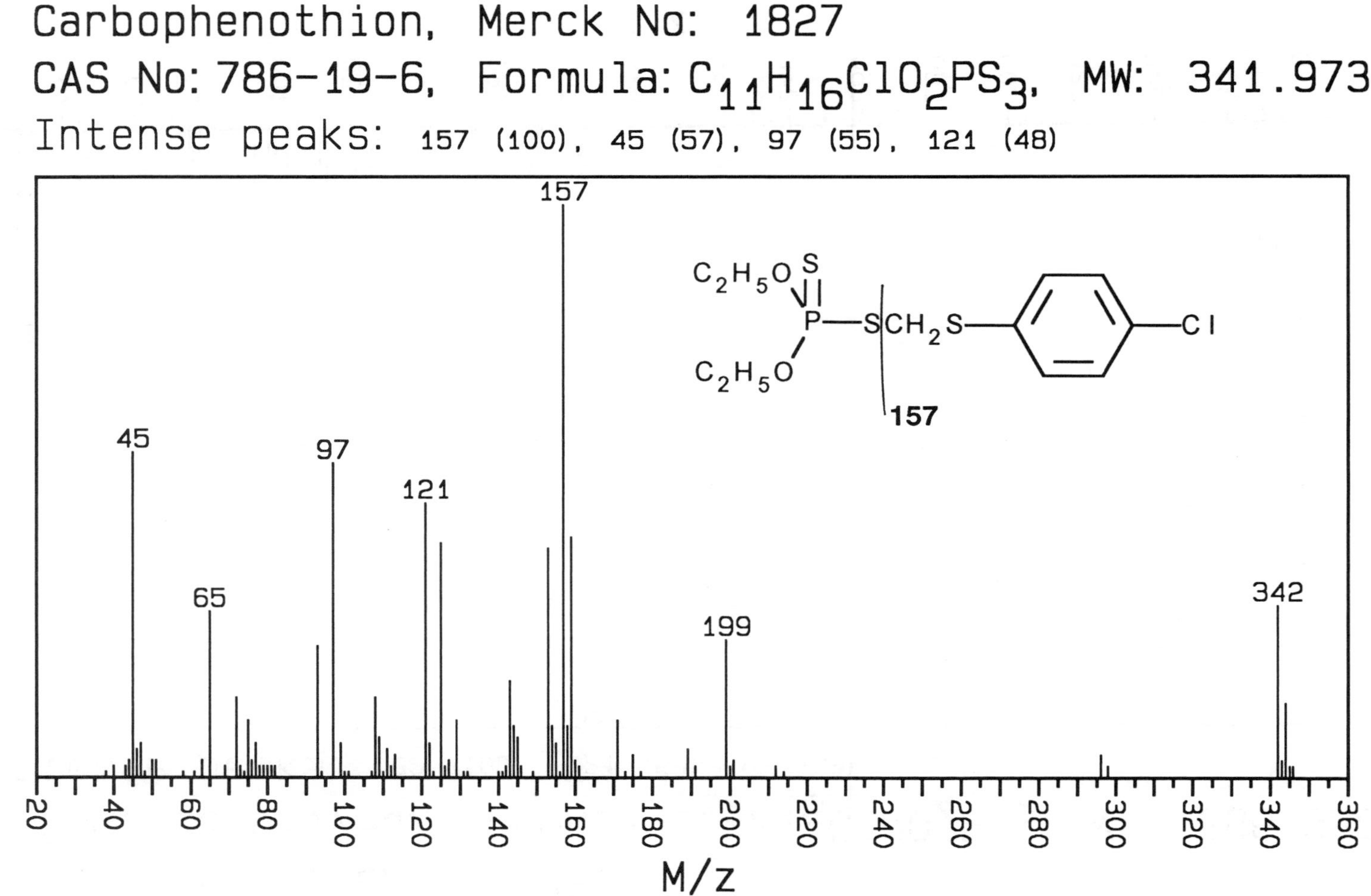

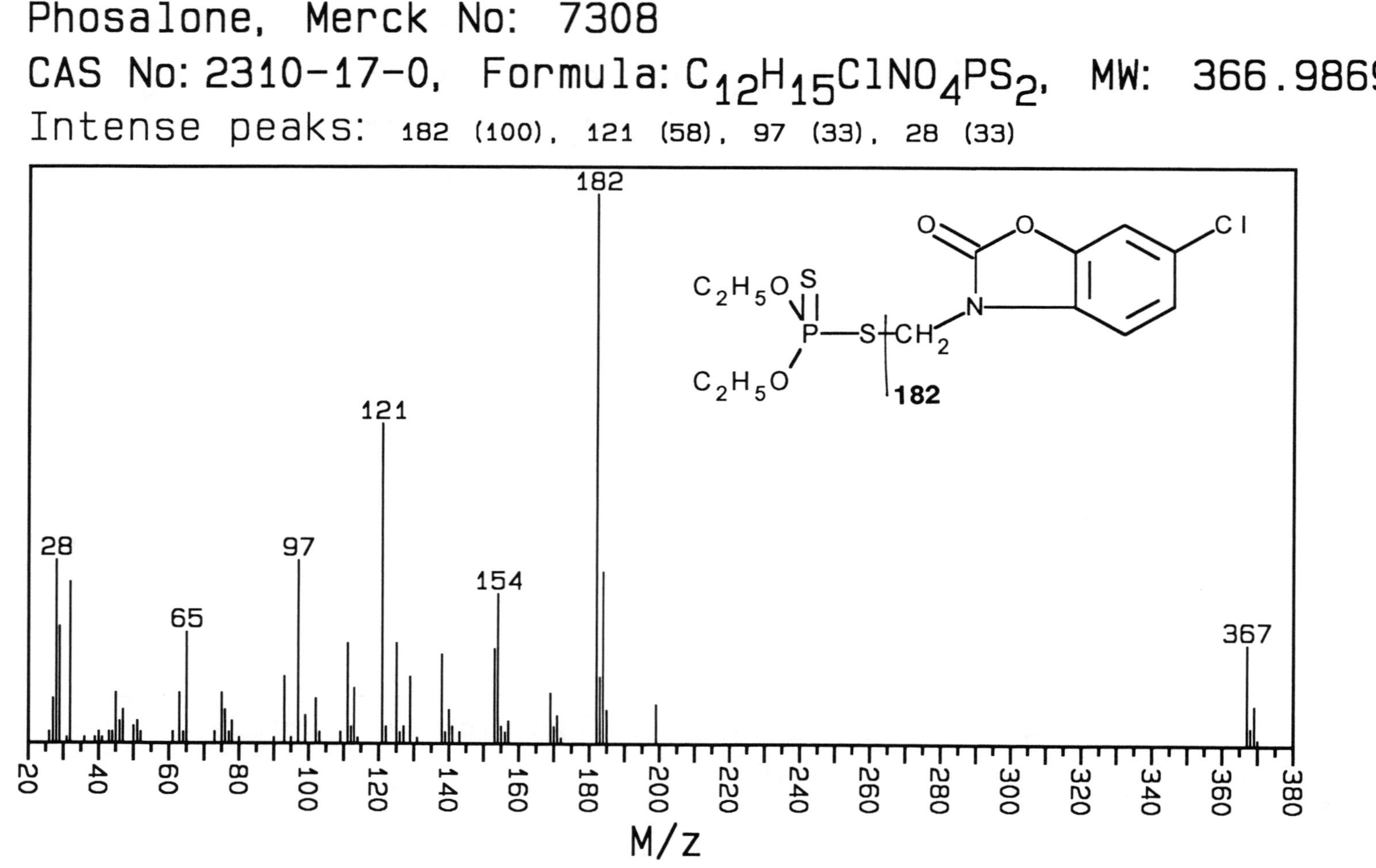
Phosalone, Merck No: 7308
CAS No: 2310-17-0, Formula: $C_{12}H_{15}ClNO_4PS_2$, MW: 366.9869
Intense peaks: 182 (100), 121 (58), 97 (33), 28 (33)
182
121
28
97
65
154
367
O
O
Cl
S
$C_2H_5O$
N
P
S
$CH_2$
$C_2H_5O$
182
20
40
60
80
100
120
140
160
180
200
220
240
260
280
300
320
340
360
380
M/Z

Dialifor, Merck No: 2949 MW: 393.0025

CAS No: 10311-84-9, Formula: $C_{14}H_{17}ClNO_4PS_2$

Intense peaks: 208 (100), 210 (42), 40 (21), 94 (19)

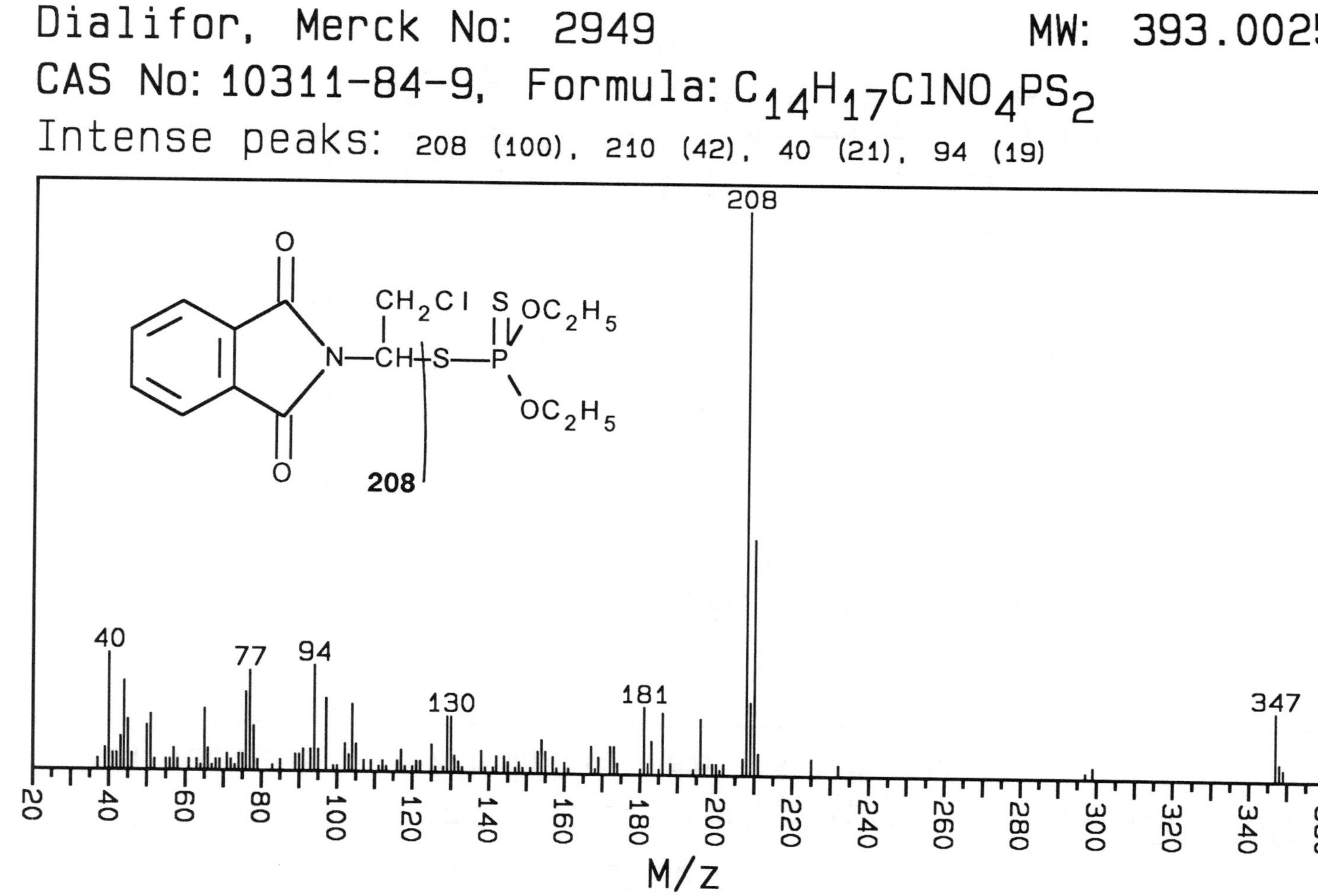

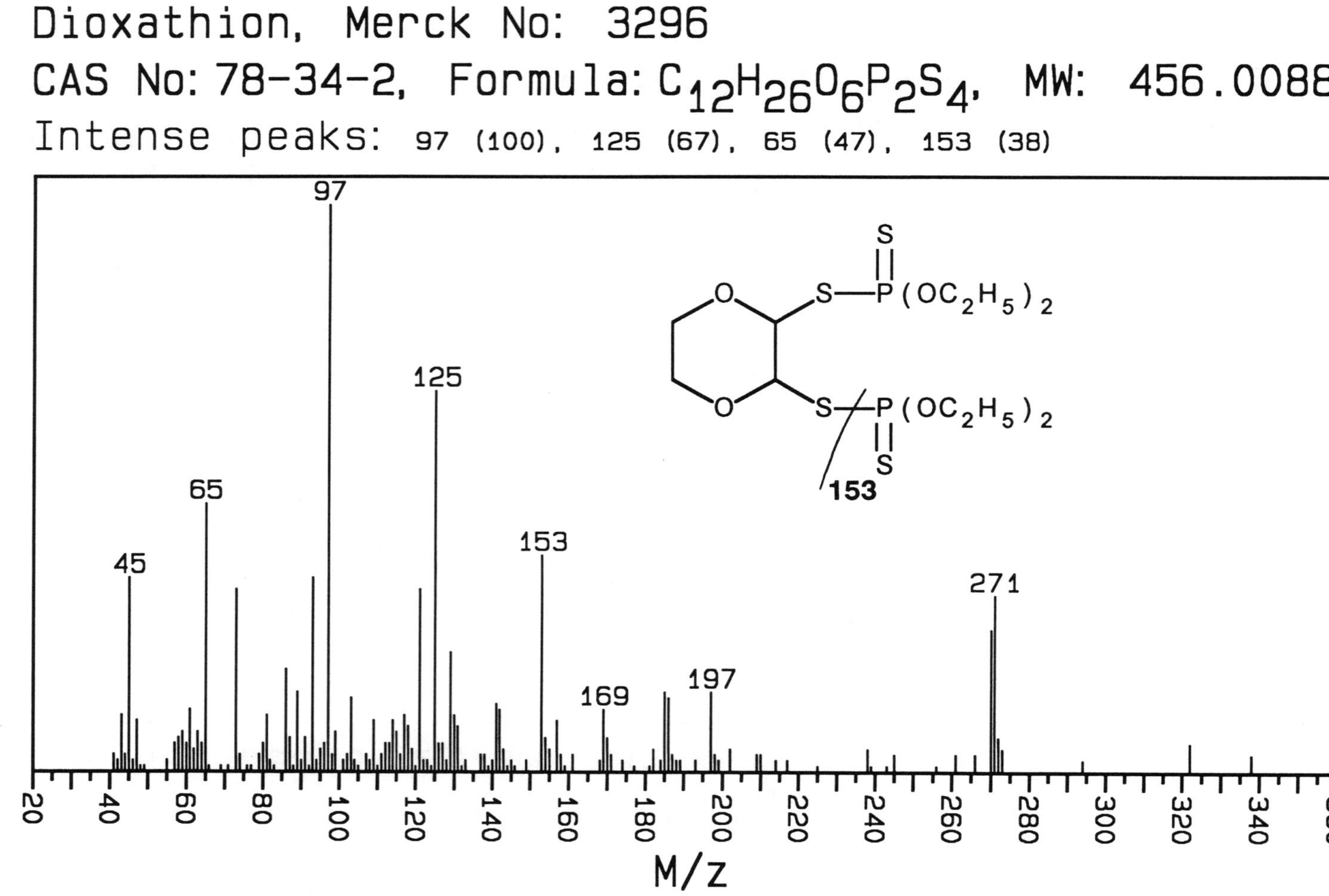
Dioxathion, Merck No: 3296
CAS No: 78-34-2, Formula: $C_{12}H_{26}O_6P_2S_4$, MW: 456.0088
Intense peaks: 97 (100), 125 (67), 65 (47), 153 (38)
97
125
65
153
45
271
169
197
153
M/Z

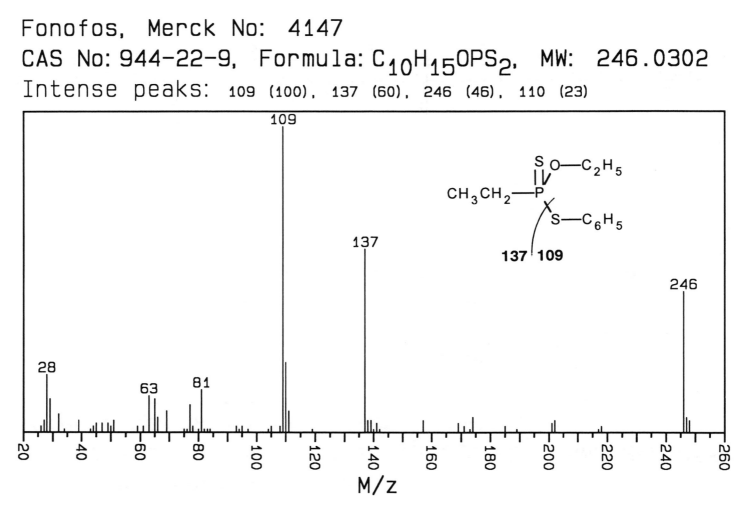

Fonofos, Merck No: 4147

CAS No: 944-22-9, Formula: $C_{10}H_{15}OPS_2$, MW: 246.0302

Intense peaks: 109 (100), 137 (60), 246 (46), 110 (23)

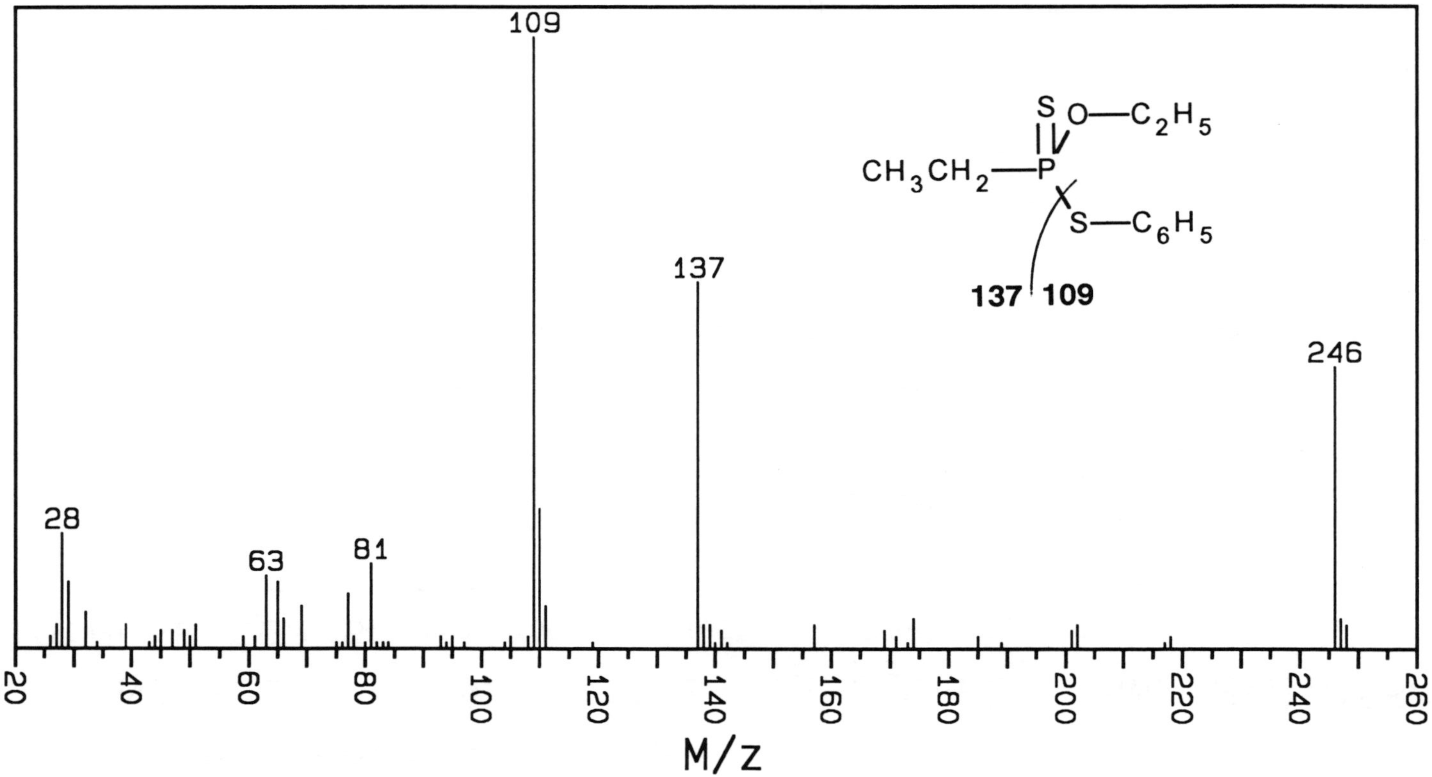

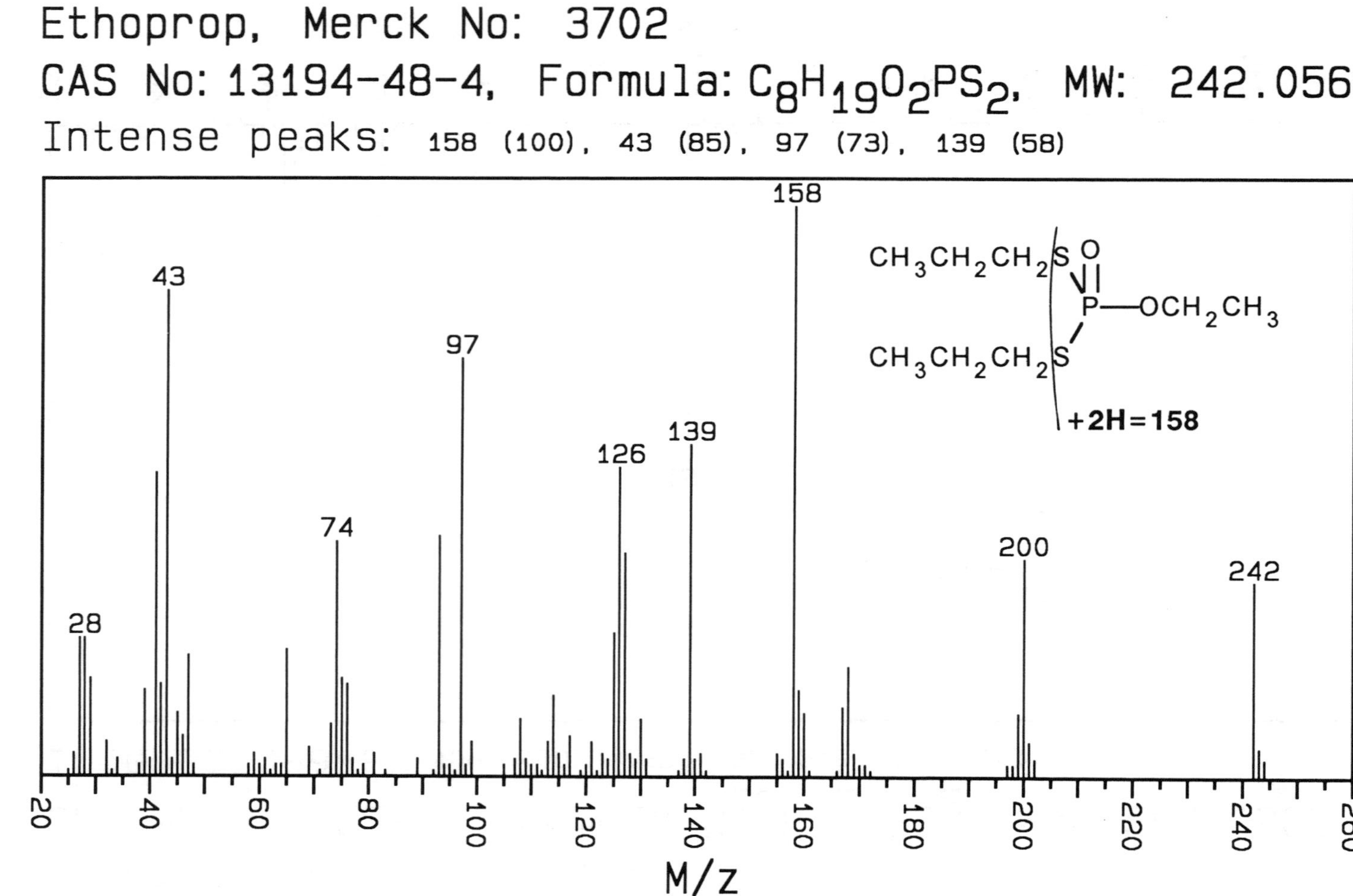

Ethoprop, Merck No: 3702
CAS No: 13194-48-4, Formula: $C_8H_{19}O_2PS_2$, MW: 242.0564
Intense peaks: 158 (100), 43 (85), 97 (73), 139 (58)
$CH_3CH_2CH_2S$
$CH_3CH_2CH_2S$
O
P
$OCH_2CH_3$
+2H=158
28
43
74
97
126
139
158
200
242
20
40
60
80
100
120
140
160
180
200
220
240
260
M/Z

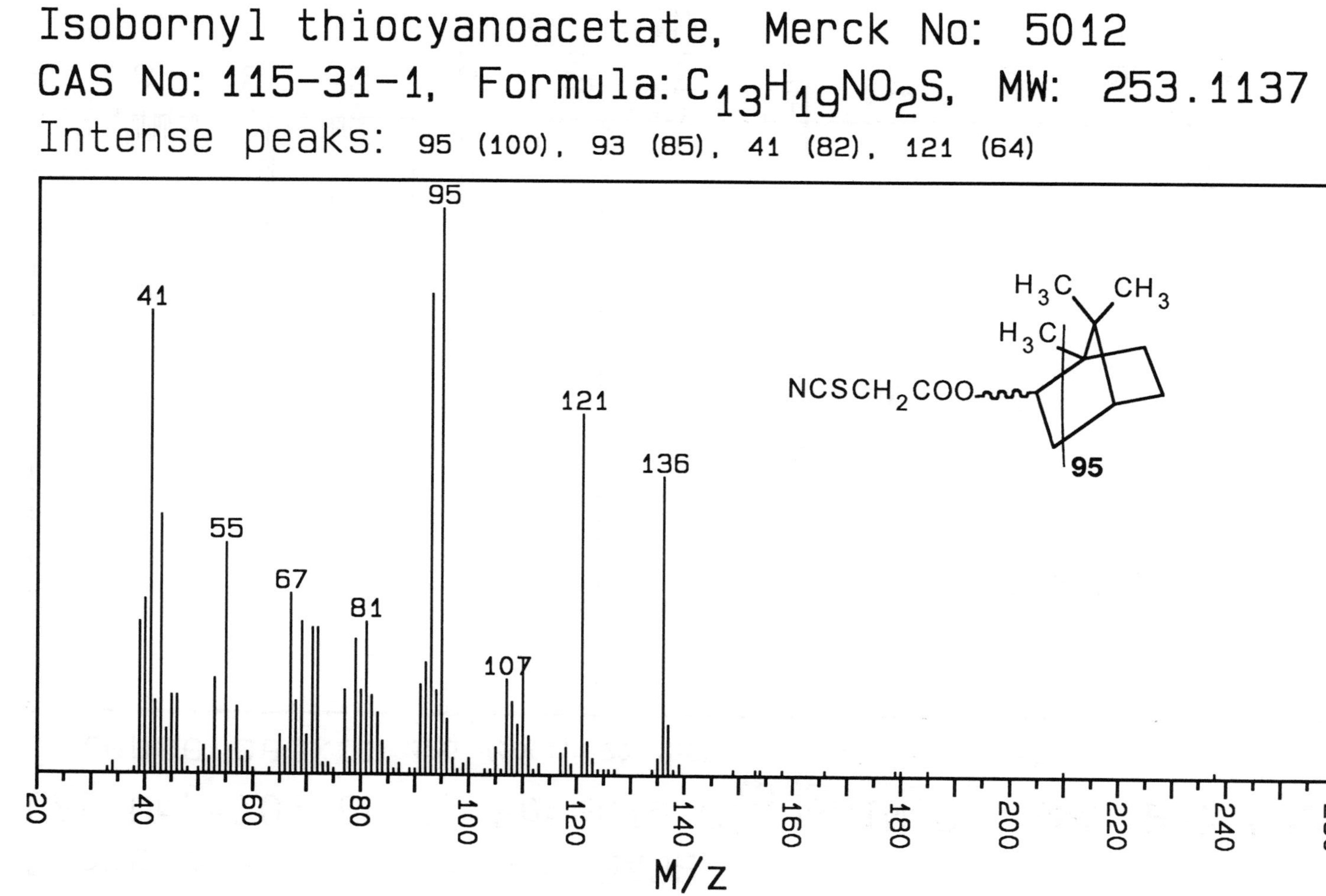
Isobornyl thiocyanoacetate, Merck No: 5012
CAS No: 115-31-1, Formula: C13H19NO2S, MW: 253.1137
Intense peaks: 95 (100), 93 (85), 41 (82), 121 (64)
41
55
67
81
95
107
121
136
H3C
CH3
H3C
NCSCH2COO
95
20
40
60
80
100
120
140
160
180
200
220
240
260
M/Z

Drazoxolon, Merck No: 3432
CAS No: 5707-69-7, Formula: $C_{10}H_8ClN_3O_2$, MW: 237.0305
Intense peaks: 125 (100), 90 (55), 63 (34), 127 (31)

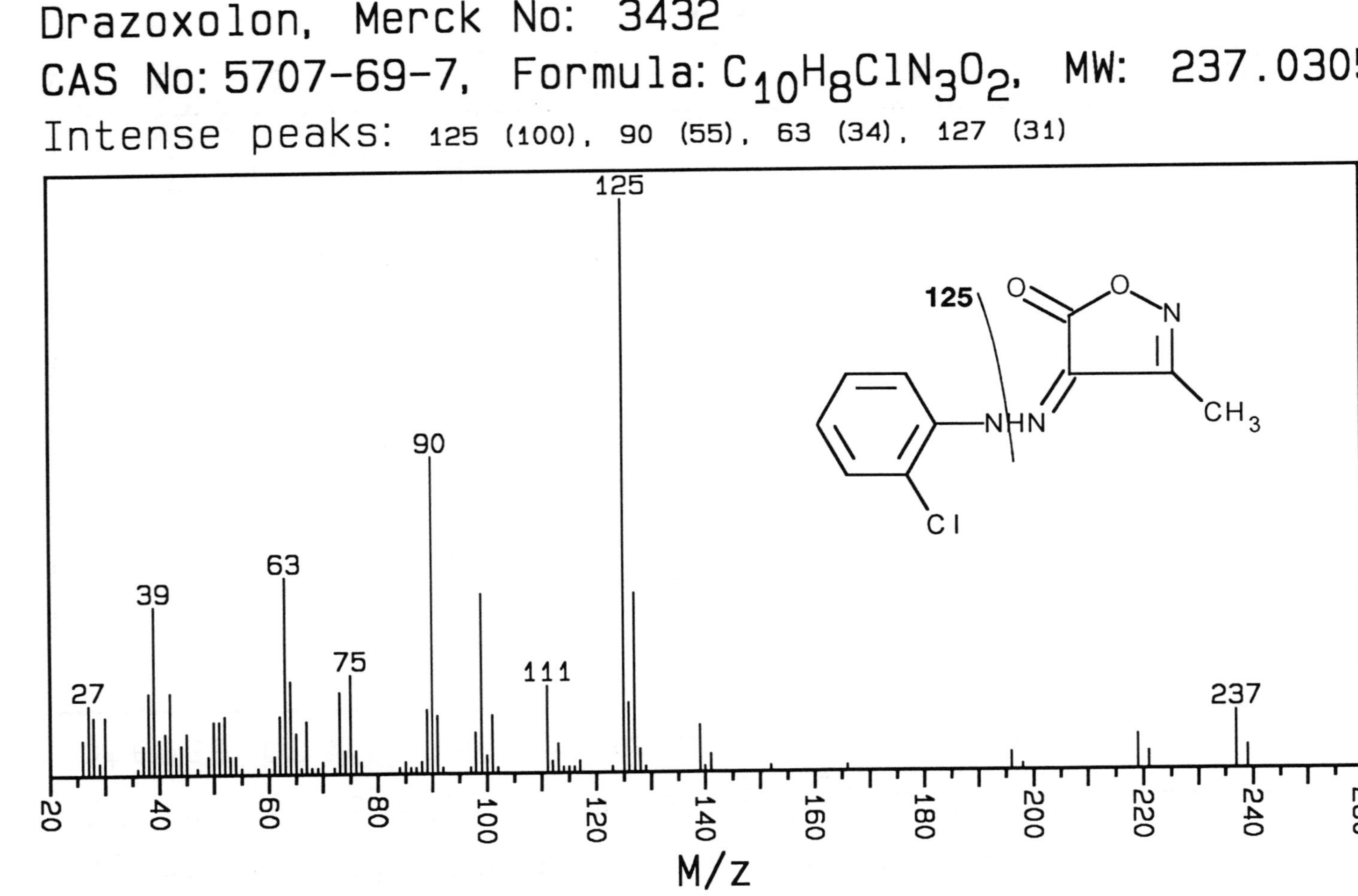

Ethoxyquin, Merck No: 3710

CAS No: 91-53-2, Formula: $C_{14}H_{19}NO$, MW: 217.1467

Intense peaks: 202 (100), 108 (53), 174 (48), 28 (40)

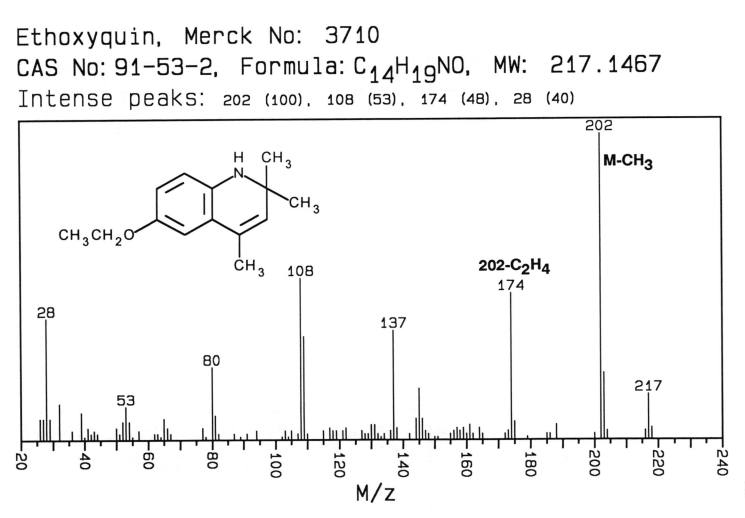

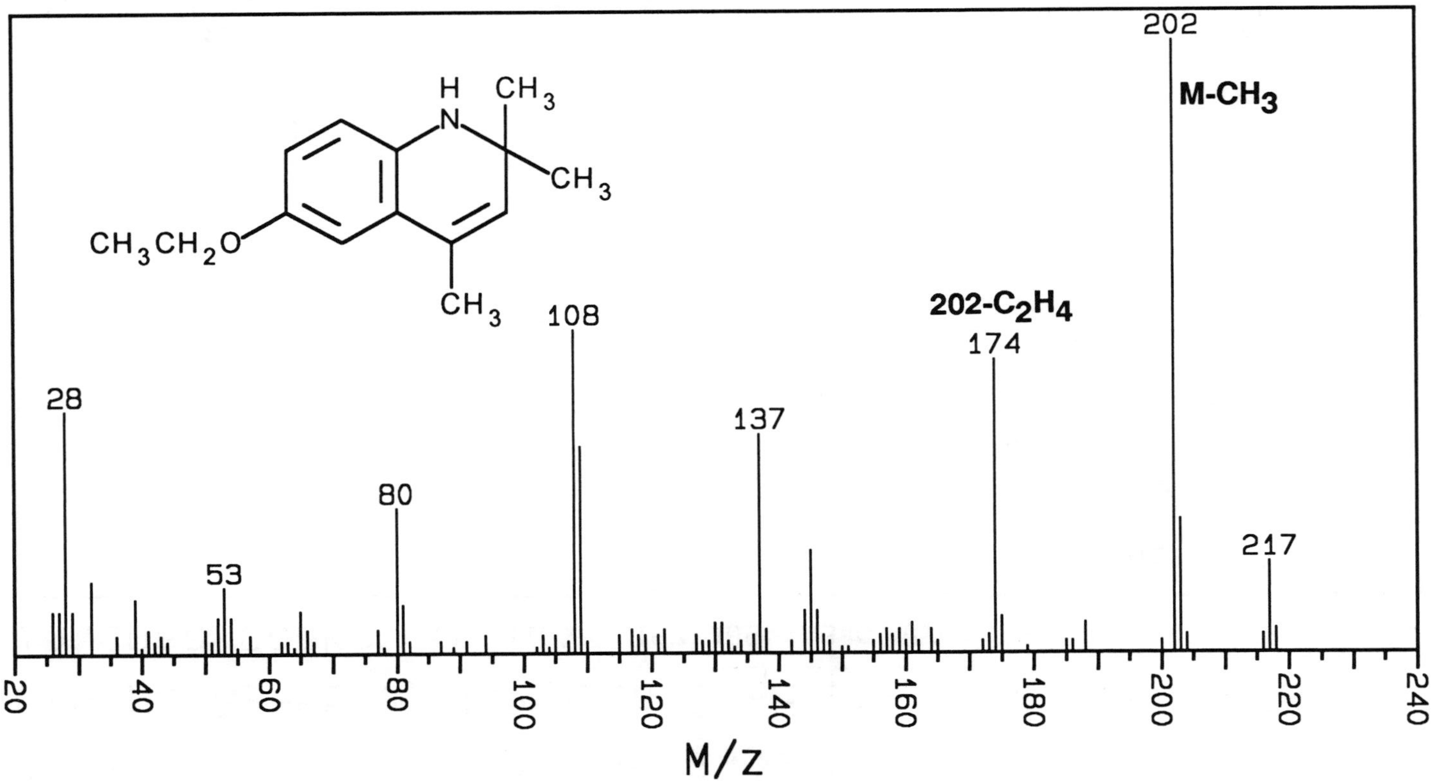

$CH_3CH_2O$ H N $CH_3$ $CH_3$ $CH_3$

202 **M-$CH_3$**

**202-$C_2H_4$** 174

108 137 80 53 28 217

20 40 60 80 100 120 140 160 180 200 220 240

M/Z

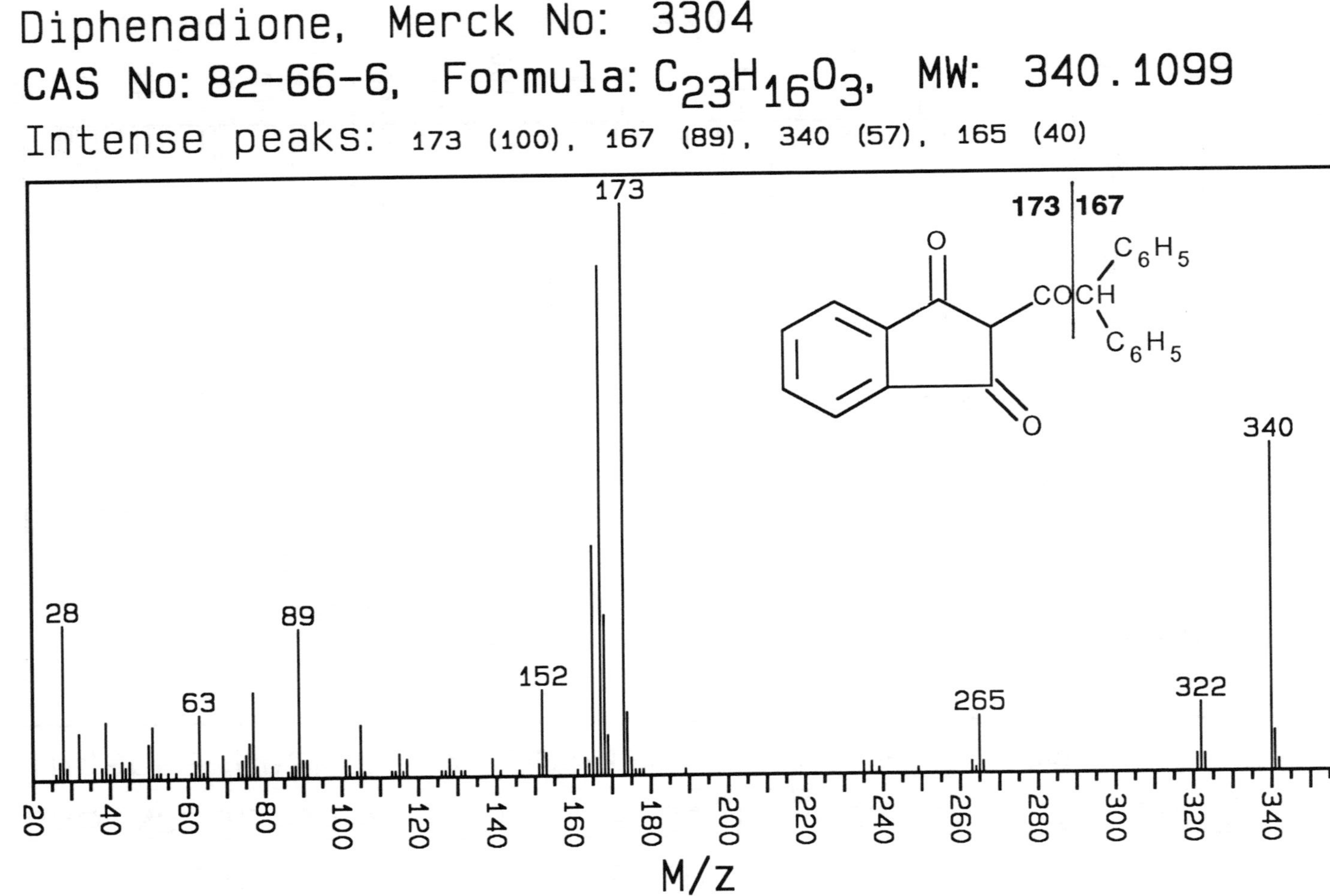
Diphenadione, Merck No: 3304
CAS No: 82-66-6, Formula: $C_{23}H_{16}O_3$, MW: 340.1099
Intense peaks: 173 (100), 167 (89), 340 (57), 165 (40)
173 | 167
O
COCH
$C_6H_5$
$C_6H_5$
O
28
63
89
152
173
265
322
340
20
40
60
80
100
120
140
160
180
200
220
240
260
280
300
320
340
360
M/Z

Ethofumesate, Merck No: 3697

CAS No: 26225-79-6, Formula: $C_{13}H_{18}O_5S$, MW: 286.0875

Intense peaks: 43 (100), 29 (91), 161 (85), 207 (83)

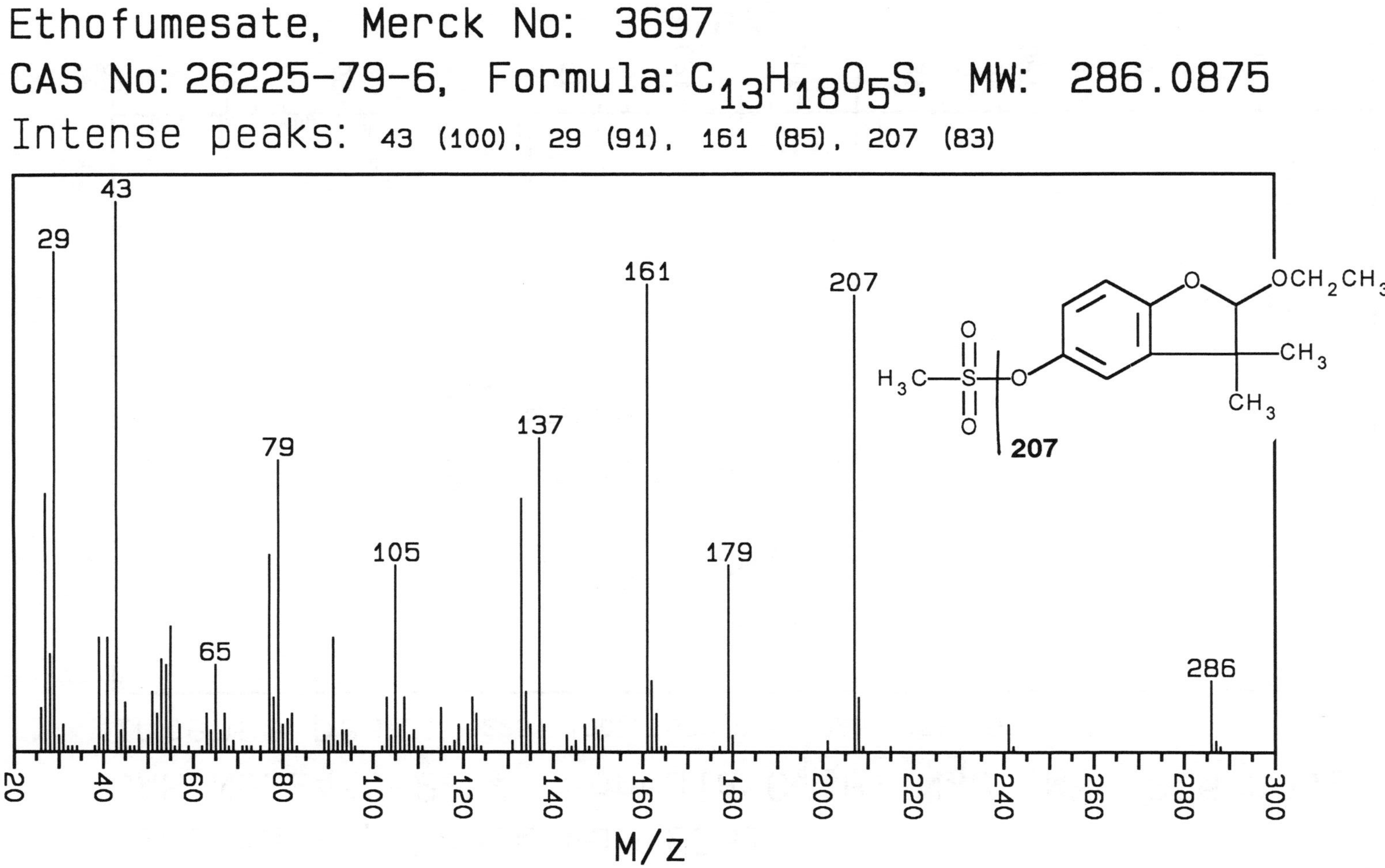

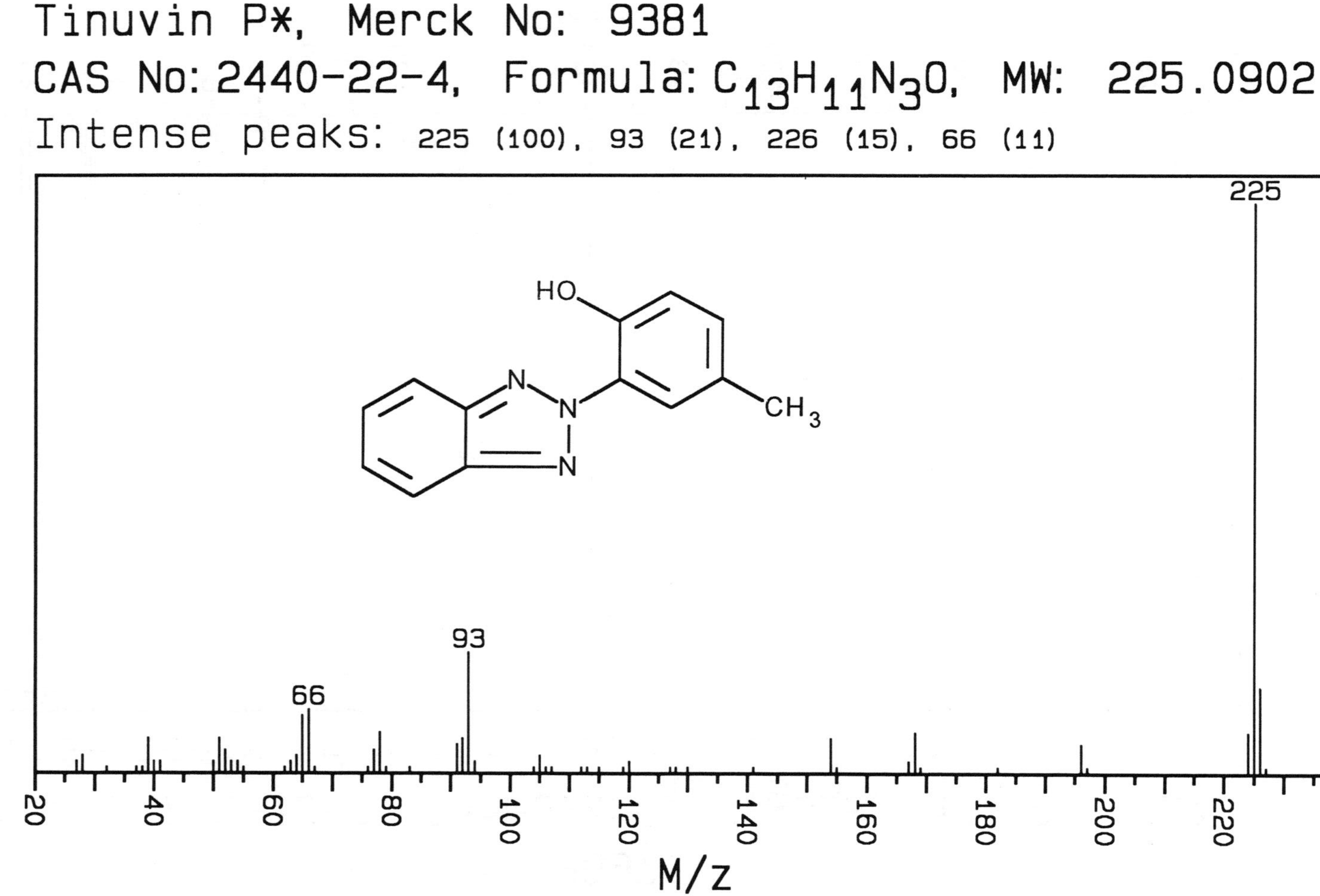
Tinuvin P*, Merck No: 9381
CAS No: 2440-22-4, Formula: $C_{13}H_{11}N_3O$, MW: 225.0902
Intense peaks: 225 (100), 93 (21), 226 (15), 66 (11)
225
93
66
HO
N
N
N
$CH_3$
20
40
60
80
100
120
140
160
180
200
220
240
M/Z

Benzothiazole, 2-mercapto-, Merck No: 5759

CAS No: 149-30-4, Formula: $C_7H_5NS_2$, MW: 166.9863

Intense peaks: 167 (100), 69 (33), 45 (23), 63 (20)

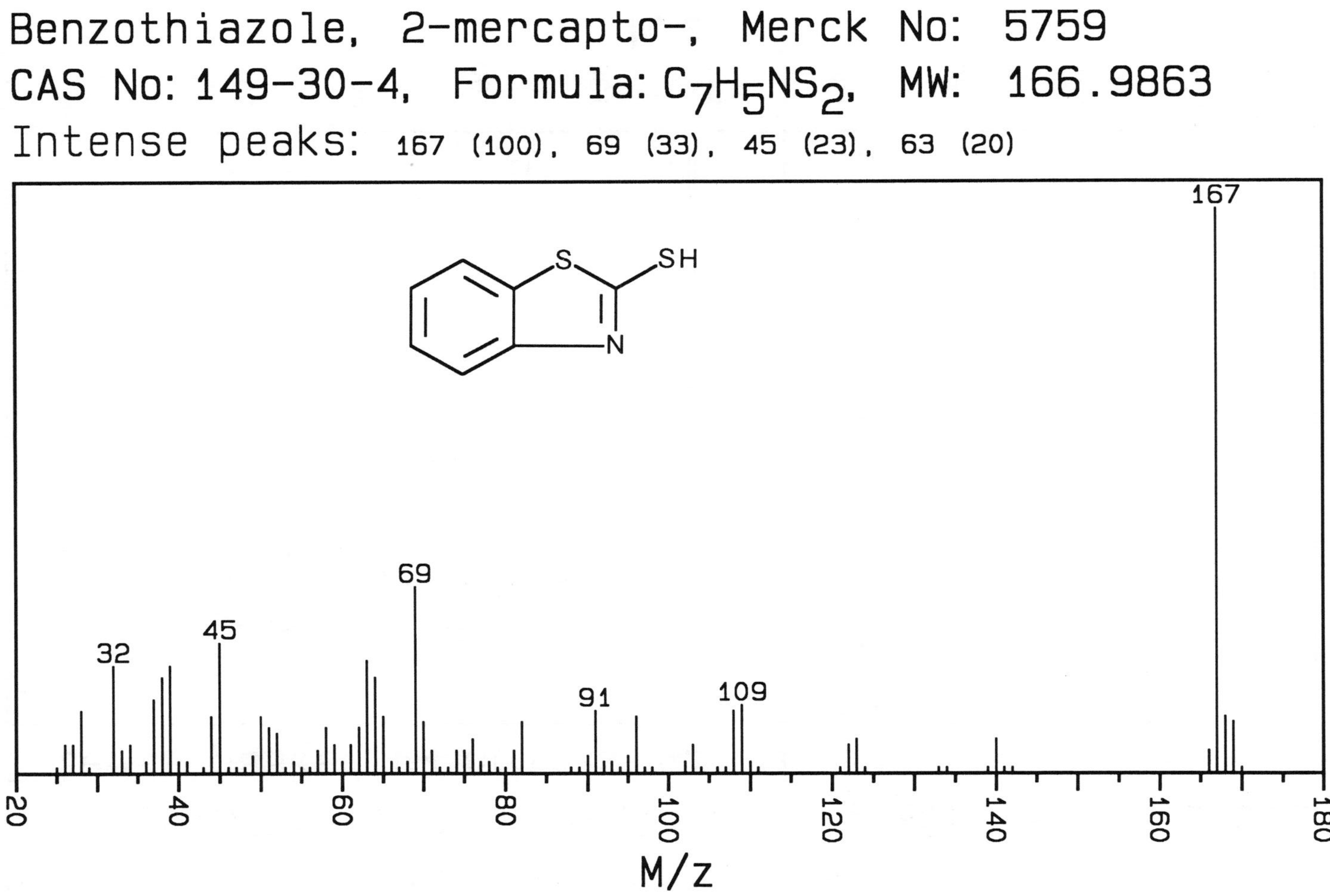

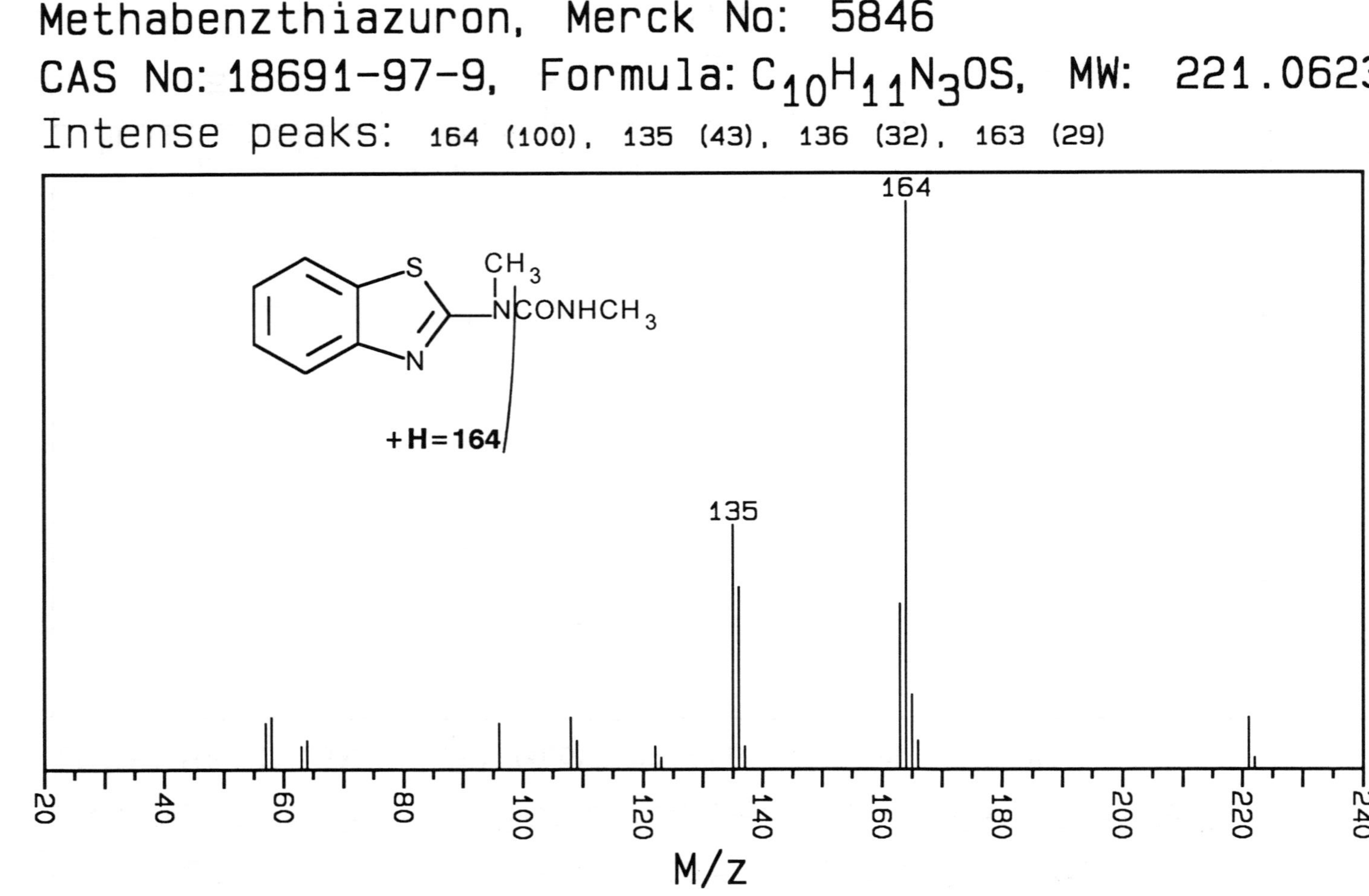
Methabenzthiazuron, Merck No: 5846
CAS No: 18691-97-9, Formula: C10H11N3OS, MW: 221.0623
Intense peaks: 164 (100), 135 (43), 136 (32), 163 (29)
164
135
CH3
NCONHCH3
S
N
+H=164
20
40
60
80
100
120
140
160
180
200
220
240
M/Z

Caffeine, Merck No: 1635

CAS No: 58-08-2, Formula: $C_8H_{10}N_4O_2$, MW: 194.0804

Intense peaks: 194 (100), 109 (88), 67 (88), 55 (71)

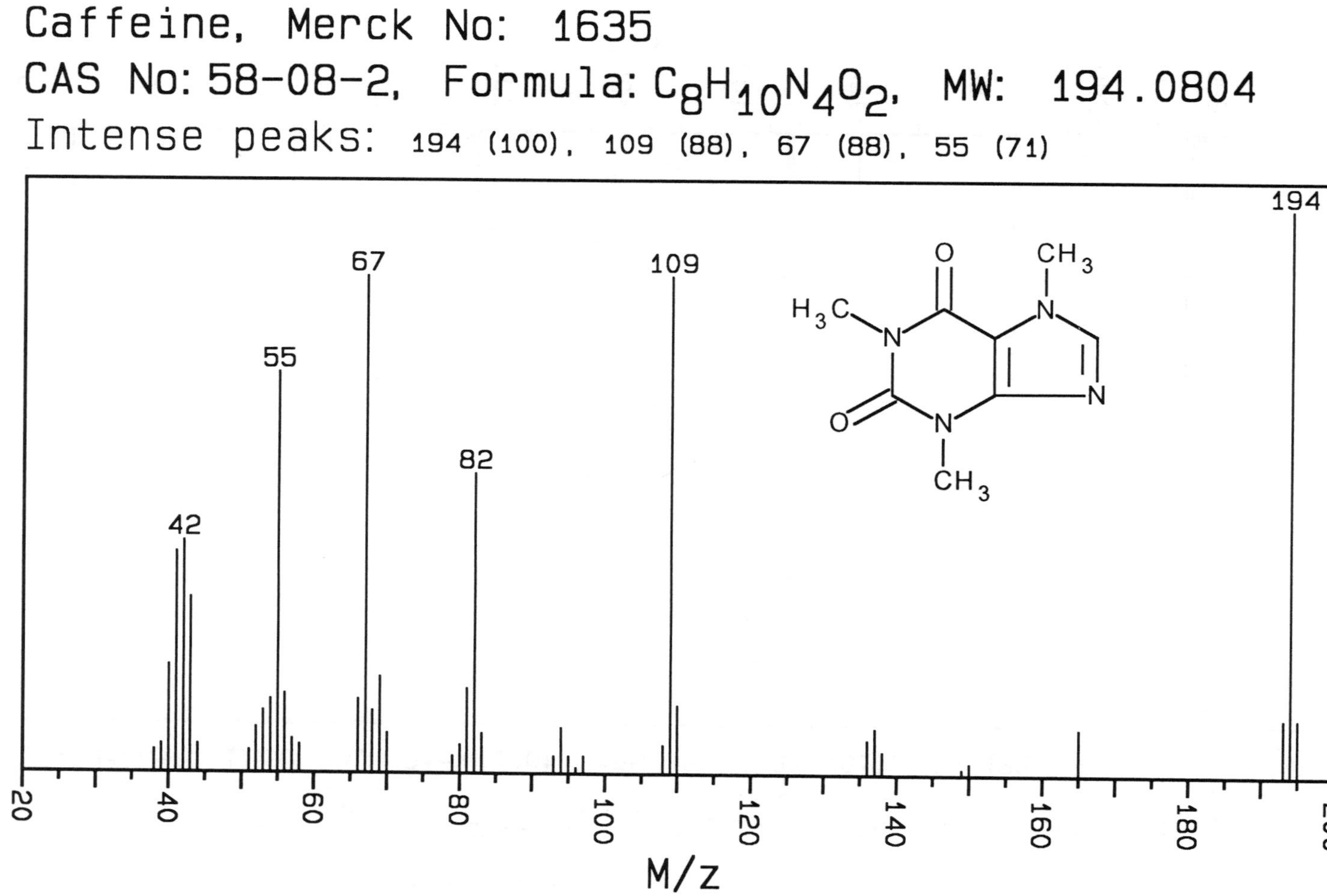

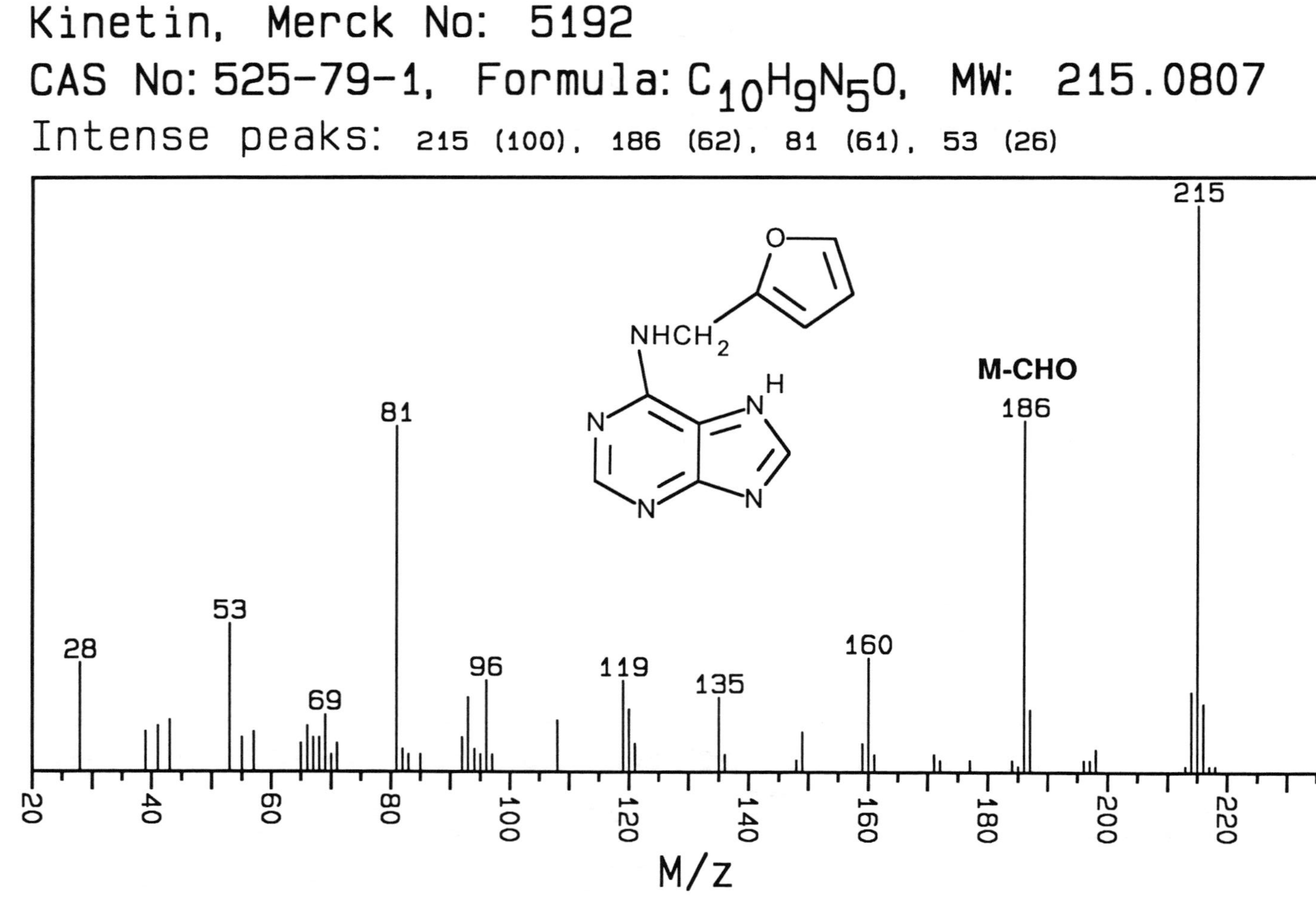

Kinetin, Merck No: 5192
CAS No: 525-79-1, Formula: $C_{10}H_9N_5O$, MW: 215.0807
Intense peaks: 215 (100), 186 (62), 81 (61), 53 (26)
215
M-CHO
186
81
160
135
119
96
69
53
28
NHCH$_2$
O
N
H
M/Z
20
40
60
80
100
120
140
160
180
200
220
240

Iridomyrmecin, Merck No: 4972

CAS No: 485-43-8, Formula: $C_{10}H_{16}O_2$, MW: 168.1151

Intense peaks: 95 (100), 67 (87), 41 (77), 81 (76)

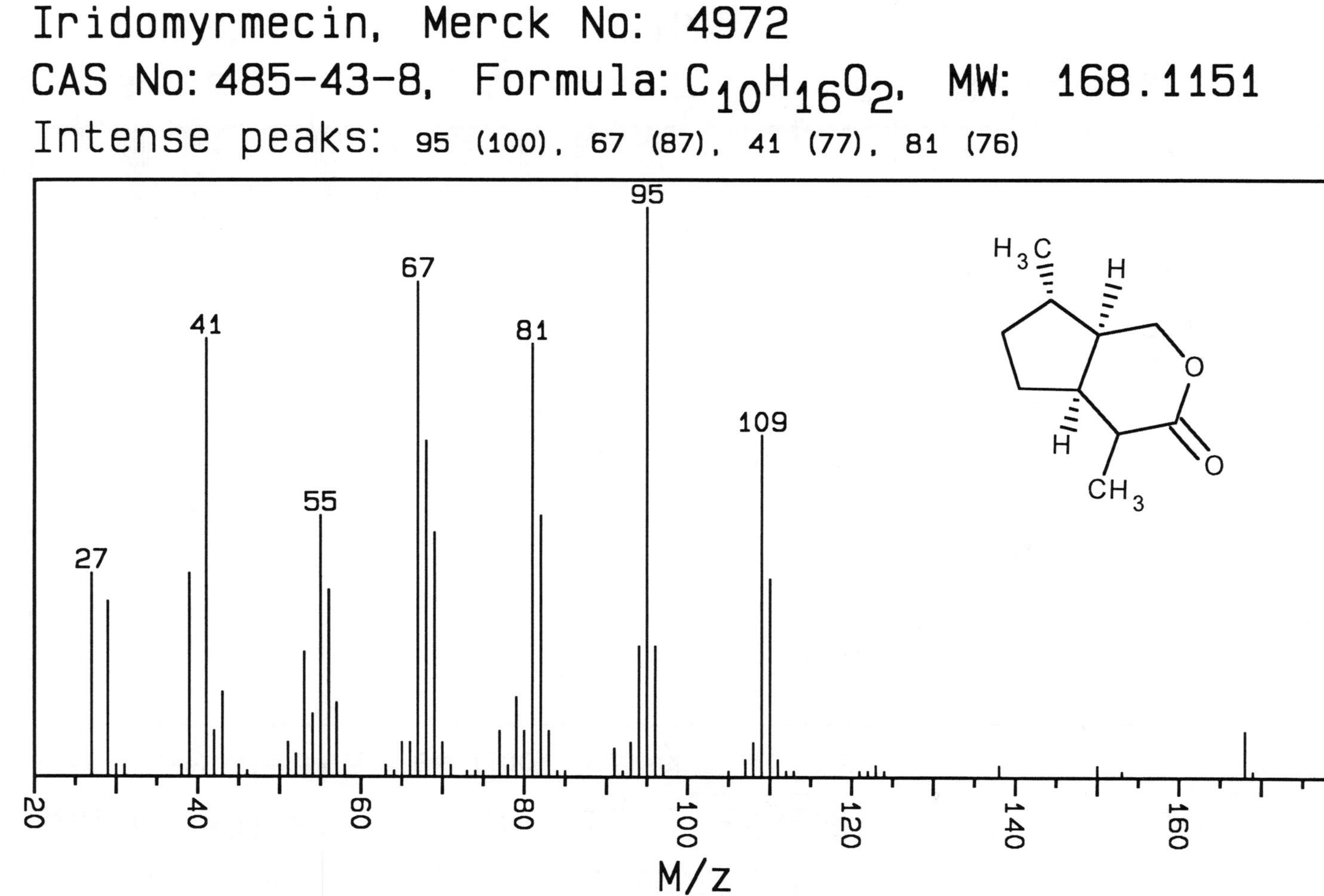

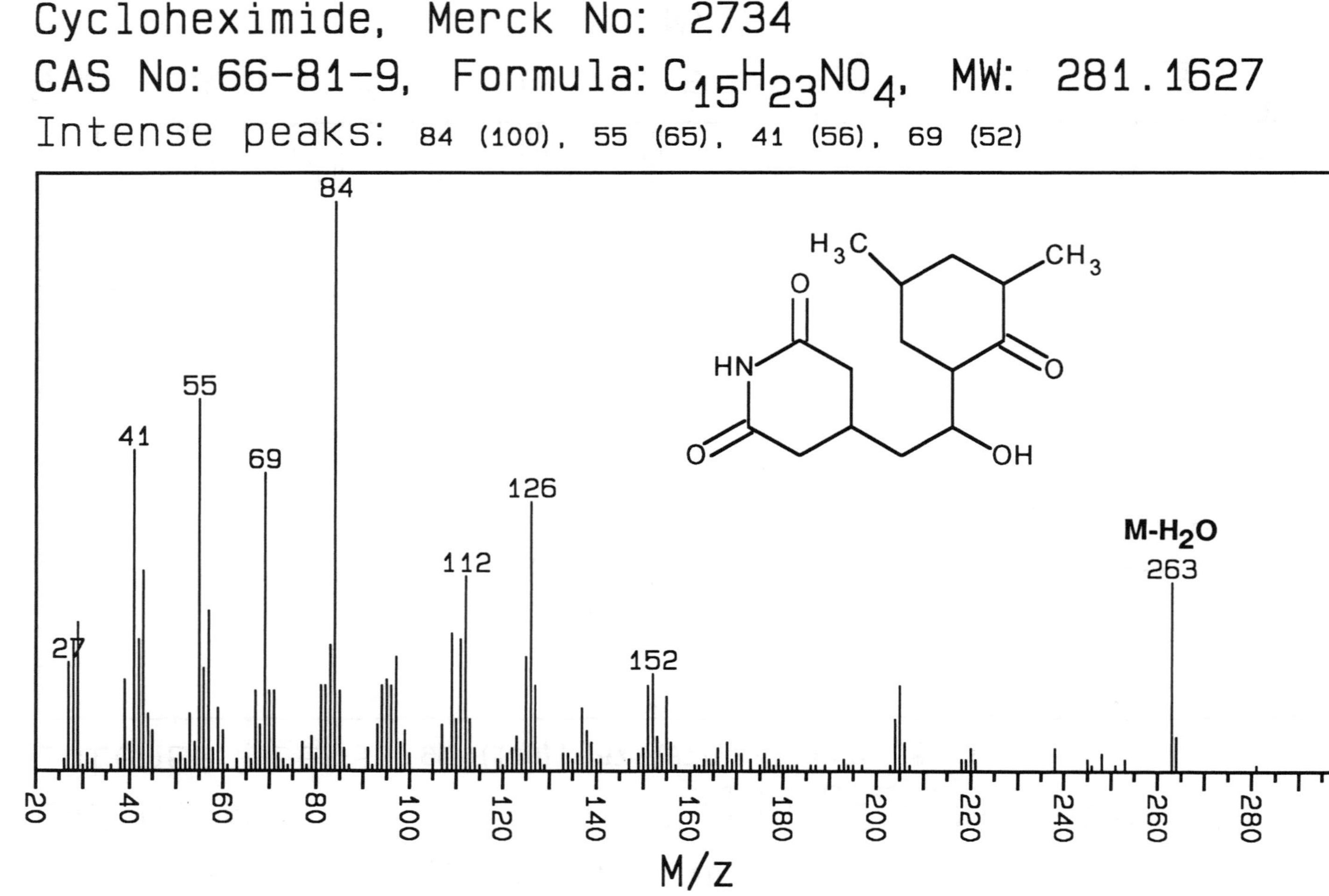
Cycloheximide, Merck No: 2734
CAS No: 66-81-9, Formula: $C_{15}H_{23}NO_4$, MW: 281.1627
Intense peaks: 84 (100), 55 (65), 41 (56), 69 (52)
$H_3C$
$CH_3$
O
HN
O
O
OH
84
55
41
69
126
112
27
152
M-$H_2O$
263
20
40
60
80
100
120
140
160
180
200
220
240
260
280
300
M/Z

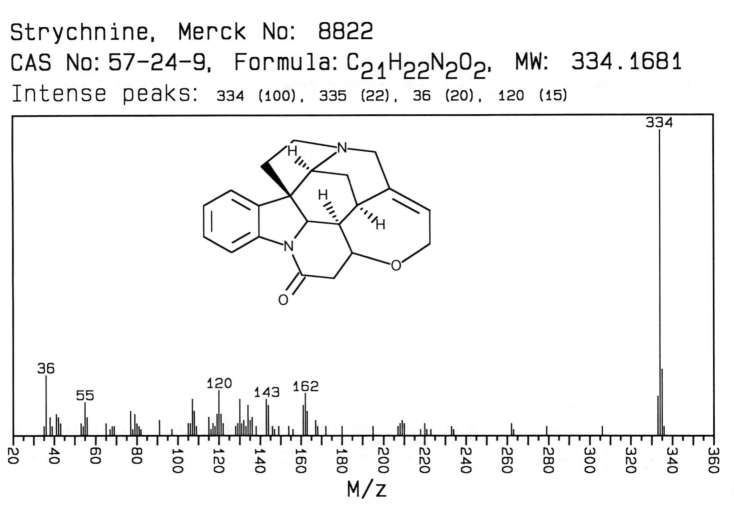

Strychnine, Merck No: 8822

CAS No: 57-24-9, Formula: $C_{21}H_{22}N_2O_2$, MW: 334.1681

Intense peaks: 334 (100), 335 (22), 36 (20), 120 (15)

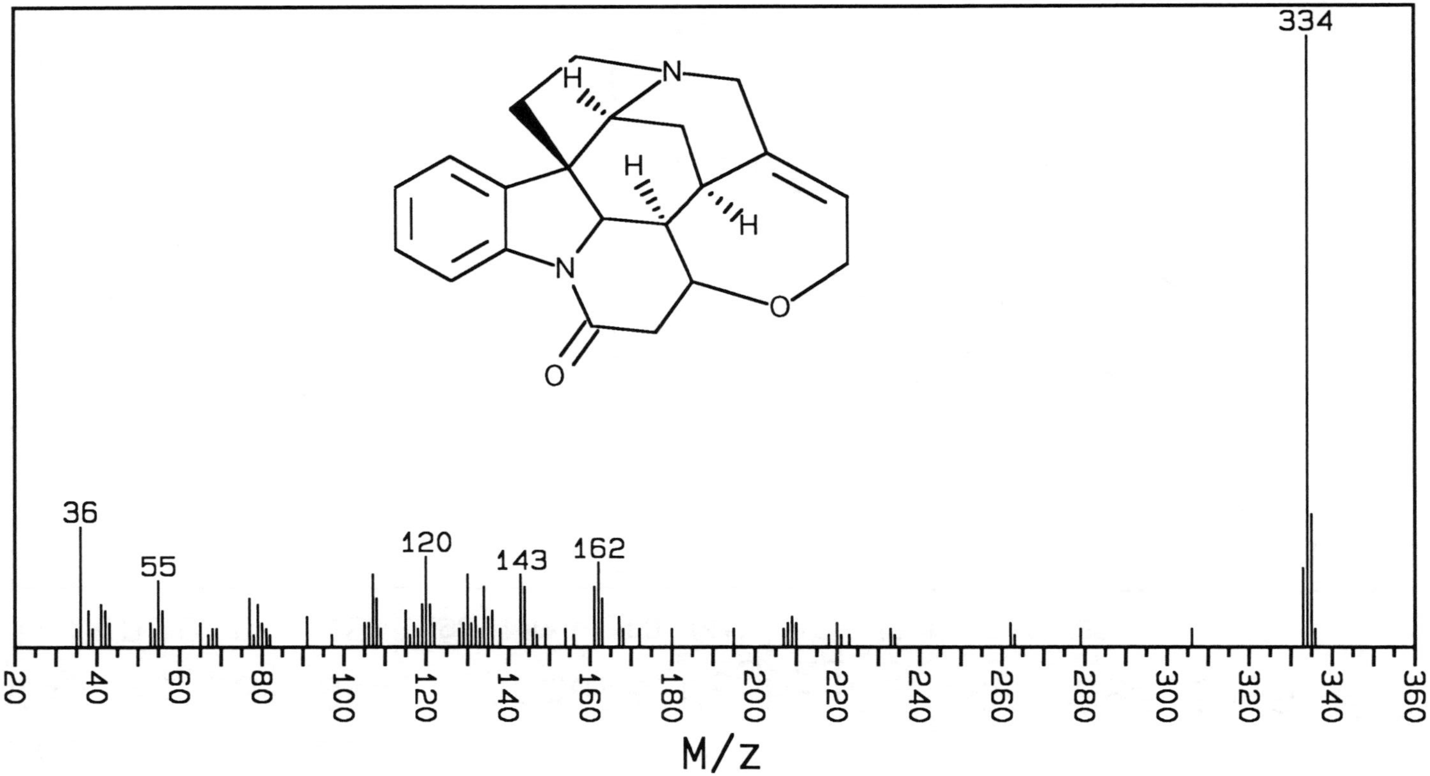

334

36

55

120

143

162

20 40 60 80 100 120 140 160 180 200 220 240 260 280 300 320 340 360

M/Z

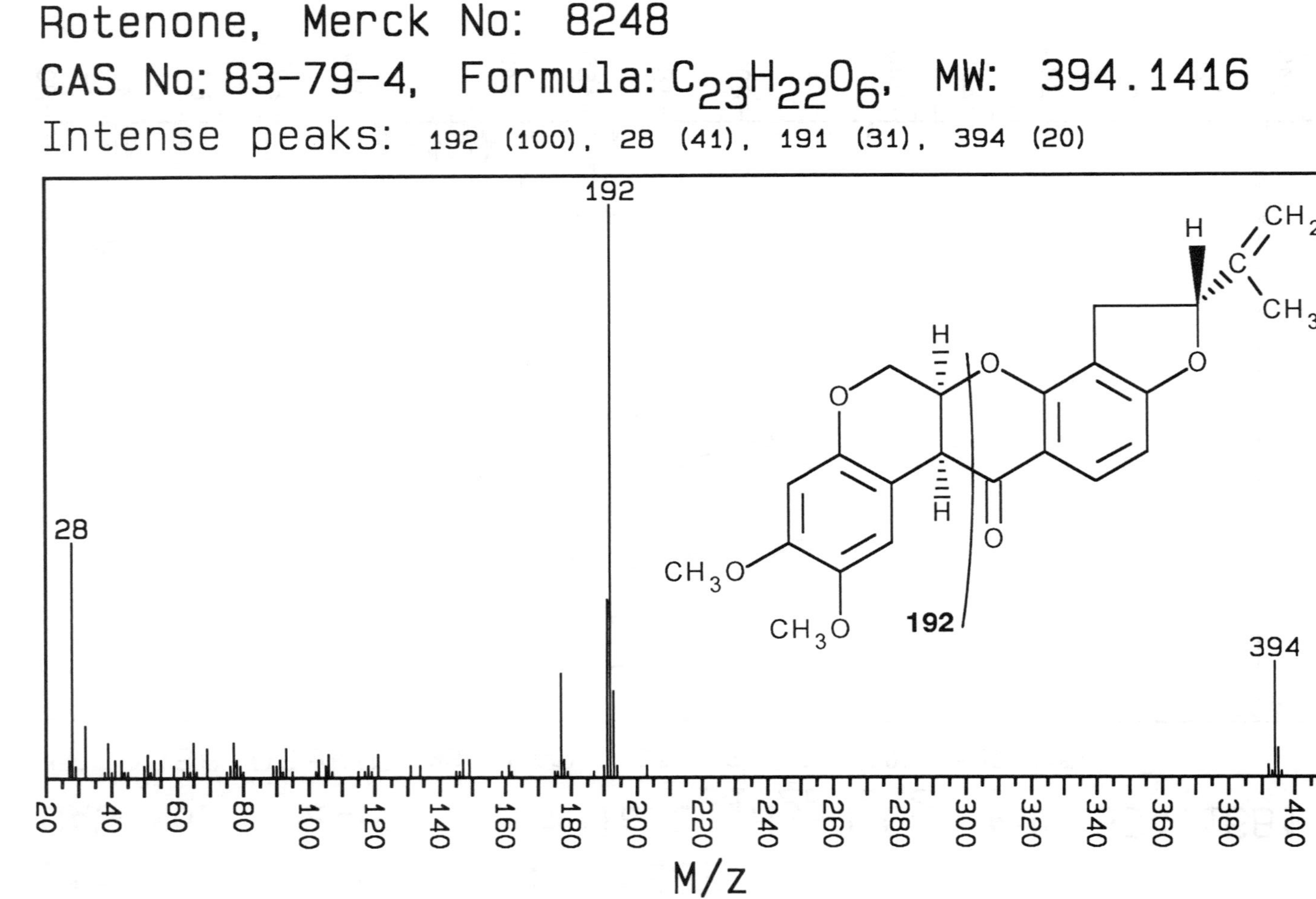
Rotenone, Merck No: 8248
CAS No: 83-79-4, Formula: $C_{23}H_{22}O_6$, MW: 394.1416
Intense peaks: 192 (100), 28 (41), 191 (31), 394 (20)
192
28
394
H
CH2
C
CH3
H
O
O
O
H
O
CH3O
CH3O
192
20
40
60
80
100
120
140
160
180
200
220
240
260
280
300
320
340
360
380
400
420
M/Z

Deguelin, Merck No: 2854

CAS No: 522-17-8, Formula: $C_{23}H_{22}O_6$, MW: 394.1416

Intense peaks: 192 (100), 394 (35), 392 (23), 191 (23)

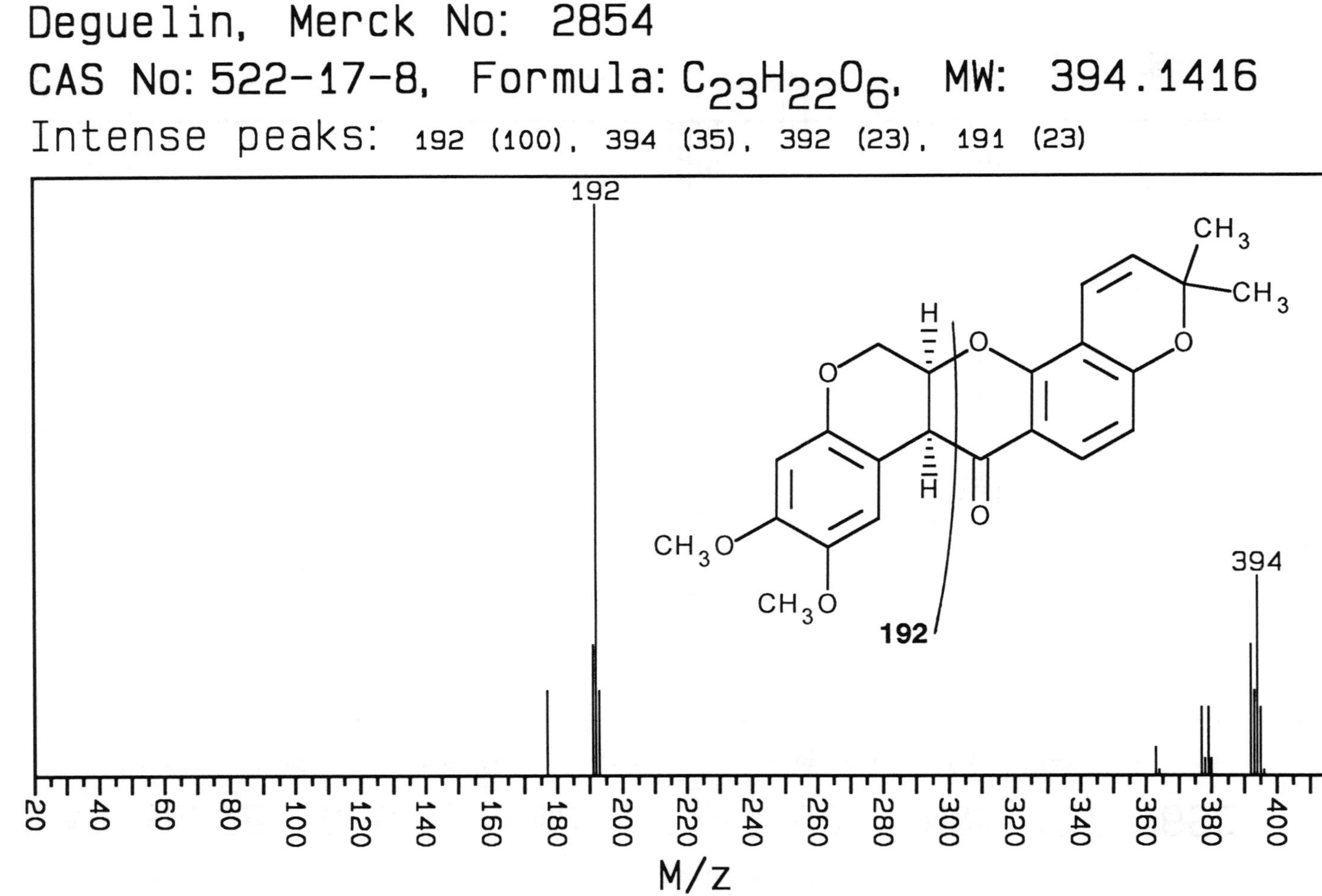

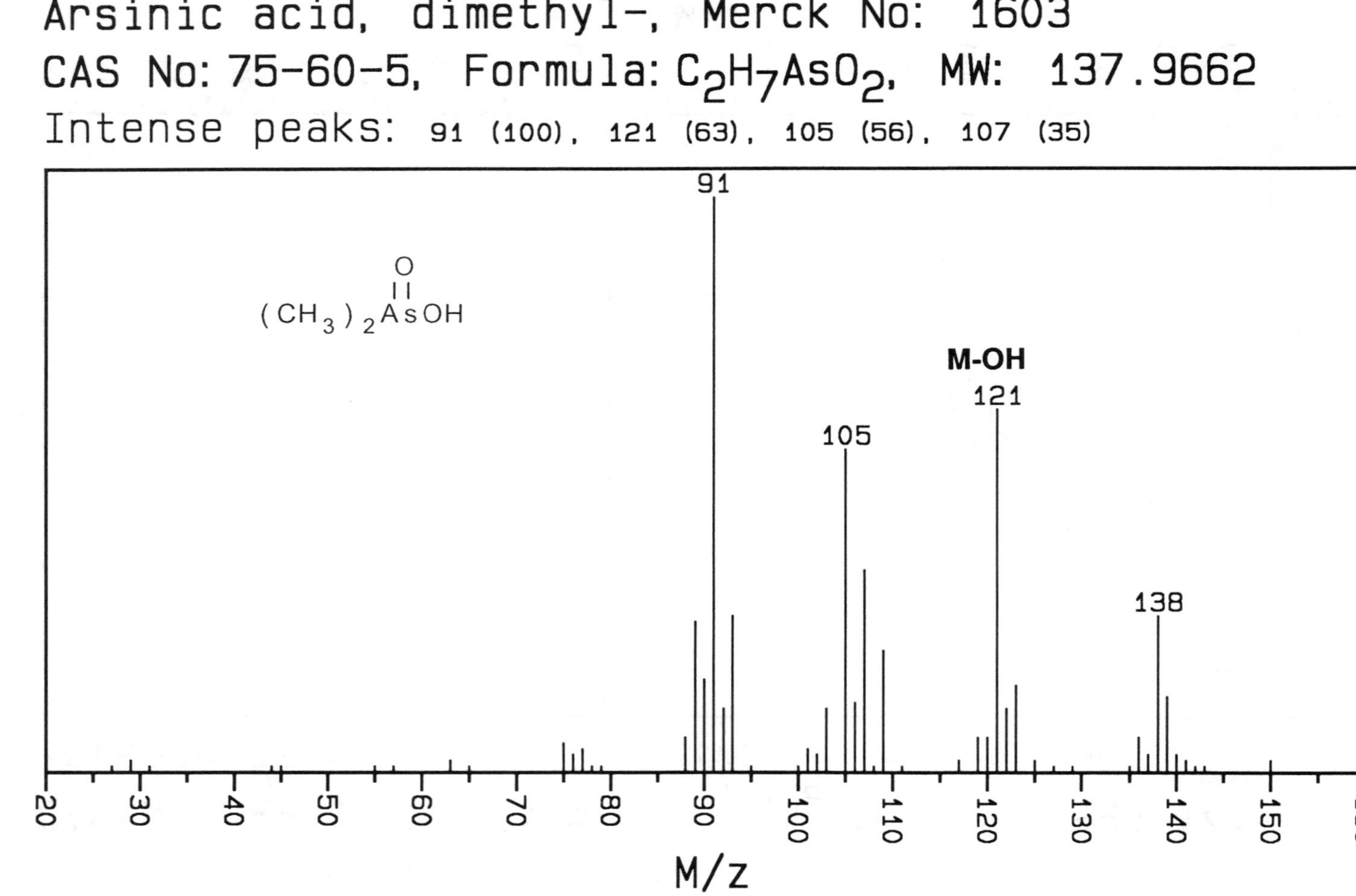

Arsinic acid, dimethyl-, Merck No: 1603
CAS No: 75-60-5, Formula: $C_2H_7AsO_2$, MW: 137.9662
Intense peaks: 91 (100), 121 (63), 105 (56), 107 (35)
$(CH_3)_2As(=O)OH$
91
105
M-OH
121
138
20
30
40
50
60
70
80
90
100
110
120
130
140
150
160
M/Z

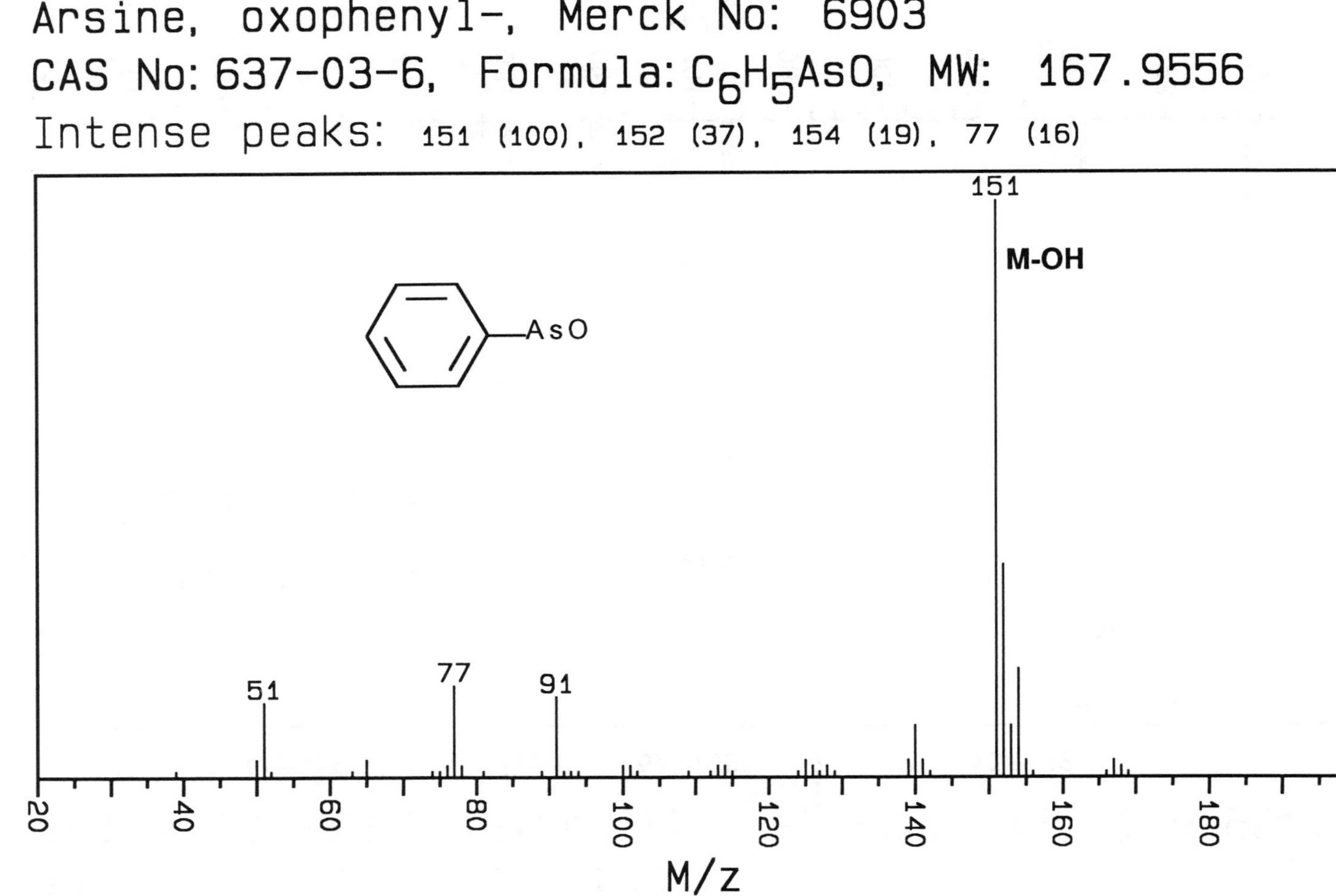
Arsine, oxophenyl-, Merck No: 6903
CAS No: 637-03-6, Formula: $C_6H_5AsO$, MW: 167.9556
Intense peaks: 151 (100), 152 (37), 154 (19), 77 (16)
151
M-OH
AsO
51
77
91
20
40
60
80
100
120
140
160
180
200
M/Z

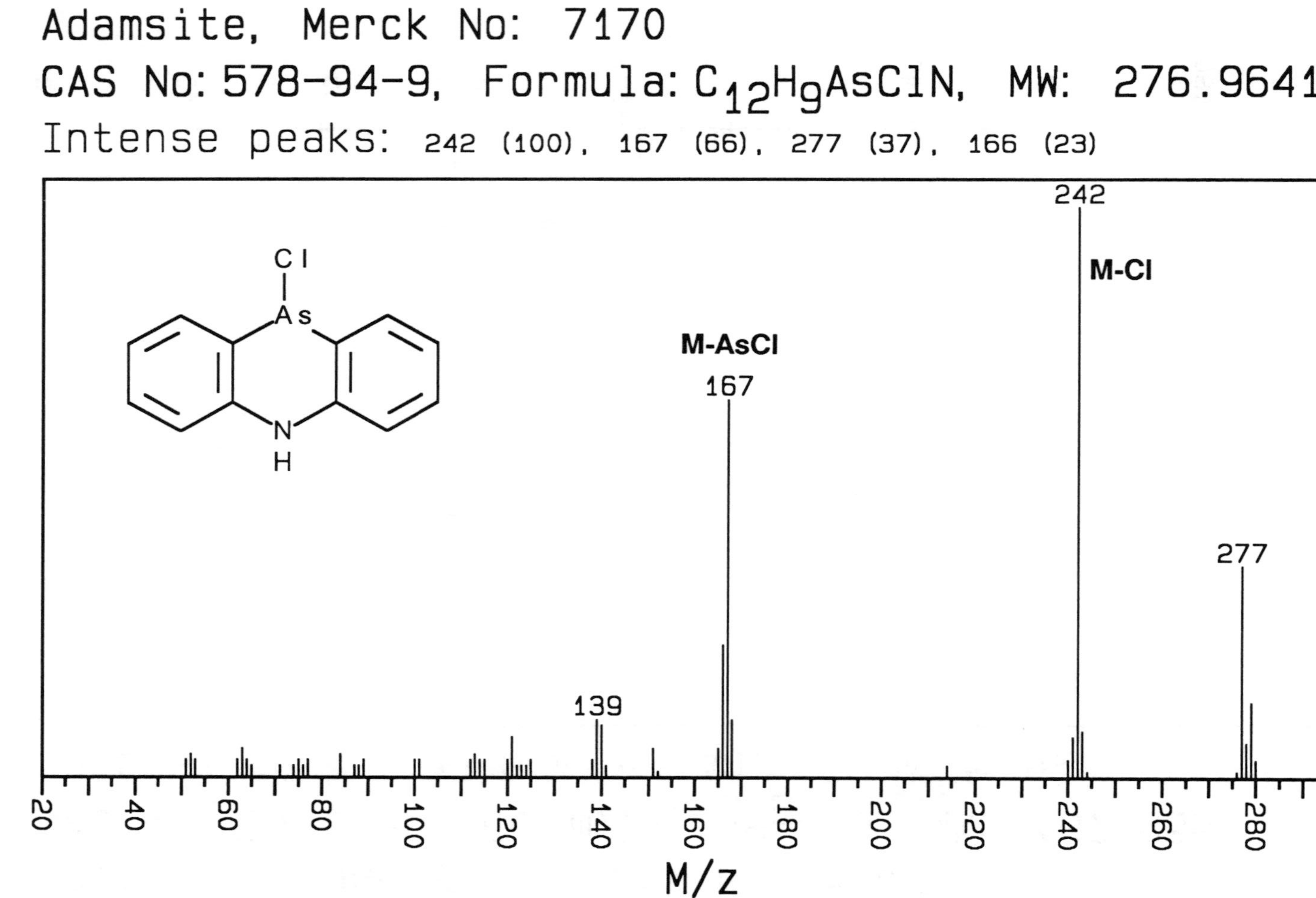
Adamsite, Merck No: 7170
CAS No: 578-94-9, Formula: C12H9AsClN, MW: 276.9641
Intense peaks: 242 (100), 167 (66), 277 (37), 166 (23)
Cl
As
N
H
242
M-Cl
M-AsCl
167
277
139
20
40
60
80
100
120
140
160
180
200
220
240
260
280
300
M/Z

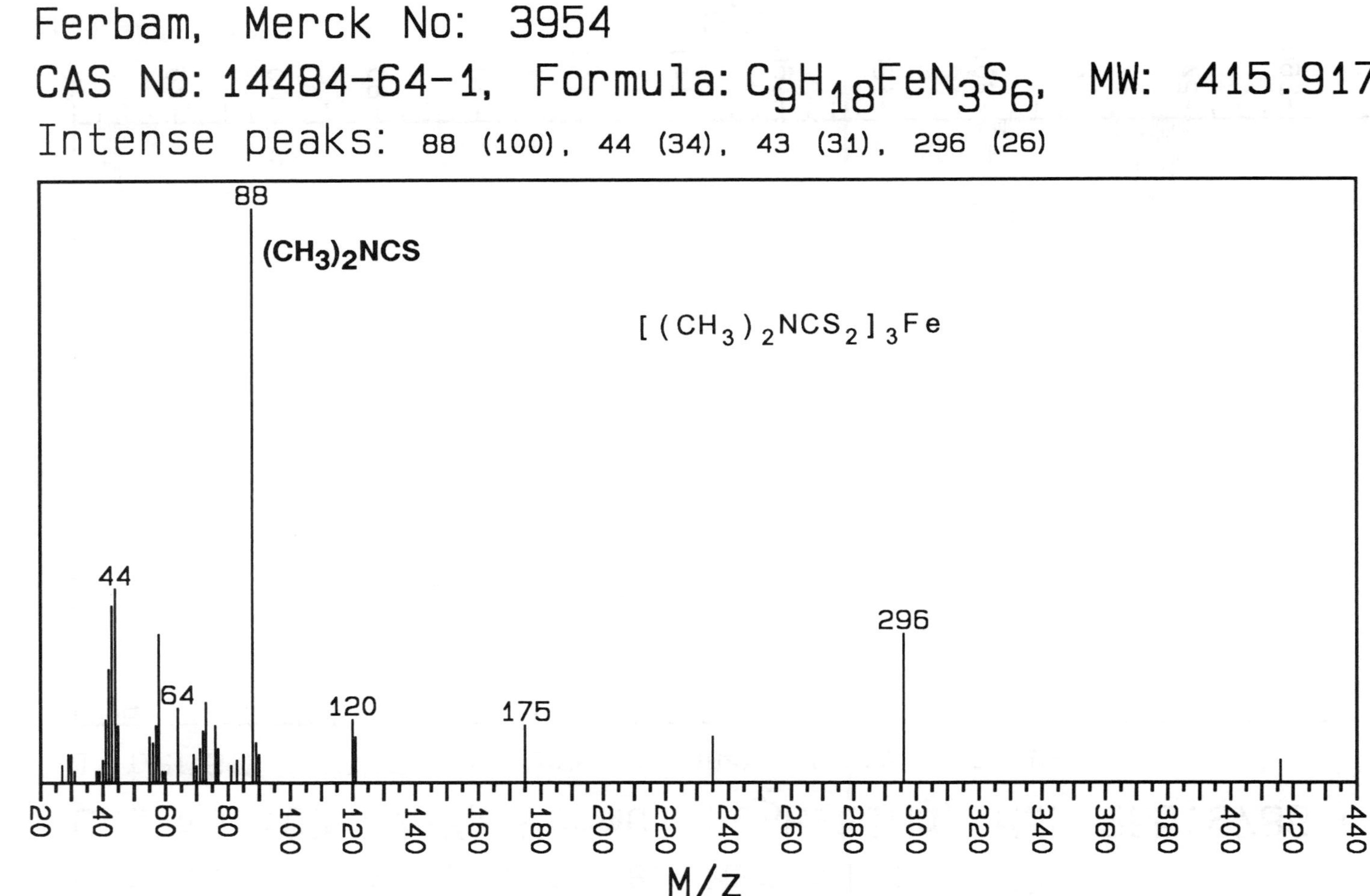

Ferbam, Merck No: 3954
CAS No: 14484-64-1, Formula: $C_9H_{18}FeN_3S_6$, MW: 415.9174
Intense peaks: 88 (100), 44 (34), 43 (31), 296 (26)
88
$(CH_3)_2NCS$
$[(CH_3)_2NCS_2]_3Fe$
44
64
120
175
296
20
40
60
80
100
120
140
160
180
200
220
240
260
280
300
320
340
360
380
400
420
440
M/Z

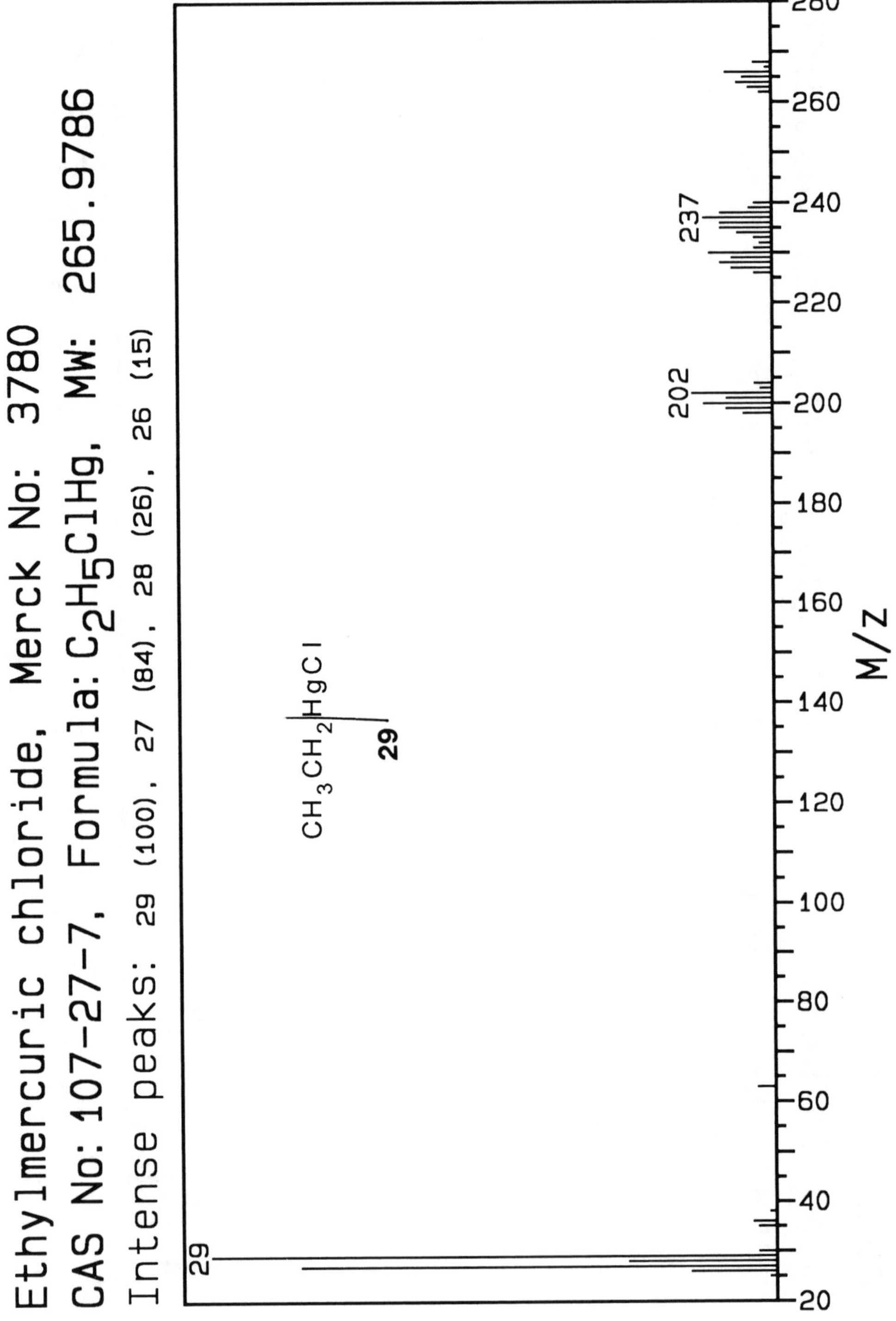
Ethylmercuric chloride, Merck No: 3780
CAS No: 107-27-7, Formula: $C_2H_5ClHg$, MW: 265.9786
Intense peaks: 29 (100), 27 (84), 28 (26), 26 (15)
$CH_3CH_2HgCl$
29
202
237
29
M/Z

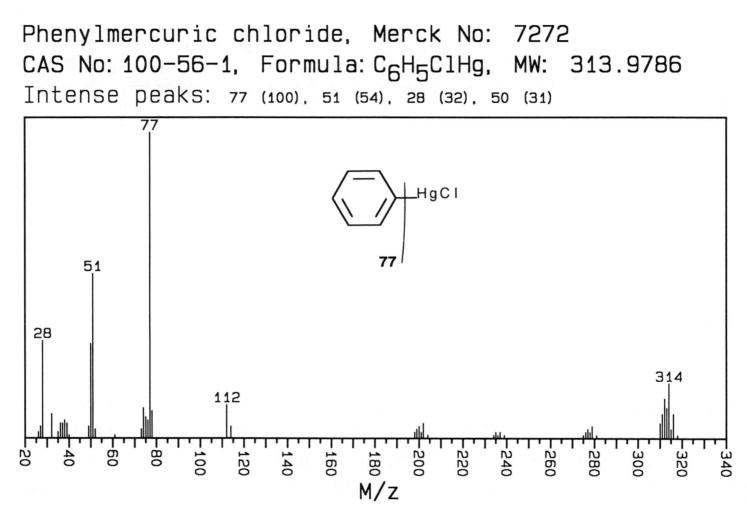

Phenylmercuric chloride, Merck No: 7272

CAS No: 100-56-1, Formula: $C_6H_5ClHg$, MW: 313.9786

Intense peaks: 77 (100), 51 (54), 28 (32), 50 (31)

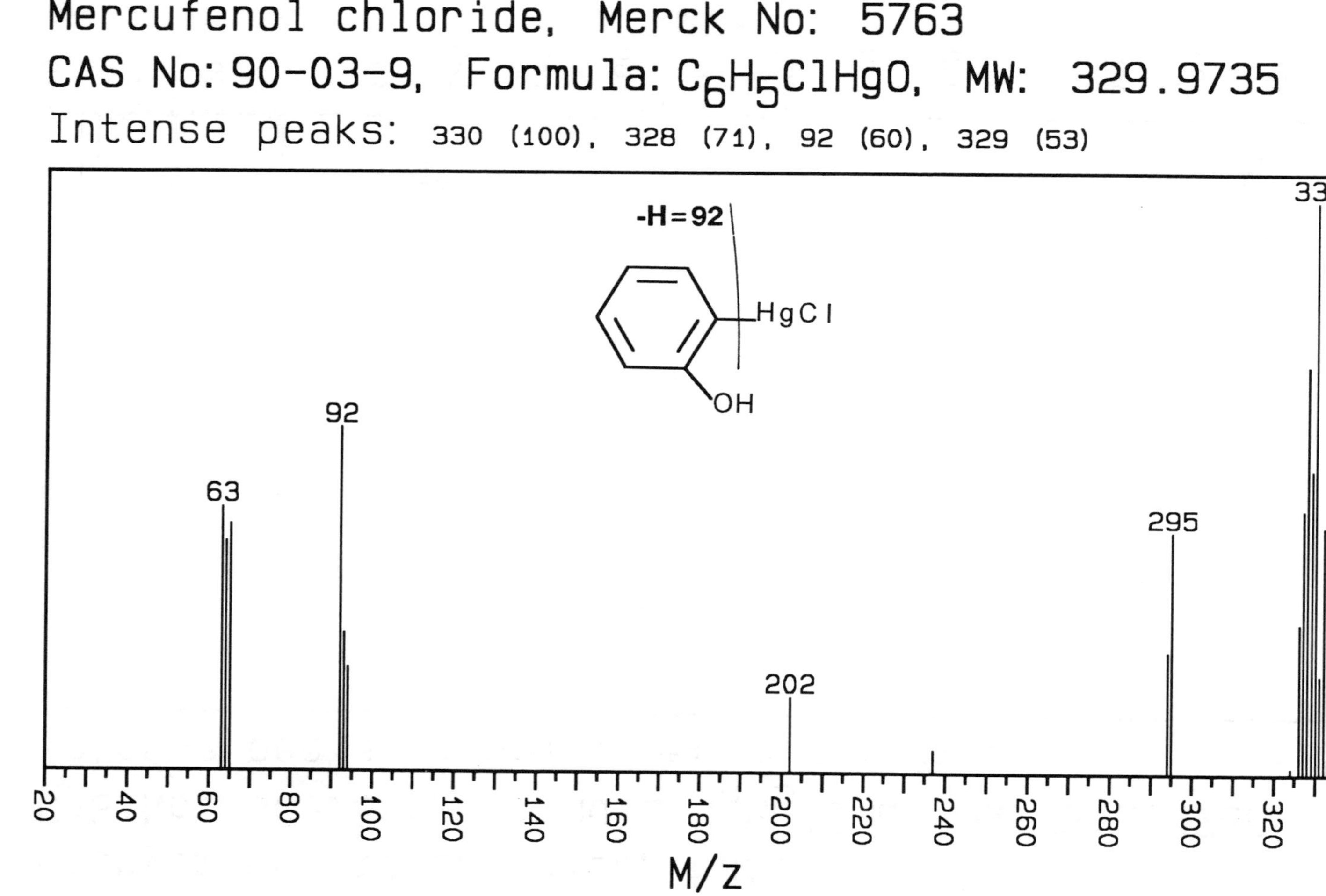

Mercufenol chloride, Merck No: 5763
CAS No: 90-03-9, Formula: C6H5ClHgO, MW: 329.9735
Intense peaks: 330 (100), 328 (71), 92 (60), 329 (53)
-H=92
HgCl
OH
63
92
202
295
330
20
40
60
80
100
120
140
160
180
200
220
240
260
280
300
320
340
M/z

Phenylmercuric acetate, Merck No: 7271

CAS No: 62-38-4, Formula: $C_8H_8HgO_2$, MW: 338.0231

Intense peaks: 77 (100), 51 (62), 50 (28), 279 (15)

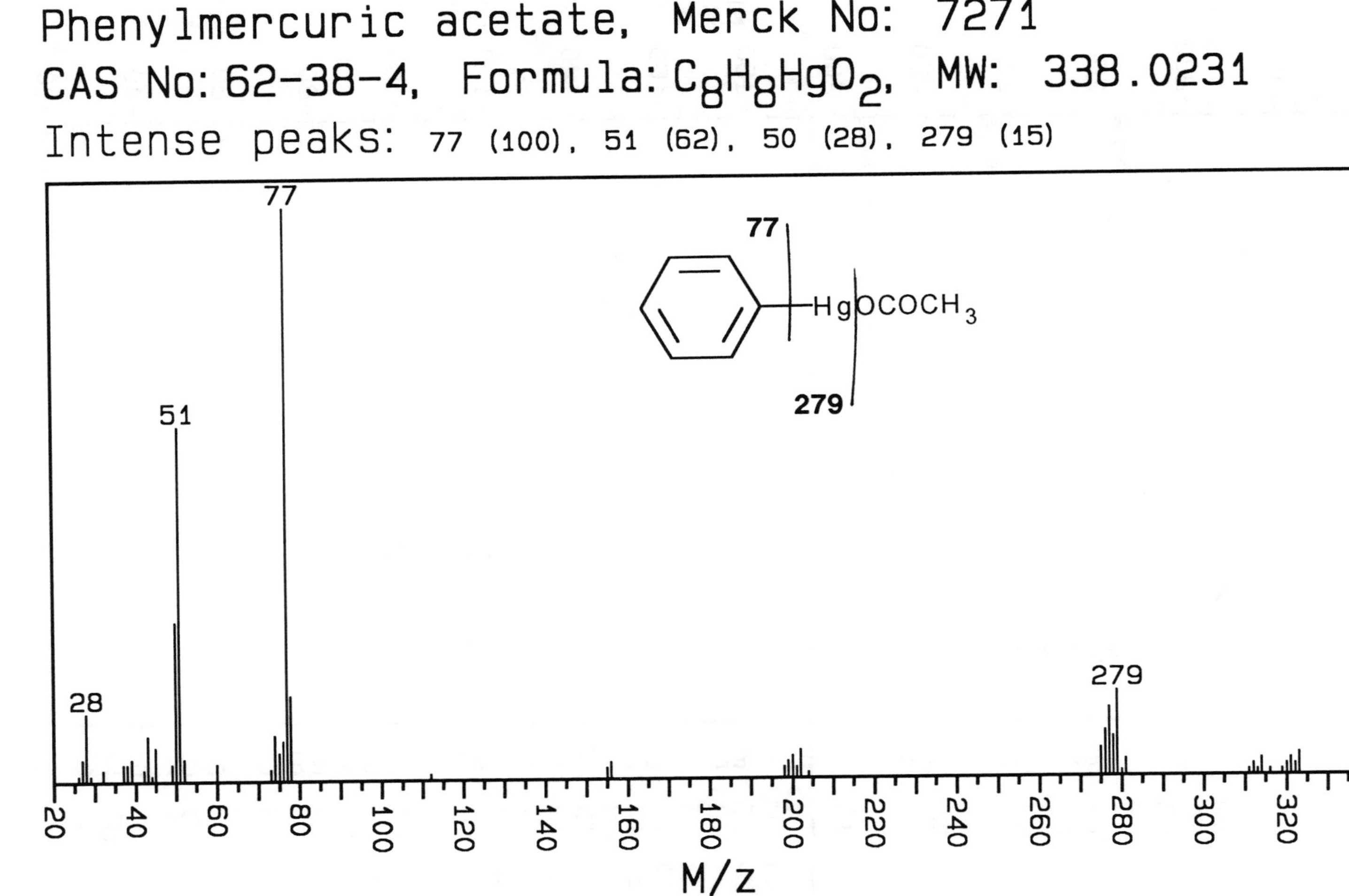

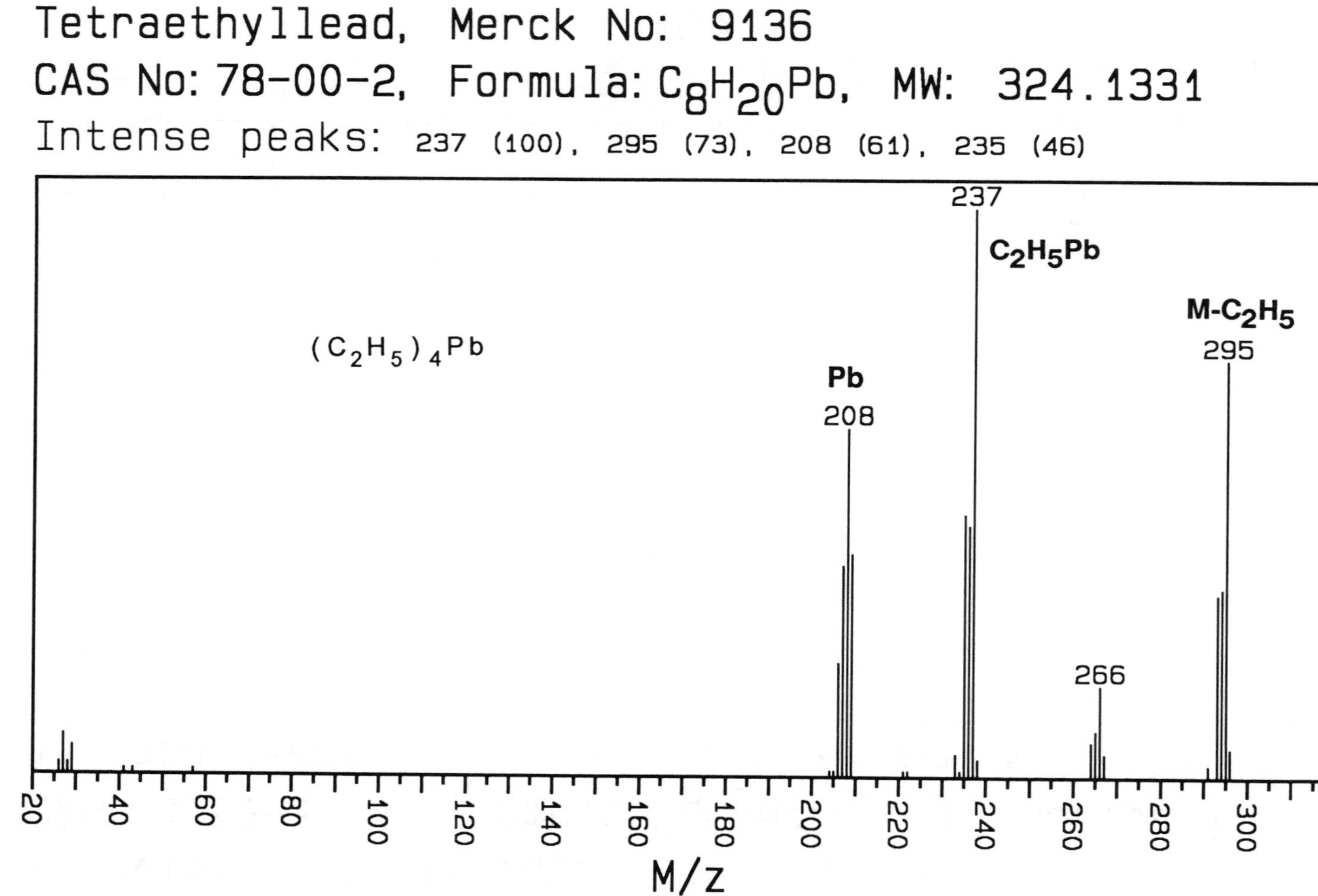
Tetraethyllead, Merck No: 9136
CAS No: 78-00-2, Formula: $C_8H_{20}Pb$, MW: 324.1331
Intense peaks: 237 (100), 295 (73), 208 (61), 235 (46)
$(C_2H_5)_4Pb$
Pb
208
237
$C_2H_5Pb$
M-$C_2H_5$
295
266
20
40
60
80
100
120
140
160
180
200
220
240
260
280
300
320
M/Z

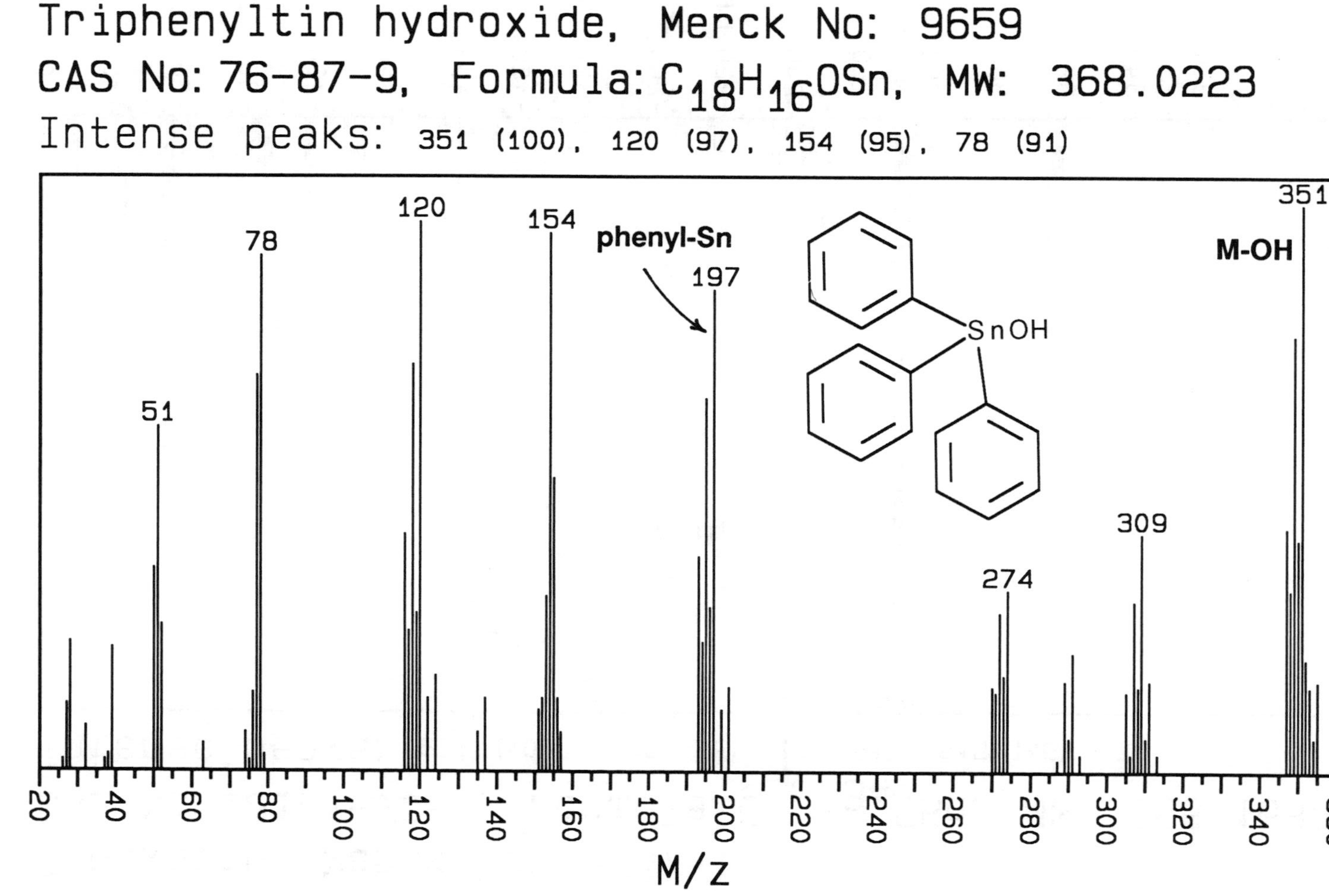
Triphenyltin hydroxide, Merck No: 9659
CAS No: 76-87-9, Formula: C18H16OSn, MW: 368.0223
Intense peaks: 351 (100), 120 (97), 154 (95), 78 (91)
351
M-OH
309
274
197
phenyl-Sn
154
120
78
51
SnOH
20
40
60
80
100
120
140
160
180
200
220
240
260
280
300
320
340
360
M/Z

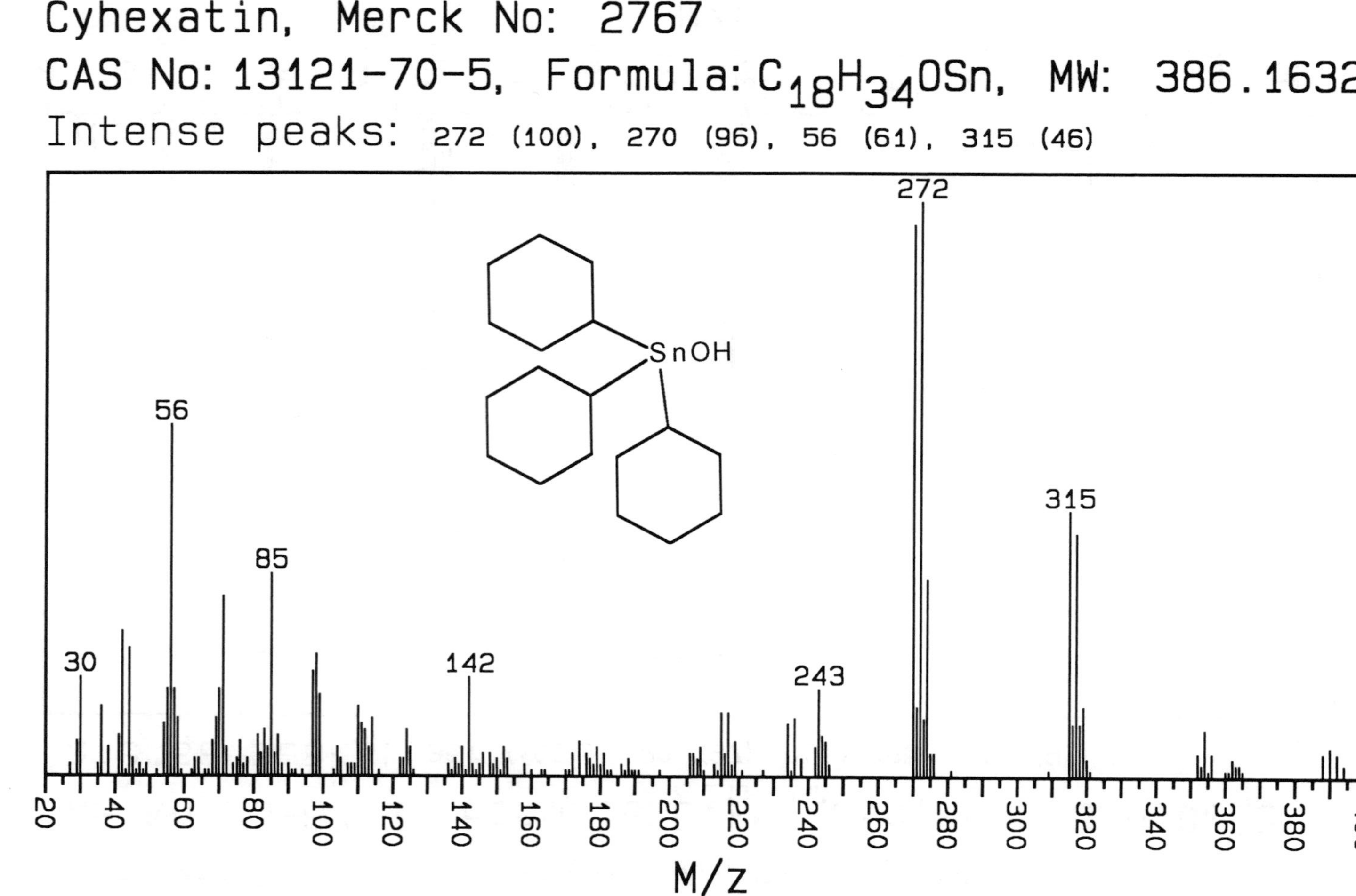
Cyhexatin, Merck No: 2767
CAS No: 13121-70-5, Formula: $C_{18}H_{34}OSn$, MW: 386.1632
Intense peaks: 272 (100), 270 (96), 56 (61), 315 (46)
SnOH
272
56
315
85
30
142
243
20
40
60
80
100
120
140
160
180
200
220
240
260
280
300
320
340
360
380
400
M/Z

Ziram, Merck No: 10075

CAS No: 137-30-4, Formula: $C_6H_{12}N_2S_4Zn$, MW: 303.9175

Intense peaks: 88 (100), 44 (28), 58 (25), 43 (22)

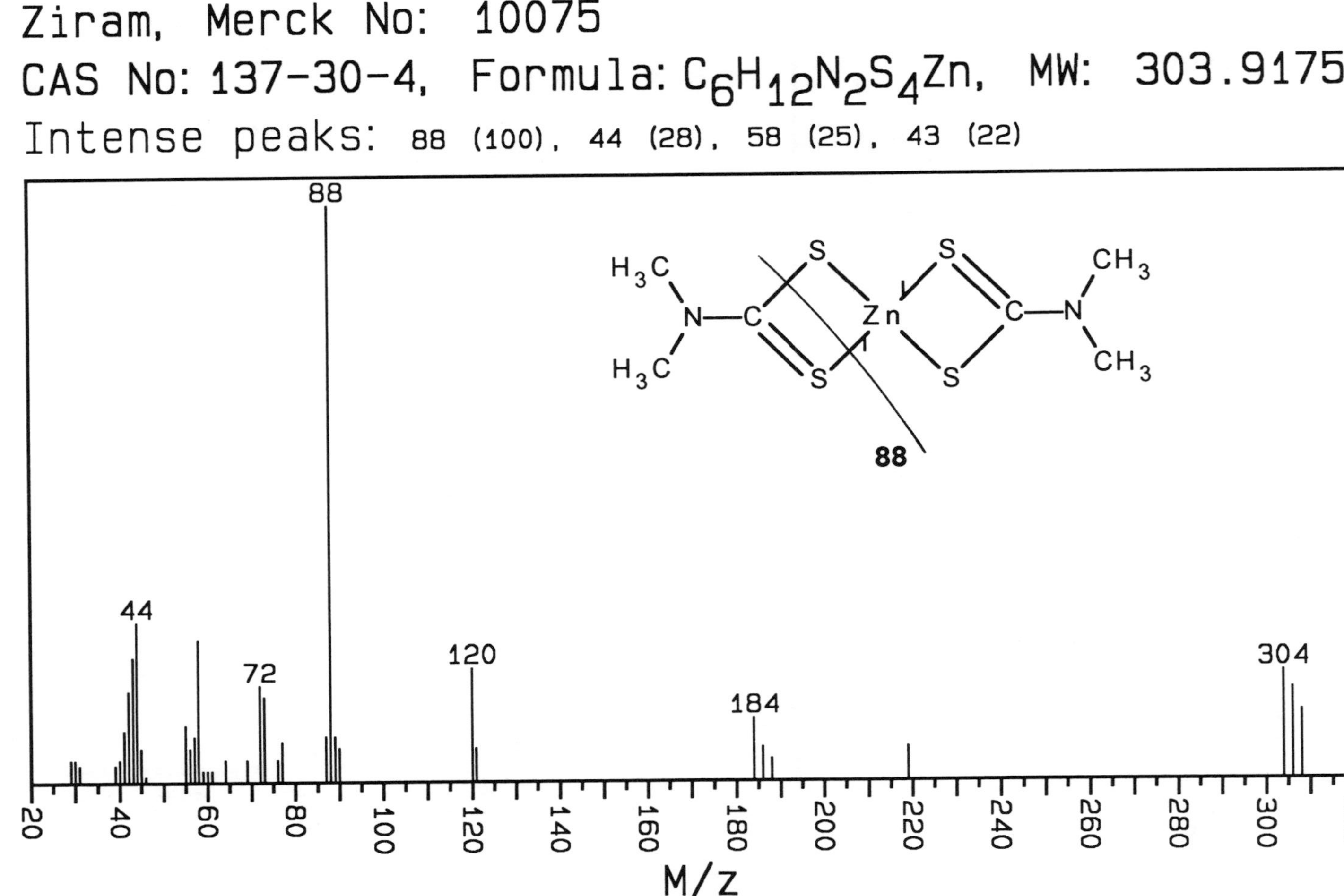

# COMMON NAME INDEX


| Common name | CAS no. | Merck no. | Page |
|---|---|---|---|
| 245-T | 93-76-5 | 8999 | 284 |
| 24-D | 94-75-7 | 2802 | 281 |
| 24-DB | 94-82-6 | 2828 | 283 |
| Acenaphthene | 83-32-9 | 23 | 158 |
| Acenaphthylene | 208-96-8 | — | 157 |
| Acephate | 30560-19-1 | 26 | 487 |
| Acetamide, 2-fluoro- | 640-19-7 | 4095 | 76 |
| Acetic acid, phenoxy- | 122-59-8 | 7223 | 278 |
| Acetic acid, trichloro- | 76-03-9 | 9539 | 50 |
| Acetone cyanohydrin | 75-86-5 | 59 | 81 |
| Acetonitrile, trichloro- | 545-06-2 | 9540 | 82 |
| Acetylacetone | 123-54-6 | 75 | 46 |
| Acridine | 260-94-6 | 117 | 163 |
| Acrolein | 107-02-8 | 122 | 47 |
| Acrylamide | 79-06-1 | 123 | 75 |
| Acrylonitrile | 107-13-1 | 125 | 80 |
| Adamsite | 578-94-9 | 7170 | 524 |
| Alachlor | 15972-60-8 | 193 | 360 |
| Aldicarb | 116-06-3 | 216 | 325 |
| Aldrin | 309-00-2 | 219 | 119 |
| Allidochlor | 93-71-0 | 250 | 357 |
| Allyl alcohol | 107-18-6 | 284 | 32 |
| Ametryn | 834-12-8 | 402 | 393 |
| Aminoazobenzene, 4- | 60-09-3 | 430 | 295 |
| Aminoazobenzene, N,N-dimethyl- | 60-11-7 | 3218 | 297 |
| Aminoazotoluene, 4- | 97-56-3 | — | 296 |
| Aminocarb | 2032-59-9 | 443 | 338 |
| Amitraz | 33089-61-1 | 503 | 354 |
| Amitrole | 61-82-5 | 506 | 108 |
| Amyl ether, normal- | 693-65-2 | 646 | 44 |
| Anabasine | 494-52-0 | 655 | 307 |
| Ancymidol | 12771-68-5 | 665 | 207 |
| Anilazine | 101-05-3 | 685 | 397 |
| Aniline | 62-53-3 | 687 | 289 |
| Aniline, 4-fluoro- | 371-40-4 | 4098 | 290 |
| Anthanthrene | 191-26-4 | — | 183 |
| Anthracene | 120-12-7 | 712 | 162 |
| Anthraquinone | 84-65-1 | 717 | 166 |
| ANTU | 86-88-4 | 755 | 324 |
| Aramite* | 140-57-8 | 794 | 426 |
| Arsine, oxophenyl- | 637-03-6 | 6903 | 523 |
| Arsinic acid, dimethyl- | 75-60-5 | 1603 | 522 |
| Atrazine | 1912-24-9 | 886 | 388 |
| Azinphos-methyl | 86-50-0 | 926 | 493 |
| Azobenzene | 103-33-3 | 930 | 294 |
| Azulene | 275-51-4 | 939 | 145 |
| Barban | 101-27-9 | 969 | 331 |
| Bendiocarb | 22781-23-3 | 1044 | 343 |
| Benfluralin | 1861-40-1 | 1048 | 257 |

| Common name | CAS no. | Merck no. | Page |
|---|---|---|---|
| Benomyl | 17804-35-2 | 1053 | 334 |
| Bentazon | 25057-89-0 | 1060 | 433 |
| Benzanthrone | 82-05-3 | 1070 | 172 |
| Benzene | 71-43-2 | 1074 | 123 |
| Benzene, chloro- | 108-90-7 | 2121 | 132 |
| Benzene, dichloro- | 95-50-1 | 3044 | 133 |
| Benzene, ethyl- | 100-41-4 | 3723 | 126 |
| Benzene, hexachloro- | 118-74-1 | 4600 | 135 |
| Benzene, nitro- | 98-95-3 | 6509 | 240 |
| Benzene, trichloro- | 87-61-6 | 9542 | 134 |
| Benzidine | 92-87-5 | 1086 | 300 |
| Benzidine, 3,3'-dichloro- | 91-94-1 | 3047 | 302 |
| Benzoguanamine | 91-76-9 | 1099 | 386 |
| Benzoic acid | 65-85-0 | 1101 | 272 |
| Benzoic acid, 2-chloro- | 118-91-2 | 2125 | 273 |
| Benzonitrile | 100-47-0 | 1107 | 310 |
| Benzophenone | 119-61-9 | 1108 | 190 |
| Benzothiazole, 2-mercapto- | 149-30-4 | 5759 | 513 |
| Benzoyl chloride | 98-88-4 | 1124 | 262 |
| Benzo[a]pyrene | 50-32-8 | 1113 | 180 |
| Benzo[c]phenanthrene | 195-19-7 | 1129 | 178 |
| Benzo[e]pyrene | 192-97-2 | 1114 | 181 |
| Benzo[ghi]fluoranthene | 203-12-3 | — | 173 |
| Benzo[ghi]perylene | 191-24-2 | — | 184 |
| Benzo[j]fluoranthene | 205-82-3 | — | 179 |
| Benzyl benzoate | 120-51-4 | 1141 | 263 |
| Benzyl chloride | 100-44-7 | 1143 | 131 |
| Benz[a]anthracene | 56-55-3 | 1069 | 175 |
| Benz[a]anthracene, 7,12-dimethyl- | 57-97-6 | 3224 | 176 |
| Bifenox | 42576-02-3 | 1228 | 250 |
| Binapacryl | 485-31-4 | 1237 | 260 |
| Biphenyl | 92-52-4 | 3314 | 159 |
| Biphenylamine, 4- | 92-67-1 | 1248 | 299 |
| Biphenyl, 2-nitro- | 86-00-0 | 6513 | 142 |
| Biphenyl, heptachloro- | 28655-71-2 | — | 141 |
| Biphenyl, hexachloro- | 26601-64-9 | — | 140 |
| Biphenyl, pentachloro- | 25429-29-2 | — | 139 |
| Biphenyl, tetrachloro- | 26914-33-0 | — | 138 |
| Biphenyl, trichloro- | 25323-68-6 | 7541 | 137 |
| Bisphenol A | 80-05-7 | 1311 | 204 |
| Bromacil | 314-40-9 | 1370 | 379 |
| Bromoform | 75-25-2 | 1407 | 8 |
| Bromophos | 2104-96-3 | 1419 | 465 |
| Bromopropylate | 18181-80-1 | 1422 | 199 |
| Bromoxynil | 1689-84-5 | 1431 | 313 |
| Bronopol | 52-51-7 | 1437 | 37 |
| Bulan* | 117-26-0 | 1470 | 201 |
| Bupirimate | 41483-43-6 | 1484 | 383 |
| Butachlor | 23184-66-9 | 1498 | 308 |
| Butralin | 33629-47-9 | 1532 | 252 |
| Butyl acetate | 123-86-4 | 1535 | 59 |

| Common name | CAS no. | Merck no. | Page |
|---|---|---|---|
| Crufomate | 299-86-5 | 2607 | 436 |
| Cumene | 98-82-8 | 2619 | 130 |
| Cyanazine | 21725-46-2 | 2692 | 392 |
| Cyanofenphos | 13067-93-1 | 2697 | 484 |
| Cyanuric acid | 108-80-5 | 2704 | 385 |
| Cyclohexane | 110-82-7 | 2729 | 84 |
| Cyclohexanecarboxylic acid | 98-89-5 | 2730 | 87 |
| Cyclohexanol | 108-93-0 | 2731 | 86 |
| Cycloheximide | 66-81-9 | 2734 | 518 |
| Cyclohexylamine | 108-91-8 | 2735 | 88 |
| Cyclopenta[cd]pyrene | 27208-37-3 | — | 174 |
| Cyhexatin | 13121-70-5 | 2767 | 532 |
| Cythioate | 115-93-5 | 2791 | 467 |
| Dactin* | 118-52-5 | 3054 | 105 |
| Dalapon | 75-99-0 | 2806 | 52 |
| Daminozide | 1596-84-5 | 2810 | 353 |
| Dazomet | 533-74-4 | 2827 | 99 |
| DBCP | 96-12-8 | 3003 | 25 |
| DCPA | 1861-32-1 | 2830 | 271 |
| DDD, o,p′- | 53-19-0 | 6134 | 195 |
| DDD, p,p′- | 72-54-8 | 3049 | 194 |
| DDE, p,p′- | 72-55-9 | — | 193 |
| DDT, p,p′- | 50-29-3 | 2832 | 192 |
| Decanol, 1- | 112-30-1 | 2847 | 29 |
| Deet | 134-62-3 | 2848 | 369 |
| Deguelin | 522-17-8 | 2854 | 521 |
| Dehydroacetic acid | 520-45-6 | 2855 | 90 |
| Dialifor | 10311-84-9 | 2949 | 503 |
| Diallate | 2303-16-4 | 2950 | 348 |
| Diaminodiphenylmethane, 4,4′- | 101-77-9 | 2958 | 298 |
| Diazinon* | 333-41-5 | 2978 | 474 |
| Dibenzofuran, octachloro- | 39001-02-0 | — | 224 |
| Dibenz[a,h]anthracene | 53-70-3 | 2989 | 185 |
| Dicamba | 1918-00-9 | 3026 | 276 |
| Dicapthon | 2463-84-5 | 3027 | 463 |
| Dichlobenil | 1194-65-6 | 3029 | 311 |
| Dichlofenthion | 97-17-6 | 3030 | 478 |
| Dichlofluanid | 1085-98-9 | 3031 | 424 |
| Dichlone | 117-80-6 | 3032 | 188 |
| Dichloroethyl ether, sym- | 111-44-4 | 3055 | 40 |
| Dichloromethyl ether, sym- | 542-88-1 | 3058 | 39 |
| Dichlorophene | 97-23-4 | 3059 | 205 |
| Dichlorprop | 120-36-5 | 3068 | 282 |
| Dichlorvos | 62-73-7 | 3069 | 440 |
| Dicofol | 115-32-2 | 3075 | 196 |
| Dicrotophos | 141-66-2 | 3077 | 444 |
| Dicryl | 2164-09-2 | 3078 | 365 |
| Dieldrin | 60-57-1 | 3093 | 121 |
| Diethanolamine | 111-42-2 | 3097 | 71 |
| Diethylamine | 109-89-7 | 3100 | 68 |
| Diflubenzuron | 35367-38-5 | 3128 | 323 |

| Common name | CAS no. | Merck no. | Page |
|---|---|---|---|
| Dimefox | 115-26-4 | 3191 | 435 |
| Dimethirimol | 5221-53-4 | 3206 | 381 |
| Dimethoate | 60-51-5 | 3209 | 489 |
| Dimethoxane | 828-00-2 | 3212 | 92 |
| Dimetilan | 644-64-4 | 3253 | 328 |
| Dinobuton | 973-21-7 | 3280 | 261 |
| Dinoseb | 88-85-7 | 3282 | 246 |
| Di-octyl adipate | 103-23-1 | — | 66 |
| Dioxathion | 78-34-2 | 3296 | 504 |
| Dioxin, 1,2,3,4-tetrachloro- | 30746-58-8 | — | 221 |
| Dioxin, 2,3,7,8-tetrachloro- | 1746-01-6 | 9052 | 222 |
| Dioxin, octachloro- | 3268-87-9 | — | 223 |
| Diphenadione | 82-66-6 | 3304 | 510 |
| Diphenamid | 957-51-7 | 3305 | 372 |
| Diphenyl sulfone | 127-63-9 | 3336 | 427 |
| Diphenylacetic acid | 117-34-0 | 3316 | 191 |
| Diphenylamine | 122-39-4 | 3317 | 293 |
| Dipropetryn | 4147-51-7 | 3349 | 396 |
| Disulfiram | 97-77-8 | 3370 | 422 |
| Disulfoton | 298-04-4 | 3371 | 497 |
| Dithianone | 3347-22-6 | 3375 | 419 |
| Diuron | 330-54-1 | 3388 | 318 |
| Dixanthogen | 502-55-6 | 3390 | 421 |
| DMPA | 299-85-4 | 3394 | 488 |
| Dodemorph | 1593-77-7 | 3405 | 96 |
| Drazoxolon | 5707-69-7 | 3432 | 508 |
| Duraset* | 85-72-3 | 6027 | 375 |
| Endosulfan | 115-29-7 | 3529 | 115 |
| Endothall | 145-73-3 | 3530 | 93 |
| Endrin | 72-20-8 | 3533 | 122 |
| Epichlorohydrin | 106-89-8 | 3563 | 38 |
| EPN | 2104-64-5 | 3576 | 485 |
| EPTC | 759-94-4 | 3580 | 345 |
| Erbon | 136-25-4 | 3587 | 286 |
| Ethalfluralin | 55283-68-6 | 3671 | 258 |
| Ethanethiol | 75-08-1 | 3680 | 28 |
| Ethane, 1,1-dichloro- | 75-34-3 | 3766 | 12 |
| Ethane, 1,1,1-trichloro- | 71-55-6 | 9549 | 15 |
| Ethane, 1,1,2-trichloro- | 79-00-5 | 9550 | 16 |
| Ethane, 1,1,2,2-tetrachloro- | 79-34-5 | 9125 | 17 |
| Ethane, hexachloro- | 67-72-1 | 4601 | 19 |
| Ethane, pentachloro- | 76-01-7 | 7058 | 18 |
| Ethanol, 2-phenoxy- | 122-99-6 | 7226 | 238 |
| Ethanol, 2-(ethylthio)- | 110-77-0 | 3813 | 33 |
| Ethene, 1,1-dichloro- | 75-35-4 | 9900 | 21 |
| Ethephon | 16672-87-0 | 3686 | 437 |
| Ethion | 563-12-2 | 3691 | 500 |
| Ethirimol | 23947-60-6 | 3695 | 382 |
| Ethofumesate | 26225-79-6 | 3697 | 511 |
| Ethohexadiol | 94-96-2 | 3699 | 35 |
| Ethoprop | 13194-48-4 | 3702 | 506 |

| Common name | CAS no. | Merck no. | Page |
|---|---|---|---|
| Ethoxyquin | 91-53-2 | 3710 | 509 |
| Ethylene dibromide | 106-93-4 | 3753 | 14 |
| Ethylene dichloride | 107-06-2 | 3754 | 13 |
| Ethylene, tetrachloro- | 127-18-4 | 9126 | 23 |
| Ethylene thiourea | 96-45-7 | 3759 | 104 |
| Ethylene, trichloro- | 79-01-6 | 9552 | 22 |
| Ethylidene diacetate | 542-10-9 | 3767 | 62 |
| Ethylmercuric chloride | 107-27-7 | 3780 | 526 |
| Eugenol | 97-53-0 | 3855 | 235 |
| Famphur | 52-85-7 | 3882 | 468 |
| Fenitrothion | 122-14-5 | 3922 | 460 |
| Fensulfothion | 115-90-2 | 3943 | 477 |
| Fenthion | 55-38-9 | 3945 | 461 |
| Fenuron | 101-42-8 | 3951 | 314 |
| Ferbam | 14484-64-1 | 3954 | 525 |
| Fluchloralin | 33245-39-5 | 4052 | 259 |
| Fluometuron | 2164-17-2 | 4080 | 317 |
| Fluoranthene | 206-44-0 | — | 170 |
| Fluorene, 9H- | 86-73-7 | 4081 | 160 |
| Fluoridamid | 47000-92-0 | 4089 | 361 |
| Folpet | 133-07-3 | 4142 | 412 |
| Fonofos | 944-22-9 | 4147 | 505 |
| Formothion | 2540-82-1 | 4160 | 490 |
| Fumaric acid | 110-17-8 | 4200 | 57 |
| Furfural | 98-01-1 | 4214 | 100 |
| Genite* | 97-16-5 | 4283 | 432 |
| Geraniol | 106-24-1 | 4298 | 31 |
| Glyphosate | 1071-83-6 | 4408 | 438 |
| Halazone | 80-13-7 | 4503 | 274 |
| Hempa | 680-31-9 | 4568 | 434 |
| Heptachlor | 76-44-8 | 4576 | 113 |
| Heptachlor epoxide | 1024-57-3 | — | 114 |
| Hexachlorocyclopentadiene | 77-47-4 | — | 109 |
| Hexachlorophene | 70-30-4 | 4602 | 206 |
| Hexachloro-p-xylene | 68-36-0 | 1313 | 136 |
| Imidazolidinone, 2- | 120-93-4 | 4830 | 103 |
| Indoleacetic acid | 87-51-4 | 4870 | 288 |
| Indole, 1H- | 120-72-9 | 4869 | 144 |
| Iodofenphos | 18181-70-9 | 4925 | 466 |
| Iprodione | 36734-19-7 | 4964 | 401 |
| Iridomyrmecin | 485-43-8 | 4972 | 517 |
| Isoamyl acetate | 123-92-2 | 4993 | 61 |
| Isobenzan | 297-78-9 | 5010 | 111 |
| Isobornyl thiocyanoacetate | 115-31-1 | 5012 | 507 |
| Isobutyl acetate | 110-19-0 | 5014 | 60 |
| Isobutyric acid | 79-31-2 | 5039 | 54 |
| Isocil | 314-42-1 | 5042 | 378 |
| Isodrin | 465-73-6 | 219 | 120 |
| Isolan* | 119-38-0 | 5063 | 326 |
| Isophorone | 78-59-1 | — | 89 |
| Isoprene | 78-79-5 | 5087 | 1 |

| Common name | CAS no. | Merck no. | Page |
|---|---|---|---|
| Naphthalene, 1-bromo- | 90-11-9 | 1413 | 148 |
| Naphthalene, chloro- | 90-13-1 | 2149 | 147 |
| Naphthalene, octachloro- | 2234-13-1 | — | 151 |
| Naphthalene, tetrachloro- | 53555-64-9 | — | 150 |
| Naphthalene, trichloro- | 55720-37-1 | — | 149 |
| Naphthylamine, 1- | 134-32-7 | 6318 | 291 |
| Naphthylamine, 2- | 91-59-8 | 6319 | 292 |
| Napropamide | 15299-99-7 | 6336 | 363 |
| Naptalam | 132-66-1 | 6338 | 374 |
| Neburon | 555-37-3 | 6354 | 321 |
| Niclosamide | 50-65-7 | 6425 | 376 |
| Nicotine | 54-11-5 | 6434 | 306 |
| Nitralin | 4726-14-1 | 6486 | 254 |
| Nitrapyrin | 1929-82-4 | 6490 | 304 |
| Nitrofen | 1836-75-5 | 6519 | 249 |
| N-Nitrosodiethylamine | 55-18-5 | 6557 | 74 |
| N-Nitrosodimethylamine | 62-75-9 | 6558 | 73 |
| N-Nitrosopyrrolidine | 930-55-2 | 6564 | 102 |
| Norea | 18530-56-8 | 6611 | 316 |
| Norflurazon | 27314-13-2 | 6618 | 398 |
| Nornicotine | 494-97-3 | 6631 | 305 |
| Octanol, 2- | 123-96-6 | 6675 | 30 |
| Octhilinone | 26530-20-1 | 6677 | 106 |
| Oryzalin | 19044-88-3 | 6840 | 255 |
| Ovex | 80-33-1 | 6856 | 431 |
| Oxadiazon | 19666-30-9 | 6860 | 399 |
| Oxamyl | 23135-22-0 | 6873 | 355 |
| Oxythioquinox | 2439-01-2 | 6933 | 417 |
| Paraoxon | 311-45-5 | 6979 | 448 |
| Parathion | 56-38-2 | 6983 | 476 |
| Pebulate | 1114-71-2 | 7007 | 350 |
| Pendimethalin | 40487-42-1 | 7026 | 251 |
| Pentac | 2227-17-0 | 3095 | 116 |
| Perfluidone | 37924-13-3 | 7113 | 429 |
| Permethrin | 52645-53-1 | 7132 | 415 |
| Perthane* | 72-56-0 | 3050 | 202 |
| Perylene | 198-55-0 | 7137 | 182 |
| Phenanthraquinone, 9-10- | 84-11-7 | 7168 | 169 |
| Phenanthrene | 85-01-8 | 7167 | 167 |
| Phenanthridine | 229-87-8 | — | 168 |
| Phenmedipham | 13684-63-4 | 7199 | 332 |
| Phenol | 108-95-2 | 7206 | 208 |
| Phenol, 2-chloro- | 95-57-8 | 2154 | 211 |
| Phenol, 4-chloro-2-methyl- | 59-50-7 | 2133 | 227 |
| Phenol, 2-cyclohexyl-4,6-dinitro- | 131-89-5 | 2739 | 247 |
| Phenol, 2-methyl- | 95-48-7 | 2580 | 225 |
| Phenol, 2-methyl-4,6-dinitro- | 534-52-1 | 3272 | 245 |
| Phenol, 2-nitro- | 88-75-5 | 6541 | 241 |
| Phenol, 2-phenyl- | 90-43-7 | 7276 | 210 |
| Phenol, 2-phenyl-6-chloro- | 85-97-2 | 7251 | 143 |
| Phenol, 2,3,4,6-tetrachloro- | 58-90-2 | — | 219 |

| Common name | CAS no. | Merck no. | Page |
|---|---|---|---|
| Phenol, 2,4-dichloro- | 120-83-2 | 3061 | 216 |
| Phenol, 2,4,5-trichloro- | 95-95-4 | 9555 | 217 |
| Phenol, 2,4,6-trichloro- | 88-06-2 | 9556 | 218 |
| Phenol, 2,6-dichloro- | 87-65-0 | 3062 | 215 |
| Phenol, 2,4-dinitro- | 51-28-5 | 3274 | 244 |
| Phenol, 3-chloro- | 108-43-0 | 2154 | 212 |
| Phenol, 3-methyl- | 108-39-4 | 2579 | 226 |
| Phenol, 4-bromo- | 106-41-2 | 1416 | 214 |
| Phenol, 4-chloro- | 106-48-9 | 2154 | 213 |
| Phenol, 4-methyl- | 106-44-5 | 2581 | 228 |
| Phenol, 4-nitro- | 100-02-7 | 6542 | 242 |
| Phenol, 4-tert-pentyl- | 80-46-6 | 7098 | 233 |
| Phenol, nonyl- | 25154-52-3 | 6599 | 237 |
| Phenol, pentachloro- | 87-86-5 | 7059 | 220 |
| Phenothiazine, 10H- | 92-84-2 | 7220 | 164 |
| Phenyl ether | 101-84-8 | 7259 | 248 |
| Phenylmercuric acetate | 62-38-4 | 7271 | 529 |
| Phenylmercuric chloride | 100-56-1 | 7272 | 527 |
| Phlorol | 90-00-6 | 7302 | 229 |
| Phorate | 298-02-2 | 7305 | 496 |
| Phosalone | 2310-17-0 | 7308 | 502 |
| Phosfolan | 947-02-4 | 7309 | 470 |
| Phosmet | 732-11-6 | 7311 | 495 |
| Phosphamidon | 13171-21-6 | 7312 | 445 |
| Phoxim | 14816-18-3 | 7341 | 472 |
| Phthalate, benzyl butyl | 85-68-7 | — | 267 |
| Phthalate, bis-(2-ethylhexyl) | 117-81-7 | 1262 | 268 |
| Phthalate, diethyl | 84-66-2 | 7345 | 265 |
| Phthalate, dimethyl | 131-11-3 | 3243 | 264 |
| Phthalate, di-n-butyl | 84-74-2 | 1586 | 266 |
| Phthalate, phenyl | 84-62-8 | 7278 | 269 |
| Picloram | 1918-02-1 | 7370 | 303 |
| Pindone | 83-26-1 | 7413 | 411 |
| Pinene, alpha- | 80-56-8 | 7414 | 2 |
| Piperine | 94-62-2 | 7442 | 403 |
| Piperonyl butoxide | 51-03-6 | 7446 | 405 |
| Pirimiphos-ethyl | 23505-41-1 | 7469 | 475 |
| Prolan | 117-27-1 | 6515 | 200 |
| Promecarb | 2631-37-0 | 7794 | 336 |
| Prometon | 1610-18-0 | 7799 | 394 |
| Prometryn | 7287-19-6 | 7800 | 395 |
| Propachlor | 1918-16-7 | 7805 | 358 |
| Propane, 1,2-dichloro- | 78-87-5 | 7867 | 24 |
| Propanil | 709-98-8 | 7814 | 362 |
| Propanoic acid | 79-09-4 | 7837 | 51 |
| Propargite | 2312-35-8 | 7818 | 425 |
| Propazine* | 139-40-2 | 7822 | 390 |
| Propene, 1,3-dichloro- | 542-75-6 | 3064 | 26 |
| Propene, 3-chloro-2-methyl- | 563-47-3 | 2148 | 27 |
| Propham | 122-42-9 | 7828 | 329 |
| Propoxur | 114-26-1 | 7849 | 340 |

| Common name | CAS no. | Merck no. | Page |
|---|---|---|---|
| Propyzamide | 23950-58-5 | 7886 | 370 |
| Pyracarbolid | 24691-76-7 | 7966 | 91 |
| Pyrazophos | 13457-18-6 | 7976 | 480 |
| Pyrene | 129-00-0 | 7977 | 171 |
| Pyridine, 2,6-dimethyl- | 108-48-5 | 5485 | 128 |
| Pyridine, 4-methyl- | 108-89-4 | 7374 | 125 |
| Pyrolan* | 87-47-8 | 8013 | 327 |
| Quinoline | 91-22-5 | 8097 | 152 |
| Quinoline, 2-methyl- | 91-63-4 | 8055 | 153 |
| Quinolinol, 8- | 148-24-3 | 4778 | 154 |
| Quintozene | 82-68-8 | 8108 | 243 |
| Resorcinol | 108-46-3 | 8158 | 209 |
| Ronnel | 299-84-3 | 8239 | 464 |
| Rotenone | 83-79-4 | 8248 | 520 |
| Salicylanilide | 87-17-2 | 8299 | 368 |
| Schradan | 152-16-9 | 8351 | 457 |
| Sesamex | 51-14-9 | 8419 | 404 |
| Siduron | 1982-49-6 | 8433 | 322 |
| Silvex | 93-72-1 | 8483 | 285 |
| Simazine | 122-34-9 | 8485 | 387 |
| Simetryne* | 1014-70-6 | 8488 | 389 |
| Solan | 2307-68-8 | 8659 | 367 |
| Squalene | 111-02-4 | 8727 | 5 |
| Stearic acid | 57-11-4 | 8761 | 55 |
| Stirofos | 22248-79-9 | 8777 | 447 |
| Strychnine | 57-24-9 | 8822 | 519 |
| Styrene | 100-42-5 | 8830 | 129 |
| Sulfallate | 95-06-7 | 8882 | 351 |
| Sulfotep | 3689-24-5 | 8945 | 499 |
| Sulfoxide | 120-62-7 | 8946 | 406 |
| Sulphenone* | 80-00-2 | 8969 | 428 |
| Tebuthiuron | 34014-18-1 | 9053 | 107 |
| Temephos | 3383-96-8 | 9075 | 469 |
| Terbacil | 5902-51-2 | 9085 | 377 |
| Terbufos | 13071-79-9 | 9088 | 498 |
| Tetradifon | 116-29-0 | 9132 | 430 |
| Tetraethyl pyrophosphate | 107-49-3 | 9138 | 456 |
| Tetraethyllead | 78-00-2 | 9136 | 530 |
| Tetralin* | 119-64-2 | 9152 | 156 |
| Tetramethrin | 7696-12-0 | 9154 | 416 |
| Thiabendazole | 148-79-8 | 9217 | 402 |
| Thionazin | 297-97-2 | 9275 | 473 |
| Thiophanate | 23564-06-9 | 9282 | 335 |
| Thioquinox | 93-75-4 | 9288 | 418 |
| Thiram | 137-26-8 | 9304 | 420 |
| Thymol | 89-83-8 | 9333 | 231 |
| Tinuvin P* | 2440-22-4 | 9381 | 512 |
| Tolidine, 2- | 119-93-7 | 9437 | 301 |
| Toluene | 108-88-3 | 9455 | 124 |
| Tranid* | 15271-41-7 | 9488 | 344 |
| Triallate | 2303-17-5 | 9510 | 349 |

# CAS REGISTRY NUMBER INDEX

| CAS no. | Common name | Page |
|---|---|---|
| 108-46-3 | Resorcinol | 209 |
| 108-48-5 | Pyridine, 2,6-dimethyl- | 128 |
| 108-80-5 | Cyanuric acid | 385 |
| 108-88-3 | Toluene | 124 |
| 108-89-4 | Pyridine, 4-methyl- | 125 |
| 108-90-7 | Benzene, chloro- | 132 |
| 108-91-8 | Cyclohexylamine | 88 |
| 108-93-0 | Cyclohexanol | 86 |
| 108-95-2 | Phenol | 208 |
| 109-89-7 | Diethylamine | 68 |
| 110-16-7 | Maleic acid | 56 |
| 110-17-8 | Fumaric acid | 57 |
| 110-19-0 | Isobutyl acetate | 60 |
| 110-75-8 | Vinyl ether, 2-chloroethyl- | 41 |
| 110-77-0 | Ethanol, 2-(ethylthio)- | 33 |
| 110-82-7 | Cyclohexane | 84 |
| 110-91-8 | Morpholine | 94 |
| 111-02-4 | Squalene | 5 |
| 111-42-2 | Diethanolamine | 71 |
| 111-44-4 | Dichloroethyl ether, sym- | 40 |
| 111-76-2 | Butyl cellosolve* | 34 |
| 112-30-1 | Decanol, 1- | 29 |
| 112-34-5 | Butyl carbitol* | 42 |
| 112-56-1 | Lethane* | 43 |
| 112-61-8 | Methyl stearate | 63 |
| 114-26-1 | Propoxur | 340 |
| 115-26-4 | Dimefox | 435 |
| 115-29-7 | Endosulfan | 115 |
| 115-31-1 | Isobornyl thiocyanoacetate | 507 |
| 115-32-2 | Dicofol | 196 |
| 115-86-6 | Triphenyl phosphate | 454 |
| 115-90-2 | Fensulfothion | 477 |
| 115-93-5 | Cythioate | 467 |
| 116-06-3 | Aldicarb | 325 |
| 116-29-0 | Tetradifon | 430 |
| 117-26-0 | Bulan* | 201 |
| 117-27-1 | Prolan | 200 |
| 117-34-0 | Diphenylacetic acid | 191 |
| 117-52-2 | Coumafuryl | 408 |
| 117-80-6 | Dichlone | 188 |
| 117-81-7 | Phthalate, bis-(2-ethylhexyl) | 268 |
| 118-52-5 | Dactin* | 105 |
| 118-74-1 | Benzene, hexachloro- | 135 |
| 118-75-2 | Chloranil | 187 |
| 118-91-2 | Benzoic acid, 2-chloro- | 273 |
| 119-38-0 | Isolan* | 326 |
| 119-61-9 | Benzophenone | 190 |
| 119-64-2 | Tetralin* | 156 |
| 119-65-3 | Isoquinoline | 155 |
| 119-93-7 | Tolidine, 2- | 301 |
| 120-12-7 | Anthracene | 162 |
| 120-36-5 | Dichlorprop | 282 |

# MOLECULAR WEIGHT INDEX


| Mol wt | CAS no. | Common name | Page |
|---|---|---|---|
| 53.0265 | 107-13-1 | Acrylonitrile | 80 |
| 56.0261 | 107-02-8 | Acrolein | 47 |
| 57.0215 | 624-83-9 | Methane, isocyanato- | 78 |
| 58.0419 | 107-18-6 | Allyl alcohol | 32 |
| 59.0735 | 75-50-3 | Trimethylamine | 69 |
| 61.9923 | 75-01-4 | Vinyl chloride | 20 |
| 62.0191 | 75-08-1 | Ethanethiol | 28 |
| 68.0626 | 78-79-5 | Isoprene | 1 |
| 70.0419 | 4170-30-3 | Crotonaldehyde | 48 |
| 71.0371 | 79-06-1 | Acrylamide | 75 |
| 72.9986 | 556-61-6 | Methane, isothiocyanato- | 79 |
| 73.0891 | 109-89-7 | Diethylamine | 68 |
| 73.0891 | 75-64-9 | Butylamine, tert- | 67 |
| 74.0368 | 79-09-4 | Propanoic acid | 51 |
| 74.0481 | 62-75-9 | N-Nitrosodimethylamine | 73 |
| 77.0277 | 640-19-7 | Acetamide, 2-fluoro- | 76 |
| 78.0469 | 71-43-2 | Benzene | 123 |
| 83.9534 | 75-09-2 | Methane, dichloro- | 6 |
| 84.0436 | 61-82-5 | Amitrole | 108 |
| 84.0939 | 110-82-7 | Cyclohexane | 84 |
| 85.0528 | 75-86-5 | Acetone cyanohydrin | 81 |
| 86.0368 | 108-05-4 | Vinyl acetate | 58 |
| 86.0481 | 120-93-4 | Imidazolidinone, 2- | 103 |
| 87.0684 | 110-91-8 | Morpholine | 94 |
| 88.0524 | 79-31-2 | Isobutyric acid | 54 |
| 88.0524 | 107-92-6 | Butyric acid | 53 |
| 89.0477 | 51-79-6 | Urethane | 77 |
| 90.0236 | 563-47-3 | Propene, 3-chloro-2-methyl- | 27 |
| 92.0029 | 106-89-8 | Epichlorohydrin | 38 |
| 92.0029 | 78-95-5 | Chloroacetone | 45 |
| 92.0626 | 108-88-3 | Toluene | 124 |
| 93.0578 | 108-89-4 | Pyridine, 4-methyl- | 125 |
| 93.0578 | 62-53-3 | Aniline | 289 |
| 94.0419 | 108-95-2 | Phenol | 208 |
| 95.9534 | 75-35-4 | Ethene, 1,1-dichloro | 21 |
| 96.0211 | 98-01-1 | Furfural | 100 |
| 97.9691 | 107-06-2 | Ethylene dichloride | 13 |
| 97.9691 | 75-34-3 | Ethane, 1,1-dichloro- | 12 |
| 98.0004 | 108-31-6 | Maleic anhydride | 101 |
| 99.1048 | 108-91-8 | Cyclohexylamine | 88 |
| 100.0524 | 123-54-6 | Acetylacetone | 46 |
| 100.0637 | 930-55-2 | N-Nitrosopyrrolidine | 102 |
| 100.0888 | 108-93-0 | Cyclohexanol | 86 |
| 101.1204 | 121-44-8 | Triethylamine | 70 |
| 102.0252 | 96-45-7 | Ethylene thiourea | 104 |
| 102.0793 | 55-18-5 | N-Nitrosodiethylamine | 74 |
| 103.0422 | 100-47-0 | Benzonitrile | 310 |
| 104.0626 | 100-42-5 | Styrene | 129 |
| 105.0791 | 111-42-2 | Diethanolamine | 71 |

| Mol wt | CAS no. | Common name | Page |
|---|---|---|---|
| 106.0185 | 110-75-8 | Vinyl ether, 2-chloroethyl- | 41 |
| 106.0452 | 110-77-0 | Ethanol, 2-(ethylthio)- | 33 |
| 106.0782 | 1330-20-7 | Xylene | 127 |
| 106.0783 | 100-41-4 | Benzene, ethyl- | 126 |
| 107.0735 | 108-48-5 | Pyridine, 2,6-dimethyl- | 128 |
| 108.0575 | 106-44-5 | Phenol, 4-methyl- | 228 |
| 108.0575 | 95-48-7 | Phenol, 2-methyl- | 225 |
| 108.0575 | 108-39-4 | Phenol, 3-methyl- | 226 |
| 109.9691 | 542-75-6 | Propene, 1,3-dichloro- | 26 |
| 110.0368 | 108-46-3 | Resorcinol | 209 |
| 111.0484 | 371-40-4 | Aniline, 4-fluoro- | 290 |
| 111.9847 | 78-87-5 | Propane, 1,2-dichloro- | 24 |
| 112.0081 | 108-90-7 | Benzene, chloro- | 132 |
| 113.9639 | 542-88-1 | Dichloromethyl ether, sym- | 39 |
| 116.0111 | 110-16-7 | Maleic acid | 56 |
| 116.0111 | 110-17-8 | Fumaric acid | 57 |
| 116,0837 | 110-19-0 | Isobutyl acetate | 60 |
| 116.0837 | 123-86-4 | Butyl acetate | 59 |
| 117.0578 | 120-72-9 | Indole, 1H- | 144 |
| 117.9144 | 67-66-3 | Chloroform | 7 |
| 118.0994 | 111-76-2 | Butyl cellosolve* | 34 |
| 119.9345 | 75-71-8 | Methane, dichlorodifluoro- | 10 |
| 120.0939 | 98-82-8 | Cumene | 130 |
| 122.0368 | 65-85-0 | Benzoic acid | 272 |
| 122.0732 | 90-00-6 | Phlorol | 229 |
| 123.0321 | 98-95-3 | Benzene, nitro- | 240 |
| 126.0236 | 100-44-7 | Benzyl chloride | 131 |
| 128.0029 | 95-57-8 | Phenol, 2-chloro- | 211 |
| 128.0029 | 108-43-0 | Phenol, 3-chloro- | 212 |
| 128.0029 | 106-48-9 | Phenol, 4-chloro- | 213 |
| 128.0626 | 91-20-3 | Naphthalene | 146 |
| 128.0626 | 275-51-4 | Azulene | 145 |
| 128.0837 | 98-89-5 | Cyclohexanecarboxylic acid | 87 |
| 129.0174 | 108-80-5 | Cyanuric acid | 385 |
| 129.0578 | 119-65-3 | Isoquinoline | 155 |
| 129.0578 | 91-22-5 | Quinoline | 152 |
| 129.9144 | 79-01-6 | Ethylene, trichloro- | 22 |
| 130.0994 | 123-92-2 | Isoamyl acetate | 61 |
| 130.1358 | 123-96-6 | Octanol, 2- | 30 |
| 131.9301 | 71-55-6 | Ethane, 1,1,1-trichloro- | 15 |
| 131.9301 | 79-00-5 | Ethane, 1,1,2-trichloro- | 16 |
| 132.0939 | 119-64-2 | Tetralin* | 156 |
| 135.9051 | 75-69-4 | Methane, trichlorofluoro- | 11 |
| 136.1252 | 80-56-8 | Pinene, alpha- | 2 |
| 137.9662 | 75-60-5 | Arsinic acid, dimethyl- | 522 |
| 138.0681 | 122-99-6 | Ethanol, 2-phenoxy- | 238 |
| 138.1045 | 78-59-1 | Isophorone | 89 |
| 139.0269 | 88-75-5 | Phenol, 2-nitro- | 241 |
| 139.0269 | 100-02-7 | Phenol, 4-nitro- | 242 |
| 140.0029 | 98-88-4 | Benzoyl chloride | 262 |
| 141.0013 | 10265-92-6 | Methamidophos | 486 |
| 141.9588 | 75-99-0 | Dalapon | 52 |

| Mol wt | CAS no. | Common name | Page |
|---|---|---|---|
| 141.9952 | 111-44-4 | Dichloroethyl ether, sym- | 40 |
| 142.0185 | 59-50-7 | Phenol, 4-chloro-2-methyl- | 227 |
| 142.9096 | 545-06-2 | Acetonitrile, trichloro- | 82 |
| 143.0735 | 91-63-4 | Quinoline, 2-methyl- | 153 |
| 143.0735 | 134-32-7 | Naphthylamine, 1- | 291 |
| 143.0735 | 91-59-8 | Naphthylamine, 2- | 292 |
| 143.9743 | 16672-87-0 | Ethephon | 437 |
| 145.0528 | 148-24-3 | Quinolinol, 8- | 154 |
| 145.9691 | 95-50-1 | Benzene, dichloro- | 133 |
| 146.0579 | 542-10-9 | Ethylidene diacetate | 62 |
| 146.1307 | 94-96-2 | Ethohexadiol | 35 |
| 148.0001 | 494-97-3 | Nornicotine | 305 |
| 149.1052 | 102-71-6 | Triethanolamine | 72 |
| 150.1045 | 89-83-8 | Thymol | 231 |
| 150.1045 | 499-75-2 | Carvacrol | 230 |
| 151.0481 | 126-11-4 | Tris(hydroxymethyl)-nitromethane | 36 |
| 151.8754 | 56-23-5 | Carbon tetrachloride | 9 |
| 152.0473 | 99-76-3 | Methylparaben | 270 |
| 152.0473 | 122-59-8 | Acetic acid, phenoxy- | 278 |
| 152.0626 | 208-96-8 | Acenaphthylene | 157 |
| 152.1201 | 76-22-2 | Camphor | 3 |
| 153.8754 | 127-18-4 | Ethylene, tetrachloro- | 23 |
| 154.0671 | 115-26-4 | Dimefox | 435 |
| 154.0782 | 92-52-4 | Biphenyl | 159 |
| 154.0783 | 83-32-9 | Acenaphthene | 158 |
| 154.1358 | 106-23-0 | Citronellal | 49 |
| 154.1358 | 106-24-1 | Geraniol | 31 |
| 155.9978 | 118-91-2 | Benzoic acid, 2-chloro- | 273 |
| 158.1671 | 693-65-2 | Amyl ether, normal- | 44 |
| 158.1671 | 112-30-1 | Decanol, 1- | 29 |
| 160.0848 | 1596-84-5 | Daminozide | 353 |
| 161.9042 | 76-03-9 | Acetic acid, trichloro- | 50 |
| 161.9639 | 87-65-0 | Phenol, 2,6-dichloro- | 215 |
| 161.9639 | 120-83-2 | Phenol, 2,4-dichloro- | 216 |
| 162.0236 | 90-13-1 | Naphthalene, chloro- | 147 |
| 162.0285 | 533-74-4 | Dazomet | 99 |
| 162.0463 | 16752-77-5 | Methomyl | 356 |
| 162.1157 | 54-11-5 | Nicotine | 306 |
| 162.1157 | 494-52-0 | Anabasine | 307 |
| 162.1256 | 112-34-5 | Butyl carbitol* | 42 |
| 162.8995 | 76-06-2 | Chloropicrin | 83 |
| 164.0837 | 97-53-0 | Eugenol | 235 |
| 164.0950 | 101-42-8 | Fenuron | 314 |
| 164.1201 | 80-46-6 | Phenol, 4-tert-pentyl- | 233 |
| 165.8911 | 79-34-5 | Ethane, 1,1,2,2-tetrachloro- | 17 |
| 166.0783 | 86-73-7 | Fluorene, 9H- | 160 |
| 166.9863 | 149-30-4 | Benzothiazole, 2-mercapto- | 513 |
| 167.0735 | 86-74-8 | Carbazole, 9H- | 161 |
| 167.9556 | 637-03-6 | Arsine, oxophenyl- | 523 |
| 168.0423 | 520-45-6 | Dehydroacetic acid | 90 |
| 168.1151 | 485-43-8 | Iridomyrmecin | 517 |
| 169.0141 | 1071-83-6 | Glyphosate | 438 |

| Mol wt | CAS no. | Common name | Page |
|---|---|---|---|
| 169.0891 | 122-39-4 | Diphenylamine | 293 |
| 169.0891 | 92-67-1 | Biphenylamine, 4- | 299 |
| 170.0732 | 101-84-8 | Phenyl ether | 248 |
| 170.0732 | 90-43-7 | Phenol, 2-phenyl- | 210 |
| 170.9643 | 1194-65-6 | Dichlobenil | 311 |
| 171.0563 | 535-89-7 | Crimidine | 380 |
| 171.9524 | 106-41-2 | Phenol, 4-bromo- | 214 |
| 173.0607 | 93-71-0 | Allidochlor | 357 |
| 174.0317 | 481-39-0 | Juglone | 189 |
| 174.0892 | 828-00-2 | Dimethoxane | 92 |
| 175.0633 | 87-51-4 | Indoleacetic acid | 288 |
| 178.0783 | 120-12-7 | Anthracene | 162 |
| 178.0783 | 85-01-8 | Phenanthrene | 167 |
| 179.0735 | 229-87-8 | Phenanthridine | 168 |
| 179.0735 | 260-94-6 | Acridine | 163 |
| 179.0946 | 122-42-9 | Propham | 329 |
| 179.1188 | 680-31-9 | Hempa | 434 |
| 179.9301 | 87-61-6 | Benzene, trichloro- | 134 |
| 180.1151 | 25013-16-5 | Butylated hydroxy anisole | 236 |
| 182.0708 | 78-40-0 | Triethyl phosphate | 451 |
| 182.0732 | 119-61-9 | Benzophenone | 190 |
| 182.0844 | 103-33-3 | Azobenzene | 294 |
| 183.0119 | 30560-19-1 | Acephate | 487 |
| 184.0121 | 51-28-5 | Phenol, 2,4-dinitro- | 244 |
| 184.0655 | 89-68-9 | Chlorothymol | 232 |
| 184.1001 | 92-87-5 | Benzidine | 300 |
| 185.8681 | 106-93-4 | Ethylene dibromide | 14 |
| 186.0528 | 145-73-3 | Endothall | 93 |
| 186.0681 | 86-87-3 | Naphthaleneacetic acid, 1- | 287 |
| 187.0858 | 91-76-9 | Benzoguanamine | 386 |
| 189.1187 | 759-94-4 | EPTC | 345 |
| 190.0776 | 116-06-3 | Aldicarb | 325 |
| 191.0582 | 555-59-9 | Maleanilic acid | 373 |
| 191.0695 | 10605-21-7 | Carbendazim | 333 |
| 191.1311 | 134-62-3 | Deet | 369 |
| 194.0579 | 131-11-3 | Phthalate, dimethyl | 264 |
| 194.0804 | 58-08-2 | Caffeine | 515 |
| 195.9250 | 95-95-4 | Phenol, 2,4,5-trichloro- | 217 |
| 195.9251 | 88-06-2 | Phenol, 2,4,6-trichloro- | 218 |
| 195.9806 | 118-52-5 | Dactin* | 105 |
| 196.0524 | 90-47-1 | Xanthone | 165 |
| 196.0767 | 6164-98-3 | Chlordimeform | 309 |
| 197.0953 | 60-09-3 | Aminoazobenzene, 4- | 295 |
| 198.0277 | 534-52-1 | Phenol, 2-methyl-4,6-dinitro- | 245 |
| 198.0561 | 150-68-5 | Monuron | 315 |
| 198.1157 | 101-77-9 | Diaminodiphenylmethane, 4,4′- | 298 |
| 198.9481 | 52-51-7 | Bronopol | 37 |
| 199.0456 | 92-84-2 | Phenothiazine, 10H- | 164 |
| 199.0633 | 86-00-0 | Biphenyl, 2-nitro- | 142 |
| 199.8521 | 76-01-7 | Ethane, pentachloro- | 18 |
| 200.0241 | 94-74-6 | Methoxone* | 279 |
| 201.0361 | 148-79-8 | Thiabendazole | 402 |

| Mol wt | CAS no. | Common name | Page |
|---|---|---|---|
| 201.0781 | 122-34-9 | Simazine | 387 |
| 201.0791 | 63-25-2 | Carbaryl | 341 |
| 202.0565 | 86-88-4 | ANTU | 324 |
| 202.0783 | 206-44-0 | Fluoranthene | 170 |
| 202.0783 | 129-00-0 | Pyrene | 171 |
| 203.0981 | 112-56-1 | Lethane* | 43 |
| 203.1311 | 483-63-6 | Crotamiton | 364 |
| 203.1344 | 1114-71-2 | Pebulate | 350 |
| 203.1344 | 1929-77-7 | Vernolate | 347 |
| 204.0342 | 85-97-2 | Phenol, 2-phenyl-6-chloro- | 143 |
| 204.9697 | 133-90-4 | Chloramben | 275 |
| 205.9731 | 90-11-9 | Naphthalene, 1-bromo- | 148 |
| 207.1259 | 2631-37-0 | Promecarb | 336 |
| 208.0524 | 84-65-1 | Anthraquinone | 166 |
| 208.0524 | 84-11-7 | Phenanthraquinone, 9-10- | 169 |
| 208.1212 | 2032-59-9 | Aminocarb | 338 |
| 209.1052 | 114-26-1 | Propoxur | 340 |
| 209.1528 | 5221-53-4 | Dimethirimol | 381 |
| 209.1528 | 23947-60-6 | Ethirimol | 382 |
| 211.0764 | 1918-16-7 | Propachlor | 358 |
| 211.1321 | 119-38-0 | Isolan* | 326 |
| 212.0837 | 117-34-0 | Diphenylacetic acid | 191 |
| 212.0837 | 120-51-4 | Benzyl benzoate | 263 |
| 212.1313 | 119-93-7 | Tolidine, 2- | 301 |
| 213.0557 | 101-21-3 | Chlorpropham | 330 |
| 213.0791 | 87-17-2 | Salicylanilide | 368 |
| 213.1048 | 1014-70-6 | Simetryne* | 389 |
| 213.1187 | 26530-20-1 | Octhilinone | 106 |
| 214.0397 | 7085-19-0 | Mecoprop | 280 |
| 214.0888 | 21087-64-9 | Metribuzin | 384 |
| 215.0807 | 525-79-1 | Kinetin | 516 |
| 215.0938 | 1912-24-9 | Atrazine | 388 |
| 216.0666 | 5902-51-2 | Terbacil | 377 |
| 217.0061 | 709-98-8 | Propanil | 362 |
| 217.1103 | 24691-76-7 | Pyracarbolid | 91 |
| 217.1467 | 91-53-2 | Ethoxyquin | 509 |
| 217.1501 | 2008-41-5 | Butylate | 346 |
| 218.0402 | 127-63-9 | Diphenyl sulfone | 427 |
| 219.0678 | 23135-22-0 | Oxamyl | 355 |
| 219.9459 | 62-73-7 | Dichlorvos | 440 |
| 219.9694 | 1918-00-9 | Dicamba | 276 |
| 219.9694 | 94-75-7 | 24-D | 281 |
| 220.1827 | 25154-52-3 | Phenol, nonyl- | 237 |
| 220.1827 | 128-37-0 | Butylated hydroxy toluene | 234 |
| 221.0623 | 18691-97-9 | Methabenzthiazuron | 514 |
| 221.1052 | 1563-66-2 | Carbofuran | 342 |
| 222.0892 | 84-66-2 | Phthalate, diethyl | 265 |
| 222.1368 | 315-18-4 | Mexacarbate | 337 |
| 222.1732 | 18530-56-8 | Norea | 316 |
| 223.0256 | 95-06-7 | Sulfallate | 351 |
| 223.0611 | 6923-22-4 | Monocrotophos | 442 |
| 223.0845 | 22781-23-3 | Bendiocarb | 343 |

| Mol wt | CAS no. | Common name | Page |
|---|---|---|---|
| 224.0451 | 7786-34-7 | Mevinphos | 443 |
| 225.0824 | 2032-65-7 | Methiocarb | 339 |
| 225.0902 | 2440-22-4 | Tinuvin P* | 512 |
| 225.1266 | 60-11-7 | Aminoazobenzene, N,N-dimethyl- | 297 |
| 225.1266 | 97-56-3 | Aminoazotoluene, 4- | 296 |
| 225.1591 | 1610-18-0 | Prometon | 394 |
| 225.9588 | 117-80-6 | Dichlone | 188 |
| 226.0783 | 27208-37-3 | Cyclopenta[cd]pyrene | 174 |
| 226.0783 | 203-12-3 | Benzo[ghi]fluoranthene | 173 |
| 227.1205 | 834-12-8 | Ametryn | 393 |
| 228.0939 | 218-01-9 | Chrysene | 177 |
| 228.0939 | 56-55-3 | Benz[a]anthracene | 175 |
| 228.0939 | 195-19-7 | Benzo[c]phenanthrene | 178 |
| 228.1045 | 34014-18-1 | Tebuthiuron | 107 |
| 228.1151 | 80-05-7 | Bisphenol A | 204 |
| 228.9021 | 1929-82-4 | Nitrapyrin | 304 |
| 228.9996 | 60-51-5 | Dimethoate | 489 |
| 229.0061 | 2164-09-2 | Dicryl | 365 |
| 229.1094 | 1912-26-1 | Trietazine | 391 |
| 229.1094 | 139-40-2 | Propazine* | 390 |
| 229.8861 | 58-90-2 | Phenol, 2,3,4,6-tetrachloro- | 219 |
| 229.9457 | 55720-37-1 | Naphthalene, trichloro- | 149 |
| 230.0201 | 8022-00-2 | Methyl demeton | 458 |
| 230.0732 | 82-05-3 | Benzanthrone | 172 |
| 230.0943 | 83-26-1 | Pindone | 411 |
| 232.0171 | 330-54-1 | Diuron | 318 |
| 232.0824 | 2164-17-2 | Fluometuron | 317 |
| 232.1576 | 1982-49-6 | Siduron | 322 |
| 233.8131 | 67-72-1 | Ethane, hexachloro- | 19 |
| 233.8447 | 96-12-8 | DBCP | 25 |
| 233.9851 | 120-36-5 | Dichlorprop | 282 |
| 233.9922 | 2439-01-2 | Oxythioquinox | 417 |
| 234.1368 | 2164-08-1 | Lenacil | 98 |
| 235.0667 | 5234-68-4 | Carboxin | 371 |
| 235.9537 | 93-75-4 | Thioquinox | 418 |
| 237.0305 | 5707-69-7 | Drazoxolon | 508 |
| 237.0766 | 141-66-2 | Dicrotophos | 444 |
| 237.9355 | 85-34-7 | Chlorfenac | 277 |
| 239.1077 | 2307-68-8 | Solan | 367 |
| 239.1311 | 957-51-7 | Diphenamid | 372 |
| 239.9261 | 1918-02-1 | Picloram | 303 |
| 239.9883 | 137-26-8 | Thiram | 420 |
| 240.0569 | 25057-89-0 | Bentazon | 433 |
| 240.0746 | 88-85-7 | Dinoseb | 246 |
| 240.0891 | 21725-46-2 | Cyanazine | 392 |
| 240.1222 | 644-64-4 | Dimetilan | 328 |
| 241.0618 | 15271-41-7 | Tranid* | 344 |
| 241.1361 | 7287-19-6 | Prometryn | 395 |
| 241.9564 | 502-55-6 | Dixanthogen | 421 |
| 242.0564 | 13194-48-4 | Ethoprop | 506 |
| 243.8653 | 118-75-2 | Chloranil | 187 |
| 245.1164 | 87-47-8 | Pyrolan* | 327 |

| Mol wt | CAS no. | Common name | Page |
|---|---|---|---|
| 246.0004 | 314-42-1 | Isocil | 378 |
| 246.0302 | 944-22-9 | Fonofos | 505 |
| 248.0007 | 94-82-6 | 24-DB | 283 |
| 248.0119 | 330-55-2 | Linuron | 319 |
| 248.0385 | 297-97-2 | Thionazin | 473 |
| 249.7629 | 75-25-2 | Bromoform | 8 |
| 252.0012 | 80-00-2 | Sulphenone* | 428 |
| 252.0221 | 91-94-1 | Benzidine, 3,3'-dichloro- | 302 |
| 252.0939 | 50-32-8 | Benzo[a]pyrene | 180 |
| 252.0939 | 192-97-2 | Benzo[e]pyrene | 181 |
| 252.0939 | 205-82-3 | Benzo[j]fluoranthene | 179 |
| 252.0939 | 198-55-0 | Perylene | 182 |
| 253.1137 | 115-31-1 | Isobornyl thiocyanoacetate | 507 |
| 253.9304 | 93-76-5 | 245-T | 284 |
| 255.0153 | 947-02-4 | Phosfolan | 470 |
| 255.0218 | 23950-58-5 | Propyzamide | 370 |
| 255.0895 | 85-72-3 | Duraset* | 375 |
| 255.1518 | 4147-51-7 | Dipropetryn | 396 |
| 255.9226 | 52-68-6 | Trichlorfon | 439 |
| 255.9613 | 25323-68-6 | Biphenyl, trichloro- | 137 |
| 256.1212 | 12771-68-5 | Ancymidol | 207 |
| 256.1252 | 57-97-6 | Benz[a]anthracene, 7,12-dimethyl- | 176 |
| 256.9945 | 2540-82-1 | Formothion | 490 |
| 257.0011 | 101-27-9 | Barban | 331 |
| 257.0819 | 51-68-3 | Meclofenoxate | 239 |
| 258.0004 | 3060-89-7 | Metobromuron | 320 |
| 259.0531 | 2533-89-3 | Karsil | 366 |
| 259.9756 | 20354-26-1 | Methazole | 400 |
| 260.0128 | 298-02-2 | Phorate | 496 |
| 260.0161 | 314-40-9 | Bromacil | 379 |
| 263.0017 | 298-00-0 | Methyl parathion | 459 |
| 263.8471 | 87-86-5 | Phenol, pentachloro- | 220 |
| 263.8816 | 1897-45-6 | Chlorothalonil | 312 |
| 263.9067 | 53555-64-9 | Naphthalene, tetrachloro- | 150 |
| 265.9786 | 107-27-7 | Ethylmercuric chloride | 526 |
| 266.0265 | 80-06-8 | Chlorfenethol | 197 |
| 266.0903 | 131-89-5 | Phenol, 2-cyclohexyl-4,6-dinitro- | 247 |
| 266.1647 | 126-73-8 | Tributyl phosphate | 452 |
| 267.9461 | 93-72-1 | Silvex | 285 |
| 267.9881 | 103-17-3 | Chlorbenside | 423 |
| 268.0058 | 97-23-4 | Dichlorophene | 205 |
| 268.9316 | 80-13-7 | Halazone | 274 |
| 269.0309 | 950-10-7 | Mephosfolan | 471 |
| 269.0408 | 2303-16-4 | Diallate | 348 |
| 269.1183 | 15972-60-8 | Alachlor | 360 |
| 269.8131 | 77-47-4 | Hexachlorocyclopentadiene | 109 |
| 271.1572 | 15299-99-7 | Napropamide | 363 |
| 273.9581 | 101-05-3 | Anilazine | 397 |
| 274.0285 | 298-04-4 | Disulfoton | 497 |
| 274.0641 | 555-37-3 | Neburon | 321 |
| 274.8582 | 1689-84-5 | Bromoxynil | 313 |
| 275.0559 | 311-45-5 | Paraoxon | 448 |

| Mol wt | CAS no. | Common name | Page |
|---|---|---|---|
| 276.0939 | 191-24-2 | Benzo[ghi]perylene | 184 |
| 276.0939 | 191-26-4 | Anthanthrene | 183 |
| 276.9639 | 578-94-9 | Adamsite | 524 |
| 277.0174 | 122-14-5 | Fenitrothion | 460 |
| 278.0201 | 55-38-9 | Fenthion | 461 |
| 278.1096 | 53-70-3 | Dibenz[a,h]anthracene | 185 |
| 278.1518 | 84-74-2 | Phthalate, di-n-butyl | 266 |
| 281.0171 | 78-57-9 | Menazon | 492 |
| 281.1376 | 40487-42-1 | Pendimethalin | 251 |
| 281.1627 | 66-81-9 | Cycloheximide | 518 |
| 281.2719 | 1593-77-7 | Dodemorph | 96 |
| 281.8131 | 118-74-1 | Benzene, hexachloro- | 135 |
| 282.9803 | 1836-75-5 | Nitrofen | 249 |
| 283.1339 | 51218-45-2 | Metolachlor | 359 |
| 284.2715 | 57-11-4 | Stearic acid | 55 |
| 285.1365 | 94-62-2 | Piperine | 403 |
| 286.0875 | 26225-79-6 | Ethofumesate | 511 |
| 286.1324 | 152-16-9 | Schradan | 457 |
| 287.8601 | 58-89-9 | Lindane | 85 |
| 288.0441 | 13071-79-9 | Terbufos | 498 |
| 289.9224 | 26914-33-0 | Biphenyl, tetrachloro- | 138 |
| 290.0684 | 107-49-3 | Tetraethyl pyrophosphate | 456 |
| 290.1379 | 17804-35-2 | Benomyl | 334 |
| 291.0331 | 56-38-2 | Parathion | 476 |
| 291.0791 | 299-86-5 | Crufomate | 436 |
| 291.0895 | 132-66-1 | Naptalam | 374 |
| 292.8372 | 82-68-8 | Quintozene | 243 |
| 293.1892 | 33089-61-1 | Amitraz | 354 |
| 294.9028 | 133-07-3 | Folpet | 412 |
| 295.1532 | 33629-47-9 | Butralin | 252 |
| 295.9714 | 3347-22-6 | Dithianone | 419 |
| 296.0443 | 47000-92-0 | Fluoridamid | 361 |
| 296.0509 | 97-77-8 | Disulfiram | 422 |
| 296.9628 | 500-28-7 | Chlorthion* | 462 |
| 296.9628 | 2463-84-5 | Dicapthon | 463 |
| 296.9895 | 115-93-5 | Cythioate | 467 |
| 297.3032 | 24602-86-6 | Tridemorph | 95 |
| 298.0541 | 14816-18-3 | Phoxim | 472 |
| 298.0841 | 117-52-2 | Coumafuryl | 408 |
| 298.1416 | 51-14-9 | Sesamex | 404 |
| 298.2872 | 112-61-8 | Methyl stearate | 63 |
| 298.9341 | 133-06-2 | Captan | 413 |
| 299.0689 | 13171-21-6 | Phosphamidon | 445 |
| 300.0939 | 191-07-1 | Coronene | 186 |
| 300.1109 | 13684-63-4 | Phenmedipham | 332 |
| 301.9571 | 97-16-5 | Genite* | 432 |
| 301.9571 | 80-33-1 | Ovex | 431 |
| 301.9619 | 950-37-8 | Methidathion | 494 |
| 303.0018 | 2303-17-5 | Triallate | 349 |
| 303.0386 | 27314-13-2 | Norflurazon | 398 |
| 303.0483 | 13067-93-1 | Cyanofenphos | 484 |
| 303.9175 | 137-30-4 | Ziram | 533 |

| Mol wt | CAS no. | Common name | Page |
|---|---|---|---|
| 304.1011 | 333-41-5 | Diazinon* | 474 |
| 306.0942 | 72-56-0 | Perthane* | 202 |
| 308.0306 | 115-90-2 | Fensulfothion | 477 |
| 308.1049 | 81-81-2 | Warfarin | 407 |
| 309.0323 | 117-27-1 | Prolan | 200 |
| 309.1689 | 33820-53-0 | Isopropalin | 253 |
| 309.8444 | 68-36-0 | Hexachloro-p-xylene | 136 |
| 310.0321 | 35367-38-5 | Diflubenzuron | 323 |
| 310.2508 | 40596-69-8 | Methoprene | 65 |
| 311.1652 | 23184-66-9 | Butachlor | 308 |
| 312.1362 | 85-68-7 | Phthalate, benzyl butyl | 267 |
| 312.9861 | 299-85-4 | DMPA | 488 |
| 313.9701 | 97-17-6 | Dichlofenthion | 478 |
| 313.9786 | 100-56-1 | Phenylmercuric chloride | 527 |
| 314.0919 | 7700-17-6 | Crotoxyphos | 446 |
| 315.9381 | 72-55-9 | DDE, p,p′- | 193 |
| 316.1569 | 41483-43-6 | Bupirimate | 383 |
| 316.9945 | 732-11-6 | Phosmet | 495 |
| 317.0058 | 86-50-0 | Azinphos-methyl | 493 |
| 317.9537 | 72-54-8 | DDD, p,p′- | 194 |
| 317.9537 | 53-19-0 | DDD, o,p′- | 195 |
| 318.0892 | 84-62-8 | Phthalate, phenyl | 269 |
| 319.8966 | 1746-01-6 | Dioxin, 2,3,7,8-tetrachloro- | 222 |
| 319.8966 | 30746-58-8 | Dioxin, 1,2,3,4-tetrachloro- | 221 |
| 319.8997 | 299-84-3 | Ronnel | 464 |
| 322.0227 | 3689-24-5 | Sulfotep | 499 |
| 322.1165 | 485-31-4 | Binapacryl | 260 |
| 322.3601 | 27519-02-4 | Muscalure | 4 |
| 323.0381 | 2104-64-5 | EPN | 485 |
| 323.0481 | 117-26-0 | Bulan* | 201 |
| 323.8834 | 25429-29-2 | Biphenyl, pentachloro- | 139 |
| 324.0321 | 510-15-6 | Chlorobenzilate | 198 |
| 324.1331 | 78-00-2 | Tetraethyllead | 530 |
| 324.1759 | 120-62-7 | Sulfoxide | 406 |
| 325.0208 | 52-85-7 | Famphur | 468 |
| 325.9861 | 50-65-7 | Niclosamide | 376 |
| 326.0708 | 115-86-6 | Triphenyl phosphate | 454 |
| 326.1114 | 973-21-7 | Dinobuton | 261 |
| 329.0334 | 36734-19-7 | Iprodione | 401 |
| 329.1781 | 1420-06-0 | Trifenmorph | 97 |
| 329.9021 | 1861-32-1 | DCPA | 271 |
| 329.9735 | 90-03-9 | Mercufenol chloride | 528 |
| 330.0361 | 121-75-5 | Malathion | 491 |
| 331.1784 | 7696-12-0 | Tetramethrin | 416 |
| 331.9623 | 1085-98-9 | Dichlofluanid | 424 |
| 333.0936 | 55283-68-6 | Ethalfluralin | 258 |
| 333.1276 | 23505-41-1 | Pirimiphos-ethyl | 475 |
| 334.1006 | 140-57-8 | Aramite* | 426 |
| 334.1681 | 57-24-9 | Strychnine | 519 |
| 335.1093 | 1582-09-8 | Trifluralin | 256 |
| 335.1093 | 1861-40-1 | Benfluralin | 257 |
| 338.0231 | 62-38-4 | Phenylmercuric acetate | 529 |

| Mol wt | CAS no. | Common name | Page |
|---|---|---|---|
| 405.7978 | 57-74-9 | Chlordane | 112 |
| 407.7771 | 297-78-9 | Isobenzan | 111 |
| 409.8701 | 21609-90-5 | Leptophos | 483 |
| 410.3913 | 111-02-4 | Squalene | 5 |
| 411.8352 | 18181-70-9 | Iodofenphos | 466 |
| 415.9174 | 14484-64-1 | Ferbam | 525 |
| 425.9466 | 18181-80-1 | Bromopropylate | 199 |
| 431.9248 | 26644-46-2 | Triforine | 352 |
| 439.7458 | 39001-02-0 | Dibenzofuran, octachloro- | 224 |
| 455.7407 | 3268-87-9 | Dioxin, octachloro- | 223 |
| 456.0088 | 78-34-2 | Dioxathion | 504 |
| 465.9897 | 3383-96-8 | Temephos | 469 |
| 469.6886 | 2227-17-0 | Pentac | 116 |
| 485.6835 | 143-50-0 | Kepone* | 117 |
| 539.6263 | 2385-85-5 | Mirex | 118 |
| 691.5809 | 126-72-7 | Tris-BP | 453 |

# MASS SPECTRAL PEAKS INDEX

| M1 | I1 | M2 | I2 | M3 | I3 | M4 | I4 | Common name | Page |
|---|---|---|---|---|---|---|---|---|---|
| 26 | 100 | 28 | 53 | 54 | 47 | 98 | 24 | Maleic anhydride | 101 |
| 27 | 100 | 56 | 74 | 28 | 65 | 26 | 54 | Acrolein | 47 |
| 27 | 100 | 63 | 91 | 43 | 65 | 44 | 54 | Vinyl ether, 2-chloroethyl- | 41 |
| 27 | 100 | 49 | 87 | 42 | 80 | 29 | 64 | Chloroacetone | 45 |
| 27 | 100 | 98 | 93 | 45 | 93 | 26 | 77 | Fumaric acid | 57 |
| 27 | 100 | 44 | 89 | 71 | 72 | 55 | 58 | Acrylamide | 75 |
| 27 | 100 | 62 | 77 | 26 | 34 | 64 | 24 | Vinyl chloride | 20 |
| 27 | 100 | 107 | 77 | 109 | 72 | 26 | 24 | Ethylene dibromide | 14 |
| 28 | 100 | 36 | 66 | 62 | 40 | 43 | 36 | Dalapon | 52 |
| 28 | 100 | 162 | 91 | 164 | 57 | 32 | 22 | Dichlorprop | 282 |
| 28 | 100 | 29 | 84 | 74 | 79 | 27 | 62 | Propanoic acid | 51 |
| 29 | 100 | 57 | 95 | 27 | 84 | 41 | 67 | Lethane* | 43 |
| 29 | 100 | 89 | 43 | 27 | 41 | 94 | 30 | Dixanthogen | 421 |
| 29 | 100 | 27 | 84 | 28 | 26 | 26 | 15 | Ethylmercuric chloride | 526 |
| 29 | 100 | 97 | 65 | 125 | 49 | 27 | 44 | Fensulfothion | 477 |
| 29 | 100 | 43 | 66 | 252 | 62 | 41 | 52 | Pendimethalin | 251 |
| 30 | 100 | 74 | 82 | 28 | 77 | 56 | 69 | Diethanolamine | 71 |
| 30 | 100 | 58 | 81 | 44 | 28 | 73 | 18 | Diethylamine | 68 |
| 30 | 100 | 75 | 92 | 341 | 71 | 63 | 57 | Bifenox | 250 |
| 31 | 100 | 44 | 91 | 45 | 80 | 43 | 69 | Urethane | 77 |
| 39 | 100 | 96 | 55 | 95 | 52 | 38 | 38 | Furfural | 100 |
| 40 | 100 | 114 | 75 | 38 | 48 | 29 | 37 | Neburon | 321 |
| 41 | 100 | 69 | 84 | 55 | 53 | 39 | 36 | Citronellal | 49 |
| 41 | 100 | 39 | 87 | 56 | 85 | 28 | 68 | Allidochlor | 357 |
| 41 | 100 | 100 | 93 | 42 | 57 | 43 | 43 | N-Nitrosopyrrolidine | 102 |
| 41 | 100 | 43 | 71 | 175 | 60 | 57 | 59 | Oxadiazon | 399 |
| 41 | 100 | 57 | 90 | 55 | 73 | 43 | 62 | Di-octyl adipate | 66 |
| 42 | 100 | 113 | 36 | 76 | 33 | 147 | 22 | Dactin* | 105 |
| 42 | 100 | 136 | 99 | 43 | 75 | 47 | 60 | Acephate | 487 |
| 42 | 100 | 44 | 42 | 43 | 35 | 57 | 28 | Dazomet | 99 |
| 43 | 100 | 41 | 42 | 27 | 40 | 73 | 22 | Isobutyric acid | 54 |
| 43 | 100 | 56 | 34 | 41 | 17 | 27 | 16 | Butyl acetate | 59 |
| 43 | 100 | 87 | 11 | 28 | 9 | 42 | 5 | Ethylidene diacetate | 62 |
| 43 | 100 | 41 | 64 | 274 | 45 | 316 | 39 | Nitralin | 254 |
| 43 | 100 | 86 | 56 | 128 | 48 | 40 | 47 | Vernolate | 347 |
| 43 | 100 | 56 | 26 | 73 | 15 | 41 | 10 | Isobutyl acetate | 60 |
| 43 | 100 | 73 | 84 | 60 | 81 | 57 | 76 | Stearic acid | 55 |
| 43 | 100 | 41 | 66 | 27 | 55 | 317 | 38 | Oryzalin | 255 |
| 43 | 100 | 85 | 31 | 100 | 20 | 27 | 12 | Acetylacetone | 46 |
| 43 | 100 | 70 | 49 | 55 | 38 | 61 | 15 | Isoamyl acetate | 61 |
| 43 | 100 | 27 | 78 | 58 | 47 | 26 | 10 | Acetone cyanohydrin | 81 |
| 43 | 100 | 93 | 89 | 41 | 57 | 179 | 36 | Propham | 329 |
| 43 | 100 | 75 | 98 | 41 | 59 | 27 | 58 | Bromopropylate | 199 |
| 43 | 100 | 141 | 89 | 71 | 81 | 41 | 47 | Solan | 367 |
| 43 | 100 | 28 | 45 | 42 | 26 | 44 | 24 | Vinyl acetate | 58 |
| 43 | 100 | 58 | 75 | 68 | 43 | 27 | 43 | Dipropetryn | 396 |
| 43 | 100 | 280 | 86 | 41 | 84 | 27 | 69 | Isopropalin | 253 |
| 43 | 100 | 55 | 93 | 56 | 88 | 70 | 79 | Decanol, 1- | 29 |
| 43 | 100 | 211 | 45 | 41 | 15 | 163 | 9 | Dinobuton | 261 |
| 43 | 100 | 45 | 33 | 42 | 22 | 71 | 19 | Dimethoxane | 92 |

| M1 | I1 | M2 | I2 | M3 | I3 | M4 | I4 | Common name | Page |
|---|---|---|---|---|---|---|---|---|---|
| 43 | 100 | 127 | 81 | 41 | 61 | 213 | 58 | Chlorpropham | 330 |
| 43 | 100 | 128 | 76 | 86 | 54 | 29 | 22 | EPTC | 345 |
| 43 | 100 | 143 | 96 | 87 | 80 | 235 | 33 | Carboxin | 371 |
| 43 | 100 | 29 | 91 | 161 | 85 | 207 | 83 | Ethofumesate | 511 |
| 44 | 100 | 77 | 61 | 33 | 12 | 42 | 6 | Acetamide, 2-fluoro- | 76 |
| 44 | 100 | 171 | 88 | 142 | 79 | 156 | 59 | Crimidine | 380 |
| 44 | 100 | 42 | 37 | 45 | 26 | 28 | 26 | Dimefox | 435 |
| 44 | 100 | 83 | 73 | 85 | 47 | 36 | 32 | Acetic acid, trichloro- | 50 |
| 44 | 100 | 201 | 78 | 186 | 51 | 43 | 51 | Simazine | 387 |
| 44 | 100 | 135 | 73 | 45 | 67 | 42 | 25 | Hempa | 434 |
| 45 | 100 | 138 | 37 | 72 | 31 | 137 | 27 | Sesamex | 404 |
| 45 | 100 | 55 | 23 | 43 | 19 | 41 | 14 | Octanol, 2- | 30 |
| 45 | 100 | 160 | 38 | 188 | 30 | 146 | 13 | Alachlor | 360 |
| 49 | 100 | 84 | 64 | 86 | 39 | 51 | 31 | Methane, dichloro- | 6 |
| 53 | 100 | 26 | 85 | 52 | 79 | 51 | 34 | Acrylonitrile | 80 |
| 54 | 100 | 105 | 74 | 40 | 41 | 42 | 39 | Methomyl | 356 |
| 55 | 100 | 43 | 94 | 57 | 89 | 83 | 87 | Muscalure | 4 |
| 55 | 100 | 39 | 53 | 29 | 31 | 90 | 30 | Propene, 3-chloro-2-methyl- | 27 |
| 55 | 100 | 43 | 33 | 276 | 24 | 56 | 24 | Ethalfluralin | 258 |
| 55 | 100 | 73 | 88 | 83 | 66 | 41 | 54 | Cyclohexanecarboxylic acid | 87 |
| 56 | 100 | 43 | 23 | 28 | 17 | 99 | 10 | Cyclohexylamine | 88 |
| 56 | 100 | 43 | 93 | 58 | 61 | 314 | 41 | Iprodione | 401 |
| 56 | 100 | 84 | 71 | 41 | 70 | 27 | 37 | Cyclohexane | 84 |
| 56 | 100 | 55 | 71 | 41 | 60 | 43 | 55 | Ethohexadiol | 35 |
| 57 | 100 | 29 | 57 | 41 | 44 | 103 | 35 | Terbufos | 498 |
| 57 | 100 | 45 | 94 | 29 | 37 | 41 | 34 | Butyl carbitol* | 42 |
| 57 | 100 | 29 | 80 | 31 | 80 | 41 | 26 | Tris(hydroxymethyl)-nitromethane | 36 |
| 57 | 100 | 56 | 40 | 28 | 24 | 55 | 9 | Methane, isocyanato- | 78 |
| 57 | 100 | 29 | 45 | 146 | 43 | 156 | 41 | Butylate | 346 |
| 57 | 100 | 128 | 76 | 72 | 71 | 41 | 48 | Pebulate | 350 |
| 57 | 100 | 27 | 96 | 29 | 71 | 31 | 39 | Epichlorohydrin | 38 |
| 57 | 100 | 45 | 38 | 29 | 35 | 41 | 31 | Butyl cellosolve* | 34 |
| 57 | 100 | 29 | 98 | 87 | 69 | 28 | 69 | Morpholine | 94 |
| 57 | 100 | 176 | 68 | 160 | 59 | 188 | 29 | Butachlor | 308 |
| 57 | 100 | 98 | 85 | 171 | 70 | 41 | 55 | Tebuthiuron | 107 |
| 57 | 100 | 31 | 34 | 29 | 32 | 28 | 31 | Allyl alcohol | 32 |
| 57 | 100 | 44 | 68 | 67 | 18 | 82 | 16 | Cyclohexanol | 86 |
| 58 | 100 | 210 | 91 | 225 | 78 | 168 | 60 | Prometon | 394 |
| 58 | 100 | 41 | 21 | 42 | 15 | 30 | 8 | Butylamine, tert- | 67 |
| 58 | 100 | 42 | 8 | 71 | 6 | 111 | 5 | Meclofenoxate | 239 |
| 58 | 100 | 214 | 84 | 229 | 53 | 43 | 52 | Propazine* | 390 |
| 58 | 100 | 41 | 99 | 86 | 97 | 89 | 75 | Aldicarb | 325 |
| 58 | 100 | 59 | 47 | 30 | 29 | 42 | 26 | Trimethylamine | 69 |
| 58 | 100 | 39 | 48 | 184 | 27 | 54 | 25 | Tranid* | 344 |
| 59 | 100 | 60 | 57 | 43 | 30 | 45 | 28 | Daminozide | 353 |
| 60 | 100 | 27 | 50 | 73 | 27 | 42 | 25 | Butyric acid | 53 |
| 61 | 100 | 46 | 29 | 248 | 11 | 160 | 8 | Linuron | 319 |
| 61 | 100 | 46 | 22 | 258 | 14 | 91 | 14 | Metobromuron | 320 |
| 61 | 100 | 96 | 61 | 98 | 38 | 63 | 32 | Ethene, 1,1-dichloro | 21 |
| 62 | 100 | 27 | 91 | 49 | 40 | 64 | 32 | Ethylene dichloride | 13 |
| 62 | 100 | 29 | 85 | 47 | 77 | 27 | 67 | Ethanethiol | 28 |
| 63 | 100 | 27 | 71 | 65 | 31 | 26 | 19 | Ethane, 1,1-dichloro- | 12 |

| M1 | I1 | M2 | I2 | M3 | I3 | M4 | I4 | Common name | Page |
|---|---|---|---|---|---|---|---|---|---|
| 63 | 100 | 43 | 92 | 27 | 81 | 306 | 52 | Fluchloralin | 259 |
| 63 | 100 | 62 | 71 | 27 | 57 | 41 | 49 | Propane, 1,2-dichloro- | 24 |
| 65 | 100 | 77 | 83 | 326 | 67 | 325 | 62 | Triphenyl phosphate | 454 |
| 65 | 100 | 121 | 66 | 50 | 49 | 137 | 46 | Halazone | 274 |
| 65 | 100 | 44 | 97 | 66 | 94 | 105 | 71 | Phthalate, phenyl | 269 |
| 66 | 100 | 79 | 43 | 91 | 34 | 263 | 32 | Aldrin | 119 |
| 67 | 100 | 68 | 85 | 53 | 61 | 39 | 34 | Isoprene | 1 |
| 68 | 100 | 44 | 83 | 225 | 64 | 43 | 60 | Cyanazine | 392 |
| 68 | 100 | 100 | 79 | 69 | 42 | 39 | 36 | Endothall | 93 |
| 69 | 100 | 41 | 97 | 39 | 49 | 63 | 12 | Dicryl | 365 |
| 69 | 100 | 41 | 65 | 68 | 20 | 29 | 10 | Geraniol | 31 |
| 69 | 100 | 120 | 74 | 188 | 54 | 91 | 48 | Crotamiton | 364 |
| 70 | 100 | 41 | 89 | 39 | 58 | 69 | 49 | Crotonaldehyde | 48 |
| 71 | 100 | 43 | 92 | 29 | 43 | 70 | 40 | Amyl ether, normal- | 44 |
| 72 | 100 | 232 | 38 | 234 | 26 | 44 | 26 | Diuron | 318 |
| 72 | 100 | 44 | 79 | 32 | 51 | 30 | 45 | Oxamyl | 355 |
| 72 | 100 | 26 | 81 | 45 | 78 | 27 | 65 | Maleic acid | 56 |
| 72 | 100 | 40 | 21 | 198 | 19 | 28 | 15 | Monuron | 315 |
| 72 | 100 | 153 | 55 | 89 | 38 | 45 | 37 | Norea | 316 |
| 72 | 100 | 164 | 26 | 44 | 25 | 65 | 22 | Fenuron | 314 |
| 72 | 100 | 28 | 39 | 240 | 23 | 73 | 16 | Dimetilan | 328 |
| 72 | 100 | 128 | 63 | 100 | 40 | 29 | 27 | Napropamide | 363 |
| 72 | 100 | 232 | 25 | 44 | 22 | 28 | 12 | Fluometuron | 317 |
| 72 | 100 | 77 | 37 | 39 | 21 | 51 | 20 | Pyrolan* | 327 |
| 72 | 100 | 28 | 7 | 45 | 6 | 27 | 6 | Isolan* | 326 |
| 73 | 100 | 72 | 48 | 45 | 28 | 44 | 12 | Methane, isothiocyanato- | 79 |
| 73 | 100 | 110 | 33 | 69 | 26 | 81 | 25 | Methoprene | 65 |
| 74 | 100 | 87 | 69 | 44 | 33 | 42 | 25 | Methyl stearate | 63 |
| 74 | 100 | 42 | 60 | 43 | 35 | 44 | 18 | N-Nitrosodimethylamine | 73 |
| 75 | 100 | 47 | 77 | 29 | 43 | 106 | 43 | Ethanol, 2-(ethylthio)- | 33 |
| 75 | 100 | 121 | 32 | 260 | 18 | 97 | 18 | Phorate | 496 |
| 75 | 100 | 39 | 55 | 77 | 32 | 49 | 26 | Propene, 1,3-dichloro- | 26 |
| 77 | 100 | 51 | 57 | 182 | 37 | 105 | 23 | Azobenzene | 294 |
| 77 | 100 | 51 | 62 | 50 | 28 | 279 | 15 | Phenylmercuric acetate | 529 |
| 77 | 100 | 141 | 67 | 51 | 61 | 63 | 27 | Genite* | 432 |
| 77 | 100 | 97 | 74 | 129 | 62 | 157 | 48 | Phoxim | 472 |
| 77 | 100 | 51 | 59 | 123 | 42 | 50 | 25 | Benzene, nitro- | 240 |
| 77 | 100 | 51 | 54 | 28 | 32 | 50 | 31 | Phenylmercuric chloride | 527 |
| 78 | 100 | 77 | 20 | 52 | 19 | 51 | 17 | Benzene | 123 |
| 79 | 100 | 80 | 29 | 77 | 21 | 78 | 18 | Captafol | 414 |
| 79 | 100 | 82 | 42 | 81 | 35 | 108 | 21 | Dieldrin | 121 |
| 79 | 100 | 49 | 62 | 81 | 32 | 29 | 23 | Dichloromethyl ether, sym- | 39 |
| 79 | 100 | 77 | 42 | 80 | 25 | 149 | 21 | Captan | 413 |
| 79 | 100 | 109 | 96 | 110 | 72 | 139 | 55 | Trichlorfon | 439 |
| 81 | 100 | 69 | 91 | 137 | 38 | 136 | 29 | Squalene | 5 |
| 81 | 100 | 263 | 98 | 265 | 65 | 261 | 65 | Endrin | 122 |
| 81 | 100 | 353 | 94 | 355 | 72 | 351 | 48 | Heptachlor epoxide | 114 |
| 81 | 100 | 89 | 57 | 109 | 55 | 346 | 50 | Coroxon | 450 |
| 82 | 100 | 39 | 28 | 138 | 17 | 27 | 17 | Isophorone | 89 |
| 82 | 100 | 81 | 55 | 109 | 41 | 27 | 31 | Ethephon | 437 |
| 83 | 100 | 55 | 50 | 84 | 16 | 82 | 14 | Binapacryl | 260 |
| 83 | 100 | 85 | 63 | 95 | 11 | 87 | 10 | Ethane, 1,1,2,2-tetrachloro- | 17 |

| M1 | I1 | M2 | I2 | M3 | I3 | M4 | I4 | Common name | Page |
|---|---|---|---|---|---|---|---|---|---|
| 83 | 100 | 85 | 64 | 47 | 35 | 35 | 19 | Chloroform | 7 |
| 84 | 100 | 29 | 74 | 57 | 54 | 44 | 27 | Amitrole | 108 |
| 84 | 100 | 133 | 31 | 162 | 30 | 161 | 20 | Nicotine | 306 |
| 84 | 100 | 105 | 58 | 106 | 45 | 133 | 42 | Anabasine | 307 |
| 84 | 100 | 55 | 65 | 41 | 56 | 69 | 52 | Cycloheximide | 518 |
| 85 | 100 | 87 | 33 | 50 | 12 | 101 | 9 | Methane, dichlorodifluoro- | 10 |
| 85 | 100 | 145 | 89 | 93 | 39 | 58 | 39 | Methidathion | 494 |
| 86 | 100 | 70 | 50 | 234 | 45 | 128 | 40 | Diallate | 348 |
| 86 | 100 | 30 | 92 | 28 | 39 | 29 | 24 | Imidazolidinone, 2- | 103 |
| 86 | 100 | 30 | 68 | 58 | 37 | 28 | 24 | Triethylamine | 70 |
| 86 | 100 | 268 | 22 | 128 | 18 | 270 | 15 | Triallate | 349 |
| 87 | 100 | 93 | 82 | 125 | 68 | 47 | 33 | Dimethoate | 489 |
| 87 | 100 | 246 | 53 | 244 | 41 | 209 | 37 | Chloranil | 187 |
| 88 | 100 | 120 | 24 | 44 | 16 | 240 | 14 | Thiram | 420 |
| 88 | 100 | 44 | 28 | 58 | 25 | 43 | 22 | Ziram | 533 |
| 88 | 100 | 60 | 49 | 57 | 27 | 29 | 24 | Methyl demeton | 458 |
| 88 | 100 | 44 | 34 | 43 | 31 | 296 | 26 | Ferbam | 525 |
| 88 | 100 | 89 | 37 | 60 | 24 | 61 | 23 | Disulfoton | 497 |
| 88 | 100 | 62 | 49 | 61 | 45 | 53 | 35 | Bromoxynil | 313 |
| 91 | 100 | 106 | 31 | 51 | 14 | 39 | 10 | Benzene, ethyl- | 126 |
| 91 | 100 | 106 | 57 | 105 | 25 | 77 | 13 | Xylene | 127 |
| 91 | 100 | 92 | 73 | 39 | 20 | 65 | 14 | Toluene | 124 |
| 91 | 100 | 121 | 63 | 105 | 56 | 107 | 35 | Arsinic acid, dimethyl- | 522 |
| 91 | 100 | 126 | 20 | 65 | 14 | 39 | 9 | Benzyl chloride | 131 |
| 92 | 100 | 140 | 61 | 196 | 51 | 60 | 49 | Phosfolan | 470 |
| 92 | 100 | 197 | 53 | 65 | 50 | 77 | 34 | Aminoazobenzene, 4- | 295 |
| 93 | 100 | 28 | 22 | 55 | 17 | 56 | 16 | Siduron | 322 |
| 93 | 100 | 63 | 74 | 27 | 38 | 95 | 32 | Dichloroethyl ether, sym- | 40 |
| 93 | 100 | 121 | 29 | 213 | 21 | 32 | 21 | Salicylanilide | 368 |
| 93 | 100 | 125 | 83 | 126 | 47 | 79 | 34 | Formothion | 490 |
| 93 | 100 | 66 | 32 | 65 | 16 | 39 | 13 | Aniline | 289 |
| 93 | 100 | 39 | 44 | 66 | 42 | 65 | 26 | Pyridine, 4-methyl- | 125 |
| 93 | 100 | 92 | 35 | 41 | 33 | 77 | 30 | Pinene, alpha- | 2 |
| 93 | 100 | 191 | 33 | 99 | 14 | 146 | 11 | Maleanilic acid | 373 |
| 94 | 100 | 77 | 31 | 39 | 16 | 51 | 16 | Ethanol, 2-phenoxy- | 238 |
| 94 | 100 | 66 | 25 | 39 | 25 | 65 | 21 | Phenol | 208 |
| 94 | 100 | 95 | 62 | 141 | 50 | 64 | 24 | Methamidophos | 486 |
| 95 | 100 | 130 | 90 | 132 | 85 | 60 | 65 | Ethylene, trichloro- | 22 |
| 95 | 100 | 67 | 87 | 41 | 77 | 81 | 76 | Iridomyrmecin | 517 |
| 95 | 100 | 93 | 85 | 41 | 82 | 121 | 64 | Isobornyl thiocyanoacetate | 507 |
| 95 | 100 | 81 | 66 | 152 | 50 | 108 | 45 | Camphor | 3 |
| 97 | 100 | 83 | 95 | 99 | 62 | 85 | 60 | Ethane, 1,1,2-trichloro- | 16 |
| 97 | 100 | 125 | 67 | 65 | 47 | 153 | 38 | Dioxathion | 504 |
| 97 | 100 | 291 | 98 | 109 | 91 | 137 | 61 | Parathion | 476 |
| 97 | 100 | 197 | 97 | 199 | 94 | 314 | 64 | Chlorpyrifos | 479 |
| 97 | 100 | 99 | 64 | 61 | 58 | 26 | 31 | Ethane, 1,1,1-trichloro- | 15 |
| 99 | 100 | 155 | 96 | 43 | 68 | 127 | 57 | Tetraethyl pyrophosphate | 456 |
| 99 | 100 | 155 | 32 | 211 | 30 | 57 | 15 | Tributyl phosphate | 452 |
| 99 | 100 | 81 | 71 | 155 | 56 | 82 | 45 | Triethyl phosphate | 451 |
| 100 | 100 | 65 | 44 | 272 | 40 | 274 | 32 | Heptachlor | 113 |
| 101 | 100 | 103 | 66 | 66 | 13 | 105 | 11 | Methane, trichlorofluoro- | 11 |
| 101 | 100 | 41 | 94 | 115 | 75 | 102 | 74 | Octhilinone | 106 |

| M1 | I1 | M2 | I2 | M3 | I3 | M4 | I4 | Common name | Page |
|---|---|---|---|---|---|---|---|---|---|
| 102 | 100 | 44 | 87 | 42 | 83 | 29 | 64 | N-Nitrosodiethylamine | 74 |
| 102 | 100 | 34 | 82 | 65 | 54 | 44 | 51 | Glyphosate | 438 |
| 102 | 100 | 30 | 89 | 73 | 35 | 45 | 34 | Ethylene thiourea | 104 |
| 103 | 100 | 40 | 64 | 311 | 51 | 313 | 45 | Isobenzan | 111 |
| 103 | 100 | 76 | 34 | 50 | 13 | 104 | 9 | Benzonitrile | 310 |
| 104 | 100 | 103 | 41 | 78 | 32 | 51 | 28 | Styrene | 129 |
| 104 | 100 | 132 | 53 | 91 | 43 | 51 | 17 | Tetralin* | 156 |
| 105 | 100 | 91 | 42 | 77 | 28 | 51 | 12 | Benzyl benzoate | 263 |
| 105 | 100 | 122 | 89 | 77 | 73 | 51 | 37 | Benzoic acid | 272 |
| 105 | 100 | 120 | 25 | 77 | 13 | 51 | 12 | Cumene | 130 |
| 105 | 100 | 77 | 49 | 182 | 32 | 51 | 15 | Benzophenone | 190 |
| 105 | 100 | 77 | 67 | 51 | 38 | 50 | 26 | Benzoyl chloride | 262 |
| 107 | 100 | 164 | 44 | 135 | 38 | 95 | 29 | Phenol, 4-tert-pentyl- | 233 |
| 107 | 100 | 108 | 91 | 77 | 28 | 79 | 21 | Phenol, 4-methyl- | 228 |
| 107 | 100 | 39 | 73 | 228 | 71 | 121 | 64 | Ancymidol | 207 |
| 107 | 100 | 77 | 93 | 152 | 81 | 94 | 24 | Acetic acid, phenoxy- | 278 |
| 107 | 100 | 135 | 69 | 137 | 66 | 123 | 58 | Bronopol | 37 |
| 107 | 100 | 106 | 77 | 76 | 65 | 104 | 61 | Duraset* | 375 |
| 107 | 100 | 39 | 39 | 106 | 29 | 66 | 22 | Pyridine, 2,6-dimethyl- | 128 |
| 107 | 100 | 122 | 36 | 77 | 27 | 39 | 17 | Phlorol | 229 |
| 107 | 100 | 96 | 98 | 106 | 73 | 97 | 64 | Thionazin | 473 |
| 108 | 100 | 107 | 85 | 79 | 35 | 39 | 31 | Phenol, 3-methyl- | 226 |
| 108 | 100 | 110 | 64 | 82 | 16 | 73 | 15 | Acetonitrile, trichloro- | 82 |
| 108 | 100 | 107 | 75 | 77 | 34 | 79 | 33 | Phenol, 2-methyl- | 225 |
| 109 | 100 | 145 | 53 | 79 | 39 | 47 | 33 | Naled | 441 |
| 109 | 100 | 137 | 60 | 246 | 46 | 110 | 23 | Fonofos | 505 |
| 109 | 100 | 81 | 93 | 149 | 64 | 99 | 60 | Paraoxon | 448 |
| 109 | 100 | 28 | 47 | 185 | 31 | 79 | 21 | Dichlorvos | 440 |
| 109 | 100 | 125 | 97 | 79 | 43 | 297 | 40 | Chlorthion* | 462 |
| 109 | 100 | 125 | 97 | 277 | 67 | 260 | 42 | Fenitrothion | 460 |
| 109 | 100 | 79 | 50 | 329 | 33 | 331 | 29 | Stirofos | 447 |
| 110 | 100 | 279 | 87 | 281 | 49 | 152 | 42 | DMPA | 488 |
| 110 | 100 | 152 | 18 | 27 | 13 | 111 | 10 | Propoxur | 340 |
| 110 | 100 | 82 | 12 | 81 | 11 | 69 | 9 | Resorcinol | 209 |
| 111 | 100 | 175 | 75 | 75 | 53 | 99 | 51 | Ovex | 431 |
| 111 | 100 | 84 | 41 | 83 | 31 | 57 | 16 | Aniline, 4-fluoro- | 290 |
| 112 | 100 | 77 | 63 | 114 | 33 | 51 | 29 | Benzene, chloro- | 132 |
| 115 | 100 | 143 | 90 | 202 | 39 | 116 | 36 | ANTU | 324 |
| 116 | 100 | 88 | 49 | 44 | 43 | 148 | 40 | Disulfiram | 422 |
| 117 | 100 | 90 | 33 | 89 | 21 | 118 | 9 | Indole, 1H- | 144 |
| 117 | 100 | 119 | 97 | 82 | 32 | 121 | 31 | Chloropicrin | 83 |
| 117 | 100 | 119 | 98 | 121 | 31 | 82 | 24 | Carbon tetrachloride | 9 |
| 118 | 100 | 56 | 69 | 45 | 60 | 42 | 56 | Triethanolamine | 72 |
| 119 | 100 | 70 | 80 | 147 | 34 | 120 | 33 | Nornicotine | 305 |
| 119 | 100 | 190 | 46 | 91 | 41 | 191 | 17 | Deet | 369 |
| 119 | 100 | 198 | 88 | 92 | 34 | 161 | 33 | Bentazon | 433 |
| 120 | 100 | 77 | 43 | 93 | 33 | 43 | 31 | Propachlor | 358 |
| 120 | 100 | 225 | 73 | 77 | 58 | 106 | 30 | Aminoazobenzene, N,N-dimethyl- | 297 |
| 121 | 100 | 53 | 96 | 39 | 93 | 52 | 79 | Phenol, 2-methyl-4,6-dinitro- | 245 |
| 121 | 100 | 152 | 80 | 93 | 70 | 28 | 68 | Methylparaben | 270 |
| 121 | 100 | 254 | 50 | 43 | 46 | 296 | 20 | Fluoridamid | 361 |
| 123 | 100 | 77 | 44 | 167 | 34 | 92 | 33 | Dichlofluanid | 424 |

| M1 | I1 | M2 | I2 | M3 | I3 | M4 | I4 | Common name | Page |
|---|---|---|---|---|---|---|---|---|---|
| 125 | 100 | 173 | 98 | 93 | 96 | 158 | 54 | Malathion | 491 |
| 125 | 100 | 235 | 80 | 241 | 59 | 276 | 56 | Bulan* | 201 |
| 125 | 100 | 90 | 55 | 63 | 34 | 127 | 31 | Drazoxolon | 508 |
| 125 | 100 | 159 | 44 | 77 | 44 | 51 | 38 | Sulphenone* | 428 |
| 125 | 100 | 43 | 90 | 55 | 44 | 97 | 44 | Pyracarbolid | 91 |
| 125 | 100 | 285 | 97 | 287 | 64 | 109 | 43 | Ronnel | 464 |
| 125 | 100 | 127 | 34 | 89 | 19 | 63 | 15 | Chlorbenside | 423 |
| 125 | 100 | 77 | 80 | 51 | 75 | 218 | 62 | Diphenyl sulfone | 427 |
| 125 | 100 | 77 | 37 | 51 | 32 | 97 | 21 | Perfluidone | 429 |
| 125 | 100 | 109 | 94 | 79 | 59 | 47 | 51 | Cythioate | 467 |
| 127 | 100 | 105 | 80 | 104 | 37 | 40 | 34 | Crotoxyphos | 446 |
| 127 | 100 | 72 | 61 | 264 | 46 | 138 | 35 | Phosphamidon | 445 |
| 127 | 100 | 67 | 66 | 58 | 36 | 97 | 34 | Monocrotophos | 442 |
| 127 | 100 | 192 | 41 | 67 | 19 | 70 | 15 | Mevinphos | 443 |
| 127 | 100 | 67 | 76 | 44 | 63 | 43 | 53 | Dicrotophos | 444 |
| 128 | 100 | 51 | 16 | 102 | 14 | 127 | 12 | Azulene | 145 |
| 128 | 100 | 65 | 35 | 130 | 33 | 64 | 20 | Phenol, 3-chloro- | 212 |
| 128 | 100 | 141 | 62 | 268 | 32 | 152 | 32 | Dichlorophene | 205 |
| 128 | 100 | 64 | 52 | 130 | 32 | 63 | 26 | Phenol, 2-chloro- | 211 |
| 128 | 100 | 51 | 13 | 129 | 11 | 64 | 11 | Naphthalene | 146 |
| 128 | 100 | 129 | 9 | 115 | 8 | 202 | 6 | Tridemorph | 95 |
| 128 | 100 | 130 | 31 | 65 | 29 | 64 | 14 | Phenol, 4-chloro- | 213 |
| 129 | 100 | 102 | 21 | 51 | 17 | 128 | 16 | Quinoline | 152 |
| 129 | 100 | 102 | 26 | 51 | 20 | 128 | 18 | Isoquinoline | 155 |
| 129 | 100 | 43 | 59 | 44 | 56 | 86 | 15 | Cyanuric acid | 385 |
| 130 | 100 | 175 | 85 | 77 | 31 | 131 | 27 | Indoleacetic acid | 288 |
| 132 | 100 | 77 | 91 | 160 | 45 | 104 | 33 | Azinphos-methyl | 493 |
| 135 | 100 | 81 | 63 | 173 | 36 | 39 | 31 | Propargite | 425 |
| 135 | 100 | 150 | 38 | 91 | 15 | 136 | 10 | Thymol | 231 |
| 135 | 100 | 150 | 70 | 91 | 25 | 58 | 23 | Promecarb | 336 |
| 135 | 100 | 150 | 27 | 91 | 14 | 136 | 10 | Carvacrol | 230 |
| 139 | 100 | 156 | 68 | 50 | 51 | 75 | 47 | Benzoic acid, 2-chloro- | 273 |
| 139 | 100 | 111 | 36 | 141 | 29 | 250 | 27 | Dicofol | 196 |
| 139 | 100 | 65 | 87 | 39 | 71 | 109 | 31 | Phenol, 4-nitro- | 242 |
| 139 | 100 | 65 | 36 | 64 | 22 | 63 | 22 | Phenol, 2-nitro- | 241 |
| 140 | 100 | 106 | 75 | 196 | 68 | 74 | 66 | Mephosfolan | 471 |
| 141 | 100 | 186 | 49 | 115 | 29 | 142 | 17 | Naphthaleneacetic acid, 1- | 287 |
| 141 | 100 | 200 | 75 | 77 | 63 | 143 | 33 | Methoxone* | 279 |
| 142 | 100 | 107 | 80 | 144 | 32 | 77 | 24 | Phenol, 4-chloro-2-methyl- | 227 |
| 142 | 100 | 77 | 78 | 214 | 75 | 107 | 71 | Mecoprop | 280 |
| 143 | 100 | 115 | 68 | 116 | 33 | 89 | 12 | Naphthylamine, 1- | 291 |
| 143 | 100 | 128 | 21 | 115 | 21 | 142 | 16 | Quinoline, 2-methyl- | 153 |
| 143 | 100 | 115 | 38 | 144 | 14 | 116 | 13 | Naphthylamine, 2- | 292 |
| 143 | 100 | 76 | 68 | 104 | 61 | 115 | 61 | Naptalam | 374 |
| 144 | 100 | 115 | 44 | 116 | 37 | 145 | 14 | Carbaryl | 341 |
| 145 | 100 | 117 | 78 | 90 | 30 | 89 | 29 | Quinolinol, 8- | 154 |
| 145 | 100 | 102 | 84 | 303 | 39 | 42 | 34 | Norflurazon | 398 |
| 146 | 100 | 148 | 64 | 111 | 38 | 75 | 23 | Benzene, dichloro- | 133 |
| 149 | 100 | 107 | 80 | 121 | 27 | 55 | 19 | Phenol, nonyl- | 237 |
| 149 | 100 | 177 | 28 | 150 | 13 | 176 | 9 | Phthalate, diethyl | 265 |
| 149 | 100 | 91 | 61 | 206 | 27 | 104 | 27 | Phthalate, benzyl butyl | 267 |
| 149 | 100 | 57 | 32 | 167 | 29 | 71 | 21 | Phthalate, bis-(2-ethylhexyl) | 268 |

| M1 | I1 | M2 | I2 | M3 | I3 | M4 | I4 | Common name | Page |
|---|---|---|---|---|---|---|---|---|---|
| 149 | 100 | 86 | 18 | 57 | 18 | 223 | 17 | Phthalate, di-n-butyl | 266 |
| 151 | 100 | 152 | 37 | 154 | 19 | 77 | 16 | Arsine, oxophenyl- | 523 |
| 151 | 100 | 166 | 69 | 126 | 61 | 58 | 60 | Bendiocarb | 343 |
| 151 | 100 | 150 | 64 | 136 | 44 | 28 | 19 | Aminocarb | 338 |
| 152 | 100 | 151 | 20 | 76 | 17 | 153 | 14 | Acenaphthylene | 157 |
| 152 | 100 | 115 | 93 | 171 | 46 | 182 | 38 | Biphenyl, 2-nitro- | 142 |
| 153 | 100 | 154 | 16 | 77 | 14 | 94 | 14 | Lenacil | 98 |
| 153 | 100 | 141 | 84 | 155 | 46 | 63 | 38 | Diflubenzuron | 323 |
| 153 | 100 | 92 | 93 | 135 | 91 | 44 | 90 | Schradan | 457 |
| 154 | 100 | 55 | 34 | 41 | 24 | 141 | 19 | Dodemorph | 96 |
| 154 | 100 | 153 | 31 | 152 | 28 | 76 | 17 | Biphenyl | 159 |
| 154 | 100 | 153 | 86 | 152 | 42 | 76 | 23 | Acenaphthene | 158 |
| 155 | 100 | 154 | 56 | 157 | 31 | 156 | 29 | Niclosamide | 376 |
| 156 | 100 | 43 | 25 | 125 | 24 | 93 | 20 | Menazon | 492 |
| 157 | 100 | 45 | 57 | 97 | 55 | 121 | 48 | Carbophenothion | 501 |
| 157 | 100 | 75 | 88 | 155 | 85 | 28 | 69 | DBCP | 25 |
| 157 | 100 | 63 | 77 | 77 | 70 | 29 | 65 | Cyanofenphos | 484 |
| 158 | 100 | 43 | 85 | 97 | 73 | 139 | 58 | Ethoprop | 506 |
| 159 | 100 | 111 | 77 | 227 | 51 | 229 | 49 | Tetradifon | 430 |
| 159 | 100 | 124 | 97 | 260 | 71 | 161 | 71 | Methazole | 400 |
| 160 | 100 | 77 | 34 | 161 | 33 | 28 | 31 | Phosmet | 495 |
| 160 | 100 | 236 | 77 | 102 | 41 | 75 | 14 | Thioquinox | 418 |
| 161 | 100 | 160 | 72 | 41 | 63 | 56 | 55 | Terbacil | 377 |
| 161 | 100 | 29 | 79 | 163 | 71 | 57 | 68 | Propanil | 362 |
| 161 | 100 | 163 | 65 | 259 | 35 | 261 | 23 | Karsil | 366 |
| 162 | 100 | 41 | 85 | 29 | 82 | 43 | 68 | Sulfoxide | 406 |
| 162 | 100 | 164 | 63 | 63 | 50 | 98 | 39 | Phenol, 2,4-dichloro- | 216 |
| 162 | 100 | 45 | 76 | 41 | 33 | 238 | 26 | Metolachlor | 359 |
| 162 | 100 | 164 | 33 | 127 | 30 | 126 | 16 | Naphthalene, chloro- | 147 |
| 162 | 100 | 164 | 66 | 87 | 33 | 43 | 29 | 24-DB | 283 |
| 162 | 100 | 164 | 64 | 63 | 21 | 98 | 15 | Phenol, 2,6-dichloro- | 215 |
| 162 | 100 | 164 | 69 | 220 | 61 | 222 | 39 | 24-D | 281 |
| 162 | 100 | 121 | 94 | 132 | 74 | 147 | 64 | Amitraz | 354 |
| 163 | 100 | 77 | 31 | 76 | 17 | 50 | 15 | Phthalate, dimethyl | 264 |
| 164 | 100 | 149 | 36 | 131 | 27 | 77 | 24 | Eugenol | 235 |
| 164 | 100 | 149 | 66 | 122 | 20 | 123 | 18 | Carbofuran | 342 |
| 164 | 100 | 135 | 43 | 136 | 32 | 163 | 29 | Methabenzthiazuron | 514 |
| 164 | 100 | 41 | 52 | 123 | 47 | 79 | 45 | Tetramethrin | 416 |
| 165 | 100 | 137 | 53 | 180 | 47 | 41 | 16 | Butylated hydroxy anisole | 236 |
| 165 | 100 | 164 | 67 | 150 | 66 | 134 | 30 | Mexacarbate | 337 |
| 166 | 100 | 164 | 82 | 131 | 71 | 129 | 71 | Ethylene, tetrachloro- | 23 |
| 166 | 100 | 165 | 80 | 167 | 15 | 163 | 12 | Fluorene, 9H- | 160 |
| 166 | 100 | 32 | 18 | 209 | 14 | 96 | 13 | Dimethirimol | 381 |
| 166 | 100 | 96 | 34 | 209 | 18 | 55 | 14 | Ethirimol | 382 |
| 167 | 100 | 165 | 52 | 152 | 29 | 168 | 18 | Diphenylacetic acid | 191 |
| 167 | 100 | 133 | 96 | 135 | 64 | 104 | 53 | Phenmedipham | 332 |
| 167 | 100 | 165 | 91 | 117 | 90 | 119 | 89 | Ethane, pentachloro- | 18 |
| 167 | 100 | 166 | 14 | 44 | 13 | 168 | 12 | Carbazole, 9H- | 161 |
| 167 | 100 | 72 | 95 | 165 | 44 | 239 | 28 | Diphenamid | 372 |
| 167 | 100 | 69 | 33 | 45 | 23 | 63 | 20 | Benzothiazole, 2-mercapto- | 513 |
| 168 | 100 | 85 | 80 | 153 | 64 | 125 | 33 | Dehydroacetic acid | 90 |
| 168 | 100 | 153 | 69 | 109 | 21 | 91 | 16 | Methiocarb | 339 |

| M1 | I1 | M2 | I2 | M3 | I3 | M4 | I4 | Common name | Page |
|---|---|---|---|---|---|---|---|---|---|
| 169 | 100 | 168 | 35 | 170 | 25 | 167 | 22 | Biphenylamine, 4- | 299 |
| 169 | 100 | 171 | 39 | 184 | 31 | 186 | 10 | Chlorothymol | 232 |
| 169 | 100 | 168 | 47 | 167 | 28 | 51 | 14 | Diphenylamine | 293 |
| 169 | 100 | 97 | 74 | 171 | 68 | 45 | 64 | Erbon | 286 |
| 170 | 100 | 169 | 69 | 141 | 35 | 115 | 26 | Phenol, 2-phenyl- | 210 |
| 170 | 100 | 141 | 39 | 51 | 36 | 77 | 35 | Phenyl ether | 248 |
| 171 | 100 | 100 | 71 | 173 | 69 | 75 | 56 | Dichlobenil | 311 |
| 171 | 100 | 377 | 62 | 28 | 60 | 375 | 45 | Leptophos | 483 |
| 172 | 100 | 174 | 99 | 65 | 31 | 93 | 20 | Phenol, 4-bromo- | 214 |
| 173 | 100 | 175 | 85 | 255 | 37 | 145 | 31 | Propyzamide | 370 |
| 173 | 100 | 171 | 50 | 175 | 49 | 93 | 22 | Bromoform | 8 |
| 173 | 100 | 220 | 92 | 175 | 70 | 191 | 65 | Dicamba | 276 |
| 173 | 100 | 167 | 89 | 340 | 57 | 165 | 40 | Diphenadione | 510 |
| 173 | 100 | 174 | 81 | 146 | 41 | 89 | 32 | Pindone | 411 |
| 173 | 100 | 374 | 19 | 165 | 19 | 174 | 18 | Chlorophacinone | 410 |
| 174 | 100 | 118 | 46 | 120 | 33 | 92 | 32 | Juglone | 189 |
| 176 | 100 | 177 | 41 | 194 | 26 | 57 | 19 | Piperonyl butoxide | 405 |
| 178 | 100 | 176 | 17 | 179 | 16 | 89 | 14 | Anthracene | 162 |
| 178 | 100 | 176 | 25 | 179 | 23 | 177 | 15 | Phenanthrene | 167 |
| 179 | 100 | 180 | 14 | 178 | 14 | 89 | 12 | Acridine | 163 |
| 179 | 100 | 180 | 25 | 178 | 24 | 76 | 16 | Phenanthridine | 168 |
| 179 | 100 | 137 | 94 | 152 | 85 | 304 | 64 | Diazinon* | 474 |
| 180 | 100 | 152 | 50 | 208 | 23 | 151 | 23 | Phenanthraquinone, 9-10- | 169 |
| 180 | 100 | 182 | 97 | 184 | 31 | 145 | 26 | Benzene, trichloro- | 134 |
| 181 | 100 | 183 | 98 | 109 | 98 | 51 | 91 | Lindane | 85 |
| 182 | 100 | 121 | 58 | 97 | 33 | 28 | 33 | Phosalone | 502 |
| 183 | 100 | 163 | 16 | 165 | 12 | 184 | 9 | Permethrin | 415 |
| 184 | 100 | 92 | 24 | 185 | 13 | 91 | 13 | Benzidine | 300 |
| 184 | 100 | 154 | 83 | 63 | 52 | 107 | 45 | Phenol, 2,4-dinitro- | 244 |
| 186 | 100 | 256 | 92 | 258 | 88 | 75 | 47 | Biphenyl, trichloro- | 137 |
| 187 | 100 | 186 | 29 | 103 | 24 | 104 | 21 | Benzoguanamine | 386 |
| 188 | 100 | 72 | 33 | 29 | 33 | 44 | 30 | Sulfallate | 351 |
| 191 | 100 | 105 | 39 | 159 | 38 | 132 | 33 | Carbendazim | 333 |
| 191 | 100 | 57 | 89 | 185 | 88 | 63 | 84 | Aramite* | 426 |
| 191 | 100 | 226 | 75 | 163 | 55 | 228 | 49 | Dichlone | 188 |
| 191 | 100 | 159 | 99 | 40 | 46 | 105 | 44 | Benomyl | 334 |
| 192 | 100 | 28 | 41 | 191 | 31 | 394 | 20 | Rotenone | 520 |
| 192 | 100 | 394 | 35 | 392 | 23 | 191 | 23 | Deguelin | 521 |
| 193 | 100 | 195 | 97 | 66 | 80 | 263 | 37 | Isodrin | 120 |
| 194 | 100 | 196 | 95 | 198 | 34 | 133 | 17 | Nitrapyrin | 304 |
| 194 | 100 | 109 | 88 | 67 | 88 | 55 | 71 | Caffeine | 515 |
| 195 | 100 | 241 | 80 | 197 | 77 | 237 | 70 | Endosulfan | 115 |
| 196 | 100 | 44 | 90 | 181 | 65 | 117 | 64 | Chlordimeform | 309 |
| 196 | 100 | 198 | 95 | 97 | 40 | 200 | 29 | Silvex | 285 |
| 196 | 100 | 198 | 97 | 97 | 41 | 200 | 32 | Phenol, 2,4,5-trichloro- | 217 |
| 196 | 100 | 198 | 98 | 209 | 58 | 211 | 56 | Hexachlorophene | 206 |
| 196 | 100 | 198 | 95 | 161 | 40 | 200 | 31 | Picloram | 303 |
| 196 | 100 | 198 | 96 | 200 | 31 | 132 | 28 | Phenol, 2,4,6-trichloro- | 218 |
| 196 | 100 | 168 | 26 | 197 | 20 | 139 | 16 | Xanthone | 165 |
| 196 | 100 | 198 | 97 | 254 | 50 | 256 | 49 | 245-T | 284 |
| 198 | 100 | 41 | 52 | 57 | 46 | 28 | 40 | Metribuzin | 384 |
| 198 | 100 | 106 | 33 | 182 | 19 | 199 | 14 | Diaminodiphenylmethane, 4,4′- | 298 |

| M1 | I1 | M2 | I2 | M3 | I3 | M4 | I4 | Common name | Page |
|---|---|---|---|---|---|---|---|---|---|
| 199 | 100 | 167 | 55 | 200 | 21 | 198 | 20 | Phenothiazine, 10H- | 164 |
| 200 | 100 | 186 | 61 | 229 | 60 | 214 | 59 | Trietazine | 391 |
| 200 | 100 | 58 | 78 | 215 | 57 | 44 | 48 | Atrazine | 388 |
| 201 | 100 | 115 | 92 | 285 | 65 | 173 | 42 | Piperine | 403 |
| 201 | 100 | 199 | 53 | 203 | 48 | 119 | 43 | Tris-BP | 453 |
| 201 | 100 | 117 | 76 | 119 | 75 | 203 | 64 | Ethane, hexachloro- | 19 |
| 201 | 100 | 174 | 69 | 202 | 15 | 175 | 9 | Thiabendazole | 402 |
| 202 | 100 | 203 | 19 | 200 | 17 | 101 | 14 | Fluoranthene | 170 |
| 202 | 100 | 203 | 26 | 200 | 21 | 101 | 21 | Pyrene | 171 |
| 202 | 100 | 108 | 53 | 174 | 48 | 28 | 40 | Ethoxyquin | 509 |
| 203 | 100 | 55 | 83 | 201 | 77 | 303 | 71 | Triforine | 352 |
| 203 | 100 | 159 | 91 | 193 | 86 | 195 | 85 | Chlorfenac | 277 |
| 204 | 100 | 139 | 55 | 141 | 42 | 69 | 38 | Phenol, 2-phenyl-6-chloro- | 143 |
| 204 | 100 | 206 | 96 | 205 | 45 | 163 | 42 | Isocil | 378 |
| 205 | 100 | 207 | 63 | 188 | 28 | 63 | 19 | Chloramben | 275 |
| 205 | 100 | 220 | 27 | 57 | 27 | 206 | 15 | Butylated hydroxy toluene | 234 |
| 205 | 100 | 207 | 98 | 41 | 68 | 29 | 34 | Bromacil | 379 |
| 206 | 100 | 29 | 78 | 133 | 69 | 150 | 56 | Thiophanate | 335 |
| 206 | 100 | 127 | 97 | 208 | 95 | 126 | 31 | Naphthalene, 1-bromo- | 148 |
| 208 | 100 | 180 | 96 | 152 | 75 | 151 | 38 | Anthraquinone | 166 |
| 208 | 100 | 210 | 42 | 40 | 21 | 94 | 19 | Dialifor | 503 |
| 211 | 100 | 163 | 40 | 147 | 23 | 117 | 19 | Dinoseb | 246 |
| 212 | 100 | 106 | 83 | 213 | 16 | 211 | 16 | Tolidine, 2- | 301 |
| 213 | 100 | 228 | 26 | 119 | 25 | 214 | 14 | Bisphenol A | 204 |
| 213 | 100 | 68 | 74 | 71 | 59 | 170 | 58 | Simetryne* | 389 |
| 215 | 100 | 186 | 62 | 81 | 61 | 53 | 26 | Kinetin | 516 |
| 216 | 100 | 368 | 90 | 97 | 89 | 125 | 76 | Coumithoate | 482 |
| 218 | 100 | 93 | 68 | 125 | 61 | 44 | 45 | Famphur | 468 |
| 221 | 100 | 97 | 33 | 232 | 31 | 373 | 22 | Pyrazophos | 480 |
| 222 | 100 | 51 | 74 | 87 | 67 | 143 | 40 | Barban | 331 |
| 223 | 100 | 224 | 19 | 179 | 9 | 167 | 9 | Perthane* | 202 |
| 225 | 100 | 134 | 41 | 79 | 38 | 107 | 28 | Aminoazotoluene, 4- | 296 |
| 225 | 100 | 93 | 21 | 226 | 15 | 66 | 11 | Tinuvin P* | 512 |
| 226 | 100 | 113 | 35 | 224 | 30 | 227 | 28 | Benzo[ghi]fluoranthene | 173 |
| 226 | 100 | 227 | 20 | 224 | 17 | 225 | 12 | Cyclopenta[cd]pyrene | 174 |
| 227 | 100 | 212 | 64 | 58 | 47 | 44 | 41 | Ametryn | 393 |
| 227 | 100 | 228 | 23 | 152 | 10 | 138 | 10 | Methoxychlor | 203 |
| 228 | 100 | 229 | 20 | 226 | 20 | 43 | 18 | Benz[a]anthracene | 175 |
| 228 | 100 | 226 | 21 | 229 | 20 | 114 | 15 | Chrysene | 177 |
| 228 | 100 | 226 | 45 | 227 | 34 | 113 | 31 | Benzo[c]phenanthrene | 178 |
| 229 | 100 | 227 | 75 | 272 | 74 | 237 | 69 | Chlorbicyclen | 110 |
| 230 | 100 | 232 | 96 | 160 | 32 | 234 | 30 | Naphthalene, trichloro- | 149 |
| 230 | 100 | 202 | 39 | 101 | 25 | 231 | 20 | Benzanthrone | 172 |
| 231 | 100 | 153 | 87 | 97 | 83 | 125 | 67 | Ethion | 500 |
| 232 | 100 | 230 | 82 | 166 | 67 | 168 | 65 | Phenol, 2,3,4,6-tetrachloro- | 219 |
| 234 | 100 | 206 | 96 | 116 | 54 | 174 | 41 | Oxythioquinox | 417 |
| 235 | 100 | 237 | 64 | 165 | 39 | 236 | 15 | DDD, o,p'- | 195 |
| 235 | 100 | 237 | 65 | 165 | 41 | 236 | 17 | DDD, p,p'- | 194 |
| 235 | 100 | 237 | 68 | 165 | 38 | 236 | 16 | DDT, p,p'- | 192 |
| 237 | 100 | 239 | 64 | 235 | 62 | 241 | 20 | Pentac | 116 |
| 237 | 100 | 239 | 64 | 235 | 63 | 95 | 41 | Hexachlorocyclopentadiene | 109 |
| 237 | 100 | 295 | 73 | 208 | 61 | 235 | 46 | Tetraethyllead | 530 |

| M1 | I1 | M2 | I2 | M3 | I3 | M4 | I4 | Common name | Page |
|---|---|---|---|---|---|---|---|---|---|
| 239 | 100 | 241 | 66 | 178 | 35 | 143 | 33 | Anilazine | 397 |
| 241 | 100 | 58 | 84 | 184 | 72 | 226 | 62 | Prometryn | 395 |
| 242 | 100 | 240 | 79 | 170 | 51 | 244 | 48 | Hexachloro-p-xylene | 136 |
| 242 | 100 | 167 | 66 | 277 | 37 | 166 | 23 | Adamsite | 524 |
| 243 | 100 | 165 | 62 | 244 | 34 | 166 | 11 | Trifenmorph | 97 |
| 246 | 100 | 318 | 79 | 248 | 66 | 316 | 61 | DDE, p,p′- | 193 |
| 251 | 100 | 139 | 88 | 43 | 86 | 253 | 62 | Chlorfenethol | 197 |
| 251 | 100 | 139 | 97 | 253 | 65 | 111 | 35 | Chlorobenzilate | 198 |
| 252 | 100 | 250 | 24 | 253 | 21 | 125 | 16 | Benzo[e]pyrene | 181 |
| 252 | 100 | 253 | 23 | 126 | 22 | 250 | 21 | Benzo[j]fluoranthene | 179 |
| 252 | 100 | 254 | 66 | 253 | 16 | 126 | 16 | Benzidine, 3,3′-dichloro- | 302 |
| 252 | 100 | 126 | 23 | 253 | 21 | 250 | 16 | Benzo[a]pyrene | 180 |
| 252 | 100 | 126 | 26 | 253 | 21 | 250 | 19 | Perylene | 182 |
| 255 | 100 | 43 | 71 | 121 | 64 | 298 | 38 | Coumafuryl | 408 |
| 256 | 100 | 241 | 40 | 239 | 37 | 240 | 24 | Benz[a]anthracene, 7,12-dimethyl- | 176 |
| 256 | 100 | 108 | 80 | 276 | 71 | 182 | 65 | Crufomate | 436 |
| 260 | 100 | 104 | 94 | 130 | 93 | 76 | 86 | Folpet | 412 |
| 262 | 100 | 264 | 78 | 235 | 77 | 125 | 76 | Prolan | 200 |
| 262 | 100 | 125 | 64 | 79 | 35 | 47 | 23 | Dicapthon | 463 |
| 263 | 100 | 125 | 99 | 109 | 99 | 79 | 39 | Methyl parathion | 459 |
| 265 | 100 | 43 | 46 | 121 | 38 | 187 | 27 | Warfarin | 407 |
| 266 | 100 | 231 | 97 | 55 | 81 | 41 | 79 | Phenol, 2-cyclohexyl-4,6-dinitro- | 247 |
| 266 | 100 | 224 | 19 | 267 | 16 | 220 | 15 | Butralin | 252 |
| 266 | 100 | 268 | 70 | 264 | 68 | 165 | 54 | Phenol, pentachloro- | 220 |
| 266 | 100 | 264 | 78 | 268 | 48 | 194 | 28 | Naphthalene, tetrachloro- | 150 |
| 266 | 100 | 264 | 85 | 268 | 52 | 109 | 30 | Chlorothalonil | 312 |
| 267 | 100 | 323 | 76 | 269 | 63 | 325 | 50 | Chlorfenvinphos | 449 |
| 272 | 100 | 270 | 96 | 56 | 61 | 315 | 46 | Cyhexatin | 532 |
| 272 | 100 | 274 | 79 | 270 | 51 | 237 | 45 | Kepone* | 117 |
| 272 | 100 | 274 | 75 | 270 | 54 | 237 | 46 | Mirex | 118 |
| 273 | 100 | 208 | 73 | 108 | 52 | 96 | 45 | Bupirimate | 383 |
| 276 | 100 | 138 | 24 | 277 | 23 | 274 | 16 | Anthanthrene | 183 |
| 276 | 100 | 138 | 37 | 137 | 28 | 277 | 25 | Benzo[ghi]perylene | 184 |
| 278 | 100 | 125 | 42 | 109 | 41 | 168 | 34 | Fenthion | 461 |
| 278 | 100 | 279 | 24 | 139 | 24 | 276 | 16 | Dibenz[a,h]anthracene | 185 |
| 279 | 100 | 223 | 50 | 281 | 42 | 162 | 42 | Dichlofenthion | 478 |
| 283 | 100 | 285 | 75 | 202 | 65 | 139 | 41 | Nitrofen | 249 |
| 284 | 100 | 286 | 82 | 282 | 52 | 288 | 35 | Benzene, hexachloro- | 135 |
| 285 | 100 | 340 | 86 | 56 | 53 | 267 | 45 | Butyl stearate | 64 |
| 292 | 100 | 220 | 89 | 290 | 78 | 222 | 57 | Biphenyl, tetrachloro- | 138 |
| 292 | 100 | 41 | 39 | 43 | 38 | 264 | 29 | Benfluralin | 257 |
| 295 | 100 | 249 | 85 | 237 | 81 | 297 | 64 | Quintozene | 243 |
| 296 | 100 | 76 | 64 | 240 | 47 | 104 | 47 | Dithianone | 419 |
| 299 | 100 | 43 | 46 | 121 | 41 | 301 | 34 | Coumachlor | 409 |
| 300 | 100 | 150 | 64 | 149 | 46 | 301 | 25 | Coronene | 186 |
| 301 | 100 | 299 | 80 | 303 | 48 | 332 | 32 | DCPA | 271 |
| 306 | 100 | 264 | 97 | 43 | 95 | 41 | 35 | Trifluralin | 256 |
| 322 | 100 | 320 | 81 | 324 | 51 | 257 | 23 | Dioxin, 1,2,3,4-tetrachloro- | 221 |
| 322 | 100 | 202 | 53 | 97 | 48 | 266 | 39 | Sulfotep | 499 |
| 322 | 100 | 320 | 79 | 324 | 48 | 257 | 25 | Dioxin, 2,3,7,8-tetrachloro- | 222 |
| 323 | 100 | 185 | 98 | 157 | 76 | 77 | 67 | EPN | 485 |
| 326 | 100 | 328 | 66 | 324 | 66 | 254 | 57 | Biphenyl, pentachloro- | 139 |

| M1 | I1 | M2 | I2 | M3 | I3 | M4 | I4 | Common name | Page |
|---|---|---|---|---|---|---|---|---|---|
| 330 | 100 | 328 | 71 | 92 | 60 | 329 | 53 | Mercufenol chloride | 528 |
| 331 | 100 | 125 | 79 | 329 | 75 | 47 | 42 | Bromophos | 465 |
| 333 | 100 | 318 | 82 | 57 | 60 | 304 | 57 | Pirimiphos-ethyl | 475 |
| 334 | 100 | 335 | 22 | 36 | 20 | 120 | 15 | Strychnine | 519 |
| 351 | 100 | 120 | 97 | 154 | 95 | 78 | 91 | Triphenyltin hydroxide | 531 |
| 359 | 100 | 361 | 82 | 289 | 69 | 394 | 61 | Biphenyl, heptachloro- | 141 |
| 360 | 100 | 325 | 88 | 290 | 83 | 362 | 80 | Biphenyl, hexachloro- | 140 |
| 362 | 100 | 109 | 95 | 226 | 78 | 97 | 74 | Coumaphos | 481 |
| 368 | 100 | 367 | 60 | 91 | 35 | 165 | 30 | Tritolyl phosphate | 455 |
| 375 | 100 | 373 | 94 | 377 | 46 | 237 | 45 | Chlordane | 112 |
| 377 | 100 | 125 | 44 | 379 | 37 | 42 | 22 | Iodofenphos | 466 |
| 404 | 100 | 402 | 85 | 406 | 61 | 332 | 41 | Naphthalene, octachloro- | 151 |
| 444 | 100 | 442 | 89 | 446 | 73 | 440 | 39 | Dibenzofuran, octachloro- | 224 |
| 460 | 100 | 458 | 90 | 28 | 85 | 462 | 63 | Dioxin, octachloro- | 223 |
| 466 | 100 | 125 | 51 | 93 | 38 | 47 | 35 | Temephos | 469 |